instructors ... get resources to help you teach

Visit *www.mhhe.com/biosci/genbio/maderhuman7*

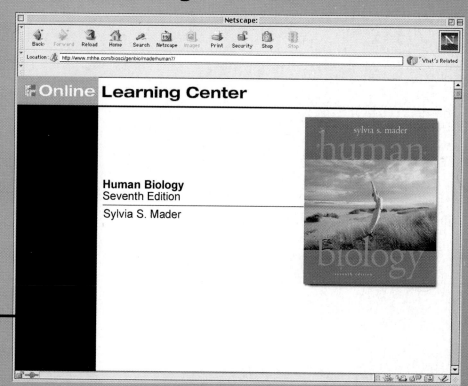

Online Learning Center

A multitude of tools awaits you on the Online Learning Center. You'll want to take advantage of our electronic illustrations and photographs from the text; classroom activities; lecture outlines; and access to PageOut™: Course Website Development Center—all available anytime you want them.

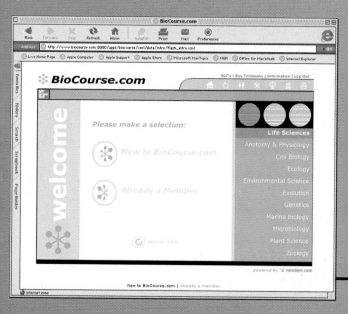

BioCourse.com

BioCourse.com is an electronic meeting place for students and instructors. It provides a comprehensive set of resources in one place that is up-to-date and easy to navigate. You can access BioCourse.com from *Human Biology*'s Online Learning Center. There, you will find resources including The Faculty Club, The Student Center, The Briefing Room, BioLabs, The Quad, and The R & D Center.

McGraw Hill

For more information visit *www.mhhe.com*.

For My Family

human biology

seventh edition

sylvia s. mader

Boston Burr Ridge, IL Dubuque, IA Madison, WI New York
San Francisco St. Louis Bangkok Bogotá Caracas Kuala Lumpur
Lisbon London Madrid Mexico City Milan Montreal New Delhi
Santiago Seoul Singapore Sydney Taipei Toronto

McGraw-Hill Higher Education

A Division of The McGraw-Hill Companies

HUMAN BIOLOGY, SEVENTH EDITION

Published by McGraw-Hill, a business unit of The McGraw-Hill Companies, Inc.,
1221 Avenue of the Americas, New York, NY 10020. Copyright © 2002, 2000, 1998, 1995, 1992, 1990, 1988 by The McGraw-Hill Companies, Inc.
All rights reserved. No part of this publication may be reproduced or distributed in any form or by any means,
or stored in a database or retrieval system, without the prior written consent of The McGraw-Hill Companies, Inc.,
including, but not limited to, in any network or other electronic storage or transmission, or broadcast for distance learning.

Some ancillaries, including electronic and print components, may not be available to customers outside the United States.

 This book is printed on recycled, acid-free paper containing 10% postconsumer waste.

International   2 3 4 5 6 7 8 9 0 QPD/QPD 0 9 8 7 6 5 4 3 2 1
Domestic   2 3 4 5 6 7 8 9 0 QPD/QPD 0 9 8 7 6 5 4 3 2

ISBN 0–07–232481–3
ISBN 0–07–112245–1 (ISE)

Sponsoring editor: *Patrick E. Reidy*
Developmental editor: *Anne L. Melde*
Senior marketing manager: *Lisa L. Gottschalk*
Senior project manager: *Jayne Klein*
Senior production supervisor: *Sandy Ludovissy*
Designer: *K. Wayne Harms*
Cover designer: *Kaye Farmer*
Cover image: *PhotoDisc Volume 38 Weekend Living*
Senior photo research coordinator: *Lori Hancock*
Photo researcher: *Connie Mueller*
Supplement producer: *Tammy Juran*
Executive producer: *Linda Meehan Avenarius*
Compositor: *GTS Graphics, Inc.*
Typeface: *10/12 Palatino*
Printer: *Quebecor World Dubuque, IA*

The credits section for this book begins on page C-1 and is considered
an extension of the copyright page.

Library of Congress Cataloging-in-Publication Data

Mader, Sylvia S.
    Human biology / Sylvia S. Mader. — 7th ed.
        p.   cm.
    Includes bibliographical references and index.
    ISBN 0–07–232481–3
    1. Human biology. I. Title.
    QP36 .M2   2002
    612—dc21   00–136453
                                                          CIP

INTERNATIONAL EDITION ISBN 0–07–112245–1
Copyright © 2002. Exclusive rights by The McGraw-Hill Companies, Inc., for manufacture and export.
This book cannot be re-exported from the country to which it is sold by McGraw-Hill.
The International Edition is not available in North America.

www.mhhe.com

Brief Contents

Contents

Readings

Preface

Human Biology introduces students to the anatomy and physiology of the human body. All systems of the body are represented and each system has its own chapter. The text can also be used to help students understand the role that humans play in the biosphere. All of us need to realize how human activities threaten ecosystems, and seek ways to lessen our impact on the biosphere.

The application of biological principles to practical human concerns is now widely accepted as a suitable approach to the study of biology because it fulfills a great need. All students should leave college with a firm grasp of how their bodies normally function, and how the human population can become more fully integrated into the biosphere. We are frequently called upon to make health and environmental decisions. Wise decisions require adequate knowledge and can help assure our continued survival as individuals and as a species.

In this edition, as in previous editions, each chapter presents the topic clearly, simply, and distinctly so that students will feel capable of achieving an adult level of understanding. Detailed, high-level scientific data and terminology are not included because I believe that true knowledge consists of working concepts rather than technical facility.

Pedagogical Features

Human Biology excels in pedagogical features. Each chapter begins with an integrated chapter outline that lists the chapter's concepts according to numbered sections of the chapter. This numbering system is continued in the chapter and summary so that instructors can assign just certain portions of the chapter, if they like.

The text is paged so that major sections start at the top of the page and illustrations are on the same or facing page to their reference. The illustrations are visually motivating, and the art program has many features that students will find helpful. Color coordination includes assigning colors to the various classes of organic molecules and to the different human tissues and organs. Visual focus illustrations give a conceptual overview that relates structure to function.

The questions at the end of the chapter are of both the essay and objective type. Studying the Concepts reviews the content of the chapter and requires that students write out their answers. Testing Your Knowledge of the Concepts includes multiple choice questions, fill in the blanks, and true-false questions. Understanding Key Terms lists the major terms in the chapter and page references the term to where it is defined in the chapter.

Revised Chapters

Every chapter in *Human Biology* has been revised or is new. The systems chapters have been fine-tuned and the illustrations in these chapters have been improved to better present the concepts. Students should have no difficulty in following the text, understanding the concepts, and applying them to their everyday lives.

Part VII of the text contains new chapters. Students learn best when the content applies to themselves, and these chapters are faithful to this educational maxim. Chapter 23 is now entitled "Human Evolution." This unique chapter teaches the principles of evolution, while at the same time reviewing human evolution from the origin of the first cell(s) to the rise of modern humans. Chapter 24, which is called "Ecosystems and Human Interferences" introduces the basics of ecology and shows how human activities have altered biogeochemical cycles to our own detriment. Chapter 25 is a new chapter entitled "Conservation of Biodiversity." We all need to be aware that other living things are valuable to the human species and to recognize that our activities threaten their very existence. In preserving other species we are ultimately preserving our own species.

Focus Readings

Health and ecological concerns are carried through the text by Health Focus readings, which help students cope with common health problems, and Ecology Focus readings, which draw attention to a particular environmental problem.

As in the previous edition, students are asked to apply the concepts to the many and varied perplexing bioethical issues that face us every day. In this edition, each bioethical issue is featured in a Bioethical Focus box which asks students to develop a point of view by answering a series of questions on such topics as genetic disease testing, modern reproductive technologies, human cloning, AIDS vaccine trials, animal rights, and fetal research. The Online Learning Center will help students fine-tune their opinion with these activities:

Taking Sides. Students answer a series of questions and their answers are tallied so that their original position is revealed.

Further Debate. Students are directed to read articles on both sides of the issue.

Explain Your Position. Students are asked to defend their position in writing. They can e-mail their essay to their professor.

Homeostasis

This edition of *Human Biology* again places an emphasis on homeostasis. An icon ⚖ calls attention to those portions of each chapter that discuss homeostasis. The chapter entitled "Organization and Regulation of Body Systems" discusses the principles of homeostasis and the contributions of the various systems to keeping the internal environment relatively constant. Well-designed illustrations, especially in the endocrine chapter, show how negative feedback control is essential to homeostasis. The Human Systems Work Together box in each systems chapter describes how that organ system works with other systems to achieve homeostasis.

Applications

Each chapter begins with a short story that applies chapter material to real-life situations. The readings stress applications and so does the running text material. This edition features expanded treatment of such topics as eating disorders, allergies, pulmonary disorders, hepatitis infections, modern reproductive technologies, the human genome project, and gene therapy. Other topics such as the cloning of humans and xenotransplantation are also included.

Technology

There are many resources that students can utilize in order to understand the content of this textbook. In addition to the end of the chapter questions and printed study guide, the Online Learning Center at www.mhhe.com/biosci/gen-bio/maderhuman7 contains readings, quizzes, animations, and other activities to help students master the concepts. New to this edition there is more integration between text material and technology. For example, Bioethical Focus boxes and Human Systems Work Together boxes have an associated online exercise that helps students make better use of these stimulating features.

Also, new to this edition, each chapter ends with an e-Learning Connection page. This page organizes the relevant technological material by major sections, helping to create a stronger association between available study activities and text material. Because this design is mimicked on the Online Learning Center the student can now easily find the appropriate learning experience.

A complete explanation of the technology package available with this textbook for students and instructors, is explained fully on pages xvi through xx of the preface.

An Overview of This Edition

- Part VII contains new chapters. "Human Evolution" traces human evolution from the origin of cell(s) to the evolution of modern humans. This unique chapter presents the principles of evolution, while at the same time reviewing human evolution. "Ecosystems and Human Interferences" presents the principles of ecology and shows how human activities disrupt biogeochemical cycles, leading to untoward effects including pollution. "Conservation of Biodiversity" explores the current biodiversity crisis and shows how the loss of so many species can be detrimental to humankind.

- Technology aids are organized according to the major sections on the e-Learning Connection page found at the end of each chapter. Students can easily determine the available resources to help explain difficult concepts. The same design is utilized for the Online Learning Center, so that students can quickly find an activity of interest. Other activities help students make full use of the Bioethical Focus boxes and Human Systems Work Together boxes.

- Relevancy of the text is enhanced due to the inclusion of such topics as sexually transmitted diseases, eating disorders, allergies, pulmonary disorders, hepatitis infections, xenotransplantation, modern reproductive technologies, human cloning, the human genome project, and gene therapy to treat cancer.

- Health Focus and Ecology Focus readings support the two major themes of the text; the study of human anatomy and physiology and the role of humans in the biosphere. A new Bioethical Focus box found throughout the text introduces students to many of the bioethical questions that face us every day. Challenging questions are provided that can be used as a basis for class discussion. The Online Learning Center allows students to further explore these issues by taking a quiz, reading articles, and writing an essay explaining their point of view.

- The vibrant art program adds vitality to illustrations and enhances the appeal of the text. Micrographs are integrated into illustrations and provide realism. Visual focus illustrations give a pictorial overview of key topics. Color coding is used both for biological molecules and for human tissues and organs.

- Homeostasis is again emphasized in this edition. An icon ⚖ calls attention to those portions of the text that discuss homeostasis. Each systems chapter has a major section that discusses how that system works with other systems of the body to achieve homeostasis. This section is supported by a Human Systems Work Together box, which also shows how that organ system works with the other systems making homeostasis possible.

Acknowledgments

To produce a text requires a concerted effort by many and it is a pleasure to thank everyone who made this edition of *Human Biology* so special. My editor Patrick Reidy and my developmental editor Anne Melde fulfilled every expectation. They planned well and supplied creativity, advice, and support whenever it was needed.

Jayne Klein, the project manager, although new to the book team, stepped right in and made the project move along smoothly. Kennie Harris did a superb job as the copy editor; Lori Hancock and Connie Mueller found just the right photographs. Again, Wayne Harms developed a design that is both beautiful and useful to students.

In my office Jo Hebert has consistently provided support and was just as diligent working on this edition as the others. I also want to take this opportunity to thank my husband and children for their continual patience and encouragement.

The Reviewers

Many instructors have contributed not only to this edition of *Human Biology* but also to previous editions. I am extremely thankful to each one, for they have all worked diligently to remain true to our calling to provide a product that will be the most useful to our students.

It is appropriate to acknowledge the help of the following individuals for the seventh edition.

David H. Arnold
University of Texas-Austin
Amir M. Assadi-Rad
San Joaquin Delta College
Ellen Baker
Santa Monica College
Linda M. Banta
Sierra College
Angela Bauer-Dantoin
University of Wisconsin-Green Bay
Cindy Beck
The Evergreen State College
George C. Boone
Susquehanna University
Judith Byrnes-Enoch
SUNY - Empire State College
Judson J. Calhoun
Mohave Community College
Joseph P. Caruso
University of the South
Debra Chapman
Wilkes University
Richard Connett
Monroe Community College
David Constantinos
Savannah State University
Charles J. Dick
Pasco-Hernando Community College

Marirose T. Ethington
Genesee Community College
Richard H. Falk
University of California-Davis
Steve Fields
Winthrop University
Dalia Giedrimiene
Saint Joseph College
Mary Louise Greeley
Salve Regina University
Kenneth W. Gregg
Winthrop University
Ryan M. Harden
Central Lakes College
Janice L. Haws
Delaware Valley College
Chris G. Haynes
Shelton State Community College
Danette I. Haynes
Shelton State Community College
Clare Hays
Metropolitan State College of Denver
Janice Ito
Leeward Community College
Dennis Jackson
Mount Aloysius College
Pushkar N. Kaul
Clark Atlantic University
Suzanne Kempke
Armstrong Atlantic State University
Michelle Kettler
University of Wisconsin-Eau Claire
Patricia Klopfenstein
Edison Community College
Troy A. Ladine
Bethel College
Thomas M. Lancraft
St. Petersburg Junior College
Michael Lentz
University of North Florida
Mary Katherine K. Lockwood
University of New Hampshire
William J. Mackay
Edinboro University of Pennsylvania
Morris V. Maniscalco
LeTourneau University
Joseph A. Mannino
University of Wisconsin-Green Bay
Vicki J. Martin
Appalachian State University
Patricia Matthews
Grand Valley State University
Elizabeth McMahon
Warren County Community College
Tekie Mehary
University of Washington
William Millington
Albany College of Pharmacy of Union University
Aaron J. Moe
Concordia University

Michele Morek
Brescia University
Robert A. Morgan
Southwestern University
Richard Mortensen
Albion College
Don Naber
University College, University of Maine
Jon R. New
Vernon Regional Junior College
Emily C. Oaks
SUNY at Oswego
Richard O'Lander
St. John's University
Sidney L. Palmer
Ricks College
Jeff Parmelee
Simpson College
Mason Posner
Ashland University
Darryl Ritter
Okaloosa-Walton Community College
Connie Rizzo
Pace University
Michael W. Ruhl
Vernon Regional Junior College
Mary Ellen St. John
Central Ohio Technical College
Don Sanders
Harding Academy of Memphis
Soma Sanyal
Penn State Altoona College
John W. Sherman
Erie Community College-North Campus
Kristin Siewert
Des Moines Area Community College at Newton
Jeff S. Simpson
Metropolitan State College of Denver

Timothy A. Stabler
Indiana University Northwest
W. Robert Stamper
Muhlenberg College
Steve K. Stocking
San Joaquin Delta Community College
Robert S. Sullivan
Marist College
Jacqueline Tanaka
University of Pennsylvania
William J. Tarutis, Jr.
Lackawanna Junior College
Kent R. Thomas
Wichita State University
Timothy S. Wakefield
Auburn University
Curt Walker
Dixie State College
Rod Waltermyer
York College of Pennsylvania
Linda Wells
California State University-Bakersfield
Frank P. Wray
University of Cincinnati
Joyce A. Wren
Surry Community College
Kathryn Yarkany
Villa Julie College
Martin D. Zahn
Thomas Nelson Community College
Nina C. Zanetti
Siena College
Henry H. Ziller
Southeastern Louisiana University
Brenda Zink
Northeastern Junior College

THE LEARNING SYSTEM

Chapter 21

DNA and Biotechnology

Chapter Concepts

21.1 DNA and RNA Structure and Function
- DNA is the genetic material, and therefore, its structure and function constitute the molecular basis of inheritance. 422
- When DNA replicates, two exact copies result. RNA structure is similar to, but also different from, that of DNA. RNA occurs in three forms, each with a specific function. 424–25

21.2 Gene Expression
- Proteins are composed of amino acids, and they function, in particular, as enzymes and as structural elements of membranes and organelles in cells. 426
- DNA's genetic information codes for the sequence of amino acids in a protein during protein synthesis. 426
- There are various levels of genetic control in human cells. 431

21.3 Biotechnology
- Recombinant DNA technology is the basis for biotechnology, an industry that produces many products. 432
- Modern-day biotechnology is expected to produce many products and to revolutionize agriculture and animal husbandry. It also permits gene therapy and the mapping of human chromosomes. 434

Figure 21.1   Cindy Cutshall.
Cindy was sick and couldn't play baseball until she underwent gene therapy. She doesn't like to talk about the procedure. She says, "It's too weird."

Cindy Cutshall was born with a rare immune disorder called SCID (severe combined immunodeficiency). Her white blood cells were ineffective because they couldn't produce a critical enzyme known as ADA. She was in and out of the hospital with infections, and her parents feared she would die young.

And so, in 1993, the Cutshalls agreed to let Cindy undergo gene therapy, sponsored by the National Institutes of Health (NIH), the top biomedical funding agency in the United States. Gene therapy is based on the idea that it is possible to supply a patient with a missing or defective gene. NIH researchers took white blood stem cells from Cindy, added the gene that specifies ADA, and injected the improved cells back into her. Stem cells for white blood cells were chosen because they reside in the bone marrow where they continually produce more white blood cells. Combined with drug treatment, the therapy apparently worked. Today, Cindy is well, and her white blood cells contain functioning ADA genes.

Like Cindy, over 600 patients have now received some form of gene therapy in treatment for AIDS, cancer, cystic fibrosis, and other diseases. In many cases, the patient's cells have not received an adequate number of working genes to make a difference. But once efficient

421

Each chapter in *Human Biology* is constructed of basic features that serve as the pedagogical framework for the chapter. Before you begin reading the text, spend a little time looking over these pages. They provide a quick guide to the learning tools found throughout the text that have been designed to enhance your understanding of biology.

Chapter Concepts

The chapter begins with an integrated outline that numbers the major sections of the chapter and lists the concepts that support each section.

Homeostasis Icon

Human Biology emphasizes homeostasis through Working Together boxes, separate discussion in each human system chapter, and through the use of an icon ⚓. The homeostasis icon has been placed adjacent to text material that discusses homeostasis.

Internal Summary Statements

Internal summaries stress the chapter's key concepts. These appear at the ends of major sections and help focus students' study efforts on the most important concepts.

---

134   Part 2   Maintenance of the Human Body

7.4 The Vascular Pathways

The cardiovascular system, which is represented in Figure 7.12, includes two circuits: the **pulmonary circuit**, which circulates blood through the lungs, and the **systemic circuit**, which serves the needs of body tissues. Both circuits, we shall see, are necessary to homeostasis. ⚓

The Pulmonary Circuit

The path of blood through the lungs can be traced as follows. Blood from all regions of the body first collects in the right atrium and then passes into the right ventricle, which pumps it into the pulmonary trunk. The pulmonary trunk divides into the right and left pulmonary arteries, which branch as they approach the lungs. The arterioles take blood to the pulmonary capillaries, where carbon dioxide is given off and oxygen is picked up. Blood then passes through the pulmonary venules, which lead to the four pulmonary veins that enter the left atrium. Since blood in the pulmonary arteries is O_2-poor but blood in the pulmonary veins is O_2-rich, it is not correct to say that all arteries carry blood that is high in oxygen and all veins carry blood that is low in oxygen. It is just the reverse in the pulmonary circuit.

> The pulmonary arteries take O_2-poor blood to the lungs, and the pulmonary veins return blood that is O_2-rich to the heart.

The Systemic Circuit

The systemic circuit includes all of the arteries and veins shown in Figure 7.13. The largest artery in the systemic circuit is the **aorta**, and the largest veins are the **superior** and **inferior venae cavae**. The superior vena cava collects blood from the head, the chest, and the arms, and the inferior vena cava collects blood from the lower body regions. Both enter the right atrium. The aorta and the venae cavae serve as the major pathways for blood in the systemic circuit.

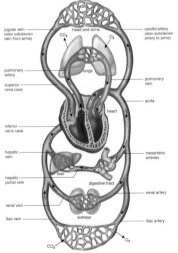

Figure 7.12   Cardiovascular system diagram.
The blue-colored vessels carry O_2-poor blood, and the red-colored vessels carry O_2-rich blood; the arrows indicate the flow of blood. Compare this diagram, useful for learning to trace the path of blood, to Figure 7.13 to realize that arteries and veins go to all parts of the body. Also, there are capillaries in all parts of the body. No cell is located far from a capillary.

The path of systemic blood to any organ in the body begins in the left ventricle, which pumps blood into the aorta. Branches from the aorta go to the organs and major body regions. For example, this is the path of blood to and from the lower legs:

Summarizing the Concepts

The numbered concepts, introduced on the chapter-opening page and explained in the body of the chapter, form the basis for the summary. This repetition helps reinforce key concepts for the student.

Summarizing the Concepts

1.1 Biologically Speaking

Human beings, like other living things, are highly organized. Cells form tissues, which form organs that function in organ systems. Human beings come into existence through reproduction, growth, and development.

Unlike other living things, human beings have a cultural heritage that sometimes hinders the realization that they are the product of an evolutionary process. Human beings are vertebrates closely related to the apes. Human beings are also a part of the biosphere where populations interact with the physical environment and with one another.

Human activities threaten the existence of ecosystems like tropical rain forests. Biodiversity is now being reduced at a rapid rate.

1.2 The Process of Science

When studying the world of living things, biologists, like other scientists, use the scientific method, which consists of these steps: making an observation, formulating a hypothesis, carrying out an experiment or simply making further observations, and coming to a conclusion.

1.3 Science and Social Responsibility

It is the responsibility of all to make ethical and moral decisions about how best to make use of the results of scientific investigations.

Studying the Concepts

This page-referenced question set reviews the concepts presented in the chapter.

Studying the Concepts

1. Name five characteristics of human beings, and discuss each one. 2–4
2. What is homeostasis, and how is it maintained? Choose one organ system and tell how it helps maintain homeostasis. 2
3. Give evidence that human beings are related to all other living things. 2
4. Describe the five-kingdom system of classification, and name types of organisms in each kingdom. 3
5. Human beings are dependent upon what services performed by plants? 4
6. Discuss the importance of scientific theory, and name several theories that are basic to understanding biological principles. 8
7. Name the steps of the scientific method, and discuss each one. 8
8. How do you recognize a control group, and what is its purpose in an experiment? 10
9. What is our social responsibility in regard to scientific findings? 11

Understanding Key Terms

Key terms are listed with page references that indicate where the term is defined in the chapter.

Understanding Key Terms

biodiversity 5
biosphere 2
cell 2
conclusion 8
control group 10
data 8
ecosystem 4
evolution 2
experiment 10
homeostasis 2
hypothesis 8
kingdom 2

organ 2
organ system 2
population 4
principle 8
reproduce 2
science 8
scientific method 8
scientific theory 8
tissue 2
variable 10
vertebrate 2

Testing Your Knowledge of the Concepts

In questions 1–4, match the human characteristics to the descriptions below.

Human beings:
a. are organized.
b. reproduce and grow.
c. have a cultural heritage.
d. are the product of evolutionary process.
e. are a part of the biosphere.

_____ 1. Humans are related to all other living things.

_____ 2. The human population encroaches on natural habitats.

_____ 3. Like cells form tissues in the human body.

_____ 4. We learn how to behave from our elders.

In questions 5–7, indicate whether the statement is true (T) or false (F).

_____ 5. Once a scientist formulates a hypothesis, he or she tests it by observation and/or experimentation.

_____ 6. The theory of evolution is so poorly supported that many scientists feel it should be discarded.

_____ 7. When an experiment has a control group, it lends validity to the resulting data.

In questions 8–10, fill in the blanks.

8. To reproduce is to make a _____ of one's self.

9. _____ has a responsibility to decide how scientific knowledge should be used.

10. _____ is a concept consistent with conclusions based on a large number of experiments and observations.

Testing Your Knowledge of the Concepts

This section consists of objective questions that test the student's ability to answer recall-based questions. Answers to these questions are given in Appendix A.

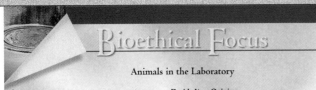

Animals in the Laboratory

Some people believe that animals should be protected in every way, and should not be used in laboratory research. In our society as a whole, the trend is toward a growing recognition of what is generally referred to as animal rights. In 1985, 63% of Americans polled agreed that "scientists should be allowed to do research that causes pain and injury to animals like dogs and chimpanzees if it produces new information about human health problems." That consensus dropped to 53% in 1995. Psychologists with Ph.D.s earned in the 1990s are half as likely to express strong support for animal research as those who earned their Ph.D.s before 1970.

Those who approve of laboratory research involving animals give examples to show that even today it would be difficult to develop new vaccines and medicines against infectious diseases, new surgical techniques for saving human lives, or new treatments for spinal cord injuries without the use of animals. Even so, most scientists today are in favor of what is now called the "three Rs": replacement of animals by in vitro, or test-tube, methods whenever possible; reduction of the numbers of animals used in experiments; and refinement of experiments to cause less suffering to animals. In the Netherlands, all scientists starting research that involves animals are well trained in the three Rs. After designing an experiment that uses animals, they are asked to find ways to answer the same questions without using animals.

F. Barbara Orlans of the Kennedy Institute of Ethics at Georgetown University says, "It is possible to be both pro research and pro reform." She feels that animal activists need to accept that sometimes animal research is beneficial to humans, and all scientists need to consider the ethical dilemmas that arise when animals are used for laboratory research.

Decide Your Opinion

To develop your opinion, answer either question 1 or question 2, and then question 3.

1. Are you opposed to the use of animals in laboratory experiments? Always or under certain circumstances? Explain.
2. Do you favor using animals in the laboratory? Always or under certain circumstances? Explain.
3. Do you feel that it would be possible for animal activists and scientists to find a compromise they could both accept? Discuss.

Figure 1C Caged animals.
These animals are used in laboratory research to help develop an AIDs vaccine for humans.

Looking at Both Sides   www.mhhe.com/biosci/genbio/maderhuman7/

Every bioethical issue has at least two sides. Even if you already have an opinion, it is important to explore the opposite opinion before finalizing your position. The Online Learning Center at www.mhhe.com/biosci/genbio/maderhuman7/ will help you fine-tune your initial opinion, explore both sides, and finalize your position. You may acquire new arguments for your original opinion, or you may even change your opinion. Be sure to complete these activities in sequence:

Taking Sides Decide your initial opinion by answering a series of questions. Then see if your opinion changes after completing the next two activities.

Further Debate Read opposing articles that give you further information on this particular bioethical issue.

Explain Your Position Answer another series of questions and then defend your original or changed opinion. You can e-mail your position to your instructor if he or she wishes.

12

Technology has become an increasingly potent force in teaching and learning. For that reason the seventh edition of *Human Biology* puts an even greater emphasis on technology by integrating it more fully with text material. Students can access the material described on these pages by going to www.mhhe.com/biosci/genbio/maderhuman7.

Bioethical Focus Boxes

The popular Bioethical Focus boxes introduced in the sixth edition have now been expanded into a full page feature. To help students further explore the complicated issues discussed in the Bioethical Focus boxes, an online feature called Looking at Both Sides has been added. Students go to the Online Learning Center where they find the following activities:

Taking Sides is a short quiz that helps students decide which side of the issue they identify with at the outset.

Further Debate facilitates the students' continued investigation of the issue by providing websites for further study. Students are asked questions that help them analyze the information and arguments provided.

Explain Your Position requires that students express and defend their position in writing. Responses can be e-mailed to the instructor if he or she wishes.

Human Systems Work Together Boxes

These helpful boxes were developed to illustrate for students how the systems in their own bodies are working together to achieve homeostasis. An online component has been developed to further emphasize this vital concept:

Systems Scramble is a matching exercise based on the Human Systems Work Together box.

Watch It Happen shows the student an animation of a process occurring in the highlighted system.

Systems Review follows up the animation with questions that require students to integrate what they have learned about the body systems.

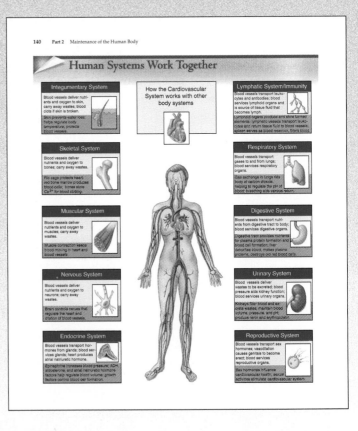

e-Learning Connection

These pages, found at the end of each chapter, tie technology directly to the major sections found within the chapter. Students are shown which McGraw-Hill study aids are available on the Online Learning Center to help them understand the concepts in each section. An icon tells the student at a glance what type of resource is being cited.

Online Learning Center

The e-Learning Connection page is duplicated on the Online Learning Center where it serves as navigation for the online chapter content. The student need only click on a link to go to the resource that is listed. This helps students find the information they need more quickly, increasing the effectiveness of their study time.
Go to www.mhhe.com/biosci/genbio/maderhuman7 to see the many resources available for students on the *Human Biology* Online Learning Center.

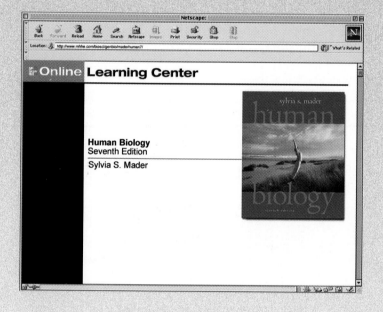

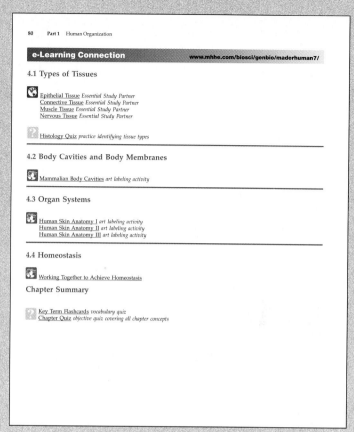

e-Learning Connection   www.mhhe.com/biosci/genbio/maderhuman7/

4.1 Types of Tissues

Epithelial Tissue *Essential Study Partner*
Connective Tissue *Essential Study Partner*
Muscle Tissue *Essential Study Partner*
Nervous Tissue *Essential Study Partner*

Histology Quiz *practice identifying tissue types*

4.2 Body Cavities and Body Membranes

Mammalian Body Cavities *art labeling activity*

4.3 Organ Systems

Human Skin Anatomy I *art labeling activity*
Human Skin Anatomy II *art labeling activity*
Human Skin Anatomy III *art labeling activity*

4.4 Homeostasis

Working Together to Achieve Homeostasis

Chapter Summary

Key Term Flashcards *vocabulary quiz*
Chapter Quiz *objective quiz covering all chapter concepts*

Icons

These icons are used on the e-Learning Connection page and the Online Learning Center to denote types of content.

 Activity — Activities are hands-on exercises that engage students in a learning experience.

 Animation — Animations help students visualize how a process occurs.

 Essential Study Partner Module — Modules combine text screens and activities to help students master difficult concepts.

 Reading — Readings give students the opportunity to explore a topic further.

 Quiz — Quizzes allow students to test themselves on the topics presented in the chapter.

Increasingly, instructors are demanding visual resources and the versatility to use them according to their needs. By adopting *Human Biology* for use in their course, instructors gain access to technological resources that can revolutionize the way information is presented to their students.

Incredible Online Resources

At the *Human Biology* Online Learning Center, you will have access to images from the textbook, case studies, chapter outlines, the Instructor's Manual, and text-specific PowerPoint presentations.

www.mhhe.com/biosci/genbio/maderhuman7

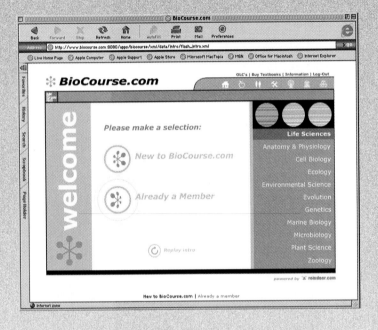

BioCourse.com is a new resource for instructors using a McGraw-Hill textbook in their course. BioCourse.com offers instructors over 10,000 images, animations, and case studies for use in their course. Visitors to BioCourse.com can check out the latest science headlines, read commentaries from McGraw-Hill authors and guests, take part in discussion boards, find relevant materials for lab courses, and much more.

Wake Up Your Lectures

Create dynamic PowerPoint presentations using art from the *Human Biology* textbook. Approximately 800 labeled and unlabeled images, including all of the illustrations and photos from the text, are available for your use on CD-ROM or by using the *Human Biology* Online Learning Center. The Visual Resource Library allows you to search and sort through a catalog of images by chapter, or by using keywords, and then place the images into your own lecture notes.

Human Biology
Visual Resource Library CD-ROM

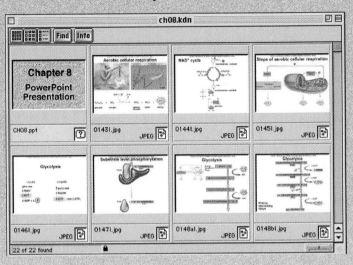

Life Science Animations
Visual Resource Library CD-ROM

The Life Science Animations 2.0 CD-ROM contains over 300 animations of complex biological processes. Help your students visualize the mechanisms at work within their own cells by incorporating these animations into your lecture presentations. Step-by-step instructions are provided to help you get the most out of this powerful resource.

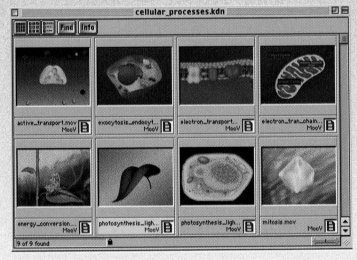

Need a Course Website?

PageOut™ makes it easy to provide class materials online by creating your own course website. Using PageOut™ you can post an interactive syllabus containing class notes, practice exercises, helpful figures, and links to relevant McGraw-Hill web content. Student registration and grade book features assist with course management.

If your time is limited you can simply copy a site from the McGraw-Hill PageOut™ library, or let the McGraw-Hill service team do the work for you. They will call you for a 30-minute consultation, create your PageOut™ website, and provide training to get you up and running.

For more information on PageOut™ contact your McGraw-Hill sales representative or go to www.pageout.net.

You Can Have What You Want

Since 1990 Primis Custom Publishing has made it possible for instructors to create their ideal textbooks. Now, in addition to printed texts, Primis custom textbooks can be made available to your students online as eBooks.

Select from over 35,000 pages of online content, then move, add, or delete materials until you are happy with your text. Once you submit your new book to the eBookstore, your students only need to go to the site to buy their download of the text. By adding a link to the eBookstore from your PageOut™ website, students have everything they need for your course in one place. Best of all, your students save money since they are only paying for the content you choose to include in the course textbook.

To find out more about customizing *Human Biology* or any other McGraw-Hill text please contact your McGraw-Hill sales representative or visit www.mhhe.com/primis/online.

New Technology

Human Biology Online Learning Center

McGraw-Hill text-specific websites allow students and instructors from all over the world to communicate. By visiting this site, students can access additional study aids, including quizzes and animations, explore links to other relevant biology sites, and catch up on current information. Log on today!

www.mhhe.com/biosci/genbio/maderhuman7

BioCourse.com

BioCourse.com provides a comprehensive set of resources in one place that is up-to-date and easy to navigate. Here is what you will find:

- The **Faculty Club** includes teaching tips, classroom activities, reference searches, presentation tools, and much more.
- The **Student Center** contains a wide range of materials to help biology students improve their study skills and achieve success in college and beyond.
- The **Briefing Room** offers instructors and students up-to-date news articles, a selection of background readings, and links to journal search tools and biology magazines.
- **BioLabs** features materials for lab students and instructors, including equipment tutorials, lab support, and simulations.
- The **Quad** is a powerful indexing tool and heirarchical out line of content resources for searching by students and faculty.
- The **R & D Center** features our newest simulations, animations, and other teaching tools.

Primis Custom Publishing

Now it is incredibly easy to create a content-rich textbook or lab manual tailored to your exact needs. Choose from our library of content, your materials, or both. Your customized course materials are delivered to your students immediately in color, online, and at a substantial savings. Customization allows you to tailor your classroom tools to meet your course needs, follow your syllabus, and teach like you do. All you need to do is log on to **www.mhhe.com/primis/online** and go to your discipline area. Simply view, select, review, and submit. Students can then easily and conveniently access the site on the Internet, reference their school and course, and purchase their "e-Book" online. Immediately, the electronic book is downloaded to their computer's hard drive.

PageOut™ Put together your own customized website with the use of PageOut™, a program designed specifically for instructors wanting to put course information on the web. No experience in web publishing is necessary; just choose from a collection of templates to create your class website.

Visual Resource Library CD-ROM

This helpful CD-ROM contains approximately 800 images, including all of the photos and line art from the text, that can be easily imported into PowerPoint to create multimedia presentations. Or, you may use the already prepared PowerPoint presentations.

Life Science Animations CD-ROM 2.0

This two CD-ROM set contains more than 200 animations of important biological concepts and processes. These animations can be imported into your PowerPoint presentations.

Life Science Animations Videotape Series

Animations of key biological processes are available on seven videotapes. The animations bring visual movement to biological processes that are difficult to understand on the text page.

Classroom Testing Software (MicroTest III)

This helpful testing software provides well-written and researched book-specific questions featured in the Test Item File.

Microbes in Motion CD-ROM, Version 2.0, by Delisle and Tomalty

This interactive CD-ROM allows students to actively explore microbial structure and function. Great for self-study, preparation for class or exams, or for classroom presentations.

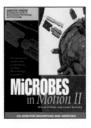

The Dynamic Human CD-ROM, Version 2.0

This guide to anatomy and physiology interactively illustrates the complex relationships between anatomical structures and their functions in the human body. Realistic, three-dimensional visuals are the premier feature of this exciting learning tool.

Dynamic Human Videodisc

Enhance your classroom presentations with movement, sound, and motion of internal organs, cells, and systems. More than 80 premier 3-D animations covering all body systems from the outstanding *Dynamic Human CD-ROM* are included.

Other Available Supplements

Instructor's Manual

The *Instructor's Manual* is designed to assist instructors as they plan and prepare for classes using **Human Biology**. The *Instructor's Manual* contains both an extended lecture outline and lecture enrichment ideas, which together review in detail the contents of the text chapter. The technology section lists relevant assets from McGraw-Hill.

Student Study Guide

To ensure close coordination with the text, Dr. Sylvia S. Mader has written the *Student Study Guide* that accompanies the text. Each text chapter has a corresponding study guide chapter that includes a listing of objectives, study questions, and a chapter test. Answers to the study questions and the chapter tests are provided to give students immediate feedback.

The concepts in the study guide are the same as those in the text, and the questions in the study guide are sequenced according to these concepts. Instructors who make their choice of concepts known to the students can thereby direct student learning in an efficient manner. Students who make use of the *Student Study Guide* should find that performance increases dramatically.

Laboratory Manual

Dr. Mader has also written the *Laboratory Manual* to accompany **Human Biology**. With few exceptions, each chapter in the text has an accompanying laboratory exercise in the manual (some chapters have more than one accompanying exercise). In this way, instructors are better able to emphasize particular portions of the curriculum. The 19 laboratory sessions in the manual are designed to further help students appreciate the scientific method and to learn the fundamental concepts of biology and the specific content of each chapter. All exercises have been tested for student interest, preparation time, and feasibility. This lab manual can be customized to fit your lab course. Contact your McGraw-Hill representative for details.

Transparencies

This set of transparency acetates to accompany the text has been expanded to 400 full-color acetates, including all of the art from the textbook.

Micrograph Slides

This ancillary provides a boxed set of 100 color slides of photomicrographs and electron micrographs from the text.

HealthQuest CD-ROM
ISBN 0-697-29723-3 (Windows)
ISBN 0-07-039335-4 (Macintosh)

Virtual Biology Laboratory CD-ROM
ISBN 0-697-37991-4

Life Science Living Lexicon CD-ROM
ISBN 0-697-37993-0

How to Study Science, Third Edition, by Fred Drewes
ISBN 0-697-36051-2

Basic Chemistry for Biology, Second Edition, by Carolyn Chapman
ISBN 0-697-36087-3

Schaum's Outlines: Biology
ISBN 0-07-022405-6

Understanding Evolution, by Volpe and Rosenbaum
ISBN 0-697-05137-4

AIDS Booklet, by Frank Cox
ISBN 0-697-29428-5

The Internet Primer
ISBN 0-07-303203-4

Critical Thinking Case Study Workbook, by Robert Allen
ISBN 0-697-14556-5

How Scientists Think, by George Johnson
ISBN 0-697-27875-1

To order any of these study tools, contact your bookstore manager or call McGraw-Hill Customer Service at 800-338-3987.

BioCourse.com

The number one source for your biology course.

BioCourse.com is an electronic meeting place for students and instructors. It provides a comprehensive set of resources in one place that is up-to-date and easy to navigate. You can access **BioCourse.com** from *Human Biology's* Online Learning Center.

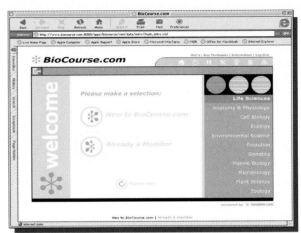

Here is what you will find at BioCourse.com:

Faculty Club is an array of information and links to related sites for instructors. Resources that you will find include:
- Teaching tips and basic information on pedagogy, assessment, etc.
- Suggestions for classroom and lecture activities.
- Reference searches and literature for faculty.
- Presentation tools.
- Test bank.
- Help for new instructors and teaching assistants.
- Information on available jobs, grant writing, and available funding.
- Case studies.

Student Center contains a wide range of materials to help biology students improve their study skills and achieve success in college and beyond. Examples of materials that will be available:
- Study aids.
- Résumé writing and information on jobs and internships.
- Graduate school options.
- Information for MCAT and other tests.
- Links to content websites by topic.

Briefing Room offers instructors and students up-to-date news articles, a selection of background readings and links to journal search tools and biology magazines. Users can e-mail articles to others, link to search engines, and read primary sources online.

BioLabs features materials for lab students and instructors. Some tools you will find include:
- For students:
  - Dissection techniques.
  - Equipment tutorials.
  - Safety and setup procedures.
- For instructors:
  - Lab preparations.
  - Lab support.
  - Simulations.

The Quad is a powerful indexing tool and hierarchical outline of content resources for searching by students and faculty. Users can search by topic through a "content warehouse" featuring text material, activities, visuals, and animations to learn more about a selected topic.

R & D Center features our newest simulations, animations, and other teaching and learning tools. This portion of our site will allow faculty members and students to try out our materials as they are being developed.

Contact your McGraw-Hill sales representative for more information or visit *w w w.mhhe.com*.

PageOut

Proven. Reliable. Class-tested.

Tens of thousands of professors have chosen PageOut to create course websites. And for good reason: PageOut offers powerful features, yet is incredibly easy to use.

Now you can be the first to use an even better version of PageOut. Through class-testing and customer feedback, we have made key improvements to the grade book, as well as the quizzing and discussion areas.

Customize the site to coincide with your lectures.

Complete the PageOut templates with your course information and you will have an interactive syllabus online. This feature lets you post content to coincide with your lectures. When students visit your PageOut website, your syllabus will direct them to components of McGraw-Hill web content germane to your text or specific material of your own.

New Features:

- Specific question selection for quizzes.
- Ability to copy your course and share it with colleagues or use as a foundation for a new semester.
- Enhanced grade book with reporting features.
- Ability to use the PageOut discussion area, or add your own third-party discussion tool.
- Password-protected courses.

Short on time? Let us do the work.

Send your course materials to our McGraw-Hill service team. They will call you for a 30-minute consultation. A team member will then create your PageOut website and provide training to get you up and running. Contact your McGraw-Hill Representative for details.

Contact your McGraw-Hill sales representative for more information or visit *www.mhhe.com*.

Visual Resource Library CD-ROMs

These two CD-ROM products contain textual images and life science animations that instructors can use to enhance their lectures. View, sort, search, and print catalog images, play chapter-specific slideshows using PowerPoint, or create customized presentations when you:

- Find and sort thumbnail image records by name, type, location, and user-defined keywords.
- Search using keywords or terms.
- View multiple images at the same time with the Small Gallery View.
- Select and view images at full size.
- Display all the important file information for easy file identification.
- Drag and place or copy and paste into virtually any graphics, desktop publishing, presentation, or multimedia application.

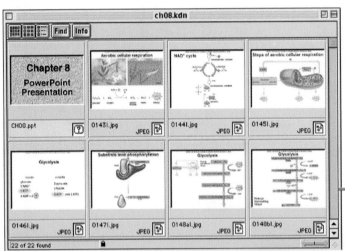

Visual Resource Library CD-ROM
(also available on the Online Learning Center)

This helpful CD-ROM contains more than 800 photographs and illustrations from the text. You'll be able to create interesting multimedia presentations with the use of these images, and students will have the ability to easily access the same images in their texts to later review the content covered in class.

Life Science Animations Visual Resource Library CD-ROM, 2.0

This instructor's tool, containing more than 300 animations of important biological concepts and processes—found in the *Essential Study Partner* and *Dynamic Human CD-ROMs*—is perfect to support your lecture. The animations contained in this library are not limited to subjects covered in the text, but include a variety of general life science topics.

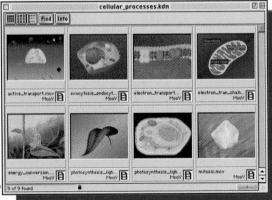

Contact your McGraw-Hill sales representative for more information or visit *www.mhhe.com*.

Chapter *1*

A Human Perspective

Chapter Concepts

Figure 1.1   Living things are alike.
The bird and the boy have many characteristics in common. Their cells, tissues, and organ systems function similarly.

Young Billy Hanson was quite excited. Today, he would let loose the black-capped chickadee he had cared for after it fell from the nest several weeks before. He had kept it warm and well fed, and now it was ready to take flight. He knew from hours of watching birds that black-capped chickadees commonly flit from one tree to another, even though they can fly long distances. Billy was hopeful this one would stay close to home, especially since he had a bird-feeder. He yearned for what he thought of as their "friendship" to continue.

We know that Billy, and yes, the bird and the trees surrounding Billy's home, are alive. By what criteria do we make this judgment? All living things respond to stimuli as they grow and reproduce. They take materials and energy from the environment to keep on living. And all living things are adapted to their way of life—the bird flies, Billy walks. Even so, they both have the same organ systems. Both have a heart, a liver, intestines, and so forth. There is a unity of life that goes beyond its diversity. This chapter discusses the characteristics of human beings—how they are like other living things and how they are different from other living things.

Biology is the scientific study of life. When biologists do their work, they attempt to answer specific questions. How do genes control characteristics in both humans and birds? What impact do humans have on their environment and the environment of all living things? Biologists use the scientific method, which is explained more fully in this chapter, to come to conclusions that they share with others. This text shares their knowledge with you.

The text as a whole has two main goals. The first goal is to explore human anatomy and physiology so you will know how the body functions. The second goal is to look at human evolution and ecology so you will better understand the place of humans in nature. Both the human body and the environment are self-regulating systems that can be thrown out of kilter by misuse and mismanagement. An appreciation of the delicate balance present in both systems provides the perspective from which future decisions can be made. It is hoped that adequate information will better enable you to keep your body and the environment healthy.

1.1 Biologically Speaking

You are about to launch on a study of human biology. Before you begin, it is appropriate to define who humans are and how they fit into the world of living things.

Who Are We?

Certain characteristics tell us who human beings are biologically speaking.

Human beings are highly organized. A **cell** is the basic unit of life, and human beings are multicellular since they are composed of many types of cells. Like cells form tissues, and **tissues** make up organs. Each type of **organ** is a part of an **organ system.** The different systems perform the specific functions listed in Table 1.1. Together, the organ systems maintain **homeostasis,** an internal environment for cells that varies only within certain limits. The text emphasizes how all the systems of the human body help maintain homeostasis. The digestive system takes in nutrients, and the circulatory system distributes these to the cells. The waste products are excreted by the urinary system. The work of the nervous and endocrine systems is critical because they coordinate the functions of the other systems.

Table 1.1   Human Organ Systems

System	Function
Digestive	Converts food particles to nutrient molecules
Cardiovascular	Transports nutrients to and wastes from cells
Lymphatic and immune	Defends against disease
Respiratory	Exchanges gases with the environment
Urinary	Eliminates metabolic wastes
Nervous	Regulates systems and internal environment
Musculoskeletal	Supports and moves organism
Endocrine	Regulates systems and internal environment
Reproductive	Produces offspring

Human beings reproduce and grow. Reproduction and growth are fundamental characteristics of all living things. Just as cells come only from preexisting cells, so living things have parents. When living things **reproduce,** they create a copy of themselves and assure the continuance of the species. (A species is a type of living thing.) Human reproduction requires that a sperm contributed by the male fertilize an egg contributed by the female. Growth occurs as the resulting cell develops into the newborn. Development includes all the changes that occur from the time the egg is fertilized until death and, therefore, all the changes that occur during childhood, adolescence, and adulthood.

Human beings have a cultural heritage. We are born without knowledge of civilized ways of behavior, and we gradually acquire these by adult instruction and imitation of role models. It is our cultural inheritance that makes us think we are separate from nature. But actually we are a product of **evolution,** a process of change that has resulted in the diversity of life, and we are a part of the **biosphere,** a network of life that spans the surface of the earth.

> Like other living things, humans are composed of cells, and when they reproduce, growth and development occur. Unlike other living things, humans have a cultural heritage.

How Do We Fit In?

Certain characteristics tell us how human beings fit into the world of living things.

Human beings are a product of an evolutionary process. Life has a history that began with the evolution of the first cell(s) about 3.5 billion years ago. It is possible to trace human ancestry from the first cell through a series of prehistoric ancestors until the evolution of modern-day humans. The presence of the same types of chemicals tells us that human beings are related to all other living things. DNA is the genetic material, and ATP is the energy currency in all cells, including human cells. It is even possible to do research with bacteria and apply the results to human beings.

The classification of living things mirrors their evolutionary relationships. It is common practice to classify living things into five major groups called **kingdoms** (Fig. 1.2). Humans are vertebrates in the animal kingdom. **Vertebrates** have a nerve cord that is protected by a vertebral column whose repeating units (the vertebrae) indicate that we and other vertebrates are segmented animals. Among the vertebrates, we are most closely related to the apes, specifically the chimpanzee, from whom we are distinguished by our highly developed brains, completely upright stance, and the power of creative language.

Human beings are a part of the biosphere. All living things are a part of the biosphere, where they live in the air,

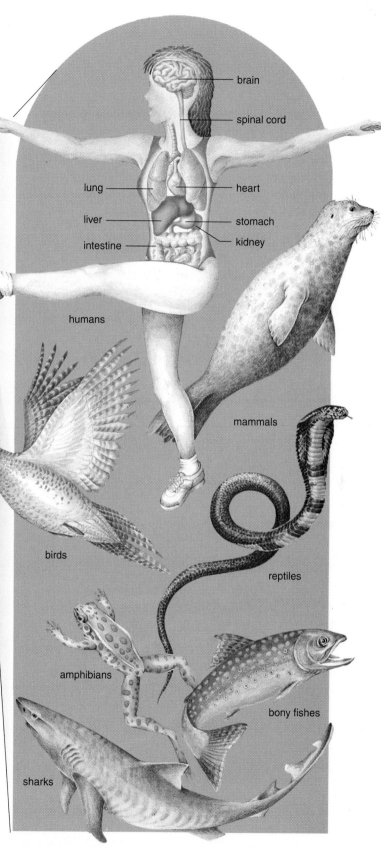

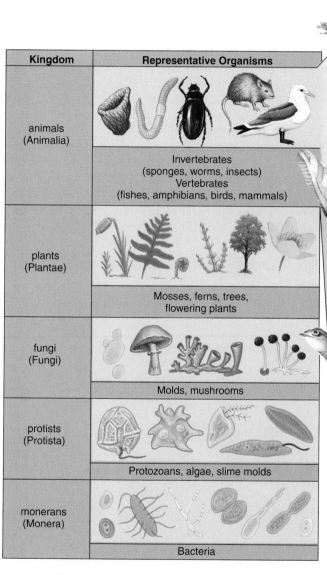

Kingdom	Representative Organisms
animals (Animalia)	Invertebrates (sponges, worms, insects) Vertebrates (fishes, amphibians, birds, mammals)
plants (Plantae)	Mosses, ferns, trees, flowering plants
fungi (Fungi)	Molds, mushrooms
protists (Protista)	Protozoans, algae, slime molds
monerans (Monera)	Bacteria

a. Survey of living things

Figure 1.2   **Classification and evolution of humans.**
a. Living things are classified into five kingdoms, and humans are in the animal kingdom. **b.** Human beings are most closely related to the other vertebrates shown. The evolutionary tree of life has many branches; the vertebrate line of descent is just one of many.

b. Survey of vertebrates

in the sea, and on land. In any portion of the biosphere, such as a forest or a pond, organisms of one type belong to a **population.** The various populations interact with one another and with the physical environment to form an **ecosystem** in which nutrients cycle and energy flows. In Figure 1.3, nutrient cycling is represented by blue arrows, and energy flow is represented by yellow to red arrows. Plants use the energy of the sun and inorganic nutrients to produce organic nutrients (food) for themselves and all living things. Animals that feed on plants may be food for other animals. A major part of the interactions between populations pertains to who eats whom. Eventually decomposers make inorganic nutrients, such as carbon dioxide and nitrates, available to plants once more.

When populations feed on one another, some of the organic nutrients are broken down to acquire energy, and eventually this energy dissipates as heat. Thus, energy flows and does not cycle in ecosystems. Notice that because energy flows through an ecosystem, it is dependent on a constant supply of solar energy. However, if an ecosystem is large enough, it needs no raw materials from the outside. A big ecosystem just keeps cycling the same inorganic nutrients.

Although ecosystems constantly undergo changes, ordinarily they still endure. Certain species may die out and new ones may come, but the ecosystem remains recognizable year after year. Ecosystems can endure and even flourish when disturbances are minor; major disturbances however, can cause irreparable damage.

Humans Threaten the Biosphere

Human populations tend to modify existing ecosystems for their own purposes. For example, humans clear forests or grasslands in order to grow crops; later, they build houses on what was once farmland; and, finally, they convert small towns into cities. Human populations ever increase in size and require greater amounts of material goods and energy input each year (Fig. 1.4). With each step, fewer and fewer original organisms remain, until at last ecosystems are completely altered. If this continues, only humans and their domesticated plants and animals will largely exist where once there were many diverse populations.

More and more ecosystems are threatened as the human population increases in size. As discussed in the Ecology Focus on pages 6 and 7, presently there is great concern among scientists and laypersons about the destruction of the

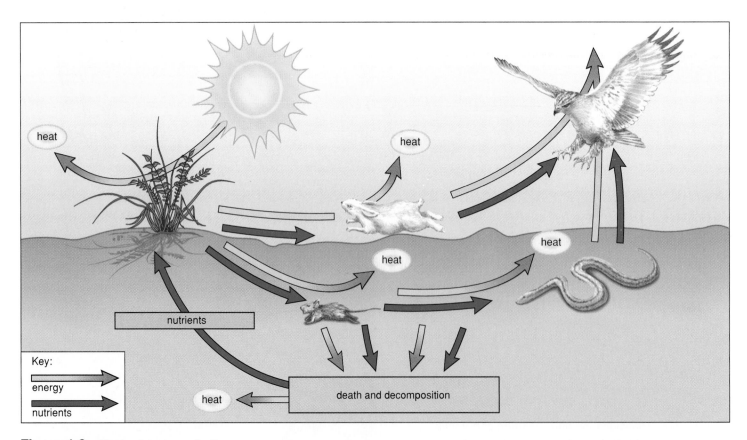

Figure 1.3   Ecosystem organization.
Within an ecosystem, plants use solar energy and inorganic nutrients to produce organic nutrients (food). Organic nutrients cycle through all the populations until decomposers make inorganic nutrients available to plants once more. Energy flows from the sun through all populations. Eventually the energy of the sun is converted to heat, which dissipates.

world's rain forests due to logging and the large numbers of persons who are starting to live and to farm there. We are beginning to realize how dependent we are on intact ecosystems and the services they perform for us. For example, the tropical rain forests act like a giant sponge, which absorbs carbon dioxide, a pollutant that pours into the atmosphere from the burning of fossil fuels like oil and coal. An increased amount of carbon dioxide in the atmosphere is expected to have many adverse effects, such as an increase in the average daily temperature.

An ever-increasing human population size is a threat to the continued existence of our species when it means that the dynamic balance of the biosphere is upset. The recognition that the workings of the biosphere need to be preserved is one of the most important developments of our new ecological awareness.

Biodiversity: Going, Going, Gone When humans modify existing ecosystems, they reduce biodiversity. **Biodiversity** is the total number of species, the variability of their genes, and the ecosystems in which they live. The present biodiversity of our planet has been estimated to be as high as 15 million species, and so far, under 2 million have been identified and named. Extinction, the death of a species, occurs when a species is unable to adapt to a change in environmental conditions. It's estimated that presently we are losing as many as 400 species a day due to human activities. For example, the existence of the species featured in the Ecology Focus on pages 6 and 7 is threatened because tropical rain forests are being reduced in size. As another example, because of seaside development, pollution, and overfishing, 14 of the most valuable finfishes are becoming commercially extinct, meaning that too few remain to justify the cost of catching them.

Most biologists are alarmed over the present rate of extinction and believe the rate may eventually rival that of the five mass extinctions that have occurred during our planet's history. The dinosaurs became extinct during the last mass extinction, 65 million years ago. Everyone needs to realize that humans are totally dependent on other species for food, clothing, medicines, and various raw materials. Therefore, it is very shortsighted of us to allow other species to become extinct. Ecosystems and the species living in them should be preserved because only then can the human species continue to exist. And it takes from 2,000 to 10,000 generations for new species to evolve and to replace the ones that have died out. Because we are dependent upon the normal function and the present biodiversity of the biosphere, the existing species should be preserved.

Humans belong to the world of living things and are vertebrates. They have modified existing ecosystems to the point that they must now be seriously concerned about the continued existence of the biosphere.

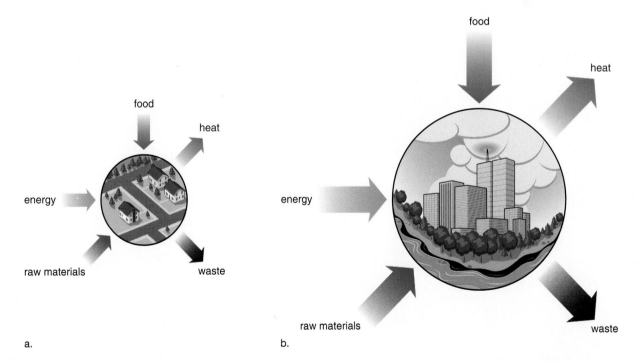

Figure 1.4 **Humans threaten the biosphere.**
As cities grow in size, more materials, fuel, and food are taken from the environment, and more heat and waste are returned to the environment. **a.** Small town. **b.** Large city.

Ecology Focus

Tropical Rain Forests: Can We Live Without Them?

Figure 1A **Tropical rain forest inhabitants.**
These animals and plants make their homes in the Amazon basin.

The tropics are home for 66% of the plant species, 90% of the nonhuman primates, 40% of the birds of prey, and 90% of the insects that have been identified thus far. Of the 15 million species estimated to exist many have not yet been discovered. These are also believed to live in the tropical rain forests that occur in a green belt spanning the planet on both sides of the equator.

If tropical rain forests are preserved, the rich diversity of plants and animals will continue to exist for scientific and pharmacological study (Fig. 1A). One-fourth of the medicines we currently use come from tropical rain forests. For example, the rosy periwinkle from Madagascar has produced two potent drugs for use against Hodgkin disease, leukemia, and other blood cancers. It is hoped that many of the still-unknown plants will provide medicines for other human ills.

Tropical forests used to cover 6–7% of the total land surface of the earth—an area roughly equivalent to our contiguous 48 states. Every year humans destroy an area of forest equivalent to the size of Oklahoma (Fig. 1B). At this rate, these forests and the species they contain will disappear completely in just a few more decades. Even if the forest areas now legally protected survive, 58–72% of all tropical forest species would still be lost.

The loss of tropical rain forests results from an interplay of social, economic, and political pressures. Many people already live in the forest, and as their numbers increase, more of the land is cleared for farming. People move to the forests because internationally financed projects build roads and open up the forests for exploitation. Small-scale farming accounts for about 60% of tropical deforestation, and decreasing percentages are due to commercial logging, cattle ranching, and mining. International demand for timber promotes destructive logging of rain forests in Southeast Asia and South America. The market for low-grade beef encourages their conversion to pastures for cattle. The lure of gold draws miners to rain forests in Costa Rica and Brazil.

The destruction of tropical rain forests gives only short-term benefits but is expected to cause long-term problems. The forests act like a giant sponge, soaking up rainfall during the wet season and releasing it during the dry season. Without them, a regional yearly regime of flooding followed by drought is expected to destroy property and reduce agricultural harvests. Worldwide, there could be changes in climate that would affect the entire human race.

However, some studies show that if the forests were used as a sustainable source of nonwood products, such as nuts, fruits, and latex rubber, they would generate as much or more revenue while continuing to perform their various ecological functions, and biodiversity could still be preserved. Brazil is exploring the concept of "extractive reserves," in which plant and animal products are harvested, but the forest itself is not cleared. Ecologists have also proposed "forest farming" systems, which mimic the natural forest as much as possible while providing abundant yields. But for such plans to work maximally, the human population size and the resource consumption per person must be stabilized.

Preserving tropical rain forests is a wise investment. Such action promotes the survival of most of the world's species—indeed, the human species, too.

Figure 1B Burning of trees in a tropical rain forest.
It is estimated that an area the size of Oklahoma is being lost each year. While trees ordinarily take up carbon dioxide, burning releases carbon dioxide to the atmosphere.

1.2 The Process of Science

Science helps human beings understand the natural world. Science aims to be objective rather than subjective even though it is very difficult to make objective observations and to come to objective conclusions because we are often influenced by our own particular prejudices. Still, scientists strive for objective observations and conclusions. We also keep in mind that scientific conclusions are subject to change whenever new findings so dictate. Quite often in science, new studies, which might utilize new techniques and equipment, tell us when previous conclusions need to be modified or changed entirely.

Scientific Theories in Biology

The ultimate goal of science is to understand the natural world in terms of **scientific theories,** concepts based on the conclusions of observations and experiments. In a movie, a detective might claim to have a theory about the crime, or you might say that you have a theory about the win-loss record of your favorite baseball team, but in science, the word theory is reserved for a conceptual scheme supported by a large number of observations and not yet found lacking. Some of the basic theories of biology are as follows:

Name of Theory	Explanation
Cell	All organisms are composed of cells.
Biogenesis	Life comes only from life.
Evolution	All living things have a common ancestor, but each is adapted to a particular way of life.
Gene	Organisms contain coded information that dictates their form, function, and behavior.

Evolution is the unifying concept of biology because it pertains to various aspects of living things. For example, the theory of evolution enables scientists to understand the history of life, the variety of living things, and the anatomy, physiology, and development of organisms—even their behavior. Because the theory of evolution has been supported by so many observations and experiments for over a hundred years, some biologists refer to the **principle** of evolution. They believe this is the appropriate terminology for theories that are generally accepted as valid by an overwhelming number of scientists.

The Scientific Method Has Steps

Scientists, including biologists, employ an approach to gathering information that is known as the **scientific method.** The approach of individual scientists to their

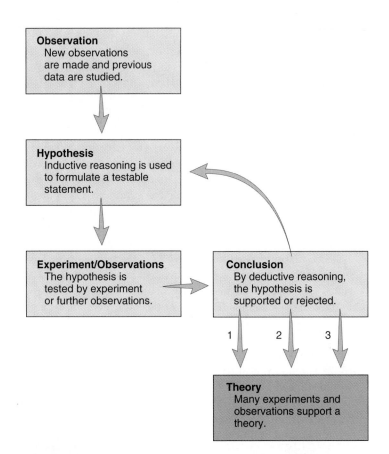

Figure 1.5   Flow diagram for the scientific method.
On the basis of observations and previous data, a scientist formulates a hypothesis. The hypothesis is tested by further observations or a controlled experiment, and new data either support or falsify the hypothesis. The return arrow indicates that a scientist often chooses to retest the same hypothesis or to test a related hypothesis. Conclusions from many different but related experiments may lead to the development of a scientific theory. For example, studies in biology of development, anatomy, and fossil remains all support the theory of evolution.

work is as varied as they themselves; still, for the sake of discussion, it is possible to speak of the scientific method as consisting of certain steps (Fig. 1.5). After making initial observations, a scientist will most likely study any previous **data,** which are facts pertinent to the matter at hand. Imagination and creative thinking also help a scientist formulate a **hypothesis** that becomes the basis for more observation and/or experimentation. The new data help a scientist come to a **conclusion** that either supports or does not support the hypothesis. Because hypotheses are always subject to modification, they can never be proven true; however, they can be proven untrue—that is, hypotheses are falsifiable. When the hypothesis is not supported by the data, it must be rejected; therefore, some think of the body of science as what is left after alternative hypotheses have been rejected.

The Discovery of Lyme Disease

In order to examine the scientific method in more detail, we will relate how scientists discovered the cause of Lyme disease, a debilitating illness that affects the whole body.

Observation

When Allen C. Steere began his work on Lyme disease in 1975, a number of adults and children in the city of Lyme, Connecticut, had been diagnosed as having rheumatoid arthritis. Steere knew that children rarely get rheumatoid arthritis, so this made him suspicious and he began to make observations. He found that (1) most victims lived in heavily wooded areas, (2) the disease was not contagious—that is, whole groups of people did not come down with Lyme disease, (3) symptoms first appeared in the summer, and (4) several victims remembered a strange bull's-eye rash occurring several weeks before the onset of symptoms.

Hypothesis

Inductive reasoning occurs when you generalize from assorted facts. Steere used inductive reasoning; that is, he put the pieces together to formulate the hypothesis that Lyme disease was caused by a pathogen most likely transmitted by the bite of an insect or a tick (Fig. 1.6).

Deductive reasoning helps scientists decide what further observations and experimentations they will make to test the hypothesis. Deductive reasoning utilizes an "if . . . then" statement: If Lyme disease is caused by the bite of a tick, then it should be possible to show that a tick carries the pathogen and that the pathogen is in the blood of those who have the disease. However, when Steere tested the blood of Lyme disease victims for the presence of infectious microbes, not a single test was positive. Finally, in 1977, one victim saved the tick that bit him, and it was identified as *Ixodes dammini*, the deer tick. Then Willy Burgdorfer, an authority on tick-borne diseases, was able to isolate a spirochete (spiral bacterium) from deer ticks, and he also found this microbe in the blood of Lyme disease victims. The new spirochete was named *Borrelia burgdorferi*, after Burgdorfer.

Conclusion

The new data collected when Burgdorfer applied deductive reasoning supported the hypothesis and allowed scientists to conclude that Lyme disease is caused by the bacterium *Borrelia burgdorferi* transmitted by the bite of the deer tick.

Even though the scientific method is quite variable, it is possible to point out certain steps that characterize it: making observations, formulating a hypothesis, testing it, and coming to a conclusion.

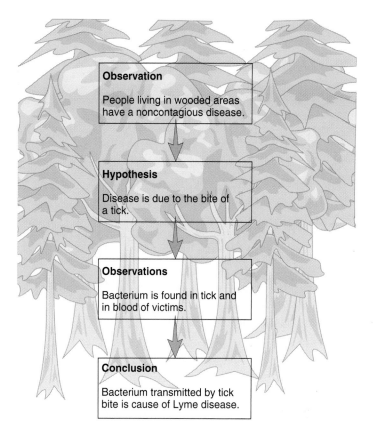

Figure 1.6 **Flow diagram for Lyme disease study.**

Reporting the Findings

It is customary to report findings in a scientific journal so that the design and the results of the experiment are available to all. For example, data about tick-borne diseases are often reported in the journal *Clinical Microbiology Review*. It is necessary to give other researchers details on how experiments were conducted because results must be repeatable; that is, other scientists using the same procedures must get the same results. Otherwise, the hypothesis is no longer supported.

Often authors of a report suggest what other types of experiments might clarify or broaden the understanding of the matter under study. People reading the report may think of other experiments to do, also. In our example, the bull's-eye rash was later found to be due to the Lyme disease spirochete.

Observations and the results of experiments are published in a journal, where they can be examined. These results are expected to be repeatable; that is, they will be obtained by anyone following the same procedure.

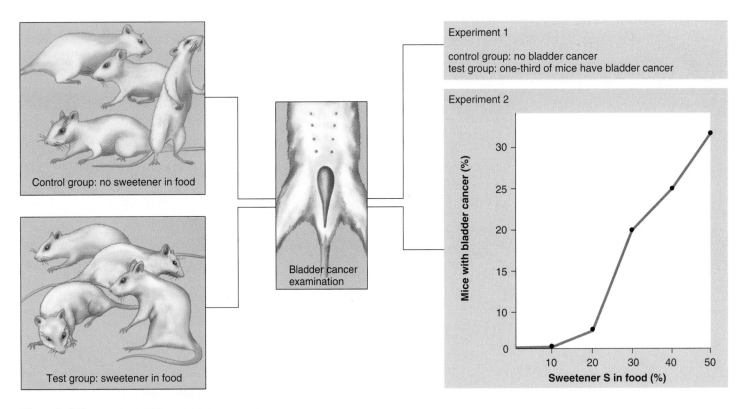

Figure 1.7   **Design of a controlled experiment.**
Genetically similar mice are randomly divided into a control group and one or more test groups that contain 100 mice each. All groups are exposed to the same conditions, such as cage setup, temperature, and water supply. The control group is not subjected to sweetener S in the food. At the end of the experiment, all mice are examined for bladder cancer. The results of experiment 1 and experiment 2 are shown on the far right.

Scientists Use Controlled Experiments

When scientists are studying a phenomenon, they often perform **experiments** in a laboratory where extraneous variables can be eliminated. A **variable** is a factor that can cause an observable change during the progress of an experiment. Experiments are considered more rigorous when they include a control group. A **control group** goes through all the steps of an experiment but lacks the factor or is not exposed to the factor being tested.

Designing the Experiment

Suppose, for example, physiologists want to determine if sweetener S is a safe food additive. On the basis of available information, they formulate a hypothesis that sweetener S is a safe food additive even when it composes up to 50% of dietary intake. Next, they design the experiment described in Figure 1.7 to test the hypothesis.

Test group: 50% of diet is sweetener S

Control group: diet contains no sweetener S

The researchers first place a certain number of randomly chosen inbred (genetically identical) mice into the various groups—say, 100 mice per group. If any of the

mice are different from the others, it is hoped random selection has distributed them evenly among the groups. The researchers also make sure that all conditions, such as availability of water, cage setup, and temperature of the surroundings, are the same for both groups. The food for each group is exactly the same except for the amount of sweetener S.

At the end of Experiment 1 in Figure 1.7, both groups of mice are to be examined for bladder cancer. Let's suppose that one-third of the mice in the test group are found to have bladder cancer, while none in the control group have bladder cancer. The results of this experiment do not support the hypothesis that sweetener S is a safe food additive when it composes up to 50% of dietary intake.

Continuing the Experiment

Science is ongoing, and one experiment frequently leads to another. Physiologists might now wish to hypothesize that sweetener S is safe if the diet contains a limited amount of sweetener S. They feed sweetener S to groups of mice at ever-greater concentrations:

Group 1: diet contains no sweetener S (the control)

Group 2: 5% of diet is sweetener S

Group 3: 10% of diet is sweetener S
↓
Group 11: 50% of diet is sweetener S

Usually, data obtained from experiments such as this are presented in the form of a table or a graph (see Experiment 2, Fig. 1.7). Researchers might run a statistical test to determine if the difference in the number of cases of bladder cancer among the various groups is significant. After all, if a significant number of mice in the control group develop cancer, the results are invalid. Scientists prefer mathematical data because such information lends itself to objectivity.

On the basis of the data, the experimenters try to develop a recommendation concerning the safety of sweetener S in the food of humans. They might caution, for example, that the intake of sweetener S beyond 10% of the diet is associated with too great a risk of bladder cancer.

Scientists often do controlled experiments in the laboratory. The use of a control sample gives assurance that the results of the experiment are due to the variable being tested.

We have seen that scientists often use the scientific method to study the natural world. A particular observation backed up by data collected previously helps them formulate a hypothesis that is then tested. Testing consists of carrying out an experiment or simply making further observations. Particularly, if the experiment is performed in the laboratory, it should contain a control sample. The control sample goes through all the steps of the experiment but lacks the factor or is not exposed to the factor being tested. In this way scientists know that their results are not due to a chance event that has nothing to do with the variable being tested. Finally, scientists come to a conclusion that either supports or rejects the hypothesis. Scientists report their findings in journals that are read by other scientists who also make similar observations or carry on the same experiment. If experiments and observations are not repeatable, the hypothesis is subject to rejection. If use of the scientific method results in conclusions that repeatedly support the same hypothesis, a theory may result.

As time goes by, it is possible that a hypothesis/theory previously accepted by the scientific community will be modified in the light of new investigations. Still, there are certain theories, such as the theory of evolution, that have stood the test of time and are generally accepted as valid.

Scientists ask questions and carry on investigations that pertain to the natural world. The conclusions of these investigations are tentative and subject to change. Eventually, it may be possible to arrive at a theory that is generally accepted by all.

1.3 Science and Social Responsibility

Science is objective and not subjective. Scientists assume that each person is capable of collecting data and seeing natural events in the same objective way. They also assume the same theories and principles are applicable to past, present, and future events. Therefore, science seeks a natural cause for the origin and history of life. Doctrines of creation that have a mythical, philosophical, or theological basis are not a part of science because they are not subject to objective observations and experimentation by all. Many cultures have their own particular set of supernatural beliefs, and various religions within a culture differ as to the application of these beliefs. Such approaches to understanding the world are not within the province of science. Similarly, scientific creationism, which states that God created all species as they are today, cannot be considered science because explanations based on supernatural rather than natural causes involve faith rather than data from experiments.

There are many ways in which science has improved our lives. The discovery of antibiotics, such as penicillin, and of the polio, measles, and mumps vaccines has increased our life span by decades. Cell biology research is helping us understand the causes of cancer. Genetic research has produced new strains of agricultural plants that have eased the burden of feeding our burgeoning world population.

Science also has effects we may find disturbing. For example, it sometimes fosters technologies that can be ecologically disastrous if not controlled properly. Too often we blame science for these developments and think that scientists are duty bound to pursue only those avenues of research that are consistent with our present system of values. But making value judgments is not a part of science. Ethical and moral decisions must be made by all people. The responsibility for how we use the fruits of science, including a given technology, must rest with people from all walks of life, not with scientists alone.

Scientists should provide the public with as much information as possible when such issues as the use of atomic energy, fetal research, and genetic engineering are being debated. Then they, along with other citizens, can help make decisions about the future role of these technologies in our society. This text, while covering all aspects of biology, focuses on human biology. It is hoped that your study of biology will enable you to make wise decisions regarding your own individual well-being and the well-being of all species, including our own.

It is the task of all persons to use scientific information as they make value judgments about their own lives and about the environment.

Bioethical Focus

Animals in the Laboratory

Some people believe that animals should be protected in every way, and should not be used in laboratory research. In our society as a whole, the trend is toward a growing recognition of what is generally referred to as animal rights. In 1985, 63% of Americans polled agreed that "scientists should be allowed to do research that causes pain and injury to animals like dogs and chimpanzees if it produces new information about human health problems." That consensus dropped to 53% in 1995. Psychologists with Ph.D.s earned in the 1990s are half as likely to express strong support for animal research as those who earned their Ph.D.s before 1970.

Those who approve of laboratory research involving animals give examples to show that even today it would be difficult to develop new vaccines and medicines against infectious diseases, new surgical techniques for saving human lives, or new treatments for spinal cord injuries without the use of animals. Even so, most scientists today are in favor of what is now called the "three Rs": replacement of animals by in vitro, or test-tube, methods whenever possible; reduction of the numbers of animals used in experiments; and refinement of experiments to cause less suffering to animals. In the Netherlands, all scientists starting research that involves animals are well trained in the three Rs. After designing an experiment that uses animals, they are asked to find ways to answer the same questions without using animals.

F. Barbara Orlans of the Kennedy Institute of Ethics at Georgetown University says, "It is possible to be both pro research and pro reform." She feels that animal activists need to accept that sometimes animal research is beneficial to humans, and all scientists need to consider the ethical dilemmas that arise when animals are used for laboratory research.

Decide Your Opinion

To develop your opinion, answer either question 1 or question 2, and then question 3.

1. Are you opposed to the use of animals in laboratory experiments? Always or under certain circumstances? Explain.
2. Do you favor using animals in the laboratory? Always or under certain circumstances? Explain.
3. Do you feel that it would be possible for animal activists and scientists to find a compromise they could both accept? Discuss.

Figure 1C   Caged animals.
These animals are used in laboratory research to help develop an AIDs vaccine for humans.

Looking at Both Sides   www.mhhe.com/biosci/genbio/maderhuman7/

Every bioethical issue has at least two sides. Even if you already have an opinion, it is important to explore the opposite opinion before finalizing your position. The Online Learning Center at www.mhhe.com/biosci/genbio/maderhuman7/ will help you fine-tune your initial opinion, explore both sides, and finalize your position. You may acquire new arguments for your original opinion, or you may even change your opinion. Be sure to complete these activities in sequence:

Taking Sides   Decide your initial opinion by answering a series of questions. Then see if your opinion changes after completing the next two activities.

Further Debate   Read opposing articles that give you further information on this particular bioethical issue.

Explain Your Position   Answer another series of questions and then defend your original or changed opinion. You can e-mail your position to your instructor if he or she wishes.

Summarizing the Concepts

1.1 Biologically Speaking
Human beings, like other living things, are highly organized. Cells form tissues, which form organs that function in organ systems. Human beings come into existence through reproduction, growth, and development.

Unlike other living things, human beings have a cultural heritage that sometimes hinders the realization that they are the product of an evolutionary process. Human beings are vertebrates closely related to the apes. Human beings are also a part of the biosphere where populations interact with the physical environment and with one another.

Human activities threaten the existence of ecosystems like tropical rain forests. Biodiversity is now being reduced at a rapid rate.

1.2 The Process of Science
When studying the world of living things, biologists, like other scientists, use the scientific method, which consists of these steps: making an observation, formulating a hypothesis, carrying out an experiment or simply making further observations, and coming to a conclusion.

1.3 Science and Social Responsibility
It is the responsibility of all to make ethical and moral decisions about how best to make use of the results of scientific investigations.

Studying the Concepts

1. Name five characteristics of human beings, and discuss each one. 2–4
2. What is homeostasis, and how is it maintained? Choose one organ system and tell how it helps maintain homeostasis. 2
3. Give evidence that human beings are related to all other living things. 2
4. Describe the five-kingdom system of classification, and name types of organisms in each kingdom. 3
5. Human beings are dependent upon what services performed by plants? 4
6. Discuss the importance of scientific theory, and name several theories that are basic to understanding biological principles. 8
7. Name the steps of the scientific method, and discuss each one. 8
8. How do you recognize a control group, and what is its purpose in an experiment? 10
9. What is our social responsibility in regard to scientific findings? 11

Understanding Key Terms

biodiversity 5	organ 2
biosphere 2	organ system 2
cell 2	population 4
conclusion 8	principle 8
control group 10	reproduce 2
data 8	science 8
ecosystem 4	scientific method 8
evolution 2	scientific theory 8
experiment 10	tissue 2
homeostasis 2	variable 10
hypothesis 8	vertebrate 2
kingdom 2	

Testing Your Knowledge of the Concepts

In questions 1–4, match the human characteristics to the descriptions below.

Human beings:
a. are organized.
b. reproduce and grow.
c. have a cultural heritage.
d. are the product of evolutionary process.
e. are a part of the biosphere.

_____ 1. Humans are related to all other living things.

_____ 2. The human population encroaches on natural habitats.

_____ 3. Like cells form tissues in the human body.

_____ 4. We learn how to behave from our elders.

In questions 5–7, indicate whether the statement is true (T) or false (F).

_____ 5. Once a scientist formulates a hypothesis, he or she tests it by observation and/or experimentation.

_____ 6. The theory of evolution is so poorly supported that many scientists feel it should be discarded.

_____ 7. When an experiment has a control group, it lends validity to the resulting data.

In questions 8–10, fill in the blanks.

8. To reproduce is to make a _____ of one's self.

9. _____ has a responsibility to decide how scientific knowledge should be used.

10. _____ is a concept consistent with conclusions based on a large number of experiments and observations.

1.1 Biologically Speaking

Life Characteristics *Essential Study Partner*
Human Organization *Essential Study Partner*

Levels of Biological Organization I *art labeling activity*
Levels of Biological Organization II *art labeling activity*

1.2 The Process of Science

Critical Thinking *Essential Study Partner*

Scientific Method *art quiz*

1.3 Science and Social Responsibility

Looking at Both Sides *critical thinking activity*

Cloning *case study*

Chapter Summary

Key Term Flashcards *vocabulary quiz*
Chapter Quiz *objective quiz*

I

Human Organization

The human body is composed of cells, the smallest units of life. An understanding of cell structure, physiology, and biochemistry serves as a foundation for understanding how the human body functions.

Principles of inorganic and organic chemistry are discussed before a study of cell structure is undertaken. The human cell is bounded by a membrane and contains organelles, which are also membranous. Membranes regulate entrance and exit of molecules and help cellular organelles carry out their functions.

The many cells of the body are specialized into tissues that are found within the organs of the various systems of the body. All body systems help maintain homeostasis, a dynamic equilibrium of the internal environment, so that proper physical conditions exist for each cell.

Chapter 2

Chemistry of Life

Chapter Concepts

2.1 Elements and Atoms
- All matter is composed of elements, each having one type of atom. 16

2.2 Molecules and Compounds
- Atoms react with one another, forming ions, molecules, and compounds. 19

2.3 Water and Living Things
- The existence of living things is dependent on the characteristics of water. 21
- The hydrogen ion concentration in water changes when acids or bases are added to water. 23

2.4 Molecules of Life
- Macromolecules are polymers that arise when their specific monomers (unit molecules) join together. 26
- The molecules found in cells are carbohydrates, lipids, proteins, and nucleic acids. 26

2.5 Carbohydrates
- Carbohydrates function as a ready source of energy in most organisms. 27
- Glucose is a simple sugar; starch, glycogen, and cellulose are polymers of glucose. 27
- Cellulose lends structural support to plant cell walls. 28

2.6 Lipids
- Lipids are varied molecules. 29
- Fats and oils, which function in long-term energy storage, are composed of glycerol and three fatty acids. 29
- Sex hormones are derived from cholesterol, a complex ring compound. 30

2.7 Proteins
- Proteins help form structures (e.g., muscles and membranes) and function as enzymes. 31
- Proteins are polymers of amino acids. 32

2.8 Nucleic Acids
- Nucleic acids are polymers of nucleotides. 35
- Genes are composed of DNA. RNA serves as an intermediary during protein synthesis. 35

Figure 2.1   Chemicals and the body.
The human body is affected by the chemicals we breathe and consume as nutrients. Even the sweeteners used in soft drinks can influence body metabolism, as people with a metabolic disorder, such as phenylketonuria, know all too well.

Glance at the back of any diet soda can, and you'll find an important warning: "Contains Phenylalanine." For most of us, this chemical—one of 20 naturally occurring amino acids—is harmless. That's because our bodies contain a liver enzyme that converts phenylalanine into a useful chemical for the body.

But people with PKU, or phenylketonuria, lack the necessary enzyme. In these people, phenylalanine builds up, and dangerous by-products flood the bloodstream. Brain damage or other problems follow. Thus, doctors keep PKU sufferers on a strict diet, limiting their intake of phenylalanine. Soda companies help by labeling products containing the chemical (Fig. 2.1).

PKU is just one of hundreds of disorders caused by malfunctioning or absent enzymes. From simple inorganic molecules to complex organic macromolecules, such as enzymes, chemicals are essential to our being. As people with PKU know, lacking just one of these precious chemicals can be disastrous.

This chapter reviews the structure of atoms and how they join to form both inorganic and organic chemicals. Inorganic chemicals like salts have significant functions in the human body just as organic chemicals like proteins do. All chemicals, whether inorganic or organic, are composed of elements.

2.1 Elements and Atoms

Matter is anything that takes up space and has weight. All matter, both nonliving and living, is composed of certain basic substances called **elements.** Considering the variety of living and nonliving things in the world, it's quite remarkable that there are only 92 naturally occurring elements. It is even more surprising that over 90% of the human body is composed of just three elements: carbon, oxygen, and hydrogen.

Every element has a name and a symbol; for example, calcium has been assigned the atomic symbol Ca (Fig. 2.2a). Some of the symbols we use for elements are derived from Latin. For example, the symbol for sodium is Na (*natrium* in Latin means sodium).

Common Elements in Living Things				
Element	Atomic Symbol	Atomic Number	Atomic Weight	Comment
hydrogen	H	1	1	These
carbon	C	6	12	elements
nitrogen	N	7	14	make up
oxygen	O	8	16	most
phosphorus	P	15	31	biological
sulfur	S	16	32	molecules.
sodium	Na	11	23	These
magnesium	Mg	12	24	elements
chlorine	Cl	17	35	occur mainly
potassium	K	19	39	as dissolved
calcium	Ca	20	40	salts.

a.

p = protons
n = neutrons
⬤ = electrons

6p
6n

Carbon
$^{12}_{6}C$

b.

Figure 2.2   Elements and atoms.
a. The atomic symbol, atomic number, and atomic weight are given for the common elements in living things. **The atomic symbol for carbon is C, the atomic number is 6, and the atomic weight is 12. b.** An atom contains the subatomic particles called protons (p) and neutrons (n) in the nucleus (colored pink) and electrons (colored blue) in shells about the nucleus.

Atoms

An **atom** is the smallest unit of an element that still retains the chemical and physical properties of an element. While it is possible to split an atom by physical means, an atom is the smallest unit to enter into chemical reactions. For our purposes, it is satisfactory to think of each atom as having a central nucleus, where subatomic particles called **protons** and **neutrons** are located, and shells, which are pathways about the nucleus where **electrons** orbit (Fig. 2.2b). Most of an atom is empty space. If we could draw an atom the size of a football stadium, the nucleus would be like a gumball in the center of the field, and the electrons would be tiny specks whirling about in the upper stands.

Two important features of protons, neutrons, and electrons are their charge and weight:

Name	Charge	Weight
Electron	One negative unit	Almost no mass
Proton	One positive unit	One atomic mass unit
Neutron	No charge	One atomic mass unit

The atomic number of an atom tells you how many protons (+) and therefore how many electrons (−) an atom has when it is electrically neutral. For example, the atomic number of calcium is 20; therefore, when calcium is neutral, it has 20 protons and 20 electrons. How many electrons are there in each shell of an atom? The inner shell has the lowest energy level and can hold only two electrons; after that, each shell for the atoms noted in Figure 2.2a can hold up to eight electrons. Using this information, calculation determines that calcium has four shells and the outer shell has two electrons. As we shall see, an atom is most stable when the outer shell has eight electrons. (Hydrogen, with only one shell, is an exception to this statement. Atoms with only one shell are stable when this shell contains two electrons.)

The subatomic particles are so light that their weight is indicated by special designations called atomic mass units. Notice in the chart above that protons and neutrons each have a weight of one atomic mass unit and electrons have almost no mass. Therefore, the atomic weight generally tells you the number of protons plus the number of neutrons. How could you calculate that carbon (C) has six neutrons? Carbon's atomic weight is 12, and you know from its atomic number that it has six protons. Therefore, carbon has six neutrons (Fig. 2.2b).

As shown in Figure 2.2b, the atomic number of an atom is often written as a subscript to the lower left of the atomic symbol. The atomic weight is often written as a superscript to the upper left of the atomic symbol. Therefore, carbon can be designated in this way:

$$^{12}_{6}C$$

All matter is composed of elements, each containing particles called atoms. Atoms have an atomic symbol, atomic number (number of protons), and atomic weight (number of protons and neutrons).

Isotopes

The atomic weights given in the Periodic Table of the Elements (Appendix C) are the average weight for each kind of atom. Actually, atoms of the same type may differ in the number of neutrons and therefore weight. Atoms that have the same atomic number and differ only in the number of neutrons are called **isotopes**. Isotopes of carbon can be written in the following manner, where the subscript stands for the atomic number and the superscript stands for the atomic weight:

$$^{12}_{6}C \qquad ^{13}_{6}C \qquad ^{14}_{6}C$$

Carbon 12 has six neutrons, carbon 13 has seven neutrons, and carbon 14, which has eight neutrons, is radioactive.

Isotopes have many uses. Each type of food has its own proportion of isotopes, and this information allows biologists to study mummified or fossilized human tissue to know what ancient peoples ate. Most isotopes are stable, but radioactive isotopes break down and emit radiation in the form of radioactive particles or radiant energy. Because carbon 14 breaks down at a known rate, the amount of carbon 14 remaining is often used to determine the age of fossils. Radioactive isotopes are widely used in biological and medical research; for example, because the thyroid gland uses iodine (I), it is possible to administer a dose of radioactive iodine and then observe later that the thyroid has taken it up (Fig. 2.3).

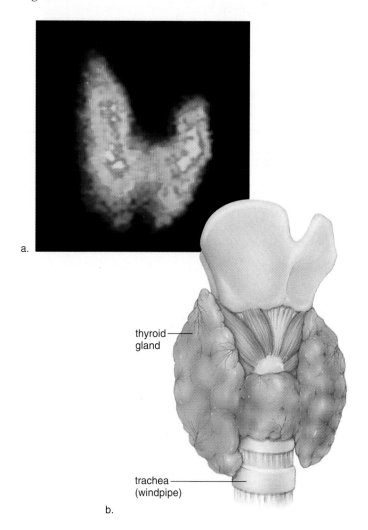

a.

thyroid
gland

trachea
(windpipe)

b.

Figure 2.3   **Use of radioactive iodine.**
a. A scan of the thyroid gland 24 hours after the patient was administered radioactive iodine. The scan resembles the shape of the thyroid shown in (**b**).

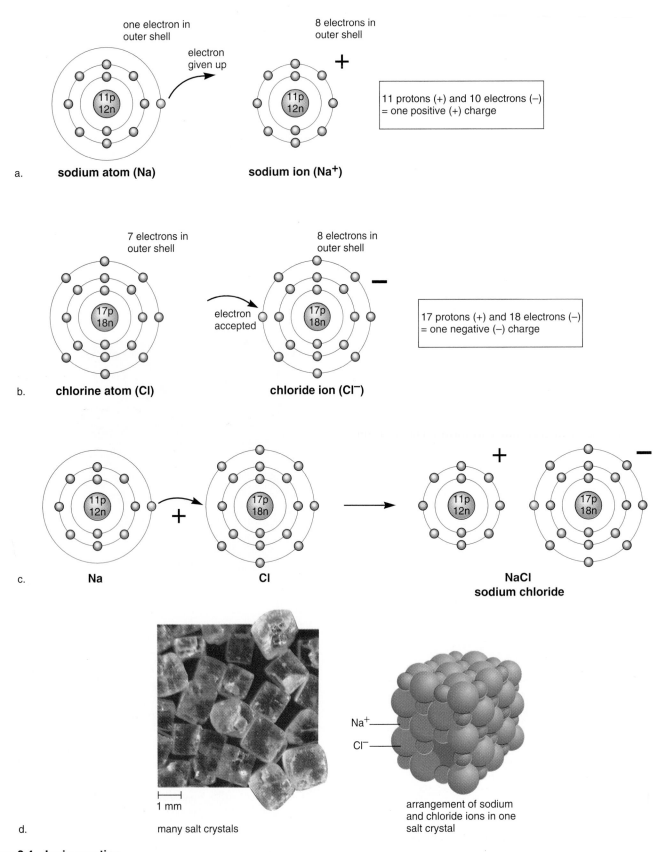

Figure 2.4 Ionic reaction.
a. When a sodium atom gives up an electron, it becomes a positive ion. **b.** When a chlorine atom gains an electron, it becomes a negative ion.
c. When sodium reacts with chlorine, the compound sodium chloride (NaCl) results. In sodium chloride, an ionic bond exists between the positive Na$^+$ and the negative Cl$^-$ ions. **d.** In a sodium chloride crystal, the ionic bonding between Na$^+$ and Cl$^-$ causes ions to form a three-dimensional lattice in which each sodium ion is surrounded by six chlorine ions, and each chlorine ion is surrounded by six sodium ions.

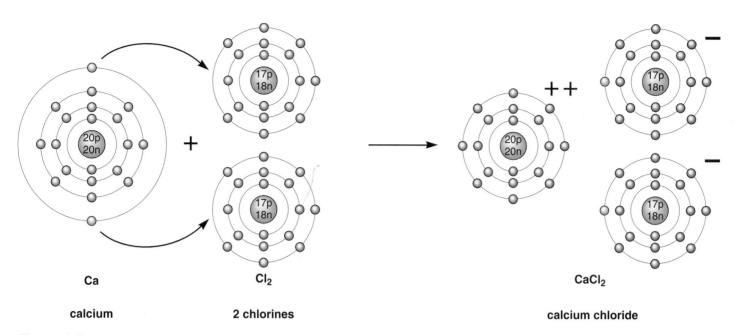

| Ca | | Cl$_2$ | | | CaCl$_2$ |
| calcium | | 2 chlorines | | | calcium chloride |

Figure 2.5   Ionic reaction.
The calcium atom gives up two electrons, one to each of two chlorine atoms. In the compound calcium chloride (CaCl$_2$), the calcium ion is attracted to two chloride ions.

2.2 Molecules and Compounds

Atoms often bond with each other to form a chemical unit called a **molecule.** A molecule can contain atoms of the same kind, as when an oxygen atom joins with another oxygen atom to form oxygen gas. Or the atoms can be different, as when an oxygen atom joins with two hydrogen atoms to form water. When the atoms are different, a compound results.

Two types of bonds join atoms: the ionic bond and the covalent bond.

Ionic Reactions

Recall that atoms (with more than one shell) are most stable when the outer shell contains eight electrons. During an ionic reaction, atoms give up or take on an electron(s) in order to achieve a stable outer shell.

Figure 2.4 depicts a reaction between a sodium (Na) and a chlorine (Cl) atom in which chlorine takes an electron from sodium. **Ions** are particles that carry either a positive ($+$) or negative ($-$) charge. The sodium ion carries a positive charge because it now has one more proton than electrons, and the chloride ion carries a negative charge because it now has one fewer proton than electrons. The attraction between oppositely charged sodium ions and chloride ions forms an **ionic bond.** The resulting compound, sodium chloride, is table salt, which we use to enliven the taste of foods.

Figure 2.5 shows an ionic reaction between a calcium atom and two chlorine atoms. Notice that calcium with two electrons in the outer shell reacts with two chlorine atoms. Why? Because with seven electrons already, each chlorine requires only one more electron to have a stable outer shell. The resulting salt is called calcium chloride.

Significant ions in the human body are listed in Table 2.1. The balance of these ions in the body is important to our health. Too much sodium in the blood can cause high blood pressure; not enough calcium leads to rickets (a bowing of the legs) in children; too much or too little potassium results in heartbeat irregularities. Bicarbonate, hydrogen, and hydroxide ions are all involved in maintaining the acid-base balance of the body. If the blood is too acidic or too basic, the body's cells cannot function properly.

An ionic bond is the attraction between oppositely charged ions.

Table 2.1	Significant Ions in the Body	
Name	**Symbol**	**Special Significance**
Sodium	Na$^+$	Found in body fluids; important in muscle contraction and nerve conduction
Chloride	Cl$^-$	Found in body fluids
Potassium	K$^+$	Found primarily inside cells; important in muscle contraction and nerve conduction
Phosphate	PO$_4$$^{3-}$	Found in bones, teeth, and the high-energy molecule ATP
Calcium	Ca^{2+}	Found in bones and teeth; important in muscle contraction
Bicarbonate	HCO$_3$$^-$	Important in acid-base balance
Hydrogen	H$^+$	Important in acid-base balance
Hydroxide	OH$^-$	Important in acid-base balance

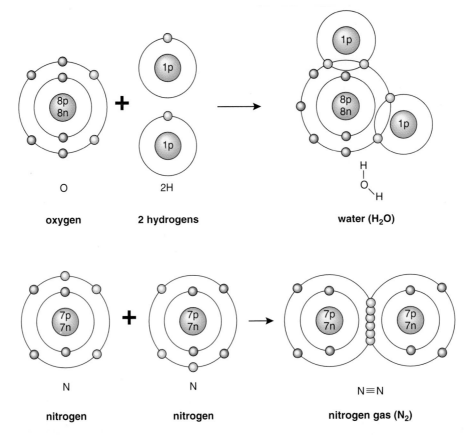

Figure 2.6   Covalent reactions.
After a covalent reaction, each atom will have filled its outer shell by sharing electrons. To determine this, it is necessary to count the shared electrons as belonging to both bonded atoms. Oxygen and nitrogen are most stable with eight electrons in the outer shell; hydrogen is most stable with two electrons in the outer shell.

Covalent Reactions

In covalent reactions, the atoms share electrons in **covalent bonds** instead of losing or gaining them. Covalent bonds can be represented in a number of ways. The overlapping outermost shells in Figure 2.6 indicate that the atoms are sharing electrons. Just as two hands participate in a handshake, each atom contributes one electron to the pair that is shared. These electrons spend part of their time in the outer shell of each atom; therefore, they are counted as belonging to both bonded atoms.

Structural formulas use straight lines to show the covalent bonds between the atoms. Each line represents a pair of shared electrons. Molecular formulas indicate only the number of each type of atom making up a molecule.

Structural formula: Cl—Cl

Molecular formula: Cl_2

Double and Triple Bonds

Besides a single bond, in which atoms share only a pair of electrons, a double or a triple bond can form. In a double bond, atoms share two pairs of electrons, and in a triple bond, atoms share three pairs of electrons between them. For example, in Figure 2.6, each nitrogen atom (N) requires three electrons to achieve a total of eight electrons in the outer shell. Notice that six electrons are placed in the outer overlapping shells in the diagram and that three straight lines are in the structural formula for nitrogen gas (N_2).

A covalent bond arises when atoms share electrons. In double covalent bonds, atoms share two pairs of electrons, and in triple covalent bonds, atoms share three pairs of electrons.

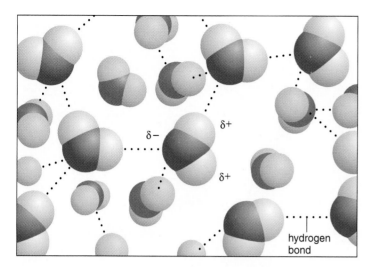

Figure 2.7 **Hydrogen bonding between water molecules.**
The polarity of the water molecules allows hydrogen bonds (dotted lines) to form between the molecules.

2.3 Water and Living Things

Water is the most abundant molecule in living organisms, and it makes up about 60–70% of the total body weight of most organisms. We will see that the physical and chemical properties of water make life possible as we know it.

Water is a polar molecule; the oxygen end of the molecule has a slight negative charge (δ^-), and the hydrogen end has a slight positive charge (δ^+):

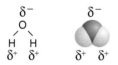

The diagram on the left shows the structural formula of the molecule, and the one on the right shows the space-filling model of the molecule.

In polar molecules, covalently bonded atoms share electrons unevenly; that is, the electrons spend more time circling the nucleus of one atom than circling the other. In water, the electrons spend more time circling the larger oxygen (O) than the smaller hydrogen (H) atoms.

In water, the negative ends and positive ends of the molecules attract one another. Each oxygen forms loose bonds to hydrogen atoms of two other water molecules (Fig. 2.7). These bonds are called hydrogen bonds. A **hydrogen bond** occurs whenever a covalently bonded hydrogen is positive and attracted to a negatively charged atom some distance away. A hydrogen bond is represented by a dotted line in Figure 2.7 because it is relatively weak and can be broken rather easily.

Properties of Water

Because of their polarity and hydrogen bonding, water molecules are cohesive and cling together. Polarity and hydrogen bonding cause water to have many characteristics beneficial to life.

1. Water is a liquid at room temperature. Therefore, we are able to drink it, cook with it, and bathe in it.

Compounds with low molecular weights are usually gases at room temperature. For example, oxygen (O_2) with a molecular weight of 32 is a gas, but water with a molecular weight of 18 is a liquid. The hydrogen bonding between water molecules keeps water a liquid and not a gas at room temperature. Water does not boil and become a gas until 100°C, one of the reference points for the Celsius temperature scale. Without hydrogen bonding between water molecules, our body fluids and indeed our bodies would be gaseous!

2. Water is the universal solvent for polar (charged) molecules and thereby facilitates chemical reactions both outside of and within our bodies.

When a salt such as sodium chloride (NaCl) is put into water, the negative ends of the water molecules are attracted to the sodium ions, and the positive ends of the water molecules are attracted to the chloride ions. This causes the sodium ions and the chloride ions to separate and to dissolve in water:

The salt NaCl dissolves in water.

When ions and molecules disperse in water, they move about and collide, allowing reactions to occur. Therefore, water is a solvent that facilitates chemical reactions.

Ions and molecules that interact with water are said to be **hydrophilic.** Nonionized and nonpolar molecules that do not interact with water are said to be **hydrophobic.**

3. Water molecules are cohesive, and therefore liquids fill vessels, such as blood vessels.

Water molecules cling together because of hydrogen bonding, and yet, water flows freely. This property allows dissolved and suspended molecules to be evenly distributed throughout a system. Therefore, water is an excellent transport medium. Within our bodies, the blood that fills our arteries and veins is 92% water. Blood transports oxygen and nutrients to the cells and removes wastes such as carbon dioxide.

a.   b.   c.

Figure 2.8   **Characteristics of water.**
a. Water boils at 100°C. If it boiled and was a gas at a lower temperature, life could not exist. **b.** It takes much body heat to vaporize sweat, which is mostly liquid water, and this helps keep bodies cool when the temperature rises. **c.** Ice is less dense than water, and it forms on top of water, making skate sailing possible.

4. The temperature of liquid water rises and falls slowly, preventing sudden or drastic changes.

The many hydrogen bonds that link water molecules cause water to absorb a great deal of heat before it boils (Fig. 2.8a). A **calorie** of heat energy raises the temperature of one gram of water 1°C. This is about twice the amount of heat required for other covalently bonded liquids. On the other hand, water holds heat, and its temperature falls slowly. Therefore, water protects us and other organisms from rapid temperature changes and helps us maintain our normal internal temperature. This property also allows great bodies of water, such as oceans, to maintain a relatively constant temperature. Water is a good temperature buffer.

5. Water has a high heat of vaporization, keeping the body from overheating.

It takes a large amount of heat to change water to steam (Fig. 2.8a). (Converting one gram of the hottest water to steam requires an input of 540 calories of heat energy.) This property of water helps moderate the earth's temperature so that life can continue to exist. Also, in a hot environment, animals sweat and the body cools as body heat is used to

evaporate sweat, which is mostly liquid water (Fig. 2.8b). Body heat is also given off when certain blood vessels dilate, bringing more blood to the surface of the body. The control of body temperature is an example of homeostasis, which is the maintenance of the internal environment within normal limits. ⚖

6. Frozen water is less dense than liquid water so that ice floats on water.

As water cools, the molecules come closer together. They are densest at 4°C, but they are still moving about. At temperatures below 4°C, there is only vibrational movement, and hydrogen bonding becomes more rigid but also more open. This makes ice less dense. Bodies of water always freeze from the top down, making skate sailing possible (Fig. 2.8c). When a body of water freezes on the surface, the ice acts as an insulator to prevent the water below it from freezing. Aquatic organisms are protected, and they have a better chance of surviving the winter.

Because of its polarity and hydrogen bonding, water has many characteristics that benefit life.

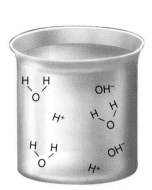

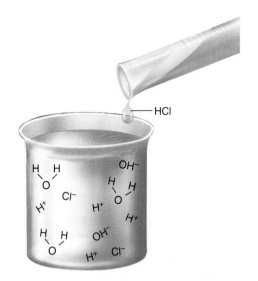

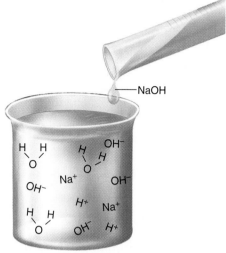

Figure 2.9   Dissociation of water molecules.

Dissociation produces an equal number of hydrogen ions (H$^+$) and hydroxide ions (OH$^-$). (These illustrations are not meant to be mathematically accurate.)

Figure 2.10   Addition of hydrochloric acid (HCl).

HCl releases hydrogen ions (H$^+$) as it dissociates. The addition of HCl to water results in a solution with more H$^+$ than OH$^-$.

Figure 2.11   Addition of sodium hydroxide (NaOH), a base.

NaOH releases OH$^-$ as it dissociates. The addition of NaOH to water results in a solution with more OH$^-$ than H$^+$.

Acidic and Basic Solutions

When water dissociates (breaks up), it releases an equal number of hydrogen ions (H$^+$) and hydroxide ions (OH$^-$).

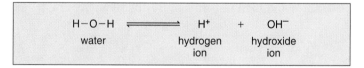

$$H-O-H \rightleftharpoons H^+ + OH^-$$

water   hydrogen   hydroxide
                    ion   ion

Only a few water molecules at a time are dissociated (Fig. 2.9). The actual number of ions is 10^{-7} moles/liter. A mole is a unit of scientific measurement for atoms, ions, and molecules.[1]

Acidic Solutions

Lemon juice, vinegar, tomato juice , and coffee are all familiar acidic solutions. What do they have in common? Acidic solutions have a sharp or sour taste, and therefore we sometimes associate them with indigestion. To a chemist, **acids** are molecules that dissociate in water, releasing hydrogen ions (H$^+$). For example, an important acid in the laboratory is hydrochloric acid (HCl), which dissociates in this manner:

$$HCl \rightarrow H^+ + Cl^-$$

Dissociation is almost complete; therefore, this is called a strong acid. When hydrochloric acid is added to a beaker of water, the number of hydrogen ions increases (Fig. 2.10).

Basic Solutions

Milk of magnesia and ammonia are common bases that most people have heard of. Bases have a bitter taste and feel slippery when in water. To a chemist, **bases** are molecules that either take up hydrogen ions (H$^+$) or release hydroxide ions (OH$^-$). For example, an important inorganic base is sodium hydroxide (NaOH), which dissociates in this manner:

$$NaOH \rightarrow Na^+ + OH^-$$

Dissociation is almost complete; therefore, sodium hydroxide is called a strong base. If sodium hydroxide is added to a beaker of water, the number of hydroxide ions increases (Fig. 2.11).

It is not recommended that you taste a strong acid or base, because they are quite destructive to cells. Any container of household cleanser, such as ammonia, has a poison symbol and carries a strong warning not to ingest the product.

The Litmus Test

A simple laboratory test for acids and bases is called the litmus test. Litmus is a vegetable dye that changes color from blue to red in the presence of an acid and from red to blue in the presence of a base. The litmus test has become a common figure of speech, as when you hear a commentator say, "The litmus test for a Republican is . . . ."

[1]A mole is the same amount of atoms, molecules, or ions as the number of atoms in exactly 12 grams of ^{12}C.

The pH Scale

The **pH scale**[2] indicates the acidity and basicity (alkalinity) of a solution. There are normally few hydrogen ions (H⁺) in a solution, and the pH scale was devised to eliminate the use of cumbersome numbers. For example, the possible hydrogen ion concentrations of a solution (in moles per liter) are on the left of this listing and the pH is on the right:

$$1 \times 10^{-6}\ [H^+] = pH\ 6$$
$$1 \times 10^{-7}\ [H^+] = pH\ 7\ (neutral)$$
$$1 \times 10^{-8}\ [H^+] = pH\ 8$$

Pure water contains only 10^{-7} moles per liter of both hydrogen ions and hydroxide ions. Therefore a pH of exactly 7 is neutral pH.

To further illustrate the relationship between hydrogen ion concentration and pH, consider the following question. Which of the pH values listed above indicates a higher hydrogen ion concentration [H⁺] than pH 7, and therefore would be an acidic solution? A number with a smaller negative exponent indicates a greater quantity of hydrogen ions than one with a larger negative exponent. Therefore pH 6 is an acidic solution.

As discussed on the previous page, basic solutions have fewer hydrogen ions compared to water. Which of the values listed is a basic solution? pH 8 is a basic solution because it indicates a lower hydrogen concentration (greater hydroxide ion concentration) than pH 7.

The pH scale (Fig. 2.12) ranges from 0 to 14. A pH of 0 to 7 is an acidic solution and a pH of 7 to 14 is a basic solution. Further, as we move down the pH scale from pH 14 to pH 0, each unit has 10 times the [H⁺] of the previous unit. As we move up the scale from 0 to 14, each unit has 10 times the [OH⁻] of the previous unit.

As discussed in the Ecology Focus on page 25, there have been detrimental environmental consequences to non-living and living things as rain and snow have become more acidic. In humans, pH needs to be maintained within a narrow range or there are health consequences. The pH of blood is around 7.4, and it is buffered in the manner described next to keep the pH within a normal range.

Buffers and pH

A **buffer** is a chemical or a combination of chemicals that keeps pH within normal limits. Many commercial products like Bufferin, shampoos, or deodorants are buffered as an added incentive to have us buy them. Buffers resist pH changes because they can take up excess hydrogen ions (H⁺) or hydroxide ions (OH⁻).

The pH of our blood is usually about 7.4, in part because it contains a combination of carbonic acid and bicarbonate

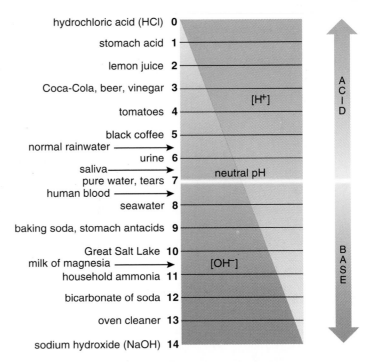

Figure 2.12   The pH scale.
The diagonal line indicates the proportionate concentration of hydrogen ions (H⁺) to hydroxide ions (OH⁻) at each pH value. Any pH value above 7 is basic, while any pH value below 7 is acidic.

ions. Carbonic acid (H_2CO_3) is a weak acid that minimally dissociates and then re-forms in the following manner:

H_2CO_3	dissociates ⇌ re-forms	H^+	+	HCO_3^-
carbonic acid		hydrogen ion		bicarbonate ion

When hydrogen ions (H⁺) are added to blood, the following reaction occurs:

$$H^+ + HCO_3^- \rightarrow H_2CO_3$$

When hydroxide ions (OH⁻) are added to blood, this reaction occurs:

$$OH^- + H_2CO_3 \rightarrow HCO_3^- + H_2O$$

These reactions prevent any significant change in blood pH.

Acids have a pH that is less than 7, and bases have a pH that is greater than 7. Buffers, which can combine with both hydrogen ions and hydroxide ions, resist pH changes.

---

[2]pH is defined as the negative logarithm of the molar concentration of the hydrogen ion [H⁺].

The Harm Done by Acid Deposition

Normally, rainwater has a pH of about 5.6 because the carbon dioxide in the air combines with water to give a weak solution of carbonic acid. Rain falling in the northeastern United States and southeastern Canada now has a pH of between 5.0 and 4.0. We have to remember that a pH of 4 is ten times more acidic than a pH of 5 to comprehend the increase in acidity this represents.

There is very strong evidence that this observed increase in rainwater acidity is a result of the burning of fossil fuels, such as coal and oil, as well as gasoline derived from oil. When fossil fuels are burned, sulfur dioxide and nitrogen oxides are produced, and they combine with water vapor in the atmosphere to form the acids sulfuric acid and nitric acid. These acids return to earth contained in rain or snow, a process properly called wet deposition, but more often called acid rain. During dry deposition, dry particles of sulfate and nitrate salts descend from the atmosphere.

Unfortunately, regulations that require the use of tall smokestacks to reduce local air pollution only cause pollutants to be carried far from their place of origin. Acid deposition in southeastern Canada is due to the burning of fossil fuels in factories and power plants in the Midwest. Acid deposition adversely affects lakes, particularly in areas where the soil is thin and lacks limestone (calcium carbonate, $CaCO_3$), a buffer to acid deposition. It leaches aluminum from the soil, carries aluminum into the lakes, and converts mercury deposits in lake bottom sediments to soluble and toxic methyl mercury. Lakes not only become more acidic, but they also show accumulation of toxic substances. In Norway and Sweden, at least 16,000 lakes contain no fish, and an additional 52,000 lakes are threatened. In Canada, some 14,000 lakes are almost fishless, and an additional 150,000 are in peril because of excess acidity. In the United States, about 9,000 lakes (mostly in the Northeast and upper Midwest) are threatened, one-third of them seriously.

In forests, acid deposition weakens trees because it leaches away nutrients and releases aluminum. By 1988, most spruce, fir, and other conifers atop North Carolina's Mt. Mitchell were dead from being bathed in ozone and acid fog for years. The soil was so acidic that new seedlings could not survive. Nineteen countries in Europe have reported woodland damage, ranging from 5 to 15% of the forested area in Yugoslavia and Sweden to 50% or more in the Netherlands, Switzerland, and the former West Germany. More than one-fifth of Europe's forests are now damaged.

These aren't the only effects of acid deposition. Reduction of agricultural yields, damage to marble and limestone monuments and buildings, and even illnesses in humans have been reported. Acid deposition has been implicated in the increased incidence of lung cancer and possibly colon cancer in residents of the North American East Coast. Tom McMillan, Canadian Minister of the Environment, says that acid rain is "destroying our lakes, killing our fish, undermining our tourism, retarding our forests, harming our agriculture, devastating our heritage, and threatening our health."

There are, of course, things that can be done. We could

a. use alternative energy sources, such as solar, wind, hydropower, and geothermal energy, whenever possible.
b. use low-sulfur coal or remove the sulfur impurities from coal before it is burned.
c. require factories and power plants to use scrubbers, which remove sulfur emissions.
d. require people to use mass transit rather than driving their own automobiles.
e. reduce our energy needs through other means of energy conservation.

Figure 2A   **Effects of acid deposition.**
Statues deteriorate and trees die due to the burning of fossil fuels. The combustion of fossil fuels results in atmospheric acids that return to the earth as acid deposition.

2.4 Molecules of Life

Inorganic molecules constitute nonliving matter, but even so, inorganic molecules like salts (e.g., NaCl) and water play important roles in living things. The molecules of life are organic molecules. **Organic molecules** always contain carbon (C) and hydrogen (H). The chemistry of carbon accounts for the formation of the very large variety of organic molecules found in living things. A carbon atom has four electrons in the outer shell. In order to achieve eight electrons in the outer shell, a carbon atom shares electrons covalently with as many as four other atoms. Methane is a molecule in which a carbon atom shares electrons with four hydrogen atoms:

$$
\begin{array}{c}
\text{H} \\
| \\
\text{H}-\text{C}-\text{H} \\
| \\
\text{H}
\end{array}
$$

A carbon atom can share with another carbon atom, and in so doing, a long hydrocarbon chain can result:

$$
\text{H}-\text{C}-\text{C}-\text{C}-\text{C}-\text{C}-\text{C}-\text{C}-\text{C}-\text{C}-\text{H}
$$

A hydrocarbon chain can also turn back on itself to form a ring compound:

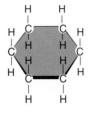

So-called functional groups can be attached to carbon chains. A **functional group** is a particular cluster of atoms that always behaves in a certain way. One functional group of interest is the acidic (carboxyl) group —COOH because it can give up a hydrogen ion (H^+) and ionize to —COO⁻.

hydrocarbon
(hydrophobic)

acid in ionized form
(hydrophilic)

Whereas a hydrocarbon chain is *hydrophobic* (not attracted to water) because it is nonpolar, a hydrocarbon chain with an attached ionized group is *hydrophilic* (attracted to water) because it is polar.

The molecules of life are divided into four classes: carbohydrates, lipids, proteins, and nucleic acids. Carbohydrates, lipids, and proteins are very familiar to you because certain foods are known to be rich in these molecules, as illustrated in Figures 2.13–2.15. The nucleic acid DNA makes up our genes, which are hereditary units that control our cells and the structure of our bodies.

Many molecules of life are macromolecules. Just as atoms can join to form a molecule, so molecules can join to form a macromolecule. The smaller molecules are called monomers, and the macromolecule is called a polymer. A polymer is a chain of monomers.

Polymer	Monomer
polysaccharide	monosaccharide
protein	amino acid
nucleic acid	nucleotide

Figure 2.13   Carbohydrate foods.
Breads, pasta, rice, corn, and oats all contain complex carbohydrates.

Figure 2.14   Foods rich in lipids.
Butter and oils contain fat, the most familiar of the lipids.

Figure 2.15   Foods rich in proteins.
Meat, eggs, cheese, and beans have a high content of protein.

2.5 Carbohydrates

Carbohydrates first and foremost function for quick and short-term energy storage in all organisms, including humans. Carbohydrate molecules are characterized by the presence of the atomic grouping H—C—OH, in which the ratio of hydrogen atoms (H) to oxygen atoms (O) is approximately 2:1. Since this ratio is the same as the ratio in water, the name "hydrates of carbon" seems appropriate.

Simple Carbohydrates

If the number of carbon atoms in a molecule is low (from three to seven), then the carbohydrate is a simple sugar, or **monosaccharide.** The designation **pentose** means a 5-carbon sugar, and the designation **hexose** means a 6-carbon sugar. Ribose and deoxyribose are two pentoses of significance because they are found respectively in the nucleic acids RNA and DNA. RNA and DNA are discussed later in the chapter. **Glucose,** a hexose, is blood sugar (Fig. 2.16); our bodies use glucose as an immediate source of energy. Other common hexoses are fructose, found in fruits, and galactose, a constituent of milk. These three hexoses

(glucose, fructose, and galactose) all occur as ring structures with the molecular formula $C_6H_{12}O_6$, but the exact shape of the ring differs, as does the arrangement of the hydrogen (—H) and hydroxide groups (—OH) attached to the ring. A larger sugar called a **disaccharide** (*di*, two; *saccharide*, sugar) contains two monosaccharides.

Organisms have a common way of joining small molecules to build larger molecules. **Condensation synthesis** of a larger molecule is so called because synthesis means "making of" and condensation means that water has been removed. Breakdown of the larger molecule is a **hydrolysis** reaction because water is used to split its bonds. Figure 2.17 shows how condensation synthesis results in a disaccharide called maltose and how hydrolysis of the maltose results in two glucose molecules again. Maltose is a disaccharide of interest because it is found in our digestive tract as a result of starch digestion.

When glucose and fructose join, the disaccharide sucrose forms. Sucrose, which is ordinarily derived from sugarcane and sugar beets, is commonly known as table sugar. When we eat a candy bar, our digestive enzymes hydrolyze sucrose into glucose and fructose.

As stated, a small organic molecule can also be a unit of a larger organic molecule often called a macromolecule. In that case, the unit is called a monomer, and the macromolecule is called a polymer. Glucose is a monomer for larger carbohydrates like glycogen, starch, and cellulose, which are discussed on the next page. Polymers are synthesized and broken down in this manner:

$$\text{monomers} \rightleftharpoons \text{polymer} + H_2O \text{ molecules}$$

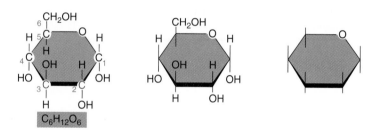

Figure 2.16   Three ways to represent the structure of glucose.
The *far left* structure shows the carbon atoms; $C_6H_{12}O_6$ is the molecular formula for glucose. The *far right* structure is the simplest way to represent glucose.

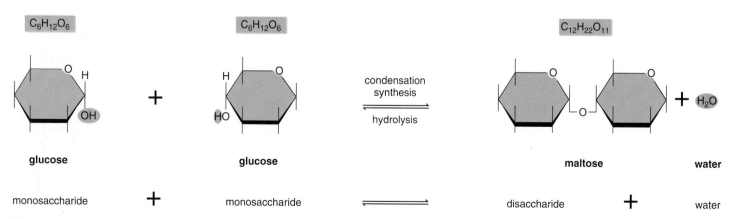

Figure 2.17   Condensation synthesis and hydrolysis of maltose, a disaccharide.
During condensation synthesis of maltose, a bond forms between the two glucose molecules, and the components of water are removed. During hydrolysis, the components of water are added, and the bond is broken.

Starch and Glycogen

Starch and glycogen are ready storage forms of glucose in plants and animals, respectively. Starch and glycogen are **polysaccharides;** that is, they are polymers of glucose formed just as a necklace might be made using only one type of bead. The following equation shows how starch is synthesized:

glucose molecules   condensation
(monomers) $\xrightarrow[\text{hydrolysis}]{\text{synthesis}}$ starch + H_2O
(polymer)

Some of the polymers in starch are long chains of up to 4,000 glucose units. Others are branched, as is glycogen (Fig. 2.18). Starch has fewer side branches, or chains of glucose that branch off from the main chain, than does glycogen.

Starch is the storage form of glucose inside plant cells. Flour, which we usually acquire by grinding wheat and use to bake bread and rolls, is high in starch. **Glycogen** is the storage form of glucose in humans. Figure 2.18 includes a micrograph of glycogen granules inside the liver.

After we eat starchy foods like bread, potatoes, and cake, starch is hydrolyzed to glucose. Then the bloodstream carries excess glucose to the liver where it is stored as glycogen. In between eating, the liver releases glucose so that the blood glucose concentration is always about 0.1%. As mentioned in chapter 4, this function of the liver is an example of homeostasis. ⚖

Cellulose

The polysaccharide **cellulose** is found in plant cell walls, and this accounts, in part, for the strong nature of these walls. In cellulose (Fig. 2.19), the glucose units are joined by a slightly different type of linkage than that in starch or glycogen. (Observe the alternating position of the oxygen atoms in the linked glucose units.) While this might seem to be a technicality, actually it is important because we are unable to digest foods containing this type of linkage; therefore, cellulose largely passes through our digestive tract as fiber, or roughage. It is believed that fiber in the diet is necessary to good health and some have suggested it may help prevent colon cancer.

Cells usually use the monosaccharide glucose as an energy source. The polysaccharides starch and glycogen are storage compounds in plant and animal cells, respectively. The polysaccharide cellulose is found in plant cell walls.

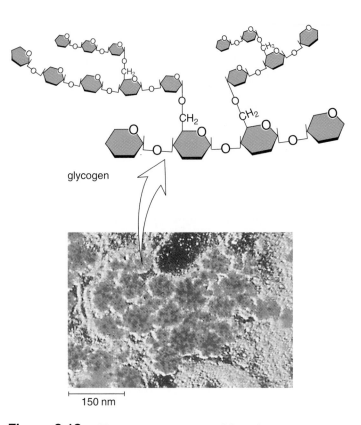

glycogen

150 nm

Figure 2.18   Glycogen structure and function.
Glycogen is a highly branched polymer of glucose molecules. The electron micrograph shows glycogen granules in liver cells. Glycogen is the storage form of glucose in animals.

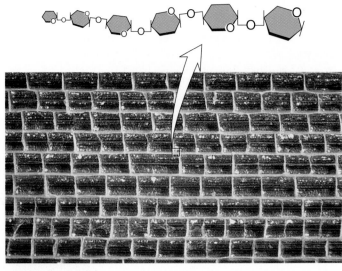

cattail leaf cell walls

Figure 2.19   Cellulose structure and function.
Cellulose contains a slightly different type of linkage between glucose molecules than that in starch or glycogen. Plant cell walls contain cellulose, and the rigidity of the cell walls permits nonwoody plants to stand upright as long as they receive an adequate supply of water.

Figure 2.20   **Condensation synthesis and hydrolysis of a fat molecule.**
Fatty acids can be saturated (no double bonds between carbon atoms) or unsaturated (have double bonds, colored yellow, between carbon atoms). When a fat molecule forms, three fatty acids combine with glycerol, and three water molecules are produced.

2.6 Lipids

Lipids are diverse in structure and function, but they have a common characteristic: they do not dissolve in water.

Fats and Oils

The most familiar lipids are those found in fats and oils. **Fats,** which are usually of animal origin (e.g., lard and butter), are solid at room temperature. **Oils,** which are usually of plant origin (e.g., corn oil and soybean oil), are liquid at room temperature. Fat has several functions in the body: it is used for long-term energy storage, it insulates against heat loss, and it forms a protective cushion around major organs.

Fats and oils form when one glycerol molecule reacts with three fatty acid molecules. A fat is sometimes called a **triglyceride** because of its three-part structure, and the term neutral fat is sometimes used because the molecule is nonpolar (Fig. 2.20).

Saturated and Unsaturated Fatty Acids

A **fatty acid** is a hydrocarbon chain that ends with the acidic group —COOH (Fig. 2.20). Most of the fatty acids in cells contain 16 or 18 carbon atoms per molecule, although smaller ones with fewer carbons are also known.

Fatty acids are either saturated or unsaturated. **Saturated fatty acids** have no double bonds between carbon atoms. The carbon chain is saturated, so to speak, with all the hydrogens it can hold. Saturated fatty acids account for the solid nature at room temperature of butter and lard, which are derived from animal sources. **Unsaturated fatty acids** have double bonds between carbon atoms wherever the number of hydrogens is less than two per carbon atom. Unsaturated fatty acids account for the liquid nature of vegetable oils at room temperature. Hydrogenation of vegetable oils can convert them to margarine and products such as Crisco.

Emulsifiers

Molecules of fat tend to clump together, and even liquid oils do not disperse in water. Almost everyone has observed the top layer of oil that accumulates in a salad dressing container.

Unlike fats, emulsifiers have a polar end that is hydrophilic in addition to the nonpolar end that is hydrophobic (the hydrocarbon chain represented by *R*). Therefore, an emulsifier does mix with water. When emulsifiers are added to oils, the oils, too, mix with water because an emulsifier positions itself about an oil droplet so that its nonpolar ends project into the fat droplet, while its polar ends project outward.

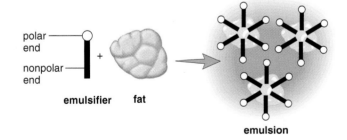

Now the droplet disperses in water, and it is said that **emulsification** has occurred. Emulsification occurs when dirty clothes are washed with soaps or detergents. Also, prior to the digestion of fatty foods, fats are emulsified by bile. A person who has had the gallbladder removed may have trouble digesting fatty foods because this organ stores bile for emulsifying fats prior to the digestive process.

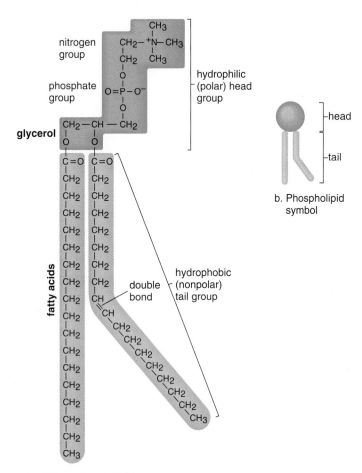

a. Lecithin, a phospholipid

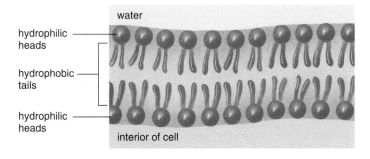

c. Phospholipid bilayer in plasma membrane

Figure 2.21 Phospholipid structure and shape.

a. Phospholipids are constructed like fats, except that they contain a phosphate group. This phospholipid also includes an organic group that contains nitrogen. **b.** The hydrophilic portion of the phospholipid molecule (head) is soluble in water, whereas the two hydrocarbon chains (tails) are not. **c.** This causes the molecule to arrange itself as shown when exposed to water.

Figure 2.22 Steroid diversity.

a. Cholesterol, like all steroid molecules, has four adjacent rings, but the effects of steroids on the body largely depend on the attached groups indicated in red. **b.** Testosterone is the male sex hormone.

Phospholipids

Phospholipids, as their name implies, contain a phosphate group (Fig. 2.21). Essentially, they are constructed like fats, except that in place of the third fatty acid, there is a phosphate group or a grouping that contains both phosphate and nitrogen. These molecules are not electrically neutral, as are fats, because the phosphate and nitrogenous groups are ionized. They form the so-called hydrophilic head of the molecule, while the rest of the molecule becomes the hydrophobic tails. The plasma membrane that surrounds cells is a phospholipid bilayer in which the heads face outward into a watery medium and the tails face each other because they are water repelling.

Steroids

Steroids are lipids having a structure that differs entirely from that of fats. Steroid molecules have a backbone of four fused carbon rings, but each one differs primarily by the arrangement of the atoms in the rings and the type of functional groups attached to them. Cholesterol is a component of an animal cell's plasma membrane and is the precursor of several other steroids, such as the sex hormones estrogen and testosterone (Fig. 2.22).

We know that a diet high in saturated fats and cholesterol can lead to circulatory disorders. This type of diet causes fatty material to accumulate inside the lining of blood vessels, therefore reducing blood flow. As discussed in the Health Focus on page 34, nutrition labels are now required to list the calories from fat per serving and the percent daily value from saturated fat and cholesterol.

Lipids include fats and oils for long-term energy storage and steroids. Phospholipids, unlike other lipids, are soluble in water because they have a hydrophilic group.

Name	Structural Formula		R Group

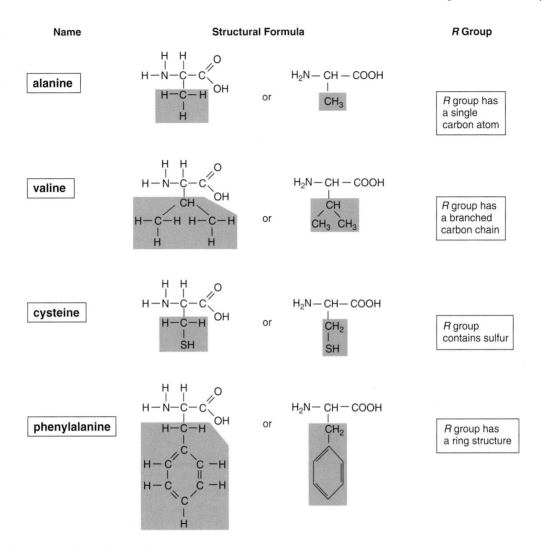

alanine — or — *R* group has a single carbon atom

valine — or — *R* group has a branched carbon chain

cysteine — or — *R* group contains sulfur

phenylalanine — or — *R* group has a ring structure

Figure 2.23   Representative amino acids.
Amino acids differ from one another by their *R* group; the simplest *R* group is a single hydrogen atom (H). The *R* groups (red) that contain carbon vary as shown.

2.7 Proteins

Proteins perform many functions. Some proteins like keratin, which makes up hair and nails, and collagen, which lends support to ligaments, tendons, and skin, are structural proteins. Many hormones, which are messengers that influence cellular metabolism, are proteins. The proteins actin and myosin account for the movement of cells and the ability of our muscles to contract. Some proteins transport molecules in the blood; hemoglobin is a complex protein in our blood that transports oxygen. Antibodies in blood and other body fluids are proteins that combine with foreign substances preventing them from destroying cells and upsetting homeostasis.

Proteins in the plasma membrane of our cells have various functions; some form channels that allow substances to enter and exit cells. Some are carriers that transport mole-

cules into and out of the cell and some here and elsewhere in the cell are enzymes. Enzymes are necessary contributors to the chemical workings of the cell and therefore of the body. **Enzymes** speed chemical reactions; they work so quickly that a reaction that normally takes several hours or days without an enzyme takes only a fraction of a second with an enzyme.

Proteins are polymers with amino acid monomers. An **amino acid** has a central carbon atom bonded to a hydrogen atom and three groups. The name of the molecule is appropriate because one of these groups is an amino group ($-NH_2$) and another is an acidic group ($-COOH$). The other group is called an *R* group because it is the *Remainder* of the molecule. Amino acids differ from one another by their *R* group; the *R* group varies from a single hydrogen (H) to a complicated ring (Fig. 2.23).

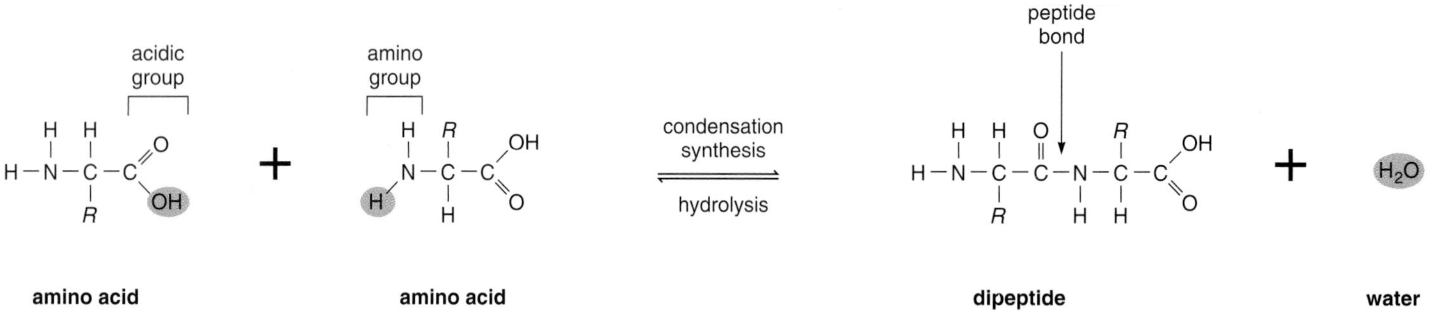

Figure 2.24   **Condensation synthesis and hydrolysis of a dipeptide.**
The two amino acids on the left-hand side of the equation differ by their *R* groups. As these amino acids join, a peptide bond forms, and a water molecule is produced. During hydrolysis, water is added, and the peptide bond is broken.

Peptides

Figure 2.24 shows that a condensation synthesis reaction between two amino acids results in a dipeptide and a molecule of water. A bond that joins two amino acids is called a **peptide bond.** The atoms associated with a peptide bond—oxygen (O), carbon (C), nitrogen (N), and hydrogen (H)—share electrons in such a way that the oxygen has a partial negative charge (δ^-) and the hydrogen has a partial positive charge (δ^+).

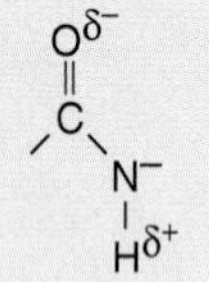

Therefore, the peptide bond is polar, and hydrogen bonding is possible between the C=O of one amino acid and the N—H of another amino acid in a polypeptide. A **polypeptide** is a single chain of amino acids.

Levels of Protein Organization

The structure of a protein has at least three levels of organization (Fig. 2.25). The first level, called the *primary structure*, is the linear sequence of the amino acids joined by peptide bonds. Polypeptides can be quite different from one another. You will recall that the structure of a polysaccharide can be likened to a necklace that contains a single type of "bead," namely, glucose. Polypeptides can make use of 20 different possible types of amino acids or "beads." Each particular polypeptide has its own sequence of amino acids. It can be said that each polypeptide differs by the sequence of its *R* groups and the number of amino acids in the sequence.

The *secondary structure* of a protein comes about when the polypeptide takes on a particular orientation in space. A coiling of the chain results in an alpha (α) helix, or a right-handed spiral, and a folding of the chain results in a pleated sheet. Hydrogen bonding between peptide bonds holds the shape in place.

The *tertiary structure* of a protein is its final three-dimensional shape. In muscles, the helical chains of myosin form a rod shape that ends in globular (globe-shaped) heads. In enzymes, the helix bends and twists in different ways. Invariably, the hydrophobic portions are packed mostly on the inside, and the hydrophilic portions are on the outside where they can make contact with water. The tertiary shape of a polypeptide is maintained by various types of bonding between the *R* groups; covalent, ionic, and hydrogen bonding all occur. One common form of covalent bonding between *R* groups is disulfide (S—S) linkages between two cysteine amino acids.

Some proteins have only one polypeptide, and some others have more than one polypeptide chain, each with its own primary, secondary, and tertiary structures. These separate polypeptides are arranged to give some proteins a fourth level of structure, termed the *quaternary structure.* Hemoglobin is a complex protein having a quaternary structure; most enzymes also have a quaternary structure.

The final shape of a protein is very important to its function. As we will discuss in chapter 3, for example, enzymes cannot function unless they have their usual shape. When proteins are exposed to extremes in heat and pH, they undergo an irreversible change in shape called **denaturation.** For example, we are all aware that the addition of acid to milk causes curdling and that heating causes egg white, which contains a protein called albumin, to coagulate. Denaturation occurs because the normal bonding between the *R* groups has been disturbed. Once a protein loses its normal shape, it is no longer able to perform its usual function.

Proteins, which contain covalently linked amino acids, are important in the structure and the function of cells. Some proteins are enzymes, which speed chemical reactions.

Visual Focus

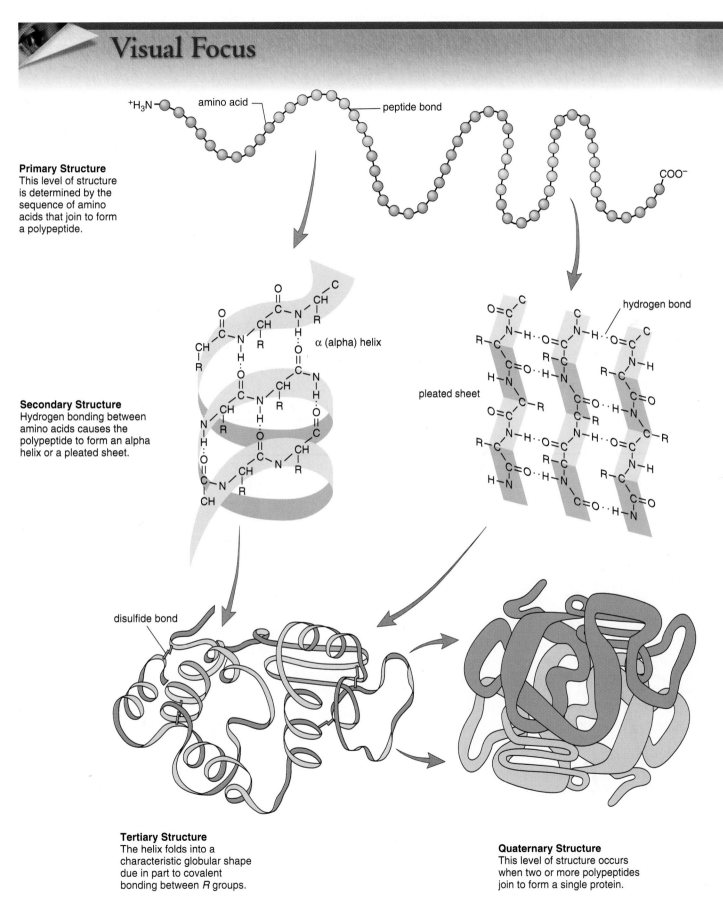

Primary Structure
This level of structure is determined by the sequence of amino acids that join to form a polypeptide.

amino acid   peptide bond

+H₃N—   COO⁻

Secondary Structure
Hydrogen bonding between amino acids causes the polypeptide to form an alpha helix or a pleated sheet.

α (alpha) helix

pleated sheet   hydrogen bond

disulfide bond

Tertiary Structure
The helix folds into a characteristic globular shape due in part to covalent bonding between *R* groups.

Quaternary Structure
This level of structure occurs when two or more polypeptides join to form a single protein.

Figure 2.25   Levels of protein organization.

Nutrition Labels

Packaged foods now have a nutrition label like the one depicted in Figure 2B. The nutrition information given on this label is based on the serving size (that is, 1¼ cup, 57 grams) of the cereal. A Calorie* is a measurement of energy. One serving of the cereal provides 220 Calories, of which 20 are from fat. At the bottom of the label, the recommended amounts of nutrients are based on a typical diet of 2,000 Calories for women and 2,500 Calories for men.

Fats are the nutrient with the highest energy content: 9 Cal/g compared to 4 Cal/g for carbohydrates and proteins. The body stores fat under the skin and around the organs for later use. A 2,000-Calorie diet should contain no more than 65 g (585 Calories) of fat. Dietary fat has been implicated in cancer of the colon, pancreas, ovary, prostate, and breast. Although saturated fat and cholesterol are essential nutrients, dietary consumption of saturated fats and cholesterol in particular should be controlled. Cholesterol and saturated fat contribute to the formation of deposits called plaques, which clog arteries and lead to cardiovascular disease, including high blood pressure.

For these reasons, it is important to know how a serving of the cereal will contribute to the maximum daily recommended amount of fat, saturated fat, and cholesterol. You can find this out by looking at the listing under % Daily Value: the total fat in one serving of the cereal provides 3% of the daily recommended amount of fat. How much will a serving of the cereal contribute to the maximum recommended daily amount of saturated fat? Of cholesterol?

Carbohydrates (sugars and polysaccharides) are the quickest, most readily available source of energy for the body. Because carbohydrates aren't usually associated with health problems, they should compose the largest proportion of the diet. Breads and cereals containing complex carbohydrates are preferable to candy and ice cream containing simple carbohydrates because they are likely to contain dietary fiber (nondigestible plant material). Insoluble fiber has a laxative effect and may reduce the risk of colon cancer; soluble fiber combines with the cholesterol in food and prevents the cholesterol from entering the body proper.

The body does not store amino acids for the production of proteins, which are found particularly in muscles but also in all cells of the body. A woman should have about 44 g of protein per day, and a man should have about 56 g of protein a day. Red meat is rich in protein, but it is usually also high in saturated fat. Therefore, it is considered good health sense to rely on protein from plant origins (e.g., whole-grain cereals, dark breads, legumes) more than is customary in the United States. Legume is a botanical term that includes peas and beans—a combination of beans and rice can provide all of the various amino acids you need to build cellular proteins.

The amount of dietary sodium (as in table salt) is of concern because excessive sodium intake has been linked to high blood pressure in some people. It is recommended that the intake of sodium be no more than 2,400 mg per day. A serving of the cereal pictured here provides what percent of this maximum amount?

Vitamins are essential requirements needed in small amounts in the diet. Each vitamin has a recommended daily intake, and the food label tells what percent of the recommended amount is provided by a serving of this cereal.

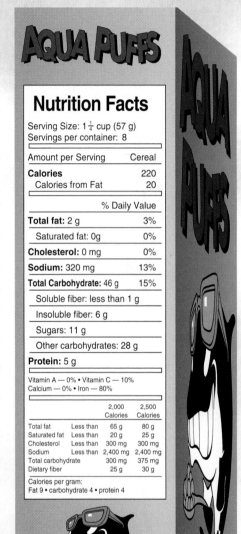

Figure 2B **Nutrition label on side panel of cereal box.**

---

*A calorie is the amount of heat required to raise the temperature of one gram of water one degree centigrade. A Calorie (capital C) is 1,000 calories.

2.8 Nucleic Acids

Nucleic acids, such as **DNA (deoxyribonucleic acid)** and **RNA (ribonucleic acid),** are huge polymers of nucleotides. Every **nucleotide** is a molecular complex of three types of subunit molecules—phosphate (phosphoric acid), a pentose sugar, and a nitrogen-containing base:

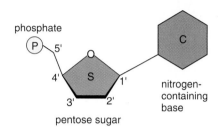

DNA makes up the genes and stores information regarding its own replication and the order in which amino acids are to be joined to form a protein. RNA is an intermediary in the process of protein synthesis, conveying information from DNA regarding the amino acid sequence in a protein.

The nucleotides in DNA contain the sugar deoxyribose, and in RNA they contain the sugar ribose; this difference accounts for their respective names (Table 2.2). As indicated in Figure 2.26, there are four different types of bases in DNA: A = adenine, T = thymine, G = guanine, and C = cytosine. The base can have two rings (adenine or guanine) or one ring (thymine or cytosine). These structures are called bases because their presence raises the pH of a solution. In RNA, the base uracil replaces the base thymine.

Although the sequence can vary between molecules, any particular DNA or RNA has a definite sequence. The nucleotides form a linear molecule called a strand, which has a

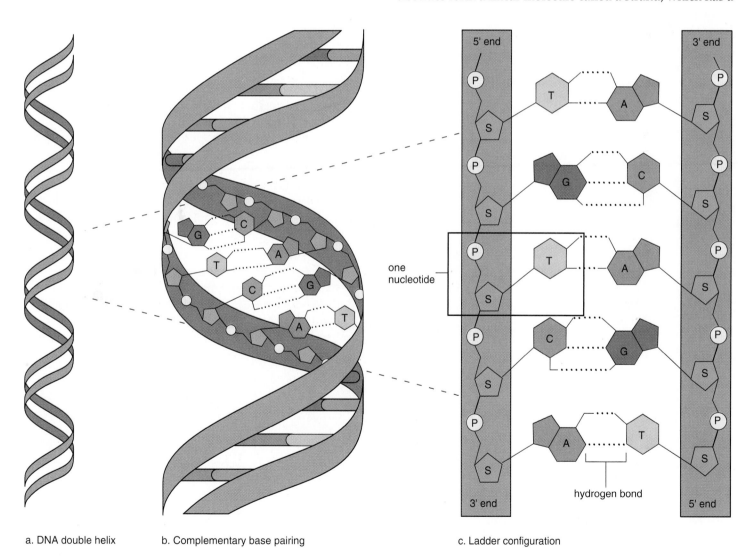

a. DNA double helix　　b. Complementary base pairing　　c. Ladder configuration

Figure 2.26　Overview of DNA structure.
a. Double helix. **b.** Complementary base pairing between strands. **c.** Ladder configuration. Notice that the uprights are composed of phosphate and sugar molecules and that the rungs are complementary paired bases.

backbone made up of phosphate-sugar-phosphate-sugar, with the bases projecting to one side of the backbone. Since the nucleotides occur in a definite order, so do the bases.

RNA is usually single stranded, while DNA is usually double stranded, with the two strands twisted about each other in the form of a double helix. In DNA, the two strands are held together by hydrogen bonds between the bases. When unwound, DNA resembles a stepladder. The sides of the ladder are made entirely of phosphate and sugar molecules, and the rungs of the ladder are made only of complementary paired bases. Thymine (T) always pairs with adenine (A), and guanine (G) always pairs with cytosine (C) (Fig. 2.26). Complementary bases have shapes that fit together.

We shall see that complementary base pairing allows DNA to replicate in a way that assures the sequence of bases will remain the same. The sequence of the bases is the genetic information that specifies the sequence of amino acids in the proteins of the cell.

DNA has a structure like a twisted ladder: sugar and phosphate molecules make up the sides, and hydrogen-bonded bases make up the rungs of the ladder.

Table 2.2	DNA Structure Compared to RNA Structure	
	DNA	RNA
SUGAR	Deoxyribose	Ribose
BASES	Adenine, guanine, thymine, cytosine	Adenine, guanine, uracil, cytosine
STRANDS	Double stranded with base pairing	Single stranded
HELIX	Yes	No

ATP (Adenosine Triphosphate)

Individual nucleotides can have metabolic functions in cells. When adenosine (adenine plus ribose) is modified by the addition of three phosphate groups, it becomes **ATP (adenosine triphosphate),** the primary energy carrier in cells.

Glucose contains too much energy to be used in cellular reactions, and cells use the energy within a glucose molecule to build ATP molecules. The amount of energy in ATP makes it useful to supply energy for chemical reactions in cells. As an analogy, consider that 20 dollar bills (i.e., ATP) are more useful for everyday purchases than 100 dollar bills (i.e., glucose).

ATP is called the energy currency of cells because when cells require energy, they often "spend" ATP. Cells use ATP for the synthesis of macromolecules like carbohydrates and proteins. In muscle cells, the energy within an ATP molecule is used for muscle contraction, and in nerve cells, it is used for the conduction of nerve impulses.

ATP is sometimes called a high-energy molecule because the last two phosphate bonds are unstable and are easily broken. Usually in cells, the terminal phosphate bond is hydrolyzed, leaving the molecule **ADP (adenosine diphosphate)** and a molecule of inorganic phosphate Ⓟ (Fig. 2.27). The terminal bond is sometimes called a high-energy bond, symbolized by a wavy line. But this terminology is misleading—the breakdown of ATP releases energy because the products of hydrolysis (ADP and Ⓟ) are more stable than ATP.

After ATP breaks down, it is rebuilt by the addition of Ⓟ to ADP (Fig. 2.27). There is enough energy in one glucose molecule to build 36 ATP molecules in this way. Homeostasis is only possible when cells continually produce and use ATP molecules. ⚖

ATP is a high-energy molecule. When ATP breaks down to ADP + Ⓟ, releasing energy, this energy is used for all metabolic work done in a cell.

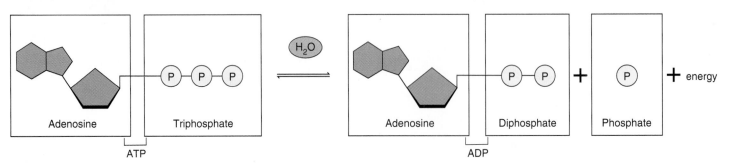

Figure 2.27   ATP reaction.
ATP, the universal energy currency of cells, is composed of adenosine and three phosphate groups. When cells require energy, ATP usually becomes ADP + Ⓟ, with the release of energy.

Organic Pollutants

Organic compounds include the carbohydrates, proteins, lipids, and nucleic acids that make up our bodies. Modern industry also uses all sorts of organic compounds that are synthetically produced. Indeed, our modern way of life wouldn't be possible without synthetic organic compounds.

Pesticides, herbicides, disinfectants, plastics, and textiles contain organic substances that are termed pollutants when they enter the natural environment and cause harm to living things. Global use of pesticides has increased dramatically since the 1950s, and modern pesticides are ten times more toxic than those of the 1950s. The Centers for Disease Control and Prevention in Atlanta, Georgia, report that 40% of children working in agricultural fields now show signs of pesticide poisoning. The U.S. Geological Survey estimates that 32 million people in urban areas and 10 million people in rural areas are using groundwater that contains organic pollutants. J. Charles Fox, an official of the Environmental Protection Agency, says that "over the life of a person ingestion of these chemicals has been shown to have adverse health effects such as cancer, reproductive problems, and developmental effects."

At one time, people failed to realize that everything in the environment is connected to everything else. In other words, they didn't know that an organic chemical can wander far from the site of its entry into the environment and that eventually these chemicals can enter our own bodies and cause harm. Now that we are aware of this outcome, we have to decide as a society how to proceed. We might decide to do nothing if the percentage of people dying from exposure to organic pollutants is small. Or we might decide to regulate the use of industrial compounds more strictly than has been done in the past. We could also decide that we need better ways of purifying public and private water supplies so that they do not contain organic pollutants.

Decide Your Opinion

1. Are you in favor of reducing the level of organic pollutants in the environment? Even if it reduces productivity and has adverse economic consequences? Explain.
2. Are you willing to stop using pesticides on your own lawn in order to prevent pollution of the water supply? Discuss.
3. Should the government regulate the production, use, and cleanup of synthetic organic compounds?
4. Are you willing to devote time and energy to promoting such government regulations? If so, to what degree?

Figure 2C   **Drinking water.**
Small traces of synthetic organic chemicals used in industry are in most public water supplies today.

Looking at Both Sides   www.mhhe.com/biosci/genbio/maderhuman7/

Every bioethical issue has at least two sides. Even if you already have an opinion, it is important to explore the opposite opinion before finalizing your position. The Online Learning Center at www.mhhe.com/biosci/genbio/maderhuman7/ will help you fine-tune your initial opinion, explore both sides, and finalize your position. You may acquire new arguments for your original opinion, or you may even change your opinion. Be sure to complete these activities in sequence:

Taking Sides   Decide your initial opinion by answering a series of questions. Then see if your opinion changes after completing the next two activities.

Further Debate   Read opposing articles that give you further information on this particular bioethical issue.

Explain Your Position   Answer another series of questions and then defend your original or changed opinion. You can e-mail your position to your instructor if he or she wishes.

Summarizing the Concepts

2.1 Elements and Atoms
All matter is composed of some 92 elements. Each element is made up of just one type of atom. An atom has a weight, which is dependent on the number of protons and neutrons in the nucleus, and its chemical properties are dependent on the number of electrons in the outer shell.

2.2 Molecules and Compounds
Atoms react with one another by forming ionic bonds or covalent bonds. Ionic bonds are an attraction between charged ions. Atoms share electrons in covalent bonds, which can be single, double, or triple bonds.

2.3 Water and Living Things
Water, acids, and bases are important inorganic molecules. The polarity of water accounts for it being the universal solvent; hydrogen bonding accounts for it boiling at 100°C and freezing at 0°C. Because it is slow to heat up and slow to freeze, water is liquid at the temperature of living things.

Pure water has a neutral pH; acids increase the hydrogen ion concentration [H$^+$] but decrease the pH, and bases decrease the hydrogen ion concentration [H$^+$] but increase the pH of water.

2.4 Molecules of Life
The chemistry of carbon accounts for the chemistry of organic compounds. Carbohydrates, lipids, proteins, and nucleic acids are molecules with specific functions in cells (Table 2.3).

2.5 Carbohydrates
Glucose is the 6-carbon sugar most utilized by cells for "quick" energy. Like the rest of the macromolecules to be studied, condensation synthesis joins two or more sugars, and a hydrolysis reaction splits the bond. Plants store glucose as starch, and animals store glucose as glycogen. Humans cannot digest cellulose, which forms plant cell walls.

2.6 Lipids
Lipids are varied in structure and function. Fats and oils, which function in long-term energy storage, contain glycerol and three fatty acids. Fatty acids can be saturated or unsaturated. Plasma membranes contain phospholipids that have a polarized end. Certain hormones are derived from cholesterol, a complex ring compound.

2.7 Proteins
The primary structure of a polypeptide is its own particular sequence of the possible 20 types of amino acids. The secondary structure is often an alpha (α) helix. The tertiary structure occurs when a polypeptide bends and twists into a three-dimensional shape. A protein can contain several polypeptides, and this accounts for a possible quaternary structure.

2.8 Nucleic Acids
Nucleic acids are polymers of nucleotides. Each nucleotide has three components: a sugar, a base, and phosphate (phosphoric acid). DNA, which contains the sugar deoxyribose, is the genetic material that stores information for its own replication and for the order in which amino acids are to be sequenced in proteins. DNA, with the help of RNA, specifies protein synthesis.

ATP, with its unstable phosphate bonds, is the energy currency of cells. Hydrolysis of ATP to ADP + Ⓟ releases energy that is used by the cell to do metabolic work.

Table 2.3	Organic Compounds Associated with Living Things	
	Unit Molecules	**Function**
Proteins	Amino acids	Enzymes speed up chemical reactions; structural components (e.g., muscle and membrane proteins)
CARBOHYDRATES		
Starch	Glucose	Energy storage in plants
Glycogen	Glucose	Energy storage in animals
Cellulose	Glucose	Plant cell walls
LIPIDS		
Fats and Oils	Glycerol, 3 fatty acids	Long-term energy storage
Phospholipids	Glycerol, 2 fatty acids, phosphate group	Plasma membrane structure
NUCLEIC ACIDS		
DNA	Nucleotides with deoxyribose sugar	Genetic material
RNA	Nucleotides with ribose sugar	Protein synthesis

Studying the Concepts

1. Name the subatomic particles of an atom; describe their charge, weight, and location in the atom. 16–17
2. Give an example of an ionic reaction, and explain it. 18–19
3. Diagram the atomic structure of calcium, and explain how it can react with two chlorine atoms. 19
4. Give an example of a covalent reaction, and explain it. 20
5. Relate the characteristics of water to its polarity and hydrogen bonding between water molecules. 21–22
6. On the pH scale, which numbers indicate a basic solution? An acidic solution? Why? 24
7. What are buffers, and why are they important to life? 24
8. Relate the variety of organic molecules to the bonding capabilities of carbon. 26
9. Name the four classes of organic molecules in cells. Which ones are polymers? Why? 26
10. Name some monosaccharides, disaccharides, and polysaccharides, and state some general functions for each. What is the most common monomer for polysaccharides? 27–28
11. How is a triglyceride synthesized? What is a saturated fatty acid? An unsaturated fatty acid? What is the function of fats? 29
12. Relate the structure of a phospholipid to that of a neutral fat. What is the function of a phospholipid? 30
13. What is the general structure and significance of cholesterol? 30
14. What are some functions of proteins? What is a peptide bond, a dipeptide, and a polypeptide? 31–34
15. Discuss the primary, secondary, and tertiary structures of proteins. 32
16. Discuss the structure and function of the nucleic acids DNA and RNA. 34–35

Testing Your Knowledge of the Concepts

In questions 1–4, match the molecule to the functions below.
a. carbohydrates
b. lipids
c. proteins
d. nucleic acids

_____ 1. long-term energy storage

_____ 2. most enzymes

_____ 3. immediate source of energy

_____ 4. hereditary units called genes

In questions 5–7, indicate whether the statement is true (T) or false (F).

_____ 5. The higher the pH, the higher the H^+ concentration.

_____ 6. Ionic bonds share electrons, and covalent bonds are an attraction between charges.

_____ 7. Protons are located in the nucleus, while electrons are located in shells about the nucleus.

In questions 8 and 9, fill in the blanks.
8. Fats and oils contain the molecules _____ and _____.

9. A nucleotide contains a _____ sugar, a _____ group, and a nitrogen-containing _____.

10. Label this diagram using these terms: *condensation, hydrolysis, monomer, polymer.*

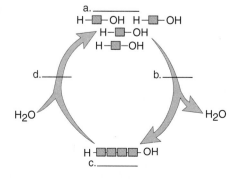

Understanding Key Terms

acid 23
ADP (adenosine diphosphate) 36
amino acid 31
atom 16
ATP (adenosine triphosphate) 36
base 23
buffer 24
calorie 22
carbohydrate 27
cellulose 28
condensation synthesis 27
covalent bond 20
denaturation 32
disaccharide 27
DNA (deoxyribonucleic acid) 35
electron 16
element 16
emulsification 29
enzyme 31
fat 29
fatty acid 29
functional group 26
glucose 27
glycogen 28
hexose 27
hydrogen bond 21

hydrolysis 27
hydrophilic 21
hydrophobic 21
inorganic molecule 26
ion 19
ionic bond 19
isotope 17
lipid 29
matter 16
molecule 19
monosaccharide 27
neutron 16
nucleotide 35
oil 29
organic molecule 26
pentose 27
peptide bond 32
pH scale 24
phospholipid 30
polypeptide 32
polysaccharide 28
protein 31
proton 16
RNA (ribonucleic acid) 35
saturated fatty acid 29
starch 28
steroid 30
triglyceride 29
unsaturated fatty acid 29

e-Learning Connection

2.1 Elements and Atoms

 <u>Atoms</u> *Essential Study Partner*

 <u>Periodic Table</u> *art quiz*
<u>Electron Energy Levels</u> *art quiz*
<u>Atomic Structure</u> *animation activity*

2.2 Molecules and Compounds

 <u>Bonds</u> *Essential Study Partner*

 <u>Covalent Bond</u> *animation activity*
<u>Ionic Bond</u> *animation activity*

2.3 Water and Living Things

 <u>Water</u> *Essential Study Partner*
<u>pH</u> *Essential Study Partner*

 <u>Molecular Structure of Water</u> *art quiz*
<u>Water as a Solvent</u> *art quiz*
<u>pH Scale</u> *art quiz*
<u>Formation of Acid Precipitation</u> *art quiz*

2.4 Molecules of Life

 <u>Organic Chemistry</u> *Essential Study Partner*

2.5 Carbohydrates

 <u>Carbohydrates</u> *Essential Study Partner*

 <u>Disaccharides</u> *art quiz*

2.6 Lipids

 <u>Lipids</u> *Essential Study Partner*

 <u>Unsaturated Fat</u> *art quiz*

2.7 Proteins

 <u>Proteins</u> *Essential Study Partner*

 <u>Peptide Bonds</u> *art quiz*

2.8 Nucleic Acids

 <u>Nucleic Acids</u> *Essential Study Partner*

 <u>DNA Base-Pairing and Hydrogen Bonds</u> *art quiz*

Chapter Summary

 <u>Key Term Flashcards</u> *vocabulary quiz*
<u>Chapter Quiz</u> *objective quiz*

Chapter 3

Cell Structure and Function

Chapter Concepts

Figure 3.1 **Racing cyclists.**
Cycling or any human activity is dependent on the functioning of skeletal muscle cells (colored red in the micrograph). Oxygen reaches muscle cells by way of the capillaries (colored blue).

Helen is nervous. But when she hops on her bike at the start of the race, her body's harmonized network of 75 trillion cells answers the challenge. Her brain sends messages along nerves to her skeletal muscles, which are fastened to her bones. When her leg muscles contract, her bones move the bike. The energy to power her muscles comes from sugars absorbed by her digestive tract. Oxygen, used to release energy from sugars, is absorbed by her lungs. Sugar and oxygen molecules are delivered to muscle cells by the circulatory system. And the waste products are expelled from Helen's body by the lungs and kidneys.

The body's organs are composed of cells. Helen can win the race because each individual muscle cell has done its job of keeping her legs moving. Cells are small and it takes a microscope to see them; therefore it is sometimes hard to imagine that individual muscle cells account for the functioning of an organ like a skeletal muscle.

Use of a microscope does show that muscles and all organs are composed of cells. The electron microscope, developed in the last century, has revealed that cells contain organelles, little bodies that are specialized in structure to carry on a particular function. Mitochondria are the organelles that oxidize sugar molecules to release energy. When you are fit, your mitochondria are conditioned to start using oxygen right away so that acids don't build up and cause fatigue. Helen trained for many months to increase her endurance. Her muscle cells have more mitochondria than those of people who have not trained, and her mitochondria are all set to help her win the race.

Despite specialization—muscle cells are specialized to contract—all cells have the same basic structure and metabolism. This chapter discusses the generalized structure of cells. It also describes the structure and function of the various organelles that carry on the activities of a cell. The chapter ends by describing the cellular reactions that provide energy for the workings of a cell.

3.1 Cell Size

All living things are made up of fundamental units called **cells.** Because cells are so small, the study of cells did not begin until the invention of the first microscope in the seventeenth century. Then the **cell theory,** which states that *all living things are composed of cells, and new cells arise only from preexisting cells,* was formulated.

Regardless of a cell's size and shape, it must carry on the functions associated with life—interacting with the environment, obtaining chemicals and energy, growing, and reproducing. A few cells, like a hen's egg or a frog's egg, are large enough to be seen by the naked eye, but most are not. This is the reason a microscope is needed to see cells. Why are cells so small (most are less than one cubic millimeter)? The small size of cells and consequently our multicellularity is explained by considering the surface/volume ratio of cells. Nutrients enter a cell and wastes exit a cell at its surface; therefore, the amount of surface represents the ability to get material in and out of the cell. A large cell requires more nutrients and produces more wastes than a small cell. In other words, the volume represents the needs of the cell. Yet, as cells get larger in volume, the proportionate amount of surface area actually decreases, as you can see by comparing these two cells:

small cell— more surface area per volume

large cell— less surface area per volume

1 mm tall cube: surface area/volume ratio = 6 : 1

2 mm tall cube: surface area/volume ratio = 3 : 1

We would expect, then, that there would be a limit to how large an actively metabolizing cell can become. Once a hen's egg is fertilized and starts actively metabolizing, it divides repeatedly without growth. Cell division restores the amount of surface area needed for adequate exchange of materials.

A cell needs a surface area that can adequately exchange materials with the environment. This explains why cells stay small.

Microscopy and Cell Structure

Three types of microscopes are most commonly used: the compound light microscope, transmission electron microscope, and scanning electron microscope. Figure 3.2 depicts these microscopes, along with a micrograph of red blood cells viewed with each one.

In a compound light microscope, light rays passing through a specimen are brought to focus by a set of glass lenses, and the resulting image is then viewed by the human eye. In the transmission electron microscope, electrons passing through a specimen are brought to focus by a set of magnetic lenses, and the resulting image is projected onto a fluorescent screen or photographic film.

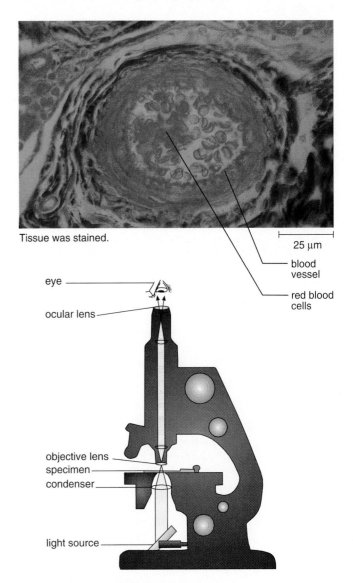

Tissue was stained.

25 μm

blood vessel

red blood cells

eye

ocular lens

objective lens

specimen

condenser

light source

Compound light microscope

Figure 3.2   Blood vessels and red blood cells viewed with three different types of microscopes.

The magnification produced by an electron microscope is much higher than that of a light microscope. Also, the ability of the electron microscope to make out detail in enlarged images is much greater. In other words, the electron microscope has a higher resolving power—that is, the ability to distinguish between two adjacent points. The following lists the resolving power of the eye, the light microscope, and the electron microscope:

Eye:	0.2 mm	= 200 μm	=	200,000 nm
Light microscope: (1,000×)	0.0002 mm	= 0.200 μm	=	200 nm
Electron microscope (50,000×):	0.00001 mm	= 0.0001 μm	=	10 nm

A scanning electron microscope provides a three-dimensional view of the surface of an object. A narrow beam of electrons is scanned over the surface of the specimen, which has been coated with a thin layer of metal. The metal gives off secondary electrons, which are collected to produce a television-type picture of the specimen's surface on a screen.

A picture obtained using a light microscope sometimes is called a photomicrograph, and a picture resulting from the use of an electron microscope is called a transmission electron micrograph (TEM) or a scanning electron micrograph (SEM), depending on the type of microscope used.

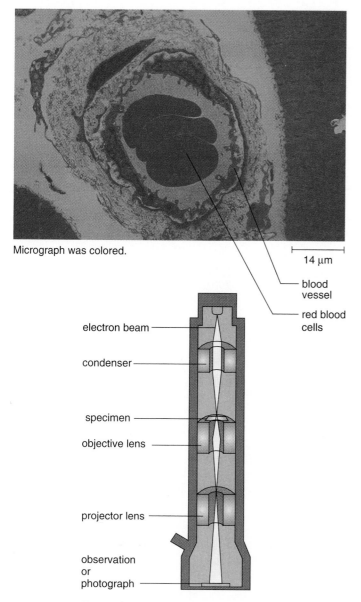

Micrograph was colored.

14 μm

blood vessel

red blood cells

Transmission electron microscope

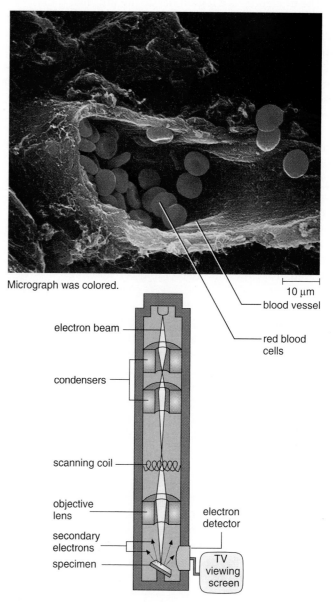

Micrograph was colored.

10 μm

blood vessel

red blood cells

Scanning electron microscope

3.2 Cellular Organization

The **plasma membrane** that surrounds and keeps the cell intact regulates what enters and exits a cell. The plasma membrane is a phospholipid bilayer that is said to be semipermeable because it allows certain molecules but not others to enter the cell. Proteins present in the plasma membrane play important roles in allowing substances to enter the cell.

The **nucleus** is a large, centrally located structure that can often be seen with a light microscope. The nucleus contains the chromosomes and is the control center of the cell. It controls the metabolic functioning and structural characteristics of the cell. The **nucleolus** is a region inside the nucleus.

The **cytoplasm** is the portion of the cell between the nucleus and the plasma membrane. The matrix of the cytoplasm is a semifluid medium that contains water and various types of molecules suspended or dissolved in the medium. The presence of proteins accounts for the semifluid nature of the matrix.

The cytoplasm contains various **organelles.** Organelles are small membranous structures that can usually only be seen with an electron microscope. Each type of organelle has a specific function. One type of organelle transports substances, for example, and another type produces ATP for the cell. Since organelles are composed of membrane, it can be seen that membrane compartmentalizes, keeping the various cellular activities separated from one another (Table 3.1 and Fig. 3.3).

Cells also have a **cytoskeleton,** a network of interconnected filaments and microtubules that occur in the cytoplasm. The name cytoskeleton is convenient in that it allows us to compare the cytoskeleton to the bones and muscles of an animal. Bones and muscle give an animal structure and produce movement. Similarly, the elements of the cytoskeleton maintain cell shape and allow the cell and its contents to move. Some cells move by using cilia and flagella, which are also made up of microtubules.

The human cell has a central nucleus and an outer plasma membrane. Various organelles are found within the cytoplasm, the portion of the cell between the nucleus and the plasma membrane.

Table 3.1   **Structures in Animal Cells**

Name	Composition	Function
Plasma membrane	Phospholipid bilayer with embedded proteins	Selective passage of molecules into and out of cell
Nucleus	Nuclear envelope surrounding nucleoplasm, chromatin, and nucleolus	Storage of genetic information
Nucleolus	Concentrated area of chromatin, RNA, and proteins	Ribosomal formation
Ribosome	Protein and RNA in two subunits	Protein synthesis
Endoplasmic reticulum (ER)	Membranous saccules and canals	Synthesis and/or modification of proteins and other substances, and transport by vesicle formation
Rough ER	Studded with ribosomes	Protein synthesis
Smooth ER	Having no ribosomes	Various; lipid synthesis in some cells
Golgi apparatus	Stack of membranous saccules	Processing, packaging, and distributing molecules
Vacuole and vesicle	Membranous sacs	Storage and transport of substances
Lysosome	Membranous vesicle containing digestive enzymes	Intracellular digestion
Mitochondrion	Inner membrane (cristae) within outer membrane	Cellular respiration
Cytoskeleton	Microtubules, actin filaments	Shape of cell and movement of its parts
Cilia and flagella	9 + 2 pattern of microtubules	Movement of cell
Centriole	9 + 0 pattern of microtubules	Formation of basal bodies

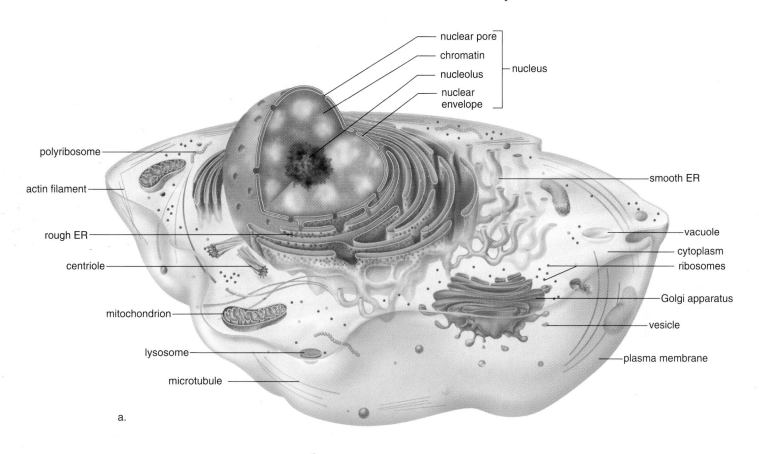

a.

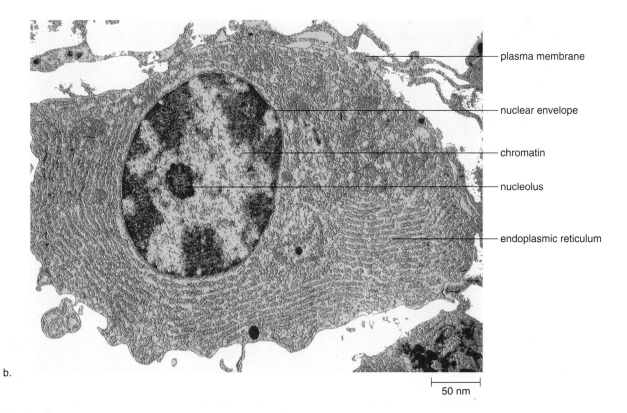

b.

50 nm

Figure 3.3   Animal cell.

a. Generalized drawing. **b.** Transmission electron micrograph. SeeTable 3.1 for a description of these structures, along with a listing of their functions.

The Plasma Membrane

An animal cell is surrounded by an outer plasma membrane. The plasma membrane marks the boundary between the outside of the cell and the inside of the cell. Plasma membrane integrity and function are necessary to the life of the cell.

The plasma membrane is a phospholipid bilayer with attached or embedded proteins. The structure of a phospholipid is such that the molecule has a polar head and nonpolar tails (Fig. 3.4). The polar heads, being charged, are hydrophilic (water loving) and face outward, toward the cytoplasm on one side and the tissue fluid on the other side, where they will encounter a watery environment. The nonpolar tails are hydrophobic (not attracted to water) and face inward toward each other, where there is no water. When phospholipids are placed in water, they naturally form a circular bilayer because of the chemical properties of the heads and the tails. At body temperature, the phospholipid bilayer is a liquid; it has the consistency of olive oil, and the proteins are able to change their position by moving laterally. The fluid-mosaic model, a working description of membrane structure, says that the protein molecules have a changing pattern (form a mosaic) within the fluid phospholipid bilayer (Fig. 3.4). Cholesterol lends support to the membrane.

Short chains of sugars are attached to the outer surface of some protein and lipid molecules (called glycoproteins and glycolipids, respectively). It is believed that these carbohydrate chains, specific to each cell, help mark it as belonging to a particular individual. They account for why people have different blood types, for example. Other glycoproteins have a special configuration that allows them to act as a receptor for a chemical messenger like a hormone. Some plasma membrane proteins form channels through which certain substances can enter cells; others are carriers involved in the passage of molecules through the membrane.

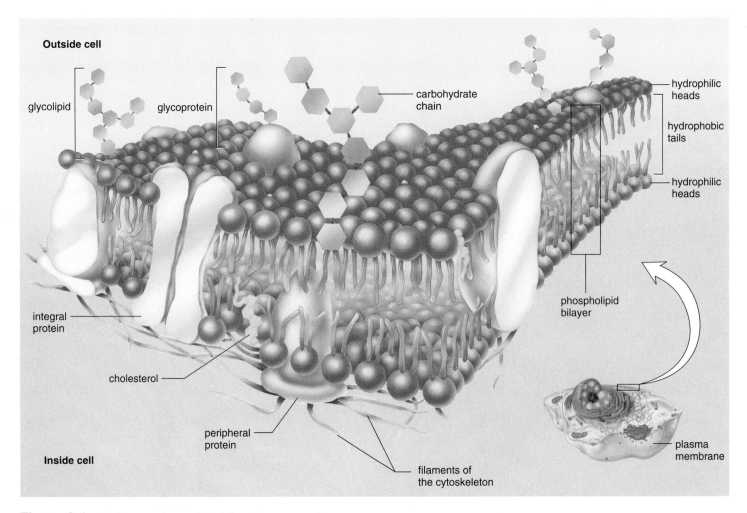

Figure 3.4 **Fluid-mosaic model of the plasma membrane.**
The membrane is composed of a phospholipid bilayer. The polar heads of the phospholipids are at the surfaces of the membrane; the nonpolar tails make up the interior of the membrane. Proteins are embedded in the membrane. Some of these function as receptors for chemical messengers, as conductors of molecules through the membrane, and as enzymes in metabolic reactions.

Plasma Membrane Functions

The plasma membrane keeps a cell intact. It allows only certain molecules and ions to enter and exit the cytoplasm freely; therefore, the plasma membrane is said to be **selectively permeable**. Small molecules that are lipid soluble, such as oxygen and carbon dioxide, can pass through the membrane easily. Certain other small molecules, like water, are not lipid soluble, but they still freely cross the membrane. Still other molecules and ions require the use of a carrier to enter a cell.

plasma membrane

Animal Cells

Under isotonic conditions, there is no net movement of water.

In a hypotonic environment, water enters the cell, which may burst.

In a hypertonic environment, water leaves the cell, which shrivels.

Figure 3.5   Tonicity.
The arrows indicate the movement of water.

The plasma membrane, composed of phospholipid and protein molecules, is selectively permeable and regulates the entrance and exit of molecules and ions into and out of the cell.

Diffusion   Diffusion is the random movement of molecules from the area of higher concentration to the area of lower concentration until they are equally distributed. To illustrate diffusion, imagine opening a perfume bottle in the corner of a room. The smell of the perfume soon permeates the room because the molecules that make up the perfume move to all parts of the room. Another example is putting a tablet of dye into water. The water eventually takes on the color of the dye as the dye molecules diffuse.

The chemical and physical properties of the plasma membrane allow only a few types of molecules to enter and exit a cell simply by diffusion. Lipid-soluble molecules such as alcohols can diffuse through the membrane because lipids are the membrane's main structural components. Gases can also diffuse through the lipid bilayer; this is the mechanism by which oxygen enters cells and carbon dioxide exits cells. As an example, consider the movement of oxygen from the alveoli (air sacs) of the lungs to blood in the lung capillaries. After inhalation (breathing in), the concentration of oxygen in the alveoli is higher than that in the blood; therefore, oxygen diffuses into the blood.

When molecules simply diffuse down their concentration gradients across plasma membranes, no cellular energy is involved.

Molecules diffuse down their concentration gradients. A few types of small molecules can simply diffuse through the plasma membrane, and no carrier protein or cellular energy is involved.

Osmosis   Osmosis is the diffusion of water across a plasma membrane. It occurs whenever there is an unequal concentration of water on either side of a selectively permeable membrane. Normally, body fluids are isotonic to cells (Fig. 3.5)—that is, there is an equal concentration of substances (solutes) and water (solvent) on both sides of the plasma membrane, and cells maintain their usual size and shape. Intravenous solutions medically administered usually have this tonicity. **Tonicity** is the degree to which a solution's concentration of solute versus water causes water to move into or out of cells.

Solutions that cause cells to swell or even to burst due to an intake of water are said to be hypotonic solutions. If red blood cells are placed in a hypotonic solution, which has a higher concentration of water (lower concentration of solute) than do the cells, water enters the cells and they swell to bursting. The term lysis is used to refer to disrupted cells; hemolysis, then, is disrupted red blood cells.

Solutions that cause cells to shrink or to shrivel due to a loss of water are said to be hypertonic solutions. If red blood cells are placed in a hypertonic solution, which has a lower concentration of water (higher concentration of solute) than do the cells, water leaves the cells and they shrink. The term crenation refers to red blood cells in this condition.

These changes have occurred due to osmotic pressure. Osmotic pressure is the force exerted on a selectively permeable membrane because water has moved from the area of higher to lower concentration of water (higher concentration of solute).

In an isotonic solution, a cell neither gains nor loses water. In a hypotonic solution, a cell gains water. In a hypertonic solution, a cell loses water and the cytoplasm shrinks.

Transport by Carriers   Most solutes do not simply diffuse across a plasma membrane; rather, they are transported by means of protein carriers within the membrane. During

facilitated transport, a molecule (e.g., an amino acid or glucose) is transported across the plasma membrane from the side of higher concentration to the side of lower concentration. The cell does not need to expend energy for this type of transport because the molecules are moving down their concentration gradient.

During **active transport,** a molecule is moving contrary to the normal direction—that is, from lower to higher concentration (Fig. 3.6). For example, iodine collects in the cells of the thyroid gland; sugar is completely absorbed from the gut by cells that line the digestive tract; and sodium (Na^+) is sometimes almost completely withdrawn from urine by cells lining kidney tubules. Active transport requires a protein carrier and the use of cellular energy obtained from the breakdown of ATP. When ATP is broken down, energy is released, and in this case the energy is used by a carrier to carry out active transport. Therefore, it is not surprising that cells involved in active transport, such as kidney cells, have a large number of mitochondria near the membrane at which active transport is occurring.

Proteins involved in active transport often are called pumps because just as a water pump uses energy to move water against the force of gravity, proteins use energy to move substances against their concentration gradients. One type of pump that is active in all cells but is especially associated with nerve and muscle cells moves sodium ions (Na^+) to the outside of the cell and potassium ions (K^+) to the inside of the cell.

The passage of salt (NaCl) across a plasma membrane is of primary importance in cells. First, sodium ions are pumped across a membrane; then, chloride ions simply diffuse through channels that allow their passage. Chloride ion channels malfunction in persons with cystic fibrosis, and this leads to the symptoms of this inherited (genetic) disorder.

Endocytosis and Exocytosis   During endocytosis, a portion of the plasma membrane invaginates to envelop a substance, and then the membrane pinches off to form an intracellular vesicle. Digestion may be required before molecules can cross a vesicle membrane to enter the cytoplasm. During exocytosis, the vesicle often formed at the Golgi apparatus fuses with the plasma membrane as secretion occurs. This is the way insulin leaves insulin-secreting cells, for instance. Table 3.2 summarizes the various ways molecules get into and out of cells.

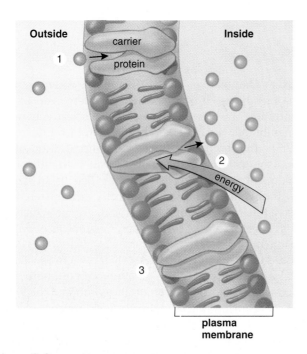

Figure 3.6   Active transport through a plasma membrane.
Active transport allows a solute to cross the membrane from lower solute concentration to higher solute concentration. ① Molecule enters carrier. ② Chemical energy of ATP is needed to transport the molecule, which exits inside of cell. ③ Carrier returns to its inactive state.

Table 3.2	Passage of Molecules into and out of Cells			
	Name	**Direction**	**Requirement**	**Examples**
PASSIVE TRANSPORT	Diffusion	Toward lower concentration	Concentration gradient	Lipid-soluble molecules, water, and gases
	Facilitated Transport	Toward lower concentration	Carrier and concentration gradient	Sugars and amino acids
ACTIVE TRANSPORT	Active Transport	Toward greater concentration	Carrier plus energy ions	Sugars, amino acids, and
	Endocytosis	Toward inside	Vesicle formation	Macromolecules
	Exocytosis	Toward outside	Vesicle fuses with plasma membrane	Macromolecules

The Nucleus

The nucleus, which has a diameter of about 5 μm, is a prominent structure in the eukaryotic cell. The nucleus is of primary importance because it stores genetic information that determines the characteristics of the body's cells and their metabolic functioning. Every cell contains a complex copy of genetic information, but each cell type has certain genes, or segments of DNA, turned on, and others turned off. Activated DNA, with RNA acting as an intermediary, specifies the sequence of amino acids during protein synthesis. The proteins of a cell determine its structure and the functions it can perform.

When you look at the nucleus, even in an electron micrograph, you cannot see DNA molecules but you can see chromatin (Fig. 3.7). **Chromatin** looks grainy, but actually it is a threadlike material that undergoes coiling into rodlike structures called **chromosomes** just before the cell divides. Chemical analysis shows that chromatin, and therefore chromosomes, contains DNA and much protein, as well as some RNA. Chromatin is immersed in a semifluid medium called the **nucleoplasm.** A difference in pH between the nucleoplasm and the cytoplasm suggests that the nucleoplasm has a different composition.

Most likely, too, when you look at an electron micrograph of a nucleus, you will see one or more regions that look darker than the rest of the chromatin. These are nucleoli (sing., nucleolus) where another type of RNA, called ribosomal RNA (rRNA), is produced and where rRNA joins with proteins to form the subunits of ribosomes. (Ribosomes are small bodies in the cytoplasm that contain rRNA and proteins.)

The nucleus is separated from the cytoplasm by a double membrane known as the **nuclear envelope,** which is continuous with the endoplasmic reticulum discussed on the next page. The nuclear envelope has **nuclear pores** of sufficient size (100 nm) to permit the passage of proteins into the nucleus and ribosomal subunits out of the nucleus.

The structural features of the nucleus include the following.

Chromatin:	DNA and proteins
Nucleolus:	Chromatin and ribosomal subunits
Nuclear envelope:	Double membrane with pores

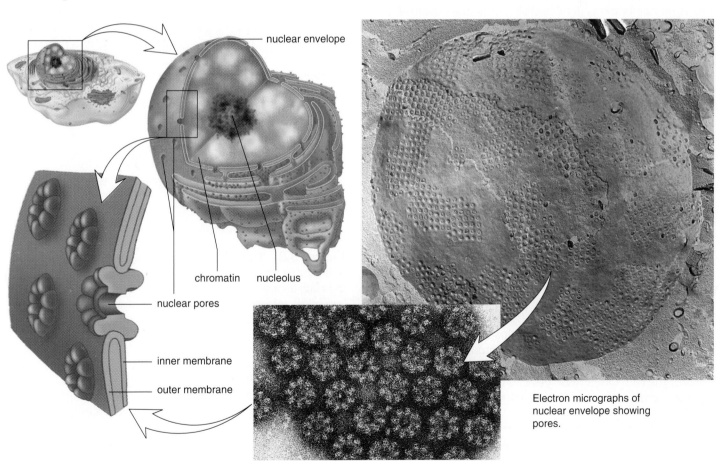

nuclear envelope

chromatin   nucleolus

nuclear pores

inner membrane

outer membrane

Electron micrographs of nuclear envelope showing pores.

Figure 3.7 The nucleus and the nuclear envelope.
The nucleus contains chromatin. Chromatin has a special region called the nucleolus, which is where rRNA is produced and ribosomal subunits are assembled. The nuclear envelope contains pores, as shown in this micrograph of a freeze-fractured nuclear envelope. Each pore is lined by a complex of eight proteins.

Ribosomes

Ribosomes are composed of two subunits, one large and one small. Each subunit has its own mix of proteins and rRNA. Protein synthesis occurs at the ribosomes. Ribosomes are found free within the cytoplasm either singly or in groups called **polyribosomes.** Ribosomes are often attached to the endoplasmic reticulum, a membranous system of saccules and channels discussed in the next section. Proteins synthesized by cytoplasmic ribosomes are used inside the cell for various purposes. Those produced by ribosomes attached to endoplasmic reticulum may eventually be secreted from the cell.

> Ribosomes are small organelles where protein synthesis occurs. Ribosomes occur in the cytoplasm, both singly and in groups (i.e., polyribosomes). Numerous ribosomes are attached to the endoplasmic reticulum.

Membranous Canals and Vesicles

The endomembrane system consists of the nuclear envelope, the endoplasmic reticulum, the Golgi apparatus, and several **vesicles** (tiny membranous sacs). This system compartmentalizes the cell so that particular enzymatic reactions are restricted to specific regions. Membranes that make up the endomembrane system are connected by direct physical contact and/or by the transfer of vesicles from one part to the other.

The Endoplasmic Reticulum

The **endoplasmic reticulum (ER),** a complicated system of membranous channels and saccules (flattened vesicles), is physically continuous with the outer membrane of the nuclear envelope. Rough ER is studded with ribosomes on the side of the membrane that faces the cytoplasm (Fig. 3.8). Here proteins are synthesized and enter the ER interior where processing and modification begin. Smooth ER, which is continuous with rough ER, does not have attached ribosomes. Smooth ER synthesizes the phospholipids that occur in membranes and has various other functions depending on the particular cell. In the testes, it produces testosterone, and in the liver it helps detoxify drugs. Regardless of any specialized function, smooth ER also forms vesicles in which large molecules are transported to other parts of the cell. Often these vesicles are on their way to the plasma membrane or the Golgi apparatus.

> ER is involved in protein synthesis (rough ER) and various other processes such as lipid synthesis (smooth ER). Molecules that are produced or modified in the ER are eventually enclosed in vesicles that often transport them to the Golgi apparatus.

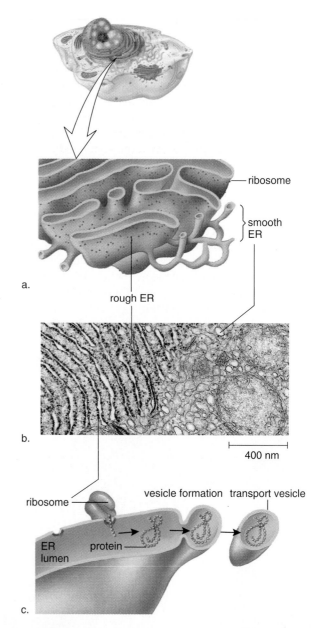

Figure 3.8   **The endoplasmic reticulum (ER).**
a. Rough ER has attached ribosomes, but smooth ER does not.
b. Rough ER appears to be flattened saccules, while smooth ER is a network of interconnected tubules. **c.** A protein made at a ribosome moves into the lumen of the system and eventually is packaged in a transport vesicle for distribution inside the cell.

The Golgi Apparatus

The **Golgi apparatus** is named for Camillo Golgi, who discovered its presence in cells in 1898. The Golgi apparatus consists of a stack of three to twenty slightly curved saccules whose appearance can be compared to a stack of pancakes (Fig. 3.9). In animal cells, one side of the stack (the inner face) is directed toward the ER, and the other side of the stack (the outer face) is directed toward the plasma membrane. Vesicles can frequently be seen at the edges of the saccules.

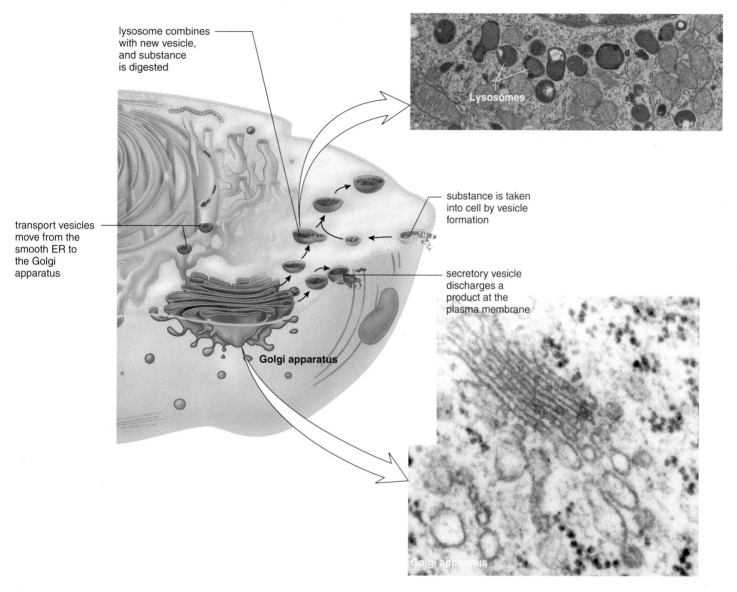

lysosome combines with new vesicle, and substance is digested

Lysosomes

transport vesicles move from the smooth ER to the Golgi apparatus

substance is taken into cell by vesicle formation

secretory vesicle discharges a product at the plasma membrane

Golgi apparatus

Golgi apparatus

Figure 3.9   The Golgi apparatus.
The Golgi apparatus modifies proteins and packages them either in vesicles for secretion from the cell or in lysosomes. Lysosomes function as digestive vesicles.

The Golgi apparatus receives protein and/or lipid-filled vesicles that bud from the ER. Some biologists believe that these fuse to form a saccule at the inner face and that this saccule remains as a part of the Golgi apparatus until the molecules are repackaged in new vesicles at the outer face. Others believe that the vesicles from the ER proceed directly to the outer face of the Golgi apparatus, where processing and packaging occur within its saccules. The Golgi apparatus contains enzymes that modify proteins and lipids. For example, it can add a chain of sugars to proteins, thereby making them glycoproteins and glycolipids, which are molecules found in the plasma membrane.

The vesicles that leave the Golgi apparatus move about the cell. Some vesicles proceed to the plasma membrane, where they discharge their contents. Because this is secretion, it is often said that the Golgi apparatus is involved in processing, packaging, and secretion. Other vesicles that leave the Golgi apparatus are lysosomes.

The Golgi apparatus processes, packages, and distributes molecules about or from the cell. It is also said to be involved in secretion.

Lysosomes

Lysosomes, vesicles produced by the Golgi apparatus, contain hydrolytic digestive enzymes. Sometimes macromolecules are brought into a cell by vesicle formation at the plasma membrane (see Fig. 3.9). When a lysosome fuses with such a vesicle, its contents are digested by lysosomal enzymes into simpler subunits that then enter the cytoplasm. Even parts of a cell are digested by its own lysosomes (called autodigestion). Normal cell rejuvenation most likely takes place in this manner, but autodigestion is also important during development. For example, when a tadpole becomes a frog, lysosomes digest away the cells of the tail. The fingers of a human embryo are at first webbed, but they are freed from one another as a result of lysosomal action.

Occasionally, a child is born with Tay-Sachs disease, a metabolic disorder involving a missing or inactive lysosomal enzyme. In these cases, the lysosomes fill to capacity with macromolecules that cannot be broken down. The cells become so full of these lysosomes that the child dies. Someday soon it may be possible to provide the missing enzyme for these children.

Lysosomes are produced by a Golgi apparatus, and their hydrolytic enzymes digest macromolecules from various sources.

Mitochondria

Most mitochondria (sing., **mitochondrion**) are between 0.5 μm and 1.0 μm in diameter and about 7 μm in length, although the size and the shape can vary. Mitochondria are bounded by a double membrane. The inner membrane is folded to form little shelves called cristae, which project into the matrix, an inner space filled with a gel-like fluid (Fig. 3.10).

Mitochondria are the site of ATP (adenosine triphosphate) production involving complex metabolic pathways. ATP molecules are the common carrier of energy in cells. A shorthand way to indicate the chemical transformation that involves mitochondria is as follows:

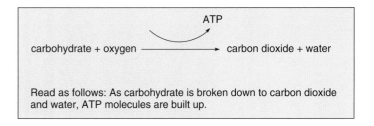

carbohydrate + oxygen ⟶ carbon dioxide + water, with ATP built up.

Read as follows: As carbohydrate is broken down to carbon dioxide and water, ATP molecules are built up.

Mitochondria are often called the powerhouses of the cell: just as a powerhouse burns fuel to produce electricity, the mitochondria convert the chemical energy of glucose products into the chemical energy of ATP molecules. In the process, mitochondria use up oxygen and give off carbon dioxide and water. The oxygen you breathe in enters cells

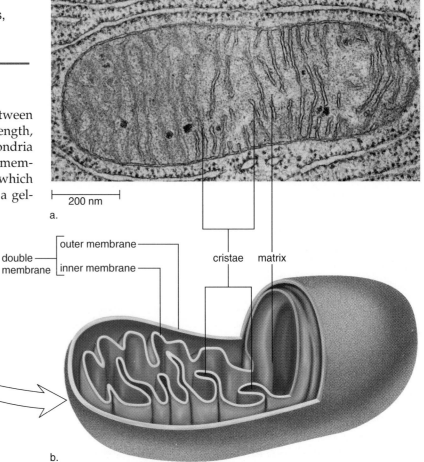

200 nm

a.

outer membrane
double membrane
inner membrane
cristae   matrix

b.

Figure 3.10 **Mitochondrion structure.**
Mitochondria are involved in cellular respiration. **a.** Electron micrograph. **b.** Generalized drawing in which the outer membrane and portions of the inner membrane have been cut away to reveal the cristae.

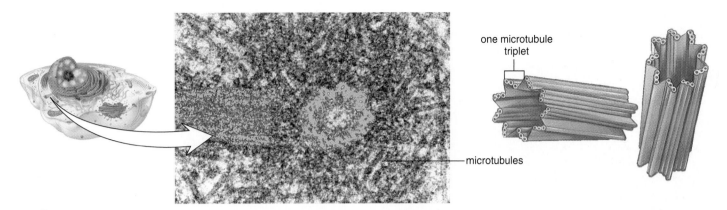

Figure 3.11   **Centrioles.**
Centrioles are composed of nine microtubule triplets. They lie at right angles to one another within the microtubule organizing center (MTOC), which is believed to assemble microtubules at the time of cell division.

and then mitochondria; the carbon dioxide you breathe out is released by mitochondria. Because oxygen is involved, it is said that mitochondria carry on cellular respiration.

The matrix of a mitochondrion contains enzymes for breaking down glucose products. ATP production then occurs at the cristae. The protein complexes that aid in the conversion of energy are located in an assembly-line fashion on these membranous shelves.

Every cell uses a certain amount of ATP energy to synthesize molecules, but many cells use ATP to carry out their specialized function. For example, muscle cells use ATP for muscle contraction, which produces movement, and nerve cells use it for the conduction of nerve impulses, which make us aware of our environment.

Mitochondria are involved in cellular respiration, a process that provides ATP molecules to the cell.

The Cytoskeleton

Several types of filamentous protein structures form a cytoskeleton that helps maintain the cell's shape and either anchors the organelles or assists their movement as appropriate. The cytoskeleton includes microtubules and actin filaments (see Fig. 3.3).

Microtubules are shaped like thin cylinders and are several times larger than actin filaments. Each cylinder contains 13 rows of tubulin, a globular protein, arranged in a helical fashion. Remarkably, microtubules can assemble and disassemble. In many cells, the regulation of microtubule assembly is under the control of a microtubule organizing center (MTOC), which lies near the nucleus. Microtubules radiate from the MTOC, helping to maintain the shape of the cell and acting as tracks along which organelles move. It is well known that during cell division, microtubules form spindle fibers, which assist the movement of chromosomes.

Actin filaments are long, extremely thin fibers that usually occur in bundles or other groupings. Actin filaments have been isolated from various types of cells, especially those in which movement occurs. Microvilli, which project from certain cells and can shorten and extend, contain actin filaments. Actin filaments, like microtubules, can assemble and disassemble.

The cytoskeleton contains microtubules and actin filaments. Microtubules (13 rows of tubulin protein molecules arranged to form a hollow cylinder) and actin filaments (thin actin strands) maintain the shape of the cell and also direct the movement of cell parts.

Centrioles

In animal cells, **centrioles** are short cylinders with a 9 + 0 pattern of microtubules. There are nine outer microtubule triplets and no center microtubules (Fig. 3.11). There is always one pair of centrioles lying at right angles to one another near the nucleus. Before a cell divides, the centrioles duplicate, and the members of the new pair are also at right angles to one another. During cell division, the pairs of centrioles separate so that each daughter cell gets one pair of centrioles.

Centrioles are part of a microtubule organizing center that also includes other proteins and substances. Microtubules begin to assemble in the center, and then they grow outward, extending through the entire cytoplasm. In addition, centrioles may be involved in other cellular processes that use microtubules, such as movement of material throughout the cell or formation of the spindle, a structure that distributes the chromosomes to daughter cells during cell division. Their exact role in these processes is uncertain, however. Centrioles also give rise to basal bodies that direct the formation of cilia and flagella.

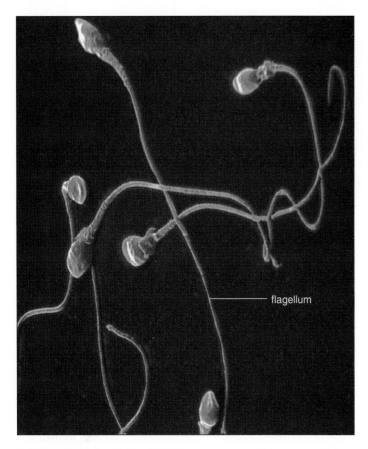

flagellum

Figure 3.12   Sperm cells.
Sperm cells use long, whiplike flagella to move about.

Cilia and Flagella

Cilia and flagella (sing., **cilium, flagellum**) are projections of cells that can move either in an undulating fashion, like a whip, or stiffly, like an oar. Cilia are short (2–10 μm) while flagella are longer (usually no longer than 200 μm). Cells that have these organelles are capable of self-movement or moving material along the surface of the cell. For example, sperm cells, carrying genetic material to the egg, move by means of flagella (Fig. 3.12). The cells that line our respiratory tract are ciliated. These cilia sweep debris trapped within mucus back up the throat, and this action helps keep the lungs clean.

Each cilium and flagellum has a basal body at its base, which lies in the cytoplasm. **Basal bodies,** like centrioles, have a 9 + 0 pattern of microtubule triplets. They are believed to organize the structure of cilia and flagella even though cilia and flagella have a 9 + 2 pattern of microtubules. In cilia and flagella, there are nine microtubule doublets surrounding two central microtubules. This arrangement is believed to be necessary to their ability to move.

Centrioles give rise to basal bodies that organize the pattern of microtubules in cilia and flagella.

3.3 Cellular Metabolism

Cellular **metabolism** includes all the chemical reactions that occur in a cell. Quite often these reactions are organized into metabolic pathways, which can be diagrammed as follows:

$$\begin{array}{ccccccccccc} & 1 & & 2 & & 3 & & 4 & & 5 & & 6 \\ A & \to & B & \to & C & \to & D & \to & E & \to & F & \to & G \end{array}$$

The letters, except A and G, are **products** of the previous reaction and the **reactants** for the next reaction. A represents the beginning reactant(s), and G represents the end product(s). The numbers in the pathway refer to different enzymes. *Every reaction in a cell requires a specific enzyme.* In effect, no reaction occurs in a cell unless its enzyme is present. For example, if enzyme 2 in the diagram is missing, the pathway cannot function; it will stop at B. Since enzymes are so necessary in cells, their mechanism of action has been studied extensively.

Most metabolic pathways are regulated by feedback inhibition: the end product of the pathway binds to a special site on the first enzyme of the pathway. This binding shuts down the pathway, and no more product is produced.

Metabolic pathways contain many enzymes that perform their reactions in a sequential order.

Enzymes and Coenzymes

When an enzyme speeds up a reaction, the reactant(s) that participate(s) in the reaction is called the enzyme's **substrate(s).** Enzymes are often named for their substrates (Table 3.3). Enzymes have a specific region, called an **active site,** where the substrates are brought together so that they can react. An enzyme's specificity is caused by the shape of the active site, where the enzyme and its substrate(s) fit together in a specific way, much as the pieces of a jigsaw puzzle fit together (Fig. 3.13). After one reaction is complete, the product or products are released, and the enzyme is ready to catalyze another reaction. This can be summarized in the following manner:

$$E + S \to ES \to E + P$$

(where E = enzyme, S = substrate, ES = enzyme-substrate complex, and P = product). The arrows in the diagram are

Table 3.3	Enzymes Named for Their Substrates
Substrate	**Enzyme**
Lipid	Lipase
Urea	Urease
Maltose	Maltase
Ribonucleic acid	Ribonuclease
Lactose	Lactase

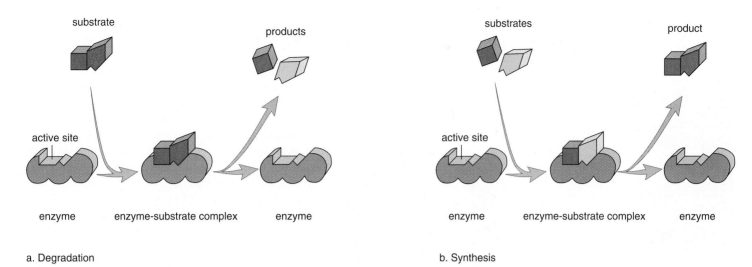

a. Degradation   b. Synthesis

Figure 3.13   **Enzymatic action.**
An enzyme has an active site where the substrates and enzyme fit together in such a way that the substrates are oriented to react. Following the reaction, the products are released and the enzyme is ready to participate in the reaction again. **a.** Some enzymes carry out degradation **b.** Some enzymes carry out synthesis.

not reversible. The energy content of the product(s) is (are) always lower than that of the substrate(s). Therefore, for the reverse reaction to occur energy has to be added.

Environmental conditions such as an incorrect pH or high temperature can cause an enzyme to become denatured. A denatured enzyme no longer has its usual shape and is therefore unable to speed up its reaction.

Many enzymes require cofactors. Some cofactors are inorganic, such as copper, zinc, or iron. Other cofactors are organic, nonprotein molecules and are called **coenzymes.** These cofactors assist the enzyme and may even accept or contribute atoms to the reaction. It is interesting that vitamins are often components of coenzymes. The vitamin niacin is a part of the coenzyme NAD, which removes hydrogen (H) atoms from substrates and therefore is called a **dehydrogenase.** Hydrogen atoms are sometimes removed by NAD as molecules are broken down. NAD that is carrying hydrogen atoms is written as $NADH_2$ because NAD removes two hydrogen atoms at a time. As we shall see, the removal of hydrogen atoms releases energy that can be used for ATP buildup.

Enzymes are specific because they have an active site that accommodates their substrates. Enzymes often have organic, nonprotein helpers called coenzymes. NAD is a dehydrogenase, a coenzyme that removes hydrogen from substrates.

Cellular Respiration

During **cellular respiration,** glucose is broken down to carbon dioxide and water. The energy released as glucose breakdown occurs is used to build up ATP molecules, the common energy carrier in cells. Even though it is possible to write an overall equation for the process, cellular respiration does not occur in one step. Glucose breakdown requires three subpathways: glycolysis, the Krebs cycle, and the electron transport system. The location of these subpathways is as follows:

- **Glycolysis** occurs in the cytoplasm, outside a mitochondrion.
- The **Krebs cycle** occurs in the matrix of a mitochondrion.
- The **electron transport system** occurs on the cristae of a mitochondrion.

Glycolysis and the Krebs cycle are a series of reactions in which the product of the previous reaction becomes the substrate for the next reaction. Every reaction that occurs during glycolysis and the Krebs cycle requires a specific enzyme. Each pathway resembles a conveyor belt in which a beginning substrate continuously enters at the start and, after a series of reactions, end products leave at the termination of the belt. It is important to realize, too, that these two pathways and the electron transport system occur at the same time. They can be compared to the inner workings of a watch, in which all parts are synchronized.

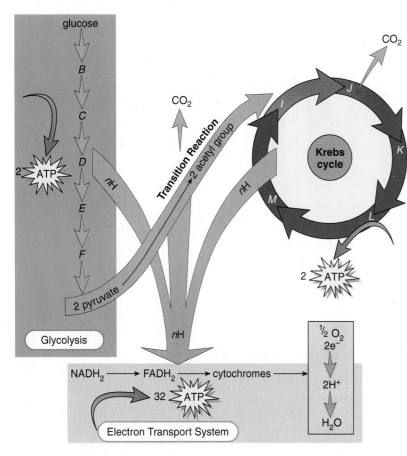

Figure 3.14   Cellular respiration.

The overall reaction shown at the top actually requires three subpathways: glycolysis, the Krebs cycle, and the electron transport system. As the reactions occur, a number of hydrogen (H) atoms and carbon dioxide (CO_2) molecules are removed from the various substrates. Oxygen (O_2) acts as the final acceptor for hydrogen atoms ($2e^- + 2H^+$) and becomes water (H_2O).

Table 3.4	Overview of Cellular Respiration
Name of Pathway	**Result**
Glycolysis	Removal of H from substrates produces 2 ATP
Transition reaction	Removal of H from substrates releases 2 CO₂
Krebs cycle	Removal of H from substrates releases 4 CO₂
	Produces 2 ATP after 2 turns
Electron transport system	Accepts H from other pathways and passes electrons on to O_2, producing H₂O
	Produces 32 ATP

The reactants of cellular respiration, namely glucose and oxygen, and the products, namely carbon dioxide and water, are related to the subpathways in the manner described next.

1. Glucose, a C_6 molecule, is to be associated with **glycolysis,** the breakdown of glucose to two molecules of pyruvate (pyruvic acid), a C_3 molecule. During glycolysis, energy is released as hydrogen (H) atoms are removed and added to NAD, forming $NADH_2$. This energy is used to form two ATP molecules (Fig. 3.14).

2. Carbon dioxide (CO_2) is to be associated with the transition reaction and the Krebs cycle, both of which occur in mitochondria. During the transition reaction, pyruvate is converted to a C_2 acetyl group after CO_2 comes off. Because the transition reaction occurs twice per glucose molecule, two molecules of CO_2 are released. Hydrogen (H) atoms are also removed at this time.

   The acetyl group enters the **Krebs cycle,** a cyclical series of reactions that gives off two CO_2 molecules and produces one ATP molecule. Since the Krebs cycle occurs twice per glucose molecule, altogether four CO_2 and two ATP are produced per glucose molecule. Hydrogen (H) atoms are removed from the substrates and added to NAD, forming $NADH_2$ as the Krebs cycle occurs.

3. Oxygen (O_2) and water (H_2O) are to be associated with the electron transport system. The **electron transport system** begins with $NADH_2$, a coenzyme that carries hydrogen (H) atoms to the system, but after that it consists of molecules that carry electrons. High-energy electrons are removed from the hydrogen atoms, leaving behind hydrogen ions (H^+), and then the electrons are passed from one molecule to another until the electrons are received by an oxygen atom. At this point, 2 H^+ combine with an oxygen to give water. As the electrons are passed down the system, their energy is released to allow the buildup of 32 ATP.

4. ATP is to be associated with glycolysis, the Krebs cycle, and the electron transport system. Altogether, 36 ATP molecules result from the breakdown of one glucose molecule (Table 3.4).

Cellular respiration requires glycolysis, which takes place in the cytoplasm; the Krebs cycle, which is located in the matrix of the mitochondria; and the electron transport system, which is located on the cristae of the mitochondria.

Fermentation

Fermentation is an anaerobic process. When oxygen is not available to cells, the electron transport system soon becomes inoperative because oxygen is not present to accept electrons. In this case, most cells have a safety valve so that some ATP can still be produced. Glycolysis operates as long as it is supplied with "free" NAD—that is, NAD that can pick up hydrogen atoms. Normally, $NADH_2$ takes hydrogens to the electron transport system and thereby becomes "free" of hydrogen atoms. However, if the system is not working due to lack of oxygen, $NADH_2$ passes its hydrogen atoms to pyruvate as shown in the following reaction:

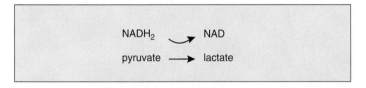

The Krebs cycle and electron transport system do not function as part of fermentation. When oxygen is available again, lactate (lactic acid) can be converted back to pyruvate, and metabolism can proceed as usual.

Fermentation takes less time than cellular respiration, but since glycolysis alone is occurring, it produces only 2 ATP per glucose molecule. Also, fermentation results in the buildup of lactate. Lactate is toxic to cells and causes muscles to cramp and fatigue. If fermentation continues for any length of time, death follows.

It is of interest to know that fermentation takes its name from yeast fermentation. Yeast fermentation produces alcohol and carbon dioxide (instead of lactate). When yeast is used to leaven bread, it is the carbon dioxide that produces the desired effect. When yeast is used to produce alcoholic beverages, it is the alcohol that humans make use of.

Fermentation is an anaerobic process, a process that does not require oxygen but produces very little ATP per glucose molecule and results in the buildup of lactate or alcohol and carbon dioxide.

Summarizing the Concepts

3.1 Cell Size
Cells are quite small, and it usually takes a microscope to see them. Small cubes, like cells, have a more favorable surface/volume ratio than do large cubes. Only inactive eggs are large enough to be seen by the naked eye; once development begins, cell division results in small-size cells.

3.2 Cellular Organization
A cell is surrounded by a plasma membrane, which regulates the entrance and exit of molecules and ions. Some molecules, such as water and gases, diffuse through the membrane. The direction in which water diffuses is dependent on its concentration within the cell compared to outside the cell.

Table 3.1 lists the cell organelles we have studied in the chapter. The nucleus is a large organelle of primary importance because it controls the rest of the cell. Within the nucleus lies the chromatin, which condenses to become chromosomes during cell division.

Proteins are made at the rough ER before being modified and packaged by the Golgi apparatus into vesicles for secretion. During secretion, a vesicle discharges its contents at the plasma membrane. Golgi-derived lysosomes fuse with incoming vesicles to digest any material enclosed within, and lysosomes also carry out autodigestion of old parts of cells.

Mitochondria are the powerhouses of the cell. During the process of cellular respiration, mitochondria convert carbohydrate energy to ATP energy.

Microtubules and actin filaments make up the cytoskeleton, which maintains the cell's shape and permits movement of cell parts. Centrioles are a part of the microtubule organizing center, which is associated with the formation of microtubules in general and the spindle that appears during cell division. Centrioles also produce basal bodies that give rise to cilia and flagella.

3.3 Cellular Metabolism
Cellular metabolism is the sum of all biochemical pathways of the cell. In a pathway, a series of reactions proceeds in an orderly step-by-step manner. Each of these reactions requires a specific enzyme. Sometimes enzymes require coenzymes, nonprotein portions that participate in the reaction. NAD is a coenzyme.

Cellular respiration (the breakdown of glucose to carbon dioxide and water) includes three pathways: glycolysis, the Krebs cycle, and the electron transport system. If oxygen is not available in cells, the electron transport system is inoperative, and fermentation (an anaerobic process) occurs. Fermentation makes use of glycolysis only, plus one more reaction in which pyruvate is reduced to lactate.

Bioethical Focus

Stem Cells

Stem cells are immature cells that develop into mature, differentiated cells that make up the adult body. For example, the red bone marrow contains stem cells for all the many different types of blood cells in the bloodstream. Embryonic cells are an even more suitable source of stem cells. The early embryo is simply a ball of cells and each of these cells has the potential to become any type of cell in the body—a muscle cell, a nerve cell, or a pancreatic cell, for example.

The use of stem cells from aborted embryos or frozen embryos left over from fertility procedures is controversial. Even though quadriplegics, like Christopher Reeve, and others with serious illnesses may benefit from this research, it is difficult to get governmental approval for use of such stem cell sources. One senator said it reminds him of the rationalization used by Nazis when they experimented on death camp inmates—"after all, they are going to be killed anyway."

Parkinson and Alzheimer are debilitating neurological disorders that people fear. It is possible that one day these disorders could be cured by supplying the patient with new nerve cells in a critical area of the brain. Suppose you had one of these disorders. Would you want to be denied a cure because the government didn't allow experimentation on human embryonic stem cells?

There are other possible sources of stem cells. It turns out that the adult body not only has blood stem cells, it also has neural stem cells in the brain. It has even been possible to coax blood stem cells and neural stem cells to become some other types of mature cells in the body. A possible source of blood stem cells is a baby's umbilical cord and it is now possible to store umbilical blood for future use. Once researchers have the know-how, it may be possible to use any type of stem cell to cure many of the afflicting human beings.

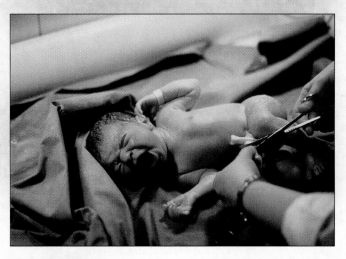

Figure 3A **Umbilical cords are valuable.**
Banking the blood from a baby's umbilical cord can be a source of blood stem cells. Investigators are learning how to convert blood stem cells to other types of mature cells aside from blood cells.

Decide Your Opinion

1. Should researchers have access to embryonic stem cells? Any source or just certain sources? Which sources and why?
2. Should an individual have access to stem cells from just his own body? Also from a relative's body? Also from a child's umbilical cord? From embryonic cells?
3. Should differentiated cells from whatever source eventually be available for sale to patients who need them? After all, you are now able to buy artificial parts, why not living parts?

Looking at Both Sides   www.mhhe.com/biosci/genbio/maderhuman7/

Every bioethical issue has at least two sides. Even if you already have an opinion, it is important to explore the opposite opinion before finalizing your position. The Online Learning Center at www.mhhe.com/biosci/genbio/maderhuman7/ will help you fine-tune your initial opinion, explore both sides, and finalize your position. You may acquire new arguments for your original opinion, or you may even change your opinion. Be sure to complete these activities in sequence:

Taking Sides   Decide your initial opinion by answering a series of questions. Then see if your opinion changes after completing the next two activities.

Further Debate   Read opposing articles that give you further information on this particular bioethical issue.

Explain Your Position   Answer another series of questions and then defend your original or changed opinion. You can e-mail your position to your instructor if he or she wishes.

Studying the Concepts

1. Describe the structure and biochemical makeup of a plasma membrane. 46
2. What are three mechanisms by which substances enter and exit cells? Define isotonic, hypertonic, and hypotonic solutions. 47
3. Describe the nucleus and its contents, including the terms DNA and RNA in your description. 49
4. Describe the structure and function of endoplasmic reticulum. Include the terms rough and smooth ER and ribosomes in your description. 50
5. Describe the structure and function of the Golgi apparatus and its relationship to vesicles and lysosomes. 50–51
6. Describe the structure of mitochondria, and relate this structure to the pathways of cellular respiration. 52–53
7. Describe the composition of the cytoskeleton. 53
8. Describe the structure and function of centrioles, cilia, and flagella. 53–54
9. Discuss and draw a diagram for a metabolic pathway. Discuss and give a reaction to describe the specificity theory of enzymatic action. Define coenzyme. 54–55
10. Name and describe the events within the three subpathways that make up cellular respiration. Why is fermentation necessary but potentially harmful to the human body? 55–57

9. Fermentation of a glucose molecule produces only _____ ATP compared to the _____ ATP produced by cellular respiration.

10. Label the parts of the cell that are involved in protein synthesis and modification. Explain your choices.

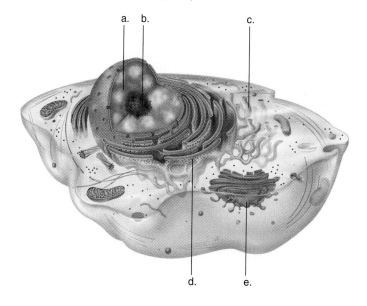

Testing Your Knowledge of the Concepts

In questions 1–4, match the organelles to the functions below.
a. mitochondria
b. nucleus
c. Golgi apparatus
d. rough ER

_____ 1. packaging and secretion

_____ 2. powerhouses of the cell

_____ 3. protein synthesis

_____ 4. control center for cell

In questions 5–7, indicate whether the statement is true (T) or false (F).

_____ 5. Microtubules and actin filaments are a part of the cytoskeleton, the framework of the cell that provides its shape and regulates movement of organelles.

_____ 6. Water enters a cell when the cell is placed in a hypertonic solution.

_____ 7. Substrates react at the active site, located on the surface of their enzyme.

In questions 8 and 9, fill in the blanks.

8. During cellular respiration, most of the ATP molecules are produced at the _____, a series of carriers located on the _____ of mitochondria.

Understanding Key Terms

active site 54
active transport 48
basal body 54
cell 42
cell theory 42
cellular respiration 55
centriole 53
chromatin 49
chromosome 49
cilium 54
coenzyme 55
cytoplasm 44
cytoskeleton 44
dehydrogenase 55
diffusion 47
electron transport system 55
endoplasmic reticulum (ER) 50
facilitated transport 48
fermentation 57
flagellum 54
glycolysis 55

Golgi apparatus 50
Krebs cycle 55
lysosome 52
metabolism 54
microtubule 53
mitochondrion 52
nuclear envelope 49
nuclear pore 49
nucleolus 44
nucleoplasm 49
nucleus 44
organelle 44
osmosis 47
plasma membrane 44
polyribosome 50
product 54
reactant 54
ribosome 50
selectively permeable 47
substrate 54
tonicity 47
vesicle 50

e-Learning Connection   www.mhhe.com/biosci/genbio/maderhuman7/

3.1 Cell Structure and Function

Cell Size *simulation*

3.2 Cellular Organization

Energy Organelles *Essential Study Partner*
Cytoskeleton *Essential Study Partner*
Membrane Structure *Essential Study Partner*
Diffusion *Essential Study Partner*
Osmosis *Essential Study Partner*
Facilitated Diffusion *Essential Study Partner*
Active Transport *Essential Study Partner*
Exo/Endocytosis *Essential Study Partner*

Animal Cell I *art labeling activity*
Animal Cell II *art labeling activity*
Anatomy of the Nucleus *art labeling activity*
Golgi Apparatus Structure *art labeling activity*
Mitochondrion Structure *art labeling activity*
The Cytoskeleton *art labeling activity*
Fluid-Mosaic Model of the Plasma Membrane Structure I
art labeling activity
Fluid-Mosaic Model of the Plasma Membrane Structure II
art labeling activity
Active Transport *simulation*

3.3 Cellular Metabolism

Enzymes *Essential Study Partner*
Pathways *Essential Study Partner*
Glycolysis *Essential Study Partner*
Krebs Cycle *Essential Study Partner*
Electron Transport System *Essential Study Partner*

Chapter Summary

Key Term Flashcards *vocabulary quiz*
Chapter Quiz *objective quiz*

Chapter 4

Organization and Regulation of Body Systems

Chapter Concepts

Figure 4.1   Human organization.
The nervous system coordinates the other systems of the body so that we respond to outside stimuli, including those from the opposite sex.

Ever wonder why a touch on the arm can send you blushing, your heart racing with swift pleasure (Fig. 4.1)? Biochemists sometimes joke that love—or the physical rush accompanying it—is really just a complex set of reactions. It's not a romantic view. Technically, though, it's true.

The coveted reactions of love begin when a light pressure on your skin is detected by sensory receptors that send messages racing to your brain. There, a cascade of messages bounce back. The brain's emotional centers can bring about all sorts of reactions—everything from a quickened heartbeat to sweaty palms. The nervous system is in communication with the endocrine system. The look a lover gives you, the sound of a lover's voice, and yes, a touch, can all start hormones racing through your body. Did you ever notice that you sometimes feel a wave of love come over you? Some investigators call one hormone, oxytocin, the love hormone and say that it is responsible for a turned-on feeling. Reactions to feelings of love by your brain can also have visible effects—your muscles can twitch or your voice can break, even as you try to act normally. When you begin to make love, oxytocin is released throughout your body, making your nerves more sensitive to pleasure and helping to account for orgasm. Some might say such a complex set of reactions caused by the nervous and endocrine systems is the best example of coordination in humans we have to offer.

In order to understand how coordination is carried out, we must first understand the organization of the body. Each system contains a number of organs, and each organ is composed of different types of tissues. This chapter reviews the structure and function of tissues and shows how they are organized within the skin. The skin is an organ system because it contains accessory organs. In a square inch of skin, there are 1,300 sensory receptors that communicate with the brain and spinal cord, enabling us to be sensitive to a lover's touch, among other stimuli.

Since the nervous and endocrine systems coordinate the other systems of the body, they play a pivotal role in maintaining **homeostasis,** the relative constancy of the internal environment. All systems of the body contribute to homeostasis, from the digestive system, which provides nutrient molecules, to the nervous and muscular system, which enables us to bring food to the mouth. ⚖

4.1 Types of Tissues

A **tissue** is composed of similarly specialized cells that perform a common function in the body. The tissues of the human body can be categorized into four major types: epithelial tissue, which covers body surfaces and lines body cavities; connective tissue, which binds and supports body parts; muscular tissue, which moves body parts; and nervous tissue, which receives stimuli and conducts impulses from one body part to another.

Cancers are classified according to the type of tissue from which they arise. **Carcinomas,** the most common type, are cancers of epithelial tissue; sarcomas are cancers arising in muscle or connective tissue (especially bone or cartilage); leukemias are cancers of the blood; and lymphomas are cancers of lymphoid tissue. The chance of developing cancer in a particular tissue shows a positive correlation to the rate of cell division; new blood cells arise at a rate of 2,500,000 cells per second, and epithelial cells also reproduce at a high rate.

Epithelial Tissue

Epithelial tissue, also called epithelium, consists of tightly packed cells that form a continuous layer or sheet lining the entire body surface and most of the body's inner cavities. On the external surface, it protects the body from injury, drying out, and possible **pathogen** (virus and bacterium) invasion. On internal surfaces, epithelial tissue may be specialized for other functions in addition to protection. For example, epithelial tissue secretes mucus along the digestive tract and sweeps up impurities from the lungs by means of cilia (sing., **cilium**). It efficiently absorbs molecules from kidney tubules and from the intestine because of minute cellular extensions called **microvilli.**

There are various types of epithelial tissue (Fig. 4.2). **Squamous epithelium** is composed of flattened cells and is found lining the lungs and blood vessels. **Cuboidal epithelium** contains cube-shaped cells and is found lining the

kidney tubules. **Columnar epithelium** has cells resembling rectangular pillars or columns, and nuclei are usually located near the bottom of each cell. This epithelium is found lining the digestive tract. Ciliated columnar epithelium is found lining the oviducts, where it propels the egg toward the uterus or womb.

An epithelium can be simple or stratified. Simple means the tissue has a single layer of cells, and stratified means the tissue has layers of cells piled one on top of the other. The walls of the smallest blood vessels, called **capillaries,** are composed of a single layer of epithelial cells. The permeability of capillaries allows exchange of substances between the blood and tissue cells. The nose, mouth, esophagus, anal canal, and vagina are all lined by stratified squamous epithelium. As we shall see, the outer layer of skin is also stratified squamous epithelium, but the cells have been reinforced by keratin, a protein that provides strength.

Pseudostratified epithelium appears to be layered; however, true layers do not exist because each cell touches the baseline. The lining of the windpipe, or trachea, is called pseudostratified ciliated columnar epithelium. A secreted covering of mucus traps foreign particles, and the upward motion of the cilia carries the mucus to the back of the throat, where it may either be swallowed or expectorated. Smoking can cause a change in mucus secretion and inhibit ciliary action, and the result is a chronic inflammatory condition called bronchitis.

A so-called **basement membrane** often joins an epithelium to underlying connective tissue. We now know that the basement membrane is glycoprotein, reinforced by fibers that are supplied by connective tissue.

An epithelium sometimes secretes a product, in which case it is described as glandular. A **gland** can be a single epithelial cell, as in the case of mucus-secreting goblet cells found within the columnar epithelium lining the digestive tract, or a gland can contain many cells. Glands that secrete their product into ducts are called exocrine glands, and those that secrete their product directly into the bloodstream are called endocrine glands. The pancreas is both an exocrine gland, because it secretes digestive juices into the small intestine via ducts, and an endocrine gland, because it secretes insulin into the bloodstream.

Epithelial tissue is named according to the shape of the cell. These tightly packed protective cells can occur in more than one layer, and the cells lining a cavity can be ciliated and/or glandular.

Visual Focus

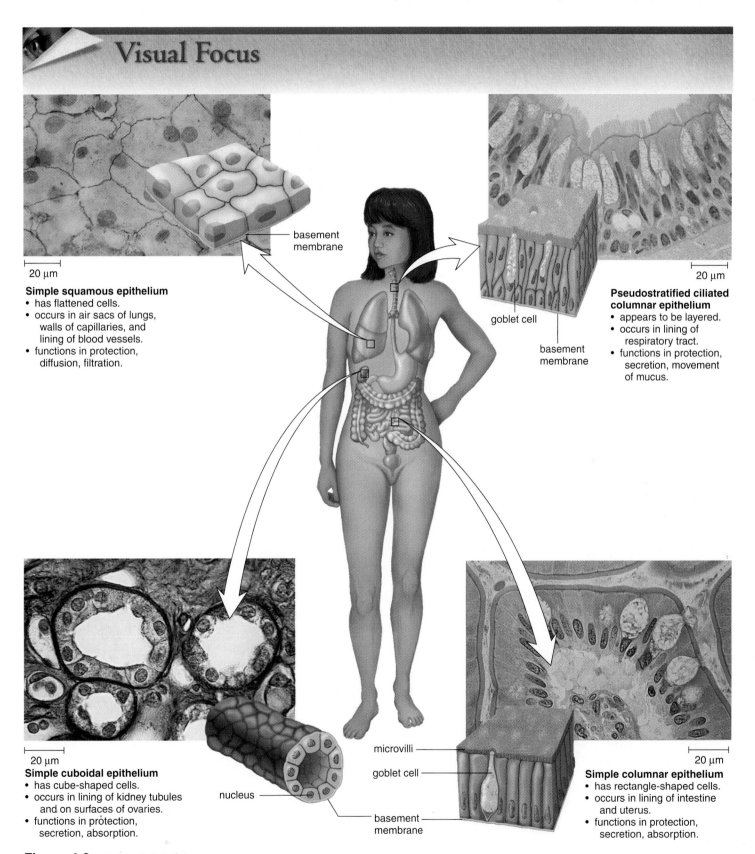

basement membrane

Simple squamous epithelium
- has flattened cells.
- occurs in air sacs of lungs, walls of capillaries, and lining of blood vessels.
- functions in protection, diffusion, filtration.

20 μm

goblet cell

basement membrane

Pseudostratified ciliated columnar epithelium
- appears to be layered.
- occurs in lining of respiratory tract.
- functions in protection, secretion, movement of mucus.

20 μm

nucleus

20 μm

Simple cuboidal epithelium
- has cube-shaped cells.
- occurs in lining of kidney tubules and on surfaces of ovaries.
- functions in protection, secretion, absorption.

microvilli

goblet cell

basement membrane

Simple columnar epithelium
- has rectangle-shaped cells.
- occurs in lining of intestine and uterus.
- functions in protection, secretion, absorption.

20 μm

Figure 4.2 Epithelial tissue.
Certain types of epithelial tissue—squamous, cuboidal, and columnar—are named for the shape of their cells. They all have a protective function, as well as the other functions noted.

Junctions Between Cells

The cells of a tissue can function in a coordinated manner when the plasma membranes of adjoining cells interact. The junctions that occur between cells help cells function as a tissue (Fig. 4.3). A **tight junction** forms an impermeable barrier because adjacent plasma membrane proteins actually join, producing a zipperlike fastening. In the intestine, the gastric juices stay out of the body, and in the kidneys, the urine stays within kidney tubules because epithelial cells are joined by tight junctions.

A **gap junction** forms when two adjacent plasma membrane channels join. This lends strength, but it also allows ions, sugars, and small molecules to pass between the two cells. Gap junctions in heart and smooth muscle ensure synchronized contraction. In an **adhesion junction** (desmosome), the adjacent plasma membranes do not touch but are held together by intercellular filaments firmly attached to buttonlike thickenings. In some organs—like the heart, stomach, and bladder, where tissues get stretched—adhesion junctions hold the cells together.

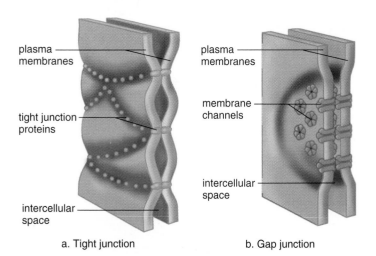

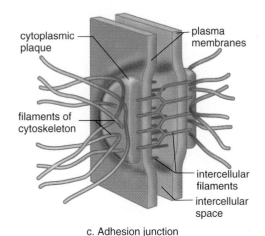

Figure 4.3   Junctions between epithelial cells.
Epithelial tissue cells are held tightly together by **a.** tight junctions that hold cells together; **b.** gap junctions that allow materials to pass from cell to cell; and **c.** adhesion junctions that allow tissues to stretch.

Connective Tissue

Connective tissue binds organs together, provides support and protection, fills spaces, produces blood cells, and stores fat. As a rule, connective tissue cells are widely separated by a **matrix,** consisting of a noncellular material that varies in consistency from solid to semifluid to fluid. The matrix may have fibers of three possible types: white **collagen fibers** contain collagen, a protein that gives them flexibility and strength. **Reticular fibers** are very thin collagen fibers that are highly branched and form delicate supporting networks. Yellow **elastic fibers** contain elastin, a protein that is not as strong as collagen but is more elastic.

Loose Fibrous and Dense Fibrous Tissues

Both loose fibrous and dense fibrous connective tissues have cells called **fibroblasts** that are located some distance from one another and are separated by a jellylike matrix containing white collagen fibers and yellow elastic fibers.

Loose fibrous connective tissue supports epithelium and also many internal organs (Fig. 4.4*a*). Its presence in lungs, arteries, and the urinary bladder allows these organs to expand. It forms a protective covering enclosing many internal organs, such as muscles, blood vessels, and nerves.

Dense fibrous connective tissue contains many collagen fibers that are packed together. This type of tissue has more specific functions than does loose connective tissue. For example, dense fibrous connective tissue is found in **tendons,** which connect muscles to bones, and in **ligaments,** which connect bones to other bones at joints.

Adipose Tissue and Reticular Connective Tissue

In **adipose tissue** (Fig. 4.4*b*), the fibroblasts enlarge and store fat. The body uses this stored fat for energy, insulation, and organ protection. Adipose tissue is found beneath the skin, around the kidneys, and on the surface of the heart. Reticular connective tissue forms the supporting meshwork of lymphoid tissue present in lymph nodes, the spleen, the thymus, and the bone marrow. All types of blood cells are produced in red bone marrow, but a certain type of lymphocyte (T lymphocyte) completes its development in the thymus. The lymph nodes store lymphocytes.

Cartilage

In **cartilage,** the cells lie in small chambers called lacunae (sing., **lacuna**), separated by a matrix that is solid yet flexible. Unfortunately, because this tissue lacks a direct blood supply, it heals very slowly. There are three types of cartilage, distinguished by the type of fiber in the matrix.

Hyaline cartilage (Fig. 4.4*c*), the most common type of cartilage, contains only very fine collagen fibers. The matrix has a white, translucent appearance. Hyaline cartilage is found in the nose and at the ends of the long bones and the ribs, and it forms rings in the walls of respiratory passages. The fetal skeleton also is made of this type of cartilage. Later, the cartilaginous fetal skeleton is replaced by bone.

Elastic cartilage has more elastic fibers than hyaline cartilage. For this reason, it is more flexible and is found, for example, in the framework of the outer ear.

Fibrocartilage has a matrix containing strong collagen fibers. Fibrocartilage is found in structures that withstand tension and pressure, such as the pads between the vertebrae in the backbone and the wedges in the knee joint.

Bone

Bone is the most rigid connective tissue. It consists of an extremely hard matrix of inorganic salts, notably calcium salts, deposited around protein fibers, especially collagen fibers. The inorganic salts give bone rigidity, and the protein fibers provide elasticity and strength, much as steel rods do in reinforced concrete.

Compact bone makes up the shaft of a long bone (Fig. 4.4*d*). It consists of cylindrical structural units called osteons (Haversian systems). The central canal of each osteon is surrounded by rings of hard matrix. Bone cells are located in spaces called lacunae between the rings of matrix. Blood vessels in the central canal carry nutrients that allow bone to renew itself. Nutrients can reach all of the bone cells because they are connected by thin processes within canaliculi (minute canals) that also reach to the central canal.

The ends of a long bone contain spongy bone, which has an entirely different structure. **Spongy bone** contains numerous bony bars and plates, separated by irregular spaces. Although lighter than compact bone, spongy bone still is designed for strength. Just as braces are used for support in buildings, the solid portions of spongy bone follow lines of stress.

Connective tissues, which bind and support body parts, differ according to the type of matrix and the abundance of fibers in the matrix.

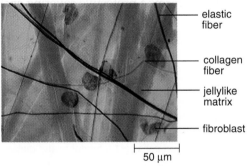

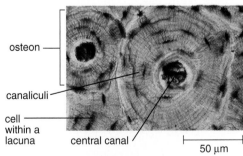

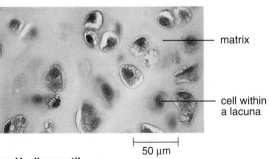

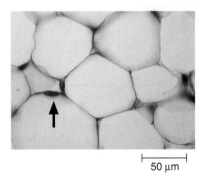

a. **Loose fibrous connective tissue**
- has space between components.
- occurs beneath skin and most epithelial layers.
- functions in support and binds organs.

b. **Adipose tissue**
- has cells filled with fat.
- occurs beneath skin, around organs including the heart.
- functions in insulation, stores fat.

c. **Hyaline cartilage**
- has cells in scattered lacunae.
- occurs in nose and walls of respiratory passages; at ends of bones including ribs.
- functions in support and protection.

d. **Compact bone**
- has cells in concentric rings of lacunae.
- occurs in bones of skeleton.
- functions in support and protection.

Figure 4.4   Connective tissue examples.
a. In loose fibrous connective tissue, cells called fibroblasts are separated by a jellylike matrix, which contains both collagen and elastic fibers. **b.** Adipose tissue cells have nuclei (arrow) pushed to one side because the cells are filled with fat. **c.** In hyaline cartilage, the flexible matrix has a white, translucent appearance. **d.** In compact bone, the hard matrix contains calcium salts. Concentric rings of cells in lacunae form an elongated cylinder called an osteon (Haversian system). An osteon has a central canal that contains blood vessels and nerve fibers.

Blood ⚖

Blood is unlike other types of connective tissue in that the matrix (i.e., plasma) is not made by the cells. Some people do not classify blood as connective tissue; instead, they suggest a separate tissue category called vascular tissue.

The internal environment of the body consists of blood and tissue fluid. The systems of the body help keep blood composition and chemistry within normal limits, and blood in turn creates tissue fluid. Blood transports nutrients and oxygen to tissue fluid and removes carbon dioxide and other wastes. It helps distribute heat and also plays a role in fluid, ion, and pH balance. Various components of blood, as discussed below, help protect us from disease, and blood's ability to clot prevents fluid loss.

If blood is transferred from a person's vein to a test tube and prevented from clotting, it separates into two layers (Fig. 4.5). The upper liquid layer, called **plasma,** represents about 55% of the volume of whole blood and contains a variety of inorganic and organic substances dissolved or suspended in water (Table 4.1). The lower layer consists of red blood cells (erythrocytes), white blood cells (leukocytes), and blood platelets (thrombocytes). Collectively, these are called the formed elements and represent about 45% of the volume of whole blood. Formed elements are manufactured in the red bone marrow of the skull, ribs, vertebrae, and ends of long bones.

The **red blood cells** are small, biconcave, disk-shaped cells without nuclei. The presence of the red pigment hemoglobin makes the cells red, and in turn, makes the blood red. Hemoglobin is composed of four units; each is composed of the protein globin and a complex iron-containing structure called heme. The iron forms a loose association with oxygen, and in this way red blood cells transport oxygen.

White blood cells may be distinguished from red blood cells by the fact that they are usually larger, have a nucleus, and without staining would appear to be translucent. White blood cells characteristically look bluish because they have been stained that color. White blood cells, which fight infection, function primarily in two ways. Some white blood cells are phagocytic and engulf infectious pathogens, while other white blood cells produce antibodies, molecules that combine with foreign substances to inactivate them.

Platelets are not complete cells; rather, they are fragments of giant cells present only in bone marrow. When a blood vessel is damaged, platelets form a plug that seals the vessel and injured tissues release molecules that help the clotting process.

Blood is a connective tissue in which the matrix is plasma.

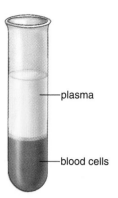

a. Blood sample

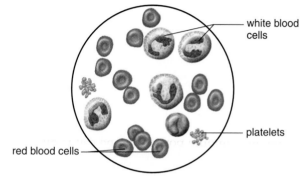

b. Blood smear

Figure 4.5   **Blood, a fluid tissue.**
a. In a test tube, a blood sample separates into its two components: blood cells and plasma. **b.** Microscopic examination of a blood smear shows that there are red blood cells, white blood cells, and platelets. Platelets are fragments of a cell. Red blood cells transport oxygen, white blood cells fight infections, and platelets are involved in blood clotting.

Table 4.1	Components of Blood Plasma
Water (92% of Total)	
Solutes (8% of Total)	
Inorganic ions (salts)	Na^+, Ca^{2+}, K^+, Mg^{2+}, Cl^-, HCO_3^-, HPO_4^{2+}, SO_4^{2+}
Gases	O_2, CO_2
Plasma proteins	Albumin, globulins, fibrinogen
Organic nutrients	Glucose, lipids, phospholipids, amino acids, etc.
Nitrogenous waste products	Urea, ammonia, uric acid
Regulatory substances	Hormones, enzymes

Muscular Tissue

Muscular (contractile) tissue is composed of cells that are called muscle fibers. Muscle fibers contain actin filaments and myosin filaments, whose interaction accounts for movement. There are three types of vertebrate muscles: skeletal, smooth, and cardiac.

Skeletal muscle, also called voluntary muscle (Fig. 4.6a), is attached by tendons to the bones of the skeleton, and when it contracts, body parts move. Contraction of skeletal muscle is under voluntary control and occurs faster than in the other muscle types. Skeletal muscle fibers are cylindrical and quite long—sometimes they run the length of the muscle. They arise during development when several cells fuse, resulting in one fiber with multiple nuclei. The nuclei are located at the periphery of the cell, just inside the plasma membrane. The fibers have alternating light and dark bands that give them a **striated** appearance. These bands are due to the placement of actin filaments and myosin filaments in the cell.

Smooth (visceral) muscle is so named because the cells lack striations. The spindle-shaped cells form layers in which the thick middle portion of one cell is opposite the thin ends of adjacent cells. Consequently, the nuclei form an irregular pattern in the tissue (Fig. 4.6b). Smooth muscle is not under voluntary control and therefore is said to be involuntary. Smooth muscle, found in the walls of viscera (intestine, stomach, and other internal organs) and blood vessels, contracts more slowly than skeletal muscle but can remain contracted for a longer time. When the smooth muscle of the intestine contracts, food moves along its lumen (central cavity). When the smooth muscle of the blood vessels contracts, blood vessels constrict, helping to raise blood pressure.

Cardiac muscle (Fig. 4.6c) is found only in the walls of the heart. Its contraction pumps blood and accounts for the heartbeat. Cardiac muscle combines features of both smooth muscle and skeletal muscle. It has striations like skeletal muscle, but the contraction of the heart is involuntary for the most part. Cardiac muscle cells also differ from skeletal muscle cells in that they have a single, centrally placed nucleus. The cells are branched and seemingly fused one with the other, and the heart appears to be composed of one large interconnecting mass of muscle cells. Actually, cardiac muscle cells are separate and individual, but they are bound end to end at **intercalated disks,** areas where folded plasma membranes between two cells contain adhesion junctions and gap junctions.

All muscular tissue contains actin filaments and myosin filaments; these form a striated pattern in skeletal and cardiac muscle, but not in smooth muscle.

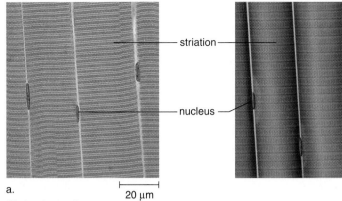

a.   20 µm

Skeletal muscle
• has striated cells with multiple nuclei.
• occurs in muscles attached to skeleton.
• functions in voluntary movement of body.
• voluntary

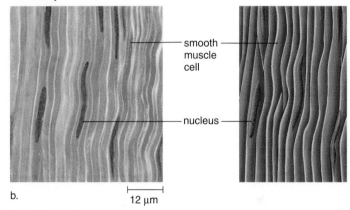

b.   12 µm

Smooth muscle
• has spindle-shaped cells, each with a single nucleus.
• cells have no striations.
• functions in movement of substances in lumens of body.
• involuntary.

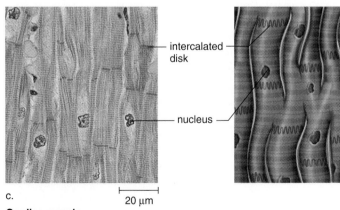

c.   20 µm

Cardiac muscle
• has branching striated cells, each with a single nucleus.
• occurs in the wall of the heart.
• functions in the pumping of blood.
• involuntary.

Figure 4.6   Muscular tissue.
a. Skeletal muscle is voluntary and striated. **b.** Smooth muscle is involuntary and nonstriated. **c.** Cardiac muscle is involuntary and striated. Cardiac muscle cells branch and fit together at intercalated disks.

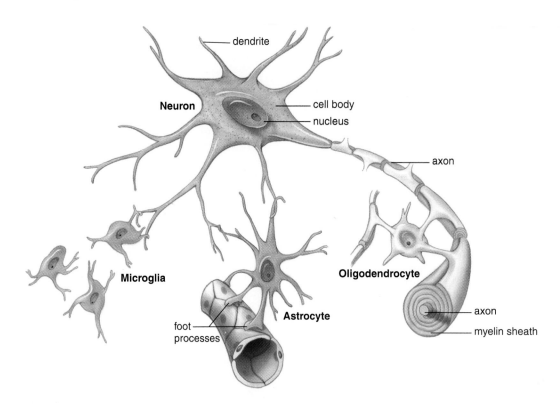

Figure 4.7   Neuron and neuroglia.
Neurons conduct nerve impulses. Microglia become mobile in response to inflammation and phagocytize debris. Microglial cells are phagocytes that clean up debris. Astrocytes lie between neurons and a capillary; therefore, substances entering neurons from the blood must first pass through astrocytes. Oligodendrocytes form the myelin sheaths around fibers in the brain and spinal cord.

Nervous Tissue

Nervous tissue, which contains nerve cells called neurons, is present in the brain and spinal cord. A **neuron** is a specialized cell that has three parts: dendrites, a cell body, and an axon (Fig. 4.7). A dendrite is a process that conducts signals toward the cell body. The cell body contains the major concentration of the cytoplasm and the nucleus of the neuron. An axon is a process that typically conducts nerve impulses away from the cell body. Long axons are covered by myelin, a white fatty substance. The term fiber[1] is used to refer to an axon along with its myelin sheath if it has one. Outside the brain and spinal cord, fibers bound by connective tissue form **nerves.**

The nervous system has just three functions: sensory input, integration of data, and motor output. Nerves conduct impulses from sensory receptors to the spinal cord and the brain where integration occurs. The phenomenon called sensation occurs only in the brain, however. Nerves also conduct nerve impulses away from the spinal cord and brain to the muscles and glands, causing them to contract and secrete, respectively. In this way, a coordinated response to the stimulus is achieved.

[1]In connective tissue, a fiber is a component of the matrix; in muscle tissue, a fiber is a muscle cell; in nervous tissue, a fiber is an axon and its myelin sheath.

In addition to neurons, nervous tissue contains neuroglial cells.

Neuroglia

There are several different types of neuroglia in the brain (Fig. 4.7), and much research is currently being conducted to determine how much "glia" contribute to the functioning of the brain. **Neuroglia** outnumber neurons nine to one and take up more than half the volume of the brain, but until recently, they were thought to merely support and nourish neurons. Three types of neuroglia are oligodendrocytes, microglia, and astrocytes. Oligodendrocytes form myelin, and microglial cells, in addition to supporting neurons, engulf bacterial and cellular debris. Astrocytes provide nutrients to neurons and produce a hormone known as glia-derived growth factor, which someday might be used as a cure for Parkinson disease and other diseases caused by neuron degeneration. Neuroglia don't have a long process, but even so, researchers are now beginning to gather evidence that they do communicate among themselves and with neurons!

Nerve cells, called neurons, have fibers (processes) called axons and dendrites. In general, neuroglia support and service neurons.

4.2 Body Cavities and Body Membranes

The internal organs are located within specific body cavities (Fig. 4.8). During human development, there is a large ventral cavity called a **coelom,** which becomes divided into the thoracic (chest) and abdominal cavities. Membranes divide the thoracic cavity into the pleural cavities, containing the right and left lungs, and the pericardial cavity, containing the heart. The thoracic cavity is separated from the abdominal cavity by a horizontal muscle called the diaphragm. The stomach, liver, spleen, gallbladder, and most of the small and large intestines are in the upper portion of the abdominal cavity. The lower portion contains the rectum, the urinary bladder, the internal reproductive organs, and the rest of the large intestine. Males have an external extension of the abdominal wall, called the scrotum, containing the testes.

The dorsal cavity also has two parts: the cranial cavity within the skull contains the brain; and the vertebral canal, formed by the vertebrae, contains the spinal cord.

Body Membranes

In this context, we are using the term *membrane* to refer to a thin lining or covering composed of an epithelium overlying a loose connective tissue layer. Body membranes line cavities and internal spaces of organs and tubes that open to the outside.

Mucous membranes line the tubes of the digestive, respiratory, urinary, and reproductive systems. The epithelium of this membrane contains goblet cells that secrete mucus. This mucus ordinarily protects the body from invasion by bacteria and viruses; hence, more mucus is secreted and expelled when a person has a cold and has to blow her/his nose. In addition, mucus usually protects the walls of the stomach and small intestine from digestive juices, but this protection breaks down when a person develops an ulcer.

Serous membranes line the thoracic and abdominal cavities and the organs they contain. They secrete a watery fluid that keeps the membranes lubricated. Serous membranes support the internal organs and compartmentalize the large thoracic and abdominal cavities. This helps hinder the spread of any infection.

The **pleural membranes** are serous membranes that line the pleural cavity and lungs. Pleurisy is a well-known infection of these membranes. The peritoneum lines the abdominal cavity and its organs. In between the organs, there is a double layer of peritoneum called mesentery. **Peritonitis,** a life-threatening infection of the peritoneum, is likely if an inflamed appendix bursts before it is removed.

Synovial membranes line freely movable joint cavities. They secrete synovial fluid into the joint cavity; this fluid lubricates the ends of the bones so that they can move freely. In

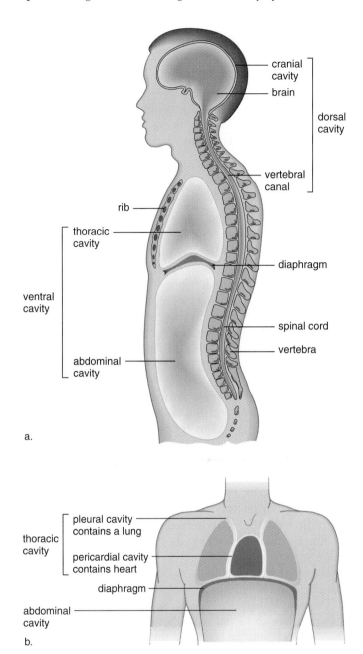

Figure 4.8   Mammalian body cavities.
a. Side view. The dorsal (toward the back) cavity contains the cranial cavity and the vertebral canal. The brain is in the cranial cavity, and the spinal cord is in the vertebral canal. There is a well-developed ventral (toward the front) cavity, which is divided by the diaphragm into the thoracic cavity and the abdominal cavity. The heart and lungs are in the thoracic cavity, and most other internal organs are in the abdominal cavity. **b.** Frontal view of the thoracic cavity.

rheumatoid arthritis, the synovial membrane becomes inflamed and grows thicker, restricting movement.

The **meninges** are membranes found within the dorsal cavity. They are composed only of connective tissue and serve as a protective covering for the brain and spinal cord. Meningitis is a life-threatening infection of the meninges.

4.3 Organ Systems

The body contains a number of systems that work together to maintain homeostasis. Page 77 introduces a feature that will be used in each organ system chapter. In this chapter, the box reviews the general functions of the body's organ systems. The corresponding boxes in other chapters will show how a particular organ system interacts with all the other systems.

Maintenance of the Body ⚖

The internal environment of the body consists of the blood within the blood vessels and the tissue fluid that surrounds the cells. Five systems (digestive, cardiovascular, lymphatic, respiratory, and urinary) add substances to and/or remove substances from the blood.

The **digestive system** consists of the mouth, esophagus, stomach, small intestine, and large intestine (colon) along with the associated organs: teeth, tongue, salivary glands, liver, gallbladder, and pancreas. This system receives food and digests it into nutrient molecules, which can enter the cells of the body.

The **cardiovascular system** consists of the heart and the blood vessels that carry blood through the body. Blood transports nutrients and oxygen to the cells, and removes their waste molecules that are to be excreted from the body. Blood also contains cells produced by the lymphatic system.

The **lymphatic system** consists of lymphatic vessels, lymph fluid, lymph nodes, and other lymphoid organs. This system protects the body from disease by purifying lymph and supporting lymphocytes, the white blood cells that produce antibodies. Lymphatic vessels absorb fat from the digestive system and collect excess tissue fluid, which is returned to the blood circulatory system.

The **respiratory system** consists of the lungs and the tubes that take air to and from them. The respiratory system brings oxygen into the body and takes carbon dioxide out of the body through the lungs.

The **urinary system** contains the kidneys and the urinary bladder. This system rids the body of nitrogenous wastes and helps regulate the fluid level and chemical content of the blood.

The digestive system, cardiovascular system, lymphatic system, respiratory system, and urinary system all perform specific processing and transporting functions to maintain the normal conditions of the body.

Support and Movement

The skeletal system and the muscular system allow the body and its parts to move. They also protect and support the body.

The skeletal system, consisting of the bones of the skeleton, protects body parts. For example, the skull forms a protective encasement for the brain, as does the rib cage for the heart and lungs. The skeleton, as a whole, serves as a place of attachment for the skeletal muscles. Contraction of muscles in the muscular system accounts for movement of body parts.

The skeletal system and the muscular system support the body and permit movement.

Coordination and Regulation of Body Systems

The **nervous system** consists of the brain, spinal cord, and associated nerves. The nerves conduct nerve impulses from receptors to the brain and spinal cord. They also conduct nerve impulses from the brain and spinal cord to the muscles and glands, allowing us to respond to both external and internal stimuli.

The **endocrine system** consists of the hormonal glands which secrete chemicals that serve as messengers between body parts. Both the nervous and endocrine systems help maintain homeostasis by coordinating and regulating the functions of the body's other systems. The endocrine system also helps maintain the proper functioning of male and female reproductive organs. ⚖

The nervous and endocrine systems coordinate and regulate the activities of the body's other systems.

Continuance of the Species

The **reproductive system** involves different organs in the male and female. The male reproductive system consists of the testes, other glands, and various ducts that conduct semen to and through the penis. The testes produce sex cells called sperm. The female reproductive system consists of the ovaries, oviducts, uterus, vagina, and external genitals. The ovaries produce sex cells called eggs. When a sperm fertilizes an egg, an offspring begins development.

The reproductive system in males and females carries out those functions that give humans the ability to reproduce.

Integumentary System

The skin and its accessory organs such as hair, nails, sweat glands, and sebaceous glands are collectively called the **integumentary system.** Skin covers the body, protecting underlying tissues from physical trauma, pathogen invasion, and water loss; it also helps regulate body temperature. Therefore, skin plays a significant role in homeostasis. The skin even synthesizes certain chemicals such as vitamin D that affect the rest of the body. Because skin contains sensory receptors, skin also helps us to be aware of our surroundings and to communicate through touch. ⚖

Regions of the Skin

The **skin** has two regions: the epidermis and the dermis (Fig. 4.9). A subcutaneous layer is found between the skin and any underlying structures, such as muscle or bone.

The **epidermis** of skin is made up of stratified squamous epithelium. New cells derived from basal cells become flattened and hardened as they push to the surface. Hardening takes place because the cells produce keratin, a waterproof protein. Dandruff occurs when the rate of keratinization in the skin of the scalp is two or three times the normal rate. A thick layer of dead keratinized cells,

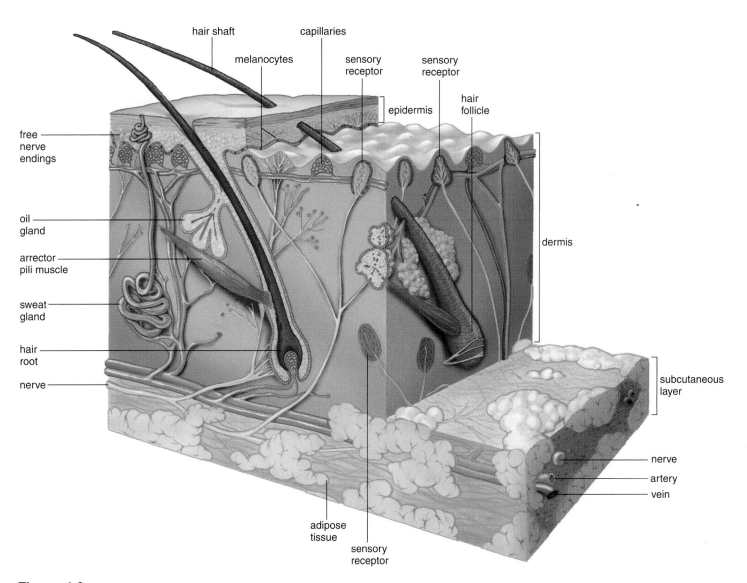

Figure 4.9   Human skin anatomy.
Skin consists of two regions, the epidermis and the dermis. A subcutaneous layer lies below the dermis.

arranged in spiral and concentric patterns, forms finger-prints and footprints. Specialized cells in the epidermis called **melanocytes** produce melanin, the pigment responsible for skin color.

The **dermis** is a region of fibrous connective tissue beneath the epidermis. The dermis contains collagen and elastic fibers. The collagen fibers are flexible but offer great resistance to overstretching; they prevent the skin from being torn. The elastic fibers maintain normal skin tension but also stretch to allow movement of underlying muscles and joints. (The number of collagen and elastic fibers decreases with exposure to the sun, and the skin becomes less supple and is prone to wrinkling.) The dermis also contains blood vessels that nourish the skin. When blood rushes into these vessels, a person blushes, and when blood is minimal in them, a person turns "blue."

Sensory receptors are specialized nerve endings in the dermis that respond to external stimuli. There are sensory receptors for touch, pressure, pain, and temperature. The fingertips contain the most touch receptors, and these add to our ability to use our fingers for delicate tasks.

The **subcutaneous layer,** which lies below the dermis, is composed of loose connective tissue and adipose tissue, which stores fat. Fat is a stored source of energy in the body. Adipose tissue helps to thermally insulate the body from either gaining heat from the outside or losing heat from the inside. A well-developed subcutaneous layer gives the body a rounded appearance and provides protective padding against external assaults. Excessive development of the subcutaneous layer accompanies obesity.

Skin has two regions: the epidermis and the dermis. A subcutaneous layer lies beneath the dermis.

Accessory Organs of the Skin

Nails, hair, and glands are structures of epidermal origin even though some parts of hair and glands are largely found in the dermis.

Nails are a protective covering of the distal part of fingers and toes. Nails can help pry open or pick up small objects. They are also used for scratching oneself or others. Nails grow from special epithelial cells at the base of the nail in the portion called the nail root. These cells become keratinized as they grow out over the nail bed. The visible portion of the nail is called the nail body. The cuticle is a fold of skin that hides the nail root. The whitish color of the half-moon-shaped base, or lunula, results from the thick layer of cells in this area (Fig. 4.10).

Hair follicles are in the dermis and continue through the epidermis where the hair shaft extends beyond the skin. Contraction of the arrector pili muscles attached to hair follicles causes the hairs to "stand on end" and causes goose bumps to develop. Epidermal cells form the root of hair, and their division causes a hair to grow. The cells become keratinized and dead as they are pushed farther from the root.

Each hair follicle has one or more **oil glands,** also called sebaceous glands, which secrete sebum, an oily substance that lubricates the hair within the follicle and the skin itself. If the sebaceous glands fail to discharge, the secretions collect and form "whiteheads" or "blackheads." The color of blackheads is due to oxidized sebum. Acne is an inflammation of the sebaceous glands that most often occurs during adolescence. Hormonal changes during this time cause the sebaceous glands to become more active.

Sweat glands, also called sudoriferous glands, are quite numerous and are present in all regions of skin. A sweat gland is a coiled tubule within the dermis, but then it straightens out near its opening. Some sweat glands open into hair follicles, but most open onto the surface of the skin. Sweat glands play a role in modifying body temperature. When body temperature starts to rise, sweat glands become active. Sweat absorbs body heat as it evaporates. Once body temperature lowers, sweat glands are no longer active.

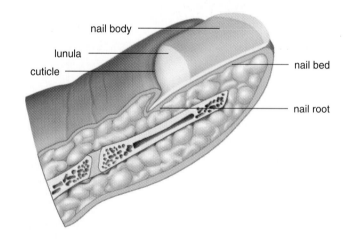

Figure 4.10   Nail anatomy.
Cells produced by the nail root become keratinized, forming the nail body.

Skin Cancer on the Rise

In the nineteenth century, and earlier, it was fashionable for Caucasian women who did not labor outdoors to keep their skin fair by carrying parasols when they went out. But early in this century, some fair-skinned people began to prefer the golden-brown look, and they took up sunbathing as a way to achieve a tan. A few hours after exposure to the sun, pain and redness due to dilation of blood vessels occur. Tanning occurs when melanin granules increase in keratinized cells at the surface of the skin as a way to prevent any further damage by ultraviolet (UV) rays. The sun gives off two types of UV rays: UV-A rays and UV-B rays. UV-A rays penetrate the skin deeply, affect connective tissue, and cause the skin to sag and wrinkle. UV-A rays are also believed to increase the effects of the UV-B rays, which are the cancer-causing rays. UV-B rays are more prevalent at midday.

Skin cancer is categorized as either nonmelanoma or melanoma. Nonmelanoma cancers are of two types. Basal cell carcinoma, the most common type, begins when UV radiation causes epidermal basal cells to form a tumor, while at the same time suppressing the immune system's ability to detect the tumor. The signs of a tumor are varied. They include an open sore that will not heal, a recurring reddish patch, a smooth, circular growth with a raised edge, a shiny bump, or a pale mark (Fig. 4A). In about 95% of patients, the tumor can be excised surgically, but recurrence is common.

Squamous cell carcinoma begins in the epidermis proper. Squamous cell carcinoma is five times less common than basal cell carcinoma, but if the tumor is not excised promptly, it is more likely to spread to nearby organs. The death rate from squamous cell carcinoma is about 1% of cases. The signs of squamous cell carcinoma are the same as for basal cell carcinoma, except that the former may also show itself as a wart that bleeds and scabs.

Melanoma that starts in pigmented cells often has the appearance of an unusual mole. Unlike a mole that is circular and confined, melanoma moles look like spilled ink spots. A variety of shades can be seen in the same mole, and they can itch, hurt,

or feel numb. The skin around the mole turns gray, white, or red. Melanoma is most apt to appear in persons who have fair skin, particularly if they have suffered occasional severe sunburns as children. The chance of melanoma increases with the number of moles a person has. Most moles appear before the age of 14, and their appearance is linked to sun exposure. Melanoma rates have risen since the turn of the century, but the incidence has doubled in the last decade. Most often, malignant moles are removed surgically; if the cancer has spread, chemotherapy and various other treatments are also available.

Since the incidence of skin cancer is related to UV exposure, scientists have developed a UV index to determine how powerful the solar rays are in different U.S. cities. In general, the more southern the city, the higher the UV index, and the greater the risk of skin cancer. Regardless of where you live, for every 10% decrease in the ozone layer, the risk of skin cancer rises 13–20%. To prevent the occurrence of skin cancer, observe the following:

- Use a broad-spectrum sunscreen, which protects you from both UV-A and UV-B radiation, with an SPF (sun protection factor) of at least 15. (This means, for example, that if you usually burn after a 20-minute exposure, it will take 15 times that long before you will burn.)
- Stay out of the sun altogether between the hours of 10 A.M. and 3 P.M. This will reduce your annual exposure by as much as 60%.
- Wear protective clothing. Choose fabrics with a tight weave and wear a wide-brimmed hat.
- Wear sunglasses that have been treated to absorb both UV-A and UV-B radiation. Otherwise, sunglasses can expose your eyes to more damage than usual because pupils dilate in the shade.
- Avoid tanning machines. Although most tanning devices use high levels of only UV-A, these rays cause the deep layers of the skin to become more vulnerable to UV-B radiation when you are later exposed to the sun.

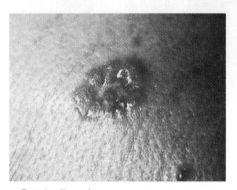

a. Basal cell carcinoma

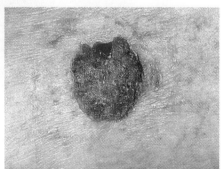

b. Squamous cell carcinoma

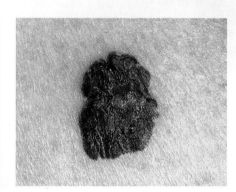

c. Melanoma

Figure 4A **Skin cancer.**
a. Basal cell carcinoma occurs when basal cells proliferate abnormally. **b.** Squamous cell carcinoma arises in epithelial cells derived from basal cells. **c.** Malignant melanoma is due to a proliferation of pigmented cells.

4.4 Homeostasis

Homeostasis is the relative constancy of the body's internal environment. Because of homeostasis, even though external conditions may change dramatically, internal conditions still stay within a narrow range (Fig. 4.11). For example, regardless of how cold or hot it gets, the temperature of the body stays around 37°C (97° to 99°F) . No matter how acidic your meal, the pH of your blood is usually about 7.4, and even if you eat a candy bar, the amount of sugar in your blood is just about 0.1%.

It is important to realize that internal conditions are not absolutely constant; they tend to fluctuate above and below a particular value. Therefore, the internal state of the body is often described as one of dynamic equilibrium. If internal conditions should change to any great degree, illness results. This makes the study of homeostatic mechanisms medically important.

Negative Feedback

A homeostatic mechanism in the body has three components: a sensor, a regulatory center, and an effector (Fig. 4.12a). The sensor detects a change in the internal environment; the regulatory center activates the effector; the effector reverses the change and brings conditions back to normal again. Now, the sensor is no longer activated.

Negative feedback is the primary homeostatic mechanism that keeps a variable, such as body temperature, close to a particular value, or set point. A home heating system illustrates how a negative feedback mechanism works (Fig. 4.12b). You set the thermostat at, say, 68°F. This is the set point. The thermostat contains a thermometer, a sensor that detects when the room temperature falls below the set point. The thermostat is also the regulatory center; it turns the furnace on. The furnace plays the role of the effector. The heat given off by the furnace raises the temperature of the room to 70°F. Now, the furnace turns off. Notice that a negative feedback mechanism prevents change in the same direction; the room does not get hotter and hotter because warmth inactivates the

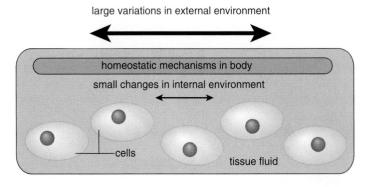

Figure 4.11   Homeostasis.
Because of homeostatic mechanisms, large external changes cause only small internal changes in such parameters as body temperature and pH of the blood.

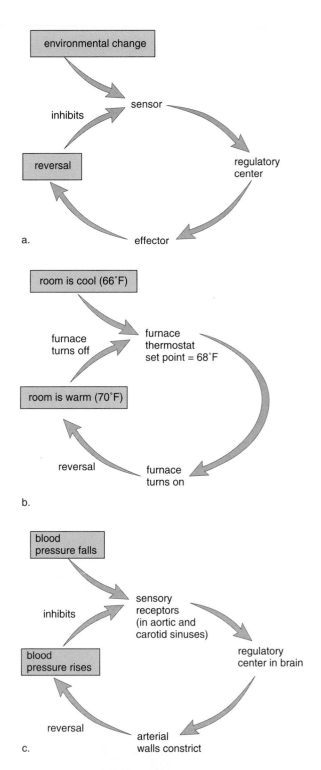

Figure 4.12   Negative feedback.
a. A sensor detects an internal environmental change and signals a regulatory center. The center activates an effector, which reverses this change. **b.** Mechanical example. When the room is cool, a thermostat that senses the room temperature signals the furnace to turn on. Once the room is warm, the furnace turns off. **c.** Biological example. When blood pressure falls, special sensory receptors in blood vessels signal a regulatory center in the brain. The brain signals the arteries to constrict, and blood pressure rises to normal. This response reverses the change.

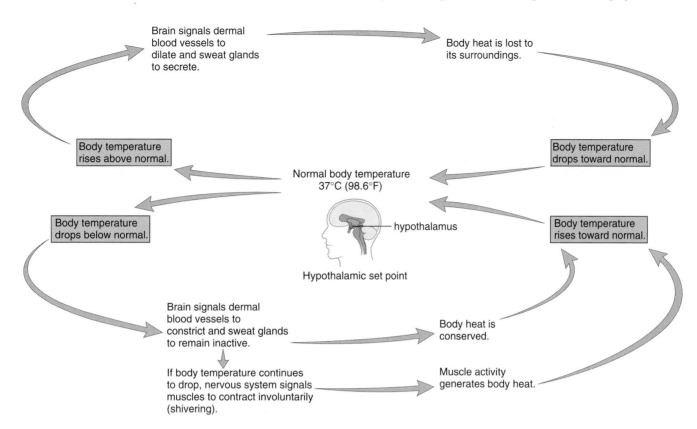

Figure 4.13   Homeostasis and body temperature regulation.
Negative feedback mechanisms control body temperature so that it remains relatively stable at 37°C. These mechanisms return the temperature to normal when it fluctuates above and below this set point.

system. Notice that negative feedback mechanisms are activated only if there is a deviation from the set point, and therefore there is a fluctuation above and below this value.

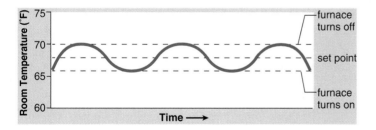

Negative feedback mechanisms in the body function similarly to this mechanical model (Fig. 4.12c). For example, when blood pressure falls, sensory receptors signal a regulatory center in the brain. This center sends out nerve impulses to the arterial walls so that they constrict. Once the blood pressure rises, the system is inactivated.

A negative feedback mechanism maintains stability by its ability to sense a change and bring about an effect that reverses that change.

Regulation of Body Temperature ⚖

The thermostat for body temperature is located in a part of the brain called the hypothalamus. When the body temperature falls below normal, the regulatory center directs (via nerve impulses) the blood vessels of the skin to constrict (Fig. 4.13). This conserves heat. If body temperature falls even lower, the regulatory center sends nerve impulses to the skeletal muscles and shivering occurs. Shivering generates heat, and gradually body temperature rises to 37°C. When the temperature rises to normal, the regulatory center is inactivated.

When the body temperature is higher than normal, the regulatory center directs the blood vessels of the skin to dilate. This allows more blood to flow near the surface of the body, where heat can be lost to the environment. In addition, the nervous system activates the sweat glands, and the evaporation of sweat helps lower body temperature. Gradually, body temperature decreases to 37°C.

The temperature of the human body is maintained at about 37°C due to activity of a regulatory center in the brain.

Positive Feedback ⚖

Positive feedback is a mechanism that brings about an ever greater change in the same direction. When a woman is giving birth, the head of the baby begins to press against the cervix, stimulating sensory receptors there. When nerve impulses reach the brain, the brain causes the pituitary gland to secrete the hormone oxytocin. Oxytocin travels in the blood and causes the uterus to contract. As labor continues, the cervix is ever more stimulated and uterine contractions become ever more severe until birth occurs.

A positive feedback mechanism can be harmful, as when a fever causes metabolic changes that push the fever still higher. Death occurs at 45°C because cellular proteins denature at this temperature and metabolism stops. Still, positive feedback loops like those involved in childbirth, blood clotting, and the stomach's digestion of protein assist the body in completing a process that has a definite cut-off point.

In contrast to negative feedback, positive feedback allows rapid change in one direction and does not achieve relative stability.

Homeostasis and Body Systems ⚖

The internal environment of the body consists of blood and tissue fluid. Tissue fluid, which bathes all the cells of the body, is refreshed when molecules such as oxygen and nutrients move into tissue fluid from the blood, and when wastes move from tissue fluid into the blood (Fig. 4.14). Tissue fluid remains constant only as long as blood composition remains constant.

As described on the next page, all systems of the body contribute toward maintaining homeostasis and therefore a relatively constant internal environment. The cardiovascular system conducts blood to and away from capillaries, where exchange occurs. The heart pumps the blood and thereby keeps it moving toward the capillaries. The formed elements also contribute to homeostasis. Red blood cells transport oxygen and participate in the transport of carbon dioxide. White blood cells fight infection, and platelets participate in the clotting process. The lymphatic system is accessory to the cardiovascular system. Lymphatic capillaries collect excess tissue fluid, and this is returned via lymphatic veins to the circulatory veins. Lymph nodes help purify lymph and keep it free of pathogens.

The digestive system takes in and digests food, providing nutrient molecules that enter the blood and replace the nutrients that are constantly being used by the body cells. The respiratory system adds oxygen to and removes carbon dioxide from the blood. The chief regulators of blood composition are the liver and the kidneys. They monitor the chemical composition of plasma and alter it as required. Immediately after glucose enters the blood, it can be removed by the liver for storage as glycogen. Later, the glycogen can

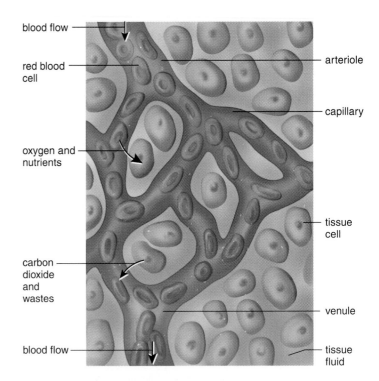

Figure 4.14   Regulation of tissue fluid composition.
Cells are surrounded by tissue fluid, which is continually refreshed because oxygen and nutrient molecules constantly exit, and carbon dioxide and waste molecules continually enter the bloodstream as shown.

be broken down to replace the glucose used by the body cells; in this way, the glucose composition of blood remains constant. The hormone insulin, secreted by the pancreas, regulates glycogen storage. The liver also removes toxic chemicals, such as ingested alcohol and other drugs. The liver makes urea, a nitrogenous end product of protein metabolism. Urea and other metabolic waste molecules are excreted by the kidneys. Urine formation by the kidneys is extremely critical to the body, not only because it rids the body of unwanted substances, but also because it offers an opportunity to carefully regulate blood volume, salt balance, and the pH of the blood.

The nervous system and the endocrine systems regulate the other systems of the body. They work together to control body systems so that homeostasis is maintained. We have already seen that in negative feedback mechanisms, sensory receptors send nerve impulses to regulatory centers in the brain, which then direct effectors to become active. Effectors can be muscles or glands. Muscles bring about an immediate change. Endocrine glands secrete hormones that bring about a slower, more lasting change that keeps the internal environment relatively stable.

All systems of the body contribute to homeostasis—that is, maintaining the relative constancy of the internal environment, blood, and tissue fluid.

Human Systems Work Together

Integumentary System

External support and protection of body; helps maintain body temperature.

Skeletal System

Internal support and protection; body movement; production of blood cells.

Muscular System

Body movement; production of heat that maintains body temperature.

Nervous System

Regulatory centers for control of all body systems; learning and memory.

Endocrine System

Secretion of hormones for chemical regulation of all body systems.

Respiratory System

Rids the blood of carbon dioxide and supplies the blood with oxygen; helps maintain the pH of the blood.

Cardiovascular System

Transport of nutrients to body cells and transport of wastes away from cells.

Lymphatic System/Immunity

Drainage of tissue fluid; purifies tissue fluid and keeps it free of pathogens.

Digestive System

Breakdown of food and absorption of nutrients into blood.

Urinary System

Maintenance of volume and chemical composition of blood.

Reproductive System

Production of sperm and egg; transfer of sperm to female system where development occurs.

Summarizing the Concepts

4.1 Types of Tissues
Human tissues are categorized into four groups. Epithelial tissue covers the body and lines its cavities. The different types of epithelial tissue (squamous, cuboidal, and columnar) can be stratified and have cilia or microvilli. Also, columnar cells can be pseudostratified. Epithelial cells sometimes form glands that secrete either into ducts or into the blood.

Connective tissues, in which cells are separated by a matrix, often bind body parts together. Connective tissues have both white and yellow fibers and may also have fat (adipose) cells. Loose fibrous connective tissue supports epthelium and encloses organs. Dense fibrous connective tissue, such as that of tendons and ligaments, contains closely packed collagen fibers. Adipose tissue stores fat. Both cartilage and bone have cells within lacunae, but the matrix for cartilage is more flexible than that for bone, which contains calcium salts. In bone, the lacunae lie in concentric circles within an osteon (or Haversian system) about a central canal. Blood is a connective tissue in which the matrix is a liquid called plasma.

Muscular tissue is of three types. Both skeletal and cardiac muscle are striated; both cardiac and smooth muscle are involuntary. Skeletal muscle is found in muscles attached to bones, and smooth muscle is found in internal organs. Cardiac muscle makes up the heart.

Nervous tissue has one main type of conducting cell, the neuron, and several types of neuroglial cells. Each neuron has dendrites, a cell body, and an axon. The brain and spinal cord contain complete neurons, while the nerves contain only neuron fibers. Axons are specialized to conduct nerve impulses.

4.2 Body Cavities and Body Membranes
The internal organs occur within cavities. The thoracic cavity contains the heart and lungs; the abdominal cavity contains organs of the digestive, urinary, and reproductive systems, among others. Membranes line body cavities and internal spaces of organs. As an example, mucous membrane lines the tubes of the digestive system, while serous membrane lines the thoracic and abdominal cavities and covers the organs they contain.

4.3 Organ Systems
The digestive, cardiovascular, lymphatic, respiratory, and urinary systems perform processing and transporting functions that maintain the normal conditions of the body. The skeletal system and muscular system support the body and permit movement. The nervous system receives sensory input from sensory receptors and directs the muscles and glands to respond to outside stimuli. The endocrine system produces hormones, some of which influence the functioning of the reproductive system, which allows humans to make more of their own kind. The skin and its accessory organs comprise the integumentary system.

The accessory organs include nails, hair, and glands. Skin protects underlying tissues from physical trauma, pathogen invasion, and water loss. Skin helps regulate body temperature, and because it contains sensory receptors, skin also helps us to be aware of our surroundings.

Skin has two regions. The epidermis contains basal cells that produce new epithelial cells that become keratinized as they move toward the surface. The dermis, a largely fibrous connective tissue, contains epidermally derived glands and hair follicles, nerve endings, and blood vessels. Sensory receptors for touch, pressure, temperature, and pain are also present in the dermis. A subcutaneous layer, which is made up of loose connective tissue containing adipose cells, lies beneath the skin.

4.4 Homeostasis
Homeostasis is the relative constancy of the internal environment. Negative feedback mechanisms keep the environment relatively stable. When a sensor detects a change above and/or below a set point, a regulatory center activates an effector that reverses the change and brings conditions back to normal again. In contrast, a positive feedback mechanism brings about rapid change in the same direction as the stimulus. Still, positive feedback mechanisms are useful under certain conditions such as when a child is born.

The internal environment consists of blood and tissue fluid. All organ systems contribute to the constancy of tissue fluid and blood. Special contributions are made by the liver, which keeps blood glucose constant, and the kidneys, which regulate the pH. The nervous and endocrine systems regulate the other systems.

Studying the Concepts

1. Name the four major types of tissues. 62
2. Name the different kinds of epithelial tissue, and give a location and function for each. 62
3. What are the functions of connective tissue? Name the different kinds, and give one location for each. 64–66
4. What are the functions of muscular tissue? Name the different kinds, and give a location for each. 67
5. Nervous tissue contains what types of cells? Which organs in the body are made up of nervous tissue? 68
6. In what cavities are the major organs located? 69
7. Distinguish between the terms plasma membrane and body membrane. 69
8. Which organ systems maintain the body, support and move the body, coordinate and regulate body systems? 70
9. Why is skin sometimes called the integumentary system? State at least two functions of skin and describe its structure. 71–72
10. Why is homeostasis sometimes defined as the dynamic equilibrium of the internal environment? 74
11. Give a mechanical and a biological example of a negative feedback mechanism. Give an example of a positive feedback mechanism. 74–76
12. After consulting the working together illustration on page 77, explain how the various systems of the body contribute to homeostasis. 76–77

Testing Your Knowledge of the Concepts

In questions 1–4, match the type of tissue to the functions listed.
  a. epithelial
  b. connective
  c. muscular
  d. nervous

_____ 1. Contracts, allowing body parts to move.

_____ 2. Supports and binds body parts to one another.

_____ 3. Lines cavities and protects surfaces.

_____ 4. Conducts messages within and to and from the brain.

In questions 5–7, indicate whether the statement is true (T) or false (F).

_____ 5. The nervous and endocrine systems coordinate the functions of the other systems in the body.

_____ 6. The lymphatic system delivers nutrients to cells and removes their wastes.

_____ 7. Regulation of body temperature involves all systems of the body except the skin.

In questions 8 and 9, fill in the blanks.

8. _____ is the relative constancy of the internal environment.

9. Smooth muscle is _____ and involuntary.

10. Give the name, the location, and the function for each of the tissues shown in the drawings below.
    a. Type of epithelial tissue
    b. Type of muscular tissue
    c. Type of connective tissue

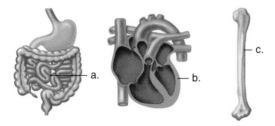

Understanding Key Terms

adhesion junction 64
adipose tissue 64
basement membrane 62
blood 66
bone 65
capillary 62
carcinoma 62
cardiac muscle 67
cardiovascular system 70
cartilage 64
cilium 62
coelom 69
collagen fiber 64
columnar epithelium 62
compact bone 65
connective tissue 64
cuboidal epithelium 62
dense fibrous connective
  tissue 64
dermis 72
digestive system 70
elastic cartilage 65
elastic fiber 64
endocrine system 70
epidermis 71
epithelial tissue 62
fibroblast 64
fibrocartilage 65
gap junction 64
gland 62
hair follicle 72
homeostasis 70
hyaline cartilage 64
integumentary system 71
intercalated disks 67
lacuna 64
ligament 64
loose fibrous connective
  tissue 64
lymphatic system 70
matrix 64
melanocyte 72
meninges 69
microvillus 62
mucous membrane 69
muscular (contractile) tissue 67
negative feedback 74
nerve 68
nervous system 70
nervous tissue 68
neuroglia 68
neuron 68
oil gland 72
pathogen 62
peritonitis 69
plasma 66
platelet 66
pleural membrane 69
positive feedback 76
red blood cell 66
reproductive system 70
respiratory system 70
reticular fiber 64
serous membrane 69
skeletal muscle 67
skin 71
smooth (visceral) muscle 67
spongy bone 65
squamous epithelium 62
striated 67
subcutaneous layer 72
sweat gland 72
synovial membrane 69
tendon 64
tight junction 64
tissue 62
urinary system 70
white blood cell 66

Further Readings

Bloomfield, L. A. April 2000. Cleaning agents. *Scientific American* 282(4):108. Article examines how soaps, detergents, bleaches, and brighteners work.

Chapman, C. 1999. *Basic chemistry for biology.* 2d ed. WCB/McGraw-Hill. The goal of this workbook is to provide a review of basic principles for biology students.

Ford, B. J. April 1998. The earliest views. *Scientific American* 278(4):50. Presents experiments of early microscopists.

Gerstein, M., and Levitt, M. November 1998. Simulating water and the molecules of life. *Scientific American* 279(5):100. Computer models show how water affects the structure and movement of proteins and other biological molecules.

Ingber, D. E. January 1998. The architecture of life. *Scientific American* 278(1):48. Simple mechanical rules may govern cell movements, tissue organization, and organ development.

Parenteau, N., and Naughton, G. April 1999. Skin: The first tissue-engineered products. *Scientific American* 280(4):83. Article discusses the procedure for tissue engineering skin.

Scerri, E. R. September 1998. The evolution of the periodic system. *Scientific American* 279(3):78. Article discusses the history and evolution of the periodic table.

Science 283(5407):1475. March 5, 1999. An entire section is devoted to topics involving mitochondria.

Scott, J. D., and Pawson, T. June 2000. Cell communication: The inside story. *Scientific American* 282(6):72. New therapies for serious disorders may be developed by mapping cell signalling networks.

e-Learning Connection   www.mhhe.com/biosci/genbio/maderhuman7/

4.1 Types of Tissues

Epithelial Tissue *Essential Study Partner*
Connective Tissue *Essential Study Partner*
Muscle Tissue *Essential Study Partner*
Nervous Tissue *Essential Study Partner*

Histology Quiz *practice identifying tissue types*

4.2 Body Cavities and Body Membranes

Mammalian Body Cavities *art labeling activity*

4.3 Organ Systems

Human Skin Anatomy I *art labeling activity*
Human Skin Anatomy II *art labeling activity*
Human Skin Anatomy III *art labeling activity*

4.4 Homeostasis

Working Together to Achieve Homeostasis

Chapter Summary

Key Term Flashcards *vocabulary quiz*
Chapter Quiz *objective quiz covering all chapter concepts*

II

Maintenance of the Human Body

All of the systems of the body help maintain homeostasis, resulting in a dynamic equilibrium of the internal environment. Our internal environment is the blood within blood vessels and the fluid that surrounds the cells of the tissues. The heart pumps the blood and sends it in vessels to the tissues, where materials are exchanged with tissue fluid. The composition of blood tends to remain relatively constant as a result of the actions of the digestive, respiratory, and urinary systems. Nutrients enter the blood at the small intestine, external gas exchange occurs in the lungs, and metabolic waste products are excreted at the kidneys. The immune system prevents pathogens from taking over the body and interfering with its proper functioning. ⚖

Chapter 5

Digestive System and Nutrition

Chapter Concepts

5.1 The Digestive System
- The human digestive system is an extended tube with specialized parts between two openings, the mouth and the anus. 82
- Food is ingested and then digested to small molecules that are absorbed. Indigestible materials are eliminated. 82

5.2 Three Accessory Organs
- The pancreas, the liver, and the gallbladder are accessory organs of digestion because their activities assist the digestive process. 90

5.3 Digestive Enzymes
- The products of digestion are small molecules, such as amino acids, fatty acids, and glucose, that can cross plasma membranes. 92
- The digestive enzymes are specific and have an optimum temperature and pH at which they function. 92

5.4 Homeostasis
- The digestive system works with the other systems of the body to maintain homeostasis. 95 ⚖

5.5 Nutrition
- Proper nutrition supplies the body with energy, nutrients, and all vitamins and minerals. 95

Has anyone in your family ever had ulcers? Aside from restricting your diet, they can be very painful. Barry Marshall believed that he knew the cause of ulcers and that, if he was correct, ulcers would be treatable and curable! One morning in 1984, he walked into his lab, stirred a beaker full of beef soup and *Helicobacter pylori,* a bacterium, and gulped the concoction. After five days, he began to vomit. His stomach grew inflamed. With further research, Marshall and others demonstrated that *H. pylori* is responsible for at least 70% of ulcers. Stress and other causes, such as prescription-drug side effects, may also play a role, but these are not usually the direct cause of ulcers.

We have only to consider the frequency of TV commercials concerned with treating gastrointestinal ills in order to conclude that the proper functioning of the digestive system is critical to our everyday lives. This chapter reviews both the anatomy and physiology of our internal tubular digestive tract and its accessory organs. The liver is an accessory organ with a myriad of functions besides its role in digestion, and we will examine many of these. Today, we recognize that in a sense "we are what we eat," and therefore a knowledge of nutrition is essential. This chapter ends with a discussion of the basic principles of nutrition.

5.1 The Digestive System

Digestion takes place within a tube called the digestive tract, which begins with the mouth and ends with the anus (Fig. 5.1). The functions of the digestive system are to ingest food, digest it to nutrients that can cross plasma membranes, absorb nutrients, and eliminate indigestible remains.

The Mouth

The mouth, which receives food, is bounded externally by the lips and cheeks. The lips extend from the base of the nose to the start of the chin. The red portion of the lips is poorly keratinized, and this allows blood to show through.

Most people enjoy eating food largely because they like its texture and taste. Sensory receptors called taste buds occur primarily on the tongue, and when these are activated by the presence of food, nerve impulses travel by way of cranial nerves to the brain. The tongue is composed of skeletal muscle whose contraction changes the shape of the tongue. Muscles exterior to the tongue cause it to move about. A fold of mucous membrane on the underside of the tongue attaches it to the floor of the oral cavity.

The roof of the mouth separates the nasal cavities from the oral cavity. The roof has two parts: an anterior (toward the front) **hard**

palate and a posterior (toward the back) **soft palate** (Fig. 5.2*a*). The hard palate contains several bones, but the soft palate is composed entirely of muscle. The soft palate ends in a finger-shaped projection called the uvula. The tonsils are in the back of the mouth, on either side of the tongue and in the nasopharynx (called adenoids). The tonsils help protect the body against infections. If the tonsils become inflamed, the person has **tonsillitis.** The infection can spread to the middle ears. If tonsillitis recurs repeatedly, the tonsils may be surgically removed (called a tonsillectomy).

Three pairs of **salivary glands** send juices (saliva) by way of ducts to the mouth. One pair of salivary glands lies at the

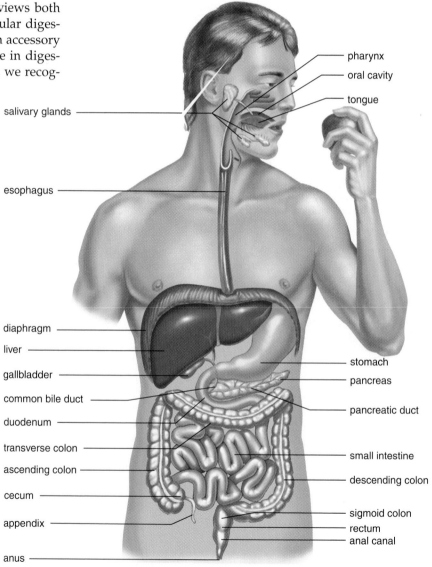

Figure 5.1 Digestive system.
Trace the path of food from the mouth to the anus. The large intestine consists of the cecum; the colon consisting of the ascending, transverse, descending, and sigmoid colon; and the rectum and anal canal. Note also the location of the accessory organs of digestion: the pancreas, the liver, and the gallbladder.

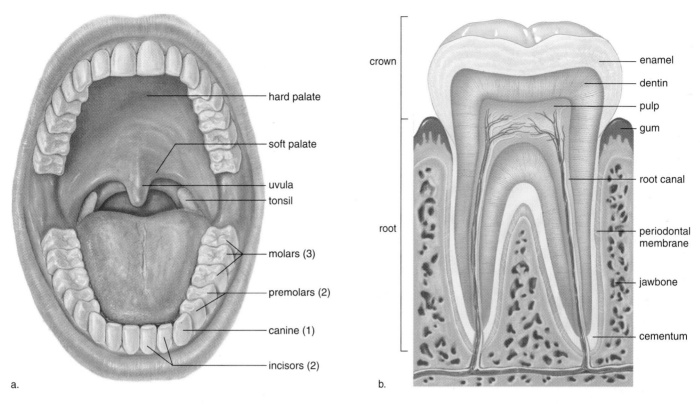

a.

b.

Figure 5.2 **Adult mouth and teeth.**
a. The chisel-shaped incisors bite; the pointed canines tear; the fairly flat premolars grind; and the flattened molars crush food. The last molar, called a wisdom tooth, may fail to erupt, or if it does, it is sometimes crooked and useless. Often dentists recommend the extraction of the wisdom teeth. **b.** Longitudinal section of a tooth. The crown is the portion that projects above the gum line and is sometimes replaced by a dentist. When a "root canal" is done, the nerves are removed. When the periodontal membrane is inflamed, the teeth can loosen.

sides of the face immediately below and in front of the ears. These glands swell when a person has the mumps, caused by a viral infection. Salivary glands have ducts that open on the inner surface of the cheek at the location of the second upper molar. Another pair of salivary glands lies beneath the tongue, and still another pair lies beneath the floor of the oral cavity. The ducts from these salivary glands open under the tongue. You can locate the openings if you use your tongue to feel for small flaps on the inside of your cheek and under your tongue. Saliva contains an enzyme called **salivary amylase** that begins the process of digesting starch.

The Teeth

With our teeth we chew food into pieces convenient for swallowing. During the first two years of life, the smaller 20 deciduous, or baby, teeth appear. These are eventually replaced by 32 adult teeth (Fig. 5.2*a*). The third pair of molars, called the wisdom teeth, sometimes fail to erupt. If they push on the other teeth and/or cause pain, they can be removed by a dentist or oral surgeon.

Each tooth has two main divisions, a crown and a root (Fig. 5.2*b*). The crown has a layer of enamel, an extremely hard outer covering of calcium compounds; dentin, a thick layer of bonelike material; and an inner pulp, which con-

tains the nerves and the blood vessels. Dentin and pulp are also found in the root.

Tooth decay, called **dental caries,** or cavities, occurs when bacteria within the mouth metabolize sugar and give off acids, which erode teeth. Two measures can prevent tooth decay: eating a limited amount of sweets and daily brushing and flossing of teeth. Fluoride treatments, particularly in children, can make the enamel stronger and more resistant to decay. Gum disease is more apt to occur with aging. Inflammation of the gums (gingivitis) can spread to the periodontal membrane, which lines the tooth socket. A person then has periodontitis, characterized by a loss of bone and loosening of the teeth so that extensive dental work may be required. Stimulation of the gums in a manner advised by your dentist is helpful in controlling this condition.

The tongue, which is composed of striated muscle and an outer layer of mucous membrane, mixes the chewed food with saliva. It then forms this mixture into a mass called a bolus in preparation for swallowing.

The salivary glands send saliva into the mouth, where the teeth chew the food and the tongue forms it into a bolus for swallowing.

Table 5.1   **Path of Food**

Organ	Function of Organ	Special Feature(s)	Function of Special Feature(s)
Oral cavity	Receives food; starts digestion of starch	Teeth Tongue	Chew food Form bolus
Esophagus	Passageway	——	——
Stomach	Storage of food; acidity kills bacteria; starts digestion of protein	Gastric glands	Release gastric juices
Small intestine	Digestion of all foods; absorption of nutrients	Intestinal glands Villi	Release fluids Absorb nutrients
Large intestine	Absorption of water; storage of indigestible remains	——	——

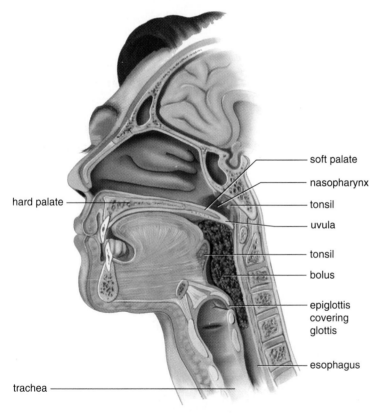

Figure 5.3   Swallowing.
When food is swallowed, the soft palate closes off the nasopharynx, and the epiglottis covers the glottis, forcing the bolus to pass down the esophagus. Therefore, you do not breathe while swallowing.

The Pharynx

The **pharynx** is a region that receives food from the mouth and air from the nasal cavities. The food passage (Table 5.1) and air passage cross in the pharynx because the trachea (windpipe) is ventral to (in front of) the esophagus, a long muscular tube that takes food to the stomach.

Swallowing, a process that occurs in the pharynx (Fig. 5.3), is a **reflex action** performed automatically, without conscious thought. Usually during swallowing, the soft palate moves back to close off the **nasopharynx,** and the trachea moves up under the **epiglottis** to cover the glottis. The **glottis** is the opening to the larynx (voice box). During swallowing, food normally enters the esophagus because the air passages are blocked. Unfortunately, we have all had the unpleasant experience of having food "go the wrong way." The wrong way may be either into the nasal cavities or into the trachea. If it is the latter, coughing will most likely force the food up out of the trachea and into the pharynx again. The up-and-down movement of the Adam's apple, the front part of the larynx, is easy to observe when a person swallows. We do not breathe when we swallow.

The air passage and the food passage cross in the pharynx. When you swallow, the air passage usually is blocked off, and food must enter the esophagus.

The Esophagus

The **esophagus** is a muscular tube that passes from the pharynx through the thoracic cavity and diaphragm into the abdominal cavity where it joins the stomach. The esophagus is ordinarily collapsed, but it opens and receives the bolus when swallowing occurs. A rhythmic contraction called **peristalsis** pushes the food along the digestive tract. Occasionally, peristalsis begins even though there is no food in the esophagus. This produces the sensation of a lump in the throat.

The esophagus plays no role in the chemical digestion of food. Its sole purpose is to conduct the food bolus from the mouth to the stomach. **Sphincters** are muscles that encircle tubes and act as valves; tubes close when sphincters contract, and they open when sphincters relax. The entrance of the esophagus to the stomach is marked by a constriction, often called a sphincter, although the muscle is not as developed as in a true sphincter. Relaxation of the sphincter allows the bolus to pass into the stomach, while contraction prevents the acidic contents of the stomach from backing up into the esophagus. **Heartburn,** which feels like a burning pain rising up into the throat, occurs when some of the stomach contents escape into the esophagus. When vomiting occurs, a contraction of the abdominal muscles and diaphragm propels the contents of the stomach upward through the esophagus.

The esophagus conducts the bolus of food from the pharynx to the stomach. Peristalsis begins in the esophagus and occurs along the entire length of the digestive tract.

The Wall of the Digestive Tract

The wall of the esophagus in the abdominal cavity is comparable to that of the digestive tract, which has these layers (Fig. 5.4):

Mucosa (mucous membrane layer) A layer of epithelium supported by connective tissue and smooth muscle lines the **lumen** (central cavity) and contains glandular epithelial cells that secrete digestive enzymes and goblet cells that secrete mucus.

Submucosa (submucosal layer) A broad band of loose connective tissue that contains blood vessels lies beneath the mucosa. Lymph nodules, called Peyer's patches, are in the submucosa. Like the tonsils, they help protect us from disease.

Muscularis (smooth muscle layer) Two layers of smooth muscle make up this section. The inner, circular layer encircles the gut; the outer, longitudinal layer lies in the same direction as the gut. (The stomach also has oblique muscles.)

Serosa (serous membrane layer) Most of the digestive tract has a serosa, a very thin, outermost layer of squamous epithelium supported by connective tissue. The serosa secretes a serous fluid that keeps the outer surface of the intestines moist so that the organs of the abdominal cavity slide against one another. The esophagus has an outer layer composed only of loose connective tissue called the adventitia.

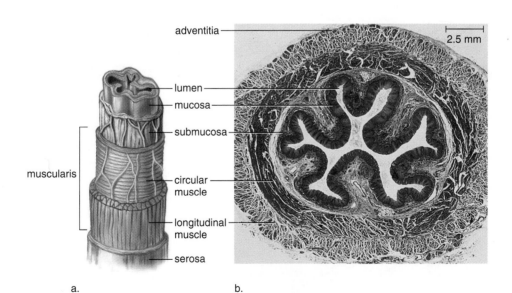

a.   b.

Figure 5.4   **Wall of the digestive tract.**
a. Several different types of tissues are found in the wall of the digestive tract. Note the placement of circular muscle inside longitudinal muscle.
b. Micrograph of the wall of the esophagus.

The Stomach

The **stomach** (Fig. 5.5) is a thick-walled, J-shaped organ that lies on the left side of the body beneath the diaphragm. The stomach is continuous with the esophagus above and the duodenum of the small intestine below. The stomach stores food and aids in digestion. The wall of the stomach has deep folds, which disappear as the stomach fills to an approximate capacity of one liter. Its muscular wall churns, mixing the food with gastric juice. The term *gastric* always refers to the stomach.

The columnar epithelial lining of the stomach (i.e., the mucosa) has millions of gastric pits, which lead into **gastric glands.** The gastric glands produce gastric juice. Gastric juice contains an enzyme called **pepsin,** which digests protein, plus hydrochloric acid (HCl) and mucus. HCl causes the stomach to have a high acidity with a pH of about 2, and this is beneficial because it kills most bacteria present in food. Although HCl does not digest food, it does break down the connective tissue of meat and activates pepsin.

The wall of the stomach is protected by a thick layer of mucus secreted by goblet cells in its lining. If, by chance, HCl penetrates this mucus, the wall can begin to break down, and an ulcer results. An **ulcer** is an open sore in the wall caused by the gradual disintegration of tissue. As discussed in the opening vignette, it now appears that most ulcers are due to a bacterial (*Helicobacter pylori*) infection that impairs the ability of epithelial cells to produce protective mucus.

Alcohol is absorbed in the stomach, but food substances are not. Normally, the stomach empties in about 2–6 hours. When food leaves the stomach, it is a thick, soupy liquid called **chyme.** Chyme leaves the stomach and enters the small intestine in squirts by way of a sphincter that repeatedly opens and closes.

The stomach can expand to accommodate large amounts of food. When food is present, the stomach churns, mixing food with acidic gastric juice.

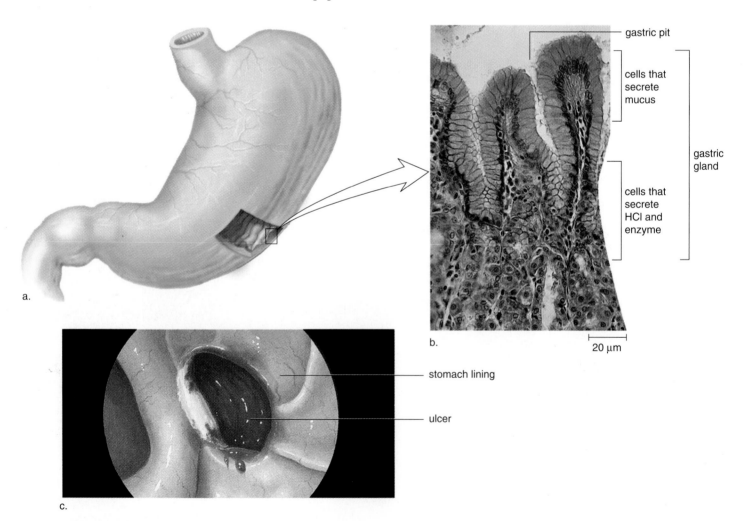

Figure 5.5   Anatomy and histology of the stomach.
a. The stomach has a thick wall with folds that allow it to expand and fill with food. **b.** The mucosa contains gastric glands, which secrete mucus and a gastric juice active in protein digestion. **c.** A bleeding ulcer viewed by using an endoscope (a tubular instrument bearing a tiny lens and a light source) that can be inserted into the abdominal cavity.

The Small Intestine

The **small intestine** is named for its small diameter (compared to that of the large intestine), but perhaps it should be called the long intestine. The small intestine averages about 6 meters (18 feet) in length, compared to the large intestine, which is about 1.5 meters (4½ ft) in length.

The first 25 cm of the small intestine is called the **duodenum.** Ducts from the liver and pancreas join to form one duct that enters the duodenum (see Fig. 5.1). The small intestine receives bile from the liver and pancreatic juice from the pancreas via this duct. **Bile** emulsifies fat—emulsification causes fat droplets to disperse in water. The intestine has a slightly basic pH because pancreatic juice contains sodium bicarbonate ($NaHCO_3$), which neutralizes chyme. The enzymes in pancreatic juice and enzymes produced by the intestinal wall complete the process of food digestion.

It's been suggested that the surface area of the small intestine is approximately that of a tennis court. What factors contribute to increasing its surface area? The wall of the small intestine contains fingerlike projections called **villi** (sing. **villus**), which give the intestinal wall a soft, velvety appearance (Fig. 5.6). A villus has an outer layer of columnar epithelial cells and each of these cells has thousands of microscopic extensions called microvilli. Collectively in electron micrographs, microvilli give the villi a fuzzy border known as a "brush border." Since the microvilli bear the intestinal enzymes, these enzymes are called brush-border enzymes. The microvilli greatly increase the surface area of the villus for the absorption of nutrients.

Nutrients are absorbed into the vessels of a villus. A villus contains blood capillaries and a small lymphatic capillary, called a **lacteal.** The lymphatic system is an adjunct to the cardiovascular system—its vessels carry a fluid called lymph to the cardiovascular veins. Sugars and amino acids enter the blood capillaries of a villus. Fats are digested to glycerol and fatty acids. These components enter the epithelial cells of the villi and within these cells they are joined and packaged as lipoprotein droplets which enter a lacteal. After nutrients are absorbed, they are eventually carried to all the cells of the body by the bloodstream.

The large surface area of the small intestine facilitates absorption of nutrients into the cardiovascular (glucose and amino acids) and lymphatic (fats) systems.

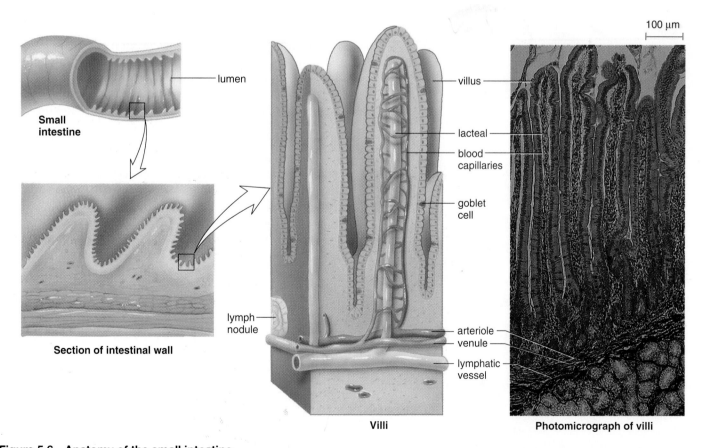

Figure 5.6 Anatomy of the small intestine.
The wall of the small intestine has folds that bear fingerlike projections called villi. The products of digestion are absorbed into the blood capillaries and the lacteals of the villi.

Regulation of Digestive Secretions

The nervous system promotes the secretion of digestive juices, but so do hormones (Fig. 5.7). A **hormone** is a substance produced by one set of cells that affects a different set of cells, the so-called target cells. Hormones are usually transported by the bloodstream. When a person has eaten a meal particularly rich in protein, the stomach produces the hormone gastrin. Gastrin enters the bloodstream, and soon the stomach is churning, and the secretory activity of gastric glands is increasing. A hormone produced by the duodenal wall, GIP (gastric inhibitory peptide), works opposite to gastrin: it inhibits gastric gland secretion.

Cells of the duodenal wall produce two other hormones that are of particular interest—secretin and CCK (cholecystokinin). Acid, especially hydrochloric acid (HCl) present in chyme, stimulates the release of secretin, while partially digested protein and fat stimulate the release of CCK. Soon after these hormones enter the bloodstream, the pancreas increases its output of pancreatic juice, which helps digest food, and the liver increases its output of bile. The gallbladder contracts to release bile. Regulation of digestive gland secretions is one of the ways hormones contribute to homeostasis. ⚖

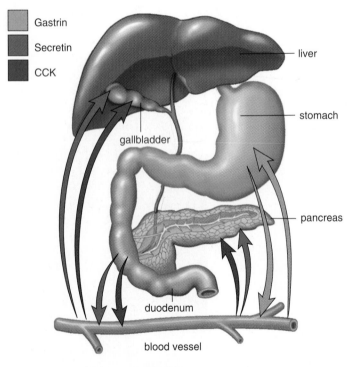

Figure 5.7   Hormonal control of digestive gland secretions.
Gastrin (blue), produced by the lower part of the stomach, enters the bloodstream and thereafter stimulates the upper part of the stomach to produce more digestive juice. Secretin (green) and CCK (purple), produced by the duodenal wall, stimulate the pancreas to secrete its digestive juice and the gallbladder to release bile.

The Large Intestine

The **large intestine,** which includes the cecum, the colon, the rectum, and the anal canal, is larger in diameter than the small intestine (6.5 cm compared to 2.5 cm), but it is shorter in length (see Fig. 5.1). The large intestine absorbs water, salts, and some vitamins. It also stores indigestible material until it is eliminated at the anus.

The **cecum,** which lies below the junction with the small intestine, is the blind end of the large intestine. The cecum has a small projection called the vermiform **appendix** (*vermiform* means wormlike) (Fig. 5.8). In humans, the appendix also may play a role in fighting infections. This organ is subject to inflammation, a condition called appendicitis. If inflamed, the appendix should be removed before the fluid content rises to the point that the appendix bursts, a situation that may cause peritonitis, a generalized infection of the lining of the abdominal cavity. Peritonitis can lead to death.

The **colon** includes the ascending colon, which goes up the right side of the body to the level of the liver; the transverse colon, which crosses the abdominal cavity just below the liver and the stomach; the descending colon, which passes down the left side of the body; and the sigmoid colon, which enters the rectum, the last 20 cm of the large intestine. The rectum opens at the **anus,** where **defecation,** the expulsion of feces, occurs. When feces are forced into the rectum by peristalsis, a defecation reflex occurs. The stretching of the rectal wall initiates nerve impulses to the spinal cord, and shortly thereafter the rectal muscles contract and the anal sphincters relax (Fig. 5.9). Ridding the body of indigestible remains is another way the digestive system helps maintain homeostasis. Feces are three-quarters water and one-quarter solids. Bacteria, fiber (indigestible remains), and other indigestible materials are in the solid portion. The brown color of feces is due to bilirubin (see page 90), and the odor is due to breakdown products as bacteria work on the nondigested remains. This bacterial action also produces gases. ⚖

For many years, it was believed that facultative bacteria (bacteria that can live with or without oxygen), such as *Escherichia coli,* were the major inhabitants of the colon, but new culture methods show that over 99% of the colon bacteria are obligate anaerobes (bacteria that die in the presence of oxygen). Not only do the bacteria break down indigestible material, they also produce some vitamins and other molecules that can be absorbed and used by our bodies. In this way, they perform a service for us.

Water is considered unsafe for swimming when the coliform (nonpathogenic intestinal) bacterial count reaches a certain number. A high count is an indication that a significant amount of feces has entered the water. The more feces present, the greater the possibility that disease-causing bacteria are also present.

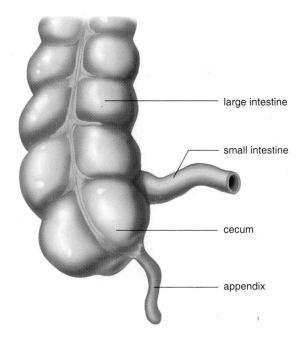

Figure 5.8 **Junction of the small intestine and the large intestine.**
The cecum is the blind end of the ascending colon. The appendix is attached to the cecum.

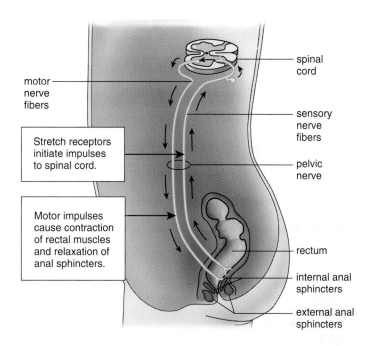

Figure 5.9 **Defecation reflex.**
The accumulation of feces in the rectum causes it to stretch, which initiates a reflex action resulting in rectal contraction and expulsion of the fecal material.

Polyps

The colon is subject to the development of **polyps,** small growths arising from the epithelial lining. Polyps, whether benign or cancerous, can be removed surgically. If colon cancer is detected while still confined to a polyp, the expected outcome is a complete cure. Some investigators believe that dietary fat increases the likelihood of colon cancer because dietary fat causes an increase in bile secretion. It could be that intestinal bacteria convert bile salts to substances that promote the development of cancer. On the other hand, fiber in the diet seems to inhibit the development of colon cancer. Dietary fiber absorbs water and adds bulk, thereby diluting the concentration of bile salts and facilitating the movement of substances through the intestine. Regular elimination reduces the time that the colon wall is exposed to any cancer-promoting agents in feces.

Diarrhea and Constipation

Two common everyday complaints associated with the large intestine are **diarrhea** and **constipation.** The major causes of diarrhea are infection of the lower intestinal tract and nervous stimulation. In the case of infection, such as food poisoning caused by eating contaminated food, the intestinal wall becomes irritated, and peristalsis increases. Water is not absorbed, and the diarrhea that results rids the body of the infectious organisms. In nervous diarrhea, the nervous system stimulates the intestinal wall, and diarrhea results.

Prolonged diarrhea can lead to dehydration because of water loss and to disturbances in the heart's contraction due to an imbalance of salts in the blood.

When a person is constipated, the feces are dry and hard. One reason for this condition is that socialized persons have learned to inhibit defecation to the point that the desire to defecate is ignored. Two components of the diet that can help prevent constipation are water and fiber. Water intake prevents drying out of the feces, and fiber provides the bulk needed for elimination. The frequent use of laxatives is discouraged. If, however, it is necessary to take a laxative, a bulk laxative is the most natural because, like fiber, it produces a soft mass of cellulose in the colon. Lubricants, like mineral oil, make the colon slippery, and saline laxatives, like milk of magnesia, act osmotically—they prevent water from being absorbed and, depending on the dosage, may even cause water to enter the colon. Some laxatives are irritants; they increase peristalsis to the degree that the contents of the colon are expelled.

Chronic constipation is associated with the development of hemorrhoids, enlarged and inflamed blood vessels at the anus.

The large intestine does not produce digestive enzymes; it does absorb water, salts, and some vitamins.

5.2 Three Accessory Organs

The pancreas, liver, and gallbladder are accessory digestive organs. Figure 5.1 shows how the pancreatic duct from the pancreas and the common bile duct from the liver and gallbladder join before entering the duodenum.

The Pancreas

The **pancreas** lies deep in the abdominal cavity, resting on the posterior abdominal wall. It is an elongated and somewhat flattened organ that has both an endocrine and an exocrine function. As an endocrine gland, it secretes insulin and glucagon, hormones that help keep the blood glucose level within normal limits. In this chapter, we are interested in its exocrine function. Most pancreatic cells produce pancreatic juice, which contains sodium bicarbonate ($NaHCO_3$) and digestive enzymes for all types of food. Sodium bicarbonate neutralizes chyme; whereas pepsin acts best in an acid pH of the stomach, pancreatic enzymes require a slightly basic pH. **Pancreatic amylase** digests starch, **trypsin** digests protein, and **lipase** digests fat. In cystic fibrosis, a thick mucus blocks the pancreatic duct, and the patient must take supplemental pancreatic enzymes by mouth for proper digestion to occur.

The Liver

The **liver,** which is the largest gland in the body, lies mainly in the upper right section of the abdominal cavity, under the diaphragm (see Fig. 5.1). The liver has two main lobes, the right lobe and the smaller left lobe, which crosses the midline and lies above the stomach. The liver contains approximately 100,000 lobules that serve as the structural and functional units of the liver (Fig. 5.10). Triads consisting of these three structures are located between the lobules: a bile duct that takes bile away from the liver; a branch of the hepatic artery that brings O_2-rich blood to the liver; and a branch of the hepatic portal vein that transports nutrients from the intestines. The central veins of lobules enter a hepatic vein. Trace the path of blood in Figure 5.11 from the intestines to the liver via the hepatic portal vein and from the liver to the inferior vena cava via the hepatic veins.

In some ways, the liver acts as the gatekeeper to the blood. As the blood from the hepatic portal vein passes through the liver, it removes poisonous substances and detoxifies them. The liver also removes nutrients and works to keep the contents of the blood constant. It removes and stores iron and the fat-soluble vitamins A, D, E, and K. The liver makes the plasma proteins from amino acids and it helps regulate the quantity of cholesterol in the blood.

The liver maintains the blood glucose level at about 100 mg/100 ml (0.1%), even though a person eats intermittently. Any excess glucose that is present in blood is

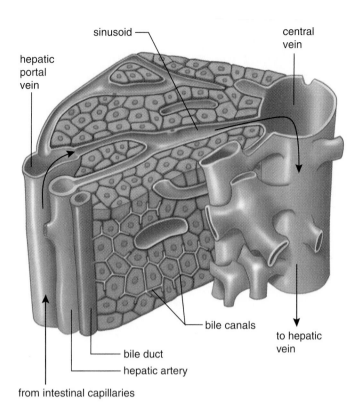

Figure 5.10 Hepatic lobules.
The liver contains over 100,000 lobules. Each lobule contains many cells that perform the various functions of the liver. They remove from and/or add materials to blood and deposit bile in bile ducts.

removed and stored by the liver as glycogen. Between meals, glycogen is broken down to glucose, which enters the hepatic veins, and in this way, the blood glucose level remains constant.

If the supply of glycogen is depleted, the liver converts glycerol (from fats) and amino acids to glucose molecules. The conversion of amino acids to glucose necessitates deamination, the removal of amino groups. By a complex metabolic pathway, the liver then combines ammonia with carbon dioxide to form urea:

$$2\,NH_3 \;+\; CO_2 \longrightarrow H_2N - \overset{\displaystyle O}{\overset{\displaystyle \|}{C}} - NH_2$$
$$\text{ammonia} \quad \text{carbon dioxide} \qquad\qquad \text{urea}$$

Urea is the usual nitrogenous waste product from amino acid breakdown in humans. After its formation in the liver, urea is excreted by the kidneys.

The liver produces bile, which is stored in the gallbladder. Bile has a yellowish-green color because it contains the bile pigment bilirubin, derived from the breakdown of hemoglobin, the red pigment of red blood cells. Bile also contains bile salts. Bile salts are derived from cholesterol and they

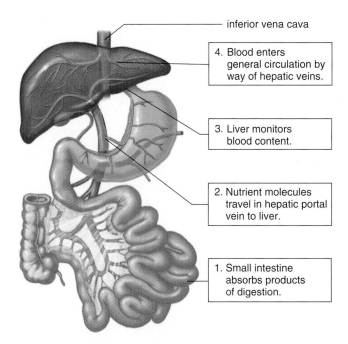

inferior vena cava

4. Blood enters general circulation by way of hepatic veins.

3. Liver monitors blood content.

2. Nutrient molecules travel in hepatic portal vein to liver.

1. Small intestine absorbs products of digestion.

Figure 5.11   Hepatic portal system.
The hepatic portal vein takes the products of digestion from the digestive system to the liver, where they are processed before entering a hepatic vein.

emulsify fat in the small intestine. When fat is emulsified, it breaks up into droplets, providing a much larger surface area, which can be acted upon by a digestive enzyme from the pancreas.

Altogether, the following are significant ways in which the liver helps maintain homeostasis.

1. Detoxifies blood by removing and metabolizing poisonous substances.
2. Stores iron (Fe^{2+}) and the fat-soluble vitamins A, D, E, K, and B_{12}.
3. Makes plasma proteins, such as albumins and fibrinogen, from amino acids.
4. Stores glucose as glycogen after eating, and breaks down glycogen to glucose to maintain the glucose concentration of blood between eating periods.
5. Produces urea after breaking down amino acids.
6. Removes bilirubin, a breakdown product of hemoglobin from the blood, and excretes it in bile, a liver product.
7. Helps regulate blood cholesterol level, converting some to bile salts.

Liver Disorders

Hepatitis and cirrhosis are two serious diseases that affect the entire liver and hinder its ability to repair itself. Therefore, they are life-threatening diseases. When a person has a liver ailment, jaundice may occur. **Jaundice** is a yellowish

tint to the whites of the eyes and also to the skin of light-pigmented persons. Bilirubin is deposited in the skin due to an abnormally large amount in the blood. In hemolytic jaundice, red blood cells have been broken down in abnormally large amounts; in obstructive jaundice, bile ducts are blocked or liver cells are damaged.

Jaundice can also result from **hepatitis,** inflammation of the liver. Viral hepatitis occurs in several forms. Hepatitis A is usually acquired from sewage-contaminated drinking water. Hepatitis B, which is usually spread by sexual contact, can also be spread by blood transfusions or contaminated needles. The hepatitis B virus is more contagious than the AIDS virus, which is spread in the same way. Thankfully, however, there is now a vaccine available for hepatitis B. Hepatitis C, which is usually acquired by contact with infected blood and for which there is no vaccine, can lead to chronic hepatitis, liver cancer, and death.

Cirrhosis is another chronic disease of the liver. First the organ becomes fatty, and liver tissue is then replaced by inactive fibrous scar tissue. Cirrhosis of the liver is often seen in alcoholics due to malnutrition and to the excessive amounts of alcohol (a toxin) the liver is forced to break down.

The liver has amazing generative powers and can recover if the rate of regeneration exceeds the rate of damage. During liver failure, however, there may not be enough time to let the liver heal itself. Liver transplantation is usually the preferred treatment for liver failure, but artificial livers have been developed and tried in a few cases. One type is a cartridge that contains liver cells. The patient's blood passes through the cellulose acetate tubing of the cartridge and is serviced in the same manner as with a normal liver. In the meantime, the patient's liver has a chance to recover.

The Gallbladder

The **gallbladder** is a pear-shaped, muscular sac attached to the surface of the liver (see Fig. 5.1). About 1,000 ml of bile are produced by the liver each day, and any excess is stored in the gallbladder. Water is reabsorbed by the gallbladder so that bile becomes a thick, mucuslike material. When needed, bile leaves the gallbladder and proceeds to the duodenum via the common bile duct.

The cholesterol content of bile can come out of solution and form crystals. If the crystals grow in size, they form gallstones. The passage of the stones from the gallbladder may block the common bile duct and cause obstructive jaundice. Then the gallbladder must be removed.

The pancreas produces pancreatic juice, which contains enzymes for the digestion of food. Among its many functions, the liver produces bile, which is stored in the gallbladder.

5.3 Digestive Enzymes

The digestive enzymes are **hydrolytic enzymes,** which break down substances by the introduction of water at specific bonds. Digestive enzymes, like other enzymes, are proteins with a particular shape that fits their substrate. They also have an optimum pH, which maintains their shape, thereby enabling them to speed up their specific reaction.

The various digestive enzymes present in the digestive juices, mentioned previously, help break down carbohydrates, proteins, nucleic acids, and fats, the major components of food. Starch is a carbohydrate, and its digestion begins in the mouth. Saliva from the salivary glands has a neutral pH and contains **salivary amylase,** the first enzyme to act on starch:

$$\text{starch} + H_2O \xrightarrow{\text{salivary amylase}} \text{maltose}$$

In this equation, salivary amylase is written above the arrow to indicate that it is neither a reactant nor a product in the reaction. It merely speeds the reaction in which its substrate, starch, is digested to many molecules of maltose, a disaccharide. Maltose molecules cannot be absorbed by the intestine; additional digestive action in the small intestine converts maltose to glucose, which can be absorbed.

Protein digestion begins in the stomach. Gastric juice secreted by gastric glands has a very low pH—about 2—because it contains hydrochloric acid (HCl). Pepsinogen, a precursor that is converted to the enzyme **pepsin** when exposed to HCl, is also present in gastric juice. Pepsin acts on protein to produce peptides:

$$\text{protein} + H_2O \xrightarrow{\text{pepsin}} \text{peptides}$$

Peptides vary in length, but they always consist of a number of linked amino acids. Peptides are usually too large to be absorbed by the intestinal lining, but later they are broken down to amino acids in the small intestine.

Starch, proteins, nucleic acids, and fats are all enzymatically broken down in the small intestine. Pancreatic juice, which enters the duodenum, has a basic pH because it contains sodium bicarbonate (NaHCO₃). Sodium bicarbonate neutralizes chyme, producing the slightly basic pH that is optimum for pancreatic enzymes. One pancreatic enzyme, **pancreatic amylase,** digests starch:

$$\text{starch} + H_2O \xrightarrow{\text{pancreatic amylase}} \text{maltose}$$

Another pancreatic enzyme, **trypsin,** digests protein:

$$\text{protein} + H_2O \xrightarrow{\text{trypsin}} \text{peptides}$$

Trypsin is secreted as trypsinogen, which is converted to trypsin in the duodenum.

Lipase, a third pancreatic enzyme, digests fat molecules in the fat droplets after they have been emulsified by bile salts:

$$\text{fat} \xrightarrow{\text{bile salts}} \text{fat droplets}$$

$$\text{fat droplets} + H_2O \xrightarrow{\text{lipase}} \text{glycerol} + \text{fatty acids}$$

The end products of lipase digestion, glycerol and fatty acid molecules, are small enough to cross the cells of the intestinal villi, where absorption takes place. As mentioned previously, glycerol and fatty acids enter the cells of the villi, and within these cells, they are rejoined and packaged as lipoprotein droplets before entering the lacteals (see Fig. 5.6).

Peptidases and **maltase,** enzymes produced by the small intestine, complete the digestion of protein to amino acids and starch to glucose, respectively. Amino acids and glucose are small molecules that cross into the cells of the villi. Peptides, which result from the first step in protein digestion, are digested to amino acids by peptidases:

$$\text{peptides} + H_2O \xrightarrow{\text{peptidases}} \text{amino acids}$$

Maltose, a disaccharide that results from the first step in starch digestion, is digested to glucose by maltase:

$$\text{maltose} + H_2O \xrightarrow{\text{maltase}} \text{glucose} + \text{glucose}$$

Other disaccharides, each of which has its own enzyme, are digested in the small intestine. The absence of any one of these enzymes can cause illness. For example, many people, including as many as 75% of African Americans, cannot digest lactose, the sugar found in milk, because they do not produce lactase, the enzyme that converts lactose to its components, glucose and galactose. Drinking untreated milk often gives these individuals the symptoms of **lactose intolerance** (diarrhea, gas, cramps), caused by a large quantity of nondigested lactose in the intestine. In most areas, it is possible to purchase milk made lactose-free by the addition of synthetic lactase or *Lactobacillus acidophilus* bacteria, which break down lactose.

Table 5.2 lists some of the major digestive enzymes produced by the digestive tract, salivary glands, or the pancreas. Each type of food is broken down by specific enzymes.

Digestive enzymes present in digestive juices help break down food to the nutrient molecules: glucose, amino acids, fatty acids, and glycerol. The first two are absorbed into the blood capillaries of the villi, and the last two re-form within epithelial cells before entering the lacteals as lipoprotein droplets.

Table 5.2 Major Digestive Enzymes

Food	Digestion	Enzyme	Optimum pH	Produced By	Site of Action
Starch	Starch + $H_2O \longrightarrow$ maltose	Salivary amylase	Neutral	Salivary glands	Mouth
		Pancreatic amylase	Basic	Pancreas	Small intestine
	Maltose + $H_2O \longrightarrow$ glucose + glucose	Maltase	Basic	Small intestine	Small intestine
Protein	Protein + $H_2O \longrightarrow$ peptides	Pepsin	Acidic	Gastric glands	Stomach
		Trypsin	Basic	Pancreas	Small intestine
	Peptide + $H_2O \longrightarrow$ amino acids	Peptidases	Basic	Small intestine	Small intestine
Nucleic acid	RNA and DNA + $H_2O \longrightarrow$ nucleotides	Nuclease	Basic	Pancreas	Small intestine
	Nucleotide + $H_2O \longrightarrow$ base + sugar + phosphate	Nucleosidases	Basic	Small intestine	Small intestine
Fat	Fat droplet + $H_2O \longrightarrow$ glycerol + fatty acids	Lipase	Basic	Pancreas	Small intestine

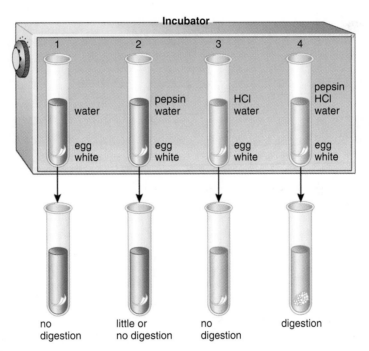

Figure 5.12 **Digestion experiment.**
This experiment is based on the optimum conditions for digestion by pepsin in the stomach. Knowing that the correct enzyme, optimum pH, optimum temperature, and the correct substrate must be present for digestion to occur, explain the results of this experiment. Colors indicate pH of test tubes (blue, basic; red, acidic).

Conditions for Digestion

Laboratory experiments can define the necessary conditions for digestion. For example, the four test tubes described in Figure 5.12 can be prepared and observed for the digestion of egg white, a protein digested in the stomach by the enzyme pepsin.

After all tubes are placed in an incubator at body temperature for at least one hour, the results depicted are observed. Tube 1 is a control tube; no digestion has occurred in this tube because the enzyme and HCl are missing. (If a control gives a positive result, then the experiment is invalidated.) Tube 2 shows limited or no digestion because HCl is missing, and therefore the pH is too high for pepsin to be effective. Tube 3 shows no digestion because although HCl is present, the enzyme is missing. Tube 4 shows the best digestive action because the enzyme is present and the presence of HCl has resulted in an optimum pH. This experiment supports the hypothesis that for digestion to occur, the substrate and enzyme must be present and the environmental conditions must be optimum. The optimal environmental conditions include a warm temperature and the correct pH.

Human Systems Work Together

Integumentary System

Digestive tract provides nutrients needed by skin.

Skin helps to protect digestive organs; helps to provide vitamin D for Ca^{2+} absorption.

Skeletal System

Digestive tract provides Ca^{2+} and other nutrients for bone growth and repair.

Bones provide support and protection; hyoid bone assists swallowing.

Muscular System

Digestive tract provides glucose for muscle activity; liver metabolizes lactic acid following anaerobic muscle activity.

Smooth muscle contraction accounts for peristalsis; skeletal muscles support and help protect abdominal organs.

Nervous System

Digestive tract provides nutrients for growth, maintenance, and repair of neurons and neuroglia.

Brain controls nerves, which innervate smooth muscle and permit tract movements.

Endocrine System

Stomach and small intestine produce hormones.

Hormones help control secretion of digestive glands and accessory organs; insulin and glucagon regulate glucose storage in liver.

How the Digestive System works with other body systems

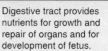

Cardiovascular System

Digestive tract provides nutrients for plasma protein formation and blood cell formation; liver detoxifies blood, makes plasma proteins, destroys old red blood cells.

Blood vessels transport nutrients from digestive tract to body; blood services digestive organs.

Lymphatic System/Immunity

Digestive tract provides nutrients for lymphoid organs; stomach acidity prevents pathogen invasion of body.

Lacteals absorb fats; Peyer's patches prevent invasion by pathogens; appendix contains lymphoid tissue.

Respiratory System

Breathing is possible through the mouth because digestive tract and respiratory tract share the pharynx.

Gas exchange in lungs provides oxygen to digestive tract and excretes carbon dioxide from digestive tract.

Urinary System

Liver synthesizes urea; digestive tract excretes bile pigments from liver and provides nutrients.

Kidneys convert vitamin D to active form needed for Ca^{2+} absorption; compensate for any water loss by digestive tract.

Reproductive System

Digestive tract provides nutrients for growth and repair of organs and for development of fetus.

Pregnancy crowds digestive organs and promotes heartburn and constipation.

5.4 Homeostasis

The illustration on the previous page tells how the digestive system works with other systems in the body to maintain homeostasis.

Within the digestive tract, the food we eat is broken down to nutrients small enough to be absorbed by the villi of the small intestine. Digestive enzymes are produced by the salivary glands, gastric glands, and intestinal glands. Three accessory organs of digestion (the pancreas, the liver, and the gallbladder) also contribute secretions that help break down food. The liver produces bile (stored by the gallbladder), which emulsifies fat. The pancreas produces enzymes for the digestion of carbohydrates, proteins, and fat. Secretions from these glands, which are sent by ducts into the small intestine, are regulated by hormones such as secretin produced by the digestive tract. Therefore, the digestive tract is also a part of the endocrine system.

Blood laden with nutrients passes from the region of the small intestine to the liver by way of the hepatic portal vein. The liver is the most important of the metabolic organs. Aside from making bile, the liver regulates the cholesterol content of the blood, makes plasma proteins, stores glucose as glycogen, produces urea, and metabolizes poisons. Because the liver is such an important organ, diseases affecting the liver, such as hepatitis and cirrhosis, are extremely dangerous.

5.5 Nutrition

The body requires three major classes of macronutrients in the diet: carbohydrate, protein, and fat. These supply the energy and the building blocks that are needed to synthesize cellular contents. Micronutrients—especially vitamins and minerals—are also required because they are necessary for optimum cellular metabolism.

Several modern nutritional studies suggest that certain nutrients can protect against heart disease, cancer, and other serious illnesses. These studies include analysis of the eating habits of healthy people in the United States and from around the world, especially those with lower rates of heart disease and cancer. The result has been the dietary recommendations illustrated by a food pyramid (Fig. 5.13).

The bulk of the diet should consist of bread, cereal, rice, and pasta as energy sources. Whole grains are preferred over those that have been milled because they contain fiber, vitamins, and minerals. Vegetables and fruits are another rich source of fiber, vitamins, and minerals. Notice, then, that a largely vegetarian diet is recommended.

Animal products, especially meat, need only be minimally included in the diet; fats and sweets should be used sparingly. Dairy products and meats tend to be high in saturated fats, and an intake of saturated fats increases the risk of cardiovascular disease (see Lipids, p. 98). Low-fat dairy products are available, but there is no way to take much of the fat out of meat. Beef, in particular, contains a relatively high fat content. Ironically, the affluence of people in the United States contributes to a poor diet and, therefore, possible illness. Only comparatively rich people can afford fatty meats from grain-fed cattle and carbohydrates that have been highly processed to remove fiber and to add sugar and salt.

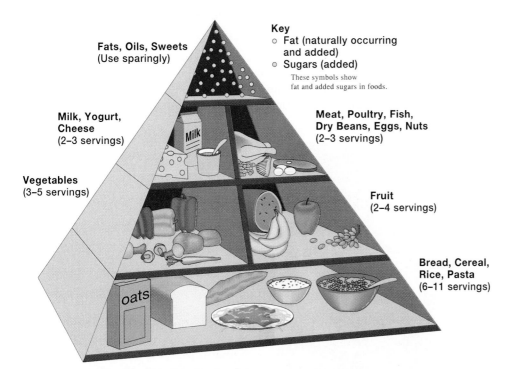

Figure 5.13 Food guide pyramid: A guide to daily food choices.
The U.S. Department of Agriculture uses a pyramid to show the ideal diet because it emphasizes the importance of including grains, fruits, and vegetables in the diet. Meats and dairy products are needed in limited amounts; fats, oils, and sweets should be used sparingly.
Source: Data from the U.S. Department of Agriculture.

Carbohydrates

The quickest, most readily available source of energy for the body is glucose. Carbohydrates are digested to simple sugars, which are or can be converted to glucose. Glucose is stored by the liver in the form of glycogen. Between eating periods, the blood glucose level is maintained at about 0.1% by the breakdown of glycogen or by the conversion of glycerol (from fats) or amino acids to glucose. If necessary, amino acids are taken from the muscles—even from the heart muscle. While body cells can utilize fatty acids as an energy source, brain cells require glucose. For this reason alone, it is necessary to include carbohydrates in the diet. According to Figure 5.13, carbohydrates should make up the bulk of the diet. Further, these should be complex and not simple carbohydrates. Complex sources of carbohydrates include preferably whole-grain pasta, rice, bread, and cereal (Fig. 5.14). Potatoes and corn, although considered vegetables, are also sources of carbohydrates.

Simple carbohydrates (e.g., sugars) are labeled "empty calories" by some dieticians because they contribute to energy needs and weight gain without supplying any other nutritional requirements. Table 5.3 gives suggestions on how to reduce dietary sugar (simple carbohydrates). In contrast to simple sugars, complex carbohydrates are likely to be accompanied by a wide range of other nutrients and by **fiber,** which is indigestible plant material.

The intake of fiber is recommended because it may decrease the risk of colon cancer, a major type of cancer, and cardiovascular disease, the number one killer in the United States. Insoluble fiber, such as that found in wheat bran, has a laxative effect and may guard against colon cancer by limiting the amount of time cancer-causing substances are in contact with the intestinal wall. Soluble fiber, such as that found in oat bran, combines with bile acids and cholesterol

Table 5.3	Reducing Dietary Sugar

To reduce dietary sugar:

1. Eat fewer sweets, such as candy, soft drinks, ice cream, and pastry.
2. Eat fresh fruits or fruits canned without heavy syrup.
3. Use less sugar—white, brown, or raw—and less honey and syrups.
4. Avoid sweetened breakfast cereals.
5. Eat less jelly, jam, and preserves.
6. Drink pure fruit juices, not imitations.
7. When cooking, use spices, such as cinnamon, instead of sugar to flavor foods.
8. Do not put sugar in tea or coffee.

in the intestine and prevents them from being absorbed. The liver then removes cholesterol from the blood and changes it to bile acids, replacing the bile acids that were lost. While the diet should have an adequate amount of fiber, a high-fiber diet can be detrimental. Some evidence suggests that the absorption of iron, zinc, and calcium is also impaired by a diet too high in fiber.

Complex carbohydrates, which contain fiber, should form the bulk of the diet.

Proteins

Foods rich in protein include red meat, fish, poultry, dairy products, legumes (i.e., peas and beans), nuts, and cereals. Following digestion of protein, amino acids enter the

Figure 5.14   Complex carbohydrates.
To meet our energy needs, dieticians recommend consuming foods rich in complex carbohydrates, like those shown here, rather than foods consisting of simple carbohydrates, like candy and ice cream. Simple carbohydrates provide monosaccharides but few other types of nutrients.

bloodstream and are transported to the tissues. Ordinarily, amino acids are not used as an energy source. Most are incorporated into structural proteins found in muscles, skin, hair, and nails. Others are used to synthesize such proteins as hemoglobin, plasma proteins, enzymes, and hormones.

Adequate protein formation requires 20 different types of amino acids. Of these, eight are required from the diet in adults (nine in children) because the body is unable to produce them. These are termed the **essential amino acids.** The body produces the other amino acids by simply transforming one type into another type. Some protein sources, such as meat, milk, and eggs, are complete; they provide all 20 types of amino acids. Vegetables and grains supply us with amino acids, but each vegetable or grain alone is an incomplete protein source because of a deficiency in at least one of the essential amino acids. Absence of one essential amino acid prevents utilization of the other 19 amino acids. Soybeans and their product, tofu, are rich in amino acids, but it is wise to combine foods to acquire all the essential amino acids. For example, the combinations of cereal with milk, or beans (a legume) with rice (a grain), will provide all the essential amino acids (Table 5.4).

Amino acids are not stored in the body, and a daily supply is needed. However, it does not take very much protein to meet the daily requirement. Two servings of meat a day (equal in size to a deck of cards) is usually enough. Some meats (e.g., hamburger) are high in protein but also high in fat. Everything considered, it is probably a good idea to depend on protein from plant origins (e.g., whole-grain cereals, dark breads, and legumes) to a greater extent than is often the custom in the United States. This can be illustrated by the health statistics of native Hawaiians who no longer eat as their ancestors did (Fig. 5.15). The modern diet depends on animal rather than plant protein and is 42% fat. A statistical study showed that the island's native peoples now have a higher than average death rate from cardiovascular disease and cancer. Diabetes is also common in persons who follow the modern diet. But the health of those who have switched back to the ancient diet has improved immensely!

Nutritionists do not recommend the use of protein and/or amino acid supplements. Protein supplements that athletes take to build muscle cost more than food and can be harmful. When excess protein is broken down more urea is excreted in the urine. The water needed for excretion of urea can cause dehydration when a person is exercising and also losing water by sweating. Also, some studies suggest that protein supplements lead to calcium loss and weakened bones. Amino acid supplements can also be dangerous to your health. Mistaken ideas abound. For example, contrary

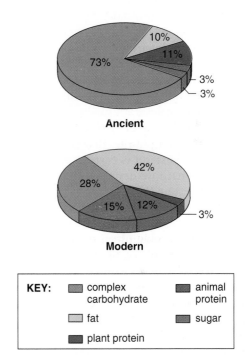

Figure 5.15   Ancient versus modern diet of native Hawaiians.
Among those native Hawaiians who have switched back to the native diet, the incidence of cardiovascular disease, cancer, and diabetes has dropped.

Table 5.4	Complementary Protein Combinations		
Combine foods from two or more of these columns to obtain complete protein.			
Grains	*Legumes*	*Seeds and Nuts*	*Vegetables*
Barley	Dried beans	Sesame seeds	Leafy greens
Bulgur	Dried lentils	Sunflower seeds	Broccoli
Cornmeal	Dried peas	Walnuts	Others
Oats	Peanuts	Cashews	
Rice	Soy products	Other nuts	
Whole-grain breads		Nut butters	
Pasta			

to popular reports, the taking of lysine does not relieve or cure herpes sores. An excess of any particular amino acid in the diet can lead to decreased absorption of other amino acids in food.

Lipids

Fat and cholesterol are both lipids. Fat is present not only in butter, margarine, and oils, but also in many foods high in animal protein.

The current guidelines suggest that fat should account for no more than 30% of our daily calories. The chief reason is that an intake of fat not only causes weight gain, but it also increases the risk of cancer and cardiovascular disease. Dietary fat may increase the risk of colon, hepatic, and pancreatic cancers. Although recent studies suggest no link between dietary fat and breast cancer, other researchers still believe that the matter deserves further investigation.

Cardiovascular disease is often due to arteries blocked by fatty deposits, called **plaque,** that contain saturated fats and cholesterol. Cholesterol is carried in the blood by two types of lipoproteins: low-density lipoprotein (LDL) and high-density lipoprotein (HDL). LDL is thought of as "bad" because it carries cholesterol from the liver to the cells, while HDL is thought of as "good" because it carries cholesterol to the liver, which takes it up and converts it to bile salts.

Saturated fatty acids have no double bonds; polyunsaturated fatty acids have many double bonds. Saturated fats, whether in butter or margarine, can raise LDL cholesterol levels, while monounsaturated (one double bond) fats and polyunsaturated (many double bonds) fats lower LDL cholesterol levels. Olive oil and canola oil contain mostly monounsaturated fats; corn oil and safflower oil contain mostly polyunsaturated fats. These oils have a liquid consistency and come from plants. Saturated fats, which are solids at room temperature, usually have an animal origin; two well-known exceptions are palm oil and coconut oil, which contain mostly saturated fats and come from the plants mentioned.

Nutritionists stress that it is more important to consume a diet low in fat rather than to be overly concerned about which type fat is in the diet. Still, polyunsaturated fats are nutritionally essential because they are the only type fat that contains linoleic acid, a fatty acid the body cannot make. Table 5.5 gives suggestions on how to reduce dietary fat.

Fake Fat

Olestra is a substance made to look, taste, and act like real fat, but the digestive system is unable to digest it. It travels down the length of the digestive system without being absorbed or contributing any calories to the day's total.

Table 5.5	**Reducing Lipids**

To reduce dietary fat:

1. Choose poultry, fish, or dry beans and peas as a protein source.
2. Remove skin from poultry before cooking, and place on a rack so that fat drains off.
3. Broil, boil, or bake rather than frying.
4. Limit your intake of butter, cream, hydrogenated oils, shortenings, and tropical oils (coconut and palm oils).*
5. Use herbs and spices to season vegetables instead of butter, margarine, or sauces. Use lemon juice instead of salad dressing.
6. Drink skim milk instead of whole milk, and use skim milk in cooking and baking.
7. Eat nonfat or low-fat foods.

To reduce dietary cholesterol:

1. Avoid cheese, egg yolks, liver, and certain shellfish (shrimp and lobster). Preferably, eat white fish and poultry.
2. Substitute egg whites for egg yolks in both cooking and eating.
3. Include soluble fiber in the diet. Oat bran, oatmeal, beans, corn, and fruits such as apples, citrus fruits, and cranberries are high in soluble fiber.

*Although coconut and palm oils are from plant sources, they are mostly saturated fats.

Therefore it is commonly known as "fake fat." Unfortunately, the fat-soluble vitamins A, D, E, and K tend to be taken up by olestra, and thereafter they are not absorbed by the body. Similarly, people using olestra have reduced amounts of carotenoids in their blood. In one study, just a handful of olestra-soaked potato chips caused a 20% decline in blood beta-carotene levels. Manufacturers fortify olestra-containing foods with the vitamins mentioned but not with carotenoids.

Fake fat has other side effects. Some people who consume olestra have developed anal leakage resulting in underwear staining. Others experience diarrhea, intestinal cramping, and gas. Presently the FDA has limited the use of olestra to potato chips and other salty snacks, but the manufacturer wants approval to add it to ice cream, salad dressings, and cheese.

Dietary protein supplies the essential amino acids; proteins from plant origins generally have less accompanying fat. A diet composed of no more than 30% fat is recommended because fat intake, particularly saturated fats, is known to be associated with various health problems.

Weight Loss the Healthy Way

People who wish to lose weight need to reduce their caloric intake and/or increase their level of exercise. For a woman 19 to 22 years of age and 5 feet 4 inches tall, who exercises lightly, the normal recommendation is 2,100 Cal* per day. For a man the same age, 5 feet 10 inches tall, who exercises lightly, the recommendation is 2,900 Cal. Exercising is a good idea, because to maintain good nutrition, the intake of calories per day should probably not go below 1,200 Cal. Also, for the reasons discussed in this chapter, carbohydrates should still make up at least 58% of these calories, proteins should be no more than 25%, and the rest can be fats. A deficit of 500 Cal a day (through intake reduction or increased exercise) is sufficient to lose a pound of body fat in a week. Once you realize that a diet needs to be judged according to the principles of adequate nutrients, of balanced carbohydrates, proteins, and fats, of moderate number of calories, and of variety of food sources, it is easy to see that many of the diets and gimmicks people use to lose weight are bad for their health. Some unhealthy approaches are described here.

Pills

The most familiar pills, and the only ones approved by the Food and Drug Administration (FDA), are those that claim to suppress the appetite. They may work at first, but the appetite soon returns to normal and the weight lost is regained. Then the user has the problem of trying to get off the drug without gaining additional weight. Other types of pills are under investigation and sometimes can be obtained illegally. But, as yet, there is no known drug that is both safe and effective for weight loss.

Liquid Diets

Despite the fact that liquid diets provide proteins and vitamins, the number of Calories is so restricted that the body cannot burn fat quickly enough to compensate, and muscle is still broken down to provide energy. A few people on this regime have died, probably because even the heart muscle was not spared by the body.

Low-Carbohydrate Diets

The dramatic weight loss that occurs with a low-carbohydrate diet is not due to a loss of fat; it is due to a loss of muscle mass and water. Glycogen and important minerals are also lost. When a normal diet is resumed, so is the normal weight.

Single-Category Diets

These diets rely on the intake of only one kind of food, either a fruit or vegetable or rice alone. However, no single type of food provides the balance of nutrients needed to maintain health. Some dieters on strange diets suffer the consequences—in one instance, an individual lost hair and fingernails.

*Cal = 1,000 calories

Questions to Ask About a Weight-Loss Diet

1. Does the diet have a reasonable number of Calories?
   (10 Cal per pound of current weight is suggested. In any case, no fewer than 1,000–1,200 Cal are recommended for a normal-sized person.)
2. Does the diet provide enough protein?
   (For a 120-lb woman, 44 grams of protein each day are recommended. For a 154-lb man, 56 grams are recommended. More than twice this amount is too much. For reference, 1 c milk and 1 oz meat each has 8 grams of protein.)
3. Does the diet provide too much fat?
   (No more than 20–30% of total Cal is recommended. For reference, a pat of butter has 45 Cal; 1 gram fat = 9 Cal.)
4. Does the diet provide enough carbohydrates?
   (100 grams = 400 Cal is the very least recommended per day; 50% of total Cal should be carbohydrates. For reference, a slice of bread contains 14 grams of carbohydrates.)
5. Does the diet provide a balanced assortment of foods?
   (The diet should include breads, cereals, legumes, vegetables (especially dark-green and yellow ones), low-fat milk products, and meats or a meat substitute.)
6. Does the diet make use of ordinary foods that are available locally?
   (Diets should not require the purchase of unusual or expensive foods.)

Figure 5A Fruit is a healthy and low-calorie snack.

Vitamins

Vitamins are organic compounds (other than carbohydrate, fat, and protein) that the body uses for metabolic purposes but is unable to produce in adequate quantity. Many vitamins are portions of coenzymes, which are enzyme helpers. For example, niacin is part of the coenzyme NAD, and riboflavin is part of another dehydrogenase, FAD. Coenzymes are needed in only small amounts because each can be used over and over again. Not all vitamins are coenzymes; vitamin A, for example, is a precursor for the visual pigment that prevents night blindness. If vitamins are lacking in the diet, various symptoms develop (Fig. 5.16). Altogether, there are 13 vitamins, which are divided into those that are fat soluble (Table 5.6) and those that are water soluble (Table 5.7).

Antioxidants

Over the past 20 years, numerous statistical studies have been done to determine whether a diet rich in fruits and vegetables is protective against cancer. Cellular metabolism generates free radicals, unstable molecules that carry an extra electron. The most common free radicals in cells are superoxide (O_2^-) and hydroxide (OH^-). In order to stabilize themselves, free radicals donate an electron to DNA, to proteins, including enzymes, or to lipids, which are found in plasma membranes. Such donations most likely damage these cellular molecules and thereby may lead to disorders, perhaps even cancer.

Vitamins C, E, and A are believed to defend the body against free radicals, and therefore they are termed antioxidants. These vitamins are especially abundant in fruits and vegetables. The dietary guidelines shown in Figure 5.13 suggest that we eat a minimum of five servings of fruits and vegetables a day. To achieve this goal, include salad greens, raw or cooked vegetables, dried fruit, and fruit juice, in addition to traditional apples and oranges and such.

Dietary supplements may provide a potential safeguard against cancer and cardiovascular disease, but nutritionists do not think it is appropriate to take supplements instead of improving intake of fruits and vegetables. There are many beneficial compounds in fruits that cannot be obtained from a vitamin pill. These compounds enhance each other's absorption or action and also perform independent biological functions.

Vitamin D

Skin cells contain a precursor cholesterol molecule that is converted to vitamin D after UV exposure. Vitamin D leaves the skin and is modified first in the kidneys and then in the liver until finally it becomes calcitriol. Calcitriol promotes the absorption of calcium by the intestines. The lack of vitamin D leads to rickets in children (see Fig. 5.16*a*). Rickets, characterized by a bowing of the legs, is caused by defective mineralization of the skeleton. Most milk today is fortified with vitamin D, which helps prevent the occurrence of rickets.

Vitamins are essential to cellular metabolism; many are protective against identifiable illnesses and conditions.

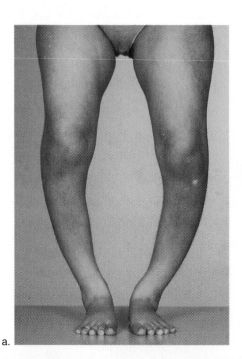

a.

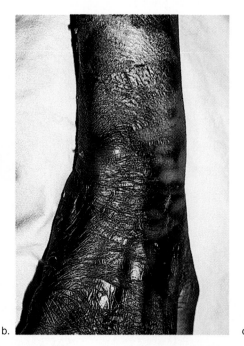

b.

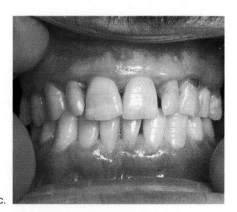

c.

Figure 5.16   Illnesses due to vitamin deficiency.
a. Bowing of bones (rickets) due to vitamin D deficiency. **b.** Dermatitis (pellagra) of areas exposed to light due to niacin (vitamin B₃) deficiency.
c. Bleeding of gums (scurvy) due to vitamin C deficiency.

Table 5.6   Fat-Soluble Vitamins

Vitamin	Functions	Food Sources	Conditions With	
			Too Little	**Too Much**
Vitamin A	Antioxidant synthesized from beta-carotene; needed for healthy eyes, skin, hair, and mucous membranes, and for proper bone growth	Deep yellow/orange and leafy, dark green vegetables, fruits, cheese, whole milk, butter, eggs	Night blindness, impaired growth of bones and teeth	Headache, dizziness, nausea, hair loss, abnormal development of fetus
Vitamin D	A group of steroids needed for development and maintenance of bones and teeth	Milk fortified with vitamin D, fish liver oil; also made in the skin when exposed to sunlight	Rickets, bone decalcification and weakening	Calcification of soft tissues, diarrhea, possible renal damage
Vitamin E	Antioxidant that prevents oxidation of vitamin A and polyunsaturated fatty acids	Leafy green vegetables, fruits, vegetable oils, nuts, whole-grain breads and cereals	Unknown	Diarrhea, nausea, headaches, fatigue, muscle weakness
Vitamin K	Needed for synthesis of substances active in clotting of blood	Leafy green vegetables, cabbage, cauliflower	Easy bruising and bleeding	Can interfere with anticoagulant medication

Table 5.7   Water-Soluble Vitamins

Vitamin	Functions	Food Sources	Conditions With	
			Too Little	**Too Much**
Vitamin C	Antioxidant; needed for forming collagen; helps maintain capillaries, bones, and teeth	Citrus fruits, leafy green vegetables, tomatoes, potatoes, cabbage	Scurvy, delayed wound healing, infections	Gout, kidney stones, diarrhea, decreased copper
Thiamine (vitamin B_1)	Part of coenzyme needed for cellular respiration; also promotes activity of the nervous system	Whole-grain cereals, dried beans and peas, sunflower seeds, nuts	Beriberi, muscular weakness, enlarged heart	Can interfere with absorption of other vitamins
Riboflavin (vitamin B_2)	Part of coenzymes, such as FAD; aids cellular respiration, including oxidation of protein and fat	Nuts, dairy products, whole-grain cereals, poultry, leafy green vegetables	Dermatitis, blurred vision, growth failure	Unknown
Niacin (nicotinic acid)	Part of coenzymes NAD and NADP; needed for cellular respiration, including oxidation of protein and fat	Peanuts, poultry, whole-grain cereals, leafy green vegetables, beans	Pellagra, diarrhea, mental disorders	High blood sugar and uric acid, vasodilation, etc.
Folacin (folic acid)	Coenzyme needed for production of hemoglobin and formation of DNA	Dark leafy green vegetables, nuts, beans, whole-grain cereals	Megaloblastic anemia, spina bifida	May mask B_{12} deficiency
Vitamin B_6	Coenzyme needed for synthesis of hormones and hemoglobin; CNS control	Whole-grain cereals, bananas, beans, poultry, nuts, leafy green vegetables	Rarely, convulsions, vomiting, seborrhea, muscular weakness	Insomnia, neuropathy
Pantothenic acid	Part of coenzyme A needed for oxidation of carbohydrates and fats; aids in the formation of hormones and certain neurotransmitters	Nuts, beans, dark green vegetables, poultry, fruits, milk	Rarely, loss of appetite, mental depression, numbness	Unknown
Vitamin B_{12}	Complex, cobalt-containing compound; part of the coenzyme needed for synthesis of nucleic acids and myelin	Dairy products, fish, poultry, eggs, fortified cereals	Pernicious anemia	Unknown
Biotin	Coenzyme needed for metabolism of amino acids and fatty acids	Generally in foods, especially eggs	Skin rash, nausea, fatigue	Unknown

Minerals

In addition to vitamins, various **minerals** are required by the body. Minerals are divided into macrominerals and microminerals. The body contains more than 5 grams of each macromineral and less than 5 grams of each micromineral (Fig. 5.17). The macrominerals are constituents of cells and body fluids and are structural components of tissues. For example, calcium (present as Ca^{2+}) is needed for the construction of bones and teeth and for nerve conduction and muscle contraction. Phosphorus (present as PO_4^{3-}) is stored in the bones and teeth and is a part of phospholipids, ATP, and the nucleic acids. Potassium (K^+) is the major positive ion inside cells and is important in nerve conduction and muscle contraction, as is sodium (Na^+). Sodium also plays a major role in regulating the body's water balance, as does chloride (Cl^-). Magnesium (Mg^{2+}) is critical to the functioning of hundreds of enzymes.

The microminerals are parts of larger molecules. For example, iron is present in hemoglobin, and iodine is a part of thyroxine and triidothyronine, hormones produced by the thyroid gland. Zinc, copper, and selenium are present in enzymes that catalyze a variety of reactions. Proteins, called zinc-finger proteins because of their characteristic shapes, bind to DNA when a particular gene is to be activated. As research continues, more and more elements are added to the list of microminerals considered essential. During the past three decades, for example, very small amounts of selenium, molybdenum, chromium, nickel, vanadium, silicon, and even arsenic have been found to be essential to good health. Table 5.8 lists the functions of various minerals and gives their food sources and signs of deficiency and toxicity.

Occasionally, individuals do not receive enough iron (especially women), calcium, magnesium, or zinc in their diets. Adult females need more iron in the diet than males (18 mg compared to 10 mg) because they lose hemoglobin each month during menstruation. Stress can bring on a magnesium deficiency, and due to its high-fiber content, a vegetarian diet may make zinc less available to the body. However, a varied and complete diet usually supplies enough of each type of mineral.

Calcium

Many people take calcium supplements to counteract **osteoporosis,** a degenerative bone disease that afflicts an estimated one-fourth of older men and one-half of older women in the United States. Osteoporosis develops because bone-eating cells called osteoclasts are more active than bone-forming cells called osteoblasts. Therefore, the bones are porous, and they break easily because they lack sufficient calcium. Due to recent studies that show consuming more calcium does slow bone loss in elderly people, the guidelines have been revised. A calcium intake of 1,000 mg a day is recommended for men and for women who are premenopausal or who use estrogen replacement therapy; 1,300 mg a day is recommended for postmenopausal women who do not use estrogen replacement therapy. To achieve this amount, supplemental calcium is most likely necessary.

Vitamin D is an essential companion to calcium in preventing osteoporosis. Other vitamins may also be helpful; for example, magnesium has been found to suppress the cycle that leads to bone loss.

Estrogen replacement therapy and exercise, in addition to adequate calcium and vitamin intake, also help prevent osteoporosis. Drinking more than nine cups of caffeinated coffee per day and smoking are risk factors for osteoporosis. Medications are also available that slow bone loss while increasing skeletal mass. These are still being studied for their effectiveness and possible side effects.

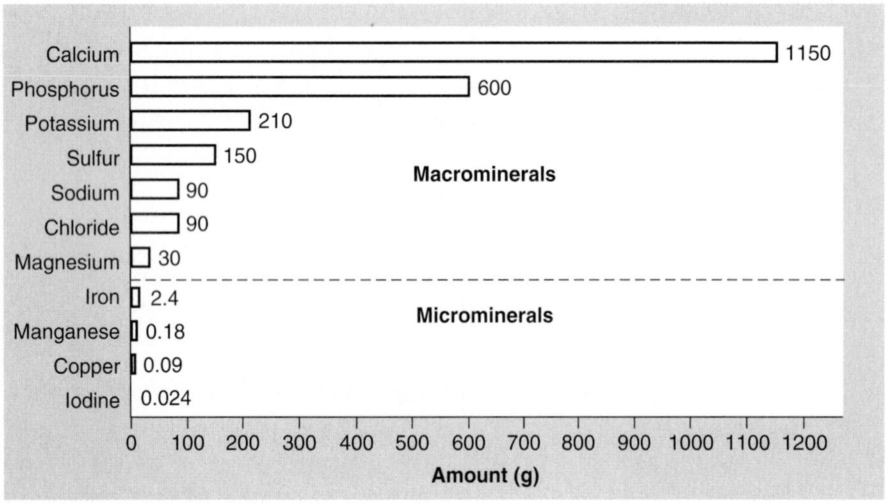

Figure 5.17   Minerals in the body.
This chart shows the usual amount of certain minerals in a 60-kilogram (135 lb) person. The macrominerals are present in amounts larger than 5 grams (about a teaspoon) and the microminerals are present in lesser amounts. The functions of these minerals are given in Table 5.8.

Presently, calcium supplements, estrogen therapy for women, and exercise are thought to be the best ways to prevent osteoporosis.

Table 5.8 Minerals

Mineral	Functions	Food Sources	Conditions With	
			Too Little	Too Much
MACROMINERALS (MORE THAN 100 MG/DAY NEEDED)				
Calcium (Ca^{2+})	Strong bones and teeth, nerve conduction, muscle contraction	Dairy products, leafy green vegetables	Stunted growth in children, low bone density in adults	Kidney stones; interferes with iron and zinc absorption
Phosphorus (PO_4^{3-})	Bone and soft tissue growth; part of phospholipids, ATP, and nucleic acids	Meat, dairy products, sunflower seeds, food additives	Weakness, confusion, pain in bones and joints	Low blood and bone calcium levels
Potassium (K^+)	Nerve conduction, muscle contraction	Many fruits and vegetables, bran	Paralysis, irregular heartbeat, eventual death	Vomiting, heart attack, death
Sodium (Na^+)	Nerve conduction, pH and water balance	Table salt	Lethargy, muscle cramps, loss of appetite	Edema, high blood pressure
Chloride (Cl^-)	Water balance	Table salt	Not likely	Vomiting, dehydration
Magnesium (Mg^{2+})	Part of various enzymes for nerve and muscle contraction, protein synthesis	Whole grains, leafy green vegetables	Muscle spasm, irregular heartbeat, convulsions, confusion, personality changes	Diarrhea
MICROMINERALS (LESS THAN 20 MG/DAY NEEDED)				
Zinc (Zn^{2+})	Protein synthesis, wound healing, fetal development and growth, immune function	Meats, legumes, whole grains	Delayed wound healing, night blindness, diarrhea, mental lethargy	Anemia, diarrhea, vomiting, renal failure, abnormal cholesterol levels
Iron (Fe^{2+})	Hemoglobin synthesis	Whole grains, meats, prune juice	Anemia, physical and mental sluggishness	Iron toxicity disease, organ failure, eventual death
Copper (Cu^{2+})	Hemoglobin synthesis	Meat, nuts, legumes	Anemia, stunted growth in children	Damage to internal organs if not excreted
Iodine (I^-)	Thyroid hormone synthesis	Iodized table salt, seafood	Thyroid deficiency	Depressed thyroid function, anxiety
Selenium (SeO_4^{2-})	Part of antioxidant enzyme	Seafood, meats, eggs	Vascular collapse, possible cancer development	Hair and fingernail loss, discolored skin

Sodium

The recommended amount of sodium intake per day is 500 mg, although the average American takes in 4,000–4,700 mg every day. In recent years, this imbalance has caused concern because high sodium intake has been linked to hypertension (high blood pressure) in some people. About one-third of the sodium we consume occurs naturally in foods; another one-third is added during commercial processing; and we add the last one-third either during home cooking or at the table in the form of table salt.

Clearly, it is possible for us to cut down on the amount of sodium in the diet. Table 5.9 gives recommendations for doing so.

Excess sodium in the diet can lead to hypertension; therefore, excess sodium intake should be avoided.

Table 5.9 Reducing Dietary Sodium

To reduce dietary sodium:

1. Use spices instead of salt to flavor foods.
2. Add little or no salt to foods at the table, and add only small amounts of salt when you cook.
3. Eat unsalted crackers, pretzels, potato chips, nuts, and popcorn.
4. Avoid hot dogs, ham, bacon, luncheon meats, smoked salmon, sardines, and anchovies.
5. Avoid processed cheese and canned or dehydrated soups.
6. Avoid brine-soaked foods, such as pickles or olives.
7. Read labels to avoid high-salt products.

Persons with obesity have

- weight 20% or more above appropriate weight for height.
- body fat content in excess of that consistent with optimal health, probably due to a diet rich in fats.
- low levels of exercise.

Figure 5.18   **Recognizing obesity.**

Eating Disorders

Authorities recognize three primary eating disorders: obesity, bulimia nervosa, and anorexia nervosa. Although they exist in a continuum as far as body weight is concerned, they all represent an inability to maintain normal body weight because of eating habits.

Obesity

As indicated in Figure 5.18, **obesity** is most often defined as a body weight 20% or more above the ideal weight for a person's height. By this standard, 28% of women and 10% of men in the United States are obese. Moderate obesity is 41–100% above ideal weight, and severe obesity is 100% or more above ideal weight.

Obesity is most likely caused by a combination of hormonal, metabolic, and social factors. It's known that obese individuals have more fat cells than normal, and when they lose weight the fat cells simply get smaller; they don't disappear. The social factors that cause obesity include the eating habits of other family members. Consistently eating fatty foods, for example, will make you gain weight. Sedentary activities, like watching television instead of exercising, also determine how much body fat you have. The risk of heart disease is higher in obese individuals, and this alone tells us that excess body fat is not consistent with optimal health.

The treatment depends on the degree of obesity. Surgery to remove body fat may be required for those who are moderately or greatly overweight. But for most people, a knowledge of good eating habits along with behavior modification may suffice, particularly if a balanced diet is accompanied by a sensible exercise program. A lifelong commitment to a properly planned program is the best way to prevent a cycle of weight gain followed by weight loss. Such a cycle is not conducive to good health. The Health Focus on page 99 discusses the proper way to lose weight.

Bulimia Nervosa

Bulimia nervosa can coexist with either obesity or anorexia nervosa, which is discussed next. People with this condition have the habit of eating to excess (called binge eating) and then purging themselves by some artificial means, such as self-induced vomiting or use of a laxative. Bulimic individuals are overconcerned about their body shape and weight, and therefore they may be on a very restrictive diet. A restrictive diet may bring on the desire to binge, and typically the person chooses to consume sweets, like cakes, cookies, and ice cream (Fig. 5.19). The amount of food consumed is far beyond the normal number of calories for one meal, and the person keeps on eating until every bit is gone. Then, a feeling of guilt most likely brings on the next phase, which is a purging of all the calories that have been taken in.

Bulimia can be dangerous to your health. Blood composition is altered, leading to an abnormal heart rhythm, and damage to the kidneys can even result in death. At the very least, vomiting may result in inflammation of the pharynx and esophagus, and stomach acids can cause the teeth to erode. The esophagus and stomach may even rupture and tear due to strong contractions during vomiting.

The most important aspect of treatment is to get the patient on a sensible and consistent diet. Again, behavioral modification is helpful, and so perhaps is psychotherapy to help the patient understand the emotional causes of the behavior. Medications, including antidepressants, have sometimes been helpful to reduce the bulimic cycle and restore normal appetite.

Obesity and bulimia nervosa have complex causes and may be damaging to health. Therefore, they require competent medical attention.

Anorexia Nervosa

In **anorexia nervosa,** a morbid fear of gaining weight causes the person to be on a very restrictive diet. Athletes such as distance runners, wrestlers, and dancers are at risk of anorexia nervosa because they believe that being thin gives them a competitive edge. In addition to eating only low-calorie foods, the person may induce vomiting and use laxatives to bring about further loss of weight. No matter how thin they have become, people with anorexia nervosa think they are overweight (Fig. 5.20). Such a distorted self-image may prevent recognition of the need for medical help.

Actually, the person is starving and has all the symptoms of starvation, such as low blood pressure, irregular heartbeat, constipation, and constant chilliness. Bone density decreases and stress fractures occur. The body begins to shut down; menstruation ceases in females; the internal organs including the brain don't function well; and the skin dries up. Impairment of the pancreas and digestive tract means that any food consumed does not provide nourishment. Death may be imminent. If so, the only recourse may be hospitalization and force-feeding. Eventually, it is necessary to use behavior therapy and psychotherapy to enlist the cooperation of the person to eat properly. Family therapy may be necessary, because anorexia nervosa in children and teens is believed to be a way for them to gain some control over their lives.

> In anorexia nervosa, the individual has a distorted body image and always feels fat. Competent medical help is often a necessity.

Persons with bulimia nervosa have

- recurrent episodes of binge eating characterized by consuming an amount of food much higher than normal for one sitting and a sense of lack of control over eating during the episode.
- an obsession about their body shape and weight, but often without excercising.
- increase in fine body hair, halitosis, and gingivitis.

Body weight is regulated by

- a restrictive diet, excessive exercise.
- purging (self-induced vomiting or misuse of laxatives).

Figure 5.19   Recognizing bulimia nervosa.

Persons with anorexia nervosa have

- a morbid fear of gaining weight; body weight no more than 85% normal.
- a distorted body image so that person feels fat even when emaciated.
- in females, an absence of a menstrual cycle for at least three months.

Body weight is kept too low by either/or

- a restrictive diet, often with excessive exercise.
- binge eating/purging (person engages in binge eating and then self-induces vomiting or misuses laxatives).

Figure 5.20   Recognizing anorexia nervosa.

Summarizing the Concepts

5.1 The Digestive System
The salivary glands send saliva into the mouth, where the teeth chew the food and the tongue forms a bolus for swallowing.

The air passage and food passage cross in the pharynx. When a person swallows, the air passage is usually blocked off, and food must enter the esophagus where peristalsis begins.

The stomach expands and stores food. While food is in the stomach, it churns, mixing food with the acidic gastric juices.

The walls of the small intestine have fingerlike projections called villi where nutrient molecules are absorbed into the cardiovascular and lymphatic systems.

The large intestine consists of the cecum, the colon which includes the ascending, transverse, descending, and sigmoid colon, and the rectum, which ends at the anus.

The large intestine does not produce digestive enzymes; it does absorb water, salts, and some vitamins.

5.2 Three Accessory Organs
Three accessory organs of digestion—the pancreas, liver, and gallbladder—send secretions to the duodenum via ducts. The pancreas produces pancreatic juice, which contains digestive enzymes for carbohydrate, protein, and fat.

The liver produces bile, which is stored in the gallbladder. The liver receives blood from the small intestine by way of the hepatic portal vein. It has numerous important functions, and any malfunction of the liver is a matter of considerable concern.

5.3 Digestive Enzymes
Digestive enzymes are present in digestive juices and break down food into the nutrient molecules glucose, amino acids, fatty acids, and glycerol (see Table 5.2). Glucose and amino acids are absorbed into the blood capillaries of the villi. Fatty acids and glycerol are rejoined and repackaged as lipoprotein droplets which enter the lacteals of the villi.

Digestive enzymes have the usual enzymatic properties. They are specific to their substrate and speed up specific reactions at optimum body temperature and pH.

5.4 Homeostasis
The digestive system works with the other systems of the body in the ways described in "Human Systems Work Together" on page 94.

5.5 Nutrition
The nutrients released by the digestive process should provide us with an adequate amount of energy, essential amino acids and fatty acids, and all necessary vitamins and minerals.

The bulk of the diet should be carbohydrates (e.g., bread, pasta, and rice) and fruits and vegetables. These are low in saturated fatty acids and cholesterol molecules, whose intake is linked to cardiovascular disease. The vitamins A, E, and C are antioxidants that protect cell contents from damage due to free radicals.

Studying the Concepts

1. List the organs of the digestive tract, and state the contribution of each to the digestive process. 82–88
2. Discuss the absorption of the products of digestion into the lymphatic and cardiovascular systems. 87
3. Name and state the functions of the hormones that assist the nervous system in regulating digestive secretions. 88
4. Name the accessory organs, and describe the part they play in the digestion of food. 90–91
5. Choose and discuss any three functions of the liver. 90–91
6. Name and discuss two serious illnesses of the liver. 91
7. Discuss the digestion of starch, protein, and fat, listing all the steps that occur with each of these. 92–93
8. How does the digestive system help maintain homeostasis? 94–95
9. How does the circulatory system assist the digestive system in maintaining homeostasis? 94–95
10. What is the chief contribution of each of these constituents of the diet: a. carbohydrates; b. proteins; c. fats; d. fruits and vegetables? 96–98, 100
11. Why should the amount of saturated fat be curtailed in the diet? 98
12. Name and discuss three eating disorders. 104–5

Testing Your Knowledge of the Concepts

In questions 1–4, match the organ to the functions listed below.
   a. stomach
   b. small intestine
   c. large intestine
   d. liver

_____ 1. makes bile

_____ 2. digests all types of food

_____ 3. absorbs water

_____ 4. produces gastric juice

In questions 5–7, indicate whether the statement is true (T) or false (F).

_____ 5. The pancreas secretes both hormones and digestive enzymes.

_____ 6. Lipase is an enzyme that digests carbohydrates.

_____ 7. Only if the diet contains meat can we be assured that the diet includes all the essential amino acids.

In questions 8 and 9, fill in the blanks.

8. The gallbladder stores _____, a substance that _____ fat.

9. Whereas a(n) _____ pH is optimum for pepsin, a(n) _____ pH is optimum for the enzymes found in pancreatic juice.

10. Label each organ indicated in the diagram (a–h). For the arrows (i–k), use either glucose, amino acids, lipids, or water.

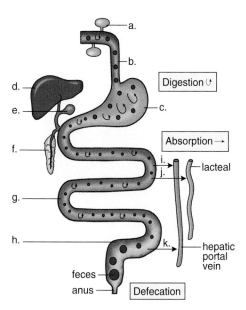

Understanding Key Terms

anorexia nervosa 105
anus 88
appendix 88
bile 87
bulimia nervosa 104
cecum 88
chyme 86
cirrhosis 91
colon 88
constipation 89
defecation 88
dental caries 83
diarrhea 89
duodenum 87
epiglottis 84
esophagus 84
essential amino acids 97
fiber 96
gallbladder 91
gastric gland 86
glottis 84
hard palate 82
heartburn 85
hepatitis 91
hormone 88
hydrolytic enzyme 92
jaundice 91
lacteal 87
lactose intolerance 92

large intestine 88
lipase 90, 92
liver 90
lumen 85
maltase 92
mineral 102
nasopharynx 84
obesity 104
osteoporosis 102
pancreas 90
pancreatic amylase 90, 92
pepsin 86, 92
peptidase 92
peristalsis 84
pharynx 84
plaque 98
polyp 89
reflex action 84
salivary amylase 83, 92
salivary gland 82
small intestine 87
soft palate 82
sphincter 85
stomach 86
tonsillitis 82
trypsin 90, 92
ulcer 86
villus 87
vitamin 100

e-Learning Connection www.mhhe.com/biosci/genbio/maderhuman7/

5.1 The Digestive System

 Human Digestion *Essential Study Partner*
Mouth and Esophagus *Essential Study Partner*
Stomach *Essential Study Partner*
Small Intestine *Essential Study Partner*
Large Intestine *Essential Study Partner*

 Human Digestion *ESP activity*
Stomach *ESP activity*
Digestive System I *art labeling activity*
Digestive System II *art labeling activity*
Digestive System III *art labeling activity*
Mouth *art labeling activity*
Teeth *art labeling activity*
Swallowing *art labeling activity*
Wall of the Digestive Tract *art labeling activity*
Anatomy of Small Intestine *art labeling activity*

5.2 Three Accessory Organs

 Accessory Organs *Essential Study Partner*

 Accessory Organs *ESP activity*
Hepatic Lobules *art labeling activity*

5.3 Digestive Enzymes

 Enzymes and Hormones *Essential Study Partner*

5.4 Homeostasis

 Working Together to Achieve Homeostasis

5.5 Nutrition

 Nutrition *Essential Study Partner*

Chapter Summary

Key Term Flashcards *vocabulary quiz*
Chapter Quiz *objective quiz*

Chapter 6

Composition and Function of the Blood

Figure 6.1 **Childhood leukemia.**
Children who are recovering from leukemia are enjoying going to camp despite temporarily losing their hair due to chemotherapy.

Johnny didn't feel well. His mother let him stay home from school because he had a fever and just wanted to lie on the couch. When he didn't seem better the next day, she took him to the doctor. The doctor was concerned because Johnny was having frequent infections and was losing weight. He suggested a complete physical examination along with blood work.

In a few days, the response came back that Johnny had acute lymphoblastic leukemia (ALL), a type of cancer that occurs in children. *Acute* means that the condition progresses rapidly; *leukemia* means there has been an abnormal increase in immature lymphocytes, one of the types of white blood cells. In Johnny's case, as in so many others, the cause of ALL was unknown.

Johnny's parents were really worried, but the doctor was very reassuring. He recommended a medical center where a combination of many different types of cancer drugs works as a cure in 75% of cases (Fig. 6.1). One year later, Johnny was so well it wasn't possible to tell that he had been through a rough time. He was playing ball and getting into mischief as usual.

Why did Johnny have so many infections? Our white blood cells fight infections, and although in ALL the lymphocyte count is high, the cells are immature and don't fight infections effectively. The other functions of blood were not affected by Johnny's illness.

FORMED ELEMENTS	Function and Description	Source
Red Blood Cells (erythrocytes) 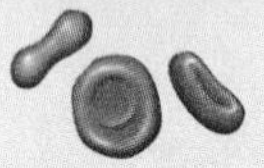 4 million–6 million per mm³ blood	Transport O_2 and help transport CO_2 7–8 μm in diameter Bright-red to dark-purple biconcave disks without nuclei	Red bone marrow
White Blood Cells (leukocytes) 4,000–11,000 per mm³ blood	Fight infection	Red bone marrow
Granular leukocytes		
• Basophils  20–50 per mm³ blood	10–12 μm in diameter Spherical cells with lobed nuclei; large, irregularly shaped, deep-blue granules in cytoplasm; release histamine which promotes blood flow to injured tissues	
• Eosinophils 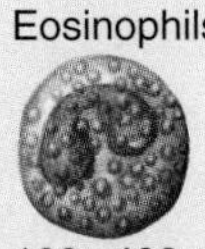 100–400 per mm³ blood	10–14 μm in diameter Spherical cells with bilobed nuclei; coarse, deep-red, uniformly sized granules in cytoplasm; phagocytize antigen-antibody complexes and allergens	
• Neutrophils  3,000–7,000 per mm³ blood	10–14 μm in diameter Spherical cells with multilobed nuclei; fine, pink granules in cytoplasm; phagocytize pathogens	
Agranular leukocytes		
• Lymphocytes 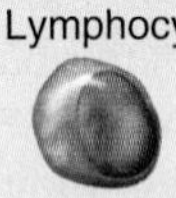 1,500–3,000 per mm³ blood	5–17 μm in diameter (average 9–10 μm) Spherical cells with large round nuclei; responsible for specific immunity	
• Monocytes 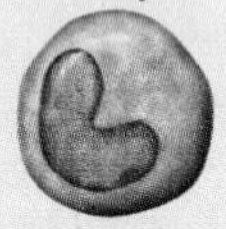 100–700 per mm³ blood	10–24 μm in diameter Large spherical cells with kidney-shaped, round, or lobed nuclei; become macrophages which phagocytize pathogens and cellular debris	
• **Platelets** (thrombocytes)  150,000–300,000 per mm³ blood	Aid clotting 2–4 μm in diameter Disk-shaped cell fragments with no nuclei; purple granules in cytoplasm	Red bone marrow

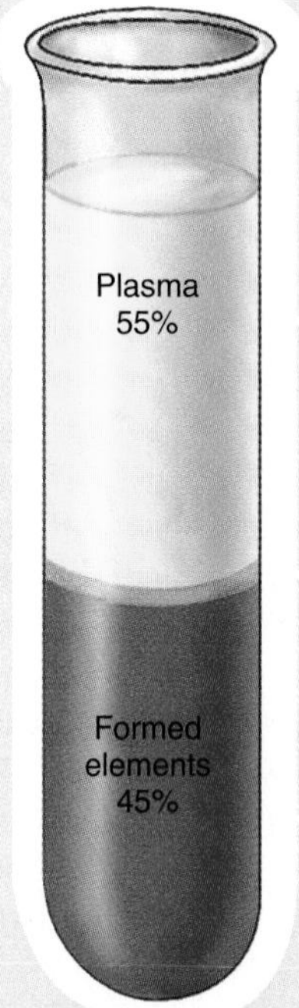

PLASMA	Function	Source
Water (90–92% of plasma)	Maintains blood volume; transports molecules	Absorbed from intestine
Plasma proteins (7–8% of plasma)	Maintain blood osmotic pressure and pH	Liver
Albumin	Maintain blood volume and pressure	
Globulins	Transport; fight infection	
Fibrinogen	Clotting	
Salts (less than 1% of plasma)	Maintain blood osmotic pressure and pH; aid metabolism	Absorbed from intestine
Gases		
Oxygen	Cellular respiration	Lungs
Carbon dioxide	End product of metabolism	Tissues
Nutrients	Food for cells	Absorbed from intestine
Lipids		
Glucose		
Amino acids		
Nitrogenous wastes	Excretion by kidneys	Liver
Urea		
Uric acid		
Other		
Hormones, vitamins, etc.	Aid metabolism	Varied

Plasma 55%

Formed elements 45%

• with Wright's stain

Figure 6.2   Composition of blood.
When blood is transferred to a test tube and is prevented from clotting, it forms two layers. The transparent, yellow top layer is plasma, the liquid portion of blood. The formed elements are in the bottom layer. This table describes these components in detail.

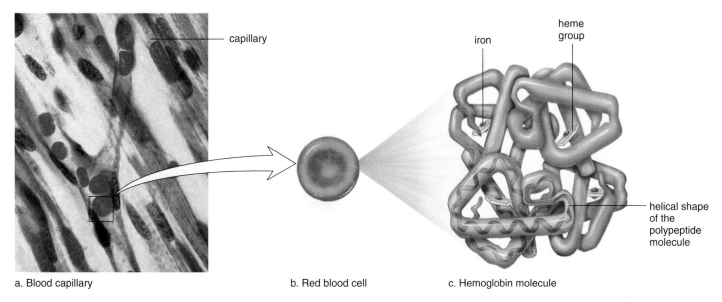

a. Blood capillary   b. Red blood cell   c. Hemoglobin molecule

Figure 6.3   Physiology of red blood cells.
a. Red blood cells move single file through the capillaries. **b.** Each red blood cell is a biconcave disk containing many molecules of hemoglobin, the respiratory pigment. **c.** Hemoglobin contains four polypeptide chains (blue). There is an iron-containing heme group in the center of each chain. Oxygen combines loosely with iron when hemoglobin is oxygenated. Oxyhemoglobin is bright red, and deoxyhemoglobin is a dark maroon color.

This chapter covers the makeup and functions of blood. If blood is transferred from a person's vein to a test tube and is prevented from clotting, it separates into two layers (Fig. 6.2). The lower layer consists of red blood cells (erythrocytes), white blood cells (leukocytes), and blood platelets (thrombocytes). Collectively, these are called the **formed elements.** Formed elements make up about 45% of the total volume of whole blood. The upper layer is plasma, which contains a variety of inorganic and organic molecules dissolved or suspended in water. Plasma accounts for about 55% of the total volume of whole blood.

The functions of blood fall into three categories: transport, defense, and regulation. Blood *transports* oxygen and nutrients to the tissues and removes wastes for excretion from the body. It also transports hormones, which help control the function of the body's organs. Blood *defends* the body against invasion by **pathogens** (microscopic infectious agents, such as bacteria and viruses), and it clots, which prevents the loss of blood. Among its various *regulatory* functions, blood helps maintain normal body temperature and the pH of body fluids.

The functions of blood contribute to homeostasis, a dynamic equilibrium of the internal environment. Cells are surrounded by tissue fluid whose composition must be kept within relatively narrow limits or the cells cease to function in an effective manner. Only if the composition of blood is within the normal range can tissue fluid also have the correct composition. All three functions of blood are necessary to keep tissue fluid relatively stable.

Blood functions to maintain homeostasis so that the environment of cells (tissue fluid) remains relatively stable.

6.1 The Red Blood Cells

Red blood cells (erythrocytes) are small, biconcave disks that lack a nucleus when mature. They occur in great quantity; there are 4 to 6 million red blood cells per mm^3 of whole blood. The absence of a nucleus provides more space for hemoglobin. **Hemoglobin** is called a respiratory pigment because it carries oxygen, and is red. A red blood cell contains about 200 million hemoglobin molecules. If this much hemoglobin were suspended within the plasma rather than enclosed within the cells, blood would be so viscous that the heart would have difficulty pumping it. In a hemoglobin molecule, the iron portion of hemoglobin carries oxygen, a molecule that cells require for cellular respiration (Fig. 6.3c). The equation for oxygenation of hemoglobin is usually written as

$$Hb + O_2 \xrightleftharpoons[\text{tissues}]{\text{lungs}} HbO_2$$

The hemoglobin on the right, which is combined with oxygen, is called oxyhemoglobin. Oxyhemoglobin, which forms in the lungs, has a bright red color. The hemoglobin on the left, which has given up oxygen to tissue fluid, is called deoxyhemoglobin. Deoxyhemoglobin is a dark maroon color. Unfortunately, as discussed in the Ecology Focus on page 114, carbon monoxide combines with hemoglobin more readily than does oxygen, and it stays combined for several hours, making hemoglobin unavailable for oxygen transport.

Visual Focus

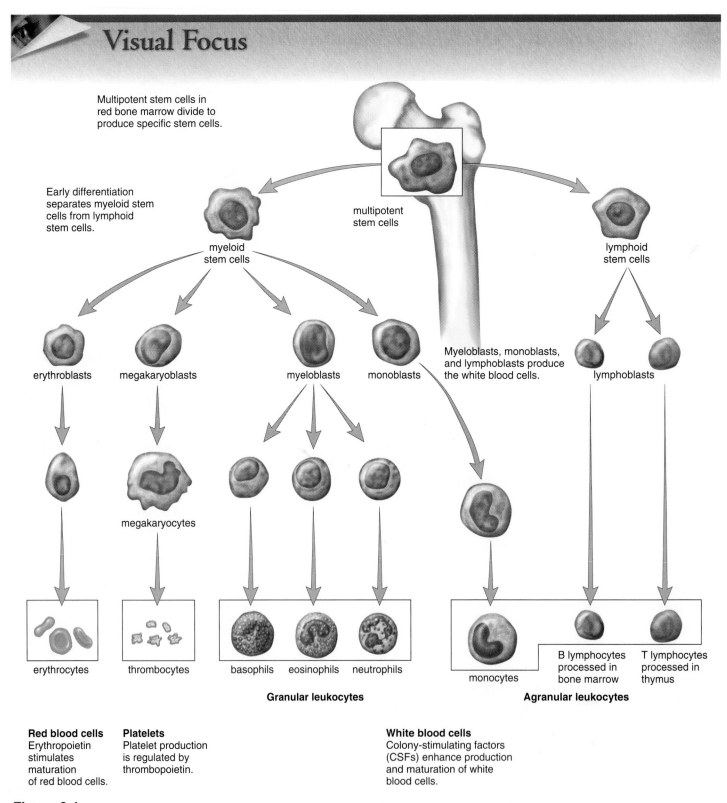

Multipotent stem cells in red bone marrow divide to produce specific stem cells.

multipotent stem cells

Early differentiation separates myeloid stem cells from lymphoid stem cells.

myeloid stem cells

lymphoid stem cells

erythroblasts

megakaryoblasts

myeloblasts

monoblasts

Myeloblasts, monoblasts, and lymphoblasts produce the white blood cells.

lymphoblasts

megakaryocytes

erythrocytes

thrombocytes

basophils   eosinophils   neutrophils

Granular leukocytes

monocytes

B lymphocytes processed in bone marrow

T lymphocytes processed in thymus

Agranular leukocytes

Red blood cells
Erythropoietin stimulates maturation of red blood cells.

Platelets
Platelet production is regulated by thrombopoietin.

White blood cells
Colony-stimulating factors (CSFs) enhance production and maturation of white blood cells.

Figure 6.4   Blood cell formation in red bone marrow.
Multipotent stem cells give rise to two specialized stem cells. The myeloid stem cell gives rise to still other cells, which become red blood cells, platelets, and all the white blood cells except lymphocytes. The lymphoid stem cell gives rise to lymphoblasts, which become lymphocytes.

The Life Cycle of Red Blood Cells

In infants, red blood cells are produced in the red bone marrow of all bones, but in adults, production primarily occurs in the red bone marrow of the skull bones, ribs, sternum, vertebrae, and pelvic bones.

All blood cells, including erythrocytes, are formed from special red bone marrow cells called stem cells (Fig. 6.4). A **stem cell** is ever capable of dividing and producing new cells that differentiate into specific types of cells. As red blood cells mature, they lose their nucleus and acquire hemoglobin. Possibly because they lack a nucleus, red blood cells live only about 120 days. As they age, red blood cells are destroyed in the liver and spleen, where they are engulfed by macrophages. It is estimated that about 2 million red blood cells are destroyed per second, and therefore an equal number must be produced to keep the red blood cell count in balance.

When red blood cells are broken down, the hemoglobin is released. The globin portion of the hemoglobin is broken down into its component amino acids, which are recycled by the body. The iron is recovered and is returned to the bone marrow for reuse. The heme portion of the molecule undergoes chemical degradation and is excreted as bile pigments by the liver into the bile. These are the bile pigments bilirubin and biliverdin, which contribute to the color of feces. Chemical breakdown of heme is also what causes a bruise of the skin to change color from red/purple to blue to green to yellow.

The number of red blood cells produced increases whenever arterial blood carries a reduced amount of oxygen, as happens when an individual first takes up residence at a high altitude, loses red blood cells, or has impaired lung function. Under these circumstances, the kidneys accelerate their release of **erythropoietin,** a hormone that is carried in blood to red bone marrow (Fig. 6.5). Once there, it speeds up the maturation of cells that are in the process of becoming red blood cells. The liver and other tissues also produce erythropoietin. Erythropoietin, now mass-produced through biotechnology, is sometimes abused by athletes in order to raise their red blood cell counts and thereby increase the oxygen-carrying capacity of their blood.

Red blood cells are produced from stem cells in red bone marrow. As red blood cells mature they lose a nucleus and gain hemoglobin, a molecule that transports oxygen. Red blood cells live only 120 days and are destroyed by phagocytic cells in the liver and spleen.

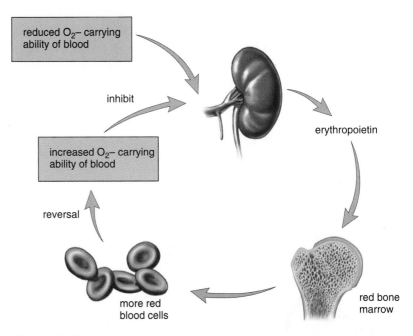

reduced O$_2$– carrying ability of blood

inhibit

erythropoietin

increased O$_2$– carrying ability of blood

reversal

more red blood cells

red bone marrow

Figure 6.5   Action of erythropoietin.
The kidneys release increased amounts of erythropoietin whenever the oxygen capacity of the blood is reduced. Erythropoietin stimulates the red bone marrow to speed up its production of red blood cells, which carry oxygen. Once the oxygen-carrying capacity of the blood is sufficient to support normal cellular activity, the kidneys cut back on their production of erythropoietin.

Anemia

When there is an insufficient number of red blood cells or the cells do not have enough hemoglobin, the individual suffers from **anemia** and has a tired, run-down feeling. In some types of anemia, only the hemoglobin level is low. It may be that the diet does not contain enough iron or folic acid. Certain foods, such as whole-grain cereals, are rich in iron and folic acid, and the inclusion of these in the diet can help prevent anemia.

In another type of anemia, called pernicious anemia, the digestive tract is unable to absorb enough vitamin B$_{12}$, found in dairy products, fish, eggs, and poultry. This vitamin is essential to the proper formation of red blood cells; without it, immature red blood cells tend to accumulate in the bone marrow in large quantities. A special diet and administration of vitamin B$_{12}$ by injection are effective treatments for pernicious anemia.

Hemolysis is the rupturing of red blood cells. In hemolytic anemia, there is an increased rate of red blood cell destruction. **Sickle-cell disease** is a hereditary condition in which the individual has sickle-shaped red blood cells that tend to rupture as they pass through the narrow capillaries. Hemolytic disease of the newborn, which is discussed at the end of this chapter (see page 121), is also a type of hemolytic anemia.

Anemia results when the blood has too few red blood cells and/or not enough hemoglobin.

Carbon Monoxide: A Deadly Poison

Carbon monoxide (CO) is an air pollutant that comes primarily from the incomplete combustion of natural gas and gasoline. Figure 6A shows that transportation contributes most of the carbon monoxide to our cities' air. But power plants, factories, waste incineration, and home heating also contribute to the carbon monoxide level. Cigarette smoke contains carbon monoxide and is delivered directly to the smoker's blood and also to nonsmokers nearby.

Because carbon monoxide is a colorless, odorless gas, people can be unaware that it is affecting their systems. But it binds to iron 200 times more tightly than oxygen. Hemoglobin contains iron, and so does cytochrome oxidase, the carrier in the electron transport system that passes electrons on to oxygen. When these molecules bind to carbon monoxide preferentially, they cannot perform their usual functions. The end result is that delivery of oxygen to mitochondria is impaired, and so is the functioning of the mitochondria.

Flushed red skin, especially on facial cheeks, is a first sign of carbon monoxide poisoning, because hemoglobin bound to carbon monoxide is brighter red than oxygenated hemoglobin. Marked euphoria, then sleepiness, coma, and death follow. Removing a person from the carbon monoxide source is not sufficient treatment because carbon monoxide, unlike oxygen, remains tightly bound to iron for many hours. A transfusion of red blood cells will help increase the carrying capacity of the blood, and pure oxygen given under pressure will displace some carbon monoxide. Despite good medical care, some people still die each year from CO poisoning.

The level of carbon monoxide in polluted air may not be sufficient to kill people, but it does interfere with the ability of the body to function properly. The elderly and those with cardiovascular disease are especially at risk. One study found that even levels once thought to be safe can cause angina patients to experience chest pains, because oxygen delivery to the heart is reduced.

We can all lessen air pollution and reduce the amount of carbon monoxide in the air by doing the following:

- Don't smoke (especially indoors).

- Walk or bicycle instead of driving.

- Use public transportation instead of driving.

- Heat your home with solar energy—not a furnace.

- Support the development of more efficient automobiles, factories, power plants, and home furnaces.

- Support the development of alternative fuels. When hydrogen gas is burned, for example, the result is water, not carbon monoxide and carbon dioxide.

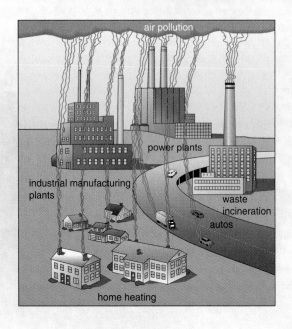

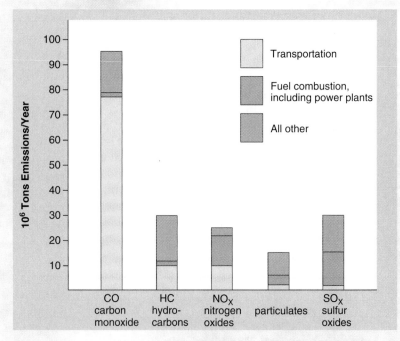

Figure 6A Air pollution.

Air pollutants (carbon monoxide, hydrocarbons, nitrogen oxides, particulates, and sulfur oxides) enter the atmosphere from the sources noted.

6.2 The White Blood Cells

White blood cells (leukocytes) differ from red blood cells in that they are usually larger, have a nucleus, lack hemoglobin, and without staining are translucent. White blood cells are not as numerous as red blood cells. There are only 5,000–11,000 per mm³ of blood. White blood cells fight infection and in this way are important contributors to homeostasis. This function of white blood cells is discussed at greater length in chapter 8, which concerns immunity. ⚖

White blood cells are derived from stem cells in the red bone marrow, and they, too, undergo several maturation stages. **Colony-stimulating factors (CSFs)** are proteins that help regulate the production of white blood cells. Researchers have shown that there is a different colony-stimulating factor for white cells derived from specific stem cells (see Fig. 6.4).

Red blood cells are confined to the blood, but white blood cells are able to squeeze through pores in the capillary wall, and therefore they are found in tissue fluid and lymph (Fig. 6.6). When there is an infection, white blood cells greatly increase in number. Many white blood cells live only a few days—they probably die while engaging pathogens. Others live months or even years.

White blood cells fight infection. They defend us against pathogens that have invaded the body.

Types of White Blood Cells

White blood cells are classified into the **granular leukocytes** and the **agranular leukocytes.** Both types of cells have granules in the cytoplasm surrounding the nucleus, but the granules are more visible upon staining in granular leukocytes. The granules contain various enzymes and proteins, which help white blood cells defend the body. There are three types of granular leukocytes and two types of agranular leukocytes. They differ somewhat by the size of the cell and the shape of the nucleus (see Fig. 6.2), and they also differ in their functions.

Granular Leukocytes

Neutrophils are the most abundant of the white blood cells. They have a multilobed nucleus joined by nuclear threads; therefore, they are also called polymorphonuclear. They have granules that do not significantly take up the stain eosin, a pink to red stain, or a basic stain that is blue to purple. (This accounts for their name, neutrophil.) Neutrophils are the first type of white blood cell to respond to an infection, and they engulf pathogens during **phagocytosis.**

Eosinophils have a bilobed nucleus, and their large, abundant granules take up eosin and become a red color. (This accounts for their name, eosinophil.) Not much is known specifically about the function of eosinophils, but

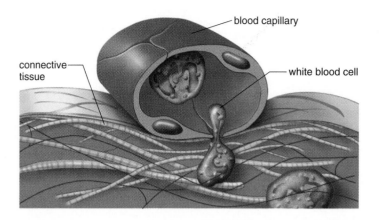

Figure 6.6   Mobility of white blood cells.
White blood cells can squeeze between the cells of a capillary wall and enter the tissues of the body.

they are known to increase in number when there is a parasitic worm infection or in the case of allergic reactions.

Basophils have a U-shaped or lobed nucleus. Their granules take up the basic stain and become a dark blue color. (This accounts for their name, basophil.) In the connective tissues, basophils, and also similar type cells called mast cells, release histamine associated with allergic reactions. Histamine dilates blood vessels and causes contraction of smooth muscle.

Agranular Leukocytes

The agranular leukocytes include monocytes, which have a kidney-shaped nucleus, and lymphocytes, which have a spherical nucleus. Lymphocytes are responsible for specific immunity to particular pathogens and their toxins (poisonous substances). Pathogens have molecules called **antigens** that allow the immune system to recognize them as foreign.

Monocytes are the largest of the white blood cells, and after taking up residence in the tissues, they differentiate into even larger macrophages. Macrophages phagocytize pathogens, old cells, and cellular debris. They also stimulate other white blood cells, including lymphocytes, to defend the body.

The **lymphocytes** are of two types, B lymphocytes and T lymphocytes. B lymphocytes protect us by producing **antibodies** that combine with antigens and thereby target pathogens for destruction. T lymphocytes, on the other hand, directly destroy any cell that has antigens. B lymphocytes and T lymphocytes are discussed more fully in chapter 8.

Leukemia

Leukemia is characterized by an abnormally large number of immature white blood cells that fill the red bone marrow and prevent red blood cell development. Anemia results, and the immature white cells offer little protection from disease. The cause of leukemia, a type of cancer, is unknown, but combined chemotherapy has been most successful, particularly in acute childhood leukemia.

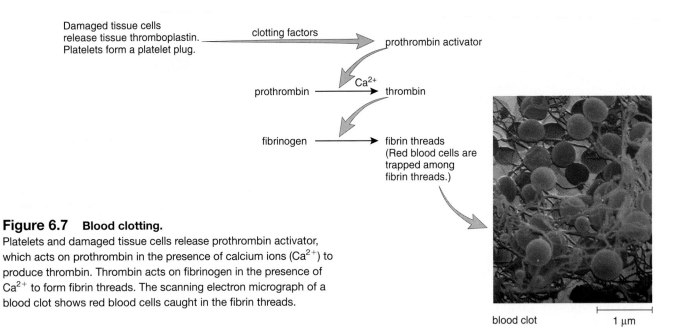

Figure 6.7 **Blood clotting.**
Platelets and damaged tissue cells release prothrombin activator,
which acts on prothrombin in the presence of calcium ions (Ca^{2+}) to
produce thrombin. Thrombin acts on fibrinogen in the presence of
Ca^{2+} to form fibrin threads. The scanning electron micrograph of a
blood clot shows red blood cells caught in the fibrin threads.

6.3 Blood Clotting

Platelets (thrombocytes) result from fragmentation of certain large cells, called **megakaryocytes,** in the red bone marrow. Platelets are produced at a rate of 200 billion a day, and
the blood contains 150,000–300,000 per mm^3. These formed
elements are involved in the process of blood **clotting,** or
coagulation.

When a blood vessel in the body is damaged, platelets
clump at the site of the puncture and seal the break if it is not
too extensive. A large break may also require a blood clot to
stop the bleeding.

Stages of Blood Clotting

There are at least 12 clotting factors that participate in the
formation of a blood clot. We will discuss the roles played
by fibrinogen, prothrombin, and thrombin. **Fibrinogen**
and **prothrombin** are proteins manufactured and deposited in blood by the liver. Vitamin K, found in green
vegetables and also formed by intestinal bacteria, is necessary for the production of prothrombin, and if by chance
this vitamin is missing from the diet, hemorrhagic disorders develop.

When a break occurs in a blood vessel, damaged tissue
releases tissue thromboplastin, a blood clotting factor that
initiates a series of reactions involving several clotting factors and calcium ions (Ca^{2+}). These reactions lead to production of **prothrombin activator** which converts
prothrombin to thrombin. This reaction also requires Ca^{2+}.
Thrombin, in turn, acts as an enzyme that severs two short
amino acid chains from each fibrinogen molecule. These
activated fragments then join end to end, forming long
threads of **fibrin.** Fibrin threads wind around the platelet

Table 6.1	Body Fluids Related to Blood
Name	**Composition**
Blood	Formed elements and plasma
Plasma	Liquid portion of blood
Serum	Plasma minus fibrinogen
Tissue fluid	Plasma minus most proteins
Lymph	Tissue fluid within lymphatic vessels

plug in the damaged area of the blood vessel and provide
the framework for the clot. Red blood cells also are trapped
within the fibrin threads; these cells make a clot appear red
(Fig. 6.7). A fibrin clot is present only temporarily. As soon
as blood vessel repair is initiated, an enzyme called plasmin destroys the fibrin network and restores the fluidity of
plasma.

If blood is allowed to clot in a test tube, a yellowish fluid
develops above the clotted material. This fluid is called
serum, and it contains all the components of plasma except
fibrinogen. Table 6.1 reviews the many different terms we
have used to refer to various body fluids related to blood.

Hemophilia

Hemophilia is an inherited clotting disorder due to a deficiency in a clotting factor. The slightest bump can cause
bleeding into the joints. Cartilage degeneration in the joints,
and resorption of underlying bone can follow. Bleeding into
muscles can lead to nerve damage and muscular atrophy.
The most frequent cause of death is bleeding into the brain
with accompanying neurological damage.

Health Focus

What to Know When Giving Blood

The Procedure

After you register to give blood, you are asked private and confidential questions about your health history and your lifestyle, and any questions you have are answered.

Your temperature, blood pressure, and pulse are checked, and a drop of your blood is tested to ensure that you are not anemic.

You will have several opportunities prior to giving blood and even afterwards to let Red Cross officials know whether your blood is safe to give to another person.

All of the supplies, including the needle, are sterile and are used only once—for YOU. You cannot get infected with HIV (the virus that causes AIDS) or any other disease by donating blood.

When the actual donation is started, you may feel a brief "sting." The procedure takes about 10 minutes, and you will have given about a pint of blood. Your body replaces the liquid part (plasma) in hours and the cells in a few weeks.

After you donate, you are given a card with a number to call if you decide after you leave that your blood may not be safe to give to another person.

An area is provided in which to relax after donating blood. Most people feel fine while they give blood and afterward. A few may have an upset stomach, a faint or dizzy feeling, or bruising, redness, and pain where the needle was inserted. Very rarely, a person may faint, have muscle spasms, and/or suffer nerve damage.

Your blood is tested for syphilis, AIDS antibodies, hepatitis, and other viruses. You are notified if tests give a positive result. Your blood won't be used if it could make someone ill.

The Cautions

\ \ \ DO NOT GIVE BLOOD / / /

if you have

> ever had hepatitis;

> had malaria, have taken drugs to prevent malaria in the last 3 years, or have traveled to a country where malaria is common;

> been treated for syphilis or gonorrhea in the last 12 months.

if you have AIDS or one of its symptoms:

> unexplained weight loss (4.5 kilograms or more in less than 2 months);

> night sweats;

> blue or purple spots on or under skin;

> long-lasting white spots or unusual sores in mouth;

> lumps in neck, armpits, or groin for over a month;

> diarrhea lasting over a month;

> persistent cough and shortness of breath;

> fever higher than 37 ℃ for more than 10 days.

if you are at risk for AIDS—that is, if you have:

> taken illegal drugs by needle, even once;

> taken clotting factor concentrates for a bleeding disorder such as hemophilia;

> tested positive for any AIDS virus or antibody;

> been given money or drugs for sex since 1977;

> had a sexual partner within the last 12 months who did any of the above things;

> (for men) had sex *even once* with another man since 1977, *or* had sex with a female prostitute within the last 12 months;

> (for women) had sex with a male or female prostitute within the last 12 months, *or* had a male sexual partner who had sex with another man *even once* since 1977.

DO NOT GIVE BLOOD to find out whether you test positive for antibodies to the viruses (HIV) that cause AIDS. Although the tests for HIV are very good, they aren't perfect. HIV antibodies may take weeks to develop after infection with the virus. If you were infected recently, you may have a negative test result yet be able to infect someone. **It is for this reason that you must not give blood if you are at risk of getting AIDS or other infectious diseases.**

Courtesy of the American Red Cross.

6.4 Plasma

Plasma is the liquid portion of blood, and about 92% of plasma is water. The remaining 8% of plasma consists of various salts (ions) and organic molecules (Table 6.2). The salts, which are simply dissolved in plasma, help maintain the pH of the blood. Small organic molecules like glucose, amino acids, and urea can also dissolve in plasma. Glucose and amino acids are nutrients for cells; urea is a nitrogenous waste product on its way to the kidneys for excretion. The large organic molecules in plasma include hormones and the plasma proteins.

The Plasma Proteins

The plasma proteins are so named because they usually remain in the plasma. The three major types of plasma proteins are the albumins, globulins, and fibrinogen. Most plasma proteins are made in the liver. An exception is the antibodies produced by B lymphocytes, which function in immunity.

The plasma proteins have many functions that help maintain homeostasis. They are able to take up and release hydrogen ions; therefore, the plasma proteins help buffer the blood and keep its pH around 7.40. They also contribute to the osmotic pressure, which helps keep water in the blood. The plasma protein **albumin** is primarily responsible for the osmotic pressure of the blood. Certain plasma proteins combine with and transport large organic molecules. For example, albumin transports the molecule bilirubin, a breakdown product of hemoglobin. Lipoproteins, whose protein portion is a globulin, transport cholesterol. There are three types of globulins, designated alpha, beta, and gamma globulins. Antibodies, which help fight infections by combining with antigens, are gamma globulins. Other plasma proteins also have specific functions. Fibrinogen is necessary to blood clotting, for example. ⚖

The numerous functions of plasma proteins contribute to maintaining homeostasis.

Table 6.2	Blood Plasma Solutes
Plasma proteins	Albumin, globulins, fibrinogen
Inorganic ions (salts)	Na^+, Ca^{2+}, K^+, Mg^{2+}, Cl^-, HCO_3^-, HPO_4^{2-}, SO_4^{2-}
Gases	O_2, CO_2
Organic nutrients	Glucose, fats, phospholipids, amino acids, etc.
Nitrogenous waste products	Urea, ammonia, uric acid
Regulatory substances	Hormones, enzymes

6.5 Capillary Exchange

The internal environment consists of blood and **tissue fluid**. The composition of tissue fluid stays relatively constant because of exchanges with blood in the region of capillaries (Fig. 6.8). Water makes up a large part of tissue fluid, and any excess is collected by lymphatic capillaries, which are always found near blood capillaries. ⚖

Blood Capillaries

A capillary has an arterial end, a midsection, and a venous end.

Arterial End of Capillary

When arterial blood enters the tissue capillaries, it is bright red because red blood cells are carrying oxygen. It is also rich in nutrients, which are dissolved in the plasma. At the arterial end of the capillary, blood pressure (30 mm Hg) is higher than the osmotic pressure of the blood (21 mm Hg). Blood pressure, you recall, is created by the pumping of the heart; the osmotic pressure is caused by the presence of salts and, in particular, by the plasma proteins that are too large to pass through the wall of the capillary. Since the blood pressure is higher than the osmotic pressure, fluid and nutrients (glucose and amino acids) exit the capillary. Red blood cells and most all plasma proteins generally remain in the capillaries, but small substances leave the capillaries. Therefore, tissue fluid, created by this process, consists of all the components of plasma except the proteins.

Midsection of Capillary

Along the length of the capillary, molecules follow their concentration gradient as diffusion occurs. Diffusion, you recall, is the movement of molecules from an area of greater concentration to an area of lesser concentration. In the tissues, the area of greater concentration for nutrients and oxygen is always blood, because after these molecules have passed into tissue fluid, they are taken up and metabolized by the tissue cells. The cells use glucose ($C_6H_{12}O_6$) and oxygen (O_2) in the process of cellular respiration, and they use amino acids for protein synthesis. Following cellular respiration, the cells give off carbon dioxide (CO_2) and water (H_2O). Carbon dioxide and other waste products of metabolism leave the cell by diffusion. Since tissue fluid is always the area of greater concentration for these waste materials, they diffuse into the capillary.

Oxygen and nutrient molecules (e.g., glucose and amino acids) exit a capillary near the arterial end; waste molecules (e.g., carbon dioxide) enter a capillary near the venous end.

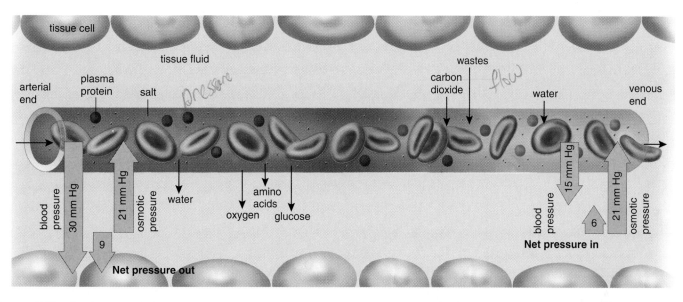

Figure 6.8   Capillary exchange.
At the arterial end of a capillary, the blood pressure is higher than the osmotic pressure; therefore, water (H_2O) tends to leave the bloodstream. In the midsection, oxygen (O_2) and carbon dioxide (CO_2) follow their concentration gradients. At the venous end of a capillary, the osmotic pressure is higher than the blood pressure; therefore, water tends to enter the bloodstream.

Venous End of Capillary

At the venous end of the capillary, blood pressure is much reduced (15 mm Hg), as shown in Figure 6.8. However, there is no reduction in osmotic pressure (21 mm Hg), which tends to pull fluid back into the capillary. As water enters a capillary, it brings with it additional waste molecules. Blood that leaves the capillaries is deep maroon in color because red blood cells contain reduced hemoglobin.

Retrieving fluid by means of osmotic pressure is not completely effective. There is always some fluid that is not picked up at the venous end. This excess tissue fluid enters the lymphatic capillaries.

Lymphatic Capillaries

Lymphatic capillaries are the smallest of the lymphatic vessels, which are a one-way system. The structure of lymphatic vessels is similar to that of cardiovascular veins, except that their walls are thinner and they have more valves. The valves prevent the backward flow of lymph as lymph flows toward the thoracic cavity. Lymphatic capillaries join to form larger vessels that merge into the lymphatic ducts (Fig. 6.9). The right lymphatic duct empties into a cardiovascular vein within the thoracic cavity.

Lymphatic vessels carry **lymph,** which has the same composition as tissue fluid. Why? Because lymphatic capillaries absorb excess tissue fluid at the blood capillaries. The lymphatic system contributes to homeostasis in several ways. One way is to maintain normal blood volume and pressure by returning excess tissue fluid to the blood. Edema is swelling that occurs when tissue fluid is not col-

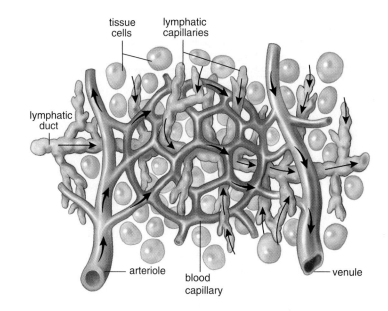

Figure 6.9   Lymphatic capillaries.
Arrows indicate that lymph is formed when lymphatic capillaries take up excess tissue fluid. Lymphatic capillaries lie near blood capillaries.

lected by the lymphatic capillaries. Edema can be due to many causes. One dramatic cause is a parasitic infection of lymphatic vessels by a small worm. An affected leg can become so large that the disease is called elephantiasis.

Exchange of nutrients for wastes occurs at the capillaries. Here lymphatic capillaries also collect excess tissue fluid.

Table 6.3	The ABO System						
Blood Type	**Antigen on Red Blood Cells**	**Antibody in Plasma**	**% U.S. African American**	**% U.S. Caucasian**	**% U.S. Asian**	**% North American Indian**	**% Americans of Chinese Descent**
A	A	Anti-B	27	41	28	8	25
B	B	Anti-A	20	9	27	1	35
AB	A, B	None	4	3	5	0	10
O	None	Anti-A and anti-B	49	47	40	92	30

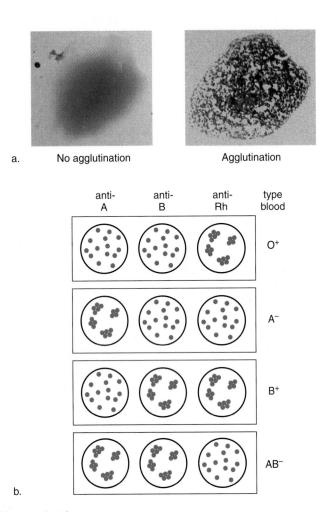

a.   No agglutination   Agglutination

Figure 6.10   Blood typing.
The standard test to determine ABO and Rh blood type consists of putting a drop of anti-A antibodies, anti-B antibodies, and anti-Rh antibodies separately on a slide. To each of these three antibody solutions, a drop of the person's blood is added. **a.** If agglutination occurs, as seen in the *top right* photo, the person has this antigen on red blood cells. **b.** Several possible results. The ABO blood type is indicated by using letters, and the Rh blood type is indicated by using the symbols (+) and (–).

6.6   Blood Typing

Blood typing involves the two types of molecules called antigens and antibodies. As mentioned on page 115, when an antigen (foreign substance) is present in the body, an antibody reacts with it.

ABO System

The most common system for typing blood is the ABO system. In the ABO system, the presence or absence of type A and type B antigens on red blood cells determines a person's blood type. For example, if a person has type A blood, the A antigen is on his or her red blood cells. This molecule is not an antigen to this individual, although it can be an antigen to a recipient who does not have type A blood.

In the simplified ABO system, there are four types of blood: A, B, AB, and O (Table 6.3). Within the plasma, there are antibodies to the antigens that are *not* present on the person's red blood cells. These antibodies are called anti-A and anti-B.

Blood Type	**Antigen on Red Blood Cells**	**Antibody in Plasma**
A	A	Anti-B
B	B	Anti-A
AB	A, B	None
O	None	Anti-A and anti-B

It is reasonable that type A blood would have anti-B and not anti-A antibodies in the plasma. If anti-A antibodies were present in plasma, **agglutination,** or clumping of red blood cells, would occur. Agglutination of red blood cells can cause blood to stop circulating in small blood vessels, and this leads to organ damage. It also is followed by hemolysis, which may cause the death of the individual.

For a recipient to receive blood from a donor, the recipient's plasma must not have an antibody that causes the donor's cells to agglutinate. For this reason, it is important

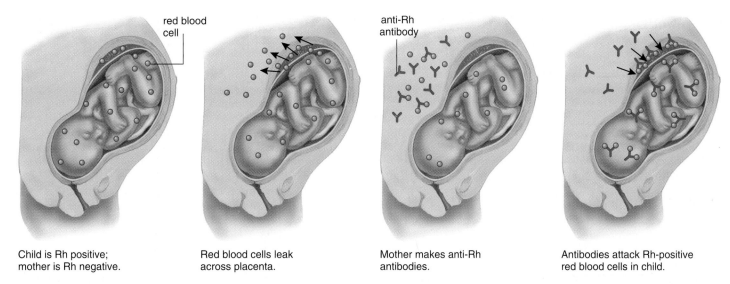

red blood cell		anti-Rh antibody	
Child is Rh positive; mother is Rh negative.	Red blood cells leak across placenta.	Mother makes anti-Rh antibodies.	Antibodies attack Rh-positive red blood cells in child.

Figure 6.11   Hemolytic disease of the newborn.
Due to a pregnancy in which the child is Rh^+, an Rh^- mother can begin to produce antibodies against Rh^+ red blood cells. In another pregnancy, these antibodies can cross the placenta and cause hemolysis of an Rh^+ child's red blood cells.

to determine each person's blood type. Figure 6.10 demonstrates a way to use the antibodies derived from plasma to determine the blood type. If clumping occurs after a sample of blood is exposed to a particular antibody, the person has that type of blood.

Today, blood transfusions are a matter of concern not only because blood types should match, but also because each person wants to receive blood that is of good quality and free of infectious agents. Blood is tested for the more serious agents such as those that cause AIDS, hepatitis, and syphilis. Donors can help protect the nation's blood supply by knowing when not to give blood. This is the topic of the Health Focus on page 117.

For the purpose of blood transfusions, the donor's blood must be compatible with the recipient's blood. The ABO system is used to determine compatibility of donor's and recipient's blood.

Rh System

Another important antigen in matching blood types is the Rh factor. Eighty-five percent of the U.S. population have this particular antigen on the red blood cells and are Rh^+ (Rh positive). Fifteen percent do not have this antigen and are Rh^- (Rh negative). Rh^- individuals normally do not have antibodies to the Rh factor, but they may make them when exposed to the Rh factor.

To test whether an individual is Rh^- or Rh^+, blood is mixed with anti-Rh antibodies. When Rh^+ blood is mixed with anti-Rh antibodies, agglutination occurs (Fig. 6.10).

The designation of blood type usually also includes whether the person has or does not have the Rh factor on the red blood cells.

During pregnancy, if the mother is Rh^- and the father is Rh^+, the child may be Rh^+ (Fig. 6.11). The Rh^+ red blood cells may begin leaking across the placenta into the mother's cardiovascular system, as placental tissues normally break down before and at birth. The presence of these Rh antigens causes the mother to produce anti-Rh antibodies. In this or a subsequent pregnancy with another Rh^+ baby, anti-Rh antibodies produced by the mother may cross the placenta and destroy the child's red blood cells. This is called hemolytic disease of the newborn (HDN) because hemolysis continues after the baby is born. Due to red blood cell destruction, excess bilirubin in the blood can lead to brain damage and mental retardation or even death.

The Rh problem is prevented by giving Rh^- women an Rh immunoglobulin injection either midway through the first pregnancy or no later than 72 hours after giving birth to an Rh^+ child. This injection contains anti-Rh antibodies that attack any of the baby's red blood cells in the mother's blood before these cells can stimulate her immune system to produce her own antibodies. This injection is not beneficial if the woman has already begun to produce antibodies; therefore, the timing of the injection is most important.

The possibility of hemolytic disease of the newborn exists when the mother is Rh^- and the father is Rh^+.

Summarizing the Concepts

Blood, which is composed of formed elements and plasma, has several functions. It transports hormones, oxygen, and nutrients to the cells and carbon dioxide and other wastes away from cells. It fights infections and has various regulatory functions. It maintains blood pressure, regulates body temperature, and keeps the pH of body fluids within normal limits. All of these functions help maintain homeostasis.

All blood cells are produced within red bone marrow from stem cells, which are ever capable of dividing and producing new cells.

6.1 The Red Blood Cells

Red blood cells are small, biconcave disks that lack a nucleus. They live about 120 days and are destroyed in the liver and spleen when they are old or abnormal. The production of red blood cells is controlled by the oxygen concentration of the blood. When the oxygen concentration decreases, the kidneys increase their production of erythropoietin, and more red blood cells are produced. Red blood cells contain hemoglobin, the respiratory pigment, which combines with oxygen and transports it to the tissues.

6.2 The White Blood Cells

White blood cells are larger than red blood cells, have a nucleus, and are translucent unless stained. Like red blood cells, they are produced in the red bone marrow. White blood cells are divided into the granular leukocytes and the agranular leukocytes. The granular leukocytes have conspicuous granules; in eosinophils, granules are red when stained with eosin, and in basophils, granules are blue when stained with a basic dye. The granules in neutrophils don't take up either dye significantly. Neutrophils are the most plentiful of the white blood cells, and they are able to phagocytize pathogens. Many neutrophils die within a few days when they are fighting an infection. The agranulocytes include the lymphocytes and the monocytes, which function in specific immunity. On occasion, the monocytes become large phagocytic cells of great significance. They engulf worn-out red blood cells and pathogens at a ferocious rate.

6.3 Blood Clotting

When there is a break in a blood vessel, the platelets clump to form a plug. Blood clotting itself requires a series of enzymatic reactions involving blood platelets, prothrombin, and fibrinogen. In the final reaction, fibrinogen becomes fibrin threads, entrapping cells. The fluid that escapes from a clot is called serum and consists of plasma minus fibrinogen.

6.4 Plasma

Plasma is mostly water (92%) and the plasma proteins (8%). The plasma proteins, most of which are produced by the liver, occur in three categories: albumins, globulins, and fibrinogen. The plasma proteins maintain osmotic pressure, help regulate pH, and transport molecules. Some plasma proteins have specific functions: the gamma globulins, which are antibodies produced by B lymphocytes, function in immunity, and fibrinogen is necessary to blood clotting.

Small organic molecules like glucose and amino acids are dissolved in plasma and serve as nutrients for cells; the gas oxygen is needed for cellular respiration, and carbon dioxide is a waste product of this process.

6.5 Capillary Exchange

At the arterial end of a cardiovascular capillary, blood pressure is greater than osmotic pressure; therefore, water leaves the capillary. In the midsection, oxygen and nutrients diffuse out of the capillary, while carbon dioxide and other wastes diffuse into the capillary. At the venous end, osmotic pressure created by the presence of proteins exceeds blood pressure, causing water to enter the capillary.

Retrieving fluid by means of osmotic pressure is not completely effective. There is always some fluid that is not picked up at the venous end of the cardiovascular capillary. This excess tissue fluid enters the lymphatic capillaries. Lymph is tissue fluid contained within lymphatic vessels. The lymphatic system is a one-way system, and lymph is returned to blood by way of a cardiovascular vein.

6.6 Blood Typing

The red blood cells of an individual are not necessarily received without difficulty by another individual. For example, the membranes of red blood cells may contain type A, B, AB, or no antigens. In the plasma, there are two possible antibodies: anti-A or anti-B. If the corresponding antigen and antibody are put together, clumping, or agglutination, occurs; in this way, the blood type of an individual may be determined in the laboratory. After determination of the blood type, it is theoretically possible to decide who can give blood to whom. For this, it is necessary to consider the donor's antigens and the recipient's antibodies.

Another important antigen is the Rh antigen. This particular antigen must also be considered in the transfusing of blood, and it is important during pregnancy because an Rh^- mother may form antibodies to the Rh antigen while carrying or after the birth of a child who is Rh^+. These antibodies can cross the placenta to destroy the red blood cells of any subsequent Rh^+ child.

Studying the Concepts

1. State the two main components of blood, and give the functions of blood.   111
2. What is hemoglobin, and how does it function?   111
3. Describe the life cycle of red blood cells, and tell how the production of red blood cells is regulated.   113
4. Name the five types of white blood cells; describe the structure and give a function for each type.   115
5. Name the steps that take place when blood clots. Which substances are present in blood at all times, and which appear during the clotting process?   116
6. Define blood, plasma, tissue fluid, lymph, and serum.   111, 116, 118, 119
7. List and discuss the major components of plasma. Name several plasma proteins, and give a function for each.   118
8. What forces operate to facilitate exchange of molecules across the capillary wall?   118–19
9. What are the four ABO blood types? For each, state the antigen(s) on the red blood cells and the antibody(ies) in the plasma.   120
10. Explain why a person with type O blood cannot receive a transfusion of type A blood.   120
11. Problems can arise during childbearing if the mother is which Rh type and the father is which Rh type? Explain why this is so.   121

Testing Your Knowledge of the Concepts

In questions 1–4, match the components of blood to the descriptions, and fill in the blanks.
  a. red blood cell
  b. white blood cells
  c. both red and white blood cells
  d. plasma

  _____ 1. Includes monocytes that _____ debris

  _____ 2. Contains _____ that transports oxygen

  _____ 3. Contains fibrinogen that is needed for _____

  _____ 4. Antigens in plasma membrane that determine the blood type

In questions 5–7, indicate whether the statement is true (T) or false (F).

  _____ 5. Carbon dioxide exits the arteriole end of the capillary and oxygen enters the venule end of the capillary.

  _____ 6. Without the work of the lymphatic system, the tissues would fill with fluid.

  _____ 7. If a woman is Rh$^+$, she doesn't have to worry about hemolytic disease of the newborn.

In questions 8 and 9, fill in the blanks.

8. B lymphocytes produce _____ that react with antigens.
9. A person with type AB blood has _____ antibodies in the plasma.
10. Match the key terms to these definitions:
    a. _____ Iron-containing protein in red blood cells that combines with and transports oxygen.
    b. _____ Clumping of cells, particularly red blood cells involved in an antigen-antibody reaction.
    c. _____ Liquid portion of blood.
    d. _____ Cell fragment that is necessary to blood clotting.
    e. _____ Plasma protein that is converted to thrombin during the steps of blood clotting.

11. Label arrows as either blood pressure or osmotic pressure.

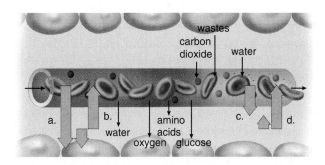

Understanding Key Terms

agglutination 120
agranular leukocyte 115
albumin 118
anemia 113
antibodies 115
antigen 115
basophil 115
clotting 116
colony-stimulating factors 115
eosinophil 115
erythropoietin 113
fibrin 116
fibrinogen 116
formed element 111
granular leukocyte 115
hemoglobin 111
hemolysis 113
hemophilia 116
leukemia 115
lymph 119

lymphocyte 115
megakaryocyte 116
monocyte 115
neutrophil 115
pathogen 111
phagocytosis 115
plasma 118
platelet (thrombocyte) 116
prothrombin 116
prothrombin activator 116
red blood cell
    (erythrocyte) 111
serum 116
sickle-cell disease 113
stem cell 113
thrombin 116
tissue fluid 118
white blood cell
    (leukocyte) 115

e-Learning Connection

6.1 The Red Blood Cells

 <u>Blood Doping in Cyclers</u> *case study*

 <u>Erythrocytes</u> *Essential Study Partner*

6.2 The White Blood Cells

 <u>Leukocytes</u> *Essential Study Partner*

6.3 Blood Clotting

 <u>Artificial Blood</u> *case study*

6.4 Plasma

 <u>Blood</u> *Essential Study Partner*

6.5 Capillary Exchange

 <u>Anatomy of a Capillary Bed</u> *art labeling activity*

6.6 Blood Typing

 <u>ABO Blood Types</u> *Essential Study Partner activity*

 <u>Blood Typing</u> *art quiz*

Chapter Summary

 <u>Key Term Flashcards</u> *vocabulary quiz*
<u>Chapter Quiz</u> *objective quiz*

Chapter 7

Cardiovascular System

Chapter Concepts

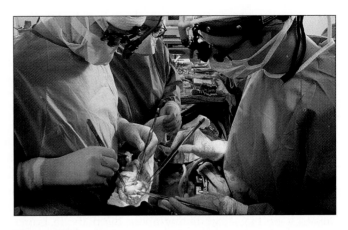

Figure 7.1   **Heart transplant operation.**
Will it be safe one day to use pigs' hearts for human transplant operations? Pigs have been genetically engineered to be immunocompatible with humans, but there is the possibility of acquiring a virus unique to pigs.

Dolores Manning, a 50-year-old mother of six, had congestive heart failure. Her heart was unable to beat effectively; blood was backing up in blood vessels, and fluids were collecting in her lungs. She might suffocate. A diuretic helped rid her body of excess fluid, but she was still sick. Dolores needed a heart transplant.

Doctors didn't know when a heart might become available, so in the meantime they gave Dolores an LVAD (left ventricular assist device). This device takes over some of the heart's pumping functions so that the heart doesn't have to work so hard.

Surgeons implanted the device in an abdominal "pocket" of skin they created. It was run by a small electric generator. The device worked so well that Dolores was able to exercise and increase her fitness. But she still felt constrained by not having a fully functioning heart. Everyone said the improvement in her general health was bound to increase the chance of a successful heart transplant.

Still there was a problem. No one knew when a heart might become available. The transplanted heart had to be immunologically compatible or it couldn't even be considered. While Dolores was waiting, she came across an article about pigs whose tissues had been genetically altered to be compatible with those of any human being (Fig. 7.1). The idea is that some day pigs will be a source of organs for transplantation into humans. Dolores was desperate to lead a normal life with her children, so she asked the doctors if she could have a pig's heart. "It's experimental right now," the doctors said. "Some investigators are afraid that the tissue of a pig might give the human recipient a virus unique to pigs." "I'm willing to take a chance," Dolores said.

As Dolores knew, the heart is a vital organ because it ordinarily keeps blood moving in the cardiovascular system. Circulation of the blood is so important that if the heart stops beating for only a few minutes, death results. This chapter reviews the cardiovascular system and how it operates to keep blood circulating about the body. In humans, the right side of the heart pumps blood to the lungs, and the left side pumps blood to the tissues. The blood never runs free, and is conducted to and from the tissues by blood vessels in these two separate vascular circuits. Among the types of blood vessels, only the capillaries have walls thin enough to allow exchange of molecules with the tissues. This exchange refreshes tissue fluid so that homeostasis is maintained. Otherwise, cells would die from the lack of nutrients and buildup of waste products.

7.1 The Blood Vessels

The cardiovascular system has three types of blood vessels: the **arteries** (and arterioles), which carry blood away from the heart to the capillaries; the **capillaries,** which permit exchange of material with the tissues; and the **veins** (and venules), which return blood from the capillaries to the heart.

The Arteries

The arterial wall has three layers (Fig. 7.2*a*). The inner layer is a simple squamous epithelium called endothelium with a connective tissue basement membrane that contains elastic fibers. The middle layer is the thickest layer and consists of smooth muscle that can contract to regulate blood flow and blood pressure. The outer layer is fibrous connective tissue near the middle layer, but it becomes loose connective tissue at its periphery. Some arteries are so large that they require their own blood vessels.

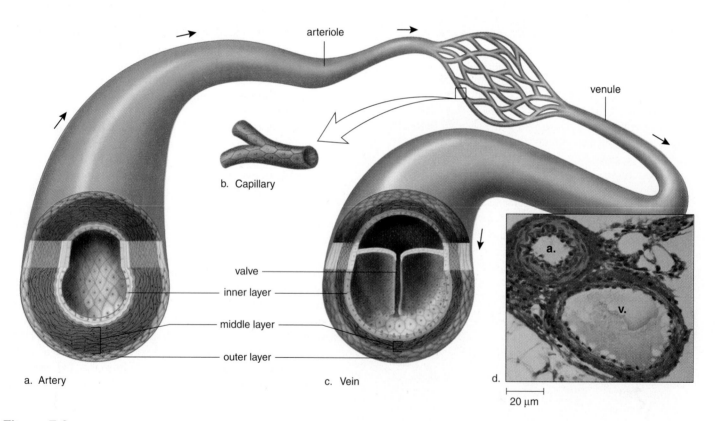

Figure 7.2 **Blood vessels.**
The walls of arteries and veins have three layers. The inner layer is composed largely of endothelium with a basement membrane that has elastic fibers; the middle layer is smooth muscle tissue; the outer layer is connective tissue (largely collagen fibers). **a.** Arteries have a thicker wall than veins because they have a larger middle layer than veins. **b.** Capillary walls are one-cell-thick endothelium. **c.** Veins are larger in diameter than arteries, so that collectively veins have a larger holding capacity than arteries. **d.** Light micrograph of an artery and a vein.

Arterioles are small arteries just visible to the naked eye. The middle layer of arterioles has some elastic tissue but is composed mostly of smooth muscle whose fibers encircle the arteriole. When these muscle fibers are contracted, the vessel has a smaller diameter (is constricted); when these muscle fibers are relaxed, the vessel has a larger diameter (is dilated). Whether arterioles are constricted or dilated affects blood pressure. The greater the number of vessels dilated, the lower the blood pressure.

The Capillaries

Arterioles branch into capillaries (Fig. 7.2*b*). Each capillary is an extremely narrow, microscopic tube with one-cell-thick walls composed only of endothelium with a basement membrane. Capillary beds (networks of many capillaries) are present in all regions of the body; consequently, a cut to any body tissue draws blood. Capillaries are a very important part of the human cardiovascular system because an exchange of substances takes place across their thin walls. Oxygen and nutrients, such as glucose, diffuse out of a capillary into the tissue fluid that surrounds cells. Wastes, such as carbon dioxide, diffuse into the capillary. The relative constancy of tissue fluid is absolutely dependent upon capillary exchange.

Only certain capillaries are open at any given time. For example, after eating, the capillaries that serve the digestive system are open and those that serve the muscles are closed. When a capillary bed is closed, the precapillary sphincters contract, and the blood moves from arteriole to venule by way of an arteriovenous shunt (Fig. 7.3).

The Veins

Venules are small veins that drain blood from the capillaries and then join to form a vein. The walls of venules (and veins) have the same three layers as arteries, but there is less smooth muscle and connective tissue (Fig. 7.2*c*). Veins often have **valves,** which allow blood to flow only toward the heart when open and prevent the backward flow of blood when closed.

Since the walls of veins are thinner, they can expand to a greater extent (Fig. 7.2*d*). At any one time, about 70% of the blood is in the veins. In this way, the veins act as a blood reservoir.

Arteries and arterioles carry blood away from the heart toward the capillaries; capillaries join arterioles to venules; veins and venules return blood from the capillaries to the heart.

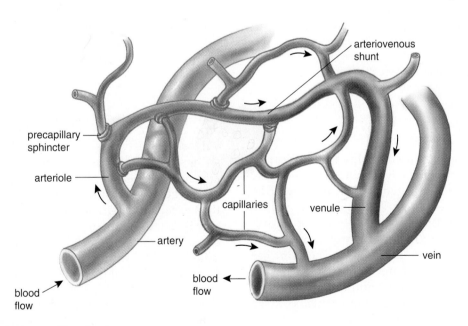

Figure 7.3 **Anatomy of a capillary bed.**
A capillary bed forms a maze of capillary vessels that lies between an arteriole and a venule. When sphincter muscles are relaxed, the capillary bed is open, and blood flows through the capillaries. When sphincter muscles are contracted, blood flows through a shunt that carries blood directly from an arteriole to a venule. As blood passes through a capillary in the tissues, it gives up its oxygen (O_2). Therefore, blood goes from being O_2-rich in the arteriole (red color) to being O_2-poor in the vein (blue color).

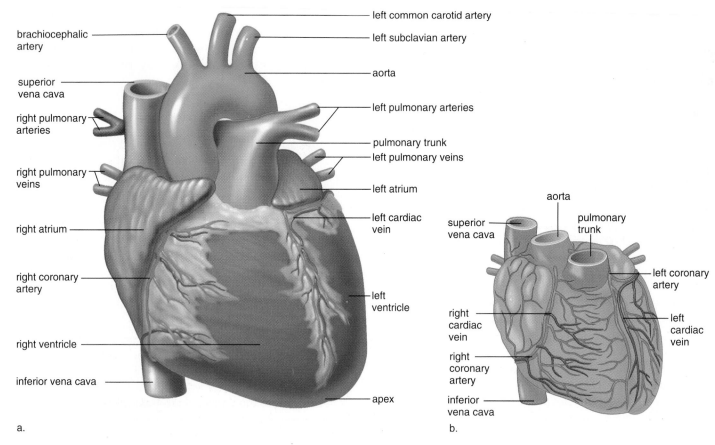

Figure 7.4   External heart anatomy.
a. The superior vena cava and the pulmonary trunk are attached to the right side of the heart. The aorta and pulmonary veins are attached to the left side of the heart. The right ventricle forms most of the ventral surface of the heart, and the left ventricle forms most of the dorsal surface.
b. The coronary arteries and cardiac veins pervade cardiac muscle. The coronary arteries bring oxygen and nutrients to cardiac cells, which derive no benefit from blood coursing through the heart.

7.2 The Heart

The **heart** is a cone-shaped, muscular organ about the size of a fist. It is located between the lungs directly behind the sternum (breastbone) and is tilted so that the apex (the pointed end) is oriented to the left. The major portion of the heart, called the **myocardium,** consists largely of cardiac muscle tissue. The muscle fibers of the myocardium are branched and tightly joined to one another. The heart lies within the **pericardium,** a thick, membranous sac that secretes a small quantity of lubricating liquid. The inner surface of the heart is lined with endocardium, which consists of connective tissue and endothelial tissue.

The heart has four chambers. The two upper, thin-walled atria (sing., **atrium**) have wrinkled protruding appendages called auricles. The two lower chambers are the thick-walled **ventricles,** which pump the blood (Fig. 7.4).

Internally, a wall called the septum separates the heart into a right side and a left side (Fig. 7.5*a*). The heart has four valves, which direct the flow of blood and prevent its backward movement. The two valves that lie between the atria and the ventricles are called the **atrioventricular valves.** These valves are supported by strong fibrous strings called **chordae tendineae.** The chordae, which are attached to muscular projections of the ventricular walls, support the valves and prevent them from inverting when the heart contracts. The atrioventricular valve on the right side is called the tricuspid valve because it has three flaps, or cusps. The atrioventricular valve on the left side is called the bicuspid (or mitral) valve because it has two flaps. The remaining two valves are the **semilunar valves,** whose flaps resemble half-moons, between the ventricles and their attached vessels. The pulmonary semilunar valve lies between the right ventricle and the pulmonary trunk. The aortic semilunar valve lies between the left ventricle and the aorta.

Humans have a four-chambered heart (two atria and two ventricles). A septum separates the right side from the left side.

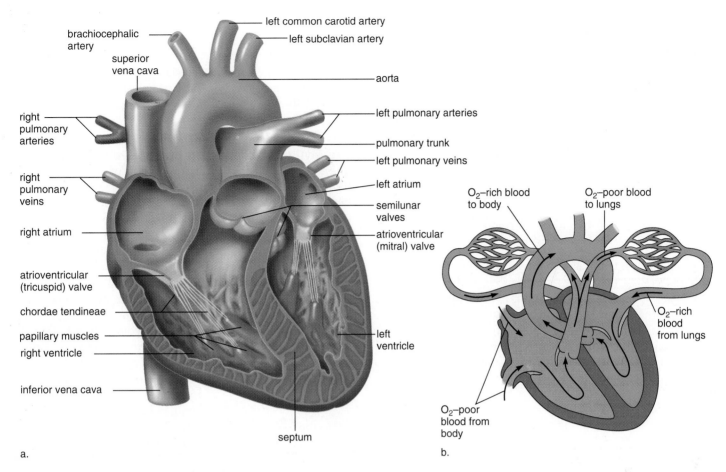

left common carotid artery

left subclavian artery

brachiocephalic artery

superior vena cava

aorta

right pulmonary arteries

left pulmonary arteries

pulmonary trunk

right pulmonary veins

left pulmonary veins

left atrium

semilunar valves

right atrium

atrioventricular (mitral) valve

atrioventricular (tricuspid) valve

chordae tendineae

papillary muscles

right ventricle

left ventricle

inferior vena cava

septum

O₂–rich blood to body

O₂–poor blood to lungs

O₂–rich blood from lungs

O₂–poor blood from body

a.

b.

Figure 7.5 **Internal view of the heart.**
a. The heart has four valves. The atrioventricular valves allow blood to pass from the atria to the ventricles, and the semilunar valves allow blood to pass out of the heart. **b.** This diagrammatic representation of the heart allows you to trace the path of the blood through the heart.

Passage of Blood Through the Heart

We can trace the path of blood through the heart (Fig. 7.5b) in the following manner:

- The superior vena cava and the inferior vena cava, which carry O₂-poor blood, enter the right atrium.
- The right atrium sends blood through an atrioventricular valve (the tricuspid valve) to the right ventricle.
- The right ventricle sends blood through the pulmonary semilunar valve into the pulmonary trunk. The pulmonary trunk divides into two **pulmonary arteries,** which go to the lungs.
- Four **pulmonary veins,** which carry O₂-rich blood, enter the left atrium.
- The left atrium sends blood through an atrioventricular valve (the bicuspid or mitral valve) to the left ventricle.
- The left ventricle sends blood through the aortic semilunar valve into the aorta to the body proper.

From this description, you can see that O₂-poor blood never mixes with O₂-rich blood and that blood must go through the lungs in order to pass from the right side to the left side of the heart. In fact, the heart is a double pump because the right ventricle of the heart sends blood through the lungs, and the left ventricle sends blood throughout the body. Since the left ventricle has the harder job of pumping blood to the entire body, its walls are thicker than those of the right ventricle, which pumps blood a relatively short distance to the lungs.

The pumping of the heart sends blood out under pressure into the arteries. Because the left side of the heart is the stronger pump, blood pressure is greatest in the aorta. Blood pressure then decreases as the cross-sectional area of arteries and then arterioles increases.

The right side of the heart pumps blood to the lungs, and the left side of the heart pumps blood throughout the body.

The Heartbeat

Each heartbeat is called a **cardiac cycle** (Fig. 7.6). When the heart beats, first the two atria contract at the same time; then the two ventricles contract at the same time. Then all chambers relax. The word **systole** refers to contraction of heart muscle, and the word **diastole** refers to relaxation of heart muscle. The heart contracts, or beats, about 70 times a minute, and each heartbeat lasts about 0.85 second.

Time	Atria	Ventricles
0.15 sec	Systole	Diastole
0.30 sec	Diastole	Systole
0.40 sec	Diastole	Diastole

A normal adult rate at rest can vary from 60 to 80 beats per minute.

When the heart beats, the familiar "lub-dup" sound occurs. The longer and lower-pitched "lub" is caused by vibrations occurring when the atrioventricular valves close due to ventricular contraction. The shorter and sharper "dup" is heard when the semilunar valves close due to back pressure of blood in the arteries. A heart murmur, or a slight slush sound after the "lub," is often due to ineffective valves, which allow blood to pass back into the atria after the atrioventricular valves have closed. Rheumatic fever resulting from a bacterial infection is one possible cause of a faulty valve, particularly the bicuspid valve. Faulty valves can be surgically corrected.

Intrinsic Control of Heartbeat

The rhythmical contraction of the atria and ventricles is due to the intrinsic conduction system of the heart. Nodal tissue, which has both muscular and nervous characteristics, is a unique type of cardiac muscle located in two regions of the heart. The **SA (sinoatrial) node** is located in the upper dorsal wall of the right atrium; the **AV (atrioventricular) node** is located in the base of the right atrium very near the septum (Fig. 7.7a). The SA node initiates the heartbeat and automatically sends out an excitation impulse every 0.85 second; this causes the atria to contract. When impulses reach the AV node, there is a slight delay that allows the atria to finish their contraction before the ventricles begin their contraction. The signal for the ventricles to contract travels from the AV node through the two branches of the **atrioventricular bundle** (AV bundle) before reaching the numerous and smaller **Purkinje fibers.** The AV bundle, its branches, and the Purkinje fibers consist of specialized cardiac muscle fibers that efficiently cause the ventricles to contract.

The SA node is called the **pacemaker** because it usually keeps the heartbeat regular. If the SA node fails to work properly, the heart still beats due to impulses generated by the AV node. But the beat is slower (40 to 60 beats per minute). To correct this condition, it is possible to implant an artificial pacemaker, which automatically gives an electrical stimulus to the heart every 0.85 second.

The intrinsic conduction system of the heart consists of the SA node, the AV node, the atrioventricular bundle, and the Purkinje fibers.

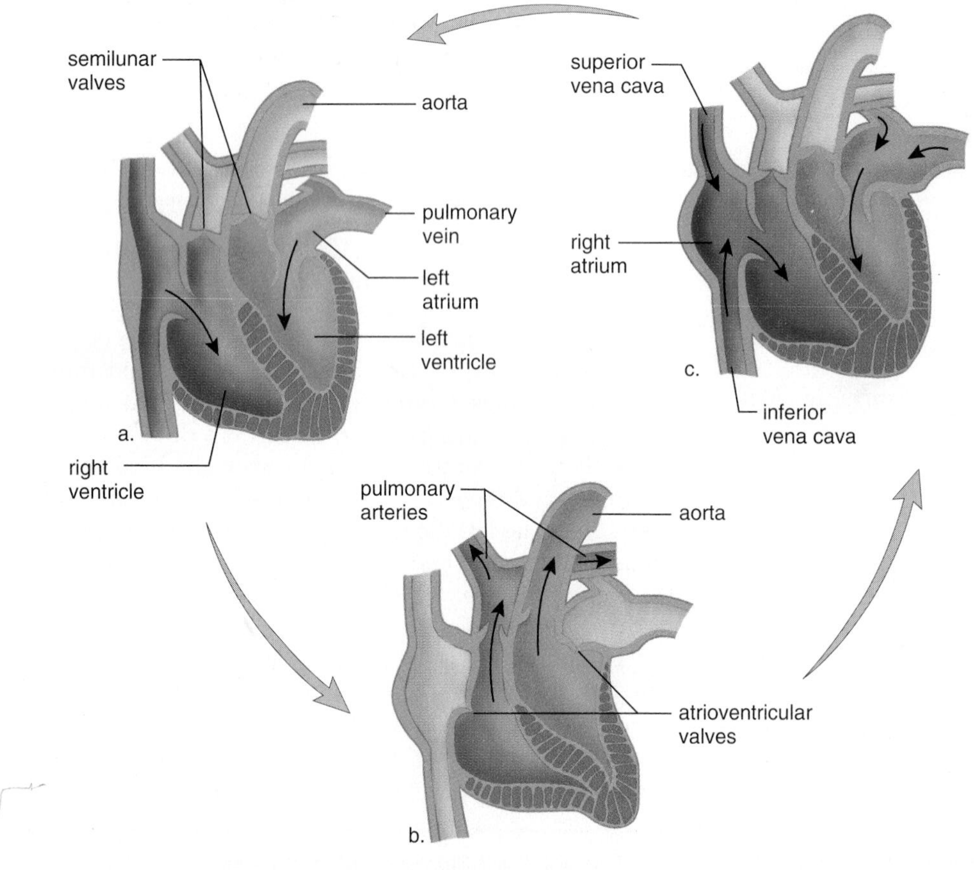

Figure 7.6   **Stages in the cardiac cycle.**
a. When the atria contract, the ventricles are relaxed and filling with blood. **b.** When the ventricles contract, the atrioventricular valves are closed, the semilunar valves are open, and the blood is pumped into the pulmonary trunk and aorta. **c.** When the heart is relaxed, both the atria and the ventricles are filling with blood.

Extrinsic Control of Heartbeat

The body has an extrinsic way to regulate the heartbeat. A cardiac control center in the medulla oblongata, a portion of the brain that controls internal organs, can alter the beat of the heart by way of the autonomic system, a division of the nervous system. The autonomic system has two subdivisions: the parasympathetic system, which promotes those functions we tend to associate with a resting state, and the sympathetic system, which brings about those responses we associate with increased activity and/or stress. The parasympathetic system decreases SA and AV nodal activity when we are inactive, and the sympathetic system increases SA and AV nodal activity when we are active or excited.

The hormones epinephrine and norepinephrine, which are released by the adrenal medulla, also stimulate the heart. During exercise, for example, the heart pumps faster and stronger due to sympathetic stimulation and due to the release of epinephrine and norepinephrine.

The body has an extrinsic way to regulate the heartbeat. The autonomic system and hormones can modify the heartbeat rate.

The Electrocardiogram

An **electrocardiogram (ECG)** is a recording of the electrical changes that occur in myocardium during a cardiac cycle. Body fluids contain ions that conduct electrical currents, and therefore the electrical changes in myocardium can be detected on the skin's surface. When an electrocardiogram is being taken, electrodes placed on the skin are connected by wires to an instrument that detects the myocardium's electrical changes. Thereafter, a pen rises or falls on a moving strip of paper. Figure 7.7b depicts the pen's movements during a normal cardiac cycle.

When the SA node triggers an impulse, the atrial fibers produce an electrical change that is called the P wave. The P wave indicates that the atria are about to contract. After that, the QRS complex signals that the ventricles are about to contract. The electrical changes that occur as the ventricular muscle fibers recover produces the T wave.

Various types of abnormalities can be detected by an electrocardiogram. One of these, called ventricular fibrillation, causes uncoordinated contraction of the ventricles (Fig. 7.7c). Ventricular fibrillation is of special interest because it can be caused by an injury or drug overdose. It is the most common cause of sudden cardiac death in a seemingly healthy person over age 35. Once the ventricles are fibrillating, they have to be defibrillated by applying a strong electrical current for a short period of time. Then the SA node may be able to reestablish a coordinated beat.

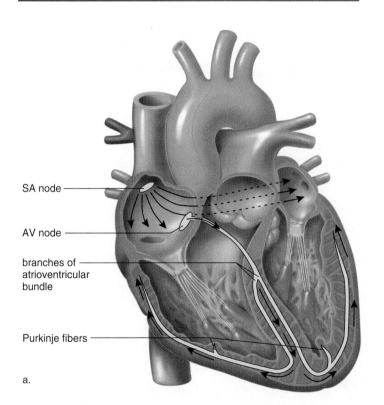
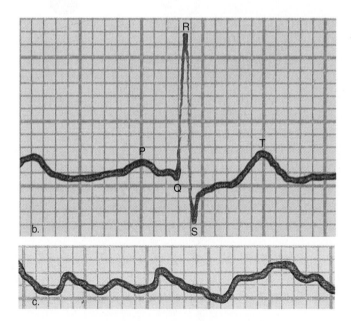

Figure 7.7 **Conduction system of the heart.**
a. The SA node sends out a stimulus, which causes the atria to contract. When this stimulus reaches the AV node, it signals the ventricles to contract. Impulses pass down the two branches of the atrioventricular bundle to the Purkinje fibers, and thereafter the ventricles contract. **b.** A normal ECG indicates that the heart is functioning properly. The P wave occurs just prior to atrial contraction; the QRS complex occurs just prior to ventricular contraction; and the T wave occurs when the ventricles are recovering from contraction. **c.** Ventricular fibrillation produces an irregular electrocardiogram due to irregular stimulation of the ventricles.

7.3 Features of the Cardiovascular System

When the left ventricle contracts, blood is sent out into the aorta under pressure. A progressive decrease in pressure occurs as blood moves through the arteries, capillaries, and finally the veins. Blood pressure is highest in the aorta and lowest in the venae cavae which enter the right atrium.

Pulse

The surge of blood entering the arteries causes their elastic walls to stretch, but then they almost immediately recoil. This alternating expansion and recoil of an arterial wall can be felt as a **pulse** in any artery that runs close to the body's surface. It is customary to feel the pulse by placing several fingers on the radial artery, which lies near the outer border of the palm side of a wrist (Fig. 7.8). A carotid artery, on either side of the trachea in the neck, is another accessible location to feel the pulse. Normally, the pulse rate indicates the rate of the heartbeat because the arterial walls pulse whenever the left ventricle contracts. The pulse is usually 70 times per minute but can vary between 60 to 80 times per minute.

Blood Flow

The beating of the heart is necessary to homeostasis because it creates the pressure that propels blood in the arteries and the arterioles. Arterioles lead to the capillaries where exchange with tissue fluid takes place.

Blood Flow in Arteries

Blood pressure is the pressure of blood against the wall of a blood vessel. A sphygmomanometer is used to measure blood pressure usually in the brachial artery of the arm (Fig. 7.9). The highest arterial pressure, called the **systolic pressure,** is reached during ejection of blood from the heart. The lowest arterial pressure is called the **diastolic pressure.** Diastolic pressure occurs while the heart ventricles are relaxing. Normal resting blood pressure for a young adult is said to be 120 mm mercury (Hg) over 80 mm Hg, or simply 120/80. The higher number is the systolic pressure, and the lower number is the diastolic pressure. Actually blood pressure varies throughout the body. As already stated, blood pressure is highest in the aorta and lowest in the venae cavae. It is customary, however, to take the blood pressure in the brachial artery of the arm where it is usually 120/80.

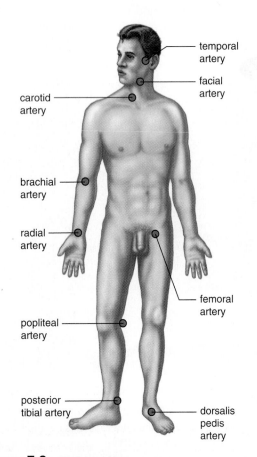

Figure 7.8   Pulse points.
The pulse can be taken at these arterial sites.

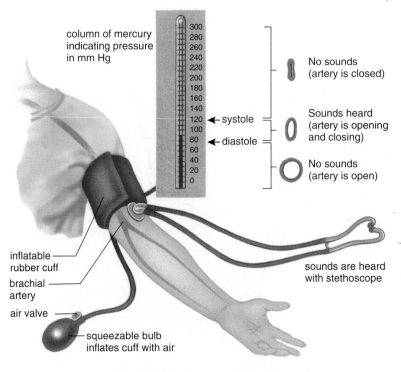

Figure 7.9   Use of a sphygmomanometer.
The technician inflates the cuff with air, gradually reduces the pressure, and listens with a stethoscope for the sounds that indicate blood is moving past the cuff in an artery. This is systolic blood pressure. The pressure in the cuff is further reduced until no sound is heard, indicating that blood is flowing freely through the artery. This is diastolic pressure.

Both systolic and diastolic blood pressure decrease with distance from the left ventricle because the total cross-sectional area of the blood vessels increases—there are more arterioles than arteries. The decrease in blood pressure causes the blood velocity to gradually decrease as it flows toward the capillaries.

Blood Flow in Capillaries

There are many more capillaries than arterioles, and blood moves slowly through the capillaries (Fig. 7.10). This is important because the slow progress allows time for the exchange of substances between the blood in the capillaries and the surrounding tissues.

Blood Flow in Veins

Blood pressure is minimal in venules and veins (20–0 mm Hg). Instead of blood pressure, venous return is dependent upon three factors: skeletal muscle contraction, the presence of valves in veins, and respiratory movements. When the skeletal muscles contract, they compress the weak walls of the veins. This causes blood to move past the next valve. Once past the valve, blood cannot flow backward (Fig. 7.11). The importance of muscle contraction in moving blood in the venous vessels can be demonstrated by forcing a person to stand rigidly still for an hour or so. Frequently, fainting occurs because blood collects in the limbs, depriving the brain of needed blood flow and oxygen. In this case, fainting is beneficial because the resulting horizontal position aids in getting blood to the head.

When inhalation occurs, the thoracic pressure falls and abdominal pressure rises as the chest expands. This also aids the flow of venous blood back to the heart because blood flows in the direction of reduced pressure. Blood velocity increases slightly in the venous vessels due to a progressive reduction in the cross-sectional area as small venules join to form veins.

Blood pressure accounts for the flow of blood in the arteries and the arterioles. Skeletal muscle contraction, valves in veins, and respiratory movements account for the flow of blood in the venules and the veins.

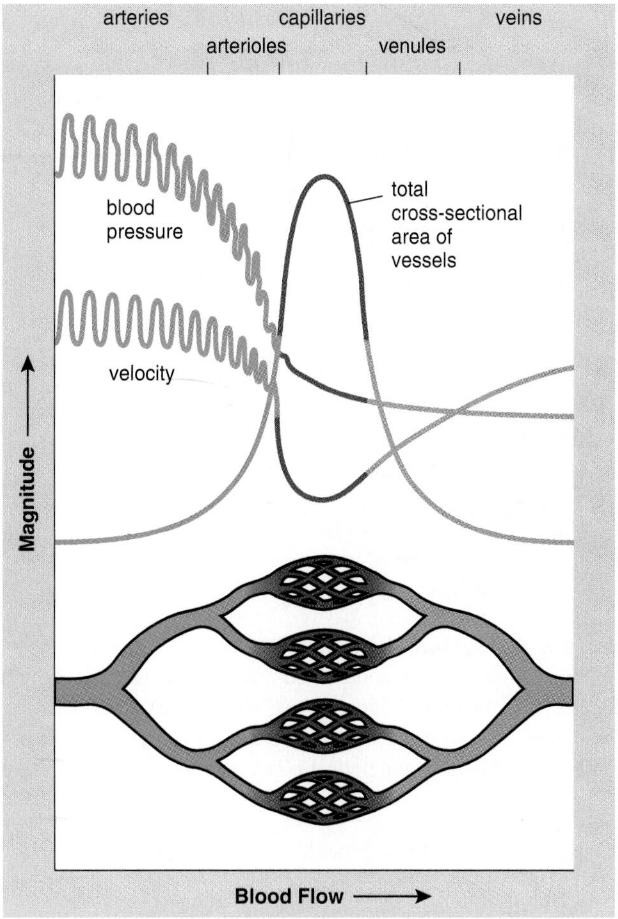

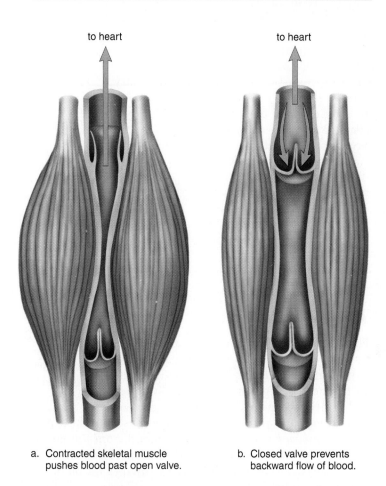

Figure 7.10 **Cross-sectional area as it relates to blood pressure and blood velocity.**
Blood pressure and blood velocity drop off in capillaries because capillaries have a greater cross-sectional area than arterioles.

a. Contracted skeletal muscle pushes blood past open valve.
b. Closed valve prevents backward flow of blood.

Figure 7.11 **Skeletal muscle contraction moves blood in veins.**
a. Muscle contraction exerts pressure against the vein, and blood moves past the valve. **b.** Blood cannot flow back once it has moved past the valve.

7.4 The Vascular Pathways

The cardiovascular system, which is represented in Figure 7.12, includes two circuits: the **pulmonary circuit,** which circulates blood through the lungs, and the **systemic circuit,** which serves the needs of body tissues. Both circuits, as we shall see, are necessary to homeostasis. ⚖

The Pulmonary Circuit

The path of blood through the lungs can be traced as follows. Blood from all regions of the body first collects in the right atrium and then passes into the right ventricle, which pumps it into the pulmonary trunk. The pulmonary trunk divides into the right and left pulmonary arteries, which branch as they approach the lungs. The arterioles take blood to the pulmonary capillaries, where carbon dioxide is given off and oxygen is picked up. Blood then passes through the pulmonary venules, which lead to the four pulmonary veins that enter the left atrium. Since blood in the pulmonary arteries is O_2-poor but blood in the pulmonary veins is O_2-rich, it is not correct to say that all arteries carry blood that is high in oxygen and all veins carry blood that is low in oxygen. It is just the reverse in the pulmonary circuit.

The pulmonary arteries take O_2-poor blood to the lungs, and the pulmonary veins return blood that is O_2-rich to the heart.

The Systemic Circuit

The systemic circuit includes all of the arteries and veins shown in Figure 7.13. The largest artery in the systemic circuit is the **aorta,** and the largest veins are the **superior** and **inferior venae cavae.** The superior vena cava collects blood from the head, the chest, and the arms, and the inferior vena cava collects blood from the lower body regions. Both enter the right atrium. The aorta and the venae cavae serve as the major pathways for blood in the systemic circuit.

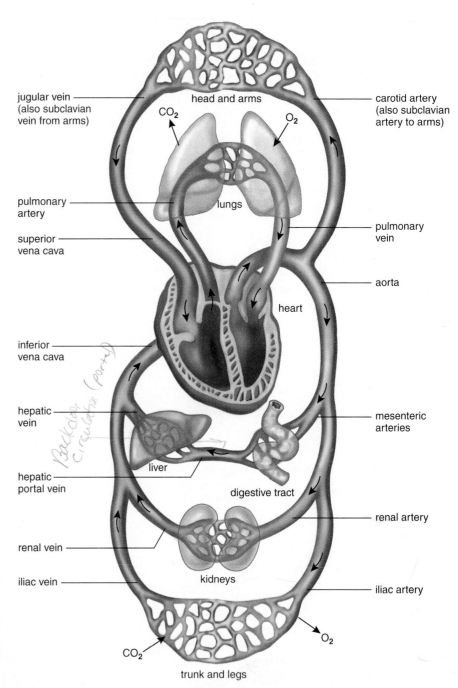

Figure 7.12 **Cardiovascular system diagram.**
The blue-colored vessels carry O_2-poor blood, and the red-colored vessels carry O_2-rich blood; the arrows indicate the flow of blood. Compare this diagram, useful for learning to trace the path of blood, to Figure 7.13 to realize that arteries and veins go to all parts of the body. Also, there are capillaries in all parts of the body. No cell is located far from a capillary.

The path of systemic blood to any organ in the body begins in the left ventricle, which pumps blood into the aorta. Branches from the aorta go to the organs and major body regions. For example, this is the path of blood to and from the lower legs:

left ventricle—aorta—common iliac artery—femoral artery—lower leg capillaries—femoral vein—common iliac vein—inferior vena cava—right atrium

Notice that when tracing blood, you need only mention the aorta, the proper branch of the aorta, the region, and the vein returning blood to the vena cava. In most instances, the artery and the vein that serve the same region are given the same name (Fig. 7.13). What happens when the blood reaches a particular region? Capillary exchange takes place, refreshing tissue fluid so that its composition stays relatively constant.

The **coronary arteries** (see Fig. 7.4) serve the heart muscle itself. (The heart is not nourished by the blood in its chambers.) The coronary arteries are the first branches off the aorta. They originate just above the aortic semilunar valve, and they lie on the exterior surface of the heart, where they divide into diverse arterioles. Because they have a very small diameter, the coronary arteries may become clogged, as discussed on page 137. The coronary capillary beds join to form venules. The venules converge to form the cardiac veins, which empty into the right atrium.

The body has a portal system called the **hepatic portal system,** which is associated with the liver. A portal system begins and ends in capillaries; in this instance, the first set of capillaries occurs at the villi of the small intestine and the second occurs in the liver. Blood passes from the capillaries of the intestinal villi into venules that join to form the **hepatic portal vein.** The hepatic portal vein connects the villi of the intestine with the liver, an organ that monitors the makeup of the blood. The **hepatic vein** leaves the liver and enters the inferior vena cava. While Figure 7.12 is helpful in tracing the path of blood, remember that all parts of the body receive both arteries and veins, as illustrated in Figure 7.13.

The systemic circuit takes blood from the left ventricle of the heart through the body proper, and back to the right atrium of the heart.

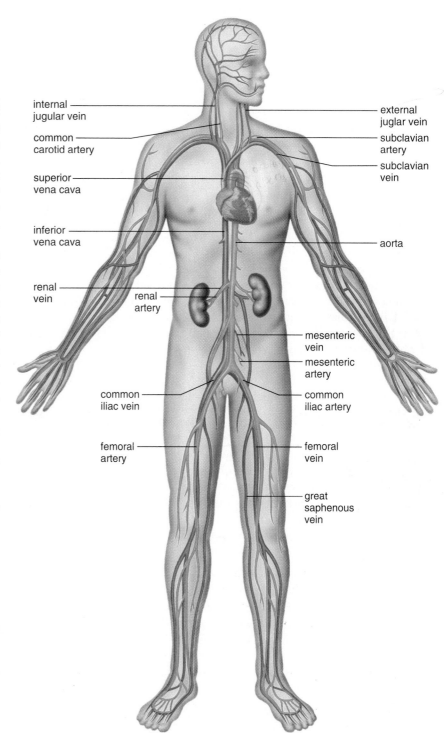

Figure 7.13   **Major arteries and veins of the systemic circuit.**
A realistic representation of major blood vessels of the systemic circuit shows how the systemic arteries and veins are actually arranged in the body. The superior and inferior venae cavae take their names from their relationship to which organ?

Health Focus

Prevention of Cardiovascular Disease

There are genetic factors that predispose an individual to cardiovascular disease, such as family history of heart attack under age 55, male gender, and ethnicity (African Americans are at greater risk). Those with one or more of these risk factors need not despair, however. It only means that they should pay particular attention to these guidelines for a heart-healthy lifestyle.

The Don'ts

Smoking

When a person smokes, the drug nicotine, present in cigarette smoke, enters the bloodstream. Nicotine causes the arterioles to constrict and the blood pressure to rise. Restricted blood flow and cold hands are associated with smoking by most people. More serious is the need for the heart to pump harder to propel the blood through the lungs at a time when the oxygen-carrying capacity of the blood is reduced.

Drug Abuse

Stimulants, such as cocaine and amphetamines, can cause an irregular heartbeat and lead to heart attacks in people who are using drugs even for the first time. Intravenous drug use may also result in a cerebral blood clot and stroke.

Too much alcohol can destroy just about every organ in the body, the heart included. But investigators have discovered that people who take an occasional drink have a 20% lower risk of heart disease than do teetotalers. Two to four drinks a week is the recommended limit for men, one to three drinks for women.

Weight Gain

Hypertension (high blood pressure) is prevalent in persons who are more than 20% above the recommended weight for their height. More tissues require servicing, and the heart sends the extra blood out under greater pressure in those who are overweight. It may be harder to lose weight once it is gained, and therefore it is recommended that weight control be a lifelong endeavor. Even a slight decrease in weight can bring with it a reduction in hypertension.

The Do's

Healthy Diet

A diet low in saturated fats and cholesterol is protective against cardiovascular disease. Cholesterol is ferried in the blood by two types of plasma proteins, called LDL (low-density lipoprotein) and HDL (high-density lipoprotein). LDL (called "bad" lipoprotein) takes cholesterol from the liver to the tissues, and HDL (called "good" lipoprotein) transports cholesterol out of the tissues to the liver. When the LDL level in blood is abnormally high or the HDL level is abnormally low, cholesterol accumulates in the cells. When cholesterol-laden cells line the arteries, plaque develops, which interferes with circulation (Fig. 7A).

It is recommended that everyone know his or her blood cholesterol level. Individuals with a high blood cholesterol level (200 mg/100 ml) should be further tested to determine their LDL-cholesterol level. The LDL-cholesterol level, together with other risk factors such as age, family history, general health, and whether the patient smokes, will determine who needs dietary therapy to lower their LDL. Eating foods high in saturated fat (red meat, cream, and butter) and foods containing so-called trans-fats (margarine, commercially baked goods, and deep-fried foods) raises the LDL-cholesterol level. Replacement of these harmful fats with healthier ones like monounsaturated fats (olive and canola oil) and polyunsaturated ones (corn, safflower, and soybean oil) is recommended.

Evidence is mounting to suggest a role for antioxidant vitamins (A, E, and C) in the prevention of cardiovascular disease. Antioxidants protect the body from free radicals that may damage HDL cholesterol through oxidation or damage the lining of an artery, leading to a blood clot that can block the vessel. Nutritionists believe that the consumption of at least five servings of fruits and vegetables a day may be protective against cardiovascular disease.

Exercise

Those who exercise are less apt to have cardiovascular disease. One study found that moderately active men who spent an average of 48 minutes a day on a leisure-time activity such as gardening, bowling, or dancing had one-third fewer heart attacks than peers who spent an average of only 16 minutes each day. Exercise which helps keep weight under control, may help minimize stress, and reduce hypertension.

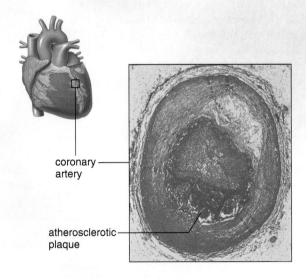

coronary artery

atherosclerotic plaque

Figure 7A Coronary arteries and plaque.
When plaque is present in a coronary artery, a heart attack is more apt to occur because of restricted blood flow.

7.5 Cardiovascular Disorders

Cardiovascular disease (CVD) is the leading cause of untimely death in the Western countries. Modern research efforts have resulted in improved diagnosis, treatment, and prevention. This section discusses the range of advances that have been made in these areas. The Health Focus on page 136 emphasizes how to prevent CVD from developing in the first place.

Hypertension

It is estimated that about 20% of all Americans suffer from **hypertension,** which is high blood pressure. Hypertension is present when the systolic blood pressure is 140 or greater or the diastolic blood pressure is 90 or greater. While both systolic and diastolic pressures are considered important, it is the diastolic pressure that is emphasized when medical treatment is being considered.

Hypertension is sometimes called a silent killer because it may not be detected until a stroke or heart attack occurs. It has long been thought that a certain genetic makeup might account for the development of hypertension. Now researchers have discovered two genes that may be involved in some individuals. One gene codes for angiotensinogen, a plasma protein that is converted to a powerful vasoconstrictor in part by the product of the second gene. Persons with hypertension due to overactivity of these genes might one day be cured by gene therapy.

At present, however, the best safeguard against the development of hypertension is to have regular blood pressure checks and to adopt a lifestyle that lowers the risk of hypertension.

Atherosclerosis

Hypertension also is seen in individuals who have **atherosclerosis,** an accumulation of soft masses of fatty materials, particularly cholesterol, beneath the inner linings of arteries. Such deposits are called **plaque.** As it develops, plaque tends to protrude into the lumen of the vessel and interfere with the flow of blood (see Fig. 7A). In certain families, atherosclerosis is due to an inherited condition. The presence of the associated mutation can be detected, and this information is helpful if measures are taken to prevent the occurrence of the disease. In most instances, atherosclerosis begins in early adulthood and develops progressively through middle age, but symptoms may not appear until an individual is 50 or older. To prevent the onset and development of plaque, the American Heart Association and other organizations recommend a diet low in saturated fat and cholesterol and rich in fruits and vegetables.

Plaque can cause a clot to form on the irregular arterial wall. As long as the clot remains stationary, it is called a **thrombus,** but when and if it dislodges and moves along with the blood, it is called an **embolus.** If **thromboembolism** (a clot that has been carried in the bloodstream but is now stationary) is not treated, complications can arise, as mentioned in the following section.

Stroke, Heart Attack, and Aneurysm

Stroke, heart attack, and aneurysm are associated with hypertension and atherosclerosis. A cerebrovascular accident (CVA), also called a **stroke,** often results when a small cranial arteriole bursts or is blocked by an embolus. A lack of oxygen causes a portion of the brain to die, and paralysis or death can result. A person is sometimes forewarned of a stroke by a feeling of numbness in the hands or the face, difficulty in speaking, or temporary blindness in one eye.

A myocardial infarction (MI), also called a **heart attack,** occurs when a portion of the heart muscle dies due to a lack of oxygen. If a coronary artery becomes partially blocked, the individual may then suffer from **angina pectoris,** characterized by a radiating pain in the left arm. Nitroglycerin or related drugs dilate blood vessels and help relieve the pain. When a coronary artery is completely blocked, perhaps because of thromboembolism, a heart attack occurs.

An **aneurysm** is a ballooning of a blood vessel, most often the abdominal artery or the arteries leading to the brain. Atherosclerosis and hypertension can weaken the wall of an artery to the point that an aneurysm develops. If a major vessel like the aorta should burst, death is likely. It is possible to replace a damaged or diseased portion of a vessel, such as an artery, with a plastic tube. Cardiovascular function is preserved, because exchange with tissue cells can still take place at the capillaries.

Dissolving Blood Clots

Medical treatment for thromboembolism includes the use of t-PA, a biotechnology drug. This drug converts plasminogen, a molecule found in blood, into plasmin, an enzyme that dissolves blood clots. In fact, t-PA, which stands for tissue plasminogen activator, is the body's own way of converting plasminogen to plasmin. t-PA is also being used for thrombolytic stroke patients but with limited success because some patients experience life-threatening bleeding in the brain. A better treatment might be new biotechnology drugs that act on the plasma membrane to prevent brain cells from releasing and/or receiving toxic chemicals caused by the stroke.

If a person has symptoms of angina or a stroke, aspirin may be prescribed. Aspirin reduces the stickiness of platelets and thereby lowers the probability that a clot will form. There is evidence that aspirin protects against first heart attacks, but there is no clear support for taking aspirin every day to prevent strokes in symptom-free people. Physicians warn that long-term use of aspirin might have harmful effects, including bleeding in the brain.

Coronary Bypass Operations

Each year thousands of persons have coronary bypass surgery because of an obstructed coronary artery. During this operation, a surgeon takes a segment from another blood vessel and stitches one end to the aorta and the other end to a coronary artery past the point of obstruction. Figure 7.14 shows a triple bypass in which three blood vessels have been used to allow blood to flow freely from the aorta to cardiac muscle by way of the coronary artery.

Gene therapy is now being used experimentally to grow new blood vessels that will carry blood to cardiac muscle. The surgeon need only make a small incision and inject many copies of the gene that codes for VEGF (vascular endothelial growth factor) between the ribs directly into the area of the heart that most needs improved blood flow. VEGF encourages new blood vessels to sprout out of an artery. If collateral blood vessels do form, they transport blood past clogged arteries, making bypass surgery unnecessary. About 60% of all patients who undergo the procedure do show signs of vessel growth within two to four weeks.

Another way to achieve the growth of new blood vessels is to drill tiny holes directly into the beating heart muscle with a laser. The holes close up immediately with the help of pressure from the surgeon's fingers. But the channels created inside the muscle remain open for a time, and apparently new blood vessels are formed.

Clearing Clogged Arteries

In **angioplasty,** a cardiologist threads a plastic tube into an artery of an arm or a leg and guides it through a major blood vessel toward the heart. When the tube reaches the region of plaque in an artery, a balloon attached to the end of the tube is inflated, forcing the vessel open (Fig. 7.15). However, the artery may not remain open because the trauma causes smooth muscle cells in the wall of the artery to proliferate and close it.

Two lines of attack are being explored. Small metal devices—either metal coils or slotted tubes, called stents—are expanded inside the artery to keep the artery open. When the stents are coated with heparin to prevent blood clotting and with chemicals to prevent arterial closing, results have been promising.

Heart Transplants and Other Treatments

Persons with weakened hearts eventually may suffer from **congestive heart failure,** meaning the heart no longer is able to pump blood adequately, and blood backs up in the heart and lungs. Sometimes it is possible to repair a weak heart. For example, a back muscle can be wrapped around a heart to strengthen it. The muscle's nerve is stimulated with a kind of pacemaker that gives a burst of stimulation every 0.85 second. One day it may be possible to use cardiac cell transplants, because researchers who have injected live cardiac muscle cells into animal hearts find that they contribute to the pumping of the heart.

On December 2, 1982, Barney Clark became the first person to receive an artificial heart that was driven by bursts of air received from a large external machine. Some clinically used artificial hearts today are driven by small battery-powered systems that can be carried about by a shoulder strap. The National Institutes of Health support the development of a hot-air engine that is fully implantable and is driven by an atomic heat source. The artificial heart is presently used only while the patient is waiting for a heart transplant. The difficulties with a heart transplant are, first, availability, and second, the tendency of the body to reject foreign organs. Recently, through genetic engineering, researchers have altered the immune system of a strain of pigs with the hope that they will soon be a source of donor hearts.

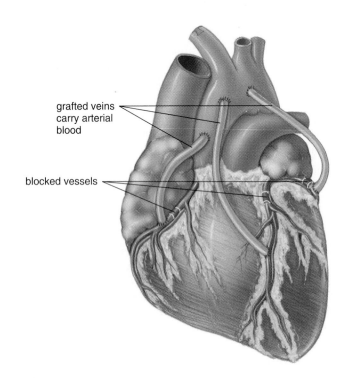

grafted veins carry arterial blood

blocked vessels

Figure 7.14 **Coronary bypass operation.**
During this operation, the surgeon grafts segments of another vessel, usually a small vein from the leg, between the aorta and the coronary vessels, bypassing areas of blockage. Patients who require surgery often receive two to five bypasses in a single operation.

Stroke, heart attack, and aneurysm are associated with both hypertension and atherosclerosis. Various treatments are available, including both medical and surgical procedures. In the end, a heart transplant may be the only recourse for an ailing heart.

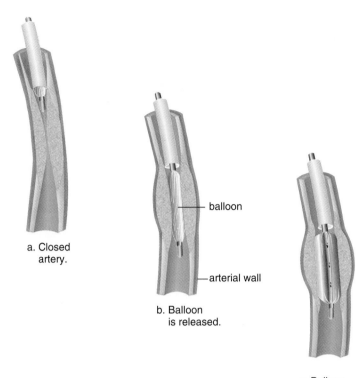

a. Closed artery.

balloon

arterial wall

b. Balloon is released.

c. Balloon is inflated.

Figure 7.15 Angioplasty.
During this procedure (**a**) a plastic tube is inserted into the coronary artery until it reaches the clogged area. **b.** A metal tip with balloon attached is pushed out the end of the plastic tube into the clogged area. **c.** When the balloon is inflated, the vessel opens. Sometimes metal coils or slotted tubes, called stents, are inserted to keep the vessel open.

Dilated and Inflamed Veins

Varicose veins develop when the valves of veins become weak and ineffective due to the backward pressure of blood. Abnormal and irregular dilations are particularly apparent in the superficial (near the surface) veins of the lower legs. Crossing the legs or sitting in a chair so that its edge presses against the back of the knees can contribute to the development of varicose veins. Varicose veins also occur in the rectum, where they are called piles, or more properly, **hemorrhoids.**

Phlebitis, or inflammation of a vein, is a more serious condition, particularly when a deep vein is involved. Blood in an unbroken but inflamed vessel may clot, and the clot may be carried in the bloodstream until it lodges in a small vessel. If a blood clot blocks a pulmonary vessel, death can result.

7.6 Homeostasis

Homeostasis is possible only if the cardiovascular system delivers oxygen and nutrients to and takes away metabolic wastes from tissue fluid that surrounds cells. The next page tells how the cardiovascular system works with the other systems of the body to maintain homeostasis.

The composition of the blood is maintained by the other systems of the body. The digestive system absorbs nutrients into the blood, and the lungs and kidneys dispose of metabolic wastes. The liver, of course, is a key regulator of blood components by producing plasma proteins, storing glucose until it is needed, transforming ammonia into urea, and changing other poisons into molecules that are also excreted.

The pumping of the heart is critical to creating the blood pressure that moves blood to the lungs and to the tissues. Sensory receptors within the aorta signal a regulatory center in the brain when blood pressure falls. This center subsequently increases heartbeat and constricts arterioles. Thereafter, blood pressure is restored. The lymphatic system collects excess tissue fluid at blood capillaries and returns it to cardiovascular veins in the thoracic cavity. In this way, the lymphatic system makes an important contribution to regulating blood volume and pressure.

The endocrine system assists the nervous system in maintaining homeostasis, so it is not surprising that hormones are also involved in regulating blood pressure. Epinephrine and norephinephrine bring about the constriction of arterioles. Other hormones not yet discussed regulate urine excretion. After all, if water is retained, blood volume and pressure will rise, and if water is excreted, blood volume and pressure will drop. In fact, some drugs prescribed for hypertension increase the amount of urine excreted.

As discussed in chapter 6, hormones regulate the manufacture of formed elements in the red bone marrow, which is a lymphoid organ. Red blood cells carry oxygen, and white blood cells fight infection. Homeostasis would not be possible without our ability to kill off pathogens.

While blood pressure accounts for the arterial flow of blood from the heart to the capillaries, venous return from the capillaries to the heart is dependent on two other systems of the body. Skeletal muscle contraction pushes blood past the valves in the veins, and breathing movements encourage the flow of blood toward the heart in the thoracic cavity.

The other systems of the body assist the cardiovascular system in carrying out its functions so that homeostasis is maintained.

Human Systems Work Together

Integumentary System

Blood vessels deliver nutrients and oxygen to skin, carry away wastes; blood clots if skin is broken.

Skin prevents water loss; helps regulate body temperature; protects blood vessels.

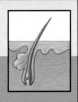

How the Cardiovascular System works with other body systems

Lymphatic System/Immunity

Blood vessels transport leukocytes and antibodies; blood services lymphoid organs and is source of tissue fluid that becomes lymph.

Lymphoid organs produce and store formed elements; lymphatic vessels transport leukocytes and return tissue fluid to blood vessels; spleen serves as blood reservoir, filters blood.

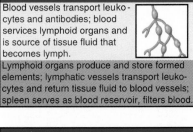

Skeletal System

Blood vessels deliver nutrients and oxygen to bones; carry away wastes.

Rib cage protects heart; red bone marrow produces blood cells; bones store Ca^{2+} for blood clotting.

Respiratory System

Blood vessels transport gases to and from lungs; blood services respiratory organs.

Gas exchange in lungs rids body of carbon dioxide, helping to regulate the pH of blood; breathing aids venous return.

Muscular System

Blood vessels deliver nutrients and oxygen to muscles; carry away wastes.

Muscle contraction keeps blood moving in heart and blood vessels.

Digestive System

Blood vessels transport nutrients from digestive tract to body; blood services digestive organs.

Digestive tract provides nutrients for plasma protein formation and blood cell formation; liver detoxifies blood, makes plasma proteins, destroys old red blood cells.

Nervous System

Blood vessels deliver nutrients and oxygen to neurons; carry away wastes.

Brain controls nerves that regulate the heart and dilation of blood vessels.

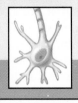

Urinary System

Blood vessels deliver wastes to be excreted; blood pressure aids kidney function; blood services urinary organs.

Kidneys filter blood and excrete wastes; maintain blood volume, pressure, and pH; produce renin and erythropoietin.

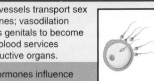

Endocrine System

Blood vessels transport hormones from glands; blood services glands; heart produces atrial natriuretic hormone.

Epinephrine increases blood pressure; ADH, aldosterone, and atrial natriuretic hormone factors help regulate blood volume; growth factors control blood cell formation.

Reproductive System

Blood vessels transport sex hormones; vasodilation causes genitals to become erect; blood services reproductive organs.

Sex hormones influence cardiovascular health; sexual activities stimulate cardiovascular system.

A Healthy Lifestyle

According to a 1993 study, about one million deaths a year in the United States could be prevented if people adopted the healthy lifestyle described in the Health Focus on page 136. Tobacco, lack of exercise, and a high-fat diet probably cost the nation about $200 billion per year in health-care costs. To what lengths should we go to prevent these deaths and reduce health-care costs?

E.A. Miller, a meat-packing entity of ConAgra in Hyrum, Utah, charges extra for medical coverage of employees who smoke. Eric Falk, Miller's director of human resources, says, "We want to teach employees to be responsible for their behavior." Anthem Blue Cross–Blue Shield of Cincinnati, Ohio, takes a more positive approach. They give insurance plan participants $240 a year in extra benefits, like additional vacation days, if they get good scores in five out of seven health-related categories. The University of Alabama, Birmingham, School of Nursing has a health-and-wellness program that councils employees about how to get into shape in order to keep their insurance coverage. Audrey Brantley, a participant in the program, has mixed feelings. She says, "It seems like they are trying to control us, but then, on the other hand, I know of folks who found out they had high blood pressure or were borderline diabetics and didn't know it."

Another question is, Does it really work? Turner Broadcasting System in Atlanta has a policy that affects all employees hired after 1986. They will be fired if caught smoking—whether at work or at home—but some admit they still manage to sneak a smoke.

Decide Your Opinion

1. Do you think employers who pay for their employees' health insurance have the right to demand, or encourage, or support a healthy lifestyle?
2. Do you think all participants in a health insurance program should qualify for the same benefits, regardless of their lifestyle?
3. What steps are ethical to encourage people to adopt a healthy lifestyle?

Figure 7B   Eating habits.

Does an employer who provides you with health insurance have the right to require you to have a healthy lifestyle?

Looking at Both Sides   www.mhhe.com/biosci/genbio/maderhuman7/

Every bioethical issue has at least two sides. Even if you already have an opinion, it is important to explore the opposite opinion before finalizing your position. The Online Learning Center at www.mhhe.com/biosci/genbio/maderhuman7/ will help you fine-tune your initial opinion, explore both sides, and finalize your position. You may acquire new arguments for your original opinion, or you may even change your opinion. Be sure to complete these activities in sequence:

Taking Sides   Decide your initial opinion by answering a series of questions. Then see if your opinion changes after completing the next two activities.

Further Debate   Read opposing articles that give you further information on this particular bioethical issue.

Explain Your Position   Answer another series of questions and then defend your original or changed opinion. You can e-mail your position to your instructor if he or she wishes.

Summarizing the Concepts

7.1 The Blood Vessels
Blood vessels include arteries (and arterioles) that take blood away from the heart; capillaries, where exchange of substances with the tissues occurs; and veins (and venules) that take blood to the heart.

7.2 The Heart
The heart has a right and left side and four chambers. On the right side, an atrium receives O_2-poor blood from the body, and a ventricle pumps it into the pulmonary circuit. On the left side, an atrium receives O_2-rich blood from the lungs, and a ventricle pumps it into the systemic circuit. During the cardiac cycle, the SA node (pacemaker) initiates the heartbeat by causing the atria to contract. The AV node conveys the stimulus to the ventricles, causing them to contract. The heart sounds, "lub-dup," are due to the closing of the atrioventricular valves, followed by the closing of the semilunar valves.

7.3 Features of the Cardiovascular System
The pulse rate indicates the heartbeat rate. Blood pressure caused by the beating of the heart accounts for the flow of blood in the arteries, but because blood pressure drops off after the capillaries, it cannot cause blood flow in the veins. Skeletal muscle contraction, the presence of valves, and respiratory movements account for blood flow in veins. The reduced velocity of blood flow in capillaries facilitates exchange of nutrients and wastes.

7.4 The Vascular Pathways
The cardiovascular system is divided into the pulmonary circuit and the systemic circuit. In the pulmonary circuit, the pulmonary trunk from the right ventricle and the two pulmonary arteries take O_2-poor blood to the lungs, and four pulmonary veins return O_2-rich blood to the left atrium. To trace the path of blood in the systemic circuit, start with the aorta from the left ventricle. Follow its path until it branches to an artery going to a specific organ. It can be assumed that the artery divides into arterioles and capillaries, and that the capillaries lead to venules. The vein that takes blood to the vena cava most likely has the same name as the artery that delivered blood to the organ. In the adult systemic circuit, unlike the pulmonary circuit, the arteries carry O_2-rich blood, and the veins carry O_2-poor blood.

7.5 Cardiovascular Disorders
Hypertension and atherosclerosis are two cardiovascular disorders that lead to stroke, heart attack, and aneurysm. Medical and surgical procedures are available to control cardiovascular disease, but the best policy is prevention by following a heart-healthy diet, getting regular exercise, maintaining a proper weight, and not smoking.

7.6 Homeostasis
Homeostasis is absolutely dependent upon the cardiovascular system because it serves the needs of the cells. However, several other body systems are critical to the functioning of the cardiovascular system. The digestive system supplies nutrients, and the respiratory system supplies oxygen and removes carbon dioxide from the blood. Like the heart, the nervous and endocrine systems are involved in maintaining the blood pressure that moves blood in the arteries and arterioles. The lymphatic system returns tissue fluid to the veins where blood is propelled by skeletal muscle contraction and breathing movements.

Studying the Concepts

1. What types of blood vessels are there? Discuss their structure and function.   126
2. Trace the path of blood through the heart, mentioning the vessels attached to, and the valves within, the heart.   129
3. Describe the cardiac cycle (using the terms systole and diastole), and explain the heart sounds.   130
4. Describe the cardiac conduction system and an ECG. Tell how an ECG is related to the cardiac cycle.   130–31
5. In what type of vessel is blood pressure highest? Lowest? Why is the slow movement of blood in capillaries beneficial?   132–33
6. What factors assist venous return of blood?   133
7. Trace the path of blood in the pulmonary circuit as it travels from and returns to the heart.   134
8. Trace the path of blood to and from the kidneys in the systemic circuit.   134–35
9. What is atherosclerosis? Name two illnesses associated with hypertension and thromboembolism.   137–38
10. Discuss the medical and surgical treatment of cardiovascular disease.   138–39
11. How does the circulatory system help maintain homeostasis?   139–40
12. How does the liver assist the circulatory system in maintaining homeostasis?   139–40

Testing Your Knowledge of the Concepts

In questions 1–4, match the circuit to the descriptions below.
a. pulmonary circuit
b. systemic circuit
c. both

_____ 1. Arteries carry O_2-rich blood.

_____ 2. Carbon dioxide leaves the capillaries, and oxygen enters the capillaries.

_____ 3. Arteries carry blood away from the heart, and veins carry blood toward the heart.

_____ 4. Contains the renal arteries and veins.

In questions 5–7, indicate whether the statement is true (T) or false (F).

_____ 5. SA node impulses and the atrioventricular bundle cause the atria to contract.

_____ 6. Venous return is dependent in part on skeletal muscle contraction.

_____ 7. The total cross-sectional area of arteries, arterioles, and capillaries affects the pressure and velocity of blood in these vessels.

In questions 8 and 9, fill in the blanks.

8. Reducing the amount of _____ and _____ in the diet reduces the chance of a heart attack.

9. A _____ occurs when a small cranial arteriole bursts or is blocked by an embolus.

10. Match the key terms to these definitions:
   a. _____ Relaxation of a heart chamber.
   b. _____ Large systemic vein that returns blood to the right atrium of the heart.
   c. _____ Obstruction of a blood vessel by a thrombus that moved from its site of formation.
   d. _____ The SA node because it initiates the heartbeat.
   e. _____ The circuit that serves body parts and does not include the gas-exchanging surfaces in the lungs.

11. Label this diagram of the heart.

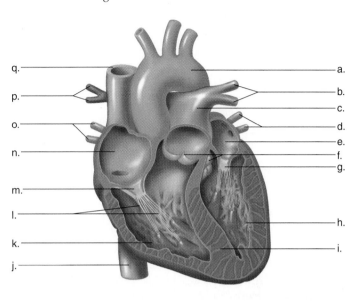

Understanding Key Terms

aneurysm 137
angina pectoris 137
angioplasty 138
aorta 134
arteriole 127
artery 126
atherosclerosis 137
atrioventricular bundle 130
atrioventricular valve 128
atrium 128
AV (atrioventricular) node 130
blood pressure 132
capillary 126
cardiac cycle 130
chordae tendineae 128
congestive heart failure 138
coronary artery 135
diastole 130
diastolic pressure 132
electrocardiogram (ECG) 131
embolus 137
heart 128
heart attack 137
hemorrhoids 139
hepatic portal system 135
hepatic portal vein 135

hepatic vein 135
hypertension 137
inferior vena cava 134
myocardium 128
pacemaker 130
pericardium 128
phlebitis 139
plaque 137
pulmonary artery 129
pulmonary circuit 134
pulmonary vein 129
pulse 132
Purkinje fibers 130
SA (sinoatrial) node 130
semilunar valve 128
stroke 137
superior vena cava 134
systemic circuit 134
systole 130
systolic pressure 132
thromboembolism 137
thrombus 137
valve 127
varicose veins 139
vein 126
ventricle 128
venule 127

e-Learning Connection   www.mmhe.com/biosci/genbio/maderhuman7/

7.1 Blood Vessels

Blood Vessels I *art labeling activity*
Blood Vessels II *art labeling activity*
Anatomy of a Capillary Bed *art labeling activity*

7.2 The Heart

Cardiac Cycle Blood Flow—Normal Speed *animation activity*
Cardiac Cycle Blood Flow—Slow Speed *animation activity*
Cardiac Cycle—Electrical *animation activity*
Cardiac Cycle—Muscular *animation activity*
External Heart Anatomy I *art labeling activity*
External Heart Anatomy II *art labeling activity*
External Heart Anatomy III *art labeling activity*
Internal View of the Heart I *art labeling activity*
Internal View of the Heart II *art labeling activity*
Internal View of the Heart III *art labeling activity*
Internal View of the Heart IV *art labeling activity*

Heart Blood Flow *Essential Study Partner*
Heart Sounds *Essential Study Partner*

7.3 Features of the Cardiovascular System

Vessels and Pressure *Essential Study Partner*

7.4 The Vascular Pathway

Portal System *animation activity*
Human Circulatory System I *art labeling activity*
Human Circulatory System II *art labeling activity*

7.5 Cardiovascular Disorders

Myocardial Infarction *animation activity*
Plaque *art labeling activity*

Heart Transplant *bioethics case study*

7.6 Homeostasis

Working Together to Achieve Homeostasis

Chapter Summary

Key Term Flashcards *vocabulary quiz*
Chapter Quiz *objective quiz covering all chapter concepts*

Chapter 8

Lymphatic and Immune Systems

Chapter Concepts

Figure 8.1   Allergies.
The immune system protects us from disease but has unwanted side effects such as reactions to environmental substances that will do no harm to the body.

Every spring, Paula C. knows what's coming. The flowers bloom. The wind blows. And she sniffles. Itching, sneezing, and rubbing her watery eyes with a tissue, Paula endures the onslaught of allergy season.

Ironically, Paula's agony stems from her immune system's admirable ability to fight off pathogens (bacteria and viruses). Allergies occur when the immune system responds to substances that most people are insensitive to. When Paula's immune system detects that these substances are foreign to her, it mounts an attack that results in the release of chemicals, like histamines. These chemicals lead to the allergy symptoms she endures (Fig. 8.1). For relief, allergy sufferers turn to antihistamines, nonprescription pills that block histamine production. And so Paula swallows the medicine, grabs a tissue, and waits—fitfully—for cold weather.

The immune system reacts to antigens, any substance, usually protein or carbohydrate, that the system is capable of recognizing as being "nonself." Get a bacterial infection, and certain lymphocytes start producing antibodies. An antibody combines with the antigen, and later the complex is phagocytized by a macrophage. In Paula's case, however, the antibodies also attach to mast cells in the tissues and they release the offending chemicals. Get the flu and other lymphocytes gear up to kill any cell that is infected with the virus. Whereas you can take an antibiotic to help cure a bacterial infection, there is nothing to do for a viral infection but wait for the immune system to win the battle.

Immunity also involves some defenses that occur immediately as a response to any type of pathogen that enters the body. Cut yourself and the skin becomes inflamed as white blood cells rush to the scene and begin their attack. Then, too, certain plasma proteins latch onto bacterial surfaces, marking them for destruction. These proteins belong to the "complement system," so called because it complements certain immune responses.

So despite the fact that Paula's immune system causes her to have allergies, it otherwise does a magnificent job of keeping her well. Unfortunately, pesticides have been found to suppress the immune system, as discussed in the Ecology Focus on page 153.

8.1 Lymphatic System

The **lymphatic system** consists of lymphatic vessels and the lymphoid organs. This system, which is closely associated with the cardiovascular system, has three main functions that contribute to homeostasis: (1) lymphatic capillaries take up excess tissue fluid and return it to the bloodstream; (2) lacteals receive lipoproteins[1] at the intestinal villi and transport them to the bloodstream (see Fig. 5.6); and (3) the lymphatic system helps defend the body against disease.

Lymphatic Vessels

Lymphatic vessels are quite extensive; most regions of the body are richly supplied with lymphatic capillaries (Fig. 8.2). The construction of the larger lymphatic vessels is similar to that of cardiovascular veins, including the presence of valves. Also, the movement of lymph within these vessels is dependent upon skeletal muscle contraction. When the muscles contract, the lymph is squeezed past a valve that closes, preventing the lymph from flowing backwards.

The lymphatic system is a one-way system that begins with lymphatic capillaries. These capillaries take up fluid that has diffused from and has not been reabsorbed by the blood capillaries. **Edema** is localized swelling caused by the accumulation of tissue fluid. This can happen if too much tissue fluid is made and/or not enough of it is drained away. Once tissue fluid enters the lymphatic vessels, it is called **lymph**. The lymphatic capillaries join to form lymphatic vessels that merge before entering one of two ducts: the thoracic duct or the right lymphatic duct. The thoracic duct is much larger than the right lymphatic duct. It serves the lower extremities, the abdomen, the left arm, and the left side of both the head and the neck. The right lymphatic duct serves the right arm, the right side of both the head and the neck, and the right thoracic area. The lymphatic ducts enter the subclavian veins, which are cardiovascular veins in the thoracic region.

Lymph flows one way from a capillary to ever-larger lymphatic vessels and finally to a lymphatic duct, which enters a subclavian vein.

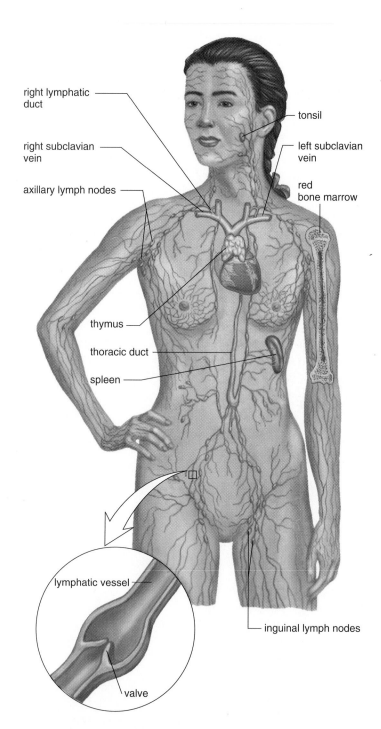

Figure 8.2 Lymphatic system.
Lymphatic vessels drain excess fluid from the tissues and return it to the cardiovascular system. The enlargement shows that lymphatic vessels, like cardiovascular veins, have valves to prevent backward flow. The tonsils, spleen, thymus gland, and red bone marrow are among those lymphoid organs that assist immunity.

[1]After glycerol and fatty acids are absorbed, they are rejoined and packaged as lipoprotein droplets which enter the lacteals.

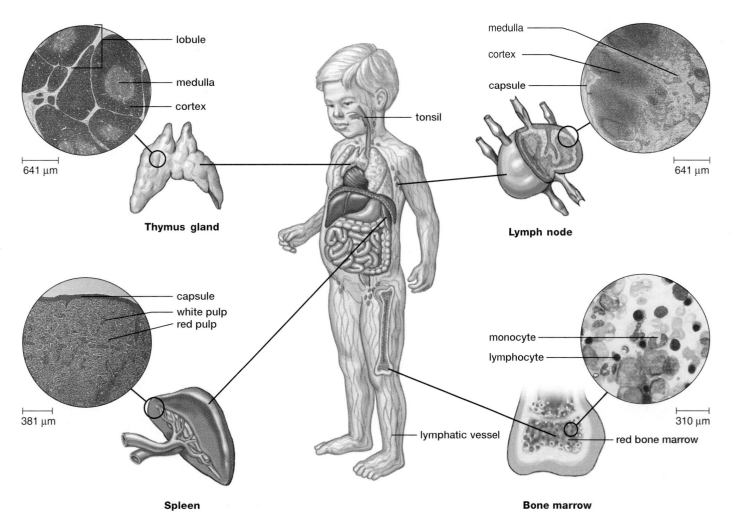

Figure 8.3   The lymphoid organs.
The lymphoid organs include the lymph nodes, tonsils (not shown), the spleen, the thymus gland, and the red bone marrow, which all contain lymphocytes.

Lymphoid Organs

The **lymphoid organs** of special interest are the lymph nodes, the tonsils, the spleen, the thymus gland, and the red bone marrow (Fig. 8.3).

Lymph nodes, which are small (about 1–25 mm in diameter) ovoid or round structures, are found at certain points along lymphatic vessels. A lymph node is composed of a capsule surrounding two distinct regions known as the cortex and medulla, which contain many lymphocytes. The cortex contains nodules where lymphocytes congregate when they are fighting off a pathogen. Macrophages, concentrated in the medulla, work to cleanse the lymph. Lymph nodes are named for their location. Inguinal nodes are in the groin, and axillary nodes are in the armpits. Physicians often feel for the presence of swollen, tender lymph nodes in the neck as evidence that the body is fighting an infection. This is a noninvasive, preliminary way to help make such a diagnosis.

The **tonsils** are patches of lymphatic tissue located in a ring about the pharynx (see Fig. 8.2). The well-known pharyngeal tonsils are also called adenoids, while the larger palatine tonsils located on either side of the posterior oral cavity are more apt to be infected. The tonsils perform the same functions as lymph nodes inside the body, but because of their location they are the first to encounter pathogens and antigens that enter the body by way of the nose and mouth.

The **spleen** is located in the upper left region of the abdominal cavity just beneath the diaphragm. It is much larger than a lymph node, about the size of a fist. Whereas the lymph nodes cleanse lymph, the spleen cleanses blood. The spleen is composed of a capsule surrounding tissue known as white pulp and red pulp. White pulp contains lymphocytes and performs the immune functions of the spleen. The red pulp contains red blood cells and plentiful macrophages. The red pulp helps purify blood that passes through the spleen by removing debris and worn-out or damaged red blood cells.

The spleen's outer capsule is relatively thin, and an infection and/or a blow can cause the spleen to burst. Although its functions are replaced by other organs, a person without a spleen is often slightly more susceptible to infections and may have to receive antibiotic therapy indefinitely.

The **thymus gland** is located along the trachea behind the sternum in the upper thoracic cavity. This gland varies in size, but it is larger in children than in adults and may disappear completely in old age. The thymus is divided into lobules by connective tissue. The T lymphocytes mature in these lobules. The interior (medulla) of the lobule, which consists mostly of epithelial cells, stains lighter than the outer layer cortex. It produces thymic hormones, such as thymosin, that are thought to aid in maturation of T lymphocytes. Thymosin may also have other functions in immunity.

Red bone marrow is the site of origination for all types of blood cells, including the five types of white blood cells pictured in Figure 6.2. The marrow contains stem cells that are ever capable of dividing and producing cells that go on to differentiate into the various types of blood cells (see Fig. 6.4). In a child, most bones have red bone marrow, but in an adult it is present only in the bones of the skull, the sternum (breastbone), the ribs, the clavicle, the pelvic bones, and the vertebral column. The red bone marrow consists of a network of connective tissue fibers, called reticular fibers, which are produced by cells called reticular cells. These and the stem cells and their progeny are packed around thin-walled sinuses filled with venous blood. Differentiated blood cells enter the bloodstream at these sinuses.

Lymph is cleansed in lymph nodes; blood is cleansed in the spleen; all blood cells are made in red bone marrow; most white blood cells mature in the marrow, but T lymphocytes mature in the thymus.

8.2   Nonspecific Defenses

The **immune system** includes the cells and tissues that are responsible for immunity. **Immunity** is the ability of the body to defend itself against infectious agents, foreign cells, and even abnormal body cells, such as cancer cells. Thereby, the internal environment has a better chance of remaining stable. Immunity includes nonspecific and specific defenses. The four types of nonspecific defenses—barriers to entry, the inflammatory reaction, natural killer cells, and protective proteins—are effective against many types of infectious agents.

Barriers to Entry

Skin and the mucous membranes lining the respiratory, digestive, and urinary tracts serve as mechanical barriers to entry by pathogens. Oil gland secretions contain chemicals that weaken or kill certain bacteria on the skin. The upper respiratory tract is lined by ciliated cells that sweep mucus and trapped particles up into the throat, where they can be swallowed or expectorated (coughed out). The stomach has an acidic pH, which inhibits the growth of or kills many types of bacteria. The various bacteria that normally reside in the intestine and other areas, such as the vagina, prevent pathogens from taking up residence.

Inflammatory Reaction

Whenever tissue is damaged, a series of events occurs that is known as the **inflammatory reaction.** The inflamed area has four outward signs: redness, heat, swelling, and pain. Figure 8.4 illustrates the participants in the inflammatory reaction. **Mast cells,** which occur in tissues, resemble basophils, one of the white cells found in the blood.

When an injury occurs, damaged tissue cells and mast cells release chemical mediators, such as **histamine** and **kinins,** which cause the capillaries to dilate and become more permeable. The enlarged capillaries cause the skin to redden, and the increased permeability allows proteins and fluids to escape into the tissues, resulting in swelling. The swollen area as well as kinins stimulate free nerve endings, causing the sensation of pain.

Neutrophils and monocytes migrate to the site of injury. They are amoeboid and can change shape to squeeze through capillary walls to enter tissue fluid. Neutrophils, and also mast cells, phagocytize pathogens. The engulfed pathogens are destroyed by hydrolytic enzymes when the endocytic vesicle combines with a lysosome, one of the cellular organelles.

As they leave the blood and enter the tissues, monocytes differentiate into **macrophages,** large phagocytic cells that are able to devour a hundred pathogens and still survive. Some tissues, particularly connective tissue, have resident macrophages, which routinely act as scavengers, devouring old blood cells, bits of dead tissue, and other debris. Macrophages can also bring about an explosive increase in the number of leukocytes by liberating colony-stimulating factors, which pass by way of blood to the red bone marrow, where they stimulate the production and the release of white blood cells, primarily neutrophils. As the infection is being overcome, some neutrophils may die. These—along with dead cells, dead bacteria, and living white blood cells—form pus, a whitish material. Pus indicates that the body is trying to overcome the infection.

When a blood vessel ruptures, the blood forms a clot to seal the break. The chemical mediators (e.g., histamine and kinins) and antigens move through the tissue fluid and lymph to the lymph nodes. Now lymphocytes are activated to react to the threat of an infection. Sometimes inflammation persists, and the result is chronic inflammation that is often treated by administering anti-inflammatory agents such as aspirin, ibuprofen, or cortisone. They act against the chemical mediators released by the white blood cells in the area.

The inflammatory reaction is a "call to arms"— it marshals phagocytic white blood cells to the site of bacterial invasion and stimulates the immune system to react against a possible infection.

Visual Focus

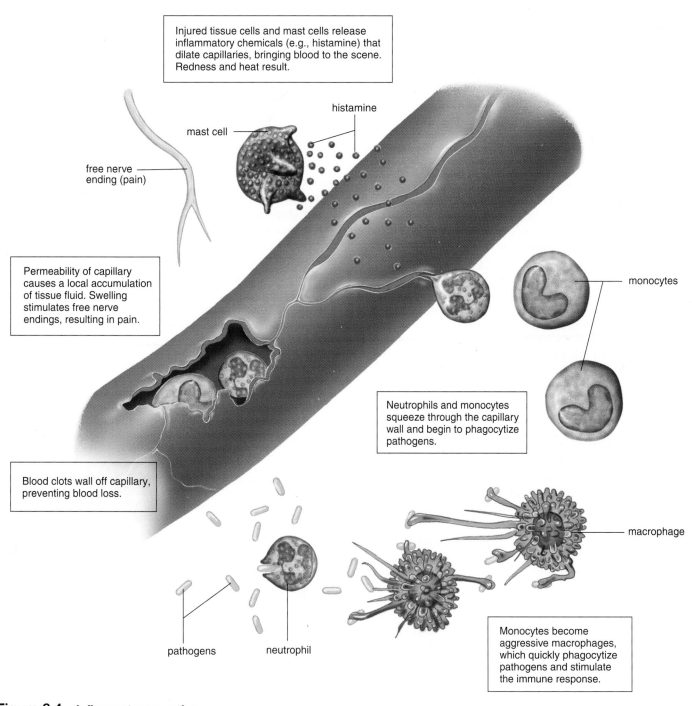

Injured tissue cells and mast cells release inflammatory chemicals (e.g., histamine) that dilate capillaries, bringing blood to the scene. Redness and heat result.

histamine

mast cell

free nerve ending (pain)

Permeability of capillary causes a local accumulation of tissue fluid. Swelling stimulates free nerve endings, resulting in pain.

Blood clots wall off capillary, preventing blood loss.

monocytes

Neutrophils and monocytes squeeze through the capillary wall and begin to phagocytize pathogens.

macrophage

Monocytes become aggressive macrophages, which quickly phagocytize pathogens and stimulate the immune response.

pathogens   neutrophil

Figure 8.4   Inflammatory reaction.

Natural Killer Cells

Natural killer (NK) cells kill virus-infected cells and tumor cells by cell-to-cell contact. They are large, granular lymphocytes. They have no specificity and no memory. Their number is not increased by immunization.

Protective Proteins

The **complement system,** often simply called complement, is a number of plasma proteins designated by the letter C and a subscript. A limited amount of activated complement protein is needed because a domino effect occurs: each activated protein in a series is capable of activating many other proteins.

Complement is activated when pathogens enter the body. It "complements" certain immune responses, which accounts for its name. For example, it is involved in and amplifies the inflammatory response because complement proteins attract phagocytes to the scene. Some complement proteins bind to the surface of pathogens already coated with antibodies, which ensures that the pathogens will be phagocytized by a neutrophil or macrophage.

Certain other complement proteins join to form a membrane attack complex that produces holes in the walls and plasma membranes of bacteria. Fluids and salts then enter the bacterial cell to the point that they burst (Fig. 8.5).

Interferon is a protein produced by virus-infected cells. Interferon binds to receptors of noninfected cells, causing them to prepare for possible attack by producing substances that interfere with viral replication. Interferon is specific to the species; therefore, only human interferon can be used in humans.

Immunity includes these nonspecific defenses: barriers to entry, the inflammatory reaction, natural killer cells, and protective proteins.

8.3   Specific Defenses

When nonspecific defenses have failed to prevent an infection, specific defenses come into play. An **antigen** is any foreign substance (often a protein or polysaccharide) that stimulates the immune system to react to it. Pathogens have antigens, but antigens can also be part of a foreign cell or a cancer cell. Because we do not ordinarily become immune to our own normal cells, it is said that the immune system is able to distinguish self from nonself. Only in this way can the immune system aid rather than counter homeostasis.

Lymphocytes are capable of recognizing an antigen because they have **antigen receptors**—plasma membrane receptor proteins whose shape allows them to combine with a specific antigen. It is often said that the receptor and the antigen fit together like a lock and a key. It is estimated that during our lifetime, we encounter a million different antigens, so we need a great diversity of lymphocytes to protect us against them. Remarkably, diversification occurs to such an extent during the maturation process that there is a lymphocyte type for any possible antigen.

Immunity usually lasts for some time. For example, once we recover from the measles, we usually do not get the illness a second time. Immunity is primarily the result of the action of the **B lymphocytes** and the **T lymphocytes.** B lymphocytes mature in the *b*one marrow,[1] and T lymphocytes mature in the *t*hymus gland. B lymphocytes, also called B cells, give rise to plasma cells, which produce **antibodies,** proteins shaped like the antigen receptor and capable of combining with and neutralizing a specific antigen. These antibodies are secreted into the blood, lymph, and other body fluids. In contrast, T lymphocytes, also called T cells, do not produce antibodies. Instead, certain T cells directly attack cells that bear nonself antigens. Other T cells regulate the immune response.

---

[1]Historically, the B stands for bursa of Fabricius, an organ in the chicken where these cells were first identified. As it turns out, however, the B can conveniently be thought of as referring to bone marrow.

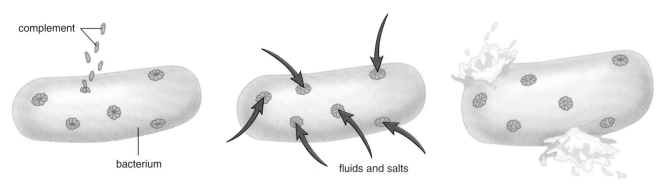

| Complement proteins form holes in the bacterial cell wall and membrane. | Holes allow fluids and salts to enter the bacterium. | Bacterium expands until it bursts. |

Figure 8.5   Action of the complement system against a bacterium.
When complement proteins in the plasma are activated by an immune reaction, they form holes in bacterial cell walls and plasma membranes, allowing fluids and salts to enter until the cell eventually bursts.

B Cells and Antibody-Mediated Immunity

A toxin is a chemical (produced by bacteria, for example) that is poisonous to other living things. When a B cell in a lymph node or the spleen encounters a bacterial cell or a toxin bearing a specific antigen, it becomes activated to divide many times. Most of the resulting cells are plasma cells. A **plasma cell** is a mature B cell that mass-produces antibodies in the lymph nodes and in the spleen.

The **clonal selection theory** states that the antigen selects which lymphocyte will undergo clonal expansion and produce more lymphocytes bearing the same type of antigen receptor (Fig. 8.6). Notice that a B cell does not divide until a specific antigen is present and binds to its receptors. B cells are stimulated to divide and become plasma cells by helper T cell secretions called cytokines, as is discussed in the next section. Some members of the clone become memory cells, which are the means by which long-term immunity is possible. If the same antigen enters the system again, **memory B cells** quickly divide and give rise to more lymphocytes capable of quickly producing antibodies.

Once the threat of an infection has passed, the development of new plasma cells ceases and those present undergo apoptosis. **Apoptosis** is a process of programmed cell death (PCD) involving a cascade of specific cellular events leading to the death and destruction of the cell. The methodology of PCD is still being worked out, but we know it is an essential physiological mechanism regulating the cell population within an organ system. PCD normally plays a central role in maintaining tissue homeostasis.

Defense by B cells is called **antibody-mediated immunity** because the various types of B cells produce antibodies. It is also called humoral immunity because these antibodies are present in blood and lymph. A humor is any fluid normally occurring in the body.

Characteristics of B Cells

- Antibody-mediated immunity against bacteria
- Produced and mature in bone marrow
- Reside in spleen and lymph nodes, circulate in blood and lymph
- Directly recognize antigen and then undergo clonal selection
- Clonal expansion produces antibody-secreting plasma cells as well as memory B cells

---

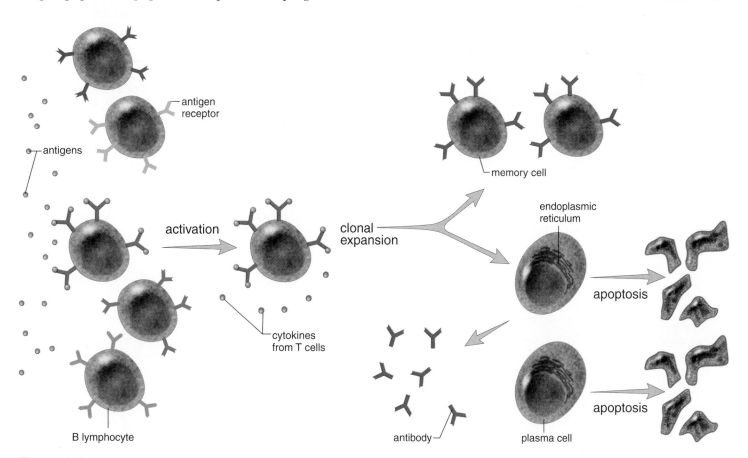

Figure 8.6 **Clonal selection theory as it applies to B cells.**

When an antigen combines with a B cell's antigen receptors, and it is stimulated by T cell secretions called cytokines, the B cell undergoes clonal expansion. The result is many plasma cells, which produce specific antibodies against this antigen and memory B cells. After the infection passes, plasma cells undergo apoptosis. Memory B cells, which keep the ability to recognize this antigen, are retained in the body.

Structure of IgG

The most common type of antibody (IgG) is a Y-shaped protein molecule with two arms. Each arm has a "heavy" (long) polypeptide chain and a "light" (short) polypeptide chain. These chains have constant regions, where the sequence of amino acids is set, and variable regions, where the sequence of amino acids varies between antibodies (Fig. 8.7). The constant regions are not identical among all the antibodies. Instead, they are almost the same within different classes of antibodies. The variable regions form an antigen-binding site, and their shape is specific to a particular antigen. The antigen combines with the antibody at the antigen-binding site in a lock-and-key manner.

The antigen-antibody reaction can take several forms, but quite often the reaction produces complexes of antigens combined with antibodies. Such antigen-antibody complexes, sometimes called immune complexes, mark the antigens for destruction. For example, an antigen-antibody complex may be engulfed by neutrophils or macrophages, or it may activate complement. Complement makes pathogens more susceptible to phagocytosis, as discussed previously.

Other Types of Antibodies

There are five different classes of circulating antibody proteins or **immunoglobulins (Igs)** (Table 8.1). IgG antibodies are the major type in blood, and lesser amounts are also found in lymph and tissue fluid. IgG antibodies bind to pathogens and their toxins. IgM antibodies are pentamers, meaning that they contain five of the Y-shaped structures shown in Figure 8.7a. These antibodies appear in blood soon after an infection begins and disappear before it is over. They are good activators of the complement system. IgA antibodies are monomers or dimers containing two Y-shaped structures. They are the main type of antibody found in bodily secretions. They bind to pathogens before they reach the bloodstream. The main function of IgD molecules seems to be to serve as antigen receptors on virgin B cells. IgE antibodies, which are responsible for immediate allergic responses, are discussed on page 158 and in the Health Focus on page 159.

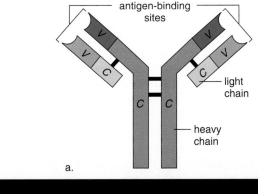

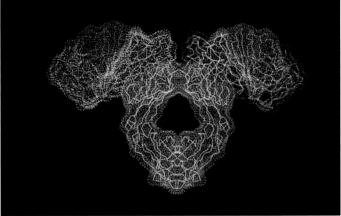

b.

Figure 8.7   **Structure of the most common antibody (IgG).**
a. An IgG antibody contains two heavy (long) polypeptide chains and two light (short) chains arranged so there are two variable regions, where a particular antigen is capable of binding with an antibody (V = variable region, C = constant region). **b.** Computer model of an antibody molecule. The antigen combines with the two side branches.

An antigen combines with an antibody at the antigen-binding site in a lock-and-key manner. The reaction can produce antigen-antibody complexes, which contain several molecules of antibody and antigen.

Table 8.1	Antibodies	
Classes	**Presence**	**Function**
IgG	Main antibody type in circulation	Binds to pathogens, activates complement, and enhances phagocytosis
IgM	Antibody type found in circulation; largest antibody	Activates complement; clumps cells
IgA	Main antibody type in secretions such as saliva and milk	Prevents pathogens from attaching to epithelial cells in digestive and respiratory tract
IgD	Antibody type found on surface of virgin B cells	Presence signifies readiness of B cell
IgE	Antibody type found as antigen receptors on basophils in blood and on mast cells in tissues	Responsible for immediate allergic response and protection against certain parasitic worms

Pesticide: An Asset and a Liability

A pesticide is any one of 55,000 chemical products used to kill insects, plants, fungi, or rodents that interfere with human activities. Increasingly, we have discovered that pesticides are harmful to the environment and humans (Fig. 8A). The effect of pesticides on the immune system is a new area of concern; no testing except for skin sensitization has thus far been required. Contact or respiratory allergic responses are among the most immediately obvious toxic effects of pesticides. But pesticides can cause suppression of the immune system, leading to increased susceptibility to infection or tumor development. Lymphocyte impairment was found following the worst industrial accident in the world, which took place in India in 1984. Nerve gas used to produce an insecticide was released into the air; 3,700 people were killed, and 30,000 were injured. Vietnam veterans claim a variety of health effects due to contact with a herbicide called Agent Orange used as a defoliant during the Vietnam War. The EPA says that pesticide residues on food possibly cause suppression of the immune system, disorders of the nervous system, birth defects, and cancer. The immune, nervous, and endocrine systems communicate with one another by way of hormones, and what affects one system can affect the other. Testicular atrophy, low sperm counts, and abnormal sperm in men may be due to the ability of pesticides to mimic the effects of estrogen, a female sex hormone.

The argument for pesticides at first seems attractive. Pesticides are meant to kill off disease-causing agents, increase yield, and work quickly with minimal risk. But over time, it has been found that pesticides do not meet these claims. Instead, pests become resistant to pesticides, which kill off natural enemies in addition to the pest. Then the pest population explodes. For example, at first DDT did a marvelous job of killing off the mosquitos that carry malaria; now malaria is as big a problem as ever. In the meantime, DDT has accumulated in the tissues of wildlife and humans, causing harmful effects. The problem was made obvious when birds of prey became unable to reproduce due to weak eggshells. The use of DDT is now banned in the United States.

There are alternatives to the use of pesticides. Integrated pest management uses a diversified environment, mechanical and physical means, natural enemies, disruption of reproduction, and resistant plants to control rather than eradicate pest populations. Chemicals are used only as a last resort. Maintaining hedgerows, weedy patches, and certain trees can provide diverse habitats and food for predators and parasites that help

Figure 8A Pesticide warning signs indicate that pesticides are harmful to our health.

control pests. The use of strip farming and crop rotation by farmers denies pests a continuous food source.

Natural enemies abound in the environment. When lacewings were released in cotton fields, they reduced the boll weevil population by 96% and increased cotton yield threefold. A predatory moth was used to reclaim 60 million acres in Australia overrun by the prickly pear cactus, and another type of moth is now used to control alligator weed in the southern United States. Also in the United States, the *Chrysolina* beetle controls Klamath weed, except in shady places where a root-boring beetle is more effective.

Sex attractants and sterile insects are used in other types of biological control. In Sweden and Norway, scientists synthesized the sex pheromone of the Ips bark beetle, which attacks spruce trees. Almost 100,000 baited traps collected females normally attracted to males emitting this pheromone. Sterile males have also been used to reduce pest populations. The screwworm fly parasitizes cattle in the United States. Flies raised in a laboratory were made sterile by exposure to radiation. The entire Southeast was freed from this parasite when female flies mated with sterile males and then laid eggs that did not hatch.

In the past, only crossbreeding could produce resistant plants. Now genetic engineering, a new technique by which certain genes are introduced into organisms, including plants, may eventually make large-scale use of pesticides unnecessary. Already, 50 types of genetically engineered plants that resist insects or viruses have entered small-scale field trials.

In general, biological control is a more sophisticated method of controlling pests than the use of pesticides. It requires an in-depth knowledge of pests and/or their life cycles. Because it does not have an immediate effect on the pest population, the effects of biological control may not be apparent to the farmer or gardener who eventually benefits from it.

Citizens too can promote biological control of pests by

- Urging elected officials to support legislation to protect humans and the environment against the use of pesticides.
- Allowing native plants to grow on all or most of the land to give natural predators a place to live.
- Cutting down on the use of pesticides, herbicides, and fertilizers for the lawn, garden, and house.
- Using alternative methods such as cleanliness and good sanitation to keep household pests under control.
- Disposing of pesticides in a safe manner.

T Cells and Cell-Mediated Immunity

When T cells leave the thymus, they have unique receptors just as B cells do. Unlike B cells, however, T cells are unable to recognize an antigen present in lymph, blood, or the tissues without help. The antigen must be presented to them by an **antigen-presenting cell (APC).** When an APC, usually a macrophage, presents a viral or cancer cell antigen the antigen is linked to a major histocompatibility complex (MHC) protein in the plasma membrane.

Human MHC proteins are called **HLA (human leukocyte-associated) antigens.** Because they mark the cell as belonging to a particular individual, HLA antigens are self-antigens. The importance of self-antigens in plasma membranes was first recognized when it was discovered that they contribute to the specificity of tissues and make it difficult to transplant tissue from one human to another. In other words, when the donor and the recipient are histo (tissue)-compatible (the same or nearly so), a transplant is more likely to be successful.

Figure 8.8 shows a macrophage presenting an antigen to a T cell. Once a T cell is activated in this manner, it undergoes clonal expansion and produces **cytokines,** chemicals that stimulate various immune cells (e.g., macrophages, B cells, and T cells) to perform their functions. Later, activated T cells destroy any cell, such as a viral infected cell or a cancer cell, that displays the same antigen. As the illness disappears, the immune reaction wanes and fewer cytokines are produced. Now, the activated T cells become susceptible to apoptosis. As mentioned previously, apoptosis is a programmed cell death that contributes to homeostasis by regulating the number of cells that are present in an organ, or in this case, in the immune system. When apoptosis does not occur as it should, T cell cancers (i.e., lymphomas and leukemias) can result.

Apoptosis also occurs in the thymus as T cells are maturing. Any T cell that has the potential to destroy the body's own cells undergoes suicide.

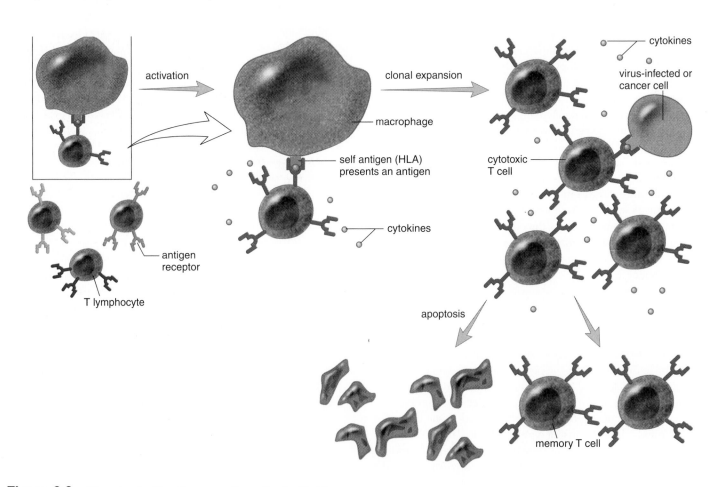

Figure 8.8   Clonal selection theory as it applies to T cells.
Each type of T cell bears a specific antigen receptor. When a macrophage presents an antigen in the groove of an HLA molecule to the correct T cell, it secretes cytokines and undergoes clonal expansion. After the immune response has been successful, the majority of T cells undergo apoptosis while a small number become memory T cells. Memory T cells provide protection should the same antigen enter the body again at a future time.

Types of T Cells

The two main types of T cells are cytotoxic T cells and helper T cells. **Cytotoxic T cells** can bring about the destruction of antigen-bearing cells, such as virus-infected or cancer cells. Cancer cells also have nonself antigens. Cytotoxic T cells have storage vacuoles containing perforin molecules. **Perforin** molecules perforate a plasma membrane, forming a pore that allows water and salts to enter. The cell then swells and eventually bursts. Cytotoxic T cells are responsible for so-called **cell-mediated immunity** (Fig. 8.9).

Helper T cells regulate immunity by secreting cytokines, the chemicals that enhance the response of other immune cells. Because HIV, the virus that causes AIDS, infects helper T cells and certain other cells of the immune system, it inactivates the immune response. For more information on an HIV infection see the AIDS supplement, page 355.

Notice in Figure 8.8 that a few of the clonally expanded T cells are labeled memory T cells. They remain in the body and can jump-start an immune reaction to an antigen previously present in the body.

Characteristics of T Cells

- Cell-mediated immunity against viruses
- Produced in bone marrow, mature in thymus
- Antigen must be presented in groove of an HLA molecule
- Cytotoxic T cells destroy nonself antigen-bearing cells
- Helper T cells secrete cytokines that control the immune response

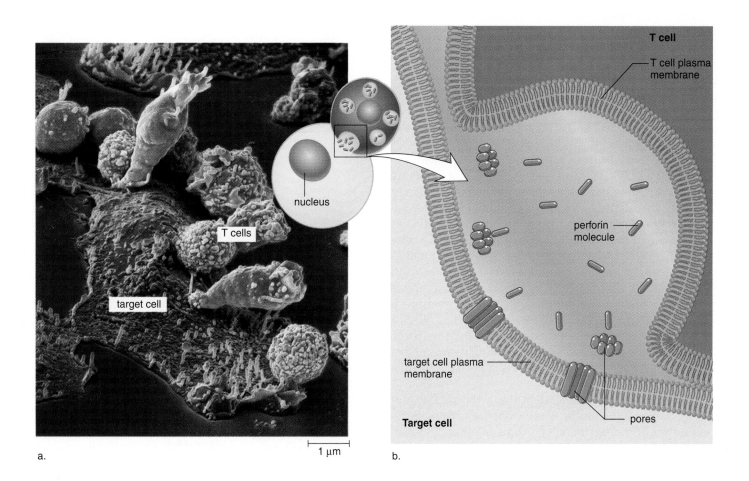

a.   1 μm

b.

Figure 8.9   Cell-mediated immunity.

a. The scanning electron micrograph shows cytotoxic T cells attacking and destroying a cancer cell. Cancer cells are subject to attack because they have acquired nonself antigens. **b.** During the killing process, the vacuoles in a cytotoxic T cell release perforin molecules. These molecules combine to form pores in the target cell plasma membrane. Thereafter, fluid and salts enter so that the target cell eventually bursts.

8.4   Induced Immunity

Immunity occurs naturally through infection or is brought about artificially by medical intervention. There are two types of induced immunity: active and passive. In active immunity, the individual alone produces antibodies against an antigen; in passive immunity, the individual is given prepared antibodies.

Active Immunity

Active immunity sometimes develops naturally after a person is infected with a pathogen. However, active immunity is often induced when a person is well so that possible future infection will not take place. To prevent infections, people can be artificially immunized against them. The United States is committed to the goal of immunizing all children against the common types of childhood diseases listed in the immunization schedule in Figure 8.10a.

Immunization involves the use of **vaccines,** substances that contain an antigen to which the immune system responds. Traditionally, vaccines are the pathogens themselves, or their products, that have been treated so they are no longer virulent (able to cause disease). Today, it is possible to genetically engineer bacteria to mass-produce a protein from pathogens, and this protein can be used as a vaccine. This method now has been used to produce a vaccine against hepatitis B, a viral disease, and is being used to prepare a vaccine against malaria, a protozoan disease.

After a vaccine is given, it is possible to follow an immune response by determining the amount of antibody present in a sample of plasma—this is called the antibody titer. After the first exposure to a vaccine, a primary response occurs. For a period of several days, no antibodies are present; then, there is a slow rise in the titer, followed by first a plateau and then a gradual decline as the antibodies bind to the antigen or simply break down (Fig. 8.10b). After a second exposure to the vaccine, a secondary response is expected. The titer rises rapidly to a plateau level much greater than before. The second exposure is called a "booster" because it boosts the antibody titer to a high level. The high antibody titer now is expected to help prevent disease symptoms even if the individual is exposed to the disease-causing antigen.

Active immunity is dependent upon the presence of memory B cells and memory T cells that are capable of responding to lower doses of antigen. Active immunity is usually long-lasting, although a booster may be required every so many years.

Active (long-lasting) immunity can be induced by the use of vaccines. Active immunity is dependent upon the presence of memory B cells and memory T cells in the body.

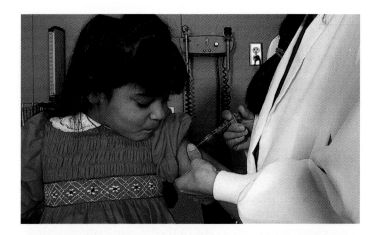

Suggested Immunization Schedule

Vaccine	Age (Months)	Age (Years)
Hepatitis B	Birth–2, 1–18	11–12
Diphtheria, tetanus, pertussis	2, 4, 6, 15–18	4–6
Tetanus only		11–12, 14–16
Haemophilus influenza, type b	2, 4, 6, 12–15	
Polio	2, 4, 6–18	4–6
Pneumococcal	2, 4, 6, 12–15	
Measles, mumps, rubella	12–15	4–6, 11–12
Varicella (chicken pox)	12–18	11–12
Hepatitis A (in selected areas)	24	4–12

a.

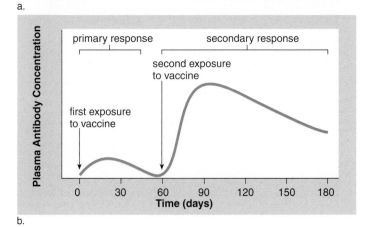

b.

Figure 8.10   Active immunity due to immunizations.
a. Suggested immunization schedule for infants and children.
b. During immunization, the primary response, after the first exposure to a vaccine, is minimal, but the secondary response, which may occur after the second exposure, shows a dramatic rise in the amount of antibody present in plasma.

Passive Immunity

Passive immunity occurs when an individual is given prepared antibodies (immunoglobulins) to combat a disease. Since these antibodies are not produced by the individual's B cells, passive immunity is temporary. For example, newborn infants are passively immune to some diseases because antibodies have crossed the placenta from the mother's blood. These antibodies soon disappear, however, so that within a few months, infants become more susceptible to infections. Breast-feeding prolongs the natural passive immunity an infant receives from the mother because antibodies are present in the mother's milk (Fig. 8.11).

Even though passive immunity does not last, it sometimes is used to prevent illness in a patient who has been unexpectedly exposed to an infectious disease. Usually, the patient receives a gamma globulin injection (serum that contains antibodies), perhaps taken from individuals who have recovered from the illness. In the past, horses were immunized, and serum was taken from them to provide the needed antibodies against such diseases as diphtheria, botulism, and tetanus. A patient who received these antibodies became ill about 50% of the time, because the serum contained proteins that the individual's immune system recognized as foreign. This was called serum sickness. But problems can still occur with products produced in other ways. An immunoglobulin intravenous product called Gammagard was withdrawn from the market because of possible implication in the transmission of hepatitis.

Passive immunity provides immediate protection when an individual is in immediate danger of succumbing to an infectious disease. Passive immunity is temporary because there are no memory cells.

Cytokines and Immunity

Cytokines are signaling molecules produced by T lymphocytes, monocytes, and other cells. Because cytokines regulate white blood cell formation and/or function, they are being investigated as possible adjunct therapy for cancer and AIDS. Both interferon and **interleukins,** which are cytokines produced by various white blood cells, have been used as immunotherapeutic drugs, particularly to enhance the ability of the individual's own T cells to fight cancer.

Interferon, discussed previously on page 150, is a substance produced by leukocytes, fibroblasts, and probably most cells in response to a viral infection. Interferon still is being investigated as a possible cancer drug, but so far it has proven to be effective only in certain patients, and the exact reasons for this as yet cannot be discerned.

When and if cancer cells carry an altered protein on their cell surface, they should be attacked and destroyed by cyto-

Figure 8.11   Passive immunity.
Breast-feeding is believed to prolong the passive immunity an infant receives from the mother because antibodies are present in the mother's milk.

toxic T cells. Whenever cancer does develop, it is possible that the cytotoxic T cells have not been activated. In that case, cytokines might awaken the immune system and lead to the destruction of the cancer. In one technique being investigated, researchers first withdraw T cells from the patient, present cancer cell antigens to them, and then activate the cells by culturing them in the presence of an interleukin. The T cells are reinjected into the patient, who is given doses of interleukin to maintain the killer activity of the T cells.

Those who are actively engaged in interleukin research believe that interleukins soon will be used as adjuncts for vaccines, for the treatment of chronic infectious diseases, and perhaps for the treatment of cancer. Interleukin antagonists also may prove helpful in preventing skin and organ rejection, autoimmune diseases, and allergies.

The interleukins and other cytokines show some promise of enhancing the individual's own immune system.

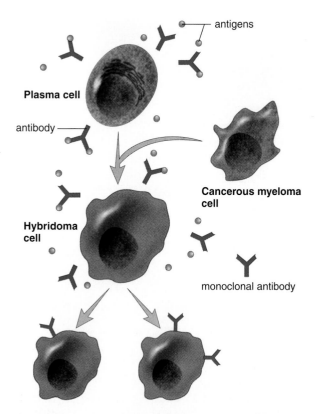

Figure 8.12   Production of monoclonal antibodies.
Plasma cells (derived from immunized mice) are fused with myeloma
(cancerous) cells, producing hybridoma cells that are "immortal."
Hybridoma cells divide and continue to produce the same type of
antibody, called monoclonal antibodies.

Monoclonal Antibodies

Every plasma cell derived from the same B cell secretes antibodies against a specific antigen. These are **monoclonal antibodies** because all of them are the same type and because
they are produced by plasma cells derived from the same B
cell. One method of producing monoclonal antibodies in
vitro (outside the body in glassware) is depicted in Figure 8.12. B lymphocytes are removed from an animal (today,
usually mice are used) and are exposed to a particular antigen. The activated B lymphocytes are fused with myeloma
cells (malignant plasma cells that live and divide indefinitely). The fused cells are called hybridomas—*hybrid-*
because they result from the fusion of two different cells,
and *-oma* because one of the cells is a cancer cell.

At present, monoclonal antibodies are being used for
quick and certain diagnosis of various conditions. For example, a particular hormone is present in the urine of a pregnant
woman. A monoclonal antibody can be used to detect this
hormone; if it is present, the woman knows she is pregnant.
Monoclonal antibodies are also used to identify infections.
And because they can distinguish between cancer and normal
tissue cells, they are used to carry radioactive isotopes or toxic
drugs to tumors so that they can be selectively destroyed.

8.5   Immunity Side Effects

The immune system usually protects us from disease
because it can distinguish self from nonself. Sometimes,
however, it responds in a manner that does harm to the
body, as when individuals develop allergies, have an autoimmune disease, or suffer tissue rejection.

Allergies

Allergies are hypersensitivities to substances such as
pollen or animal hair that ordinarily would do no harm to
the body. The response to these antigens, called **allergens,**
usually includes some degree of tissue damage. There are
four types of allergic responses, but we will consider only
two of these: immediate allergic response and delayed allergic response.

Immediate Allergic Response

An **immediate allergic response** can occur within seconds
of contact with the antigen. As discussed in the Health Focus
on page 159, coldlike symptoms are common. Anaphylactic
shock is a severe reaction characterized by a sudden and life-
threatening drop in blood pressure.

Immediate allergic responses are caused by antibodies
known as IgE (see Table 8.1). IgE antibodies are attached to
the plasma membrane of mast cells in the tissues and also
basophils in the blood. When an allergen attaches to the IgE
antibodies on these cells, they release histamine and other
substances that bring about the coldlike symptoms or,
rarely, anaphylactic shock.

Allergy shots sometimes prevent the onset of an allergic response. It's been suggested that injections of the allergen may cause the body to build up high quantities of IgG
antibodies, and these combine with allergens received
from the environment before they have a chance to reach
the IgE antibodies located in the membrane of mast cells
and basophils.

Delayed Allergic Response

Delayed allergic responses are initiated by memory T cells
at the site of allergen in the body. The allergic response is
regulated by the cytokines secreted by both T cells and
macrophages.

A classic example of a delayed allergic response is the
skin test for tuberculosis (TB). When the result of the test is
positive, the tissue where the antigen was injected becomes
red and hardened. This shows that there was prior exposure to tubercle bacilli, which cause TB. Contact dermatitis,
such as occurs when a person is allergic to poison ivy, jew-
elry, cosmetics, and so forth, is also an example of a delayed
allergic response.

Immediate Allergic Responses

The runny nose and watery eyes of hay fever are often caused by an allergic reaction to the pollen of trees, grasses, and ragweed. Worse, the airways leading to the lungs constrict if a person has asthma, resulting in difficult breathing characterized by wheezing. Windblown pollen, particularly in the spring and fall, brings on the symptoms of hay fever. Most people can inhale pollen with no ill effects. But others have developed a hypersensitivity, meaning that their immune system responds in a deleterious manner. The problem stems from a type of antibody called immunoglobulin E (IgE) that causes the release of histamine from mast cells and also basophils whenever they are exposed to an allergen. Histamine causes mucosal membranes of the nose and eyes to release fluid as a defense against pathogen invasion. But in the case of allergy, copious fluid is released although no real danger is present.

Most food allergies are also due to the presence of IgE antibodies that bind usually to a protein in the food. The symptoms, such as nausea, vomiting, and diarrhea, are due to the mode of entry of the allergen. Skin symptoms may also occur, however. Adults are often allergic to shellfish, nuts, eggs, cows' milk, fish, and soybeans. Peanut allergy is a common food allergy in the United States possibly because peanut butter is a staple in the diet. People seem to outgrow allergies to cows' milk and eggs more often than allergies to peanuts and soybeans.

Celiac disease occurs in people who are allergic to wheat, rye, barley, and sometimes oats—in short, any grain that contains gluten proteins. It is thought that the gluten proteins elicit a delayed cell-mediated immune response by T cells with the resultant production of cytokines. The symptoms of celiac disease include diarrhea, bloating, weight loss, anemia, bone pain, chronic fatigue, and weakness.

People can reduce the chances of a reaction to airborne and food allergens by avoiding the offending substances. The reaction to peanuts can be so severe that airlines are now required to have a peanut-free zone in their planes for those who are allergic. The people in Figure 8B are trying to avoid windblown allergens. The taking of antihistamines can also be helpful.

If these procedures are inadequate, patients can be tested to measure their susceptibility to any number of possible allergens. A small quantity of a suspected allergen is inserted just beneath the skin, and the strength of the subsequent reaction is noted. A wheal-and-flare response at the skin prick site demonstrates that IgE antibodies attached to mast cells have reacted to an allergen. In an immunotherapy called hyposensitization, ever-increasing doses of the allergen are periodically injected subcutaneously with the hope that the body will build up a supply of IgG. IgG, in contrast to IgE, does not cause the release of histamine after it combines with the allergen. If IgG combines first upon exposure to the allergen, the allergic response does not occur. Patients know they are cured when the allergic symptoms no longer occur. Therapy may have to continue for as long as two to three years.

Allergic-type reactions can occur without involving the immune system. Wasp and bee stings contain substances that cause swellings, even in those whose immune system is not sensitized to substances in the sting. Also, jellyfish tentacles and foods such as strawberries or fish that is not fresh contain histamine or closely related substances that can cause a reaction. Immunotherapy is also not possible in those who are allergic to penicillin and bee stings. High sensitivity has built up upon the first exposure, and when reexposed, anaphylactic shock can occur. Among its many effects, histamine causes increased permeability of the capillaries, the smallest blood vessels. In these individuals, there is a drastic decrease in blood pressure that can be fatal within a few minutes. People who know they are allergic to bee stings can obtain a syringe of epinephrine to carry with them. This medication can delay the onset of anaphylactic shock until medical help is available.

Figure 8B **Protection against allergies.**
The allergic reactions that result in hay fever and asthma attacks can have many triggers, one of which is the pollen of a variety of plants. A dramatic solution to the problem has been found by these people.

Autoimmune Diseases

When cytotoxic T cells or antibodies mistakenly attack the body's own cells as if they bear foreign antigens, the resulting condition is known as an **autoimmune disease.** Exactly what causes autoimmune diseases is not known. However, sometimes they occur after an individual has recovered from an infection.

In the autoimmune disease myasthenia gravis, neuromuscular junctions do not work properly, and muscular weakness results. In multiple sclerosis, the myelin sheath of nerve fibers breaks down, and this causes various neuromuscular disorders. A person with systemic lupus erythematosus has various symptoms prior to death due to kidney damage. In rheumatoid arthritis, the joints are affected. Researchers suggest that heart damage following rheumatic fever and type I diabetes are also autoimmune illnesses. As yet there are no cures for autoimmune diseases, but they can be controlled with drugs.

Autoimmune diseases occur when antibodies or cytotoxic T cells recognize and destroy the body's own cells.

Tissue Rejection

Certain organs, such as skin, the heart, and the kidneys, could be transplanted easily from one person to another if the body did not attempt to *reject* them. Rejection occurs because antibodies and cytotoxic T cells bring about destruction of foreign tissues in the body. When rejection occurs, the immune system is correctly distinguishing between self and nonself.

Organ rejection can be controlled by carefully selecting the organ to be transplanted and administering immunosuppressive drugs. It is best if the transplanted organ has the same type of HLA antigens as those of the recipient, because cytotoxic T cells recognize foreign HLA antigens. The immunosuppressive drug cyclosporine has been used for many years. A new drug, tacrolimus (formerly known as FK-506), shows some promise, especially in liver transplant patients. However, both drugs, which act by inhibiting the response of T cells to cytokines, are known to adversely affect the kidneys.

The hope is that tissue engineering, the production of organs that lack antigens or that can be protected in some way from the immune system, will one day do away with the problem of rejection. For example, pancreatic cells have been placed in protective capsules that are then implanted in the abdominal cavity.

When an organ is rejected, the immune system has recognized and destroyed cells that bear foreign HLA antigens.

8.6   Homeostasis

Blood and tissue fluid provide an internal environment for the body's cells. The lymphatic vessels collect excess tissue fluid and return it as lymph to cardiovascular veins in the thorax. The lymphoid organs, along with the immune system, protect us from infectious diseases.

Nonspecific ways of protecting the body from disease precede specific immunity. The skin and the mucous membranes of the respiratory tract, the digestive tract, and the urinary system all resist invasion by viruses and bacteria. If a pathogen should enter the body, the infection is localized as much as possible. During the inflammatory reaction, the phagocytic white blood cells immediately rush to the scene and engulf as many pathogens as possible. Macrophages are especially good at devouring viruses and bacteria by phagocytosis. If the infection cannot be confined and pathogens do enter the blood, complement is a series of proteins that work in diverse ways to keep the blood free of disease-causing organisms and their toxins.

Not surprisingly, specific defenses are dependent upon blood cells; the lymphocytes and macrophages play central roles. B and T cells have antigen receptors and can distinguish self from nonself. The binding of the antigen selects which specific B or T cells will undergo clonal expansion. B cells are capable of recognizing an antigen directly, but T cells must have the antigen displayed by an APC in the groove of an HLA antigen. Plasma cells (mature B cells) produce antibodies, but T cells kill virus infected and cancer cells outright.

The lymphoid organs play a central role in immunity. White blood cells are made in the red bone marrow where B cells also mature. T cells mature in the thymus. The spleen filters the blood directly. Clonal expansion of lymphocytes occurs in the lymph nodes, which also filter the lymph.

A strong connection exists between the immune, nervous, and endocrine systems. Lymphocytes have receptor proteins for a wide variety of hormones, and the thymus gland produces hormones that influence the immune response. Cytokines help the body recover from disease by affecting the brain's temperature control center. The high body temperature of a fever is thought to create an unfavorable environment for the foreign invaders. Also, cytokines bring about a feeling of sluggishness, sleepiness, and loss of appetite. These behaviors tend to make us take care of ourselves until we feel better. A close connection between the immune and endocrine systems is illustrated by the ability of cortisone to mollify the inflammatory reaction in the joints.

The lymphatic vessels collect excess tissue fluid and return it as lymph to the cardiovascular veins. The immune system normally keeps the body free of infectious diseases.

Human Systems Work Together

Integumentary System

Lymphatic vessels pick up excess tissue fluid; immune system protects against skin infections.

Skin serves as a barrier to pathogen invasion; Langerhans cells phagocytize pathogens; protects lymphatic vessels.

Skeletal System

Lymphatic vessels pick up excess tissue fluid; immune system protects against infections.

Red bone marrow produces leukocytes involved in immunity.

Muscular System

Lymphatic vessels pick up excess tissue fluid; immune system protects against infections.

Skeletal muscle contraction moves lymph; physical exercise enhances immunity.

Nervous System

Lymphatic vessels pick up excess tissue fluid; immune system protects against infections of nerves.

Microglia engulf and destroy pathogens.

Endocrine System

Lymphatic vessels pick up excess tissue fluid; immune system protects against infections.

Thymus is necessary to maturity of T lymphocytes.

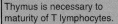

How the Lymphatic System works with other body systems

Cardiovascular System

Lymphoid organs produce and store formed elements; lymphatic vessels transport leukocytes and return tissue fluid to blood vessels; spleen serves as blood reservoir, filters blood. Blood vessels transport leukocytes and antibodies; blood services lymphoid organs and is source of tissue fluid that becomes lymph.

Respiratory System

Lymphatic vessels pick up excess tissue fluid; immune system protects against respiratory tract and lung infections.

Tonsils and adenoids occur along respiratory tract; breathing aids lymph flow; lungs carry out gas exchange.

Digestive System

Lacteals absorb fats; Peyer's patches prevent invasion of pathogens; appendix contains lymphoid tissue.

Digestive tract provides nutrients for lymphoid organs; stomach acidity prevents pathogen invasion of body.

Urinary System

Lymphatic system picks up excess tissue fluid, helping to maintain blood pressure for kidneys to function; immune system protects against infections.

Kidneys control volume of body fluids, including lymph.

Reproductive System

Immune system does not attack sperm or fetus, even though they are foreign to the body.

Sex hormones influence immune functioning; acidity of vagina helps prevent pathogen invasion of body; milk passes antibodies to newborn.

Summarizing the Concepts

8.1 Lymphatic System
The lymphatic system consists of lymphatic vessels and lymphoid organs. The lymphatic vessels receive lipoproteins at intestinal villi and excess tissue fluid at blood capillaries, and carry these to the bloodstream.

Lymphocytes are produced and accumulate in the lymphoid organs (red bone marrow, lymph nodes, tonsils, spleen, and thymus gland). Lymph is cleansed of pathogens and/or their toxins in lymph nodes, and blood is cleansed of pathogens and/or their toxins in the spleen. T lymphocytes mature in the thymus, while B lymphocytes mature in the red bone marrow where all blood cells are produced. White blood cells are necessary for nonspecific and specific defenses.

8.2 Nonspecific Defenses
Immunity involves nonspecific and specific defenses. Nonspecific defenses include barriers to entry, the inflammatory reaction, natural killer cells, and protective proteins.

8.3 Specific Defenses
Specific defenses require B lymphocytes and T lymphocytes, also called B cells and T cells. B cells undergo clonal selection with production of plasma cells and memory B cells after their antigen receptors combine with a specific antigen. Plasma cells secrete antibodies and eventually undergo apoptosis. Plasma cells are responsible for antibody-mediated immunity. IgG antibody is a Y-shaped molecule that has two binding sites for a specific antigen. Memory B cells remain in the body and produce antibodies if the same antigen enters the body at a later date.

T cells are responsible for cell-mediated immunity. The two main types of T cells are cytotoxic T cells and helper T cells. Cytotoxic T cells kill virus-infected or cancer cells on contact because they bear a nonself antigen. Helper T cells produce cytokines and stimulate other immune cells. Like B cells, each T cell bears antigen receptors. However, for a T cell to recognize an antigen, the antigen must be presented by an antigen-presenting cell (APC), usually a macrophage, along with an HLA (human leukocyte-associated antigen). Thereafter, the activated T cell undergoes clonal expansion until the illness has been stemmed. Then most of the activated T cells undergo apoptosis. A few cells remain, however, as memory T cells.

8.4 Induced Immunity
Immunity can be induced in various ways. Vaccines are available to induce long-lasting, active immunity, and antibodies sometimes are available to provide an individual with temporary, passive immunity.

Cytokines, including interferon, are used in an attempt to promote the body's ability to recover from cancer and to treat AIDS.

8.5 Immunity Side Effects
Allergic responses occur when the immune system reacts vigorously to substances not normally recognized as foreign. Immediate allergic responses, usually consisting of coldlike symptoms, are due to the activity of antibodies. Delayed allergic responses, such as contact dermatitis, are due to the activity of T cells.

8.6 Homeostasis
The lymphatic system works with the other systems of the body in the ways described in the illustration on page 161.

Studying the Concepts

1. What is the lymphatic system, and what are its three functions? 146
2. Describe the structure and the function of lymph nodes, tonsils, the spleen, the thymus, and red bone marrow. 147–48
3. What are the body's nonspecific defense mechanisms? 148–49
4. Describe the inflammatory reaction, and give a role for each type of cell and molecule that participates in the reaction. 148
5. What is the clonal selection theory as it applies to B cells? B cells are responsible for which type of immunity? 151
6. Describe the structure of an antibody, and define the terms variable regions and constant regions. 152
7. Describe the clonal selection theory as it applies to T cells. 154
8. Name the two main types of T cells, and state their functions. 155
9. How is active immunity artificially achieved? How is passive immunity achieved? 156–57
10. What are cytokines, and how are they used in immunotherapy? 157
11. How are monoclonal antibodies produced, and what are their applications? 158
12. Discuss allergies, autoimmune diseases, and tissue rejection as they relate to the immune system. 158–60
13. How do the lymphatic and immune systems help maintain homeostasis? 160–61
14. How does the skeletal system assist the immune system in maintaining homeostasis? 160–61

Testing Your Knowledge of the Concepts

In questions 1–4, match the cells to their functions. The answer can require more than one type of cell.
   a. T cells
   b. B cells
   c. macrophage
   d. neutrophils

   _____ 1. antibody-mediated immunity

   _____ 2. presents antigen to T cells

   _____ 3. cell-mediated immunity

   _____ 4. nonspecific defense only

In questions 5–7, indicate whether the statement is true (T) or false (F).

_____ 5. During the inflammatory reaction, monocytes differentiate into macrophages, cells that release histamine and cause capillary permeability.

_____ 6. The presentation of the antigen to helper T cells causes them to secrete cytokines.

_____ 7. Mast cells are involved in the inflammatory reaction and aggravate the symptoms of allergies.

In questions 8–18, fill in the blanks.

8. Lymphatic vessels take up excess _____ and return it to the _____ veins.

9. The function of lymph nodes is to _____ lymph.

10. T cells mature in the _____.

11. _____ is a group of proteins present in plasma that plays a role in destroying bacteria.

12. A stimulated B cell becomes antibody-secreting _____ cells and _____ cells, which are ready to produce the same type of antibody at a later time.

13. B lymphocytes and T lymphocytes have _____ _____ in the plasma membrane which combine with a specific antigen.

14. T cells produce _____, which are stimulatory molecules for all types of immune cells.

15. In order for a T cell to recognize an antigen, it must be presented by a(n) _____ along with an MHC protein.

16. Immunization with _____ brings about active immunity.

17. Antibodies produced by a single clone of cells are called _____ antibodies.

18. Good active immunity lasts as long as clones of _____ B cells and _____ T cells are present in the body.

19. Match the key terms to these definitions:
    a. _____ Treated antigens that promote active immunity without causing disease.
    b. _____ Abdominal organ that purifies blood.
    c. _____ Foreign substance that stimulates the immune system to react, such as to produce antibodies.
    d. _____ Process of programmed cell death.
    e. _____ Lymphocyte that matures in the thymus; one type kills antigen-bearing cells outright.

20. Label *a–c* on this IgG molecule using these terms: antigen-binding sites, light chain, heavy chain.

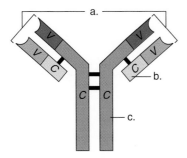

d. What does *V* stand for in the diagram?

e. What does *C* stand for in the diagram?

Understanding Key Terms

allergen 158
allergy 158
antibody 150
antibody-mediated immunity 151
antigen 150
antigen-presenting cell (APC) 154
antigen receptor 150
apoptosis 151
autoimmune disease 160
B lymphocyte 150
cell-mediated immunity 155
clonal selection theory 151
complement system 150
cytokine 154
cytotoxic T cell 155
delayed allergic response 158
edema 146
helper T cell 155
histamine 148
HLA (human leukocyte-associated) antigen 154
immediate allergic response 158
immune system 148

immunity 148
immunization 156
immunoglobulin (Ig) 152
inflammatory reaction 148
interferon 150
interleukin 157
kinin 148
lymph 146
lymph node 147
lymphatic system 146
lymphatic vessel 146
lymphoid organ 147
macrophage 148
mast cell 148
memory B cell 151
monoclonal antibody 158
natural killer (NK) cell 150
perforin 155
plasma cell 151
red bone marrow 148
spleen 147
T lymphocyte 150
thymus gland 148
tonsils 147
vaccine 156

e-Learning Connection

8.1 Lymphatic System

 Lymphatic System *Essential Study Partner*

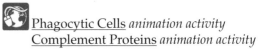 Lymphatic System *art labeling activity*
The Lymphoid Organs *art labeling activity*
Thymus Gland *art labeling activity*
Lymph Node *art labeling activity*
Spleen *art labeling activity*
Bone Marrow *art labeling activity*
Lymphatic System *animation activity*

8.2 Nonspecific Defenses

 Defense Mechanism Overview *Essential Study Partner*
Nonspecific Immunity *Essential Study Partner*

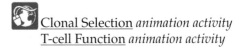 Phagocytic Cells *animation activity*
Complement Proteins *animation activity*

8.3 Specific Defenses

 Specific Immunity *Essential Study Partner*

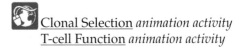 Clonal Selection *animation activity*
T-cell Function *animation activity*

8.4 Induced Immunity

 Overview *Essential Study Partner*

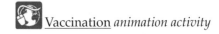 Vaccination *animation activity*

8.5 Immunity Side Effects

 Abnormalities *Essential Study Partner*

 Allergies *art quiz*

8.6 Homeostasis

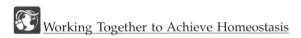 Working Together to Achieve Homeostasis

Chapter Summary

Key Term Flashcards *vocabulary quiz*
Chapter Quiz *objective quiz covering all chapter concepts*

Chapter 9

Respiratory System

Chapter Concepts

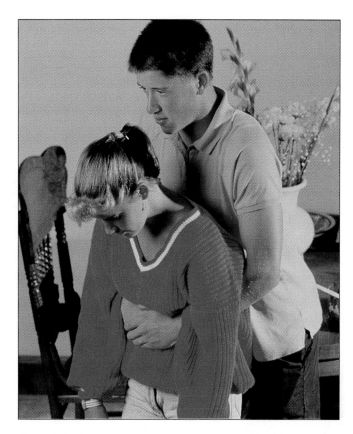

Figure 9.1   The Heimlich maneuver.
This young man shows how he saved his girlfriend from choking on a piece of candy by using the Heimlich maneuver.

At one time, Dr. Henry Heimlich thought, like many other people, that choking to death on a piece of food or some other object occurred rarely. But then he learned that choking was the sixth leading cause of accidental death in the United States—more than 20 persons a day die from choking. Many children choke on toys or fragments of balloons that explode when they are trying to blow them up.

Dr. Heimlich started doing some research. When a person chokes, quick action is needed because cessation of breathing causes damage and death within just four minutes. Back slaps don't work because they drive the offending object downward to totally block small air passages instead of sending the object upward, out of the air passages. Heimlich discovered what to do: make a fist; grab it with your other hand and press forcefully upward on the diaphragm. Standing behind someone helps, but you can even do it on yourself. The rush of air from the lungs expels the object every time. The procedure is easy to do— the man in Figure 9.1 saved his girlfriend using what is now called the Heimlich maneuver.

The air passages and the food passages cross in the pharynx, and that's why you shouldn't eat and talk at the same time. Food can go the wrong way and lodge in the windpipe. The air passages repeatedly divide, getting smaller and smaller until they reach small air sacs called alveoli in the lungs. Here gas exchange occurs. Food is stored in the body, but oxygen is not, so you have to keep on breathing to get oxygen into your system. Oxygen is transported in the blood to all the cells, where it is used in cellular respiration, the process that produces energy in the form of ATP, the common energy carrier in cells. Carbon dioxide, an end product of cellular respiration, moves in the opposite direction. Blood transports carbon dioxide from the tissues to the lungs, where it is expired. The manner in which oxygen and carbon dioxide are carried in the blood is of interest. The red blood cells and the respiratory pigment hemoglobin play a role in both processes.

As you would expect, breathing is controlled by the brain, which automatically keeps the rib cage moving up and down. But the breathing rate can be modified, particularly by the amount of carbon dioxide in the blood, and we can voluntarily alter the quantity of air we take in and release. Homeostasis is usually maintained by self-regulatory mechanisms, but it also is possible for us to consciously appraise the need for air and increase the depth of breathing. ⚖

9.1 Respiratory Tract

During **inspiration** or inhalation (breathing in) and **expiration** or exhalation (breathing out), air is conducted toward or away from the lungs by a series of cavities, tubes, and openings, illustrated in Figure 9.2.

Table 9.1 traces the path of air from the nose to the lungs. As air moves in along the airways, it is filtered, warmed, and moistened. Filtering is accomplished by coarse hairs, cilia, and mucus in the region of the nostrils and by cilia alone in the rest of the nasal cavity and the other airways of the respiratory tract. In the nose, the hairs and the cilia act as a screening device. In the trachea and other airways, the cilia beat upward, carrying mucus, dust, and occasional bits of food that "went down the wrong way" into the pharynx, where the accumulation can be swallowed or expectorated. The air is warmed by heat given off by the blood vessels lying close to the surface of the lining of the airways, and it is moistened by the wet surface of these passages.

Conversely, as air moves out during expiration, it cools and loses its moisture. As the air cools, it deposits its moisture on the lining of the windpipe and the nose, and the nose may even drip as a result of this condensation. The air

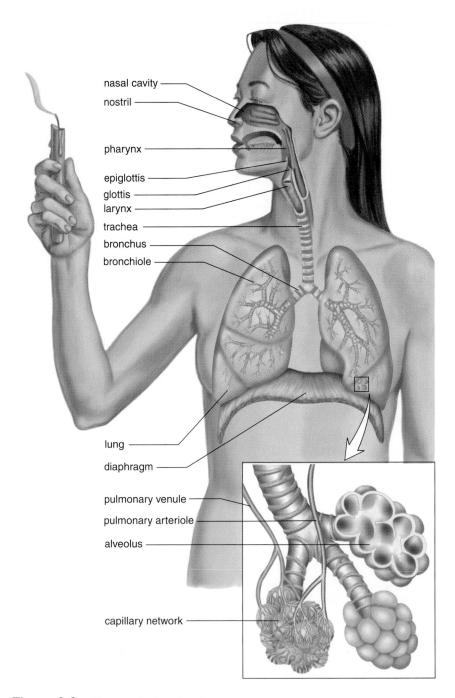

Figure 9.2   The respiratory tract.
The respiratory tract extends from the nose to the lungs, which are composed of air sacs called alveoli. Gas exchange occurs between air in the alveoli and blood within a capillary network that surrounds the alveoli. Notice that the pulmonary arteriole is colored blue—it carries O_2-poor blood away from the heart to alveoli. The pulmonary venule is colored red—it carries O_2-rich blood from alveoli toward the heart.

still retains so much moisture, however, that upon expiration on a cold day, it condenses and forms a small cloud.

Air is filtered, warmed, and moistened as it moves from the nose toward the lungs.

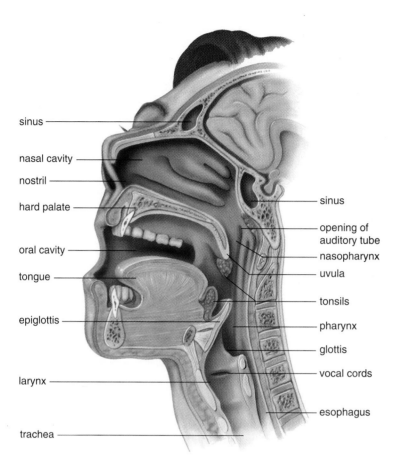

sinus

nasal cavity

nostril

hard palate

oral cavity

tongue

epiglottis

larynx

trachea

sinus

opening of
auditory tube

nasopharynx

uvula

tonsils

pharynx

glottis

vocal cords

esophagus

Figure 9.3   The path of air.
This drawing shows the path of air from the nose to the trachea.

Table 9.1	**Path of Air**	
Structure	**Description**	**Function**
Nasal cavities	Hollow spaces in nose	Filter, warm, and moisten air
Pharynx	Chamber behind oral cavity and between nasal cavity and larynx	Connection to surrounding regions
Glottis	Opening into larynx	Passage of air into larynx
Larynx	Cartilaginous organ that contains vocal cords (voice box)	Sound production
Trachea	Flexible tube that connects larynx with bronchi (windpipe)	Passage of air to bronchi
Bronchi	Divisions of the trachea that enter lungs	Passage of air to lungs
Bronchioles	Branched tubes that lead from bronchi to the alveoli	Passage of air to each alveolus
Lungs	Soft, cone-shaped organs that occupy a large portion of the thoracic cavity	Gas exchange

The Nose

The nose contains two **nasal cavities,** which are narrow canals separated from one another by a septum composed of bone and cartilage (Fig. 9.3). Special ciliated cells in the narrow upper recesses of the nasal cavities act as receptors. Nerves lead from these cells to the brain, where the impulses generated by the odor receptors are interpreted as smell.

The tear (lacrimal) glands drain into the nasal cavities by way of tear ducts. For this reason, crying produces a runny nose. The nasal cavities also communicate with the cranial sinuses, air-filled mucosa-lined spaces in the skull. If inflammation due to a cold or an allergic reaction blocks the ducts leading from the sinuses, mucus may accumulate, causing a sinus headache.

The nasal cavities empty into the nasopharynx, the upper portion of the pharynx. The auditory tubes lead from the nasopharynx to the middle ears.

The path of air starts with the nasal cavities, which open into the nasopharynx.

The Pharynx

The **pharynx** is a funnel-shaped passageway that connects the nasal and oral cavities to the larynx. Therefore, the pharynx, which is commonly referred to as the "throat," has three parts: the nasopharynx, where the nasal cavities open above the soft palate; the oropharynx, where the oral cavity opens; and the laryngopharynx, which opens into the larynx. The tonsils form a protective ring at the junction of the oral cavity and the pharynx. Being lymphoid tissue, the tonsils contain lymphocytes that protect against invasion of foreign antigens that are inhaled. In the tonsils, B cells and T cells are prepared to respond to antigens that may subsequently invade internal tissues and fluids. Therefore, the respiratroy tract assists the role the immune system plays in homeostasis.

In the pharynx, the air passage and the food passage cross because the larynx, which receives air, is ventral to the esophagus, which receives food. The larynx lies at the top of the trachea. The larynx and trachea are normally open, allowing the passage of air, but the esophagus is normally closed and opens only when swallowing occurs.

Air from either the nose or the mouth enters the pharynx, as does food. The passage of air continues in the larynx and then in the trachea.

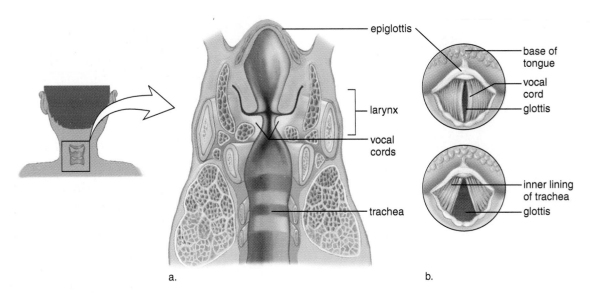

Figure 9.4   **Placement of the vocal cords.**
a. Frontal section of the larynx shows the location of the vocal cords inside the larynx. The vocal cords viewed from above are stretched across the glottis. When air passes through the glottis, the vocal cords vibrate, producing sound. **b.** The glottis is narrow when we produce a high-pitched sound *(top)* and widens as the pitch deepens *(bottom)*.

The Larynx

The **larynx** can be pictured as a triangular box whose apex, the Adam's apple, is located at the front of the neck. The Adam's apple is more prominent in men than in women. At the top of the larynx is a variable-sized opening called the **glottis.** When food is swallowed, the larynx moves upward against the **epiglottis,** a flap of tissue that prevents food from passing into the larynx. You can detect this movement by placing your hand gently on your larynx and swallowing.

The larynx is called the voice box because the vocal cords are inside the larynx. The **vocal cords** are mucosal folds supported by elastic ligaments, which are stretched across the glottis (Fig. 9.4). When air passes through the glottis, the vocal cords vibrate, producing sound. At the time of puberty, the growth of the larynx and the vocal cords is much more rapid and accentuated in the male than in the female, causing the male to have a more prominent Adam's apple and a deeper voice. The voice "breaks" in the young male due to his inability to control the longer vocal cords. These changes cause the lower pitch of the voice in males.

The high or low pitch of the voice is regulated when speaking and singing by changing the tension on the vocal cords. The greater the tension, as when the glottis becomes more narrow, the higher the pitch. When the glottis is wider, the pitch is lower (Fig. 9.4*b*). The loudness, or intensity, of the voice depends upon the amplitude of the vibrations—that is, the degree to which the vocal cords vibrate.

The Trachea

The **trachea,** commonly called the windpipe, is a tube connecting the larynx to the primary bronchi. The trachea lies ventral to the esophagus and is held open by C-shaped cartilaginous rings. The open part of the C-shaped rings faces the esophagus, and this allows the esophagus to expand when swallowing. The mucosa that lines the trachea has a layer of pseudostratified ciliated columnar epithelium. (Pseudostratified means that while the epithelium appears to be layered, actually each cell touches the basement membrane.) The cilia that pro-

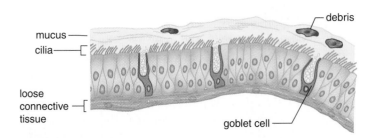

ject from the epithelium keep the lungs clean by sweeping mucus, produced by goblet cells, and debris toward the pharynx. Smoking is known to destroy the cilia, and consequently the soot in cigarette smoke collects in the lungs. Smoking is discussed more fully at the end of this chapter.

If the trachea is blocked because of illness or the accidental swallowing of a foreign object, it is possible to insert a tube by way of an incision made in the trachea. This tube acts as an artificial air intake and exhaust duct. The operation is called a **tracheostomy.**

The Bronchial Tree

The trachea divides into right and left primary bronchi (sing., **bronchus**), which lead into the right and left lungs (see Fig. 9.2). The bronchi branch into a great number of secondary bronchi that eventually lead to **bronchioles.** The bronchi resemble the trachea in structure, but as the bronchial tubes divide and subdivide, their walls become thinner, and the small rings of cartilage are no longer present. During an asthma attack, the smooth muscle of the bronchioles contracts, causing bronchiolar constriction and characteristic wheezing. Each bronchiole terminates in an elongated space enclosed by a multitude of air pockets, or sacs, called alveoli (sing., **alveolus**). The alveoli make up the lungs.

The Lungs

The **lungs** are paired, cone-shaped organs within the thoracic cavity. The right lung has three lobes, and the left lung has two lobes, allowing room for the heart, which is on the left side of the body. A lobe is further divided into lobules, and each lobule has a bronchiole serving many alveoli. The lungs lie on either side of the heart in the thoracic cavity. The base of each lung is broad and concave so that it fits the convex surface of the diaphragm. The other surfaces of the lungs follow the contours of the ribs and the diaphragm in the thoracic cavity.

The Alveoli

Each alveolar sac is made up of simple squamous epithelium surrounded by blood capillaries. Gas exchange occurs between air in the alveoli and blood in the capillaries (Fig. 9.5). Oxygen diffuses across the alveolar wall and enters the bloodstream, while carbon dioxide diffuses from the blood across the alveolar wall to enter the alveoli.

The alveoli of human lungs are lined with a surfactant, a film of lipoprotein that lowers the surface tension and prevents them from closing. The lungs collapse in some newborn babies, especially premature infants, who lack this film. The condition, called **infant respiratory distress syndrome,** is now treatable by surfactant replacement therapy.

There are altogether about 300 million alveoli, with a total cross-sectional area of 50–70 m^2. This is the surface area of a typical classroom and at least 40 times the surface area of the skin. Because of their many air spaces, the lungs are very light; normally, a piece of lung tissue dropped in a glass of water floats.

The trachea divides into the primary bronchi, which divide repeatedly to give rise to the bronchioles. The bronchioles have many branches and terminate at the alveoli, which make up the lungs.

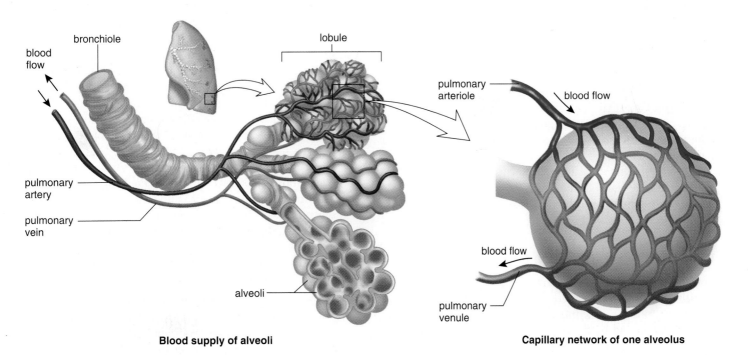

blood flow

bronchiole

lobule

pulmonary arteriole

blood flow

pulmonary artery

pulmonary vein

alveoli

blood flow

pulmonary venule

Blood supply of alveoli

Capillary network of one alveolus

Figure 9.5 **Gas exchange in the lungs.**

The lungs consist of alveoli, surrounded by an extensive capillary network. Notice that the pulmonary arteriole carries O$_2$-poor blood (colored blue) and the pulmonary venule carries O$_2$-rich blood (colored red).

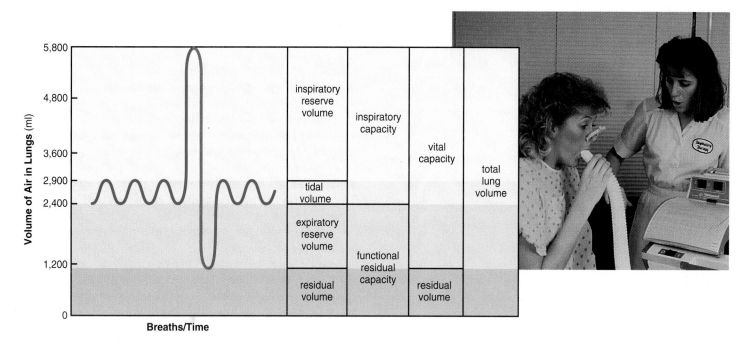

Figure 9.6   **Vital capacity.**
A spirometer measures the amount of air inhaled and exhaled with each breath. During inspiration, the pen moves up and during expiration, the pen moves down. Vital capacity (red) is the maximum amount of air a person can exhale after taking the deepest inhalation possible.

9.2 Mechanism of Breathing

The term respiration refers to the complete process of supplying oxygen to body cells for cellular respiration and the reverse process of ridding the body of carbon dioxide given off by cells. Respiration includes the following components:

1. Breathing: *inspiration* (entrance of air into the lungs) and *expiration* (exit of air from the lungs).
2. *External respiration:* exchange of the gases oxygen (O_2) and carbon dioxide (CO_2) between air and blood in the lungs.
3. *Internal respiration:* exchange of the gases O_2 and CO_2 between blood and tissue fluid.
4. *Cellular respiration:* production of ATP in cells.

Respiratory Volumes

When we breathe, the normal amount of air moved in and out with each breath is called the **tidal volume.** The tidal volume is about 500 ml, but we can increase the amount inhaled and exhaled by deep breathing. The maximum volume of air that can be moved in and out during a single breath is called the **vital capacity** (Fig. 9.6).

We can increase inspiration by as much as 2,900 ml of air by forced inspiration. This is called the **inspiratory reserve volume.** Even so, some of the inhaled air never reaches the lungs; instead, it fills the nasal cavities, trachea, bronchi, and bronchioles (see Fig. 9.2). These passages are not used for gas exchange, and therefore, they are said to contain dead space air. To ensure that inhaled air reaches the lungs, it is better to breathe slowly and deeply. Similarly, we can increase expiration by contracting the abdominal and thoracic muscles. This is called the **expiratory reserve volume,** and it measures approximately 1,400 ml of air. Vital capacity is the sum of tidal, inspiratory reserve, and expiratory reserve volumes.

Note in Figure 9.6 that even after very deep breathing, some air (about 1,000 ml) remains in the lungs; this is called the **residual volume.** This air is no longer useful for gas exchange purposes. In some lung diseases, such as emphysema (see p. 179), the residual volume builds up because the individual has difficulty emptying the lungs. This means that the vital capacity is reduced and the lungs tend to be filled with useless air.

The air used for gas exchange excludes both the air in the dead space of the respiratory tract and the residual volume in the lungs.

Ecology Focus

Photochemical Smog Can Kill

Most industrialized cities have photochemical smog at least occasionally. Photochemical smog arises when primary pollutants react with one another under the influence of sunlight to form a more deadly combination of chemicals. For example, the primary pollutants nitrogen oxides (NO_x) and hydrocarbons (HC) react with one another in the presence of sunlight to produce nitrogen dioxide (NO_2), ozone (O_3), and PAN (peroxyacetylnitrate). Ozone and PAN are commonly referred to as oxidants. Breathing oxidants affects the respiratory and nervous systems, resulting in respiratory distress, headache, and exhaustion.

Cities with warm, sunny climates that are large and industrialized, such as Los Angeles, Denver, and Salt Lake City in the United States, Sydney in Australia, Mexico City in Mexico, and Buenos Aires in Argentina, are particularly susceptible to photochemical smog. If the city is surrounded by hills, a thermal inversion may aggravate the situation. Normally, warm air near the ground rises, so that pollutants are dispersed and carried away by air currents. But sometimes during a thermal inversion, smog gets trapped near the earth by a blanket of warm air (Fig. 9A). This may occur when a cold front brings in cold air, which settles beneath a warm layer. The trapped pollutants cannot disperse, and the results can be disastrous. In 1963, about 300 people died, and in 1966, about 168 people died when air pollutants accumulated over New York City. Even worse were the events in London in 1957, when 700 to 800 people died, and in 1962, when 700 people died, due to the effects of air pollution.

Even though we have federal legislation to bring air pollution under control, more than half the people in the United States live in cities polluted by too much smog. We should place our emphasis on pollution prevention because, in the long run, prevention is usually easier and cheaper than pollution cleanup methods. Some prevention suggestions are as follows:

- Build more efficient automobiles or burn fuels that do not produce pollutants.
- Reduce the amount of waste to be incinerated by recycling materials instead.
- Reduce our energy use so that power plants need to provide less, and/or use renewable energy sources such as solar, wind, or water power.
- Require industries to meet clean air standards.

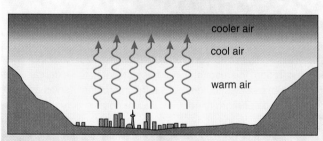

a. Normal pattern

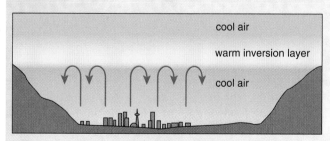

b. Thermal inversion

Figure 9A **Thermal inversion.**
a. Normally, pollutants escape into the atmosphere when warm air rises. **b.** During a thermal inversion, a layer of warm air (warm inversion layer) overlies and traps pollutants in cool air below. **c.** Los Angeles, a city of 8.5 million cars and thousands of factories, is particularly susceptible to thermal inversions, and this accounts for why this city is the "air pollution capital" of the United States.

Inspiration and Expiration

To understand **ventilation,** the manner in which air enters and exits the lungs, it is necessary to remember first that normally there is a continuous column of air from the pharynx to the alveoli of the lungs.

Second, the lungs lie within the sealed-off thoracic cavity. The **rib cage** forms the top and sides of the thoracic cavity. It contains the ribs, hinged to the vertebral column at the back and to the sternum (breastbone) at the front, and the intercostal muscles that lie between the ribs. The **diaphragm,** a dome-shaped horizontal sheet of muscle and connective tissue, forms the floor of the thoracic cavity.

The lungs are enclosed by two membranes called **pleural membranes.** An infection of the pleural membranes is called pleurisy. The parietal pleura adheres to the rib cage and the diaphragm, and the visceral pleura is fused to the lungs. The two pleural layers lie very close to one another, separated only by a small amount of fluid. Normally, the intrapleural pressure (pressure between the pleural membranes) is lower than atmospheric pressure by 4 mm Hg. The importance of the reduced intrapleural pressure is demonstrated when, by design or accident, air enters the intrapleural space. The affected lobules collapse.

The pleural membranes enclose the lungs and line the thoracic cavity. Intrapleural pressure is lower than atmospheric pressure.

Inspiration

A **respiratory center is** located in the medulla oblongata of the brain. The respiratory center consists of a group of neurons that exhibit an automatic rhythmic discharge that triggers inspiration. Carbon dioxide (CO_2) and hydrogen ions (H^+) are the primary stimuli that directly cause changes in the activity of this center. This center is not affected by low oxygen (O_2) levels. Chemoreceptors in the **carotid bodies,** located in the carotid arteries, and in the **aortic bodies,** located in the aorta, are sensitive to the level of oxygen in blood. When the concentration of oxygen decreases, these bodies communicate with the respiratory center, and the rate and depth of breathing increase.

The respiratory center sends out impulses by way of nerves to the diaphragm and the external intercostal muscles of the rib cage (Fig. 9.7). In its relaxed state, the diaphragm is dome-shaped, but upon stimulation, it contracts and lowers. Also, the external intercostal muscles contract, causing the rib cage to move upward and out-

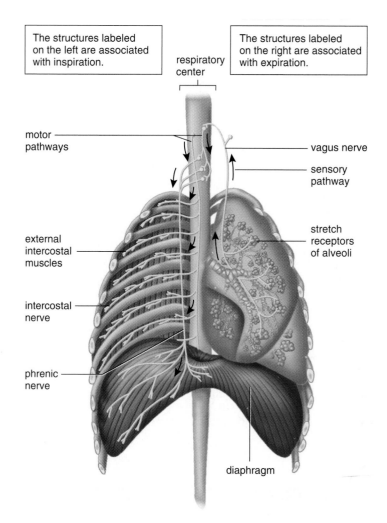

The structures labeled on the left are associated with inspiration.

The structures labeled on the right are associated with expiration.

respiratory center

motor pathways

vagus nerve

sensory pathway

external intercostal muscles

stretch receptors of alveoli

intercostal nerve

phrenic nerve

diaphragm

Figure 9.7   Nervous control of breathing.
During inspiration, the respiratory center stimulates the external intercostal (rib) muscles to contract via the intercostal nerves and stimulates the diaphragm to contract via the phrenic nerve. Should the tidal volume increase above 1.5 liters, stretch receptors send inhibitory nerve impulses to the respiratory center via the vagus nerve. In any case, expiration occurs due to a lack of stimulation from the respiratory center to the diaphragm and intercostal muscles.

ward. Now the thoracic cavity increases in size, and the lungs expand. As the lungs expand, air pressure within the enlarged alveoli lowers, and air enters through the nose or the mouth.

Inspiration is the active phase of breathing (Fig. 9.8*a*). During this time, the diaphragm and the rib muscles contract, intrapleural pressure decreases, the lungs expand, and air comes rushing in. Note that air comes in because the lungs already have opened up; air does not force the lungs open. This is why it is sometimes said that *humans breathe by negative pressure.* The creation of a partial vacuum in the alveoli causes air to enter the lungs.

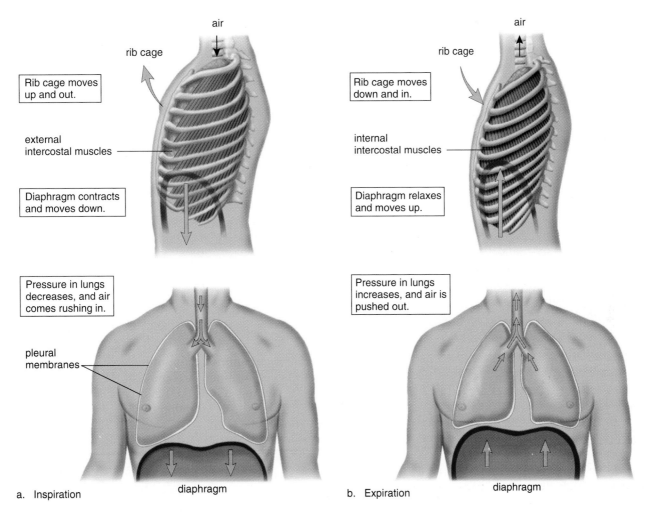

air

rib cage

Rib cage moves
up and out.

external
intercostal muscles

Diaphragm contracts
and moves down.

Pressure in lungs
decreases, and air
comes rushing in.

pleural
membranes

a. Inspiration

diaphragm

air

rib cage

Rib cage moves
down and in.

internal
intercostal muscles

Diaphragm relaxes
and moves up.

Pressure in lungs
increases, and air is
pushed out.

b. Expiration

diaphragm

Figure 9.8   Inspiration versus expiration.
a. During inspiration, the thoracic cavity and lungs expand so that air is drawn in. **b.** During expiration, the thoracic cavity and lungs resume their original positions and pressures. Now, air is forced out.

Expiration

When the respiratory center stops sending neuronal signals to the diaphragm and the rib cage, the diaphragm relaxes and resumes its dome shape. The abdominal organs press up against the diaphragm, and the rib cage moves down and inward (Fig. 9.8*b*). Now the elastic lungs recoil, and air is pushed out. The respiratory center acts rhythmically to bring about breathing at a normal rate and volume. If by chance we inhale more deeply, the lungs are expanded and the alveoli stretch. This stimulates stretch receptors in the alveolar walls, and they initiate inhibitory nerve impulses that travel from the inflated lungs to the respiratory center. This causes the respiratory center to stop sending out nerve impulses.

While inspiration is the active phase of breathing, expiration is usually passive—the diaphragm and external intercostal muscles are relaxed when expiration occurs. When breathing is deeper and/or more rapid, expiration can also be active. Contraction of internal intercostal muscles can force the rib cage to move downward and inward. Also, when the abdominal wall muscles are contracted, they push on the viscera, which push against the diaphragm, and the increased pressure in the thoracic cavity helps to expel air.

During inspiration, due to nervous stimulation, the diaphragm lowers and the rib cage lifts up and out. During expiration, due to a lack of nervous stimulation, the diaphragm rises and the rib cage lowers.

9.3 Gas Exchanges in the Body

Gas exchange is critical to homeostasis. The act of breathing brings oxygen in air to the lungs and carbon dioxide from the lungs to outside the body. Respiration includes not only the exchange of gases in the lungs, but also the exchange of gases in the tissues (Fig. 9.9). The principles of diffusion alone govern whether O_2 or CO_2 enters or leaves the blood in the lungs and in the tissues. ⚖

External Respiration

External respiration refers to the exchange of gases between air in the alveoli and blood in the pulmonary capillaries. Gases exert pressure, and the amount of pressure each gas exerts is its partial pressure, symbolized as P_{O_2} and P_{CO_2}. Blood entering the pulmonary capillaries has a higher P_{CO_2} than atmospheric air. Therefore, *CO_2 diffuses out of the blood into the lungs.* Most of the CO_2 is carried as **bicarbonate ions** (HCO_3^-). As the little remaining free CO_2 begins to diffuse out, the following reaction is driven to the right:

$$H^+ \quad + \quad HCO_3^- \longrightarrow H_2CO_3 \longrightarrow H_2O \ + \ CO_2\uparrow$$

hydrogen   bicarbonate   water   carbon
ion   ion   dioxide

"Up" arrow indicates carbon dioxide is leaving the body.

The enzyme **carbonic anhydrase,** present in red blood cells, speeds up the reaction. This reaction requires that the respiratory pigment **hemoglobin,** also present in red blood cells, gives up the hydrogen ions (H^+) it has been carrying; that is HHb becomes Hb. Hb is called deoxyhemoglobin.

The pressure pattern is the reverse for O_2. Blood entering the pulmonary capillaries is low in oxygen, and alveolar air contains a much higher partial pressure of oxygen. Therefore, *O_2 diffuses into plasma and then red blood cells in the lungs.* Hemoglobin takes up this oxygen and becomes **oxyhemoglobin** (HbO_2).

$$Hb \quad + \quad \downarrow O_2 \longrightarrow HbO_2$$

deoxyhemoglobin   oxygen   oxyhemoglobin

"Down" arrow indicates that oxygen is entering the body.

Internal Respiration

Internal respiration refers to the exchange of gases between the blood in systemic capillaries and the tissue fluid. Blood that enters the systemic capillaries is a bright red color because red blood cells contain oxyhemoglobin. Oxyhemoglobin gives up O_2, which diffuses out of blood into the tissues.

$$HbO_2 \longrightarrow Hb \quad + \quad O_2$$

oxyhemoglobin   deoxyhemoglobin   oxygen

Oxygen diffuses out of the blood into the tissues because the P_{O_2} of tissue fluid is lower than that of blood. The lower P_{O_2} is due to cells continuously using up oxygen in cellular respiration. *Carbon dioxide diffuses into the blood from the tissues* because the P_{CO_2} of tissue fluid is higher than that of blood. Carbon dioxide, produced continuously by cells, collects in tissue fluid.

After CO_2 diffuses into the blood, it enters the red blood cells, where a small amount is taken up by hemoglobin, forming **carbaminohemoglobin.** Most of the CO_2 combines with water, forming carbonic acid (H_2CO_3), which dissociates to hydrogen ions (H^+) and bicarbonate ions (HCO_3^-). The increased concentration of CO_2 in the blood drives the reaction to the right.

$$CO_2 \ + \ H_2O \xrightarrow[\text{carbonic anhydrase}]{} H_2CO_3 \rightleftharpoons H^+ \ + \ HCO_3^-$$

carbon   water   carbonic hydrogen bicarbonate
dioxide   acid   ion   ion

The enzyme carbonic anhydrase, present in red blood cells, speeds up the reaction. Bicarbonate ions diffuse out of red blood cells and are carried in the plasma. The globin portion of hemoglobin combines with excess hydrogen ions produced by the overall reaction, and Hb becomes HHb, called **reduced hemoglobin.** In this way, the pH of blood remains fairly constant. Blood that leaves the capillaries is a dark maroon color because red blood cells contain reduced hemoglobin. ⚖

External and internal respiration are the movement of gases between blood and the alveoli and between blood and the systemic capillaries, respectively. Both processes are dependent on the process of diffusion.

Visual Focus

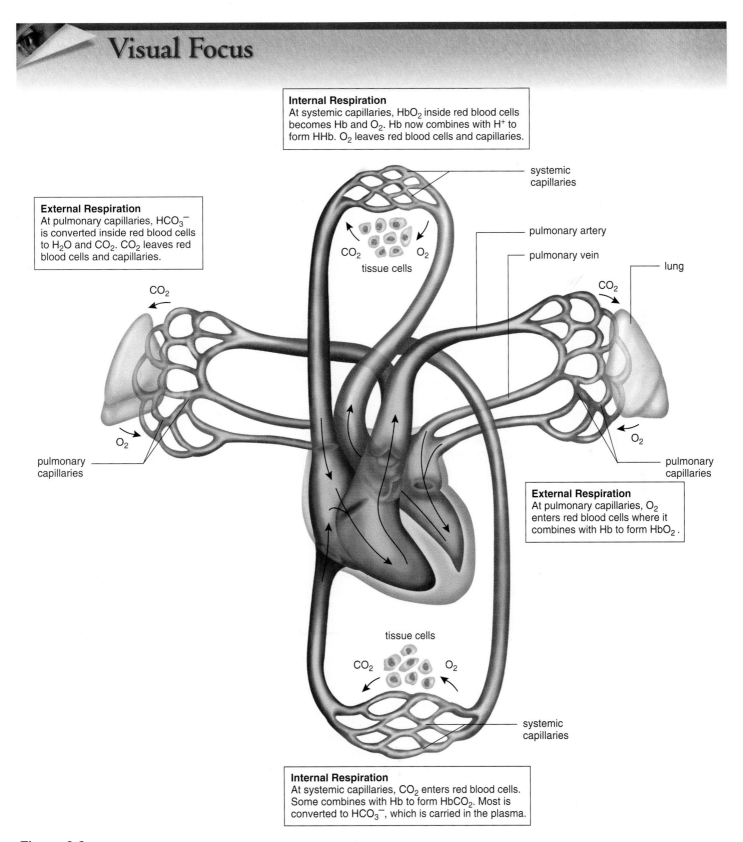

Internal Respiration
At systemic capillaries, HbO_2 inside red blood cells becomes Hb and O_2. Hb now combines with H^+ to form HHb. O_2 leaves red blood cells and capillaries.

systemic capillaries

External Respiration
At pulmonary capillaries, HCO_3^- is converted inside red blood cells to H_2O and CO_2. CO_2 leaves red blood cells and capillaries.

pulmonary artery

pulmonary vein

lung

CO_2

CO_2

CO_2

tissue cells

O_2

O_2

O_2

pulmonary capillaries

pulmonary capillaries

External Respiration
At pulmonary capillaries, O_2 enters red blood cells where it combines with Hb to form HbO_2.

tissue cells

CO_2

O_2

systemic capillaries

Internal Respiration
At systemic capillaries, CO_2 enters red blood cells. Some combines with Hb to form $HbCO_2$. Most is converted to HCO_3^-, which is carried in the plasma.

Figure 9.9   External and internal respiration.
During external respiration in the lungs, CO_2 leaves blood and O_2 enters blood. During internal respiration in the tissues, O_2 leaves blood and CO_2 enters blood.

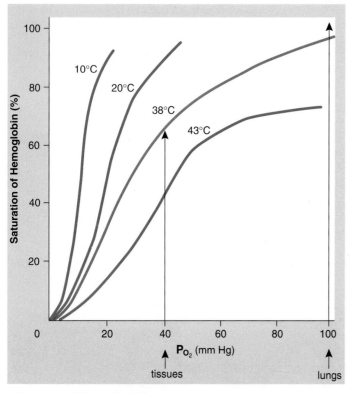

a. Saturation of Hb relative to temperature

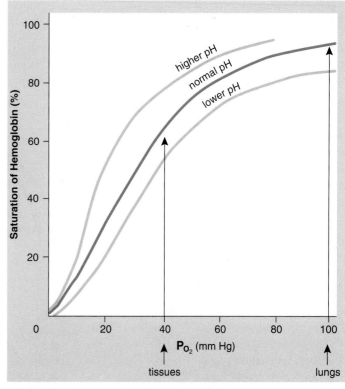

b. Saturation of Hb relative to pH

Figure 9.10   Effect of environmental conditions on hemoglobin saturation.
The partial pressure of oxygen (P_{O_2}) in pulmonary capillaries is about 98–100 mm Hg, but only about 40 mm Hg in tissue capillaries. Hemoglobin is about 98% saturated in the lungs because of P_{O_2}, and also because (**a**) the temperature is cooler and (**b**) the pH is higher in the lungs. On the other hand, hemoglobin is only about 60% saturated in the tissues because of the P_{O_2}, and also because (**a**) the temperature is warmer and (**b**) the pH is lower in the tissues.

Binding Capacity of Hemoglobin

The binding capacity of hemoglobin is also affected by partial pressures. The P_{O_2} of air entering the alveoli is about 100 mm Hg, and at this pressure the hemoglobin in the blood becomes saturated with O_2. This means that iron in hemoglobin molecules has combined with O_2. On the other hand, the P_{O_2} in the tissues is about 40 mm Hg, causing hemoglobin molecules to release O_2, and O_2 to diffuse into the tissues.

In addition to the partial pressure of O_2, temperature and pH also affect the amount of oxygen that hemoglobin can carry. The lungs have a lower temperature and a higher pH than the tissues:

	pH	Temperature
Lungs	7.40	37°C
Tissues	7.38	38°C

Figures 9.10*a* and *b* show that, as expected, hemoglobin is more saturated with O_2 in the lungs than in the tissues. This effect, which can be attributed to the difference in P_{O_2} between the lungs and tissues, is enhanced by the difference in temperature and pH between the lungs and tissues. Notice in Figure 9.10*a* that the saturation curve for hemoglobin is steeper at 10°C compared to 20°C, and so forth. Also, Figure 9.10*b* shows that the saturation curve for hemoglobin is steeper at higher pH than at lower pH.

This means that the environmental conditions in the lungs are favorable for the uptake of O_2 by hemoglobin, and the environmental conditions in the tissues are favorable for the release of O_2 by hemoglobin. Hemoglobin is about 98–100% saturated in the capillaries of the lungs and about 60–70% saturated in the tissues. During exercise, hemoglobin is even less saturated in the tissues because muscle contraction leads to higher body temperature (up to 103°F in marathoners!) and lowers the pH (due to the production of lactic acid).

The difference in P_{O_2}, temperature, and pH between the lungs and tissues causes hemoglobin to take up oxygen in the lungs and release oxygen in the tissues.

9.4 Respiration and Health

The respiratory tract is constantly exposed to environmental air. The quality of this air, as discussed in the Ecology Focus on page 171, can affect our health. The presence of a disease means that homeostasis is threatened, and if the condition is not brought under control, death is a possibility.

Upper Respiratory Tract Infections

The upper respiratory tract consists of the nasal cavities, the pharynx, and the larynx. Upper respiratory infections (URI) can spread from the nasal cavities to the sinuses, to the middle ears, and to the larynx (Fig. 9.11). Viral infections sometimes lead to secondary bacterial infections. What we call "strep throat" is a primary bacterial infection caused by *Streptococcus pyogenes* that can lead to a generalized upper respiratory infection and even a systemic (affecting the body as a whole) infection. While antibiotics have no effect on viral infections, they are successfully used for most bacterial infections, including strep throat.

Sinusitis

Sinusitis is an infection of the cranial sinuses, cavities within the facial skeleton that drain into the nasal cavities. Only about 1–3% of upper respiratory infections are accompanied by sinusitis. Sinusitis develops when nasal congestion blocks the tiny openings leading to the sinuses. Symptoms include postnasal discharge as well as facial pain that worsens when the patient bends forward. Pain and tenderness usually occur over the lower forehead or over the cheeks. If the latter, toothache is also a complaint. Successful treatment depends on restoring proper drainage of the sinuses. Even a hot shower and sleeping upright can be helpful. Otherwise, spray decongestants are preferred over oral antihistamines, which thicken rather than liquefy the material trapped in the sinuses.

Otitis Media

Otitis media is a bacterial infection of the middle ear. The middle ear is not a part of the respiratory tract, but this infection is considered here because it is a complication often seen in children who have a nasal infection. Infection can spread by way of the **auditory tube** that leads from the nasopharynx to the middle ear. Pain is the primary symptom of a middle ear infection. A sense of fullness, hearing loss, vertigo (dizziness), and fever may also be present. Antibiotics almost always bring about a full recovery, and a recurrence is probably due to a new infection. Drainage tubes (called tympanostomy tubes) are sometimes placed in the eardrums of children with multiple recurrences to help prevent the buildup of fluid in the middle ear and the possibility of hearing loss. Normally, the tubes slough out with time.

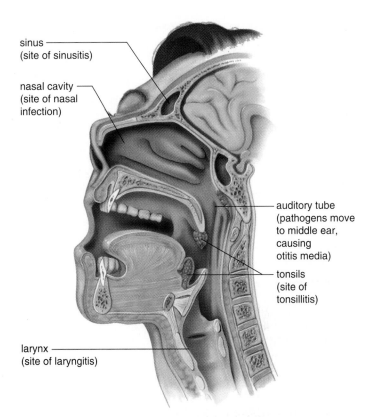

sinus
(site of sinusitis)

nasal cavity
(site of nasal
infection)

auditory tube
(pathogens move
to middle ear,
causing
otitis media)

tonsils
(site of
tonsillitis)

larynx
(site of laryngitis)

Figure 9.11   Sites of upper respiratory infections.
A nasal infection, more properly called rhinitis, is the usual symptom of a common cold due to a viral infection, but rhinitis can also be due to a bacterial infection. Secondary to a URI, the sinuses, tonsils, vocal cords, and middle ear can become infected. Allergies are also a cause of runny nose, blocked sinuses, and laryngitis.

Tonsillitis

Tonsillitis occurs when the tonsils become inflamed and enlarged. **Tonsils** are masses of lymphatic tissue in the pharynx. The tonsils in the dorsal wall of the nasopharynx are often called adenoids. If tonsillitis occurs frequently and enlargement makes breathing difficult, the tonsils can be removed surgically in a **tonsillectomy.** Fewer tonsillectomies are performed today than in the past because it is now known that the tonsils remove many of the pathogens that enter the pharynx; therefore, they are a first line of defense against invasion of the body.

Laryngitis

Laryngitis is an infection of the larynx with an accompanying hoarseness leading to the inability to talk in an audible voice. Usually laryngitis disappears with treatment of the upper respiratory infection. Persistent hoarseness without the presence of an upper respiratory infection is one of the warning signs of cancer and therefore should be looked into by a physician.

Visual Focus

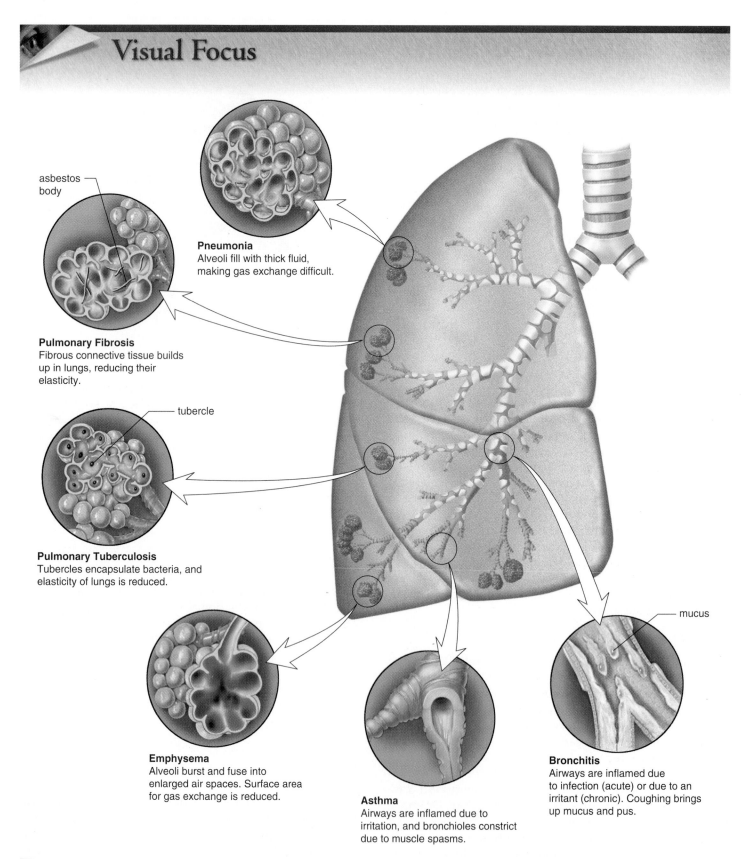

Pneumonia
Alveoli fill with thick fluid, making gas exchange difficult.

asbestos body

Pulmonary Fibrosis
Fibrous connective tissue builds up in lungs, reducing their elasticity.

tubercle

Pulmonary Tuberculosis
Tubercles encapsulate bacteria, and elasticity of lungs is reduced.

Emphysema
Alveoli burst and fuse into enlarged air spaces. Surface area for gas exchange is reduced.

Asthma
Airways are inflamed due to irritation, and bronchioles constrict due to muscle spasms.

mucus

Bronchitis
Airways are inflamed due to infection (acute) or due to an irritant (chronic). Coughing brings up mucus and pus.

Figure 9.12 **Lower respiratory tract disorders.**
Exposure to infectious pathogens and/or air pollutants, including cigarette and cigar smoke, can cause the diseases and disorders shown here.

Lower Respiratory Tract Disorders

Lower respiratory tract disorders, which are illustrated in Figure 9.12, include infections, restrictive pulmonary disorders, obstructive pulmonary disorders, and lung cancer.

Lower Respiratory Infections

Acute bronchitis, pneumonia, and tuberculosis are infections of the lower respiratory tract. **Acute bronchitis** is an infection of the primary and secondary bronchi. Usually it is preceded by a viral URI that has led to a secondary bacterial infection. Most likely, a nonproductive cough has become a deep cough that expectorates mucus and perhaps pus.

Pneumonia is a viral or bacterial infection of the lungs in which the bronchi and alveoli fill with thick fluid. Most often it is preceded by influenza. High fever and chills, with headache and chest pain, are symptoms of pneumonia. Rather than being a generalized lung infection, pneumonia may be localized in specific lobules of the lungs; obviously, the more lobules involved, the more serious the infection. Pneumonia can be caused by a bacterium that is usually held in check, but that has gained the upper hand due to stress and/or reduced immunity. AIDS patients are subject to a particularly rare form of pneumonia caused by the protozoan *Pneumocystis carinii*. Pneumonia of this type is almost never seen in individuals with a healthy immune system.

Pulmonary tuberculosis is caused by the tubercle bacillus, a type of bacterium. It is possible to tell if a person has ever been exposed to tuberculosis with a skin test in which a highly diluted extract of the bacillus is injected into the skin of the patient. A person who has never been in contact with the tubercle bacillus shows no reaction, but one who has developed immunity to the organism shows an area of inflammation that peaks in about 48 hours. When tubercle bacilli invade the lung tissue, the cells build a protective capsule about the foreigners, isolating them from the rest of the body. This tiny capsule is called a tubercle. If the resistance of the body is high, the imprisoned organisms die, but if the resistance is low, the organisms eventually can be liberated. If a chest X ray detects active tubercles, the individual is put on appropriate drug therapy to ensure the localization of the disease and the eventual destruction of any live bacteria.

Tuberculosis was a major killer in the United States before the middle of this century, after which antibiotic therapy brought it largely under control. In recent years, however, the incidence of tuberculosis is on the rise, particularly among AIDS patients, the homeless, and the rural poor. Worse, the new strains are resistant to the usual antibiotic therapy. Therefore, some physicians would like to again quarantine patients in sanitariums, as was previously done.

Restrictive Pulmonary Disorders

In restrictive pulmonary disorders, vital capacity is reduced not because air does not move freely into and out of the lungs but because the lungs have lost their elasticity. Inhaling particles such as silica (sand), coal dust, asbestos, and, now it seems, fiberglass can lead to **pulmonary fibrosis,** a condition in which fibrous connective tissue builds up in the lungs. The lungs cannot inflate properly and are always tending toward deflation. Breathing asbestos is also associated with the development of cancer. Since asbestos was formerly used so widely as a fireproofing and insulating agent, unwarranted exposure has occurred. It is projected that two million deaths caused by asbestos exposure—mostly in the workplace—will occur between 1990 and 2020.

Obstructive Pulmonary Disorders

In obstructive pulmonary disorders, air does not flow freely in the airways, and the time it takes to inhale or exhale maximally is greatly increased. Several disorders, including chronic bronchitis, emphysema, and asthma, are referred to as chronic obstructive pulmonary disorders (COPD) because they tend to recur and have flare-ups.

In **chronic bronchitis,** the airways are inflamed and filled with mucus. A cough that brings up mucus is common. The bronchi have undergone degenerative changes, including the loss of cilia and their normal cleansing action. Under these conditions, an infection is more likely to occur. Smoking cigarettes and cigars is the most frequent cause of chronic bronchitis. Exposure to other pollutants can also cause chronic bronchitis.

Emphysema is a chronic and incurable disorder in which the alveoli are distended and their walls damaged so that the surface area available for gas exchange is reduced. Emphysema is often preceded by chronic bronchitis. Air trapped in the lungs leads to alveolar damage and a noticeable ballooning of the chest. The elastic recoil of the lungs is reduced, so not only are the airways narrowed but the driving force behind expiration is also reduced. The victim is breathless and may have a cough. Because the surface area for gas exchange is reduced, oxygen reaching the heart and the brain is reduced. Even so, the heart works furiously to force more blood through the lungs, and an increased workload on the heart can result. Lack of oxygen to the brain can make the person feel depressed, sluggish, and irritable. Exercise, drug therapy, and supplemental oxygen, along with giving up smoking, may relieve the symptoms and possibly slow the progression of emphysema.

Bioethical Focus

Use of Antibiotics

Since the introduction of the first antibiotics in the 1940s, there has been a dramatic decline in deaths due to respiratory illnesses like pneumonia and tuberculosis. Strep throat and ear infections have also been brought under control with antibiotics, which are chemicals that selectively kill bacteria without harming host cells.

There are problems associated with antibiotic therapy, however. Aside from a possible allergic reaction, antibiotics not only kill off disease-causing bacteria, they also reduce the number of beneficial bacteria in the intestinal tract and other locations. These beneficial bacteria hold in check the growth of other microbes that in their absence begin to flourish. Diarrhea can result, as can a vaginal yeast infection. The use of antibiotics can also prevent natural immunity from occurring, leading to the need for recurring antibiotic therapy. Especially alarming at this time is the occurrence of resistance. Resistance takes place when vulnerable bacteria are killed off by an antibiotic, and this allows resistant bacteria to become prevalent. The bacteria that cause ear, nose, and throat infections, as well as scarlet fever and pneumonia, are becoming widely resistant because we have not been using antibiotics properly. Tuberculosis is on the rise, and the new strains are resistant to the usual combined antibiotic therapy. When a disease is caused by a resistant bacterium, it cannot be cured by the administration of any presently available antibiotic.

Although drug companies now recognize the problem and have begun to develop new antibiotics that hopefully will kill bacteria resistant to today's antibiotics, every citizen needs to be aware of our present crisis situation. Stuart Levy, a Tufts University School of Medicine microbiologist, says that we should do what is ethical for society and ourselves. Antibiotics kill bacteria, not viruses—therefore, we shouldn't take antibiotics unless we know for sure we have a bacterial infection. And we shouldn't take them prophylactically—that is, just in case we might need one. If antibiotics are taken in low dosages and intermittently, resistant strains are bound to take over. Animal and agricultural use should be pared down, and household disinfectants should no longer be spiked with antibacterial agents. Perhaps then, Levy says, vulnerable bacteria will begin to supplant the resistant ones in the population.

Decide Your Opinion

1. With regard to antibiotics, should each person think about the needs of society as well as their own needs? Why or why not?
2. Should each person do what he or she can to help prevent the growing resistance of bacteria to disease? Why or why not?
3. Should you gracefully accept a physician's decision that an antibiotic will not help an illness you may have? Why or why not?

Figure 9B   **Antibiotic therapy.**
When we misuse antibiotic therapy, resistant strains of bacteria can develop.

Looking at Both Sides   www.mhhe.com/biosci/genbio/maderhuman7

Every bioethical issue has at least two sides. Even if you already have an opinion, it is important to explore the opposite opinion before finalizing your position. The Online Learning Center at www.mhhe.com/biosci/genbio/maderhuman7/ will help you fine-tune your initial opinion, explore both sides, and finalize your position. You may acquire new arguments for your original opinion, or you may even change your opinion. Be sure to complete these activities in sequence:

Taking Sides   Decide your initial opinion by answering a series of questions. Then see if your opinion changes after completing the next two activities.

Further Debate   Read opposing articles that give you further information on this particular bioethical issue.

Explain Your Position   Answer another series of questions and then defend your original or changed opinion. You can e-mail your position to your instructor if he or she wishes.

Asthma is a disease of the bronchi and bronchioles that is marked by wheezing, breathlessness, and sometimes a cough and expectoration of mucus. The airways are unusually sensitive to specific irritants, which can include a wide range of allergens such as pollen, animal dander, dust, cigarette smoke, and industrial fumes. Even cold air can be an irritant. When exposed to the irritant, the smooth muscle in the bronchioles undergoes spasms. It now appears that chemical mediators given off by immune cells in the bronchioles cause the spasms. Most asthma patients have some degree of bronchial inflammation that reduces the diameter of the airways and contributes to the seriousness of an attack. Asthma is not curable but it is treatable. Special inhalers can control the inflammation and hopefully prevent an attack, while other types of inhalers can stop the muscle spasms should an attack occur.

Lung Cancer

Lung cancer used to be more prevalent in men than in women, but recently it has surpassed breast cancer as a cause of death in women. This can be linked to an increase in the number of women who smoke today. Autopsies on smokers have revealed the progressive steps by which the most common form of lung cancer develops. The first event appears to be thickening and callusing of the cells

lining the airways. (Callusing occurs whenever cells are exposed to irritants.) Then there is a loss of cilia so that it is impossible to prevent dust and dirt from settling in the lungs. Following this, cells with atypical nuclei appear in the callused lining. A tumor (Fig. 9.13b) consisting of disordered cells with atypical nuclei is considered to be cancer in situ (at one location). A final step occurs when some of these cells break loose and penetrate other tissues, a process called metastasis. Now the cancer has spread. The original tumor may grow until a bronchus is blocked, cutting off the supply of air to that lung. The entire lung then collapses, the secretions trapped in the lung spaces become infected, and pneumonia or a lung abscess (localized area of pus) results. The only treatment that offers a possibility of cure is to remove a lobe or the lung completely before metastasis has had time to occur. This operation is called **pneumonectomy.** If the cancer has spread, chemotherapy and radiation will also be required.

Current research indicates that secondhand smoke can also cause lung cancer and other illnesses associated with smoking. The Health Focus on the next page lists the various illnesses that are apt to occur when a person smokes. If a person stops both voluntary and involuntary smoking, and if the body tissues are not already cancerous, they may return to normal over time.

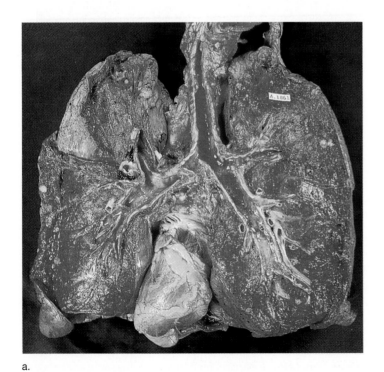

a.

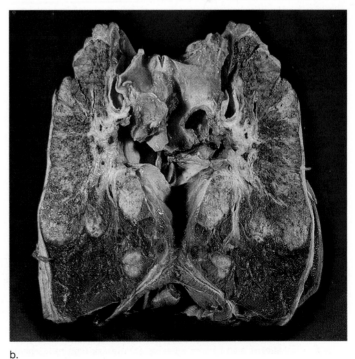

b.

Figure 9.13 Normal lung versus cancerous lung.
a. Normal lung with heart in place. Note the healthy red color. **b.** Lungs of a heavy smoker. Notice how black the lungs are except where cancerous tumors have formed.

Health Focus

The Most Often Asked Questions About Tobacco and Health

Is there a safe way to smoke?

No. All forms of tobacco can cause damage, and smoking even a small amount is dangerous. Tobacco is perhaps the only legal product whose advertised and intended use—that is, smoking it—will hurt the body.

Does smoking cause cancer?

Yes, and not only lung cancer. Besides causing lung cancer, smoking a pipe, cigarettes, or cigars is also a major cause of cancers of the mouth, larynx (voice box), and esophagus. In addition, smoking increases the risk of cancer of the bladder, kidney, pancreas, stomach, and the uterine cervix.

What are the chances of being cured of lung cancer?

Very low; the five-year survival rate is only 13%. Fortunately, lung cancer is a largely preventable disease. That is, by not smoking, it can probably be prevented.

Does smoking cause other lung diseases?

Yes. It leads to chronic bronchitis—a disease in which the airways produce excess mucus, forcing the smoker to cough frequently. Smoking is also the major cause of emphysema—a disease that slowly destroys a person's ability to breathe.

Why do smokers have "smoker's cough"?

Normally, cilia (tiny hairlike formations that line the airways) beat outwards and "sweep" harmful material out of the lungs. Smoke, however, decreases this sweeping action, so some of the poisons in the smoke remain in the lungs.

If you smoke but don't inhale, is there any danger?

Yes. Wherever smoke touches living cells, it does harm. So, even if smokers of pipes, cigarettes, and cigars don't inhale, they are at an increased risk for lip, mouth, and tongue cancer.

Does smoking affect the heart?

Yes. Smoking increases the risk of heart disease, which is the United States' number one killer. Smoking, high blood pressure, high cholesterol, and lack of exercise are all risk factors for heart disease. Smoking alone doubles the risk of heart disease.

Is there any risk for pregnant women and their babies?

Pregnant women who smoke endanger the health and lives of their unborn babies. When a pregnant woman smokes, she really is smoking for two because the nicotine, carbon monoxide, and other dangerous chemicals in smoke enter the mother's bloodstream and then pass into the baby's body. Smoking mothers have more stillbirths and babies of low birth weight than nonsmoking mothers.

Does smoking cause any special health problems for women?

Yes. Women who smoke and use the birth control pill have an increased risk of stroke and blood clots in the legs as well. In addition, women who smoke increase their chances of getting cancer of the uterine cervix.

What are some of the short-term effects of smoking cigarettes?

Almost immediately, smoking can make it hard to breathe. Within a short time, it can also worsen asthma and allergies. Only seven seconds after a smoker takes a puff, nicotine reaches the brain, where it produces a morphinelike effect.

Are there any other risks to the smoker?

Yes, there are many more risks. Smoking is a cause of stroke, which is the third leading cause of death in the United States. Chronic bronchitis and pulmonary emphysema are higher in smokers compared to nonsmokers. Smokers are more likely to have and die from stomach ulcers than nonsmokers. Smokers have a higher incidence of cancer in general. If a person smokes and is exposed to radon or asbestos, the risk for lung cancer increases dramatically.

What are the dangers of passive smoking?

Passive smoking causes lung cancer in healthy nonsmokers. Children whose parents smoke are more likely to suffer from pneumonia or bronchitis in the first two years of life than children who come from smoke-free households. Passive smokers have a 30% greater risk of developing lung cancer than nonsmokers who live in a smoke-free house.

Are chewing tobacco and snuff safe alternatives to cigarette smoking?

No, they are not. Many people who use chewing tobacco or snuff believe it can't harm them because there is no smoke. Wrong. Smokeless tobacco contains nicotine, the same addicting drug found in cigarettes and cigars. While not inhaled through the lungs, the juice from smokeless tobacco is absorbed through the lining of the mouth. There it can cause sores and white patches, which often lead to cancer of the mouth. Snuff dippers also take in an average of over ten times more cancer-causing substances than cigarette smokers.

Human Systems Work Together

Integumentary System

Gas exchange in lungs provides oxygen to skin and rids body of carbon dioxide from skin.

Skin helps protect respiratory organs and helps regulate body temperature.

Skeletal System

Gas exchange in lungs provides oxygen and rids body of carbon dioxide.

Rib cage protects lungs and assists breathing; bones provide attachment sites for muscles involved in breathing.

Muscular System

Lungs provide oxygen for contracting muscles and rid the body of carbon dioxide from contracting muscles.

Muscle contraction assists breathing; physical exercise increases respiratory capacity.

Nervous System

Lungs provide oxygen for neurons and rid the body of carbon dioxide produced by neurons.

Respiratory centers in brain regulate breathing rate.

Endocrine System

Gas exchange in lungs provides oxygen and rids body of carbon dioxide.

Epinephrine promotes ventilation by dilating bronchioles; growth factors control production of red blood cells that carry oxygen.

How the Respiratory System works with other body systems

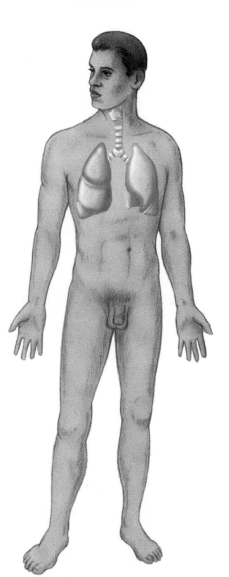

Cardiovascular System

Gas exchange in lungs rids body of carbon dioxide, helping to regulate the pH of blood; breathing aids venous return.

Blood vessels transport gases to and from lungs; blood services respiratory organs.

Lymphatic System/Immunity

Tonsils and adenoids occur along respiratory tract; breathing aids lymph flow; lungs carry out gas exchange.

Lymphatic vessels pick up excess tissue fluid; immune system protects against respiratory tract and lung infections.

Digestive System

Gas exchange in lungs provides oxygen to the digestive tract and excretes carbon dioxide from the digestive tract.

Breathing is possible through the mouth because digestive tract and respiratory tract share the pharynx.

Urinary System

Lungs excrete carbon dioxide, provide oxygen, and convert angiotensin I to angiotensin II, leading to kidney regulation.

Kidneys compensate for water lost through respiratory tract; work with lungs to maintain blood pH.

Reproductive System

Gas exchange increases during sexual activity.

Sexual activity increases breathing; pregnancy causes breathing rate and vital capacity to increase.

9.5 Homeostasis

The respiratory system contributes to homeostasis in two primary ways. First, the lungs perform gas exchange. Carbon dioxide, a waste molecule given off by cellular respiration, exits the body, and oxygen, a molecule needed for cellular respiration, enters the body at the lungs. Cellular respiration produces ATP, a molecule that allows the body to perform all sorts of work, including muscle contraction and nerve conduction. It is estimated that the brain uses 15–20% of the oxygen taken into the blood. Not surprisingly, a lack of oxygen affects the functioning of the brain and our judgment before it affects other organs.

Second, the respiratory system is involved in regulating the pH of the blood. In the tissues, carbon dioxide enters the blood and red blood cells where this reaction occurs. The bicarbonate ion (HCO_3^-) diffuses out of the red blood cells to be carried in the plasma.

$$CO_2 \ + \ H_2O \ \overset{\text{tissues}}{\underset{\text{lungs}}{\rightleftharpoons}} \ H_2CO_3 \ \overset{\text{tissues}}{\underset{\text{lungs}}{\rightleftharpoons}} \ H^+ \ + \ HCO_3^-$$

This reaction lowers the blood pH because it gives off H^+. When carbon dioxide starts to diffuse out of the blood in the lungs, the reaction occurs in the reverse direction. Now, the blood pH rises.

What happens to blood pH if you hypoventilate—that is, breathe at a low rate? A low blood pH, called acidosis, results because hydrogen ions are being held in the body. Any condition, such as emphysema, that hinders the passage of carbon dioxide out of the blood also results in acidosis. What happens to blood pH if you hyperventilate—that is, breathe at a high rate? A high blood pH, called alkalosis, results because carbon dioxide is leaving the body at a high rate. Severe anxiety can cause a person to hyperventilate.

The illustration on the previous page tells how the respiratory system depends on and assists other systems of the body. The brain controls the respiratory rate; hypoventilation occurs if the respiratory center is depressed, and hyperventilation occurs if this center is stimulated.

The cardiovascular system transports oxygen from the lungs to the tissues and carbon dioxide from the tissues to the lungs. As mentioned in chapter 7, expansion of the chest during inspiration causes a reduced pressure that promotes the flow of blood toward the thoracic cavity and the heart. Therefore, the act of breathing assists the return of blood to the heart and the transport of carbon dioxide to the lungs.

The respiratory tract assists immunity. The cilia that line the trachea sweep impurities toward the throat, for example. Also, we now know that the tonsils serve as a location where T cells are presented with antigens before they enter the body as a whole. This action helps the body prepare to respond to an antigen before it enters the bloodstream! The contributions of the respiratory system to homeostasis cannot be overemphasized.

Summarizing the Concepts

9.1 Respiratory Tract
The respiratory tract consists of the nose (nasal cavities), the nasopharynx, the pharynx, the larynx (which contains the vocal cords), the trachea, the bronchi, the bronchioles, and the alveoli. The bronchi, along with the pulmonary arteries and veins, enter the lungs, which consist of the alveoli, air sacs surrounded by a capillary network.

9.2 Mechanism of Breathing
Inspiration begins when the respiratory center in the medulla oblongata sends excitatory nerve impulses to the diaphragm and the muscles of the rib cage. As they contract, the diaphragm lowers, and the rib cage moves upward and outward; the lungs expand, creating a partial vacuum, which causes air to rush in. The respiratory center now stops sending impulses to the diaphragm and muscles of the rib cage. As the diaphragm relaxes, it resumes its dome shape, and as the rib cage retracts, air is pushed out of the lungs during expiration.

9.3 Gas Exchanges in the Body
External respiration occurs when CO_2 leaves blood via the alveoli and O_2 enters blood from the alveoli. Oxygen is transported to the tissues in combination with hemoglobin as oxyhemoglobin (HbO_2). Internal respiration occurs when O_2 leaves blood and CO_2 enters blood at the tissues. Carbon dioxide is mainly carried to the lungs within the plasma as the bicarbonate ion (HCO_3^-). Hemoglobin combines with hydrogen ions and becomes reduced (HHb).

9.4 Respiration and Health
A number of illnesses are associated with the respiratory tract. These disorders are divided into those that affect the upper respiratory tract and those that affect the lower respiratory tract. Infections of the nasal cavities, sinuses, throat, tonsils, and larynx are all well known. In addition, infections can spread from the nasopharynx to the ears.

The lower respiratory tract is also subject to infections such as acute bronchitis, pneumonia, and pulmonary tuberculosis. In restrictive pulmonary disorders, exemplified by pulmonary fibrosis, the lungs lose their elasticity. In obstructive pulmonary disorders, exemplified by chronic bronchitis, emphysema, and asthma, the bronchi (and bronchioles) do not effectively conduct air to and from the lungs. Smoking, which is associated with chronic bronchitis and emphysema, can eventually lead to lung cancer.

9.5 Homeostasis
The respiratory system works with the other systems of the body in the ways described in the illustration on page 183.

Studying the Concepts

1. List the parts of the respiratory tract. What are the special functions of the nasal cavity, the larynx, and the alveoli? 167–69
2. Name and explain the four parts of respiration. 170
3. What is the difference between tidal volume and vital capacity? Of the air we breathe, what part is not used for gas exchange? 170
4. What are the steps in inspiration and expiration? How is breathing controlled? 172–73
5. Discuss the events of external respiration, and include two pertinent equations in your discussion. 174
6. What two equations pertain to the exchange of gases during internal respiration? 174
7. State two factors that influence hemoglobin's O2- binding capacity, and relate them to the environmental conditions in the lungs and tissues. 176
8. Name four respiratory tract disorders other than cancer, and explain why breathing is difficult with these conditions. 178
9. What are emphysema and pulmonary fibrosis, and how do they affect a person's health? 179
10. List the steps by which lung cancer develops. 181

Testing Your Knowledge of the Concepts

In questions 1–4, match the organ to its description and function.
a. larynx
b. bronchi
c. tonsils
d. lungs

_____ 1. airways to the lungs supported by C-shaped cartilaginous rings

_____ 2. contains vocal cords and produces the voice

_____ 3. lymphoid tissue that helps fight infections

_____ 4. perform gas exchange in alveoli

In questions 5–7, indicate whether the statement is true (T) or false (F).

_____ 5. In tracing the path of air, the bronchioles immediately follow the trachea.

_____ 6. Reduced hemoglobin becomes oxyhemoglobin in the lungs.

_____ 7. The most frequent cause of chronic bronchitis is a strep infection.

In questions 8 and 9, fill in the blanks.

8. Air enters the lungs after they have _____.

9. The hydrogen ions (H^+) given off when carbonic acid (H_2CO_3) dissociates are carried by _____.

10. Label this diagram of the human respiratory tract.

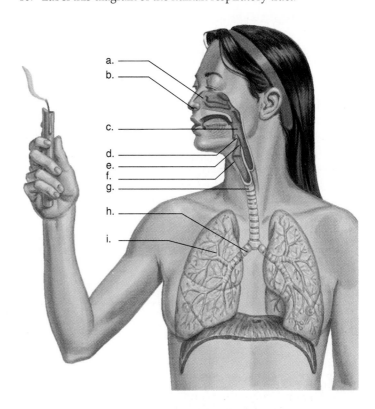

a. _____
b. _____
c. _____
d. _____
e. _____
f. _____
g. _____
h. _____
i. _____

Understanding Key Terms

acute bronchitis 179
alveolus 169
aortic bodies 172
asthma 181
auditory tube 177
bicarbonate ion 174
bronchiole 169
bronchus 169
carbaminohemoglobin 174
carbonic anhydrase 174
carotid bodies 172
chronic bronchitis 179
diaphragm 172
emphysema 179
epiglottis 168
expiration 166
expiratory reserve
    volume 170
external respiration 174
glottis 168
hemoglobin 174
infant respiratory distress
    syndrome 169
inspiration 166
inspiratory reserve
    volume 170
internal respiration 174

laryngitis 177
larynx 168
lung cancer 181
lungs 169
nasal cavity 167
otitis media 177
oxyhemoglobin 174
pharynx 167
pleural membrane 172
pneumonectomy 181
pneumonia 179
pulmonary fibrosis 179
pulmonary tuberculosis 179
reduced hemoglobin 174
residual volume 170
respiratory center 172
rib cage 172
sinusitis 177
tidal volume 170
tonsillectomy 177
tonsillitis 177
tonsils 177
trachea 168
tracheostomy 168
ventilation 172
vital capacity 170
vocal cords 168

e-Learning Connection

www.mhhe.com/biosci/genbio/maderhuman7/

9.1 Respiratory Tract

<u>Human Breathing</u> *Essential Study Partner*

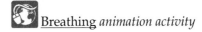

<u>The Respiratory Tract</u> *art labeling activity*
<u>The Respiratory Tract II</u> *art labeling activity*
<u>The Upper Respiratory Tract</u> *art labeling activity*
<u>The Upper Respiratory Tract II</u> *art labeling activity*
<u>Section of Larynx</u> *art labeling activity*
<u>Glottis Function</u> *art labeling activity*

9.2 Mechanism of Breathing

<u>Measuring Function</u> *Essential Study Partner*

<u>Breathing</u> *animation activity*

9.3 Gas Exchanges in the Body

<u>Gas Exchange</u> *Essential Study Partner*

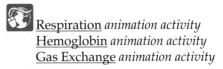

<u>Respiration</u> *animation activity*
<u>Hemoglobin</u> *animation activity*
<u>Gas Exchange</u> *animation activity*

9.4 Respiration and Health

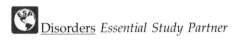

<u>Disorders</u> *Essential Study Partner*

<u>Asthma: The New Worldwide Epidemic</u> *case study*

<u>Smoking Risks</u> *animation activity*
<u>Asthma</u> *animation activity*
<u>Bronchial and Pulmonary Disorders</u> *art labeling activity*

9.5 Homeostasis

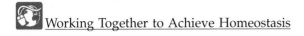

<u>Working Together to Achieve Homeostasis</u>

Chapter Summary

<u>Key Term Flashcards</u> *vocabulary quiz*
<u>Chapter Quiz</u> *objective quiz covering all chapter concepts*

Chapter *10*

Urinary System and Excretion

Chapter Concepts

Figure 10.1 Seawater.
Drinking seawater raises the osmolarity of the blood and causes the kidneys to excrete more water than usual to rid the body of the excess salt.

On a mild Saturday morning, Sarah heads off in her dad's rowboat. Once out in the peaceful deep, she stops rowing, lays back, and falls asleep. Three hours later, she wakes up and finds herself far from shore and lost. Scared, Sarah starts rowing furiously and tires before land is in sight.

Sarah decides to drop anchor, stay put, and wait for help. She sure is thirsty but brought nothing to drink. Staring at the ocean, Sarah crouches down in the boat and scoops up seawater. She gulps it. That's better. She gets more.

Fortunately for Sarah, her dad came looking for her, or else she could have died from dehydration. Ironically, the culprit would probably have been the water she had kept drinking. Seawater is far too salty for the human body to process. In trying to flush away the excess salt, Sarah's kidneys would have to excrete more liquid than she consumed. If you're ever in Sarah's position—or even if you just participate in a long bike ride or other sports activities—make sure you bring water. Your kidneys can't function without it.

When we are able to drink fresh water as we should, the kidneys maintain the water-salt balance and the acid-base balance of blood within normal limits. The kidneys also excrete the nitrogenous end products urea, uric acid, and creatinine. **Excretion** rids the body of metabolic wastes, which come from the breakdown of substances that have been metabolized in the body's cells. The kidneys are not the only organs that excrete substances. The skin contains sweat glands that excrete water, salt, and urea. The lungs excrete carbon dioxide, and the liver excretes bile pigments. But none of these organs can make up for the work of the kidneys. Excretion should not be confused with secretion, which is the release of a substance useful to the body. Also, defecation is not a form of excretion. Defecation refers only to the elimination of feces from the digestive tract. ⚖⚖

The correct functioning of the kidneys is essential to our good health, yet it is something that most of us take for granted until an illness strikes. Urinary tract infections and kidney stones cause much pain and may even result in permanent damage to the kidneys. Nationwide, many persons are on kidney machines or have received a transplanted kidney.

10.1 Urinary System

The urinary system consists of the organs labeled in Figure 10.2. This figure also traces the path of urine. This section discusses the organs of the urinary system, urination, and the functions of the urinary system.

Urinary Organs

The kidneys are found on either side of the vertebral column, just below the diaphragm. They lie in depressions against the deep muscles of the back beneath the peritoneum, the lining of the abdominal cavity. Although they are somewhat protected by these muscles and by the lower rib cage, the kidneys can be damaged by blows to the back. As you probably know, so-called "kidney punches" are not allowed in boxing for this reason.

The **kidneys** are bean-shaped and reddish-brown in color. The fist-sized organs are covered by a tough capsule of fibrous connective tissue overlaid by adipose tissue. The concave side of a kidney has a depression called the hilum. The **renal artery** enters and the **renal vein** and ureters exit a kidney at the hilum.

The **ureters,** which extend from the kidneys to the bladder, are small muscular tubes about 25 cm long. Peristalsis moves urine within the ureters, and peristaltic contractions cause urine to enter the bladder at a rate of about five jets per minute.

The **urinary bladder,** which can hold up to 600 ml of urine, is a hollow, muscular organ that gradually expands as urine enters. A sphincter is a circular muscle that encloses a tube. Two sphincters are found in close proximity where the urethra exits the bladder. When these sphincters are closed, urination does not take place.

The **urethra,** which extends from the urinary bladder to an external opening, has a different length in females and males. In females, the urethra is only about 4 cm long. As mentioned in the Health Focus on page 190, the short length of the female urethra makes bacterial invasion easier and helps explain why females are more prone to urinary tract infections than males. In males, the urethra averages 20 cm when the penis is flaccid (limp, nonerect). As the urethra leaves the male urinary bladder, it is encircled by the prostate gland. In older men, enlargement of the prostate gland can restrict urination. A surgical procedure can usually correct the condition and restore a normal flow of urine.

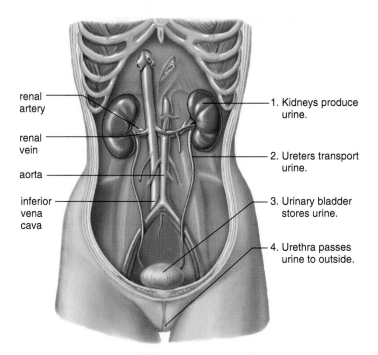

renal artery
renal vein
aorta
inferior vena cava

1. Kidneys produce urine.
2. Ureters transport urine.
3. Urinary bladder stores urine.
4. Urethra passes urine to outside.

Figure 10.2   The urinary system.
Urine is found only within the kidneys, the ureters, the urinary bladder, and the urethra.

In females, there is no connection between the reproductive and urinary systems. In males, the urethra carries urine during urination and sperm during ejaculation. This double function of the urethra in males does not alter the path of urine (Fig. 10.2).

Only the urinary system, consisting of the kidneys, the urinary bladder, the ureters, and the urethra, holds urine.

Urination

When the urinary bladder fills to about 250 ml of urine, stretch receptors send sensory nerve impulses to the spinal cord. Subsequently, motor nerve impulses from the spinal cord cause the urinary bladder to contract and the sphincters to relax so that urination is possible (Fig. 10.3). In older children and adults, the brain controls this reflex, delaying urination until a suitable time.

Functions of the Urinary System ⚖⚖

The function of the urinary system is to produce urine and conduct it to outside the body. The kidneys produce urine, and the other organs of the system store or conduct urine toward the outside of the body.

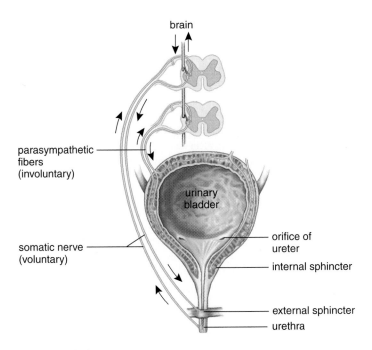

Figure 10.3   Urination.
As the bladder fills with urine, sensory impulses go to the spinal cord and then to the brain. The brain can override the urge to urinate. When urination occurs, motor nerve impulses cause the bladder to contract and an internal sphincter to open. Nerve impulses also cause an external sphincter to open.

The kidneys are the primary organs of the urinary system, and they play a central role in homeostasis by regulating the composition of blood, and therefore tissue fluid. As urine is being produced, the kidneys (1) carry out the excretion of metabolic wastes, particularly nitrogenous wastes; (2) maintain the normal water-salt balance of the blood and, as a consequence, the normal blood volume and blood pressure; and (3) maintain the acid-base balance of blood.

The kidneys also have a hormonal function, as discussed in the next section.

Excretion of Metabolic Wastes

The kidneys excrete metabolic wastes, notably nitrogenous wastes. Urea is the primary nitrogenous end product of metabolism in human beings, but humans also excrete some ammonium, creatinine, and uric acid.

Urea is a by-product of amino acid metabolism. The breakdown of amino acids in the liver releases ammonia, which the liver combines with carbon dioxide to produce urea. Ammonia is very toxic to cells, and urea is much less toxic. Because it is less toxic, less water is required to excrete urea.

The metabolic breakdown of creatine phosphate results in **creatinine.** Creatine phosphate is a high-energy phosphate reserve molecule in muscles.

The breakdown of nucleotides, such as those containing adenine and thymine, produces **uric acid.** Uric acid is rather insoluble. If too much uric acid is present in blood, crystals form and precipitate out. Crystals of uric acid sometimes collect in the joints, producing a painful ailment called gout.

Maintenance of Water-Salt Balance

A principal function of the kidneys is to maintain the appropriate water-salt balance of the blood. As we shall see, blood volume is intimately associated with the salt balance of the body. As you know, salts, such as NaCl, have the ability to cause osmosis, the diffusion of water—in this case, into the blood. The more salts there are in the blood, the greater the blood volume and the greater the blood pressure. In this way, the kidneys are involved in regulating blood pressure.

The kidneys also maintain the appropriate level of other ions, such as potassium ions (K^+), bicarbonate ions (HCO_3^-), and calcium ions (Ca^{2+}), in the blood.

Maintenance of Acid-Base Balance

The kidneys regulate the acid-base balance of the blood. In order for us to remain healthy, the blood pH should be just about 7.4. The kidneys monitor and control blood pH, mainly by excreting hydrogen ions (H^+) and reabsorbing the bicarbonate ion (HCO_3^-) as needed. Urine usually has a pH of 6 or lower because our diet often contains acidic foods.

Secretion of Hormones

The kidneys assist the endocrine system. The kidneys release renin, a substance that leads to the secretion of the hormone aldosterone from the adrenal cortex, the outer portion of the adrenal glands, which lie atop the kidneys. As described later in this chapter, aldosterone promotes the reabsorption of sodium ions (Na^+) by the kidneys.

Whenever the oxygen-carrying capacity of the blood is reduced, the kidneys secrete the hormone **erythropoietin,** which stimulates red blood cell production.

The kidneys also help activate vitamin D from the skin. Vitamin D is the precursor of the hormone calcitriol, which promotes calcium (Ca^{2+}) reabsorption from the digestive tract.

The kidneys are the primary organs of excretion, particularly of nitrogenous wastes. The kidneys are also major organs of homeostasis because they regulate the water-salt balance and the acid-base balance of the blood.

Urinary Tract Infections Require Attention

Although males can get a urinary tract infection, the condition is 50 times more common in women. The explanation lies in a comparison of male and female anatomy (Fig. 10A). The female urethral and anal openings are closer together, and the shorter urethra makes it easier for bacteria from the bowels to enter and start an infection. Although it is possible to have no outward signs of an infection, usually urination is painful, and patients often describe a burning sensation. The urge to pass urine is frequent, but it may be difficult to start the stream. Chills with fever, nausea, and vomiting may be present.

Urinary tract infections can be confined to the urethra, in which case urethritis is present. If the bladder is involved, it is called cystitis. Should the infection reach the kidneys, the person has pyelonephritis. *Escherichia coli* (*E. coli*), a normal bacterial resident of the large intestine, is usually the cause of infection. Since the infection is caused by a bacterium, it is curable by antibiotic therapy. The problem is, however, that reinfection is possible as soon as antibiotic therapy is finished.

It makes sense to try to prevent infection in the first place. These tips might help.

Men and women should drink lots of water. Try to drink from 2 to 2.5 liters of liquid a day. Try to avoid caffeinated drinks, which may be irritating. Cranberry juice is recommended because it contains a substance that stops bacteria from sticking to the bladder wall once an infection has set in. If an attack occurs, testing and antibiotic therapy may be in order. Keep in mind that sexually transmitted diseases such as gonorrhea, chlamydia, or herpes can cause urinary tract infections. All personal behaviors should be examined carefully, and suitable adjustments should be made to avoid urinary tract infections.

Women may have a urinary tract infection for the first time shortly after they become sexually active. "Honeymoon cystitis" was coined because of the common association of urinary tract infections with sexual intercourse. Washing the genitals before having sex and being careful not to introduce bacteria from the anus into the urethra are recommended. Also, urinating immediately before and after sex will help flush out any bacteria that are present. A diaphragm may press on the urethra and prevent adequate emptying of the bladder, and estrogen, such as in birth control pills, can increase the risk of cystitis. A sex partner may have an asymptomatic (no symptoms) urinary infection that causes a woman to become infected repeatedly.

Women should wipe from the front to the back after using the toilet. Perfumed toilet paper and any other perfumed products that come in contact with the genitals may be irritating. Wearing loose clothing and cotton underwear discourages the growth of bacteria, while tight clothing, such as jeans and panty hose, provides an environment for the growth of bacteria.

Personal hygiene is especially important at the time of menstruation. Hands should be washed before and after changing napkins and/or tampons. Superabsorbent tampons that are changed infrequently may encourage the growth of bacteria. Also, sexual intercourse may cause menstrual flow to enter the urethra.

In males, the prostate is a gland that surrounds the urethra just below the bladder (Fig. 10A). The prostate contributes secretions to semen whenever semen enters the urethra prior to ejaculation. An infection of the prostate, called prostatitis, is often accompanied by a urinary tract infection. Fever is present, and the prostate is tender and inflamed. The patient may have to be hospitalized and treated with a broad-spectrum antibiotic. Prostatitis, which in a young person is often preceded by a sexually transmitted disease, can lead to a chronic condition. Chronic prostatitis may be asymptomatic or, as is more typical, the person may experience irritation upon voiding and/or difficulty in voiding. The latter can lead to the need for surgery to remove the obstruction to urine flow.

Figure 10A   Female versus male urinary tract.
Females have a short urinary tract compared to that of males. This means that it is easier for bacteria to invade the urethra and helps explain why females are 50 times more likely than males to get a urinary tract infection.

Labels: female, male, kidney, ureter, bladder, rectum, prostate gland, pubic symphysis, vagina, urethra, penis

10.2 Kidneys

When a kidney is sliced lengthwise, it is possible to see the many branches of the renal artery and vein that reach inside the kidney (Fig. 10.4*a*). If the blood vessels are removed, it is easier to identify the three regions of a kidney. The **renal cortex** is an outer granulated layer that dips down in between a radially striated, or lined, inner layer called the renal medulla. The **renal medulla** consists of cone-shaped tissue masses called renal pyramids. The **renal pelvis** is a central space, or cavity, that is continuous with the ureter (Fig. 10.4*b*).

Microscopically, the kidney is composed of over one million **nephrons,** sometimes called renal or kidney tubules (Fig. 10.4*c*). The nephrons produce urine and are positioned so that the urine flows into a collecting duct. Several nephrons enter the same collecting duct; the collecting ducts enter the renal pelvis.

Macroscopically, a kidney has three regions: the renal cortex, the renal medulla, and the renal pelvis that is continuous with the ureter. Microscopically, a kidney contains over one million nephrons.

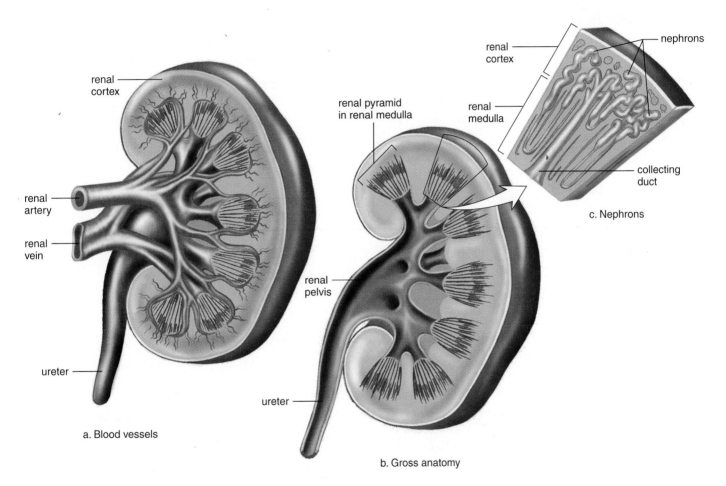

a. Blood vessels

b. Gross anatomy

c. Nephrons

Figure 10.4   Gross anatomy of the kidney.
a. A longitudinal section of the kidney showing the blood supply. Note that the renal artery divides into smaller arteries, and these divide into arterioles. Venules join to form small veins, which join to form the renal vein. **b.** The same section without the blood supply. Now it is easier to distinguish the renal cortex, the renal medulla, and the renal pelvis, which connects with the ureter. The renal medulla consists of the renal pyramids. **c.** An enlargement showing the placement of nephrons.

Anatomy of a Nephron

Each nephron has its own blood supply, including two capillary regions (Fig 10.5). From the renal artery, an afferent arteriole leads to the **glomerulus,** a knot of capillaries inside the glomerular capsule. Blood leaving the glomerulus enters the efferent arteriole. Blood pressure is higher in the glomerulus because the efferent arteriole is narrower than the afferent arteriole. The efferent arteriole takes blood to the **peritubular capillary network,** which surrounds the rest of the nephron. From there, the blood goes into a venule that joins the renal vein.

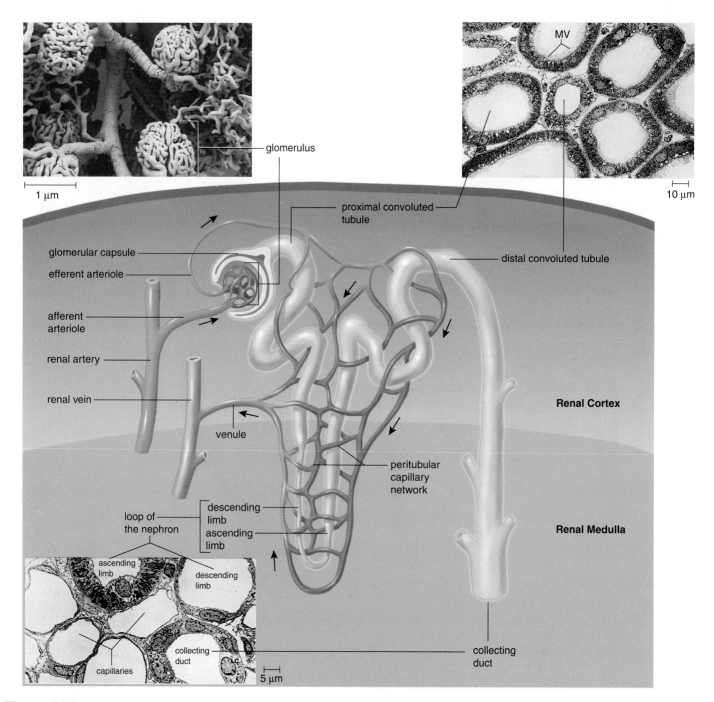

Figure 10.5   Nephron anatomy.

A nephron is made up of a glomerular capsule, the proximal convoluted tubule, the loop of the nephron, the distal convoluted tubule, and the collecting duct. The micrographs show these structures in cross section; MV = microvilli. You can trace the path of blood about the nephron by following the arrows.

Parts of a Nephron

Each nephron is made up of several parts (Fig. 10.5). The structure of each part suits its function.

First, the closed end of the nephron is pushed in on itself to form a cuplike structure called the **glomerular capsule** (Bowman's capsule). The outer layer of the glomerular capsule is composed of squamous epithelial cells; the inner layer is made up of podocytes that have long cytoplasmic processes. The podocytes cling to the capillary walls of the glomerulus and leave pores that allow easy passage of small molecules from the glomerulus to the inside of the glomerular capsule. This process, called glomerular filtration, produces a filtrate of blood.

Next, there is a **proximal** (meaning near the glomerular capsule) **convoluted tubule.** The cuboidal epithelial cells lining this part of the nephron have numerous microvilli, about 1 μm in length, that are tightly packed and form a brush border (Fig. 10.6). A brush border greatly increases the surface area for the tubular reabsorption of filtrate components. Each cell also has many mitochondria, which can supply energy for active transport of molecules from the lumen to the peritubular capillary network.

Simple squamous epithelium appears as the tube narrows and makes a U-turn called the **loop of the nephron** (loop of Henle). Each loop consists of a descending limb that allows water to leave and an ascending limb that extrudes salt (NaCl). Indeed, as we shall see, this activity facilitates the reabsorption of water by the nephron and collecting duct.

The cells of the **distal convoluted tubule** have numerous mitochondria, but they lack microvilli. This is consistent with the active role they play in moving molecules from the blood into the tubule, a process called tubular secretion. The distal convoluted tubules of several nephrons enter one collecting duct. A kidney contains many collecting ducts, which carry urine to the renal pelvis.

As shown in Figure 10.5, the glomerular capsule and the convoluted tubules always lie within the renal cortex. The loop of the nephron dips down into the renal medulla; a few nephrons have a very long loop of the nephron, which penetrates deep into the renal medulla. **Collecting ducts** are also located in the renal medulla, and they give the renal pyramids their lined appearance.

Each part of a nephron is anatomically suited to its specific function in urine formation.

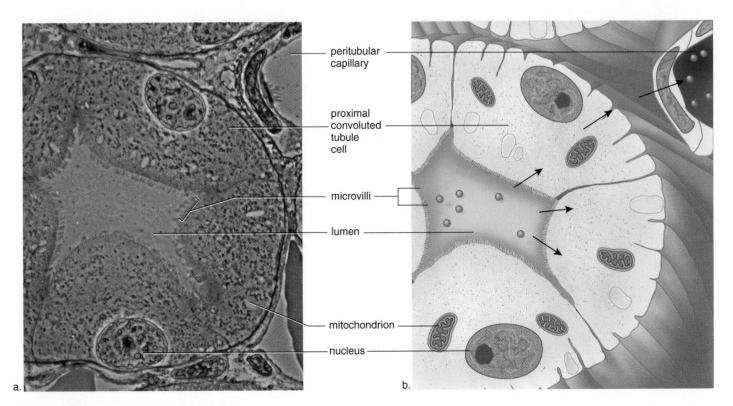

a.   b.

Figure 10.6   Proximal convoluted tubule.

a. This photomicrograph shows that the cells lining the proximal convoluted tubule have a brushlike border composed of microvilli, which greatly increase the surface area exposed to the lumen. The peritubular capillary network surrounds the cells. **b.** Diagrammatic representation of (**a**) shows that each cell has many mitochondria, which supply the energy needed for active transport, the process that moves molecules (green) from the lumen of the tubule to the capillary, as indicated by the arrows.

Visual Focus

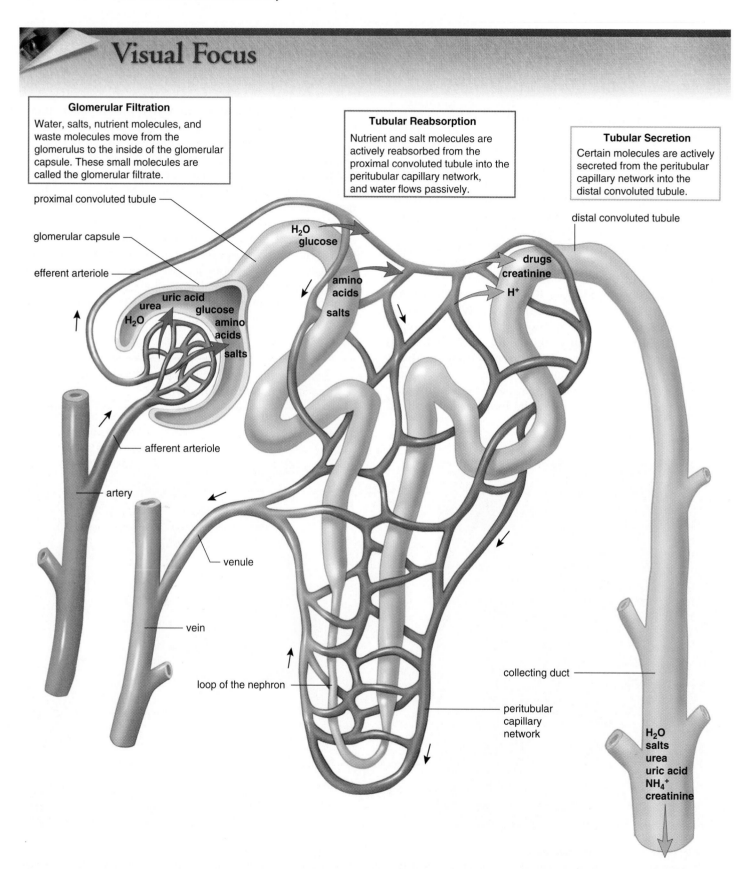

Glomerular Filtration

Water, salts, nutrient molecules, and waste molecules move from the glomerulus to the inside of the glomerular capsule. These small molecules are called the glomerular filtrate.

Tubular Reabsorption

Nutrient and salt molecules are actively reabsorbed from the proximal convoluted tubule into the peritubular capillary network, and water flows passively.

Tubular Secretion

Certain molecules are actively secreted from the peritubular capillary network into the distal convoluted tubule.

proximal convoluted tubule

glomerular capsule

efferent arteriole

distal convoluted tubule

H_2O
glucose

drugs
creatinine

H^+

amino
acids

salts

uric acid

urea
H_2O

glucose
amino
acids
salts

afferent arteriole

artery

venule

vein

loop of the nephron

collecting duct

peritubular
capillary
network

H_2O
salts
urea
uric acid
NH_4^+
creatinine

Figure 10.7   Steps in urine formation.

The three main steps in urine formation are color-coded to arrows that show the movement of molecules into or out of the nephron at specific locations. In the end, urine is composed of the substances within the collecting duct (see gray arrow).

10.3 Urine Formation

Figure 10.7 gives an overview of urine formation, which is divided into these steps: glomerular filtration, tubular reabsorption, and tubular secretion.

Glomerular Filtration

Glomerular filtration occurs when whole blood enters the afferent arteriole and the glomerulus. Due to glomerular blood pressure, water and small molecules move from the glomerulus to the inside of the glomerular capsule. This is a filtration process because large molecules and formed elements are unable to pass through the capillary wall. In effect, then, blood in the glomerulus has two portions: the filterable components and the nonfilterable components.

Filterable Blood Components	Nonfilterable Blood Components
Water	Formed elements (blood cells and platelets)
Nitrogenous wastes	Proteins
Nutrients	
Salts (ions)	

The **glomerular filtrate** contains small dissolved molecules in approximately the same concentration as plasma. Small molecules that escape being filtered and the nonfilterable components leave the glomerulus by way of the efferent arteriole.

As indicated in Table 10.1, 180 liters of water are filtered per day along with a considerable amount of small molecules (such as glucose) and ions (such as sodium). If the composition of urine were the same as that of the glomerular filtrate, the body would continually lose water, salts, and nutrients. Therefore, we can conclude that the composition of the filtrate must be altered as this fluid passes through the remainder of the tubule.

Tubular Reabsorption

Tubular reabsorption occurs as molecules and ions are both passively and actively reabsorbed from the nephron into the blood of the peritubular capillary network. The osmolarity of the blood is maintained by the presence of both plasma proteins and salt. When sodium ions (Na^+) are actively reabsorbed, chloride ions (Cl^-) follow passively. The reabsorption of salt (NaCl) increases the osmolarity of the blood compared to the filtrate, and therefore water moves passively from the tubule into the blood. About 67% of Na^+ is reabsorbed at the proximal convoluted tubule.

Nutrients such as glucose and amino acids also return to the blood at the proximal convoluted tubule. This is a selective process because only molecules recognized by carrier molecules are actively reabsorbed. Glucose is an example of

Table 10.1	Reabsorption from Nephrons		
Substance	Amount Filtered (Per Day)	Amount Excreted (Per Day)	Reabsorption (%)
Water, L	180	1.8	99.0
Sodium, g	630	3.2	99.5
Glucose, g	180	0.0	100.0
Urea, g	54	30.0	44.0

L = liters, g = grams

From A. J. Vander, et al. *Human Physiology*, 4th ed. © 1985. The McGraw-Hill Publishing Companies, Inc. All Rights Reserved. Reprinted by permission.

a molecule that ordinarily is completely reabsorbed because there is a plentiful supply of carrier molecules for it. However, every substance has a maximum rate of transport, and after all its carriers are in use, any excess in the filtrate will appear in the urine. For example, as reabsorbed levels of glucose approach 1.8–2 mg/ml plasma, the rest appears in the urine. In diabetes mellitus, excess glucose occurs in the blood, and then in the filtrate, and then in the urine, because the liver and muscles fail to store glucose as glycogen, and the kidneys cannot reabsorb all of it. The presence of glucose in the filtrate increases its osmolarity compared to that of the blood, and therefore less water is reabsorbed into the peritubular capillary network. The frequent urination and increased thirst experienced by untreated diabetics are due to the fact that water is not being reabsorbed.

We have seen that the filtrate that enters the proximal convoluted tubule is divided into two portions: components that are reabsorbed from the tubule into blood, and components that are not reabsorbed and continue to pass through the nephron to be further processed into urine.

Reabsorbed Filtrate Components	Nonreabsorbed Filtrate Components
Most water	Some water
Nutrients	Much nitrogenous waste
Required salts (ions)	Excess salts (ions)

The substances that are not reabsorbed become the tubular fluid, which enters the loop of the nephron.

Tubular Secretion

Tubular secretion is a second way by which substances are removed from blood and added to the tubular fluid. Hydrogen ions, creatinine, and drugs such as penicillin are some of the substances that are moved by active transport from blood into the distal convoluted tubule. In the end, urine contains substances that have undergone glomerular filtration but have not been reabsorbed, and substances that have undergone tubular secretion.

10.4 Maintaining Water-Salt Balance ⚖

The kidneys maintain the water-salt balance of the blood within normal limits. In this way, they also maintain the blood volume and blood pressure. Most of the water and salt (NaCl) present in the filtrate is reabsorbed across the wall of the proximal convoluted tubule.

Reabsorption of Water

The excretion of a hypertonic urine (one that is more concentrated than blood) is dependent upon the reabsorption of water from the loop of the nephron and the collecting duct.

A long loop of the nephron, which typically penetrates deep into the renal medulla, is made up of a descending limb and an ascending limb. Salt (NaCl) passively diffuses out of the lower portion of the ascending limb, but the upper, thick portion of the limb actively extrudes salt out into the tissue of the outer renal medulla (Fig. 10.8). Less and less salt is available for transport as fluid moves up the thick portion of the ascending limb. Because of these circumstances, there is an osmotic gradient within the tissues of the renal medulla: the concentration of salt is greater in the direction of the inner medulla. (Note that water cannot leave the ascending limb because the limb is impermeable to water.)

The large arrow in Figure 10.8 indicates that the innermost portion of the inner medulla has the highest concentration of solutes. This cannot be due to salt because active transport of salt does not start until fluid reaches the thick portion of the ascending limb. Urea is believed to leak from the lower portion of the collecting duct, and it is this molecule that contributes to the high solute concentration of the inner medulla.

Because of the osmotic gradient within the renal medulla, water leaves the descending limb along its entire length. This is a countercurrent mechanism: as water diffuses out of the descending limb, the remaining fluid within the limb encounters an even greater osmotic concentration of solute; therefore, water continues to leave the descending limb from the top to the bottom.

Fluid enters the collecting duct from the distal convoluted tubule. This fluid is isotonic to the cells of the cortex. This means that to this point, the net effect of reabsorption of water and salt is the production of a fluid that has the same tonicity as blood plasma. However, the filtrate within the collecting duct also encounters the same osmotic gradient mentioned earlier (Fig. 10.8). Therefore, water diffuses out of the collecting duct into the renal medulla, and the urine within the collecting duct becomes hypertonic to blood plasma.

Antidiuretic hormone (ADH) released by the posterior lobe of the pituitary plays a role in water reabsorption at the collecting duct. In order to understand the action of this hormone, consider its name. Diuresis means increased amount of urine, and antidiuresis means decreased amount of urine. When ADH is present, more water is reabsorbed (blood vol-

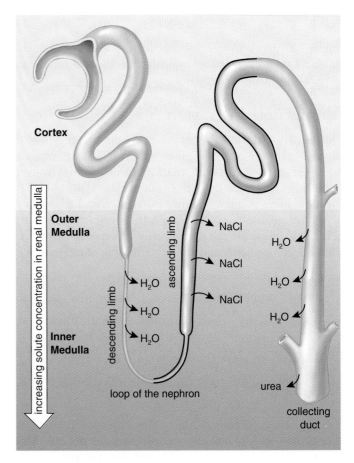

Figure 10.8   Reabsorption of water at the loop of the nephron and the collecting duct.
Salt (NaCl) diffuses and is actively transported out of the ascending limb of the loop of the nephron into the renal medulla; also, urea is believed to leak from the collecting duct and to enter the tissues of the renal medulla. This creates a hypertonic environment, which draws water out of the descending limb and the collecting duct. This water is returned to the cardiovascular system. (The thick black line means the ascending limb is impermeable to water.)

ume and pressure rise), and a decreased amount of urine results. In practical terms, if an individual does not drink much water on a certain day, the posterior lobe of the pituitary releases ADH, causing more water to be reabsorbed and less urine to form. On the other hand, if an individual drinks a large amount of water and does not perspire much, ADH is not released. In that case, more water is excreted, and more urine forms.

Reabsorption of Salt

The kidneys regulate the salt balance in blood by controlling the excretion and the reabsorption of various ions. Sodium (Na^+) is an important ion in plasma that must be regulated, but the kidneys also excrete or reabsorb other ions, such as potassium ions (K^+), bicarbonate ions (HCO_3^-), and magnesium ions (Mg^{2+}), as needed.

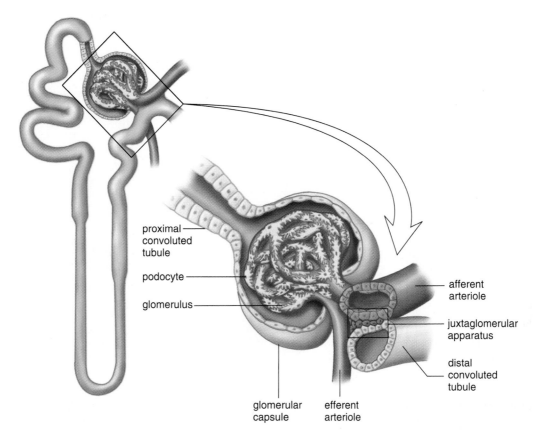

Figure 10.9 **Juxtaglomerular apparatus.**
This drawing shows that the afferent arteriole and the distal convoluted tubule usually lie next to each other. The juxtaglomerular apparatus occurs where they touch. The juxtaglomerular apparatus secretes renin, a substance that leads to the release of aldosterone by the adrenal cortex. Reabsorption of sodium ions and then water now occurs. Therefore, blood volume and blood pressure increases.

Usually, more than 99% of sodium (Na$^+$) filtered at the glomerulus is returned to the blood. Most sodium (67%) is reabsorbed at the proximal tubule, and a sizable amount (25%) is extruded by the ascending limb of the loop of the nephron. The rest is reabsorbed from the distal convoluted tubule and collecting duct.

Hormones regulate the reabsorption of sodium at the distal convoluted tubule. **Aldosterone** is a hormone secreted by the adrenal cortex. Aldosterone promotes the excretion of potassium ions (K$^+$) and the reabsorption of sodium ions (Na$^+$). The release of aldosterone is set in motion by the kidneys themselves. The **juxtaglomerular apparatus** is a region of contact between the afferent arteriole and the distal convoluted tubule (Fig. 10.9). When blood volume, and therefore blood pressure, is not sufficient to promote glomerular filtration, the juxtaglomerular apparatus secretes renin. **Renin** is an enzyme that changes angiotensinogen (a large plasma protein produced by the liver) into angiotensin I. Later, angiotensin I is converted to angiotensin II, a powerful vasoconstrictor that also stimulates the adrenal cortex to release aldosterone. The reabsorption of sodium ions is fol-lowed by the reabsorption of water. Therefore, blood volume and blood pressure increase.

Atrial natriuretic hormone (ANH) is a hormone secreted by the atria of the heart when cardiac cells are stretched due to increased blood volume. ANH inhibits the secretion of renin by the juxtaglomerular apparatus and the secretion of aldosterone by the adrenal cortex. Its effect, therefore, is to promote the excretion of Na$^+$, called natri-uresis. When Na$^+$ is excreted, so is water, and therefore blood volume and blood pressure decrease.

Diuretics

Diuretics are chemicals that increase the flow of urine. Drinking alcohol causes diuresis because it inhibits the secretion of ADH. The dehydration that follows is believed to contribute to the symptoms of a hangover. Caffeine is a diuretic because it increases the glomerular filtration rate and decreases the tubular reabsorption of Na$^+$. Diuretic drugs developed to counteract high blood pressure inhibit active transport of Na$^+$ at the loop of the nephron or at the distal convoluted tubule. A decrease in water reabsorption and a decrease in blood volume follow.

Human Systems Work Together

Integumentary System

Kidneys compensate for water loss due to sweating; activate vitamin D precursor made by skin.

Skin helps regulate water loss; sweat glands carry on some excretion.

Skeletal System

Kidneys provide active vitamin D for Ca^{2+} absorption and help maintain blood level of Ca^{2+}, needed for bone growth and repair.

Bones provide support and protection.

Muscular System

Kidneys maintain blood levels of Na^+, K^+, and Ca^{2+}, which are needed for muscle innervation, and eliminate creatinine, a muscle waste.

Smooth muscular contraction assists voiding of urine; skeletal muscles support and help protect urinary organs.

Nervous System

Kidneys maintain blood levels of Na^+, K^+, and Ca^{2+}, which are needed for nerve conduction.

Brain controls nerves, which innervate muscles that permit urination.

Endocrine System

Kidneys keep blood values within normal limits so that transport of hormones continues.

ADH and aldosterone, and atrial natriuretic hormone regulate reabsorption of Na^+ by kidneys.

How the Urinary System works with other body systems

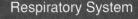

Cardiovascular System

Kidneys filter blood and excrete wastes; maintain blood volume, pressure, and pH; produce renin and erythropoietin.

Blood vessels deliver waste to be excreted; blood pressure aids kidney function; heart produces atrial natriuretic hormone.

Lymphatic System/Immunity

Kidneys control volume of body fluids, including lymph.

Lymphatic system picks up excess tissue fluid, helping to maintain blood pressure for kidneys to function; immune system protects against infections.

Respiratory System

Kidneys compensate for water lost through respiratory tract; work with lungs to maintain blood pH.

Lungs excrete carbon dioxide, provide oxygen, and convert angiotensin I to angiotensin II, leading to kidney regulation.

Digestive System

Kidneys convert vitamin D to active form needed for Ca^{2+} absorption; compensate for any water loss by digestive tract.

Liver synthesizes urea; digestive tract excretes bile pigments from liver and provides nutrients.

Reproductive System

Semen is discharged through the urethra in males; kidneys excrete wastes and maintain electrolyte levels for mother and child.

Penis in males contains the urethra and performs urination; prostate enlargement hinders urination.

10.5 Maintaining Acid-Base Balance

The bicarbonate (HCO_3^-) buffer system and breathing work together to maintain the pH of the blood. Central to the mechanism is this reaction, which you have seen before:

$$H^+ + HCO_3^- \rightleftharpoons H_2CO_3 \rightleftharpoons H_2O + CO_2$$

The excretion of carbon dioxide (CO_2) by the lungs helps keep the pH within normal limits, because when carbon dioxide is exhaled, this reaction is pushed to the right and hydrogen ions are tied up in water. Indeed, when blood pH decreases, chemoreceptors in the carotid bodies (located in the carotid arteries) and in aortic bodies (located in the aorta) stimulate the respiratory center, and the rate and depth of breathing increase. On the other hand, when blood pH begins to rise, the respiratory center is depressed, and the level of bicarbonate ions increases in the blood.

As powerful as the buffer/breathing system is, only the kidneys can rid the body of a wide range of acidic and basic substances. The kidneys are slower acting than the buffer/breathing mechanism, but they have a more powerful effect on pH. For the sake of simplicity, we can think of the kidneys as reabsorbing bicarbonate ions and excreting hydrogen ions as needed to maintain the normal pH of the blood. If the blood is acidic, hydrogen ions are excreted and bicarbonate ions are reabsorbed. If the blood is basic, hydrogen ions are not excreted and bicarbonate ions are not reabsorbed. Since the urine is usually acidic, it follows that usually an excess of hydrogen ions are excreted. Ammonia (NH_3) provides a means for buffering these hydrogen ions in urine: ($NH_3 + H^+ \rightarrow NH_4^+$). Ammonia (whose presence is quite obvious in the diaper pail or kitty litter box) is produced in tubule cells by the deamination of amino acids. Phosphate provides another means of buffering hydrogen ions in urine.

The acid-base balance of the blood is adjusted by the reabsorption of the bicarbonate ions (HCO_3^-) and the secretion of hydrogen ions (H^+) as appropriate.

10.6 Homeostasis ⚖

The kidneys are primary organs of homeostasis because they maintain the water-salt balance and the acid-base balance of the blood. The illustration on the previous page

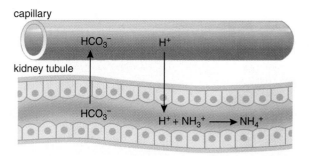

Figure 10.10   Acid-base balance.
In the kidneys, bicarbonate ions (HCO_3^-) are reabsorbed and hydrogen ions (H^+) are excreted as needed to maintain the pH of the blood. Excess hydrogen ions are buffered, for example, by ammonia (NH_3), which is produced in tubule cells by the deamination of amino acids.

tells how the urinary system works with the other systems of the body.

If blood does not have the usual osmolarity and blood pressure, exchange across capillary walls cannot take place, nor is glomerular filtration possible in the kidneys themselves. The production of renin by the kidneys and subsequently the renin-angiotensin-aldosterone sequence helps ensure that the sodium concentration of the blood stays normal. The kidneys maintain the normal concentration of sodium ions (Na^+) and also the concentration of potassium ions (K^+) and calcium ions (Ca^{2+}). All these ions are necessary to the contraction of the heart and other muscles in the body, and are also needed for nerve conduction.

The bicarbonate ion is a part of a buffer system that helps maintain the pH of the blood. The kidneys can reabsorb the bicarbonate ion and excrete hydrogen ions (H^+) as needed to maintain the pH (Fig. 10.10). The kidneys have ultimate control over the pH of the blood, and the importance of this function cannot be overemphasized. The enzymes of cells cannot continue to function if the internal environment does not have near-normal pH.

The kidneys also excrete nitrogenous wastes, which are end products of metabolism. The excretion of nitrogenous wastes may not be as critical as maintaining the water-salt and the acid-base balances, but it is still a necessity. If the urea produced day by day were not removed, in about four years we would be nothing but urea.

Finally, the kidneys assist the endocrine system. They produce erythropoietin, which stimulates red bone marrow to produce red blood cells, and they help convert vitamin D to an active hormone that stimulates the reabsorption of calcium from the digestive tract.

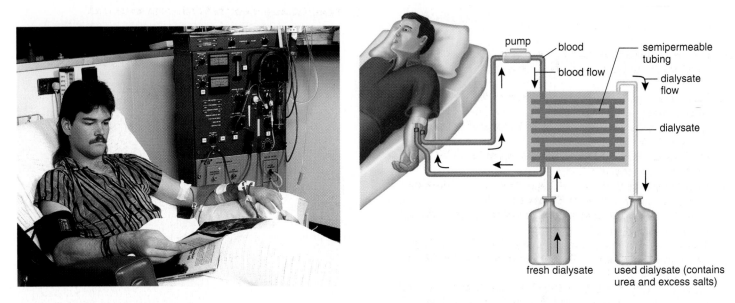

Figure 10.11 **An artificial kidney machine.**
As the patient's blood is pumped through dialysis tubing, it is exposed to a dialysate (dialysis solution). Wastes exit from blood into the solution because of a preestablished concentration gradient. In this way, blood is not only cleansed, but its water-salt and acid-base balance can also be adjusted.

10.7 Problems with Kidney Function

Many types of illnesses, especially diabetes, hypertension, and inherited conditions, cause progressive renal disease and renal failure. Urinary tract infections include urethritis, infection of the urethra; cystitis, infection of the bladder; and pyelonephritis, infection of the kidneys. The Health Focus on page 190 suggests ways to prevent urinary tract infections.

Glomerular damage sometimes leads to blockage of the glomeruli so that glomerular filtration does not occur or allows large substances to pass through. This is detected when a urinalysis is done. If the glomeruli are too permeable, albumin, white blood cells, or even red blood cells appear in the urine. A trace amount of protein in the urine is not a matter of concern, however.

When glomerular damage is so extensive that more than two-thirds of the nephrons are inoperative, waste substances accumulate in the blood. This condition is called uremia because urea is one of the substances that accumulates. Imbalance in the ionic composition of body fluids is even more serious because it can lead to loss of consciousness and to heart failure.

Hemodialysis

While waiting for a kidney transplant, it is usually necessary for the patient to undergo **hemodialysis,** utilizing either an artificial kidney machine or continuous ambulatory peritoneal (abdominal) dialysis (CAPD). Dialysis is defined as the diffusion of dissolved molecules through a semipermeable membrane, an artificial membrane with pore sizes that allow only small molecules to pass through. In an artificial kidney machine (Fig. 10.11), the patient's blood is passed through a membranous tube, which is in contact with a dialysis solution, or dialysate. Substances more concentrated in the blood diffuse into the dialysate, and substances more concentrated in the dialysate diffuse into the blood. Accordingly, the artificial kidney can be utilized either to extract substances from blood, including waste products or toxic chemicals and drugs, or to add substances to blood—for example, bicarbonate ions (HCO_3^-) if blood is acidic. In the course of a three- to six-hour hemodialysis, from 50 to 250 grams of urea can be removed from a patient, which greatly exceeds the amount excreted by normal kidneys. Therefore, a patient needs to undergo treatment only about twice a week.

In CAPD, a fresh amount of dialysate is introduced directly into the abdominal cavity from a bag attached to a permanently implanted plastic tube. Waste and salt molecules pass from the blood vessels in the abdominal wall into the dialysate before the fluid is collected four or eight hours later. Unlike hemodialysis, the individual can go about his or her normal activities during CAPD.

Replacing a Kidney

Patients with renal failure sometimes undergo a kidney transplant operation during which a functioning kidney from a donor is received. As with all organ transplants, there is the possibility of organ rejection. Receiving a kidney from a close relative has the highest chance of success. The current one-year survival rate is 97% if the kidney is received from a relative and 90% if it is received from a nonrelative.

Organ Transplants

Transplantation of the kidney, heart, liver, pancreas, lung, and other organs is now possible due to two major breakthroughs. First, solutions have been developed that preserve donor organs for several hours. This made it possible for one young boy to undergo surgery for 16 hours, during which time he received five different organs. Second, rejection of transplanted organs is now prevented by immunosuppressive drugs; therefore, organs can be donated by unrelated individuals, living or dead. After death, it is possible to give the "gift of life" to someone else—over 25 organs and tissues from one cadaver can be used for transplants. The survival rate after a transplant operation is good. So many heart recipients are now alive and healthy that they have formed basketball and softball teams, demonstrating the normalcy of their lives after surgery.

One problem persists, however, and that is the limited availability of organs for transplantation. At any one time, at least 27,000 people in the United States are waiting for a donated organ. Keen competition for organs can lead to various bioethical inequities. When the governor of Pennsylvania received a heart and lungs within a relatively short period of time, it appeared that his social status might have played a role. When Mickey Mantle received a liver transplant, people asked if it was right to give an organ to an older man who had a diseased liver due to the consumption of alcohol. If a father gives a kidney to a child, he has to undergo a major surgical operation that leaves him vulnerable to possible serious consequences in the future. If organs are taken from those who have just died, who guarantees that the individual is indeed dead? And is it right to genetically alter animals to serve as a source of organs for humans? Such organs will most likely be for sale, and does this make the wealthy more likely to receive a transplant than those who cannot pay?

Decide Your Opinion

1. Is it ethical to ask a parent to donate an organ to his or her child? Why or why not?
2. Is it ethical to put a famous person at the top of the list for an organ transplant? Why or why not?
3. Is it ethical to remove organs from a newborn who is brain dead but whose organs are still functioning? Why or why not?
4. When xenotransplants (transplants for humans from other animals) are available, should they be for sale? Why or why not?

Figure 10B **Kidney recipients.**
Sean Elliot, who returned to professional basketball after receiving a kidney transplant, was the spokesperson for the Transplant Games in which Clifford Moore received a gold medal.

Looking at Both Sides www.mhhe.com/biosci/genbio/maderhuman7/

Every bioethical issue has at least two sides. Even if you already have an opinion, it is important to explore the opposite opinion before finalizing your position. The Online Learning Center at www.mhhe.com/biosci/genbio/maderhuman7/ will help you fine-tune your initial opinion, explore both sides, and finalize your position. You may acquire new arguments for your original opinion, or you may even change your opinion. Be sure to complete these activities in sequence:

Taking Sides   Decide your initial opinion by answering a series of questions. Then see if your opinion changes after completing the next two activities.

Further Debate   Read opposing articles that give you further information on this particular bioethical issue.

Explain Your Position   Answer another series of questions and then defend your original or changed opinion. You can e-mail your position to your instructor if he or she wishes.

Summarizing the Concepts

10.1 Urinary System
The kidneys produce urine, which is conducted by the ureters to the bladder where it is stored before being released by way of the urethra.

The kidneys excrete nitrogenous wastes, including urea, uric acid, and creatinine. They maintain the normal water-salt balance and the acid-base balance of the blood.

10.2 Kidneys
Macroscopically, the kidneys are divided into the renal cortex, renal medulla, and renal pelvis. Microscopically, they contain the nephrons.

Each nephron has its own blood supply; the afferent arteriole approaches the glomerular capsule and divides to become the glomerulus, a capillary tuft. The efferent arteriole leaves the capsule and immediately branches into the peritubular capillary network.

Each region of the nephron is anatomically suited to its task in urine formation. The spaces between the podocytes of the glomerular capsule allow small molecules to enter the capsule from the glomerulus. The cuboidal epithelial cells of the proximal convoluted tubule have many mitochondria and microvilli to carry out active transport (following passive transport) from the tubule to the blood. In contrast, the cuboidal epithelial cells of the distal convoluted tubule have numerous mitochondria but lack microvilli. They carry out active transport from the blood to the tubule.

10.3 Urine Formation
Urine is composed primarily of nitrogenous waste products and salts in water.

The steps in urine formation are glomerular filtration, tubular reabsorption, and tubular secretion, as explained in Figure 10.7.

10.4 Maintaining Water-Salt Balance
The kidneys regulate the water-salt balance of the body. Water is reabsorbed from certain parts of the tubule, and the loop of the nephron establishes an osmotic gradient that draws water from the descending loop of the nephron and also from the collecting duct. The permeability of the collecting duct is under the control of the hormone ADH.

The reabsorption of salt increases blood volume and pressure because more water is also reabsorbed. Two other hormones, aldosterone and ANH, control the kidneys' reabsorption of sodium (Na^+).

10.5 Maintaining Acid-Base Balance
The kidneys keep blood pH within normal limits. They reabsorb HCO_3^- and excrete H^+ as needed to maintain the pH at about 7.4.

10.6 Homeostasis
The urinary system works with the other systems of the body to maintain homeostasis in the ways described in the illustration on page 198.

10.7 Problems with Kidney Function
Various types of problems, including repeated urinary infections, can lead to renal failure, which necessitates receiving a kidney from a donor or undergoing hemodialysis by utilizing a kidney machine or CAPD.

Studying the Concepts

1. State the path of urine and the function of each organ mentioned. 188
2. Explain how urination is controlled. 188–89
3. List and explain four functions of the urinary system. 189
4. Describe the macroscopic anatomy of a kidney. 191
5. Trace the path of blood about a nephron. 192
6. Name the parts of a nephron, and tell how the structure of the convoluted tubules suits their respective functions. 193
7. State and describe the three steps of urine formation. 194–95
8. Where in particular are water and salt reabsorbed along the length of the nephron? Describe the contribution of the loop of the nephron. 196–97
9. Name and describe the action of antidiuretic hormone (ADH), the renin-aldosterone connection, and atrial natriuretic hormone (ANH). 196–97
10. How do the kidneys maintain the pH of the blood within normal limits? 199
11. Explain how the artificial kidney machine works. 200

Testing Your Knowledge of the Concepts

In questions 1–4, match the structure to the functions below.
a. glomerulus
b. proximal convoluted tubule
c. distal convoluted tubule
d. collecting duct

_____ 1. Regulation of water reabsorption

_____ 2. Reabsorption of vital molecules

_____ 3. Formation of filtrate

_____ 4. Secretion of hydrogen ions and drugs

In questions 5–7, indicate whether the statement is true (T) or False (F).

_____ 5. The ureters conduct urine from the bladder to outside the body.
_____ 6. Amino acids are filtered, reabsorbed, and not in urine.
_____ 7. When antidiuretic hormone (ADH) is present, water is maximally reabsorbed.

In questions 8 and 9, fill in the blanks.

8. The ascending limb of the loop of the nephron extrudes _____ into the renal medulla, making the renal medulla hypertonic to fluid in the collecting duct.

9. _____ is a molecule that is found in the filtrate, is reabsorbed, and is found in urine.

10. Label this diagram of a nephron.

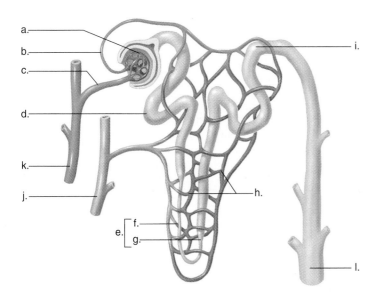

Understanding Key Terms

aldosterone 197
antidiuretic hormone
  (ADH) 196
atrial natriuretic
  hormone (ANH) 197
collecting duct 193
creatinine 189
distal convoluted
  tubule 193
diuretic 197
erythropoietin 189
excretion 187
glomerular capsule 193
glomerular filtrate 195
glomerular filtration 195
glomerulus 192
hemodialysis 200
juxtaglomerular
  apparatus 197
kidney 188

loop of the nephron 193
nephron 191
peritubular capillary
  network 192
proximal convoluted
  tubule 193
renal artery 188
renal cortex 191
renal medulla 191
renal pelvis 191
renal vein 188
renin 197
tubular reabsorption 195
tubular secretion 195
urea 189
ureter 188
urethra 188
uric acid 189
urinary bladder 188

Further Readings

Abraham, S. N. September/October 1997. Discovering the benign traits of the mast cell. *Science & Medicine* 4(5):46. Recent investigations suggest mast cells have roles in immune surveillance and control of the immune response.

Benjamini, E., and Leskowitz, S. 2000. *Immunology: A short course.* 4th ed. New York: John Wiley & Sons. Presents the essential principles of immunology.

Doyle, R. June 2000. Asthma worldwide. *Scientific American* 282(6):30. Remarkable progress has been made in the treatment of asthma with drugs such as inhaled steroids.

Glausiusz, J. September 1998. Infected hearts. *Discover* 19(9):30. Infectious bacteria may play a role in heart disease; antibiotics could prevent the need for heart surgery.

Guyton, A. C. 2000. *Textbook of medical physiology.* 10th ed. Philadelphia: W. B. Saunders Co. Presents physiological principles for those in the medical fields.

Hanson, L. A. November/December 1997. Breast feeding stimulates the infant immune system. *Science & Medicine* 4(6):12. Long-lasting protection against some infectious diseases has been reported in breast-fed infants.

Mader, S. S. 2001. *Understanding anatomy and physiology.* 4th ed. Dubuque, Iowa: The McGraw-Hill Companies. A text that emphasizes the basics for beginning allied health students.

Melton, L. July 2000. Age breakers. *Scientific American* 283(1):16. A compound is being developed that might rejuvenate hearts and muscles by breaking sugar-protein bonds that accumulate with aging.

Nature Medicine Vaccine Supplement, May 1998, Vol. 4, No. 5. Entire issue is devoted to the topic of vaccines, including history, recent developments, and research in malaria, cancer, and HIV vaccines.

Nemecek, S. March 2000. Granting immunity. *Scientific American* 282(3):15. Article discusses the safety of vaccines.

Nucci, M. L., and Abuchowski, A. February 1998. The search for blood substitutes. *Scientific American* 278(2):72. Artificial blood substitutes are being developed from synthetic chemicals, and some are based on hemoglobin.

Roitt, I., et al. 1998. *Immunology.* 5th ed. London: Mosby International, Ltd. For the advanced student, this text features a clear description of the scientific principles involved in immunology, combined with clinical examples.

Wardlaw, G., et al. 2000. *Contemporary nutrition.* 4th ed. Dubuque, Iowa: McGraw-Hill College Division. This text gives a clear understanding of nutritional information found on product labels.

Weiner, D. B., and Kennedy, R. C. July 1999. Genetic vaccines. *Scientific American* 281(1):50. Introducing bits of DNA or RNA into cells by genetic vaccines holds promise as safe preventatives and therapies for certain diseases.

West, J. B. 2000. *Respiratory physiology—The essentials.* 6th ed. Baltimore: Lippincott, Williams & Wilkins. A good reference resource that discusses all aspects of respiratory physiology, including breathing and external and internal respiration.

e-Learning Connection
www.mhhe.com/biosci/genbio/maderhuman7/

10.1 Urinary System

 Urinary System *Essential Study Partner*

 Urinary System *art labeling activity*

10.2 Kidneys

 Kidneys *Essential Study Partner*

 Kidney Function *animation activity*
Anatomy of a Kidney and Lobe *art labeling activity*
Nephron Anatomy *art labeling activity*
Nephron Anatomy II *art labeling activity*
Nephron *art quiz*
Transport Process in Mammalian Nephron *art quiz*

10.3 Urine Formation

 Urine Formation *Essential Study Partner*

10.4 Maintaining Water-Salt Balance

 Urine Concentration *Essential Study Partner*

10.6 Homeostasis

 Working Together to Achieve Homeostasis
Homeostasis of Blood Volume and Pressure *art quiz*

10.7 Problems with Kidney Function

 Looking At Both Sides *critical thinking activity*

Chapter Summary

 Key Term Flashcards *vocabulary quiz*
Chapter Quiz *objective quiz covering all chapter concepts*

III

Movement and Support in Humans

11 Skeletal System 205

The skeletal and muscular systems give the body its shape and allow it to move. The skeletal system protects and supports other organs—the skull protects the brain, and the rib cage protects the lungs and heart; the pelvis supports the organs of the abdominal cavity. The skeletal system also contributes to homeostasis because it stores minerals and produces the blood cells. ⚖

Contraction of the skeletal muscles allows the body to move but also accounts for facial expressions and our ability to speak. Homeostasis would be impossible without movement of the rib cage up and down and contraction of the heart to keep the blood moving. Smooth muscle contraction allows the other internal organs to function; it helps move food along the digestive tract and urine along the urinary tract. Indeed, life as we know it is dependent on the skeletal and muscular systems.

Chapter 11

Skeletal System

Chapter Concepts

11.1 Tissues of the Skeletal System
- Various connective tissues are necessary to the anatomy of bones, which are organs in the skeletal system.   206

11.2 Bone Growth and Repair
- Bone is a living tissue; therefore, it grows and undergoes repair.   208
- The fetal skeleton is cartilaginous and then is replaced by bone.   208
- The adult bones undergo remodeling—they are constantly being broken down and rebuilt.   209
- The mending of a fracture requires certain identifiable steps.   209

11.3 Bones of the Skeleton
- The bones of the skeleton are divided into those of the axial skeleton and those of the appendicular skeleton.   211

11.4 Articulations
- Joints are classified according to their anatomy, and only one type is freely movable.   218
- Freely movable joints are subject to arthritis, a disorder that limits their mobility.   220

11.5 Homeostasis
- The skeletal system works with the other systems of the body to maintain homeostasis.   221 ⚖

Ellen B. knew there was something different about the way she stood and walked. She leaned to one side, and her knees didn't line up properly because she had scoliosis, a sideways curvature of the spine that throws the body out of line. So, at 17, while most kids dwelt on high-school graduation, Ellen checked into the hospital. During surgery, doctors carefully hooked steel rods to vertebrae at the top and bottom of Ellen's spine, fusing the bones together with fragments taken from her hip. The fusions healed in a straightened position.

With her spine corrected, Ellen found herself two inches taller, virtually free of back pain, and physically similar to the other students in her class. In addition to being able to proceed with graduation plans, she now could go shopping for clothes without the concern that they wouldn't "hang" correctly. After the operation, Ellen knew that the skeleton ordinarily permits flexible body movement without any pain. It serves as an attachment for muscles, whose contraction makes the bones move so that we can walk, play tennis, type papers, and do all manner of activities. The bones also support and protect. For example, the large heavy bones of the legs support the entire body against the pull of gravity; the skull protects our brain; and the rib cage protects the heart and lungs. Bones have other functions too: they are the site of blood cell formation, and they store mineral salts.

This chapter reviews the structure and function of bones, which are organs in the skeletal system.

11.1 Tissues of the Skeletal System

The bones are largely composed of connective tissues—bone, cartilage, and fibrous connective tissues. Connective tissue contains cells separated by a matrix that contains fibers.

Bone

Bones are strong because their matrix contains mineral salts, notably calcium phosphate in addition to protein fibers. **Compact bone** is highly organized and composed of tubular units called osteons. In a cross section of an osteon, bone cells called **osteocytes** lie in lacunae, which are tiny chambers arranged in concentric circles around a central canal (Fig. 11.1). The matrix fills the spaces between the lacunae. Tiny canals called canaliculi (sing., canaliculus) run through the matrix, connecting the lacunae with each other and with the central canal. Central canals contain blood vessels, lymphatic vessels, and nerves. Canaliculi bring nutrients from the blood vessel in the central canal to the cells in the lacunae.

Compared to compact bone, **spongy bone** has an unorganized appearance (Fig. 11.1). It contains numerous thin plates (called trabeculae) separated by unequal spaces. Although this makes spongy bone lighter than compact bone, spongy bone is still designed for strength. Just as braces are used for support in buildings, the trabeculae follow lines of stress. The spaces of spongy bone are often filled with **red bone marrow,** a specialized tissue that produces all types of blood cells. The osteocytes of spongy bone are irregularly placed within the trabeculae, and canaliculi bring them nutrients from the red bone marrow.

Cartilage

Cartilage is not as strong as bone, but it is more flexible because the matrix is gel-like and contains many collagenous and elastic fibers. The cells, called **chondrocytes,** lie within lacunae that are irregularly grouped. Cartilage has no nerves, making it well suited for padding joints where the stresses of movement are intense. Cartilage also has no blood vessels, making it slow to heal.

The three types of cartilage differ according to the type and arrangement of fibers in the matrix. *Hyaline cartilage* is firm and somewhat flexible. The matrix appears uniform and glassy, but actually it contains a generous supply of collagenous fibers. Hyaline cartilage is found at the ends of long bones, in the nose, at the ends of the ribs, and in the larynx and trachea.

Fibrocartilage is stronger than hyaline cartilage because the matrix contains wide rows of thick, collagenous fibers. Fibrocartilage is able to withstand both tension and pressure, and is found where support is of prime importance—in the disks located between the vertebrae and also in the cartilage of the knee.

Elastic cartilage is more flexible than hyaline cartilage because the matrix contains mostly elastin fibers. This type of cartilage is found in the ear flaps and the epiglottis.

Fibrous Connective Tissue

Fibrous connective tissue contains rows of cells called fibroblasts separated by bundles of collagenous fibers. This tissue makes up the **ligaments** that connect bone to bone and the **tendons** that connect muscles to a bone at **joints,** which are also called articulations.

Structure of a Typical Bone

Figure 11.1 shows how the tissues we have been discussing are arranged in a long bone. The expanded region at the end of a long bone is called an epiphysis (pl., epiphyses). The epiphyses are composed largely of spongy bone that contains red bone marrow where blood cells are made. The epiphyses are coated with a thin layer of hyaline cartilage, which is called **articular cartilage** because it occurs at a joint.

The shaft, or main portion of the bone, is called the diaphysis. The diaphysis has a large **medullary cavity** whose walls are composed of compact bone. The medullary cavity is lined with a thin, vascular membrane (the endosteum) and is filled with fatty yellow bone marrow.

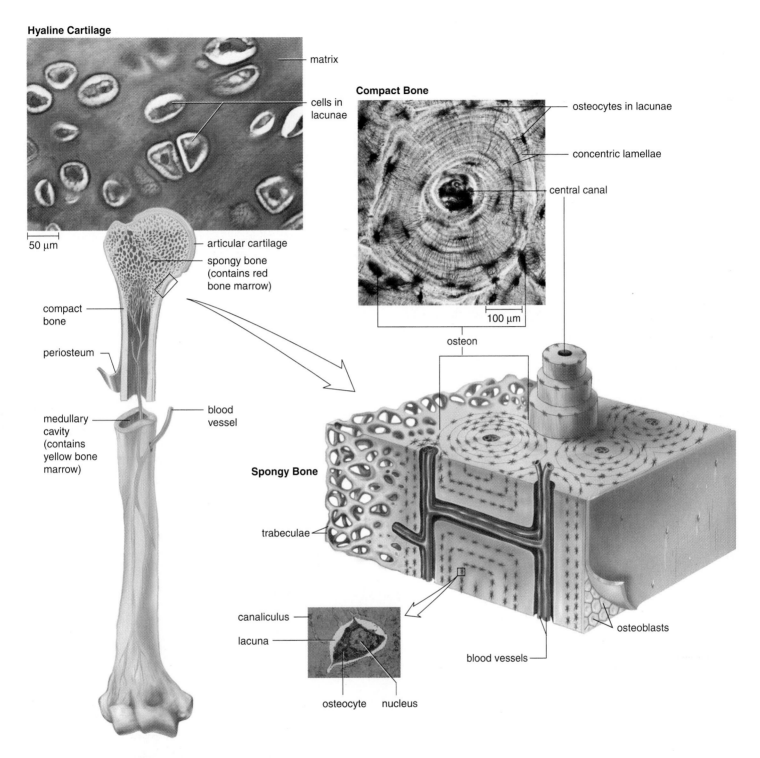

Hyaline Cartilage

matrix

cells in lacunae

50 μm

articular cartilage

spongy bone (contains red bone marrow)

compact bone

periosteum

medullary cavity (contains yellow bone marrow)

blood vessel

Compact Bone

osteocytes in lacunae

concentric lamellae

central canal

100 μm

osteon

Spongy Bone

trabeculae

osteoblasts

blood vessels

canaliculus

lacuna

osteocyte   nucleus

Figure 11.1 **Anatomy of a long bone, from the macroscopic to the microscopic level.**
A long bone is encased by the periosteum except at the ends where it is covered by hyaline (articular) cartilage (see micrograph, *top left*). Spongy bone located in each epiphysis may contain red bone marrow. The central shaft contains yellow bone marrow and is bordered by compact bone, which is shown in the enlargement and micrograph (*top right*).

Except for the articular cartilage on its ends, a long bone is completely covered by a layer of fibrous connective tissue called the **periosteum.** This covering contains blood vessels, lymphatic vessels, and nerves. Note in Figure 11.1 how a blood vessel penetrates the periosteum and enters the bone where it gives off branches within the central canals. The periosteum is continuous with ligaments and tendons that are connected to this bone.

11.2 Bone Growth and Repair

Bones are composed of living tissues, as exemplified by their ability to grow and undergo repair. Several different types of cells are involved in bone growth and repair.

Osteoprogenitor cells are unspecialized cells present in the inner portion of the periosteum, in the endosteum, and in the central canal of compact bone.

Osteoblasts are bone-forming cells derived from osteoprogenitor cells. They are responsible for secreting the matrix characteristic of bone.

Osteocytes are mature bone cells derived from osteoblasts. Once the osteoblasts are surrounded by matrix, they become the osteocytes in bone.

Osteoclasts are thought to be derived from monocytes, a type of white blood cell present in red bone marrow. Osteoclasts perform bone resorption; that is, they break down bone and deposit calcium and phosphate in the blood. The work of osteoclasts is important to the growth and repair of bone.

Bone Development and Growth

The term **ossification** refers to the formation of bone. The bones of the skeleton form during embryonic development in two distinctive ways—intramembranous ossification and endochondral ossification.

In intramembranous ossification, bones develop between sheets of fibrous connective tissue. The bones of the skull are examples of intramembranous bones. Cells, derived from connective tissue cells, become osteoblasts that lay down matrix in various directions, forming the trabeculae of spongy bone. Other osteoblasts associated with the periosteum lay down compact bone over the surface of the spongy bone. The osteoblasts become osteocytes when they are surrounded by matrix.

Most of the bones of the human skeleton form by endochondral ossification. Hyaline cartilage models which appear during fetal development are replaced by bone as development continues.

During endochondral ossification of a long bone, the cartilage begins to break down in the center of the diaphysis, which is now covered by a periosteum (Fig. 11.2). Osteoblasts invade the region and begin to lay down spongy bone in what is called a primary ossification center. Other osteoblasts lay down compact bone beneath the periosteum. As the compact bone thickens, the spongy bone of the diaphysis is broken down by osteoclasts, and the cavity created becomes the medullary cavity.

After birth, the epiphyses (ends) of developing bone continue to grow, but soon secondary ossification centers appear in these regions. Here spongy bone forms and does not break down. A band of cartilage called a **growth plate** remains between the primary ossification center and each secondary center. The limbs keep increasing in length and width as long as growth plates are still present. The rate of growth is controlled by hormones, such as growth hormones and the sex hormones. Eventually the growth plates become ossified, and the bone stops growing.

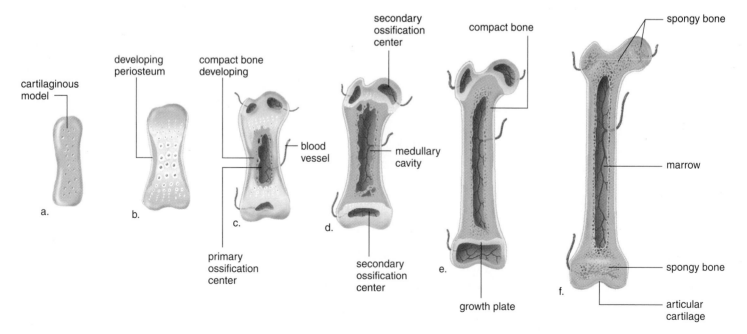

Figure 11.2　Endochondral ossification of a long bone.
a. A cartilaginous model develops during fetal development. **b.** A periosteum develops. **c.** A primary ossification center contains spongy bone surrounded by compact bone. **d.** The medullary cavity forms in the diaphysis, and secondary ossification centers develop in the epiphyses. **e.** After birth, growth is still possible as long as cartilage remains at the growth plates. **f.** When the bone is fully formed, the growth plates become a thin line.

Remodeling of Bones

In the adult, bone is continually being broken down and built up again. Osteoclasts derived from monocytes in red bone marrow break down bone, remove worn cells, and deposit calcium in the blood. After a period of about three weeks, the osteoclasts disappear, and the bone is repaired by the work of osteoblasts. As they form new bone, osteoblasts take calcium from the blood. Eventually some of these cells get caught in the matrix they secrete and are converted to osteocytes, the cells found within the lacunae of osteons.

Thus, through this process of remodeling, old bone tissue is replaced by new bone tissue. Because of continual remodeling, the thickness of bones can change. Physical use and hormone balance affect the thickness of bones. Strange as it may seem, adults apparently require more calcium in the diet (about 1,000 to 1,500 mg daily) than do children in order to promote the work of osteoblasts. Otherwise, osteoporosis, a condition in which weak and thin bones easily fracture, may develop. Osteoporosis is discussed in the Health Focus on page 210.

Bone Repair

Repair of a bone is required after it breaks or fractures. In some ways, bone repair parallels the development of a bone (Fig. 11.3) except that the first step, hematoma, indicates that injury has occurred and then fibrocartilage instead of hyaline cartilage precedes the production of compact bone.

1. *Hematoma.* Blood escapes from ruptured blood vessels and forms a hematoma (mass of clotted blood) in the space between the broken bones six to eight hours after a fracture.
2. *Fibrocartilaginous callus.* Tissue repair begins, and fibrocartilage fills the space between the ends of the broken bone for about three weeks.
3. *Bony callus.* Osteoblasts produce trabeculae of spongy bone and convert the fibrocartilage callus to a bony callus that joins the broken bones together and lasts about three to four months.
4. *Remodeling.* Osteoblasts build new compact bone at the periphery, and osteoclasts reabsorb the spongy bone, creating a new medullary cavity.

The naming of fractures tells you what kind of break occurred. A fracture is complete if the bone is broken clear through and incomplete if the bone is not separated into two parts. A fracture is simple if it does not pierce the skin and compound if it does pierce the skin. Impacted means that the broken ends are wedged into each other, and a spiral fracture occurs when there is a ragged break due to twisting of a bone.

Bone is living tissue. It develops, grows, remodels, and repairs itself. In all these processes, osteoclasts break down bone, and osteoblasts build bone.

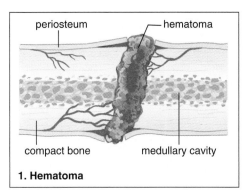

1. Hematoma

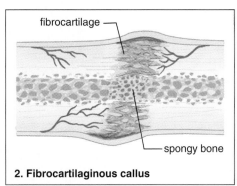

2. Fibrocartilaginous callus

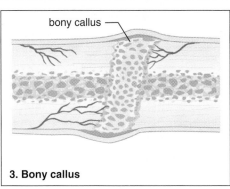

3. Bony callus

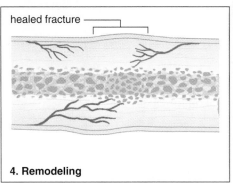

4. Remodeling

a.

b.

Figure 11.3 Bone fracture and repair.
a. Steps in the repair of a fracture. **b.** A plaster of Paris cast helps stabilize the bones while repair takes place.

Health Focus

You Can Avoid Osteoporosis

Osteoporosis is a condition in which the bones are weakened due to a decrease in the bone mass that makes up the skeleton. Throughout life, bones are continuously remodeled. While a child is growing, the rate of bone formation is greater than the rate of bone breakdown. The skeletal mass continues to increase until ages 20 to 30. After that, there is an equal rate of formation and breakdown of bone mass until ages 40 to 50. Then, reabsorption begins to exceed formation, and the total bone mass slowly decreases.

Over time, men are apt to lose 25% and women 35% of their bone mass. But we have to consider that men tend to have denser bones than women anyway, and their testosterone (male sex hormone) level generally does not begin to decline significantly until after age 65. In contrast, the estrogen (female sex hormone) level in women begins to decline at about age 45. Since sex hormones play an important role in maintaining bone strength, this difference means that women are more likely than men to suffer fractures, involving especially the hip, vertebrae, long bones, and pelvis. Although osteoporosis may at times be the result of various disease processes, it is essentially a disease of aging.

There are measures that everyone can take to avoid osteoporosis when they get older. Adequate dietary calcium throughout life is an important protection against osteoporosis. The U.S. National Institutes of Health recommend a calcium intake of 1,200–1,500 mg per day during puberty. Males and females require 1,000 mg per day until age 65 and 1,500 mg per day after age 65. In postmenopausal women not receiving estrogen replacement therapy, 1,500 mg per day is desirable.

A small daily amount of vitamin D is also necessary to absorb calcium from the digestive tract. Exposure to sunlight is required to allow skin to synthesize vitamin D. If you reside on or north of a "line" drawn from Boston to Milwaukee, to Minneapolis, to Boise, chances are you're not getting enough vitamin D during the winter months. Therefore, you should avail yourself of vitamin D present in fortified foods such as low-fat milk and cereal.

Postmenopausal women with any of the following risk factors should have an evaluation of their bone density:

- white or Asian race
- thin body type
- family history of osteoporosis
- early menopause (before age 45)
- smoking
- a diet low in calcium, or excessive alcohol consumption and caffeine intake
- sedentary lifestyle

Presently bone density is measured by a method called dual energy X-ray absorptiometry (DEXA). This test measures bone density based on the absorption of photons generated by an X-ray tube. Soon there may be a blood and urine test to detect the biochemical markers of bone loss. Then it may be possible for physicians to screen all older women and at-risk men for osteoporosis.

If the bones are thin, it is worthwhile to take other measures to gain bone density because even a slight increase can significantly reduce fracture risk. Usually, experts believe that estrogen therapy is the treatment of choice in postmenopausal women. Hormone replacement is most effective when begun at the start of menopause and continued over the long term. A combination of hormone replacement and exercise apparently yields the best results. If desired, estrogen can be replaced by a synthetic estrogen-like drug that has a selective action on the bones, rather than affecting a wide variety of organs as estrogen does.

Very inactive people, such as those confined to bed, lose bone mass 25 times faster than people who are moderately active. Thus, regular, moderate, weight-bearing exercise like walking or jogging is another good way to maintain bone strength (Fig. 11A).

a.   b.   osteoporosis

normal bone

Figure 11A   **Preventing osteoporosis.**
a. Exercise can help prevent osteoporosis, but when playing golf, you should carry your own clubs and walk instead of using a golf cart. **b.** Normal bone growth compared to bone from a person with osteoporosis.

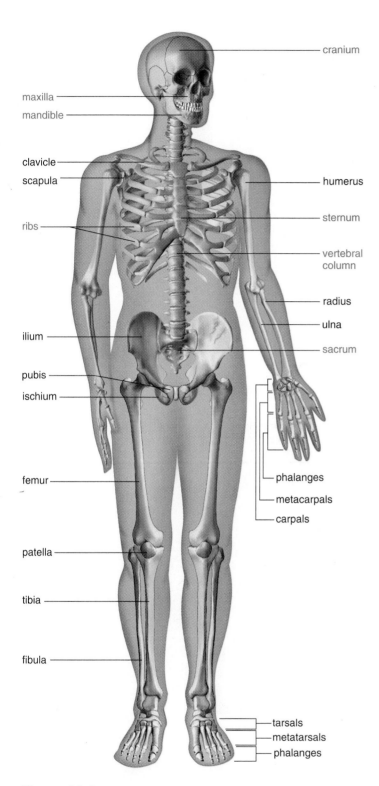

Figure 11.4 **The skeleton.**

The skeleton of a human adult contains bones that belong to the axial skeleton (red labels) and those that belong to the appendicular skeleton (black labels).

11.3 Bones of the Skeleton

Let's discuss the functions of the skeleton in relation to particular bones.

The skeleton supports the body. The bones of the legs (the femur in particular and also the tibia) support the entire body when we are standing, and the coxal bones of the pelvic girdle support the abdominal cavity.

The skeleton protects soft body parts. The bones of the skull protect the brain; the rib cage, composed of the ribs, thoracic vertebrae, and sternum, protects the heart and lungs.

The skeleton produces blood cells. All bones in the fetus have spongy bone with red bone marrow that produces blood cells. In the adult, the flat bones of the skull, ribs, sternum, clavicles, and also the vertebrae and pelvis produce blood cells. Fat is stored in yellow bone marrow.

The skeleton stores minerals and fat. All bones have a matrix that contains calcium phosphate. When bones are remodeled, osteoclasts break down bone and return calcium ions and phosphorus ions to the bloodstream.

The skeleton, along with the muscles, permits flexible body movement. While articulations (joints) occur between all the bones, we associate body movement in particular with the bones of the legs (especially the femur and tibia) and the feet (tarsals, metatarsals, and phalanges) because we use them when walking.

Classification of the Bones

The bones are classified according to their shape. Long bones, exemplified by the humerus and femur, are longer than they are wide. Short bones, such as the carpals and tarsals, are cube shaped—that is, their lengths and widths are about equal. Flat bones, like those of the skull, are platelike with broad surfaces. Round bones, exemplified by the patella, are circular in shape. Irregular bones, such as the vertebrae and facial bones, have varied shapes that permit connections with other bones.

The 206 bones of the skeleton are also classified according to whether they occur in the axial skeleton or the appendicular skeleton. The axial skeleton is in the midline of the body, and the appendicular skeleton consists of the limbs along with their girdles (Fig. 11.4).

The bones of the skeleton are not smooth; they have articulating depressions and protuberances at various joints. And they have projections, often called processes, where the muscles attach. Also, there are openings for nerves and/or blood vessels to pass through.

The skeleton is divided into the axial and appendicular skeletons. Each has different types of bones with protuberances at joints and processes where the muscles attach.

The Axial Skeleton

The **axial skeleton** lies in the mid-line of the body and consists of the skull, hyoid bone, vertebral column, and rib cage.

The Skull

The **skull** is formed by the cranium (braincase) and the facial bones. It should be noted, however, that some cranial bones contribute to the face.

The Cranium   The cranium protects the brain. In adults, it is composed of eight flat bones fitted tightly together. In newborns, certain cranial bones are not completely formed and instead are joined by membranous regions called **fontanels.** The fontanels usually close by the age of 16 months by the process of intramembranous ossification.

Some of the bones of the cranium contain the **sinuses,** air spaces lined by mucous membrane. The sinuses reduce the weight of the skull and give a resonant sound to the voice. Two sinuses called the mastoid sinuses drain into the middle ear. **Mastoiditis,** a condition that can lead to deafness, is an inflammation of these sinuses.

The major bones of the cranium have the same names as the lobes of the brain: frontal, parietal, occipital, and temporal. On the top of the cranium (Fig. 11.5*a*), the **frontal bone** forms the forehead, the **parietal bones** extend to the sides, and the **occipital bone** curves to form the base of the skull. Here there is a large opening, the **foramen magnum** (Fig. 11.5*b*), through which the spinal cord passes and becomes the brain stem. Below the much larger parietal bones, each temporal bone has an opening (external auditory canal) that leads to the middle ear.

The **sphenoid bone,** which is shaped like a bat with wings outstretched, extends across the floor of the cranium from one side to the other. The sphenoid is considered the keystone bone of the cranium because all the other bones articulate with it. The sphenoid completes the sides of the skull and also contributes to forming the orbits (eye sockets). The **ethmoid bone,** which lies in front of the sphenoid, also helps form the orbits and the nasal septum. The orbits are completed by various facial bones. The eye sockets are called orbits because of our ability to rotate the eyes.

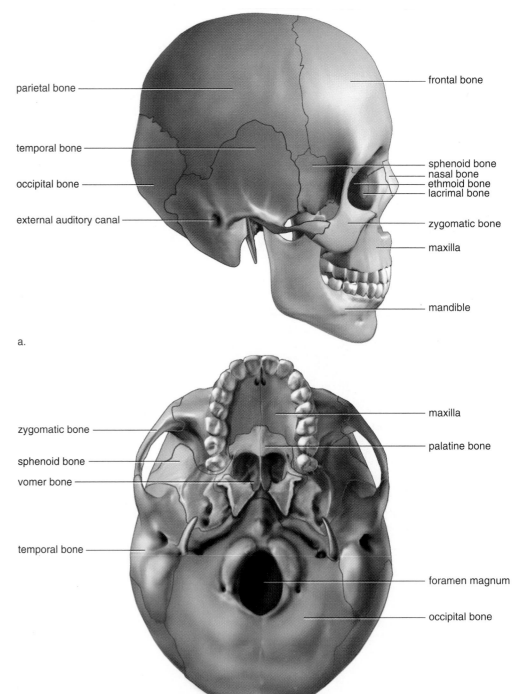

Figure 11.5   Bones of the skull.
a. Lateral view. **b.** Inferior view.

The cranium contains eight bones: the frontal, two parietal, the occipital, two temporal, the sphenoid, and the ethmoid.

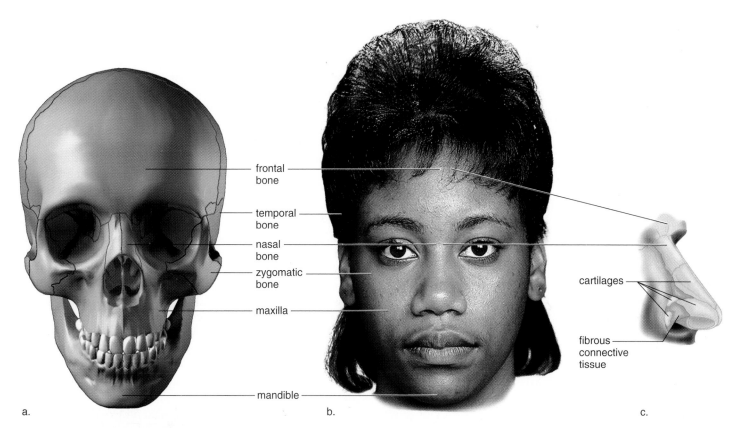

frontal
bone

temporal
bone

nasal
bone

zygomatic
bone

maxilla

mandible

cartilages

fibrous
connective
tissue

a.

b.

c.

Figure 11.6   **Bones of the face, including the nose.**

a. The frontal bone forms the forehead and eyebrow ridges; the zygomatic bones form the cheekbones; and the maxillae form the upper jaw. The maxillae are the most expansive facial bones, extending from the forehead to the lower jaw. The mandible has a projection we call the chin. **b.** The maxillae, frontal, and nasal bones help form the external nose. **c.** The rest of the nose is formed by cartilages and fibrous connective tissue.

The Facial Bones   The most prominent of the facial bones are the mandible, the maxillae (maxillary bones), the zygomatic bones, and the nasal bones.

The **mandible,** or lower jaw, is the only movable portion of the skull (Fig. 11.6*a*), and its action permits us to chew our food. It also forms the chin. Tooth sockets are located on the mandible and on the **maxillae,** the upper jaw that also forms the anterior portion of the hard palate. The palatine bones make up the posterior portion of the hard palate and the floor of the nose (see Fig. 11.5*b*).

The lips and cheeks have a core of skeletal muscle. The **zygomatic bones** are the cheekbone prominences, and the **nasal bones** form the bridge of the nose. Other bones (e.g., ethmoid and vomer) are a part of the nasal septum, which divides the interior of the nose into two nasal cavities. The lacrimal bone (see Fig. 11.5*a*) contains the opening for the nasolacrimal canal, which brings tears from the eyes to the nose.

The temporal and frontal bones are cranial bones that contribute to the face. The temporal bones account for the flattened areas we call the temples. The frontal bone forms the forehead and has supraorbital ridges where the eyebrows are located. Glasses sit where the frontal bone joins the nasal bones.

While the ears are formed only by elastic cartilage and not by bone, the nose (Fig. 11.6*c*) is a mixture of bones, cartilages, and fibrous connective tissue. The cartilages complete the tip of the nose, and fibrous connective tissue forms the flared sides of the nose.

Among the facial bones, the mandible is the lower jaw where the chin is located, the two maxillae form the upper jaw, the two zygomatic bones are the cheekbones, and the two nasal bones form the bridge of the nose.

The Hyoid Bone

Although the **hyoid bone** is not part of the skull, it will be mentioned here because it is a part of the axial skeleton. It is the only bone in the body that does not articulate with another bone. It is attached to the temporal bones by muscles and ligaments and to the larynx by a membrane. The larynx is the voice box at the top of the trachea in the neck region. The hyoid bone anchors the tongue and serves as the site for the attachment of muscles associated with swallowing.

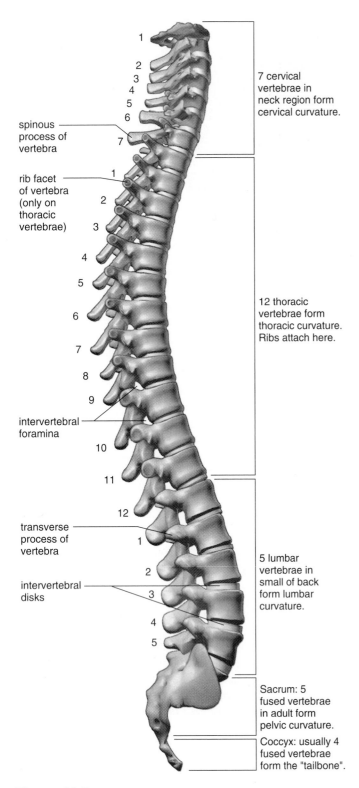

spinous process of vertebra

7 cervical vertebrae in neck region form cervical curvature.

rib facet of vertebra (only on thoracic vertebrae)

12 thoracic vertebrae form thoracic curvature. Ribs attach here.

intervertebral foramina

transverse process of vertebra

intervertebral disks

5 lumbar vertebrae in small of back form lumbar curvature.

Sacrum: 5 fused vertebrae in adult form pelvic curvature.

Coccyx: usually 4 fused vertebrae form the "tailbone".

Figure 11.7   The vertebral column.
The vertebrae are named according to their location in the vertebral column, which is flexible due to the intervertebral disks. Note the presence of the coccyx, also called the tailbone.

The Vertebral Column

The **vertebral column** consists of 33 vertebrae (Fig. 11.7). Normally, the vertebral column has four curvatures that provide more resilience and strength for an upright posture than a straight column could provide. As discussed in the introduction to this chapter, scoliosis is an abnormal lateral (sideways) curvature of the spine. There are two other well-known abnormal curvatures: kyphosis is an abnormal posterior curvature that often results in a hunchback, and lordosis is an abnormal anterior curvature resulting in a swayback.

The vertebral column forms when the vertebrae join. The spinal cord, which passes through the vertebral canal, gives off the spinal nerves at the intervertebral foramina. Spinal nerves function to control skeletal muscle contraction and the internal organs. The spinous processes of the vertebrae can be felt as bony projections along the midline of the back. The spinous processes and also the transverse processes, which extend laterally, serve as attachment sites for the muscles that move the vertebral column.

The various vertebrae are named according to their location in the vertebral column. The cervical vertebrae are located in the neck. The first cervical vertebra, called the **atlas,** holds up the head. It is so named because Atlas, of Greek mythology, held up the world. Movement of the atlas permits the "yes" motion of the head. It also allows the head to tilt from side to side. The second cervical vertebra is called the **axis** because it allows a degree of rotation as when we shake the head "no." The thoracic vertebrae have long, thin spinous processes and extra articular facets for the attachment of the ribs (Fig. 11.8a). Lumbar vertebrae have a large body and thick processes. The five sacral vertebrae are fused together in the sacrum, which is a part of the pelvic girdle. The coccyx, or tailbone, is usually composed of four fused vertebrae.

Between the vertebrae are **intervertebral disks** composed of fibrocartilage that act as a kind of padding. They prevent the vertebrae from grinding against one another and absorb shock caused by movements such as running, jumping, and even walking. The presence of the disks allows the vertebrae to move as we bend forward, backward, and from side to side. Unfortunately, these disks become weakened with age and can even slip and rupture. Pain results if a slipped disk presses against the spinal cord and/or spinal nerves. If that occurs, surgical removal of the disk may relieve the pain.

The vertebral column forms when the vertebrae join. The vertebral column supports the head and trunk, protects the spinal cord, and serves as a site for muscle attachment. The intervertebral disks provide padding and account for the flexibility of the column.

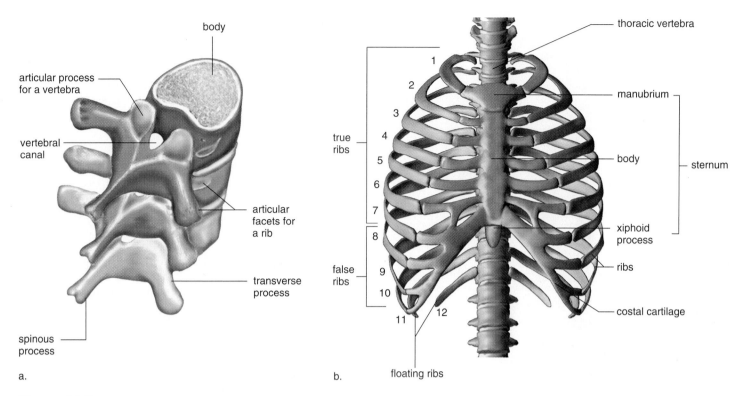

Figure 11.8 **Thoracic vertebrae and the rib cage.**
a. The thoracic vertebrae articulate with each other at the articular processes and with the ribs. A thoracic vertebra has two facets for articulation with a rib; one is on the body, and the other is on the transverse process. **b.** The rib cage consists of the thoracic vertebrae, the ribs, the costal cartilages, and the sternum.

The Rib Cage

The rib cage is composed of the thoracic vertebrae, the ribs and their associated cartilages, and the sternum (Fig. 11.8b).

The rib cage demonstrates how the skeleton is protective but also flexible. The rib cage protects the heart and lungs; yet it swings outward and upward upon inspiration and then downward and inward upon expiration.

The Ribs

There are twelve pairs of ribs. All twelve pairs connect directly to the thoracic vertebrae in the back. A rib articulates with the body and transverse process of its corresponding thoracic vertebra. Each rib curves outward and then forward and downward.

The upper seven pairs of ribs connect directly to the sternum by means of costal cartilages. These are called the "true ribs." The next three pairs of ribs do not connect directly to the sternum, and they are called the "false ribs." They attach to the sternum by means of a common cartilage. The last two pairs are called "floating ribs" because they do not attach to the sternum at all.

The Sternum

The **sternum,** or breastbone, is a flat bone that has the shape of a blade. The sternum, along with the ribs, helps protect the heart and lungs.

The sternum is composed of three bones that fuse during fetal development. These bones are the manubrium, the body, and the xiphoid process. The manubrium joins with the body of the sternum at an angle. This is an important anatomical landmark because it occurs at the level of the second rib and therefore allows the ribs to be counted. Counting the ribs is sometimes used to determine where the apex of the heart is located—usually between the fifth and sixth ribs.

The xiphoid process is the third part of the sternum. Composed of hyaline cartilage in the child, it becomes ossified in the adult. The variably shaped xiphoid process serves as an attachment site for the diaphragm, which divides the thoracic cavity from the abdominal cavity.

The rib cage, consisting of the thoracic vertebrae, the ribs, and the sternum, protects the heart and lungs in the thoracic cavity.

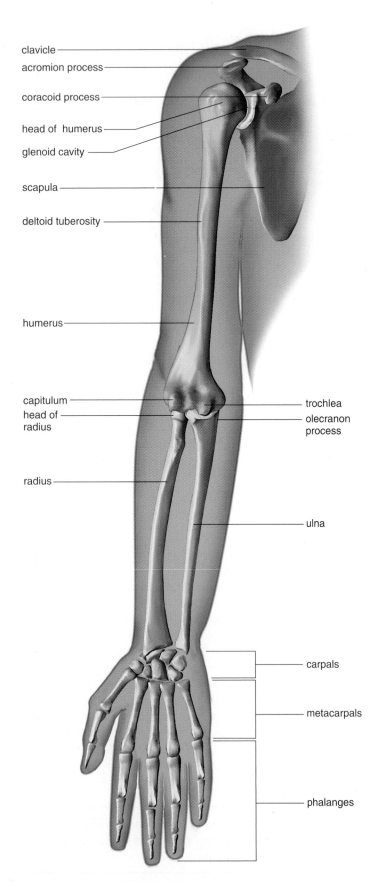

clavicle
acromion process
coracoid process
head of humerus
glenoid cavity
scapula
deltoid tuberosity
humerus
capitulum
head of radius
radius
trochlea
olecranon process
ulna
carpals
metacarpals
phalanges

Figure 11.9   **Bones of a pectoral girdle and arm.**

The Appendicular Skeleton

The **appendicular skeleton** consists of the bones within the pectoral and pelvic girdles and their attached limbs. A pectoral (shoulder) girdle and arm are specialized for flexibility; the pelvic (hip) girdle and legs are specialized for strength.

The Pectoral Girdle and Arm

A **pectoral girdle** consists of a scapula (shoulder blade) and a clavicle (collarbone) (Fig. 11.9). The **clavicle** extends across the top of the thorax; it articulates with (joins with) the sternum and the acromion process of the **scapula,** a visible bone in the back. The muscles of the arm and chest attach to the coracoid process of the scapula. The **glenoid cavity** of the scapula articulates with and is much smaller than the head of the humerus. This allows the arm to move in almost any direction, but reduces stability. This is the joint that is most apt to dislocate. Tendons that encircle and help form a socket for the humerus are collectively called the **rotator cuff.** Vigorous circular movements of the arm can lead to rotator cuff injuries.

The components of a pectoral girdle follow freely the movements of the arm, which consists of the humerus of the upper arm and the radius and ulna of the lower arm. The **humerus,** the single long bone in the upper arm, has a smoothly rounded head that fits into the glenoid cavity of the scapula as mentioned. The shaft of the humerus has a tuberosity (protuberance) where the deltoid, the prominent muscle of the chest, attaches. After death, enlargement of this tuberosity can be used as evidence that the person did a lot of heavy lifting.

The far end of the humerus has two protuberances, called the capitulum and the trochlea, which articulate respectively with the **radius** and the **ulna** at the elbow. The bump at the back of the elbow is the olecranon process of the ulna.

When the arm is held so that the palm is turned forward, the radius and ulna are about parallel to one another. When the arm is turned so that the palm is turned backward, the radius crosses in front of the ulna, a feature that contributes to the easy twisting motion of the forearm.

The hand has many bones, and this increases its flexibility. The wrist has eight **carpal** bones, which look like small pebbles. From these, five **metacarpal** bones fan out to form a framework for the palm. The metacarpal bone that leads to the thumb is opposable to the other digits. (**Digits** is a term that refers to either fingers or toes.) The knuckles are the enlarged distal ends of the metacarpals. Beyond the metacarpals are the **phalanges,** the bones of the fingers and the thumb. The phalanges of the hand are long, slender, and lightweight.

The pectoral girdle and arm are specialized for flexibility of movement.

The Pelvic Girdle and Legs

Figure 11.10 shows how a leg is attached to the pelvic girdle. The **pelvic girdle** (hip girdle) consists of two heavy, large coxal bones (hipbones). The **pelvis** is a basin composed of the pelvic girdle, sacrum, and coccyx. The pelvis bears the weight of the body, protects the organs within the pelvic cavity, and serves as the place of attachment for the legs.

Each **coxal bone** has three parts: the ilium, the ischium, and the pubis, which are fused in the adult (Fig. 11.10). The hip socket, called the acetabulum, occurs where these three bones meet. The ilium is the largest part of the coxal bones, and our hips occur where it flares out. We sit on the ischium, which has a posterior spine called the ischial spine. The pubis, from which the term pubic hair is derived, is the anterior part of a coxal bone. The two pubic bones are joined together by a fibrocartilage disk at the pubic symphysis.

The male and female pelvis differ from one another. In the female, the iliac bones are more flared; the pelvic cavity is more shallow, but the outlet is wider. These adaptations facilitate giving birth.

The **femur** (thighbone) is the longest and strongest bone in the body. The head of the femur articulates with the coxal bones at the acetabulum, and the short neck better positions the legs for walking. The femur has two large processes, the greater and lesser trochanters, which are places of attachment for the muscles of the legs and buttocks. At its distal end, the femur has medial and lateral condyles that articulate with the **tibia** of the lower leg. This is the region of the knee and the **patella,** or kneecap. The patella is held in place by the quadriceps tendon that continues as a ligament which attaches to the tibial tuberosity. At the distal end, the medial malleolus of the tibia causes the inner bulge of the ankle. The **fibula** is the more slender bone in the lower leg. The fibula has a head that articulates with the tibia and a distal lateral malleolus that forms the outer bulge of the ankle.

Each foot has an ankle, an instep, and five toes. The many bones of the foot give it considerable flexibility, especially on rough surfaces. The ankle contains seven **tarsal** bones, one of which (the talus) can move freely where it joins the tibia and fibula. Strange to say, the calcaneus, or heel bone, is also considered part of the ankle. The talus and calcaneus support the weight of the body.

The instep has five elongated **metatarsal** bones. The distal end of the metatarsals forms the ball of the foot. If the ligaments that bind the metatarsals together become weakened, flat feet are apt to result. The bones of the toes are called **phalanges,** just like those of the fingers, but in the foot, the phalanges are stout and extremely sturdy.

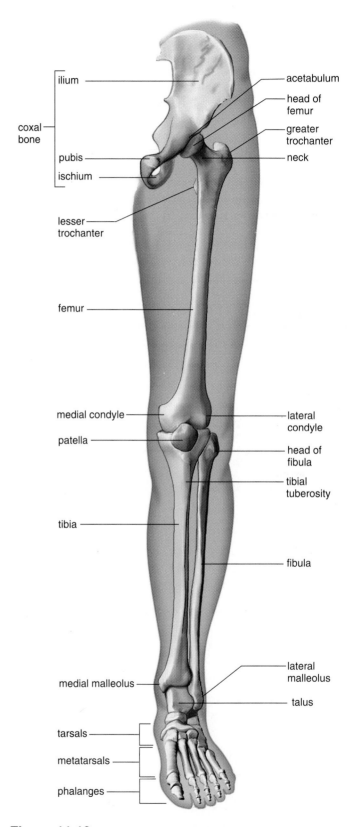

Figure 11.10   **A coxal bone and the bones of a leg.**

The pelvic girdle and leg are adapted to supporting the weight of the body. The femur is the longest and strongest bone in the body.

11.4 Articulations

Bones are joined at the joints, which are classified as fibrous, cartilaginous, and synovial. Fibrous joints, such as the **sutures** between the cranial bones, are immovable. Cartilaginous joints are connected by hyaline cartilage, as in the costal cartilages that join the ribs to the sternum, or by fibrocartilage, as in the intervertebral disks. Cartilaginous joints are slightly movable. Synovial joints are freely movable.

In **synovial joints,** the two bones are separated by a cavity. Ligaments hold the two bones in place as they form a capsule. Tendons also help stabilize the joint. The joint capsule is lined by a synovial membrane, which produces synovial fluid, a lubricant for the joint. The knee is an example of a synovial joint (Fig. 11.11). Aside from articular cartilage, the knee contains menisci (sing., **meniscus**), crescent-shaped pieces of hyaline cartilage between the bones. These give added stability and act as shock absorbers. Unfortunately, athletes often suffer injury to the menisci, known as torn cartilage. The knee joint also con-

tains 13 fluid-filled sacs called bursae (sing., **bursa**), which ease friction between the tendons and ligaments. Inflammation of the bursae is called **bursitis;** tennis elbow is a form of bursitis.

There are different types of synovial joints. The knee and elbow joints are **hinge joints** because, like a hinged door, they largely permit movement in one direction only. The joint between the radius and ulna is a pivot joint in which only rotation is possible. More movable are the **ball-and-socket joints;** for example, the ball of the femur fits into a socket on the hipbone. Ball-and-socket joints allow movement in all planes and even rotational movement. The various movements of body parts at synovial joints are depicted in Figure 11.12.

Joints are regions of articulations between bones. Synovial joints are freely movable and allow particular types of movements. Unfortunately, synovial joints are subject to various disorders.

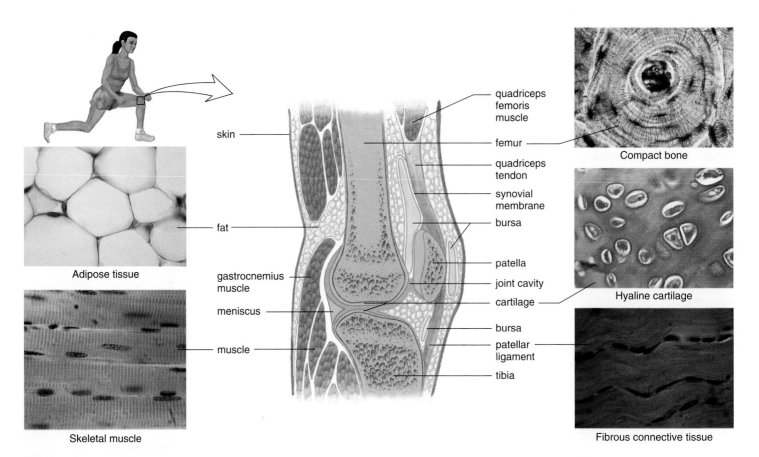

Figure 11.11 Knee joint.
The knee joint is a synovial joint. Notice the cavity between the bones, which is encased by ligaments and lined by synovial membrane. The patella (kneecap) serves to guide the quadriceps tendon over the joint when flexion or extension occurs.

Figure 11.12   Joint movements.
a. Angular movements increase or decrease
an angle between the bones of a joint.
b. Circular movements describe a circle or
part of a circle. **c.** Special movements are
unique to certain joints.

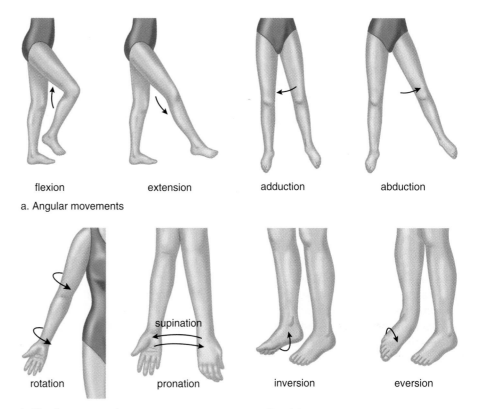

flexion   extension   adduction   abduction

a. Angular movements

rotation   pronation   inversion   eversion

supination

b. Circular movements   c. Special movements

Movements Permitted by Synovial Joints

Intact skeletal muscles are attached to bones by tendons that
span joints. When a muscle contracts, one bone moves in relation to another bone. More common types of movements
are described here:

Angular movements:

Flexion (Fig. 11.12*a*) decreases the joint angle. Flexion
of the elbow moves the forearm toward the upper
arm; flexion of the knee moves the lower leg
toward the upper leg. *Dorsiflexion* is flexion of the
foot upward, as when you stand on your heels;
plantar flexion is flexion of the foot downward, as
when you stand on your toes.

Extension increases the joint angle. Extension of the
flexed elbow straightens the arm so that there is a
180° angle at the elbow. Hyperextension occurs
when a portion of the body part is extended
beyond 180°. It is possible to hyperextend the head
and the trunk of the body.

Adduction is the movement of a body part toward the
midline. For example, adduction of the arms or
legs moves them back to the sides, toward the
body.

Abduction is the movement of a body part laterally,
away from the midline. Abduction of the arms or
legs moves them laterally, away from the body.

Circular movements:

Rotation (Fig. 11.12*b*) is the movement of a body part
around its own axis, as when the head is turned to
answer "no" or when the arm is twisted one way
and then the other.

Supination is the rotation of the lower arm so that the
palm is upward; **pronation** is the opposite—the
movement of the lower arm so that the palm is
downward.

Circumduction is the movement of a body part in a
wide circle, as when a person makes arm circles. If
the motion is observed carefully, one can see that,
because the proximal end of the arm is stationary,
the shape outlined by the arm is actually a cone.

Special movements:

Inversion and **eversion** (Fig. 11.12*c*) are terms that
apply only to the feet. Inversion is turning the foot
so that the sole is inward, and eversion is turning
the foot so that the sole is outward.

Elevation and **depression** are the lifting up and down,
respectively, of a body part, such as when you
shrug your shoulders.

Movements at joints are classified as angular,
circular, and special.

Arthritic Joints

Synovial joints are subject to **arthritis.** Inflammation of the joint leads to degenerative changes that cause pain and reduced mobility. Rheumatoid arthritis (RA) is more crippling than osteoarthritis (OA). RA is an autoimmune disease that begins after the body has brought an infection under control. An antibody known as rheumatoid factor attacks synovial membranes, and the resulting inflammatory response destroys articular cartilage. As the cartilage degenerates, the enlarged joint capsule may be invaded by fibroconnective tissue, or the bones may solidly fuse together.

OA begins in joints that are overworked. Constant compression and abrasion continually damage articular cartilage, which is repeatedly replaced. Eventually, articular cartilage softens, cracks, and wears away entirely in some areas. As the disease progresses, the exposed bone thickens and forms spurs that cause the bone ends to enlarge and joint movement to be restricted.

Weight loss can ease arthritis. Taking off 3 pounds can reduce the load on a hip or knee joint by 9–15 pounds. A sensible exercise program helps build up muscles, which stabilize joints; low-impact activities like biking and swimming are best. Pain medications and proper use of a cane or walker can also improve mobility. Replacement of damaged joints with a prosthesis (artificial substitute) may be recommended. Knee and hip replacements are common and successful (Fig. 11.13).

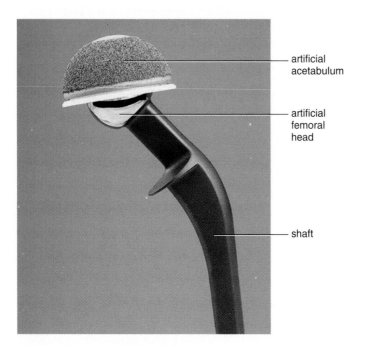

artificial acetabulum

artificial femoral head

shaft

Figure 11.13   Hip prosthesis.
The acetabulum of a coxal bone and the head of a femur are replaced by these artificial parts. The shaft is anchored in the femur.

11.5 Homeostasis

The illustration on next page tells how the skeletal system works with the other systems of the body to maintain homeostasis.

The rib cage assists the breathing process, enabling oxygen to enter the blood where it is transported by red blood cells to the tissues. Red bone marrow produces the red blood cells that transport oxygen. Without a supply of oxygen, the cells of the body could not efficiently produce ATP. ATP is needed for muscle contraction and for nerve conduction as well as for the many synthesis reactions that occur in cells.

Red bone marrow produces not only the red blood cells but also the white blood cells. The white cells are involved in defending the body against pathogens and cancerous cells. Without the ability to withstand foreign invasion, the body may soon succumb to disease and die.

The jaws contain sockets for the teeth, which chew food. Chewing breaks food into pieces small enough to be swallowed and chemically digested. Without digestion, nutrients would not enter the body to serve as building blocks for repair and a source of energy for the production of ATP.

The bones also protect the internal organs. The rib cage protects the heart and lungs; the skull protects the brain; and the vertebrae protect the spinal cord. The endocrine organs, such as the pituitary gland and pineal gland, are also protected by bone. The nervous system and the endocrine system work together to control the other organs and, ultimately, homeostasis.

The storage of calcium in the bones is under hormonal control. There is a dynamic equilibrium between the concentration of calcium in the bones and in the blood. Calcium ions play a major role in muscle contraction and nerve conduction. Calcium ions also help regulate cellular metabolism. Protein hormones, which cannot enter cells, are called the first messenger, and a second messenger such as calcium ions jump-starts cellular metabolism, directing it to proceed in a particular way.

Locomotion is efficient in human beings because they have a jointed skeleton for the attachment of muscles that move the bones. Our jointed skeleton allows us to seek out and move to a more suitable external environment in order to maintain the internal environment within reasonable limits.

The skeletal system has many diverse roles in homeostasis.

Human Systems Work Together

Integumentary System

Bones provide support for skin.

Skin protects bones; helps provide vitamin D for Ca^{2+} absorption.

Muscular System

Bones provide attachment sites for muscles; store Ca^{2+} for muscle function.

Muscular contraction causes bones to move joints; muscles help protect bones.

Nervous System

Bones protect sense organs, brain, and spinal cord; store Ca^{2+} for nerve function.

Receptors send sensory input from bones to joints.

Endocrine System

Bones provide protection for glands; store Ca^{2+} used as second messenger.

Growth hormone regulates bone development; parathyroid hormone and calcitonin regulate Ca^{2+} content.

Cardiovascular System

Rib cage protects heart; red bone marrow produces blood cells; bones store Ca^{2+} for blood clotting.

Blood vessels deliver nutrients and oxygen to bones, carry away wastes.

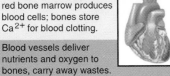

How the Skeletal System works with other body systems

Lymphatic System/Immunity

Red bone marrow produces white blood cells involved in immunity.

Lymphatic vessels pick up excess tissue fluid; immune system protects against infections.

Respiratory System

Rib cage protects lungs and assists breathing; bones provide attachment sites for muscles involved in breathing.

Gas exchange in lungs provides oxygen and rids body of carbon dioxide.

Digestive System

Jaws contain teeth that chew food; hyoid bone assists swallowing.

Digestive tract provides Ca^{2+} and other nutrients for bone growth and repair.

Urinary System

Bones provide support and protection.

Kidneys provide active vitamin D for Ca^{2+} absorption and help maintain blood level of Ca^{2+}, needed for bone growth and repair.

Reproductive System

Bones provide support and protection of reproductive organs.

Sex hormones influence bone growth and density in males and females.

Summarizing the Concepts

11.1 Tissues of the Skeletal System

A system is made up of organs, and each organ is made up of tissues. The organs known as bones are largely composed of connective tissue, particularly bone, cartilage, and fibrous connective tissue. In a long bone, hyaline (articular) cartilage covers the ends, while periosteum (fibrous connective tissue) covers the rest of the bone. Spongy bone, which may contain red bone marrow, is in the epiphyses, while yellow bone marrow is in the medullary cavity of the diaphysis. The wall of the diaphysis is compact bone.

11.2 Bone Growth and Repair

Bone is a living tissue that can grow and undergo repair. The prenatal human skeleton is at first cartilaginous, but is later replaced by a bony skeleton. During adult life, bone is constantly being broken down by osteoclasts and then rebuilt by osteoblasts that become osteocytes in lacunae. Repair of a fracture requires four steps: (1) hematoma, (2) fibrocartilaginous callus, (3) bony callus, and (4) remodeling.

11.3 Bones of the Skeleton

The skeleton supports and protects the body, permits flexible movement, produces blood cells, and serves as a storehouse for mineral salts, particularly calcium phosphate.

The axial skeleton lies in the midline of the body and consists of the skull, the hyoid bone, the vertebral column, and the rib cage. The skull contains the cranium, which protects the brain and the facial bones. On the face, the frontal bone forms the forehead; the maxillae extend from the frontal bone to the upper row of teeth; the zygomatic bones are the cheekbones; and the nasal bones form the bridge of the nose. The rest of the nose is a mixture of cartilages and fibrous connective tissue. The ears are elastic cartilage. The mandible is the lower jaw and accounts for the chin.

The appendicular skeleton consists of the bones of the pectoral girdle, arms, pelvic girdle, and legs. The pectoral girdle and arms are adapted for flexibility. The glenoid cavity of the scapula is barely large enough to receive the head of the humerus, and this means that dislocated shoulders are more likely to happen than dislocated hips. The pelvic girdle and the legs are adapted for strength; the femur is the strongest bone in the body. However, the foot, like the hand, is flexible because it contains so many bones. Both the fingers and the toes are called digits, and the term phalanges refers to the bones of each.

11.4 Articulations

There are three types of joints: fibrous joints, such as the sutures of the cranium, are immovable; cartilaginous joints, such as those between the ribs and sternum and the pubic symphysis, are slightly movable; and synovial joints, consisting of a membrane-lined (synovial membrane) capsule, are freely movable. There are different kinds of synovial joints, and the movements they permit are varied.

11.5 Homeostasis

The skeletal system works with the other systems of the body in the ways described in the illustration on page 221.

Studying the Concepts

1. Bones are organs composed of what types of tissues? What are some differences between compact bone and spongy bone? 206
2. Describe the makeup of a long bone. 206–7
3. What are the two types of ossification? Describe endochondral ossification. 208
4. How does a long bone grow in children, and how is it remodeled in all age groups? 208–9
5. What are the four steps required for fracture repair? 209
6. What are the bones of the cranium and the face? Associate the parts of a face with particular bones or cartilages. 212–13
7. What are the parts of the vertebral column, and what are its curvatures? Distinguish between the atlas, axis, sacrum, and coccyx. 214
8. What are the bones of the rib cage, and what are several functions of the rib cage? 215
9. What are the bones of the pectoral girdle? Give examples to demonstrate the flexibility of the pectoral girdle. Where does the scapula articulate with the clavicle? 216
10. What are the bones of the arm? Name several processes of the humerus. Of the elbow. 216
11. What are the bones of the pelvic girdle, and what are their functions? Give examples to demonstrate the strength and stability of the pelvic girdle. 217
12. What are the bones of the leg? Name several processes of the femur. Of the ankle. 217
13. How are joints classified? Give examples of the different types of synovial joints and the movements they permit. 218–19
14. Name the two types of arthritis and describe each. 220
15. How does the skeletal system help maintain homeostasis? 220–21
16. How does the skeletal system assist the respiratory system in maintaining homeostasis? 220–21

Testing Your Knowledge of the Concepts

In questions 1–6, match the bones to the location .
a. forehead
b. chin
c. cheekbone
d. collarbone
e. shoulder blade
f. hip
g. upper arm

_____ 1. zygomatic

_____ 2. clavicle

_____ 3. frontal bone

_____ 4. humerus

_____ 5. coxal bone

_____ 6. scapula

In questions 7–12, match the features to the bones.
a. glenoid cavity
b. olecranon process
c. acetabulum
d. spinous process
e. greater and lesser trochanter
f. xiphoid process

_____ 7. femur

_____ 8. scapula

_____ 9. ulna

_____ 10. coxal bone

_____ 11. sternum

_____ 12. vertebra

In questions 13 and 14, indicate whether the statement is true (T) or false (F).

_____ 13. The pectoral girdle is specialized for weight-bearing, while the pelvic girdle is specialized for flexibility of movement.

_____ 14. The term phalanges refers to the bones in both the fingers and the toes.

For questions 15–19, fill in the blanks.

15. Compact bone contains tubular units called _____.

16. Spongy bone often contains _____ where blood cells are produced.

17. During bone remodeling, cells called _____ break down bone and return calcium to the blood.

18. The _____ are air-filled spaces in the cranium.

19. The sacrum is a part of the _____, and the sternum is a part of the _____.

20. Label this diagram of the pelvis and leg.

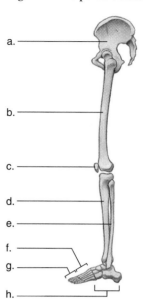

Understanding Key Terms

appendicular skeleton 216
arthritis 220
articular cartilage 206
axial skeleton 212
ball-and-socket joint 218
bursa 218
bursitis 218
cartilage 206
chondrocyte 206
compact bone 206
fibrous connective tissue 206
fontanel 212
foramen magnum 212
growth plate 208
hinge joint 218
intervertebral disk 214
joint 206
ligament 206
mastoiditis 212
medullary cavity 206
meniscus 218
ossification 208
osteoblast 208
osteoclast 208
osteocyte 206
pectoral girdle 216
pelvic girdle 217
periosteum 207
red bone marrow 206
rotator cuff 216
sinus 212
spongy bone 206
suture 218
synovial joint 218
tendon 206
vertebral column 214

e-Learning Connection   **www.mhhe.com/biosci/genbio/maderhuman7/**

11.1 Tissues of the Skeletal System

Bone Structure *Essential Study Partner*

11.2 Bone Growth and Repair

Osteoporosis *case study*

11.3 Bones of the Skeleton

Skeleton *Essential Study Partner*

Bones of the Pectoral Girdle and Arm *art labeling activity*
Bones of the Pelvic Girdle and Leg *art labeling activity*

11.4 Articulations

Joint Disorders *reading*

11.5 Homeostasis

Working Together to Achieve Homeostasis

Chapter Summary

Key Term Flashcards *vocabulary quiz*
Chapter Quiz *objective quiz covering all chapter concepts*

Chapter *12*

Muscular System

Chapter Concepts

Figure 12.1   How important is a smile?
Facial expressions are dependent on muscles that allow us to open our mouths, wink our eyes, and of course, smile.

When most kids were dreaming of new toys, sugar cookies, and other holiday treats, 7-year-old Chelsey Thomas found herself hoping for just one thing—the chance to smile (Fig. 12.1).

Chelsey was born with a rare disorder called Moebius syndrome, which causes paralysis of facial muscles. All through childhood—as her friends giggled and grinned—Chelsey could not force even the smallest of smiles. She stubbornly managed to stay happy. But it was hard. And Chelsey's parents worried that she would eventually become depressed and withdrawn.

So after much preparation and talk with doctors, Chelsey and her parents drove to a medical center in California. There, a team of surgeons began a 10-hour operation designed to give Chelsey her smile.

Moving cautiously, doctors extracted a strip of muscle, vein, and artery from Chelsey's thigh. Making an incision along her cheek, the team carefully sewed the extractions into place along Chelsey's jaw. The tiny sutures, thinner than a human hair, were sewn while viewing them under a microscope.

The painstaking surgery allowed the transplanted muscle in the left side of Chelsey's face to work. She gave a beautiful, if lopsided, smile. Then later, doctors performed the same surgery on the right side of her face. Finally, her long-awaited smile was complete. Today, Chelsey grins nonstop. Her newfound smile dazzles her family and helps her fit in with the kids at school.

Researchers are now learning to perform many kinds of muscle and nerve transplants. These operations can save or improve the use of a patient's arms or legs, for example. As knowledge of the muscular system improves, transplants will improve and become much easier and cheaper to perform. And that, say doctors, is something everyone can smile about.

All the activities of the body, from smiling to balancing on tiptoe, are dependent on muscles whose sole ability is to contract. That's why surgeons could take a muscle from the leg and transplant it to the face to allow Chelsey to smile. The location and function of certain well-known muscles are reviewed in this chapter, which concentrates on the structure of skeletal muscles from the macroscopic to the microscopic level. When muscles contract, two protein filaments slide past one another within individual muscle fibers. Much is known about the interaction of these filaments, and we will examine in detail how muscle contraction is activated.

Muscle contraction requires the presence of ATP, which is produced most efficiently when oxygen is present. However, some muscle fibers are adapted to anaerobic conditions, making them advantageous for particular sports that require power and speed, like weight lifting. Muscle anatomy and physiology are of extreme interest to sports enthusiasts, and this chapter emphasizes sports-related topics.

12.1 Skeletal Muscles ⚖

Muscles have various functions, which they are more apt to perform well if they are exercised regularly.

Skeletal muscles support the body. Skeletal muscle contraction opposes the force of gravity and allows us to remain upright.

Skeletal muscles make bones move. Muscle contraction accounts not only for the movement of arms and legs but also for movements of the eyes, facial expressions, and breathing.

Skeletal muscles help maintain a constant body temperature. Skeletal muscle contraction causes ATP to break down, releasing heat that is distributed about the body.

Skeletal muscle contraction assists movement in cardiovascular and lymphatic vessels. The pressure of skeletal muscle contraction keeps blood moving in cardiovascular veins and lymph moving in lymphatic vessels.

Skeletal muscles help protect internal organs and stabilize joints. Muscles pad the bones that protect organs, and they have tendons that help hold bones together at joints.

Cardiac and smooth muscles have specific functions. The contraction of cardiac muscle causes the heart to pump blood. Smooth muscle contraction propels food in the digestive tract and urine in the ureters.

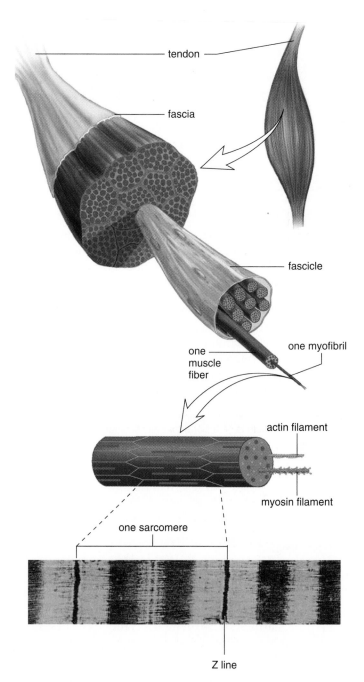

Figure 12.2   Anatomy of a muscle from the macroscopic to the microscopic level.

A fascia covers the surface of a muscle and contributes to the tendon, which attaches the muscle to a bone. Connective tissue also separates bundles of muscle fibers that contain contractile elements called myofibrils. The arrangement of myofilaments within a myofibril accounts for skeletal muscle striations. A sarcomere is a contractile unit of a myofibril.

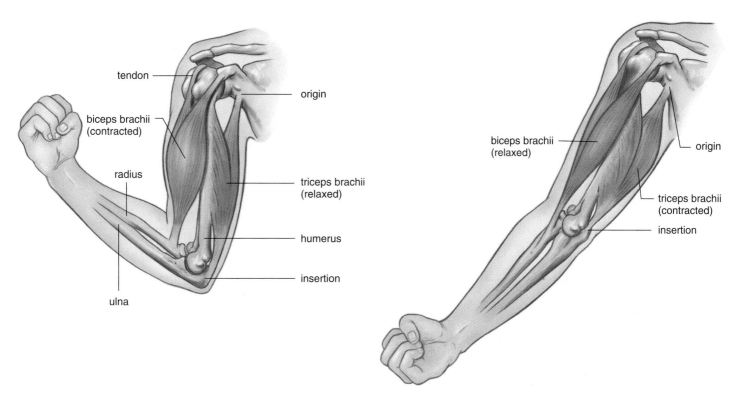

Figure 12.3   **Attachment of the skeletal muscles as exemplified by the biceps brachii and the triceps brachii.**
The origin of a muscle is on a bone that remains stationary, and the insertion of a muscle is on a bone that moves when the muscle contracts.
The muscles in this drawing are antagonistic. When the biceps brachii contracts, the lower arm flexes, and when the triceps brachii contracts, the lower arm extends.

The Structure of Muscles

This chapter is primarily concerned with skeletal muscles—those muscles that make up the bulk of the human body. Muscles are covered by several layers of fibrous connective tissue called fascia, which extends beyond the muscle to become its **tendon** (Fig. 12.2). Connective tissue also surrounds fascicles, which are bundles of muscle fibers, and separates the muscle fibers within a fascicle. A muscle fiber contains many contractile elements called myofibrils. A light microscope allows you to see muscle fibers but an electron microscope is needed to see myofibrils.

Myofibrils run the length of a muscle fiber, and the striations seen in observations of muscular tissue are due to the placement of myofilaments within sarcomeres, the structural units of a myofibril. There are thin (actin) filaments and thick (myosin) filaments in the myofibril.

Muscles Work in Pairs

Skeletal muscles are attached to the skeleton, and their contraction causes the movement of bones at a joint. When muscles contract, one bone remains fairly stationary, and the other one moves. The **origin** of a muscle is on the stationary bone, and the **insertion** of a muscle is on the bone that moves.

Frequently, a body part is moved by a group of muscles working together. Even so, one muscle does most of the work and is called the prime mover. The assisting muscles are called the synergists. When muscles contract, they shorten; therefore, muscles can only pull; they cannot push.

Most muscles have antagonists, and antagonistic pairs work opposite to one another to bring about movement in opposite directions. For example, the biceps brachii and the triceps brachii are antagonists; one flexes the forearm, and the other extends the forearm (Fig. 12.3).

Skeletal muscles have levels of organization from macroscopic to microscopic. When muscles cooperate to achieve movement, some act as prime movers, others are synergists, and still others are antagonists.

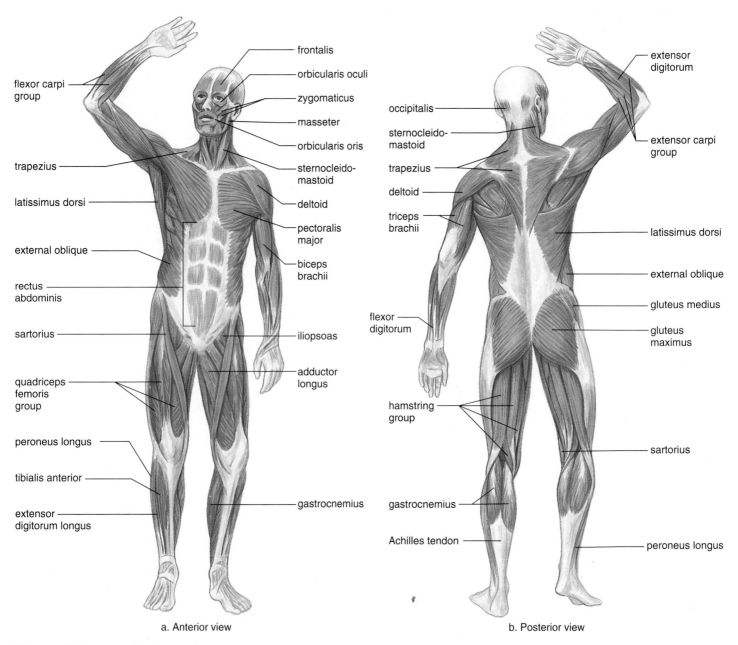

flexor carpi group

trapezius

latissimus dorsi

external oblique

rectus abdominis

sartorius

quadriceps femoris group

peroneus longus

tibialis anterior

extensor digitorum longus

frontalis

orbicularis oculi

zygomaticus

masseter

orbicularis oris

sternocleido-mastoid

deltoid

pectoralis major

biceps brachii

iliopsoas

adductor longus

gastrocnemius

a. Anterior view

occipitalis

sternocleido-mastoid

trapezius

deltoid

triceps brachii

flexor digitorum

hamstring group

gastrocnemius

Achilles tendon

extensor digitorum

extensor carpi group

latissimus dorsi

external oblique

gluteus medius

gluteus maximus

sartorius

peroneus longus

b. Posterior view

Figure 12.4   Human musculature.
Superficial skeletal muscles in (**a**) anterior and (**b**) posterior view.

Skeletal Muscles of the Body

Skeletal muscles (Fig. 12.4 and Tables 12.1 and 12.2) are given names based on the following characteristics and examples:

1. Size. The gluteus maximus that makes up the buttocks is the largest muscle.
2. Shape. The deltoid is shaped like a triangle. (The Greek letter delta has this appearance: Δ)
3. Location. The frontalis overlies the frontal bone.
4. Direction of muscle fibers. The rectus abdominis is a longitudinal muscle of the abdomen (rectus means straight).
5. Number of attachments. The biceps brachii has two attachments, or origins.
6. Action. The extensor digitorum extends the fingers and digits. Extension increases the joint angle and flexion decreases the joint angle; abduction is the movement of a body part sideways away from the midline and adduction is the movement of a body part toward the midline.

Table 12.1 Muscles (anterior view)

Name	Function
Head and neck	
Frontalis	Wrinkles forehead and lifts eyebrows
Orbicularis oculi	Closes eye (winking)
Zygomaticus	Raises corner of mouth (smiling)
Masseter	Closes jaw
Orbicularis oris	Closes and protrudes lips (kissing)
Arms and trunk	
External oblique	Compresses abdomen; rotates trunk
Rectus abdominis	Flexes spine
Pectoralis major	Flexes and adducts shoulder and arm ventrally (pulls arm across chest)
Deltoid	Abducts and raises arm at shoulder joint
Biceps brachii	Flexes forearm and supinates hand
Legs	
Adductor longus	Adducts thigh
Iliopsoas	Flexes thigh or hip joint
Sartorius	Rotates thigh (sitting cross-legged)
Quadriceps femoris group	Extends lower leg
Peroneus longus	Everts foot
Tibialis anterior	Dorsiflexes and inverts foot
Flexor digitorum longus	Flexes toes
Extensor digitorum longus	Extends toes

Table 12.2 Muscles (posterior view)

Name	Function
Head and neck	
Occipitalis	Moves scalp backward
Sternocleidomastoid	Turns head to side; flexes neck and head
Trapezius	Extends head; raises and adducts shoulders dorsally (shrugging shoulders)
Arms and trunk	
Latissimus dorsi	Extends and adducts shoulder and arm dorsally (pulls arm across back)
Deltoid	Abducts and raises arm at shoulder joint
External oblique	Rotates trunk
Triceps brachii	Extends forearm and arm
Flexor carpi group	Flexes hand
Extensor carpi group	Extends hand
Flexor digitorum	Flexes fingers
Extensor digitorum	Extends fingers
Buttocks and legs	
Gluteus medius	Abducts thigh
Gluteus maximus	Extends thigh (forms buttocks)
Hamstring group	Flexes lower leg and extends thigh
Gastrocnemius	Plantar flexes foot (tiptoeing)

Visual Focus

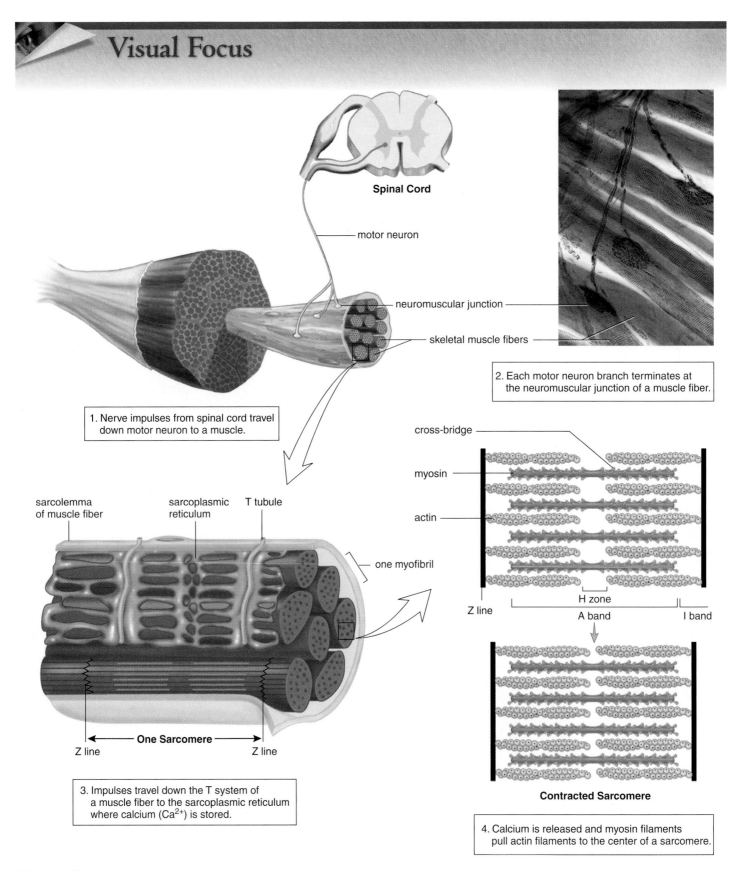

Spinal Cord

motor neuron

neuromuscular junction

skeletal muscle fibers

1. Nerve impulses from spinal cord travel down motor neuron to a muscle.

2. Each motor neuron branch terminates at the neuromuscular junction of a muscle fiber.

cross-bridge

myosin

actin

Z line

H zone

A band

I band

sarcolemma of muscle fiber

sarcoplasmic reticulum

T tubule

one myofibril

One Sarcomere

Z line   Z line

3. Impulses travel down the T system of a muscle fiber to the sarcoplasmic reticulum where calcium (Ca²⁺) is stored.

Contracted Sarcomere

4. Calcium is released and myosin filaments pull actin filaments to the center of a sarcomere.

Figure 12.5   Contraction of a muscle.

A motor neuron ends in neuromuscular junctions at a muscle fiber. At a neuromuscular junction, impulses travel down the T system of a muscle fiber, and calcium is released from the sarcoplasmic reticulum. Then the myosin filaments pull actin filaments within sarcomeres, units of a myofibril.

12.2 Mechanism of Muscle Fiber Contraction

A series of events leads up to muscle fiber contraction. After discussing these events, we will examine a neuromuscular junction and filament sliding in detail.

Overview of Muscular Contraction

Figure 12.5 shows the steps that lead to muscle contraction. Muscles are stimulated to contract by nerve impulses that begin in the brain or spinal cord. These nerve impulses travel down motor neurons and stimulate muscle fibers at neuromuscular junctions.

Each **muscle fiber** is a cell containing the usual cellular components, but special names have been assigned to some of these components. The plasma membrane is called the **sarcolemma,** the cytoplasm is the sarcoplasm, and the endoplasmic reticulum is the **sarcoplasmic reticulum.** A muscle fiber also has some unique anatomical characteristics. One feature is its T (for transverse) system; the sarcolemma forms **T (transverse) tubules** that penetrate, or dip down, into the cell so that they come into contact—but do not fuse—with expanded portions of the sarcoplasmic reticulum. The expanded portions of the sarcoplasmic reticulum contain calcium ions (Ca^{2+}), which are essential for muscle contraction. The sarcoplasmic reticulum encases hundreds and sometimes even thousands of **myofibrils,** which are the contractile portions of the muscle fibers.

Myofibrils and Sarcomeres

Myofibrils are cylindrical in shape and run the length of the muscle fiber. The light microscope shows that muscle fibers have light and dark bands called striations (Fig. 12.6). The electron microscope shows that the striations of muscle fibers are formed by the placement of myofilaments within contractile units of myofibrils called **sarcomeres** (see Fig. 12.2). A sarcomere extends between two dark lines called the Z lines. A sarcomere contains two types of protein myofilaments. The thick filaments are made up of a protein called **myosin,** and the thin filaments are made up of a protein called **actin.** Other proteins in addition to actin are also present. The I band is light colored because it contains only actin filaments attached to a Z line. The dark regions of the A band contain overlapping actin and myosin filaments, and its H zone has only myosin filaments.

Sliding Filaments

Impulses generated at a neuromuscular junction travel down a T tubule, and calcium is released from the sarcoplasmic reticulum. Now the muscle fiber contracts as the sarcomeres within the myofibrils shorten. When a sarcomere shortens, the actin (thin) filaments slide past the myosin (thick) fila-

Table 12.3	Muscle Contraction
Name	**Function**
Actin filaments	Slide past myosin, causing contraction
Ca^{2+}	Needed for myosin to bind the actin
Myosin filaments	Pull actin filaments by means of cross-bridges; are enzymatic and split ATP
ATP	Supplies energy for muscle contraction

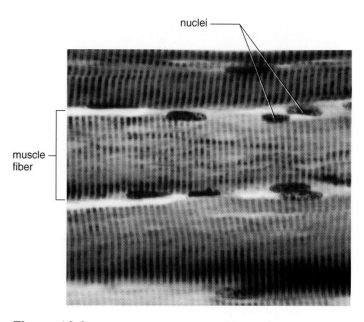

Figure 12.6 **Light micrograph of skeletal muscle.**
The striations of skeletal muscle tissue are produced by alternating dark A bands and light I bands. See the sarcomere in Figure 12.5.

ments and approach one another. This causes the I band to shorten and the H zone to almost or completely disappear. The movement of actin filaments in relation to myosin filaments is called the **sliding filament theory** of muscle contraction. During the sliding process, the sarcomere shortens even though the filaments themselves remain the same length.

The participants in muscle contraction function as shown in Table 12.3. ATP supplies the energy for muscle contraction. Although the actin filaments slide past the myosin filaments, it is the myosin filaments that do the work. Myosin filaments break down ATP and have cross-bridges that pull the actin filaments toward the center of the sarcomere.

When muscle fibers are stimulated to contract, myofilaments slide past one another, causing sarcomeres to shorten.

Muscle Innervation

Muscle fibers are innervated—that is, they are stimulated to contract by motor nerve fibers. Motor nerve fibers have several branches, each of which ends in an axon bulb that lies in close proximity to the sarcolemma of a muscle fiber. A small gap, called a synaptic cleft, separates the axon bulb from the sarcolemma. This entire region is called a **neuromuscular junction** (Fig. 12.7).

Axon bulbs contain synaptic vesicles that are filled with the neurotransmitter acetylcholine (ACh). When nerve impulses traveling down a motor neuron arrive at an axon bulb, the synaptic vesicles release ACh into the synaptic cleft. ACh quickly diffuses across the cleft and binds to receptors in the sarcolemma. Now the sarcolemma generates impulses that spread over the sarcolemma and down T tubules to the sarcoplasmic reticulum. The release of calcium from the sarcoplasmic reticulum causes the filaments within sarcomeres to slide past one another. Sarcomere contraction results in myofibril contraction, which in turn results in muscle fiber and finally muscle contraction.

At a neuromuscular junction, nerve impulses bring about the release of neurotransmitter molecules that signal a muscle fiber to contract.

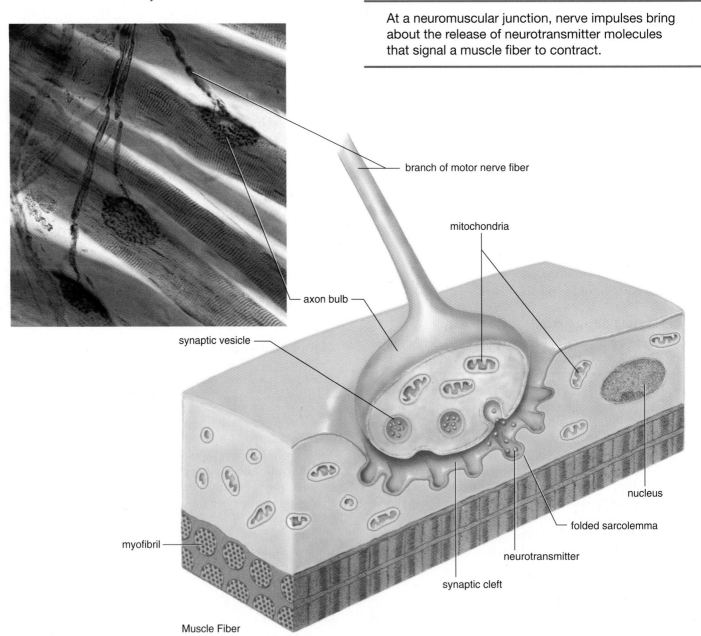

Figure 12.7 **Neuromuscular junction.**
The branch of a motor nerve fiber terminates in an axon bulb that meets but does not touch a muscle fiber. A synaptic cleft separates the axon bulb from the sarcolemma of the muscle fiber. Nerve impulses traveling down a motor fiber cause synaptic vesicles to discharge acetylcholine, which diffuses across the synaptic cleft. When the neurotransmitter is received by the sarcolemma of a muscle fiber, impulses begin and lead to muscle fiber contractions.

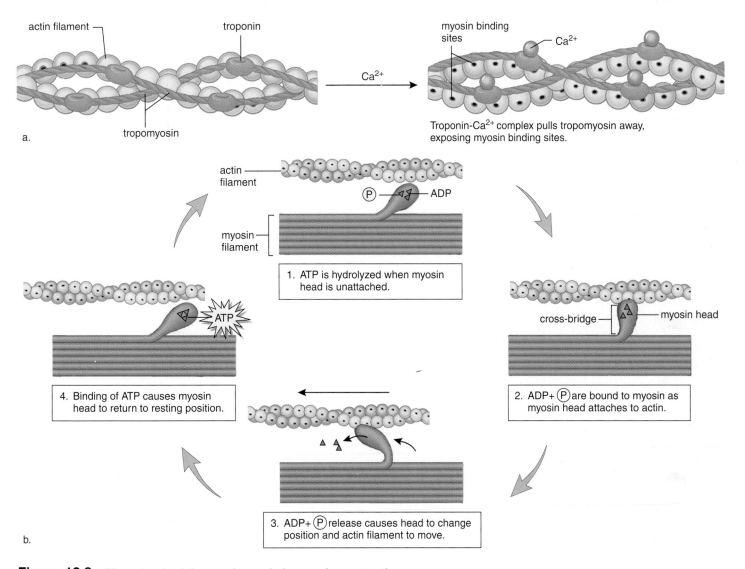

Figure 12.8 **The role of calcium and myosin in muscle contraction.**
a. Upon release, calcium binds to troponin, exposing myosin binding sites. **b.** After breaking down ATP, myosin heads bind to an actin filament, and later, a power stroke causes the actin filament to move.

Figure 12.8 shows the placement of two other proteins associated with a thin filament, which is composed of a double row of twisted actin molecules. Threads of **tropomyosin** wind about an actin filament, and **troponin** occurs at intervals along the threads. Calcium ions (Ca^{2+}) that have been released from the sarcoplasmic reticulum combine with troponin. After binding occurs, the tropomyosin threads shift their position, and myosin binding sites are exposed.

The thick filament is actually a bundle of myosin molecules, each having a double globular head with an ATP binding site. The heads function as ATPase enzymes, splitting ATP into ADP and Ⓟ. This reaction activates the head so that it will bind to actin. The ADP and Ⓟ remain on the myosin heads until the heads attach to actin, forming a cross-bridge. Now, ADP and Ⓟ are released, and this causes the cross-bridges to change their positions. This is

the power stroke that pulls the thin filaments toward the middle of the sarcomere. When another ATP molecule binds to a myosin head, the cross-bridge is broken as the head detaches from actin. The cycle begins again; the actin filaments move nearer the center of the sarcomere each time the cycle is repeated.

Contraction continues until nerve impulses cease and calcium ions are returned to their storage sacs. The membranes of the sarcoplasmic reticulum contain active transport proteins that pump calcium ions back into the sarcoplasmic reticulum.

Myosin filament heads break down ATP and then attach to an actin filament, forming cross-bridges that pull the actin filament to the center of a sarcomere.

12.3 Whole Muscle Contraction

Observation of muscle contraction in the laboratory can help us understand muscle contraction in the body.

Basic Laboratory Observations

One way to study muscle contraction in the laboratory is to remove the gastrocnemius (calf muscle) from a frog and attach it to a movable lever. The muscle is stimulated, and the mechanical force of contraction is recorded as a visual pattern called a **myogram.**

At first, the stimulus may be too weak to cause a contraction, but as soon as the strength of the stimulus reaches a threshold stimulus, the muscle contracts and then relaxes. This action—a single contraction that lasts only a fraction of a second—is called a **muscle twitch.** Figure 12.9*a* is a myogram of a twitch, which is customarily divided into three stages: the latent period, or the period of time between stimulation and initiation of contraction; the contraction period, when the muscle shortens; and the relaxation period, when the muscle returns to its former length.

Stimulation of an individual fiber within a muscle usually results in a maximal, all-or-none contraction. But the contraction of a whole muscle, as evidenced by the size of a muscle twitch, can vary in strength depending on the number of muscle fibers contracting.

If a muscle is given a rapid series of threshold stimuli, it can respond to the next stimulus without relaxing completely. In this way, muscle contraction summates until maximal sustained contraction, called **tetanus,** is achieved (Fig. 12.9*b*). The myogram no longer shows individual twitches; rather, the twitches are fused and blended completely into a straight line. Tetanus continues until the muscle fatigues due to depletion of energy reserves. Fatigue is apparent when a muscle relaxes even though stimulation continues.

Muscle Tone in the Body

Tetanic contractions ordinarily occur in the body's muscles whenever skeletal muscles are actively used. But while some muscle fibers are contracting, others are relaxing. Because of this, intact muscles rarely fatigue completely. Even when muscles appear to be at rest, they exhibit **tone** in which some of their fibers are always contracting. Muscle tone is particularly important in maintaining posture. If all the fibers within the muscles of the neck, trunk, and legs were to suddenly relax, the body would collapse.

Maintenance of the right amount of tone requires the use of special receptors called muscle spindles. A muscle spindle consists of a bundle of modified muscle fibers, with sensory nerve fibers wrapped around a short, specialized region. A muscle spindle contracts along with muscle fibers, but thereafter it sends sensory nerve impulses to the central nervous system, which then regulates muscle contraction so that tone is maintained (see Fig. 14.3).

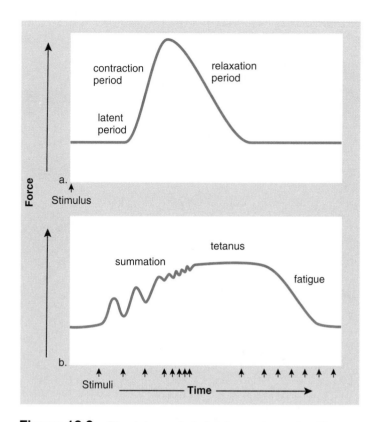

Figure 12.9   **Physiology of skeletal muscle contraction.**
These are myograms, visual representations of the contraction of a muscle that has been dissected from a frog. **a.** A simple muscle twitch is composed of three periods: latent, contraction, and relaxation. **b.** Summation and tetanus. When stimulation frequency increases, the muscle does not relax completely between stimuli, and the contraction gradually increases in intensity. The muscle becomes maximally contracted until it fatigues.

Recruitment and the Strength of Contraction

Each motor neuron innervates several muscle fibers at a time. A motor neuron together with all of the muscle fibers that it innervates is called a **motor unit.** As the intensity of nervous stimulation increases, more and more motor units are activated. This phenomenon, known as recruitment, results in stronger and stronger contractions.

Another variable of importance is the number of muscle fibers within a motor unit. In the ocular muscles that move the eyes, the innervation ratio is one motor neuron per 23 muscle fibers, while in the gastrocnemius muscle of the lower leg, the ratio is about one motor neuron per 1,000 muscle fibers. Moving the eyes requires finer control than moving the legs.

Observation of muscle contraction in the laboratory applies to muscle contraction in the body. A muscle at rest exhibits tone, and a contracting muscle exhibits degrees of contraction dependent on recruitment.

Health Focus

Exercise, Exercise, Exercise

Exercise programs improve muscular strength, muscular endurance, and flexibility. Muscular strength is the force a muscle group (or muscle) can exert against a resistance in one maximal effort. Muscular endurance is judged by the ability of a muscle to contract repeatedly or to sustain a contraction for an extended period. Flexibility is tested by observing the range of motion about a joint.

Exercise also improves cardiorespiratory endurance. The heart rate and capacity increase, and the air passages dilate so that the heart and lungs are able to support prolonged muscular activity. The blood level of high-density lipoprotein (HDL), the molecule that prevents the development of plaque in blood vessels, increases. Also, body composition, the proportion of protein to fat, changes favorably when you exercise.

Exercise also seems to help prevent certain kinds of cancer. Cancer prevention involves eating properly, not smoking, avoiding cancer-causing chemicals and radiation, undergoing appropriate medical screening tests, and knowing the early warning signs of cancer. However, studies show that people who exercise are less likely to develop colon, breast, cervical, uterine, and ovarian cancers.

Physical training with weights can improve bone density and strength and muscular strength and endurance in all adults, regardless of age. Even men and women in their eighties and nineties make substantial gains in bone and muscle strength, which can help them lead more independent lives. Exercise helps prevent osteoporosis, a condition in which the bones are weak and tend to break. Exercise promotes the activity of osteoblasts in young people as well as older people. The stronger the bones when a person is young, the less chance of osteoporosis as a person ages. Exercise helps prevent weight gain, not only because the level of activity increases but also because muscles metabolize faster than other tissues. As a person becomes more muscular, it is less likely that fat will accumulate.

Exercise relieves depression and enhances the mood. Some people report that exercise actually makes them feel more energetic, and that after exercising, particularly in the late afternoon, they sleep better that night. Self-esteem rises because of improved appearance, as well as other factors that are not well understood. For example, vigorous exercise releases endorphins, hormone-like chemicals that are known to alleviate pain and provide a feeling of tranquility.

A sensible exercise program is one that provides all the benefits without the detriments of a too strenuous program. Overexertion can actually be harmful to the body and might result in sports injuries such as a bad back or bad knees. The beneficial programs suggested in Table 12A are tailored according to age.

Dr. Arthur Leon at the University of Minnesota performed a study involving 12,000 men, and the results showed that only moderate exercise is needed to lower the risk of a heart attack by one-third. In another study conducted by the Institute for Aerobics Research in Dallas, Texas, which included 10,000 men and more than 3,000 women, even a little exercise was found to lower the risk of death from circulatory diseases and cancer. Increasing daily activity by walking to the corner store instead of driving and by taking the stairs instead of the elevator can improve your health.

Table 12A	A Checklist for Staying Fit		
Children, 7–12	**Teenagers, 13–18**	**Adults, 19–55**	**Seniors, 56 and Up**
Vigorous activity 1–2 hours daily	Vigorous activity 1 hour 3–5 days a week, otherwise 1/2 hour daily moderate activity	Vigorous activity 1 hour 3 days a week, otherwise 1/2 hour daily moderate activity	Moderate exercise 1 hour daily 3 days a week, otherwise 1/2 hour daily moderate activity
Free play	Build muscle with calisthenics	Exercise to prevent lower back pain: aerobics, stretching, yoga	Take a daily walk
Build motor skills through team sports, dance, swimming	Do aerobic exercise to control buildup of fat cells	Take active vacations: hike, bicycle, cross-country ski	Do daily stretching exercises
Encourage more exercise outside of physical education classes	Pursue tennis, swimming, horseback riding—sports that can be enjoyed for a lifetime	Find exercise partners: join a running club, bicycle club, outing group	Learn a new sport or activity: golf, fishing, ballroom dancing
Initiate family outings: bowling, boating, camping, hiking	Continue team sports, dancing, hiking, swimming		Try low-impact aerobics. Before undertaking new exercises, consult your doctor

12.4 Energy for Muscle Contraction

ATP produced previous to strenuous exercise lasts a few seconds, and then muscles acquire new ATP in three different ways: creatine phosphate breakdown, fermentation, and cellular respiration (Fig. 12.10). The first two ways are anaerobic and do not require oxygen.

Creatine phosphate is a high-energy compound built up when a muscle is resting. Creatine phosphate cannot participate directly in muscle contraction. Instead, it can regenerate ATP by the following reaction:

This reaction occurs in the midst of sliding filaments, and therefore is the speediest way to make ATP available to muscles. Creatine phosphate provides enough energy for only about eight seconds of intense activity, and then it is spent. Creatine phosphate is rebuilt when a muscle is resting by transferring a phosphate group from ATP to creatine.

Fermentation also supplies ATP without consuming oxygen. During fermentation, glucose is broken down to lactate (lactic acid):

The accumulation of lactate in a muscle fiber makes the cytoplasm more acidic, and eventually enzymes cease to function well. If fermentation continues longer than two or three minutes, cramping and fatigue set in. Cramping seems to be due to lack of ATP needed to pump calcium ions back into the sarcoplasmic reticulum and to break the linkages between the actin and myosin filaments so that muscle fibers can relax.

Fortunately, cellular respiration occurring in mitochondria usually provides most of a muscle's ATP. Glycogen and fat are stored in muscle cells. Therefore, a muscle cell can use glucose, from glycogen and fatty acids, from fat, as fuel to produce ATP if oxygen is available.

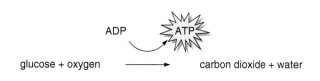

Myoglobin, an oxygen carrier similar to hemoglobin, is synthesized in muscle cells, and its presence accounts for the reddish-brown color for skeletal muscle fibers. Myoglobin has a higher affinity of oxygen than does hemoglobin. Therefore, myoglobin can pull oxygen out of blood and make it available to muscle mitochondria that are carrying on cellular respiration. Then, too, the ability of myoglobin to temporarily store oxygen reduces a muscle's immediate need for oxygen when cellular respiration begins. The end products (carbon dioxide and water) can be rapidly disposed of. The by-product heat keeps the entire body warm.

The three pathways for acquiring ATP work together during muscle contraction. But the anaerobic pathways are usually no longer needed once the body achieves an aerobic steady state. At this point, some lactate has accumulated but not enough to bring on exhaustion.

People who train rely even more heavily on cellular respiration than people who do not train. In people who train, the number of muscle mitochondria increases, and so fermentation is not needed to produce ATP. Their mitochondria can start consuming oxygen as soon as ADP concentration starts rising during muscle contraction. Because mitochondria can break down fatty acid, instead of glucose, blood glucose is spared for the activity of the brain. (The brain, unlike other organs, can only utilize glucose to produce ATP.) Because less lactate is produced in people who train, the pH of the blood remains steady, and there is less of an "oxygen debt."

Oxygen Debt

When a muscle uses the anaerobic means of supplying energy needs, it incurs an **oxygen debt.** Oxygen debt is obvious when a person continues to breathe heavily after exercising. The ability to run up an oxygen debt is one of muscle tissue's greatest assets. Brain tissue cannot last nearly as long without oxygen as muscles can.

Repaying an oxygen debt requires replenishing creatine phosphate supplies and disposing of lactic acid. Lactic acid can be changed back to pyruvic acid and metabolized completely in mitochondria, or it can be sent to the liver to reconstruct glycogen. A marathon runner who has just crossed the finish line is not exhausted due to oxygen debt. Instead, the runner has used up all the muscles', and probably the liver's, glycogen supply. It takes about two days to replace glycogen stores on a high-carbohydrate diet.

Working muscles require a supply of ATP. Anaerobic creatine phosphate breakdown and fermentation can quickly generate ATP. Cellular respiration in mitochondria is best for sustained exercise.

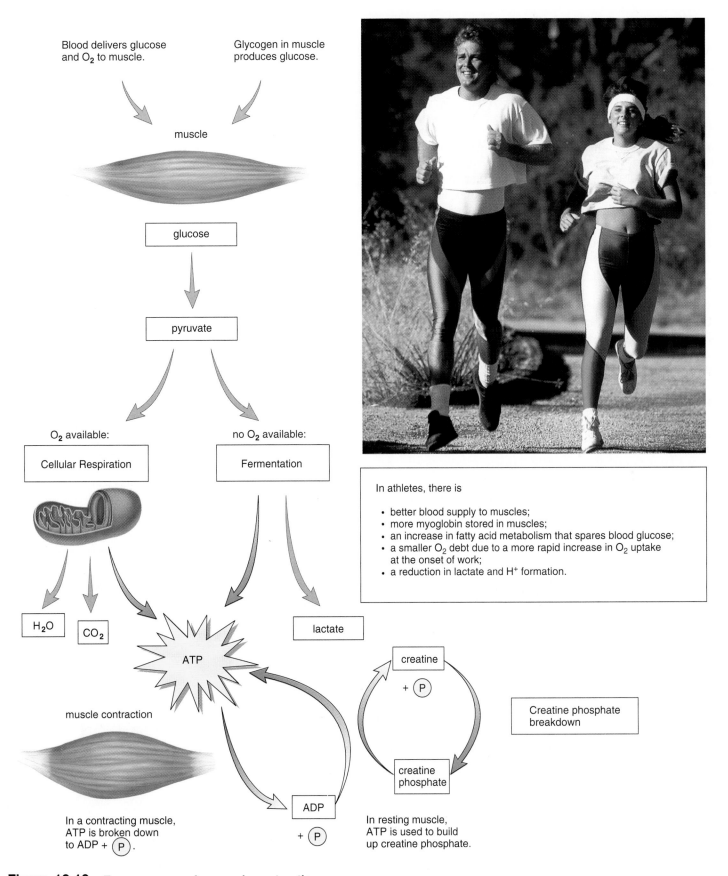

Blood delivers glucose and O_2 to muscle.

Glycogen in muscle produces glucose.

muscle

glucose

pyruvate

O_2 available:

Cellular Respiration

no O_2 available:

Fermentation

H_2O

CO_2

ATP

muscle contraction

In a contracting muscle, ATP is broken down to ADP + (P).

lactate

ADP

+ (P)

creatine

+ (P)

creatine phosphate

Creatine phosphate breakdown

In resting muscle, ATP is used to build up creatine phosphate.

In athletes, there is

- better blood supply to muscles;
- more myoglobin stored in muscles;
- an increase in fatty acid metabolism that spares blood glucose;
- a smaller O_2 debt due to a more rapid increase in O_2 uptake at the onset of work;
- a reduction in lactate and H^+ formation.

Figure 12.10   **Energy sources for muscle contraction.**

Athletics and Muscle Contraction

Athletes who excel in a particular sport, and much of the general public, are interested today in staying fit by exercising. The Health Focus on page 235 gives suggestions for exercise programs according to age.

Exercise and Size of Muscles

Muscles that are not used or that are used for only very weak contractions decrease in size, or atrophy. **Atrophy** can occur when a limb is placed in a cast or when the nerve serving a muscle is damaged. If nerve stimulation is not restored, muscle fibers gradually are replaced by fat and fibrous tissue. Unfortunately, atrophy can cause muscle fibers to shorten progressively, leaving body parts contracted in contorted positions.

Forceful muscular activity over a prolonged period causes muscle to increase in size as the number of myofibrils within the muscle fibers increases. Increase in muscle size, called **hypertrophy,** occurs only if the muscle contracts to at least 75% of its maximum tension.

Some athletes take anabolic steroids, either testosterone or related chemicals, to promote muscle growth. This practice has many undesirable side effects, such as cardiovascular disease, liver and kidney dysfunction, impotency and sterility, and even an increase in rash behavior called "roid mania."

Slow-Twitch and Fast-Twitch Muscle Fibers

We have seen that all muscle fibers metabolize both aerobically and anaerobically. Some muscle fibers, however, utilize one method more than the other to provide myofibrils with ATP. Slow-twitch fibers tend to be aerobic, and fast-twitch fibers tend to be anaerobic (Fig. 12.11).

Slow-Twitch Fibers   Slow-twitch fibers have a steadier tug and have more endurance despite motor units with a smaller number of fibers. These muscle fibers are most helpful in sports like long-distance running, biking, jogging, and swimming. Because they produce most of their energy aerobically, they tire only when their fuel supply is gone. Slow-twitch fibers have many mitochondria and are dark in color because they contain myoglobin, the respiratory pigment found in muscles. They are also surrounded by dense capillary beds and draw more blood and oxygen than fast-twitch fibers. Slow-twitch fibers have a low maximum tension, which develops slowly, but these muscle fibers are highly resistant to fatigue. Because slow-twitch fibers have a substantial reserve of glycogen and fat, their abundant mitochondria can maintain a steady, prolonged production of ATP when oxygen is available.

Fast-Twitch Fibers   Fast-twitch fibers tend to be anaerobic and seem to be designed for strength because their motor units contain many fibers. They provide explosions of energy and are most helpful in sports activities like sprinting, weight lifting, swinging a golf club, or throwing a shot. Fast-twitch fibers are light in color because they have fewer mitochondria, little or no myoglobin, and fewer blood vessels than slow-twitch fibers do. Fast-twitch fibers can develop maximum tension more rapidly than slow-twitch fibers can, and their maximum tension is greater. However, their dependence on anaerobic energy leaves them vulnerable to an accumulation of lactic acid that causes them to fatigue quickly.

Success in a particular sport is in part determined by the proportion of slow-twitch and fast-twitch muscle fibers in a person's muscles.

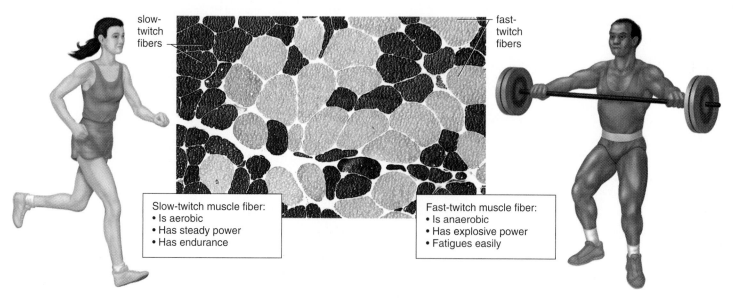

slow-twitch fibers

fast-twitch fibers

Slow-twitch muscle fiber:
- Is aerobic
- Has steady power
- Has endurance

Fast-twitch muscle fiber:
- Is anaerobic
- Has explosive power
- Fatigues easily

Figure 12.11   Slow- and fast-twitch muscle fibers.
If your muscles contain many slow-twitch fibers (dark color), you would probably do better at a sport like cross-country running. But if your muscles contain many fast-twitch fibers (light color), you would probably do better at a sport like weight lifting.

Performance-Enhancing Drugs

As we learned on page 238, athletes may be better at one sport or another, depending on whether their muscles contain fast- or slow-twitch fibers. A natural advantage of this sort does not bar an athlete from participating in and winning a medal in a particular sport at the Olympic games. Nor are athletes restricted to a certain amount of practice or required to eliminate certain foods from their diets.

Athletes cannot, however, receive Olympic medals if they have taken certain performance-enhancing drugs (Fig. 12A). There is no doubt that regular use of drugs like anabolic steroids leads to kidney disease, liver dysfunction, hypertension, and a myriad of other undesirable side effects. Even so, shouldn't individuals be allowed to take these drugs if they want to? Anabolic steroids are synthetic forms of the male sex hormone testosterone. Taking large doses along with strength training leads to much larger muscles than otherwise. Extra strength and endurance can give an athlete an advantage in sports such as racing, swimming, and weight lifting.

Should the Olympic committee outlaw the taking of anabolic steroids, and if so, on what basis? "Unfair advantage" can't be cited because some athletes naturally have an unfair advantage over other athletes. Should these drugs be outlawed on the basis of health reasons? Excessive practice alone and a purposeful decrease or increase in weight to better perform in a sport can also injure a person's health. In other words, how can you justify allowing some behaviors that enhance performance and not others?

Decide Your Opinion

1. Do you believe that the manner in which athletes wish to train and increase their performance should be regulated in any way? Why or why not?
2. Is it all right for athletes to endanger their health by practicing excessively, gaining or losing pounds, or taking drugs? Why or why not?
3. Who should be in charge of regulating the behavior of athletes so that they do not do harm to themselves?

Figure 12A   Anabolic steroid use.
In 1988, Ben Johnson won an Olympic gold medal for the 100-meter sprint, but had to return the medal a few days later because he tested positive for anabolic steroids.

Looking at Both Sides   www.mhhe.com/biosci/genbio/maderhuman7/

Every bioethical issue has at least two sides. Even if you already have an opinion, it is important to explore the opposite opinion before finalizing your position. The Online Learning Center at www.mhhe.com/biosci/genbio/maderhuman7/ will help you fine-tune your initial opinion, explore both sides, and finalize your position. Either you will acquire new arguments for your original opinion or you may even change your opinion. Be sure to complete these activities in sequence:

Taking Sides   Decide your initial opinion by answering a series of questions. Then see if your opinion changes after completing the next two activities.

Further Debate   Read opposing articles that give you further information on this particular bioethical issue.

Explain Your Position   Answer another series of questions and then defend your original or changed opinion. You can e-mail your position to your instructor if he or she wishes.

Muscular Disorders

Some muscle disorders are annoying but usually not serious. Muscular spasms and cramps are involuntary contractions that occur suddenly and cause pain. A spasm of intestinal muscles causes what is called a bellyache. Facial tics such as periodic eye blinking, head turning, or grimacing are spasms that can be controlled voluntarily but only with great effort. A leg or foot cramp can even occur when sleeping after a strenuous workout.

Tendonitis occurs when a tendon becomes painfully inflamed due to the strain of repeated athletic activity. The tendons most commonly affected are those associated with the shoulder, elbow, hip, and knee. The term myalgia refers to inflammation of muscle tissue itself.

In persons who have not been properly immunized, the toxin of the tetanus bacterium can cause muscles to lock in paralysis. A rigidly locked jaw is one of the first signs, and therefore this infection is commonly known as "lockjaw," or tetanus. Like other bacterial infections, lockjaw is curable with the administration of an antibiotic.

Muscular dystrophy is a broad term applied to a group of disorders that are characterized by a progressive degeneration and weakening of muscles. As muscle fibers die, fat and connective tissue take their place. Duchenne muscular dystrophy, the most common type, is inherited through a flawed gene carried by the mother. It is now known that the lack of a protein called dystrophin causes the condition. When dystrophin is absent, calcium leaks into the cell and activates an enzyme that dissolves muscle fibers. In an attempt to treat the condition, muscles have been injected with immature muscle cells that do produce dystrophin.

Myasthenia gravis is an autoimmune disease characterized by muscle weakness that especially affects the muscles of the eyelids, face, neck, and extremities. Muscle contraction is impaired because the immune system mistakenly produces antibodies that destroy acetylcholine receptors. In many cases, the first sign of the disease is a drooping of the eyelids and double vision (Fig. 12.12). Treatment includes drugs that are antagonistic to the enzyme acetylcholinesterase.

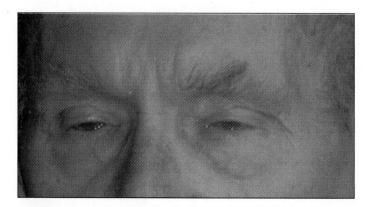

Figure 12.12   Myasthenia gravis.
The first sign of myasthenia gravis is a drooping of the eyelids.

12.5 Homeostasis

The illustration on the next page tells how the muscular system works with other systems of the body to maintain homeostasis.

Cardiac muscle contraction accounts for the heartbeat, which creates blood pressure, the force that propels blood in the arteries and arterioles. The arterioles branch into the capillaries where exchange takes place that creates and cleanses tissue fluid. Blood and tissue fluid are the internal environment of the body, and without cardiac muscle contraction, blood would never reach the capillaries for exchange to take place. Blood is returned to the heart in cardiovascular veins, and excess tissue fluid is returned to the cardiovascular system within lymphatic vessels. Skeletal muscle contraction presses on the cardiovascular veins and lymphatic vessels, and this creates the pressure that moves fluids in both types of vessels.

The flow of blood within arterioles is regulated by constriction of their smooth muscle walls, and contraction of sphincters temporarily prevents the flow of blood into a capillary. This is an important homeostatic mechanism because in times of emergency it is more important, for example, for blood to be directed to the skeletal muscles than to the tissues of the digestive tract.

Smooth muscle contraction accounts for peristalsis, the process that moves food along the digestive tract. Without this action, food would never reach the various organs of the digestive tract where digestion releases nutrients that enter the bloodstream. Smooth muscle contraction assists the voiding of urine, which is necessary for ridding the body of metabolic wastes and for regulating the blood volume, salt concentration, and pH of internal fluids.

Skeletal muscles protect internal organs, and the strength of muscles protects joints by stabilizing their movements. Skeletal muscle contraction raises the rib cage and lowers the diaphragm during the active phase of breathing. As we breathe, oxygen enters the blood and is delivered to the tissues, including the muscles where ATP is produced in mitochondria with heat as a by-product. The heat produced by skeletal muscle contraction allows the body temperature to remain within the normal range for human beings.

The overall importance of muscle contraction to the maintenance of good health cannot be overestimated. Exercise improves the functioning of the heart, increases the respiratory volume, enhances immunity, and improves the health of the individual in general.

Muscle contraction is involved in the functioning of the organs of the respiratory system, the digestive system, and the urinary system. Cardiac muscle contraction keeps the heart pumping so that blood is delivered to the tissues. Skeletal muscle contraction presses on the veins so that blood is returned to the heart.

Human Systems Work Together

Integumentary System

Muscle contraction provides heat to warm skin.

Skin protects muscles; rids the body of heat produced by muscle contraction.

How the Muscular System works with other body systems

Lymphatic System/Immunity

Skeletal muscle contraction moves lymph; physical exercise enhances immunity.

Lymphatic vessels pick up excess tissue fluid; immune system protects against infections.

Skeletal System

Muscle contraction causes bones to move joints; muscles help protect bones.

Bones provide attachment sites for muscles; store Ca^{2+} for muscle function.

Respiratory System

Muscle contraction assists breathing; physical exercise increases respiratory capacity.

Lungs provide oxygen for, and rid the body of, carbon dioxide from contracting muscles.

Nervous System

Muscle contraction moves eyes, permits speech, creates facial expressions.

Brain controls nerves that innervate muscles; receptors send sensory input from muscles to brain.

Digestive System

Smooth muscle contraction accounts for peristalsis; skeletal muscles support and help protect abdominal organs.

Digestive tract provides glucose for muscle activity; liver metabolizes lactic acid following anaerobic muscle activity.

Endocrine System

Muscles help protect glands.

Androgens promote growth of skeletal muscle; epinephrine stimulates heart and constricts blood vessels.

Urinary System

Smooth muscle contraction assists voiding of urine; skeletal muscles support and help protect urinary organs.

Kidneys maintain blood levels of Na^+, K^+, and Ca^{2+}, which are needed for muscle innervation, and eliminate creatinine, a muscle waste.

Cardiovascular System

Muscle contraction keeps blood moving in heart and blood vessels.

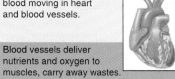

Blood vessels deliver nutrients and oxygen to muscles, carry away wastes.

Reproductive System

Muscle contraction occurs during orgasm and moves gametes; abdominal and uterine muscle contraction occurs during childbirth.

Androgens promote growth of skeletal muscle.

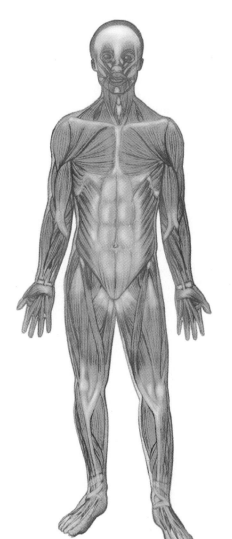
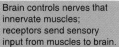
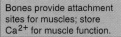
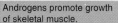

Summarizing the Concepts

12.1 Skeletal Muscles
Skeletal muscles have levels of organization. A whole muscle contains bundles of muscle fibers; muscle fibers contain myofibrils; and myofibrils contain actin and myosin filaments. Skeletal muscles are usually attached to the skeleton, where some are prime movers, some are synergists, and others are antagonists. Muscles have various functions; they provide movement and heat, help maintain posture, and protect underlying organs.

Muscles are named for their size, shape, location, direction of fibers, number of attachments, and action.

12.2 Mechanism of Muscle Fiber Contraction
Nerve impulses travel down motor neurons and stimulate muscle fibers at neuromuscular junctions. The sarcolemma of a muscle fiber forms T tubules that extend into the fiber and almost touch the sarcoplasmic reticulum, which stores calcium ions. When calcium ions are released into muscle fibers, actin filaments slide past myosin filaments within the sarcomeres of a myofibril.

At a neuromuscular junction, synaptic vesicles release acetylcholine (ACh), which binds to protein receptors on the sarcolemma, causing impulses to travel down the T tubules and calcium to leave the sarcoplasmic reticulum. Myofibril contraction follows.

Calcium ions bind to troponin and cause the tropomyosin threads that wind around actin filaments to shift their position, revealing myosin binding sites. The myosin filament is composed of many myosin molecules, each containing a head with an ATP binding site. Myosin is an ATPase, and once it breaks down ATP, the myosin head is ready to attach to actin. The release of ADP + Ⓟ causes the head to change its position. This is the power stroke that causes the actin filament to slide toward the center of a sarcomere.

When myosin catalyzes another ATP, the head detaches from actin, and the cycle begins again.

12.3 Whole Muscle Contraction
In the laboratory, muscle contraction is described in terms of a muscle twitch, summation, and tetanus. In the body, muscles exhibit tone, in which a continuous slight tension is maintained by muscle fibers that take turns contracting. The strength of muscle contraction varies according to recruitment of motor units.

12.4 Energy for Muscle Contraction
A muscle fiber has three ways to acquire ATP for muscle contraction. (1) Creatine phosphate, built up when a muscle is resting, can donate a high-energy phosphate to ADP, forming ATP. (2) Fermentation results in oxygen debt because oxygen is needed to complete the metabolism of the lactic acid that accumulates. Both of these processes quickly produce ATP for muscle contraction. (3) Cellular respiration takes longer because oxygen must be transported to mitochondria before the process can be completed.

Exercise results in muscle fiber increase, called hypertrophy. Sports like running and swimming can be associated with slow-twitch fibers, which rely on cellular respiration to acquire ATP. They have a plentiful supply of mitochondria and myoglobin, which gives them a dark color. Other sports, like weight lifting, can be associated with fast-twitch fibers, which rely on an anaerobic means of acquiring ATP. They have few mitochondria and myoglobin, and their motor units contain more muscle fibers. Fast-twitching fibers are known for their explosive power, but they fatigue quickly.

12.5 Homeostasis
The muscular system works with the other systems of the body in the ways described in the illustration on page 241.

Studying the Concepts

1. List and discuss the functions of muscles. 226
2. Describe the levels of structure of a muscle from the whole muscle level to the myofilaments within a myofibril. 227
3. Describe the steps resulting in muscle contraction by starting with the motor neuron and ending with the sliding of actin filaments. 231
4. Describe the structure and function of a neuromuscular junction. 232
5. Describe the cyclical events as myosin pulls actin toward the center of a sarcomere. 233
6. Contrast a muscle twitch with summation and tetanus. 234
7. What is tone, and how is it maintained? 234
8. By what mechanism does the strength of muscle contraction vary? 234
9. What are the three ways a muscle fiber can acquire ATP for muscle contraction? How are the three ways interrelated? 236–37
10. What is atrophy? Hypertrophy? 238
11. Contrast slow-twitch and fast-twitch fibers in as many ways as possible. 238
12. How does the muscular system help maintain homeostasis? 240–41

Testing Your Knowledge of the Concepts

In questions 1–4, match the region of the body to the muscles.
  a. head and neck
  b. trunk
  c. arm
  d. leg

_____ 1. hamstring group

_____ 2. trapezius

_____ 3. rectus abdominis

_____ 4. triceps brachii

In questions 5–7, indicate whether the statement is true (T) or false (F).

_____ 5. Myosin breaks down ATP and pulls actin filaments.

_____ 6. Slow-twitch muscle fibers are associated with weight lifting.

_____ 7. Muscle cramping appears to be due to a depletion of oxygen and a buildup of lactic acid.

In questions 8 and 9, fill in the blanks.

8. _____ stores high-energy phosphate bonds in muscle fibers.

9. _____ is the neurotransmitter released by a motor neuron at a neuromuscular junction.

10. Label this diagram of a muscle fiber, using these terms: myofibril, mitochondrion, T tubule, sarcomere, sarcolemma, sarcoplasmic reticulum.

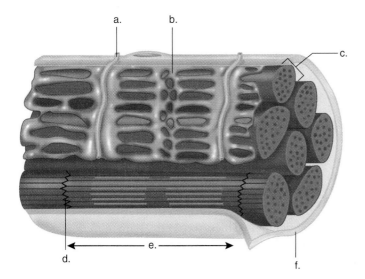

Understanding Key Terms

actin 231	origin 227
atrophy 238	oxygen debt 236
creatine phosphate 236	sarcolemma 231
hypertrophy 238	sarcomere 231
insertion 227	sarcoplasmic reticulum 231
motor unit 234	sliding filament theory 231
muscle fiber 231	T (transverse) tubule 231
muscle twitch 234	tendon 227
myofibril 231	tetanus 234
myoglobin 238	tone 234
myogram 234	tropomyosin 233
myosin 231	troponin 233
neuromuscular junction 232	

Further Readings

Anderson, J. L., et al. September 2000. Muscles, genes, and athletic performance. *Scientific American* 283(3):48. Genetics may play a role in determining the ratio of fast- to slow-twitch muscle fibers, which is important to athletic success.

Clemente, C. D. 1998. *Anatomy: A regional atlas of the human body.* 4th ed. Lippincott, Williams and Wilkins. This atlas contains both drawings and photographs of anatomical regions of the human body.

Demers, C., and Hamdy, R. C. November/December 1999. Bone morphogenetic proteins. *Science & Medicine* 6(6):8. Certain bone morphogenetic proteins may prove useful in treating difficult cases of bone damage.

Deyo, R. A. August 1998. Low-back pain. *Scientific American* 279(2):48. Treatment options for low-back pain that don't involve bed rest or surgery are improving.

Ferber, D. June 30, 2000. Cholesterol drugs show promise as bone builders. *Science* 288(5475):2297. First tested in mice, a group of drugs called statins, which are used to lower cholesterol, seems to prevent bone fractures as well as to trigger significant bone regrowth in older people.

Fox, S. I. 1998. *Human physiology.* 6th ed. Dubuque, Iowa: WCB/McGraw-Hill. This is an introductory physiology text.

Gore, R. September 2000. The unbeatable body. *National Geographic* 198(3):2. Article discusses how athletes push the limits of human performance. Steroid use is also covered.

Halstead, L. S. April 1998. Post-polio syndrome. *Scientific American* 278(4):42. Recovered polio victims are experiencing fatigue, pain, and weakness, resulting from degeneration of motor neurons.

Kunzig, R. February 1999. How to grow bones from scratch. *Discover* 20(2):18. Article discusses how researchers are growing synthetic bone on biodegradable foam to be used in bone grafts.

Mader, S. S. 2001. *Understanding human anatomy and physiology.* 4th ed. Dubuque, Iowa: The McGraw-Hill Companies, Inc. A text that emphasizes the basics for beginning allied health students.

Marieb, E. N. 1999. *Human anatomy and physiology.* 4th ed. Redwood City, Calif.: Benjamin/Cummings Publishing. A thorough anatomy and physiology text that can safely be used as a complete and accurate reference.

Marion, R. September 1999. Emergency room terror. *Discover* 20(9):44. Article discusses osteogenesis imperfecta, a disease that results in easily fractured bones in children.

Melton, L. July 2000. Age breakers. *Scientific American* 283(1):16. A compound has been developed that might rejuvenate hearts and muscles by breaking the sugar-protein bonds that accumulate with aging.

Pennisi, E. March 6, 1998. Bone marrow cells may provide muscle power. *Science* 279(5356):1456. Researchers show that bone marrow cells can move into damaged muscle and grow into new muscle fibers in mice.

Rogers, M. J. January 2000. Statins: Lower lipids and better bones? *Nature Medicine* 6(1):21. A recent study suggests that statins could be developed into new treatments for common metabolic bone diseases such as osteoporosis.

Rome, L. C. July/August 1997. Testing a muscle's design. *American Scientist* 85(4):356. Muscular systems adapt to specific functions, such as the specialized muscles found in frogs.

Science, September 1, 2000. Bone remodeling and repair. 289(5484):1497. Several articles in this section discuss bones, bone repair, and the future of tissue engineering in bone repair.

Tortora, G. J. 1996. *Atlas of the human skeleton.* New York: HarperCollins Publishers. This supplement contains carefully selected and labeled photographs of the human skeleton.

White, R. J. September 1998. Weightlessness and the human body. *Scientific American* 279(3):58. Space medicine is providing new ideas about treatment of osteoporosis and anemia.

Wolkomir, R. August 1998. Oh, my aching back. *Smithsonian* 29(5):36. Researchers try to pinpoint the source of back pain using a Virtual Corset to monitor the subject's activities.

e-Learning Connection

12.1 Skeletal Muscles

Striated Muscle *animation activity*
Attachment of Skeletal Muscles *art labeling activity*
Skeletal Muscles *art labeling activity*

12.2 Mechanism of Muscle Fiber Contraction

Muscle Cell Function *Essential Study Partner*

Action Potential *animation activity*
Striated Muscle Contraction *animation activity*
Actin-Myosin Cross-Bridges *animation activity*
Muscle Fiber *art quiz*

Interaction of Thick and Thin Filaments *art quiz*
Cross-Bridge Cycle in Muscle Contraction *art quiz*
Muscle Cell Function *Essential Study Partner activity*

12.3 Whole Muscle Contraction

Muscle Stimulation Pattern *art quiz*

12.4 Energy for Muscle Contraction

The Case of Rachel Martin and Lactic Acid *case study*

Fast-Twitch and Slow-Twitch Fibers *art quiz*

12.5 Homeostasis

Working Together to Achieve Homeostasis

Chapter Summary

Key Term Flashcards *vocabulary quiz*
Chapter Quiz *objective quiz covering all chapter concepts*

IV

Integration and Coordination in Humans

The nervous system is the ultimate coordinator of homeostasis. Nerves bring information to the brain and spinal cord from sensory receptors that detect changes both inside and outside the body. Then, nerves take the commands given by the brain and spinal cord to effectors, allowing the body to respond to these changes. For example, after low blood pressure stimulates internal receptors, the cardiovascular center in the brain sends out nerve impulses that constrict blood vessels, causing blood pressure to rise.

The endocrine system, like the nervous system, regulates other organs, but it acts more slowly and brings about a response that lasts longer. The endocrine organs secrete chemical messengers called hormones into the bloodstream. After arriving at their target organs, hormones alter cellular metabolism.

Despite their very different modes of operation, we now know that the nervous system and the endocrine system are joined in numerous ways.

Chapter 13

Nervous System

Chapter Concepts

13.1 Nervous Tissue
- The nervous system contains cells called neurons, which are specialized to carry nerve impulses. 246
- A nerve impulse is an electrochemical change that travels along the length of a neuron axon.   248
- Transmission of signals between neurons is dependent on neurotransmitter molecules. 251

13.2 The Central Nervous System
- The central nervous system is made up of the spinal cord and the brain. 252
- The spinal cord transmits messages to and from the brain and coordinates reflex responses. 253
- The parts of the brain are specialized for particular functions. 255–57
- The cerebral cortex contains motor areas, sensory areas, and association areas that communicate with each other. 255–56
- The reticular formation contains fibers that arouse the brain when they are active and account for sleep when they are inactive. 257

13.3 The Limbic System and Higher Mental Functions
- The limbic system contains cortical and subcortical areas that are involved in higher mental functions and emotional responses. 257
- Long-term memory depends upon association areas that are in contact with the limbic system. 258
- Particular areas in the left hemisphere are involved in language and speech. 259

13.4 The Peripheral Nervous System
- The peripheral nervous system contains nerves that conduct nerve impulses toward and away from the central nervous system. 260

13.5 Drug Abuse
- The use of psychoactive drugs, such as alcohol, nicotine, cocaine, heroin, and marijuana is detrimental to the body. 264

13.6 Homeostasis
- The nervous system works with the other systems of the body, particularly the endocrine system, to maintain homeostasis. 266

It was a warm spring Saturday in 1995 when a crowd gathered in Culpeper, Virginia, to enjoy a horse-riding competition. Everything was fine until—suddenly—one horse stopped dead in its tracks. The horse's rider, *Superman* star Christopher Reeve, went tumbling through the air. Hitting the ground, Reeve crushed the top two vertebrae in his neck, damaging the sensitive spinal cord underneath. In a split second, he was paralyzed. Doctors say he is lucky to be alive.

The spinal cord is a ropelike bundle of long nerve tracts that shuttle messages between the brain and the rest of the body. The nervous system contains two subdivisions. Together, the brain and spinal cord compose the **central nervous system (CNS),** which interprets sensory input before coordinating a response that helps maintain homeostasis. The **peripheral nervous system (PNS)** consists of nerves that carry sensory information to the CNS and carry motor commands from the CNS to the muscles and glands (Fig. 13.1).

When Reeve hurt his spinal cord, the CNS lost its avenue of communication with the portion of his body located below the site of damage. He receives no sensation from most of his body, nor can he command his arms and legs to move. But cranial nerves from his eyes and ears still allow him to see and hear, and his brain still enables him to have emotions, to remember, and to reason. Also, his internal organs still function normally, a sign that his injury was not as severe as it could have been. In this chapter, we will examine the structure of the nervous system and how it carries out its numerous functions.

13.1 Nervous Tissue

The nervous system contains two types of cells: neuroglia (neuroglial cells) and neurons. **Neuroglia** support and nourish **neurons,** the cells that transmit nerve impulses (see Fig. 4.7).

Neuron Structure

Neurons vary in appearance, but all of them have just three parts: dendrites, a cell body, and an axon. In Figure 13.2, the **dendrites** are the many extensions from the cell body that receive signals from other neurons and send them on to the cell body. The **cell body** contains the nucleus as well as other organelles. An **axon** conducts nerve impulses away from the cell body toward other neurons or target structures.

There are three classes of neurons: sensory neurons, motor neurons, and interneurons. Their functions are best described in relation to the CNS. A **sensory neuron** takes messages from a sensory receptor to the CNS, and a **motor neuron** takes messages away from the CNS to an effector (muscle fiber or gland) (Fig. 13.2). An **interneuron** conveys messages between neurons in the CNS. Interneurons can receive input from sensory neurons and also from other interneurons in the CNS. Thereafter, they sum up all these signals before sending commands out to the muscles and glands by way of motor neurons.

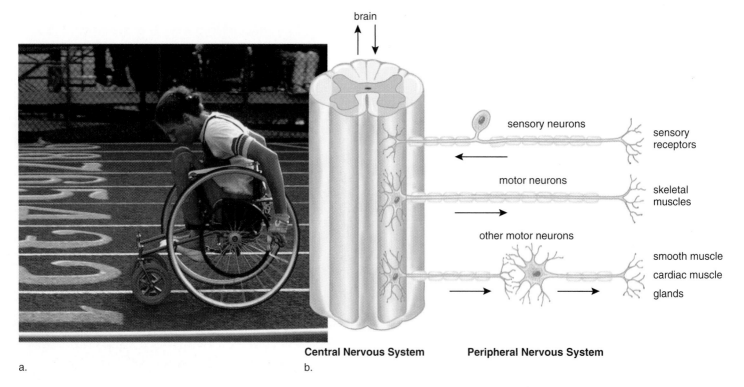

brain

sensory neurons

sensory receptors

motor neurons

skeletal muscles

other motor neurons

smooth muscle

cardiac muscle

glands

Central Nervous System   **Peripheral Nervous System**

a.   b.

Figure 13.1   Organization of the nervous system.
a. In paraplegics, messages no longer flow between the lower limbs and the central nervous system (the spinal cord and brain). **b.** The sensory neurons of the peripheral nervous system take nerve impulses from sensory receptors to the central nervous system (CNS), and motor neurons take nerve impulses from the CNS to the organs mentioned.

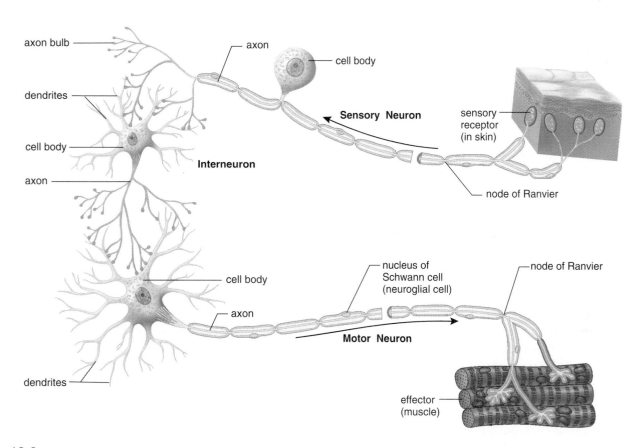

Figure 13.2   Types of neurons.

A sensory neuron, an interneuron, and a motor neuron are drawn here to show their arrangement in the body. (The breaks indicate that the fibers are much longer than shown.) How does this arrangement correlate with the function of each neuron?

Myelin Sheath

Some axons are covered by a protective **myelin sheath.** In the PNS, this covering is formed by a type of neuroglia called **Schwann cells,** which contain the lipid substance myelin in their plasma membranes. The myelin sheath develops when Schwann cells wrap themselves around an axon many times and in this way lay down several layers of plasma membrane. The myelin sheath is interrupted by gaps called **nodes of Ranvier** (Fig. 13.3). Myelin gives nerve fibers their white, glistening appearance and serves as an excellent insulator. The myelin sheath also plays an important role in nerve regeneration within the PNS. If an axon is accidentally severed, the myelin sheath remains and serves as a passageway for new fiber growth. Multiple sclerosis (MS) is a disease of the myelin sheath in the CNS. Lesions develop and become hardened scars that interfere with normal conduction of nerve impulses, and the result is various neuromuscular symptoms.

All neurons have three parts: dendrites, a cell body, and an axon. Sensory neurons take information to the CNS, and interneurons sum up sensory input before motor neurons take commands away from the CNS.

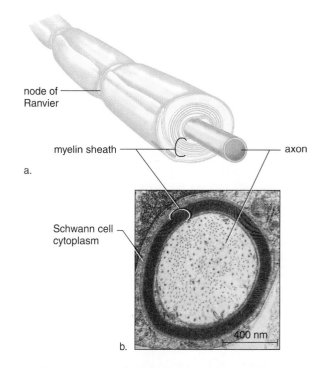

Figure 13.3   Myelin sheath.

a. In the PNS, a myelin sheath forms when Schwann cells wrap themselves around an axon. **b.** Electron micrograph of a cross section of an axon surrounded by a myelin sheath.

The Nerve Impulse

The nervous system uses the **nerve impulse** to convey information. The nature of a nerve impulse has been studied by using excised axons and a voltmeter called an **oscilloscope.** Voltage, often measured in millivolts (mV), is a measure of the electrical potential difference between two points, which in this case are the inside and the outside of the axon. Voltage is displayed on the oscilloscope screen as a trace, or pattern, over time.

Resting Potential

In the experimental setup shown in Figure 13.4a, an oscilloscope is wired to two electrodes: one electrode is placed inside an axon and the other electrode is placed outside. The axon is essentially a membranous tube filled with axoplasm (cytoplasm of the axon). When the axon is not conducting an impulse, the oscilloscope records a potential difference across a membrane equal to about -65 mV. This reading indicates that the inside of the axon is negative compared to the outside. This is called the **resting potential** because the axon is not conducting an impulse.

The existence of this polarity (charge difference) correlates with a difference in ion distribution on either side of the axomembrane (plasma membrane of the axon). As Figure 13.4a shows, the concentration of sodium ions (Na^+) is greater outside the axon than inside, and the concentration of potassium ions (K^+) is greater inside the axon than outside. The unequal distribution of these ions is due to the action of the **sodium-potassium pump,** a membrane protein that actively transports Na^+ out of and K^+ into the axon. The work of the pump maintains the unequal distribution of Na^+ and K^+ across the membrane.

The pump is always working because the membrane is somewhat permeable to these ions, and they tend to diffuse toward their lesser concentration. Since the membrane is more permeable to K^+ than to Na^+, there are always more positive ions outside the membrane than inside. This accounts for the polarity recorded by the oscilloscope. Large, negatively charged organic ions in the axoplasm also contribute to the polarity across a resting axomembrane.

Because of the sodium-potassium pump, there is a concentration of Na^+ outside an axon and K^+ inside an axon. An unequal distribution of ions causes the inside of an axon to be negative compared to the outside.

Action Potential

An **action potential** is a rapid change in polarity across an axomembrane as the nerve impulse occurs. An action potential is an all-or-none phenomenon. If a stimulus causes the axomembrane to depolarize to a certain level, called **threshold,** an action potential occurs. The strength of an action potential does not change, but an intense stimulus can cause an axon to fire (start an axon potential) more often in a given time interval than a weak stimulus.

The action potential requires two types of gated channel proteins in the membrane. One gated channel protein opens to allow Na^+ to pass through the membrane, and another opens to allow K^+ to pass through the membrane (Fig. 13.4b).

Sodium Gates Open   When an action potential occurs, the gates of sodium channels open first, and Na^+ flows into the axon. As Na^+ moves to inside the axon, the membrane potential changes from -65 mV to $+40$ mV. This is a *depolarization* because the charge inside the axon changes from negative to positive (Fig. 13.4c).

Potassium Gates Open   Second, the gates of potassium channels open, and K^+ flows to outside the axon. As K^+ moves to outside the axon, the action potential changes from $+40$ mV back to -65 mV. This is a *repolarization* because the inside of the axon resumes a negative charge as K^+ exits the axon.

The nerve impulse consists of an electrochemical change that occurs across an axomembrane. During depolarization, Na^+ moves to inside the axon, and during repolarization K^+ moves to outside the axon.

Propagation of an Action Potential

When an action potential travels down an axon, each successive portion of the axon undergoes a depolarization and then a repolarization. Like a domino effect, each preceding portion causes an action potential in the next portion of an axon.

As soon as an action potential has moved on, the previous portion of an axon undergoes a **refractory period** during which the sodium gates are unable to open. This ensures that the action potential cannot move backward and instead always moves down an axon toward its branches.

In myelinated axons, the gated ion channels that produce an action potential are concentrated at the nodes of Ranvier. Since ion exchange occurs only at the nodes, the action potential travels faster than in nonmyelinated axons. This is called saltatory conduction, meaning that the action potential "jumps" from node to node. Speeds of 200 meters per second (450 miles per hour) have been recorded.

An action potential travels along the length of an axon.

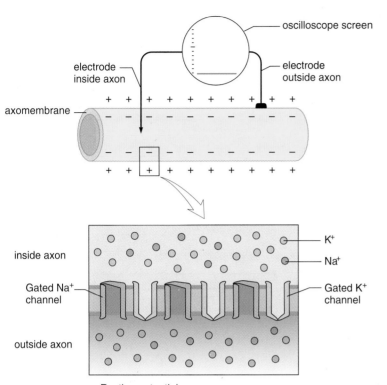

oscilloscope screen

electrode inside axon

electrode outside axon

axomembrane

K⁺

Na⁺

Gated Na⁺ channel

Gated K⁺ channel

inside axon

outside axon

a. Resting potential

direction of impulse →

Open Na⁺ channel

Open K⁺ channel

b. Action potential

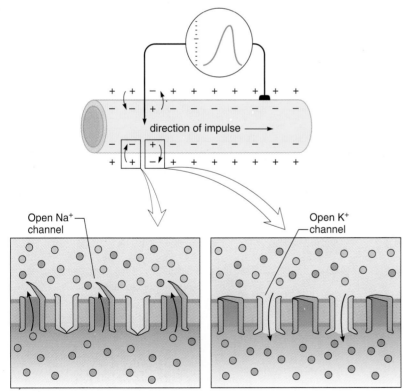

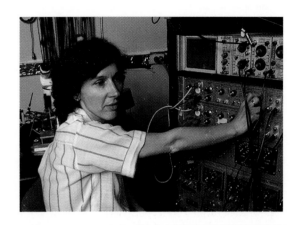

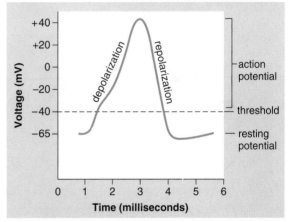

c. Enlargement of action potential

Figure 13.4 **Resting and action potential.**
a. Resting potential. An oscilloscope, an instrument that records voltage changes, records a resting potential of −65 mV. There is a preponderance of Na⁺ outside the axon and a preponderance of K⁺ inside the axon. The permeability of the membrane to K⁺ compared to Na⁺ causes the inside to be negative compared to the outside. **b.** Action potential. A depolarization occurs when Na⁺ gates open and Na⁺ moves to inside the axon; a repolarization occurs when K⁺ gates open and K⁺ moves to outside the axon. **c.** Enlargement of the action potential in (**b**), as seen by an experimenter using an oscilloscope.

Visual Focus

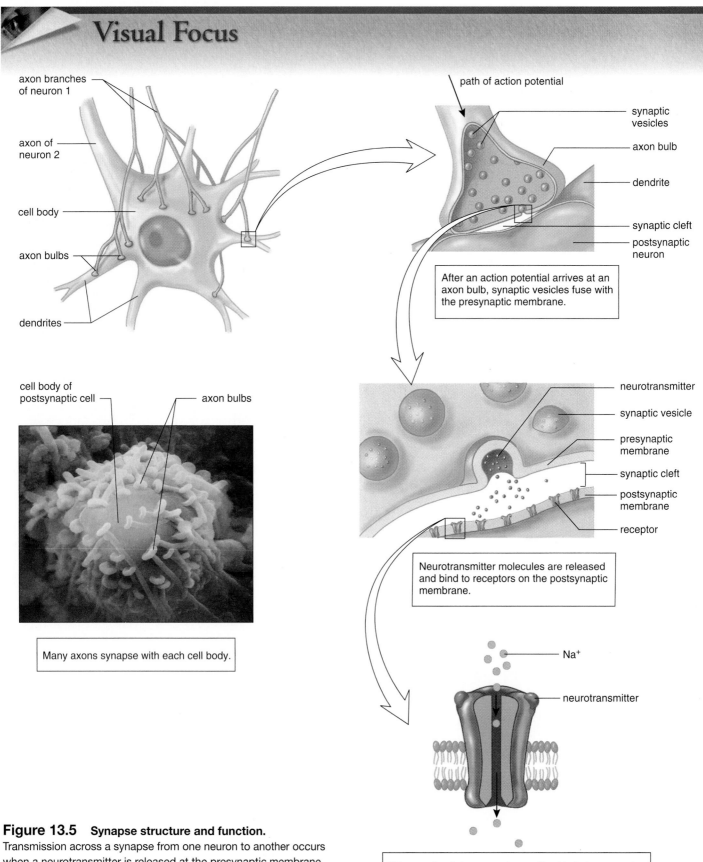

axon branches of neuron 1

axon of neuron 2

cell body

axon bulbs

dendrites

path of action potential

synaptic vesicles

axon bulb

dendrite

synaptic cleft

postsynaptic neuron

After an action potential arrives at an axon bulb, synaptic vesicles fuse with the presynaptic membrane.

cell body of postsynaptic cell

axon bulbs

Many axons synapse with each cell body.

neurotransmitter

synaptic vesicle

presynaptic membrane

synaptic cleft

postsynaptic membrane

receptor

Neurotransmitter molecules are released and bind to receptors on the postsynaptic membrane.

Na⁺

neurotransmitter

When a stimulatory neurotransmitter binds to a receptor, Na⁺ diffuses into the postsynaptic neuron.

Figure 13.5   Synapse structure and function.
Transmission across a synapse from one neuron to another occurs when a neurotransmitter is released at the presynaptic membrane, diffuses across a synaptic cleft, and binds to a receptor in the postsynaptic membrane.

Transmission Across a Synapse

Every axon branches into many fine endings, each tipped by a small swelling called an **axon bulb.** Each bulb lies very close to either the dendrite or the cell body of another neuron. This region of close proximity is called a **synapse** (Fig. 13.5). At a synapse, the membrane of the first neuron is called the *pre*synaptic membrane, and the membrane of the next neuron is called the *post*synaptic membrane. The small gap between is the **synaptic cleft.**

Transmission across a synapse is carried out by molecules called **neurotransmitters,** which are stored in synaptic vesicles in the axon bulbs. When nerve impulses traveling along an axon reach an axon bulb, gated channels for calcium ions (Ca^{2+}) open, and calcium enters the bulb. This sudden rise in Ca^{2+} stimulates synaptic vesicles to merge with the presynaptic membrane, and neurotransmitter molecules are released into the synaptic cleft. They diffuse across the cleft to the postsynaptic membrane, where they bind with specific receptor proteins.

Depending on the type of neurotransmitter and the type of receptor, the response of the postsynaptic neuron can be toward excitation or toward inhibition. Excitatory neurotransmitters that utilize gated ion channels are fast acting. Other neurotransmitters affect the metabolism of the postsynaptic cell and therefore are slower acting.

Synaptic Integration

A single neuron has many dendrites plus the cell body, and both can have synapses with many other neurons. A neuron is on the receiving end of many excitatory and inhibitory signals. An excitatory neurotransmitter produces a potential change called a *signal* that drives the neuron closer to an action potential; an inhibitory neurotransmitter produces a signal that drives the neuron farther from an action potential. Excitatory signals have a depolarizing effect, and inhibitory signals have a hyperpolarizing effect.

Neurons integrate these incoming signals. **Integration** is the summing up of excitatory and inhibitory signals (Fig. 13.6). If a neuron receives many excitatory signals (either from different synapses or at a rapid rate from one synapse), the chances are the axon will transmit a nerve impulse. On the other hand, if a neuron receives both inhibitory and excitatory signals, the summing up of these signals may prohibit the axon from firing.

Integration is the summing up of inhibitory and excitatory signals received by a postsynaptic neuron.

Neurotransmitter Molecules

At least 25 different neurotransmitters have been identified, but two very well-known ones are **acetylcholine (ACh)** and **norepinephrine (NE).**

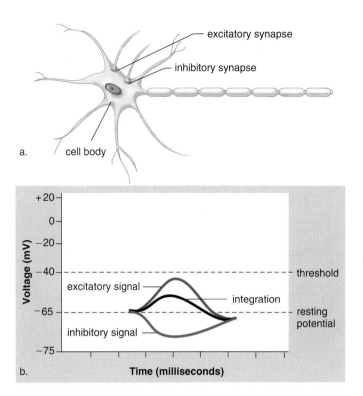

Figure 13.6   Integration.
a. Inhibitory signals and excitatory signals are summed up in the dendrite and cell body of the postsynaptic neuron. Only if the combined signals cause the membrane potential to rise above threshold does an action potential occur. **b.** In this example, threshold was not reached.

Once a neurotransmitter has been released into a synaptic cleft and has initiated a response, it is removed from the cleft. In some synapses, the postsynaptic membrane contains enzymes that rapidly inactivate the neurotransmitter. For example, the enzyme **acetylcholinesterase (AChE)** breaks down acetylcholine. In other synapses, the presynaptic membrane rapidly reabsorbs the neurotransmitter, possibly for repackaging in synaptic vesicles or for molecular breakdown. The short existence of neurotransmitters at a synapse prevents continuous stimulation (or inhibition) of postsynaptic membranes.

It is of interest to note here that many drugs that affect the nervous system act by interfering with or potentiating the action of neurotransmitters. As described in Figure 13.18, drugs can enhance or block the release of a neurotransmitter, mimic the action of a neurotransmitter or block the receptor, or interfere with the removal of a neurotransmitter from a synaptic cleft.

Transmission across a synapse is dependent on the release of neurotransmitters, which diffuse across the synaptic cleft from one neuron to the next.

13.2 The Central Nervous System

The central nervous system (CNS) consists of the spinal cord and the brain, where sensory information is received and motor control is initiated. Figure 13.7 illustrates how the CNS relates to the PNS. Both the spinal cord and the brain are protected by bone; the spinal cord is surrounded by vertebrae, and the brain is enclosed by the skull. Also, both the spinal cord and the brain are wrapped in protective membranes known as **meninges** (sing., meninx). The spaces between the meninges are filled with **cerebrospinal fluid,** which cushions and protects the CNS. A small amount of this fluid is sometimes withdrawn from around the cord for laboratory testing when a spinal tap (lumbar puncture) is performed. Meningitis is an infection of the meninges.

Cerebrospinal fluid is also contained within the ventricles of the brain and in the central canal of the spinal cord. The brain's **ventricles** are interconnecting cavities that produce and serve as a reservoir for cerebrospinal fluid. Normally, any excess cerebrospinal fluid drains away into the circulatory system. However, blockages can occur. In an infant, the brain can enlarge due to cerebrospinal fluid accumulation, resulting in a condition called hydrocephalus ("water on the brain"). If cerebrospinal fluid collects in an adult, the brain cannot enlarge, and instead is pushed against the skull, possibly becoming injured.

The CNS is composed of two types of nervous tissue—gray matter and white matter. **Gray matter** is gray because it contains cell bodies and short, nonmyelinated fibers. **White matter** is white because it contains myelinated axons that run together in bundles called **tracts.**

The CNS, which lies in the midline of the body and consists of the brain and the spinal cord, receives sensory information and initiates motor control.

The Spinal Cord

The **spinal cord** extends from the base of the brain through a large opening in the skull called the foramen magnum and into the vertebral canal formed by openings in the vertebrae.

Structure of the Spinal Cord

Figure 13.8*a* shows how an individual vertebra protects the spinal cord. The spinal nerves project from the cord between the vertebrae which make up the vertebral column. Intervertebral disks separate the vertebrae and if one slips a bit and presses on the spinal cord, pain will result.

A cross section of the spinal cord shows a central canal, gray matter, and white matter (Fig. 13.8*b,c*). The central canal contains cerebrospinal fluid, as do the meninges that

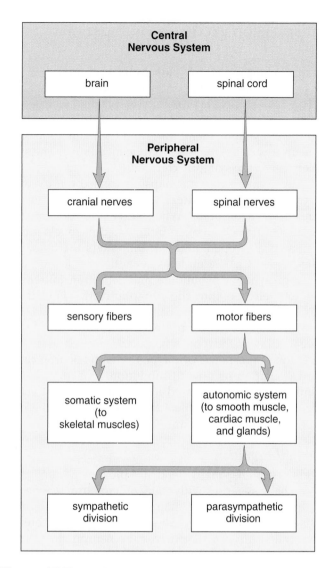

Figure 13.7   Organization of the nervous system.
The CNS, composed of the spinal cord and brain, communicates with the PNS, which contains nerves. In the somatic system, nerves conduct impulses from sensory receptors to the CNS and motor impulses from the CNS to the skeletal muscles. In the autonomic system, consisting of the sympathetic and parasympathetic divisions, motor impulses travel to smooth muscle, cardiac muscle, and the glands.

protect the spinal cord. The gray matter is centrally located and shaped like the letter H. Portions of sensory neurons and motor neurons are found there, as are interneurons that communicate with these two types of neurons. The dorsal root of a spinal nerve contains sensory fibers entering the gray matter, and the ventral root of a spinal nerve contains motor fibers exiting the gray matter. The dorsal and ventral roots join before the spinal nerve leaves the vertebral canal. Spinal nerves are a part of the PNS.

The white matter of the spinal cord occurs in areas around the gray matter. The white matter contains ascending tracts taking information to the brain (primarily located dorsally) and descending tracts taking information from the brain (primarily located ventrally). Because the tracts cross just after they enter and exit the brain, the left side of the brain controls the right side of the body, and the right side of the brain controls the left side of the body.

> The spinal cord extends from the base of the brain into the vertebral canal formed by the vertebrae. A cross section shows that the spinal cord has a central canal, gray matter, and white matter.

Functions of the Spinal Cord

The spinal cord provides a means of communication between the brain and the peripheral nerves that leave the cord. When someone touches your hand, sensory receptors generate nerve impulses that pass through sensory fibers to the spinal cord and up ascending tracts to the brain. When we voluntarily move our limbs, motor impulses originating in the brain pass down descending tracts to the spinal cord and out to our muscles by way of motor fibers. Therefore, if the spinal cord is severed, we suffer a loss of sensation and a loss of voluntary control—that is, paralysis. If the cut occurs in the thoracic region, the lower body and legs are paralyzed, a condition known as paraplegia. If the injury is in the neck region, all four limbs are usually affected, a condition called quadriplegia.

We will see that the spinal cord is also the center for thousands of reflex arcs. A stimulus causes sensory receptors to generate nerve impulses that travel in sensory axons to the spinal cord. Interneurons integrate the incoming data and relay signals to motor neurons. A response to the stimulus occurs when motor axons cause skeletal muscles to contract. Each interneuron in the spinal cord has synapses with many other neurons, and therefore they send signals to several other interneurons and motor neurons.

The spinal cord plays a similar role for the internal organs. For example, when blood pressure falls, internal receptors in the carotid arteries and aorta generate nerve impulses that pass through sensory fibers to the cord and then up an ascending tract to a cardiovascular center in the brain. Thereafter, nerve impulses pass down a descending tract to the spinal cord. Motor impulses then cause blood vessels to constrict so that the blood pressure rises.

> The spinal cord serves as a means of communication between the brain and much of the body. The spinal cord is also a center for reflex actions.

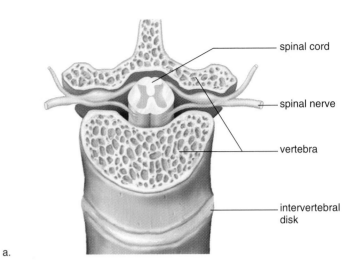

a.

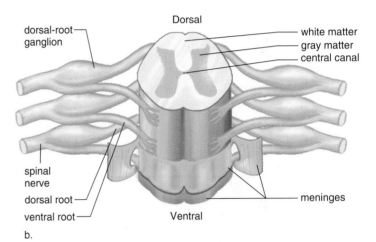

b.

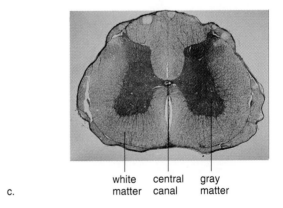

c.

Figure 13.8 Spinal cord.
a. The spinal cord passes through the vertebral canal formed by the vertebrae. **b.** The spinal cord has a central canal filled with cerebrospinal fluid, gray matter in an H-shaped configuration, and white matter around the outside. The white matter contains tracts that take nerve impulses to and from the brain. **c.** Photomicrograph of a cross section of the spinal cord.

The Brain

The human **brain** has been called the last great frontier of biology. The goal of modern neuroscience is to understand the structure and function of the brain's various parts so well that it will be possible to prevent or correct the thousands of mental disorders that rob human beings of a normal life. This section gives only a glimpse of what is known about the brain and the modern avenues of research.

We will discuss the parts of the brain with reference to the cerebrum, the diencephalon, the cerebellum, and the brain stem. The brain has four ventricles called, in turn, the two lateral ventricles, the third ventricle, and the fourth ventricle. It may be helpful to you to associate the cerebrum with the two lateral ventricles, the diencephalon with the third ventricle, and the brain stem and the cerebellum with the fourth ventricle (Fig. 13.9*a*).

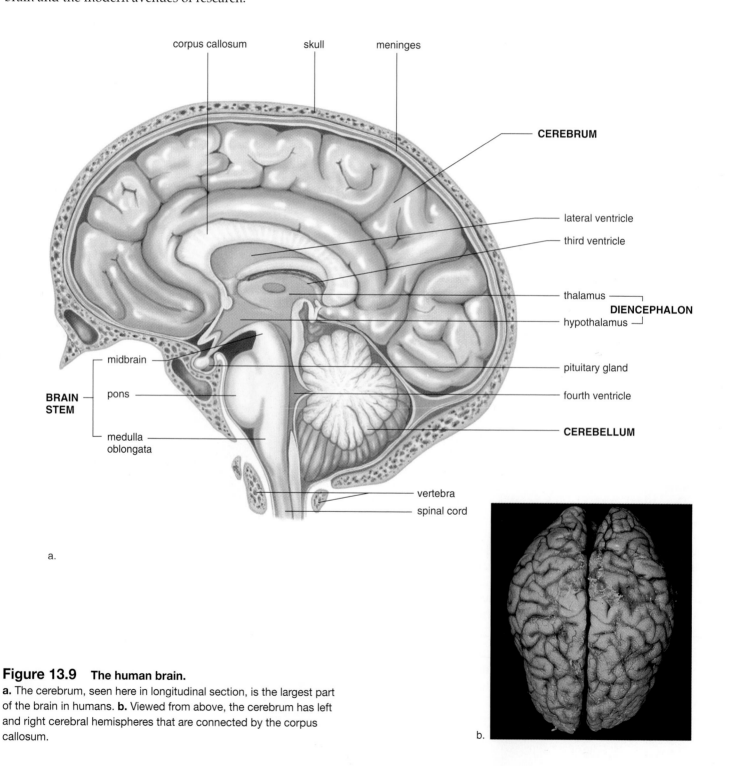

a.

b.

Figure 13.9   The human brain.
a. The cerebrum, seen here in longitudinal section, is the largest part of the brain in humans. **b.** Viewed from above, the cerebrum has left and right cerebral hemispheres that are connected by the corpus callosum.

The Cerebrum

The **cerebrum,** also called the telencephalon, is the largest portion of the brain in humans. The cerebrum is the last center to receive sensory input and carry out integration before commanding voluntary motor responses. It communicates with and coordinates the activities of the other parts of the brain. As we shall see, the cerebrum carries out higher thought processes required for learning and memory and for language and speech.

Just as the human body has two halves, so does the cerebrum. These halves are called the left and right **cerebral hemispheres** (see Fig. 13.9b). A deep groove called the longitudinal fissure divides the left and right cerebral hemispheres. Still, the two cerebral hemispheres are connected by a bridge of tracts within the corpus callosum.

Shallow grooves called sulci (sing., sulcus) divide each hemisphere into lobes (Fig. 13.10). The *frontal lobe* is toward the front of a cerebral hemisphere, and the *parietal lobe* is toward the back. The *occipital lobe* is dorsal to (behind) the parietal lobe, and the *temporal lobe* lies below the frontal and parietal lobes.

The Cereal Cortex The **cerebral cortex** is a thin but highly convoluted outer layer of gray matter that covers the cerebral hemispheres. The cerebral cortex contains over one billion cell bodies and is the region of the brain that accounts for sensation, voluntary movement, and all the thought processes we associate with consciousness.

The cerebral cortex contains motor areas and sensory areas as well as association areas. The **primary motor area** is in the frontal lobe just ventral to (before) the central sulcus. Voluntary commands to skeletal muscles begin in the primary motor area, and each part of the body is controlled by a certain section. For example, our versatile hand takes up an especially large portion of the primary motor area. Ventral to the primary motor area is a premotor area. The *premotor area* organizes motor functions for skilled motor activities, and then the primary motor area sends signals to the cerebellum, which integrates them. The unique ability of humans to speak is partially dependent upon *Broca's area,* a motor speech area in the left frontal lobe. Signals originating here pass to the premotor area before reaching the primary motor area.

The **primary somatosensory area** is just dorsal to the central sulcus. Sensory information from the skin and skeletal muscles arrives here, where each part of the body is sequentially represented. A primary visual area in the occipital lobe receives information from our eyes, and a primary

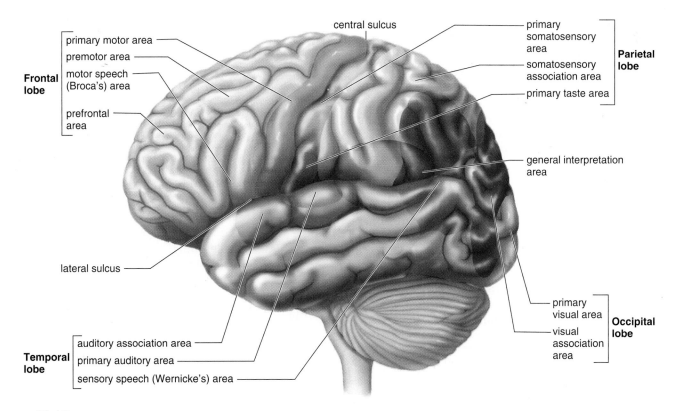

Figure 13.10 **The lobes of a cerebral hemisphere.**
Each cerebral hemisphere is divided into four lobes: frontal, parietal, temporal, and occipital. The cerebral cortex of a frontal lobe has motor areas and an association area called the prefrontal area. The cerebral cortex of the other lobes both have sensory areas and association areas.

auditory area in the temporal lobe receives information from our ears. A primary taste area in the parietal lobe accounts for taste sensations.

Association areas are places where integration occurs. For example, the *somatosensory association area,* located just dorsal to the primary somatosensory area, processes and analyzes sensory information from the skin and muscles. The *visual association area* in the occipital lobe associates new visual information with previously received visual information. It might "decide," for example, if we have seen this face or tool or whatever before. The *auditory association area* in the temporal lobe performs the same functions with regard to sounds. These association areas meet near the dorsal end of the lateral sulcus. This region is called the general interpretation area because it receives information from all the sensory association areas and allows us to quickly integrate incoming signals and send them on to the prefrontal area so that an immediate response is possible. This is the part of the brain that is in operation when people are able to quickly assess a situation and take actions that save others from danger. The **prefrontal area,** an association area in the frontal lobe, receives information from the other association areas and uses this information to reason and plan our actions. Integration in this area accounts for our most cherished human abilities to think critically and to formulate appropriate behaviors.

White Matter   Much of the rest of the cerebrum is composed of white matter. As you know, white matter in the CNS consists of long myelinated axons organized into tracts. Descending tracts from the primary motor area communicate with lower brain centers, and ascending tracts from lower brain centers send sensory information up to the primary somatosensory area. Because the tracts cross over in the medulla, the left side of the cerebrum controls the right side of the body and vice versa. Tracts within the cerebrum also take information between the different sensory, motor, and association areas pictured in Figure 13.10. As previously mentioned, the corpus callosum contains tracts that join the two cerebral hemispheres.

Basal Nuclei   While the bulk of the cerebrum is composed of tracts, there are masses of gray matter located deep within the white matter. These so-called **basal nuclei** (formerly termed basal ganglia) integrate motor commands, ensuring that proper muscle groups are activated or inhibited. Huntington disease and Parkinson disease, which are both characterized by uncontrollable movements, are believed to be due to malfunctioning of the basal nuclei.

The gray matter of the cerebrum consists of the cerebral cortex and the basal nuclei. The white matter consists of tracts.

The Diencephalon

The hypothalamus and the thalamus are in the **diencephalon,** a region that encircles the third ventricle. The **hypothalamus** forms the floor of the third ventricle. The hypothalamus is an integrating center that helps maintain homeostasis by regulating hunger, sleep, thirst, body temperature, and water balance. The hypothalamus controls the pituitary gland and thereby serves as a link between the nervous and endocrine systems.

The **thalamus** consists of two masses of gray matter located in the sides and roof of the third ventricle. The thalamus is on the receiving end for all sensory input except for smell. Visual, auditory, and somatosensory information arrives at the thalamus via the cranial nerves and tracts from the spinal cord. The thalamus integrates this information and sends it on to the appropriate portions of the cerebrum. The thalamus is involved in arousal of the cerebrum, and it also participates in higher mental functions such as memory and emotions.

The pineal gland, which secretes the hormone melatonin, is located in the diencephalon. Presently there is much popular interest in the role of melatonin in our daily rhythms; some researchers believe it can help ameliorate jet lag or insomnia. Scientists are also interested in the possibility that the hormone may regulate the onset of puberty.

The Cerebellum

The **cerebellum** is separated from the brain stem by the fourth ventricle. The cerebellum has two portions that are joined by a narrow median portion. Each portion is primarily composed of white matter, which in longitudinal section has a treelike pattern. Overlying the white matter is a thin layer of gray matter that forms a series of complex folds.

The cerebellum receives sensory input from the eyes, ears, joints, and muscles about the present position of body parts, and it also receives motor output from the cerebral cortex about where these parts should be located. After integrating this information, the cerebellum sends motor impulses by way of the brain stem to the skeletal muscles. In this way, the cerebellum maintains posture and balance. It also ensures that all of the muscles work together to produce smooth, coordinated voluntary movements. The cerebellum assists the learning of new motor skills like playing the piano or hitting a baseball.

The Brain Stem

The brain stem contains the midbrain, the pons, and the medulla oblongata (see Fig. 13.6*a*). The **midbrain** acts as a relay station for tracts passing between the cerebrum and the spinal cord or cerebellum. It also has reflex centers for visual, auditory, and tactile responses. The word **pons** means "bridge" in Latin, and true to its name, the pons contains bundles of axons traveling between the cerebellum and the rest of the CNS. In addition, the pons functions with the medulla oblongata to regulate breathing rate and has reflex centers concerned with head movements in response to visual and auditory stimuli.

The **medulla oblongata** contains a number of reflex centers for regulating heartbeat, breathing, and vasoconstriction (blood pressure). It also contains the reflex centers for vomiting, coughing, sneezing, hiccuping, and swallowing. The medulla oblongata lies just superior to the spinal cord and it contains tracts that ascend or descend between the spinal cord and higher brain centers.

The Reticular Formation   The **reticular formation** is a complex network of nuclei (masses of gray matter) and fibers that extend the length of the brain stem (Fig. 13.11). The reticular formation receives sensory signals, which it sends up to higher centers, and motor signals, which it sends to the spinal cord.

One portion of the reticular formation, called the reticular activating system (RAS), arouses the cerebrum via the thalamus and causes a person to be alert. It is believed to filter out unnecessary sensory stimuli; this may explain why you can study with the TV on. An inactive reticular formation results in sleep, and a severe injury to the RAS can cause a person to be comatose.

The other portions of the brain work with the cerebrum and they are essential to maintaining homeostasis. ♊

13.3 The Limbic System and Higher Mental Functions

The limbic system is intimately involved in our emotions and higher mental functions. After a short description, we will discuss the functions of the limbic system.

Limbic System

The **limbic system** is a complex network of tracts and nuclei that incorporates medial portions of the cerebral lobes, the basal nuclei, and the diencephalon (Fig. 13.12). The limbic system blends primitive emotions and higher mental functions into a united whole. It accounts for why activities like sexual behavior and eating seem pleasurable and also for why, say, mental stress can cause high blood pressure.

Two significant structures within the limbic system are the hippocampus and the amygdala, which are essential for learning and memory. The hippocampus is well situated in the brain to make the prefrontal area aware of past experiences stored in association areas. The amygdala, in particular, can cause these experiences to have emotional overtones.

The prefrontal area consults the hippocampus in order to use memories to modify our behavior. However, the inclusion of the frontal lobe in the limbic system means that reason can keep us from acting out strong feelings.

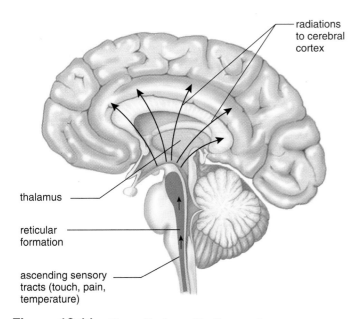

Figure 13.11   **The reticular activating system.**
The reticular formation receives and sends on motor and sensory information to various parts of the CNS. One portion, the reticular activating system (RAS; see arrows), arouses the cerebrum and in this way controls alertness versus sleep.

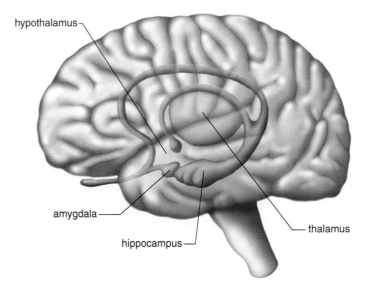

Figure 13.12   **The limbic system.**
Structures deep within each cerebral hemisphere and surrounding the diencephalon join higher mental functions like reasoning with more primitive feelings like fear and pleasure.

Higher Mental Functions

As in other areas of biological research, brain research has progressed due to technological breakthroughs. Neuroscientists now have a wide range of techniques at their disposal for studying the human brain, including modern technologies that allow us to record its functioning.

Memory and Learning

Just as the connecting tracts of the corpus callosum are evidence that the two cerebral hemispheres work together, so the limbic system indicates that cortical areas may work with lower centers to produce learning and memory. **Memory** is the ability to hold a thought in mind or to recall events from the past, ranging from a word we learned only yesterday to an early emotional experience that has shaped our lives. **Learning** takes place when we retain and utilize past memories.

Types of Memory   We have all tried to remember a seven-digit telephone number for a short period of time. If we say we are trying to keep it in the forefront of our brain, we are exactly correct. The prefrontal area, which is active during **short-term memory,** lies just dorsal to our forehead! There are some telephone numbers that we have memorized; in other words, they have gone into **long-term memory.** Think of a telephone number you know by heart and try to bring it to mind without also thinking about the place or person associated with that number. Most likely you cannot, because typically long-term memory is a mixture of what is called **semantic memory** (numbers, words, etc.) and **episodic memory** (persons, events, etc.). Due to brain damage, some people lose one type of memory but not the other. For example, without a working episodic memory, they can carry on a conversation but have no rec- ollection of recent events. If you are talking to them and then leave the room, they don't remember you when you come back!

Skill memory is another type of memory that can exist independent of episodic memory. Skill memory is involved in performing motor activities like riding a bike or playing ice hockey. When a person first learns a skill, more areas of the cerebral cortex are involved than after the skill is perfected. In other words, you have to think about what you are doing when you learn a skill, but later the actions become automatic. Skill memory involves all the motor areas of the cerebrum below the level of consciousness.

Long-term Memory Storage and Retrieval   The first step toward curing memory disorders is to know what parts of the brain are functioning when we remember something. Investigators have been able to work it out pretty well. Our long-term memories are stored in bits and pieces throughout the sensory association areas of the cerebral cortex. Visions are stored in the vision association area, sounds are stored in the auditory association area, and so forth. The **hippocampus,** a seahorse-shaped structure deep in the temporal lobe, serves as a bridge between the sensory association areas where memories are stored and the prefrontal area where memories are utilized (Fig. 13.13). The prefrontal area communicates with the hippocampus when memories are stored and when these memories are brought to mind. Why are some memories so emotionally charged? The **amygdala** seems to be responsible for fear conditioning and associating danger with sensory stimuli received from both the diencephalon and the cortical sensory areas.

Long-term Potentiation   While it is helpful to know the memory functions of various portions of the brain, an important step toward curing mental disorders is understanding

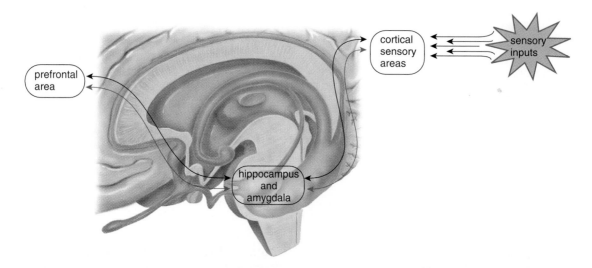

Figure 13.13   **Long-term memory circuits.**
The hippocampus and amygdala are believed to be involved in the storage and retrieval of memories. Semantic memory (red arrows) and episodic memory (black arrows) are stored separately, and therefore you can lose one without losing the other.

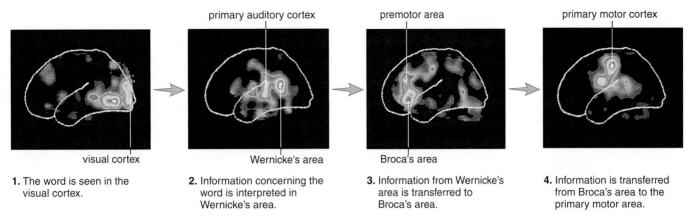

1. The word is seen in the visual cortex.

2. Information concerning the word is interpreted in Wernicke's area.

3. Information from Wernicke's area is transferred to Broca's area.

4. Information is transferred from Broca's area to the primary motor area.

Figure 13.14 Language and speech.
These functional images were captured by a high-speed computer during PET (positron emission tomography) scanning of the brain. A radioactively labeled solution is injected into the subject, and then the subject is asked to perform certain activities. Cross-sectional images of the the brain generated by the computer reveal where activity is occurring because the solution is preferentially taken up by active brain tissue and not by inactive brain tissue. These PET images show the cortical pathway for reading words and then speaking them. Red indicates the most active areas of the brain, and blue indicates the least active areas.

memory on the cellular level. **Long-term potentiation (LTP)** is an enhanced response at synapses within the hippocampus. LTP is probably essential to memory storage, but unfortunately, it sometimes causes a postsynaptic neuron to become so excited that it undergoes apoptosis, a form of cell death. This phenomenon, called excitotoxicity, may develop due to a mutation. (The longer we live, the more likely it is that any particular mutation will occur.) Excitotoxicity is due to the action of the neurotransmitter glutamate, which is active in the hippocampus. When glutamate binds with a specific type of receptor in the postsynaptic membrane, calcium (Ca^{2+}) may rush in too fast; this influx is lethal to the cell. A gradual extinction of brain cells in the hippocampus and other parts of the brain occurs in persons with Alzheimer disease (AD).

Language and Speech

Language is dependent upon semantic memory. Therefore, we would expect some of the same areas in the brain to be involved in both memory and language. Any disruption of these pathways could very well contribute to an inability to comprehend our environment and use speech correctly.

Seeing and hearing words depends on sensory centers in the occipital and temporal lobes, respectively. Damage to a sensory speech area in the temporal lobe called Wernicke's area results in the inability to comprehend speech.

Generating and speaking words depends on motor centers in the frontal lobe. Studies of patients with speech disorders have shown that damage to the motor speech area called Broca's area results in the inability to speak. Broca's area is located just in front of the primary motor area for speech musculature (lips, tongue, larynx, and so forth) (see Fig. 13.10).

The functions of the visual cortex, Wernicke's area, and Broca's area are shown in Figure 13.14.

One interesting aside pertaining to language and speech is the recognition that the left brain and the right brain have different functions. For example, only the left hemisphere and not the right contains a Broca's area and a Wernicke's area! Indeed, the left hemisphere plays a role of great importance in language functions in general and not just in speech. In an attempt to cure epilepsy in the early 1940s, the corpus callosum was surgically severed in some patients. Later studies showed that these split-brain patients could only name objects if seen by the left hemisphere. If viewed only by the right hemisphere, a split-brain patient could choose the proper object for a particular use but was unable to name it. Based on these and other types of studies, the left brain/right brain hypothesis says that the left brain can be contrasted with the right brain along these lines:

Left Hemisphere	Right Hemisphere
Verbal	Nonverbal, visuo-spatial
Logical, analytical	Intuitive
Rational	Creative

Further, it became generally thought that one hemisphere was dominant in each person, accounting in part for personality traits. But researchers now believe that the hemispheres process the same information differently. The left hemisphere is more global, whereas the right hemisphere is more specific in its approach.

Memory has been studied at various levels—behavioral, structural, and cellular. Special areas in the left hemisphere help account for our ability to comprehend and use speech.

13.4 The Peripheral Nervous System

The peripheral nervous system (PNS) lies outside the central nervous system and is composed of nerves and ganglia. **Nerves** are bundles of axons; axons are called nerve fibers when they occur in nerves.

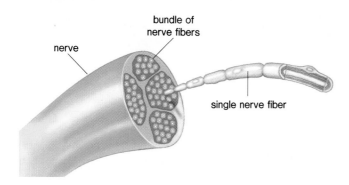

Sensory fibers carry information to the CNS, and motor fibers carry information away from the CNS. Ganglia (sing., **ganglion**) are swellings associated with nerves that contain collections of cell bodies.

Humans have 12 pairs of **cranial nerves** attached to the brain. By convention, the pairs of cranial nerves are referred to by roman numerals (Fig. 13.15*a*). Some of these are sensory nerves—that is, they contain only sensory fibers; some are motor nerves that contain only motor fibers; and others are mixed nerves that contain both sensory and motor fibers. Cranial nerves are largely concerned with the head, neck, and facial regions of the body. However, the vagus nerve (**X**) has branches not only to the pharynx and larynx, but also to most of the internal organs.

The **spinal nerves** of humans emerge in 31 pairs from either side of the spinal cord. Each spinal nerve originates when two short branches, or roots, join together (Fig. 13.15*b*). The dorsal root (at the back) contains sensory fibers that conduct impulses inward (toward the spinal cord) from sensory receptors. The cell body of a sensory neuron is in a **dorsal-root ganglion.** The ventral root (at the front) contains motor fibers that conduct impulses outward (away from the cord) to effectors. Notice, then, that all spinal nerves are mixed nerves that contain many sensory and motor fibers. Each spinal nerve serves the particular region of the body in which it is located. The intercostal muscles of the rib cage are innervated by thoracic nerves, for example.

In the PNS, cranial nerves take impulses to and/or from the brain, and spinal nerves take impulses to and from the spinal cord.

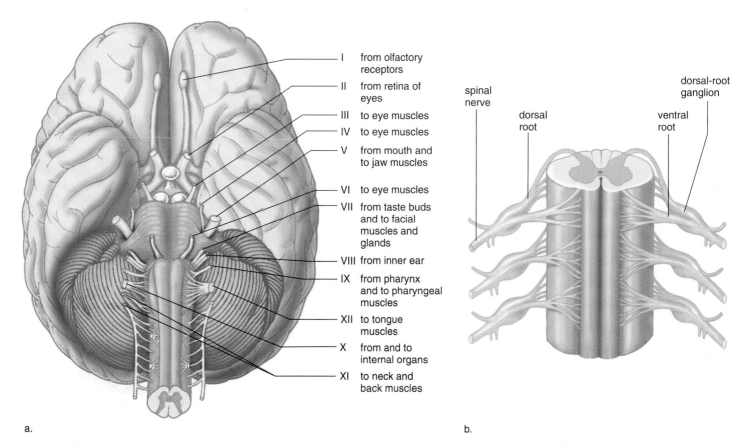

I   from olfactory receptors
II   from retina of eyes
III to eye muscles
IV   to eye muscles
V   from mouth and to jaw muscles
VI   to eye muscles
VII from taste buds and to facial muscles and glands
VIII from inner ear
IX   from pharynx and to pharyngeal muscles
XII to tongue muscles
X   from and to internal organs
XI   to neck and back muscles

a.

b.

Figure 13.15   Cranial and spinal nerves.
a. Ventral surface of the brain showing the attachment of the 12 pairs of cranial nerves. **b.** Cross section of the spinal cord, showing 3 pairs of spinal nerves. The human body has 31 pairs of spinal nerves, and each spinal nerve has a dorsal root and a ventral root attached to the spinal cord.

Somatic System

The PNS is subdivided into the somatic system and the autonomic system. The **somatic system** serves the skin, skeletal muscles, and tendons. It includes nerves that take sensory information from external sensory receptors to the CNS and motor commands away from the CNS to the skeletal muscles. Some actions in the somatic system are due to **reflexes,** automatic responses to a stimulus. A reflex occurs quickly, without us even having to think about it. Other actions are voluntary, and these always originate in the cerebral cortex as when we decide to move a limb.

The Reflex Arc

Figure 13.16 illustrates the path of a reflex that involves only the spinal cord. If your hand touches a sharp pin, sensory receptors in the skin generate nerve impulses that move along sensory fibers through the dorsal root ganglia toward the spinal cord. Sensory neurons that enter the cord dorsally pass signals on to many interneurons. Some of these interneurons synapse with motor neurons whose short dendrites and cell bodies are in the spinal cord. Nerve impulses travel along these motor fibers to an effector, which brings about a response to the stimulus. In this case, the effector is a muscle, which contracts so that you withdraw your hand from the pin. Various other reactions are also possible—you will most likely look at the pin, wince, and cry out in pain. This whole series of responses occurs because some of the interneurons involved carry nerve impulses to the brain. The brain makes you aware of the stimulus and directs these other reactions to it. Pain is not felt until the brain receives the information and interprets it.

In the somatic system, nerves take information from external sensory receptors to the CNS and take motor commands to the skeletal muscles. Involuntary reflexes allow us to respond rapidly to external stimuli.

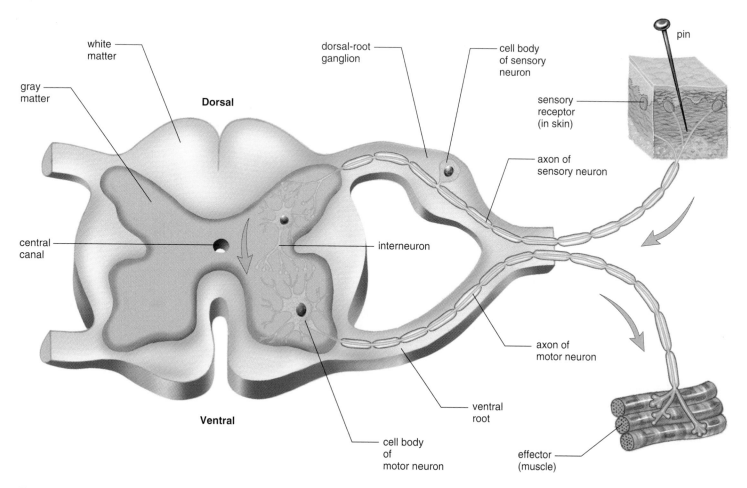

Figure 13.16 **A reflex arc showing the path of a spinal reflex.**
A stimulus (e.g., a pinprick) causes sensory receptors in the skin to generate nerve impulses that travel in sensory axons to the spinal cord. Interneurons integrate data from sensory neurons and then relay signals to motor neurons. Motor axons convey nerve impulses from the spinal cord to a skeletal muscle, which contracts. Movement of the hand away from the pin is the response to the stimulus.

Sympathetic Division

Parasympathetic Division

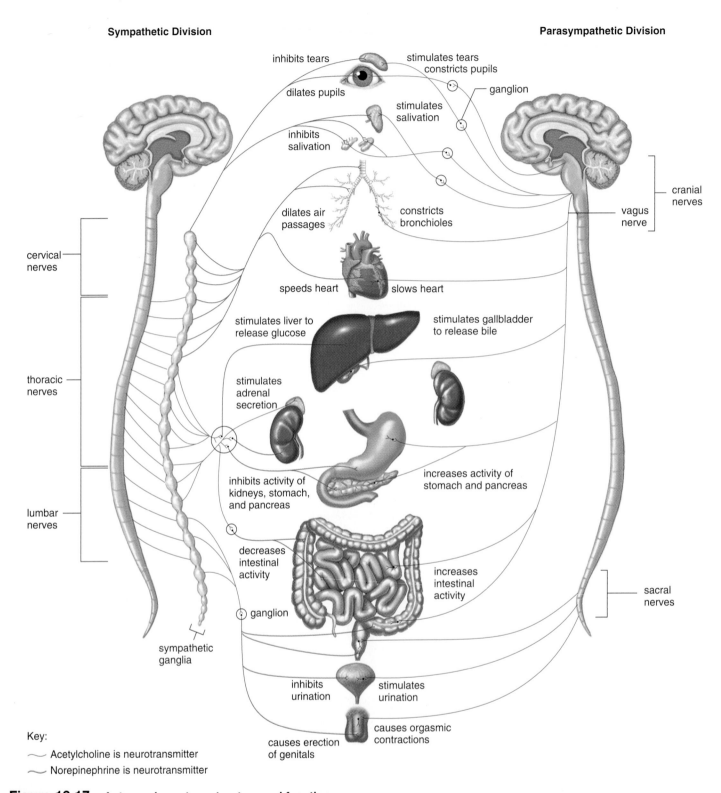

Key:

~ Acetylcholine is neurotransmitter

~ Norepinephrine is neurotransmitter

Figure 13.17   Autonomic system structure and function.
Sympathetic preganglionic fibers *(left)* arise from the cervical, thoracic, and lumbar portions of the spinal cord; parasympathetic preganglionic fibers *(right)* arise from the cranial and sacral portions of the spinal cord. Each system innervates the same organs but has contrary effects.

Autonomic System

The **autonomic system** of the PNS regulates the activity of cardiac and smooth muscle and glands. The system is divided into the sympathetic and parasympathetic divisions (Fig. 13.17). These two divisions have several features in common: (1) They function automatically and usually in an involuntary manner; (2) they innervate all internal organs; and (3) they utilize two neurons and one ganglion for each impulse. The first neuron has a cell body within the CNS and a preganglionic fiber. The second neuron has a cell body within the ganglion and a postganglionic fiber.

Reflex actions, such as those that regulate the blood pressure and breathing rate, are especially important to the maintenance of homeostasis. These reflexes begin when the sensory neurons in contact with internal organs send information to the CNS. They are completed by motor neurons within the autonomic system.

Sympathetic Division

Most preganglionic fibers of the **sympathetic division** arise from the middle, or thoracic-lumbar, portion of the spinal cord and almost immediately terminate in ganglia that lie near the cord. Therefore, in this division, the preganglionic fiber is short, but the postganglionic fiber that makes contact with an organ is long.

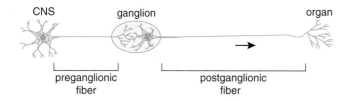

The sympathetic division is especially important during emergency situations when you might be required to fight or take flight. It accelerates the heartbeat and dilates the bronchi; active muscles, after all, require a ready supply of glucose and oxygen. On the other hand, the sympathetic division inhibits the digestive tract—digestion is not an immediate necessity if you are under attack. The neurotransmitter released by the postganglionic axon is primarily norepinephrine (NE). The structure of NE is like that of epinephrine (adrenaline), an adrenal medulla hormone that usually increases heart rate and contractility.

The sympathetic division brings about those responses we associate with "fight or flight."

Parasympathetic Division

The **parasympathetic division** includes a few cranial nerves (e.g., the vagus nerve) as well as fibers that arise from the sacral (bottom) portion of the spinal cord. Therefore, this division is often referred to as the craniosacral portion of the autonomic system. In the parasympathetic division, the preganglionic fiber is long, and the postganglionic fiber is short because the ganglia lie near or within the organ.

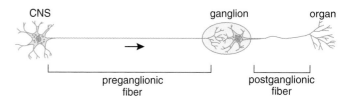

The parasympathetic division, sometimes called the housekeeper division, promotes all the internal responses we associate with a relaxed state; for example, it causes the pupil of the eye to contract, promotes digestion of food, and retards the heartbeat. The neurotransmitter utilized by the parasympathetic division is acetylcholine (ACh).

Table 13.1 summarizes the features and functions of the motor divisions of the somatic and autonomic systems.

The parasympathetic division brings about the responses we associate with a relaxed state.

Table 13.1	Comparison of Somatic Motor and Autonomic Motor Pathways		
	Somatic Motor Pathway	**Autonomic Motor Pathways**	
		Sympathetic	**Parasympathetic**
Type of control	Voluntary/involuntary	Involuntary	Involuntary
Number of neurons per message	One	Two (preganglionic shorter than postganglionic)	Two (preganglionic longer than postganglionic)
Location of motor fiber	Most cranial nerves and all spinal nerves	Thoracolumbar spinal nerves	Cranial (e.g., vagus) and sacral spinal nerves
Neurotransmitter	Acetylcholine	Norepinephrine	Acetylcholine
Effectors	Skeletal muscles	Smooth and cardiac muscle, glands	Smooth and cardiac muscle, glands

Figure 13.18   Drug actions at a synapse.

A drug can affect a neurotransmitter in these ways: (**a**) cause leakage out of a synaptic vesicle into the axon bulb; (**b**) prevent release of the neurotransmitter into the synaptic cleft; (**c**) promote release of the neurotransmitter into the synaptic cleft; (**d**) prevent reuptake by the presynaptic membrane; (**e**) block the enzyme that causes breakdown of the neurotransmitter; or (**f**) bind to a receptor, mimicking the action or preventing the uptake of a neurotransmitter.

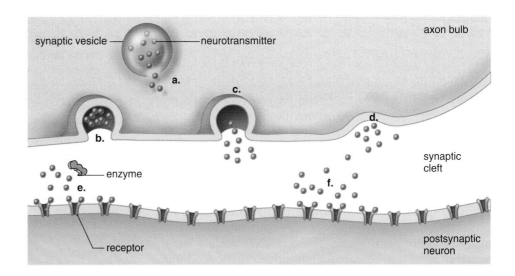

13.5 Drug Abuse

A wide variety of drugs affect the nervous system and can alter the mood and/or emotional state. Such drugs have two general effects: (1) they impact the limbic system, and (2) they either promote or decrease the action of a particular neurotransmitter (Fig. 13.18). Stimulants are drugs that increase the likelihood of neuron excitation, and depressants decrease the likelihood of excitation. Increasingly, researchers believe that dopamine, among other neurotransmitters in the brain, is responsible for mood. Cocaine is known to potentiate the effects of dopamine by interfering with its uptake from synaptic clefts. Many of the new medications developed to counter drug dependence and mental illness affect the release, reception, or breakdown of dopamine.

Drug abuse is apparent when a person takes a drug at a dose level and under circumstances that increase the potential for a harmful effect (Fig. 13.19). Drug abusers are apt to display a psychological and/or physical dependence on the drug. Dependence has developed when the person spends much time thinking about the drug or arranging to get it, and often takes more of the drug than was intended. With physical dependence, formerly called "addiction," the person has become tolerant to the drug—that is, more of the drug is needed to get the same effect, and withdrawal symptoms occur when he or she stops taking the drug.

Drugs that affect the nervous system can cause physical dependence and withdrawal symptoms.

Alcohol

It is possible that alcohol influences the action of GABA, an inhibiting transmitter, or glutamate, an excitatory neurotransmitter. Once imbibed, alcohol is primarily metabolized in the liver, where it disrupts the normal workings of this organ so that fats cannot be broken down. Fat accumulation, the first stage of liver deterioration, begins after only a single night of heavy drinking. If heavy drinking continues, fibrous scar tissue appears during a second stage of deterioration. If heavy drinking stops, the liver can still recover and become normal once again. If not, the final and irrevocable stage, cirrhosis of the liver, occurs: liver cells die, harden, and turn orange (cirrhosis means orange).

Alcohol is used by the body as an energy source, but it lacks the vitamins, minerals, essential amino acids, and fatty acid the body needs to stay healthy. Many alcoholics are undernourished and prone to illness for this reason.

The surgeon general recommends that pregnant women drink no alcohol at all. Alcohol crosses the placenta freely and causes fetal alcohol syndrome, which is characterized by mental retardation and various physical defects in newborns.

Nicotine

Nicotine, an alkaloid derived from tobacco, is a widely used neurological agent. When a person smokes a cigarette, nicotine is quickly distributed to the central and peripheral nervous systems. In the central nervous system, nicotine causes neurons to release the neurotransmitter dopamine. The excess dopamine has a reinforcing effect that leads to dependence on the drug. In the peripheral nervous system, nicotine stimulates the same postsynaptic receptors as acetylcholine and leads to increased activity of the skeletal muscles. It also increases the heart rate and blood pressure, as well as digestive tract mobility.

Many cigarette and cigar smokers find it difficult to give up the habit because nicotine induces both physiological and psychological dependence. Withdrawal symptoms include headache, stomach pain, irritability, and insomnia. Cigarette smoking in young women who are sexually active is most unfortunate because if they become pregnant, nicotine, like other psychoactive drugs, adversely affects a developing embryo and fetus.

Cocaine

Cocaine is an alkaloid derived from the shrub *Erythroxylon coca*. It is sold in powder form and as crack, a more potent extract. Cocaine prevents the synaptic uptake of dopamine, and this causes the user to experience a rush sensation. The epinephrine-like effects of dopamine account for the state of arousal that lasts for several minutes after the rush experience.

A cocaine binge can go on for days, after which the individual suffers a crash. During the binge period, the user is hyperactive and has little desire for food or sleep but has an increased sex drive. During the crash period, the user is fatigued, depressed, and irritable, has memory and concentration problems, and displays no interest in sex. Indeed, men are often impotent.

Cocaine causes extreme physical dependence. With continued cocaine use, the body begins to make less dopamine to compensate for a seemingly excess supply. The user, therefore, experiences tolerance, withdrawal symptoms, and an intense craving for cocaine. These are indications that the person is highly dependent upon the drug. Overdosing on cocaine can cause seizures and cardiac and respiratory arrest. It is possible that long-term cocaine abuse causes brain damage. Babies born to addicts suffer withdrawal symptoms and may suffer neurological and developmental problems.

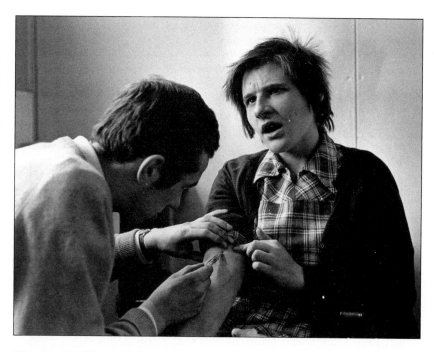

Figure 13.19   **Drug use.**
Blood-borne diseases such as AIDS and hepatitis B pass from one drug abuser to another when they share needles.

Heroin

Heroin is derived from morphine, an alkaloid of opium. Once it is injected into a vein, a feeling of euphoria, along with relief of any pain, occurs within 3 to 6 minutes. Side effects can include nausea, vomiting, dysphoria, and respiratory and circulatory depression.

Heroin binds to receptors meant for the endorphins, the special neurotransmitters that kill pain and produce a feeling of tranquility. With time, the body's production of endorphins decreases. Tolerance develops so that the user needs to take more of the drug just to prevent withdrawal symptoms. The euphoria originally experienced upon injection is no longer felt.

Heroin withdrawal symptoms include perspiration, dilation of pupils, tremors, restlessness, abdominal cramps, gooseflesh, vomiting, and increase in systolic blood pressure and respiratory rate. People who are excessively dependent may experience convulsions, respiratory failure, and death. Infants born to women who are physically dependent also experience these withdrawal symptoms.

Marijuana

The dried flowering tops, leaves, and stems of the Indian hemp plant *Cannabis sativa* contain and are covered by a resin that is rich in THC (tetrahydrocannabinol). The names *cannabis* and *marijuana* apply to either the plant or THC. Usually marijuana is smoked in a cigarette form called a "joint."

The occasional marijuana user reports experiencing a mild euphoria along with alterations in vision and judgment, which result in distortions of space and time. Motor incoordination, including the inability to speak coherently, takes place. Heavy use can result in hallucinations, anxiety, depression, rapid flow of ideas, body image distortions, paranoid reactions, and similar psychotic symptoms. The terms cannabis psychosis and cannabis delirium refer to such reactions. Craving and difficulty in stopping usage can occur as a result of regular use.

Recently, researchers have found that marijuana binds to a receptor for anandamide, a normal molecule in the body. Some researchers believe that long-term marijuana use leads to brain impairment. Fetal cannabis syndrome, which resembles fetal alcohol syndrome, has been reported. Some psychologists believe that marijuana use among adolescents is a way to avoid dealing with the personal problems that often develop during that stage of life.

Neurological drugs either potentiate or dampen the effect of the body's neurotransmitters.

13.6 Homeostasis ⚖

The nervous system (together with the endocrine system) coordinates the functioning of the other systems in the body. The governance of internal organs and the regulation of the composition of blood and tissue fluid usually take place below the level of consciousness. Subconscious control is dependent on reflex actions that involve the hypothalamus and the medulla oblongata. The hypothalamus and the medulla oblongata act through the autonomic nervous system to control such important parameters as the heart rate, the constriction of the blood vessels, and the breathing rate. The cardiovascular system works harder to carry oxygen to the skeletal muscles when we are in a "fight or flight" mode than when we are in a relaxed state.

The illustration on page 267 tells how the nervous system works with other systems in the body to maintain homeostasis. The hypothalamus works closely with the endocrine system and even produces the hormone ADH, which causes the kidneys to reabsorb water. The kidneys are under hormonal control when they help regulate blood volume and pressure.

You might think that voluntary movements don't play a role in homeostasis, but actually we usually modify our behavior to stay in as moderate an environment as possible. Otherwise, we are testing the ability of the nervous system to maintain homeostasis despite extreme conditions.

Degenerative Nervous System Diseases

Homeostasis is not possible when degenerative nervous system diseases occur.

Alzheimer Disease

Alzheimer disease (AD) is characterized by a gradual loss of reason that begins with memory lapses and ends with the inability to perform any type of daily activity. Signs of mental disturbance eventually appear, and patients gradually become bedridden and die of a complication, such as pneumonia.

In AD patients, abnormal neurons are present throughout the brain but especially in the hippocampus and amygdala. These neurons have two abnormalities: (1) plaques, containing a protein called beta amyloid, envelop the axon, and (2) neurofibrillary tangles are in the axons and surround the nucleus (Fig. 13.20). Researchers are racing furiously to discover the cause of these abnormalities. Several genes that predispose a person to AD have been identified. One of them, called $APOE_4$, is found in 65% of persons with AD. But it is unknown why inheritance of this gene leads to AD neurons.

Researchers do know that plasma membrane enzymes called secretases snip beta amyloid from a larger molecule even in healthy cells. People with AD produce too much beta amyloid, particularly one very sticky type. The beta

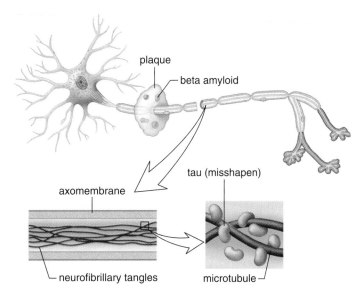

Figure 13.20   Alzheimer disease.
Some of the neurons of AD patients have neurofibrillary tangles and beta amyloid plaques. AD neurons are present throughout the brain but concentrated in the hippocampus and amygdala.

amyloid plaques grow so dense that they trigger an inflammatory reaction that ends in neuron death.

Some researchers believe that the protein-synthesizing machinery in AD neurons is faulty. The neurofibrillary tangles arise because a protein called tau no longer has the correct shape. Tau holds microtubules in place so that they support the structure of neurons. When tau changes shape, it grabs onto other tau molecules, and the tangles result.

Although researchers don't have all the answers, clinical trials are being planned. In one trial, patients will be vaccinated with a fragment of beta amyloid, and in another researchers will try to block a type of secretase enzyme. Still other efforts are concentrating on trying to prevent excitotoxicity (see page 259) in the hippocampus.

Parkinson Disease

Parkinson disease is characterized by a wide-eyed, unblinking expression, involuntary tremors of the fingers and thumbs, muscular rigidity, and a shuffling gait. In Parkinson patients, the basal nuclei (see page 256) are overactive because of the degeneration of dopamine-releasing neurons in the brain. Without dopamine, which is an inhibitory neurotransmitter, the basal nuclei send out excess excitatory signals that result in the symptoms of Parkinson disease.

Unfortunately, it is not possible to give Parkinson patients dopamine because of the impermeability of the capillaries serving the brain. However, symptoms can be alleviated by giving patients L-dopa, a chemical that can be changed to dopamine in the body until too few cells are left to do the job. Then patients must turn to a number of controversial surgical procedures.

Human Systems Work Together

Integumentary System

Brain controls nerves that regulate size of cutaneous blood vessels, activate sweat glands and arrector pili muscles.

Skin protects nerves, helps regulate body temperature; skin receptors send sensory input to brain.

Skeletal System

Receptors send sensory input from bones and joints to brain.

Bones protect sense organs, brain, and spinal cord; store Ca^{2+} for nerve function.

Muscular System

Brain controls nerves that innervate muscles; receptors send sensory input from muscles to brain.

Muscle contraction moves eyes, permits speech, and creates facial expressions.

Endocrine System

Hypothalamus is part of endocrine system; nerves innervate certain glands of secretion.

Sex hormones affect development of brain.

Cardiovascular System

Brain controls nerves that regulate the heart and dilation of blood vessels.

Blood vessels deliver nutrients and oxygen to neurons, carry away wastes.

How the Nervous System works with other body systems

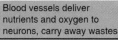

Lymphatic System/Immunity

Microglia engulf and destroy pathogens.

Lymphatic vessels pick up excess tissue fluid; immune system protects against infections of nerves.

Respiratory System

Respiratory centers in brain regulate breathing rate.

Lungs provide oxygen for neurons and rid the body of carbon dioxide produced by neurons.

Digestive System

Brain controls nerves that innervate smooth muscle and permit digestive tract movements.

Digestive tract provides nutrients for growth, maintenance, and repair of neurons and neuroglial cells.

Urinary System

Brain controls nerves that innervate muscles that permit urination.

Kidneys maintain blood levels of Na^+, K^+, and Ca^{2+}, which are needed for nerve conduction.

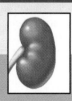

Reproductive System

Brain controls onset of puberty; nerves are involved in erection of penis and clitoris, contraction of ducts that carry gametes, and contraction of uterus.

Sex hormones masculinize or feminize the brain, exert feedback control over the hypothalamus, and influence sexual behavior.

Summarizing the Concepts

13.1 Nervous Tissue

Nervous tissue contains neurons, which are the cells that transmit nerve impulses, and neuroglia, which service neurons. Sensory neurons take information from sensory receptors to the CNS; interneurons occur within the CNS; and motor neurons take information from the CNS to effectors (muscles or glands). A neuron is composed of dendrites, a cell body, and an axon. Long axons are covered by a myelin sheath.

When an axon is not conducting a nerve impulse, the inside of the axon is negative (−65 mV) compared to the outside. The sodium-potassium pump actively transports Na^+ out of an axon and K^+ to the inside of an axon. The resting potential is due to the leakage of K^+ to the outside of the neuron. When an axon is conducting a nerve impulse (action potential), Na^+ first moves into the axoplasm, and then K^+ moves out of the axoplasm.

Transmission of a nerve impulse from one neuron to another takes place when a neurotransmitter molecule is released into a synaptic cleft. The binding of the neurotransmitter to receptors in the postsynaptic membrane causes excitation or inhibition. Integration is the summing of excitatory and inhibitory signals. Neurotransmitter molecules are removed from the cleft by enzymatic breakdown or by reabsorption.

13.2 The Central Nervous System

The CNS consists of the spinal cord and brain, which are both protected by bone. The CNS receives and integrates sensory input and formulates motor output. The gray matter of the spinal cord contains neuron cell bodies; the white matter consists of myelinated axons that occur in bundles called tracts. The spinal cord sends sensory information to the brain, receives motor output from the brain, and carries out reflex actions. Because tracts cross over, the left side of the brain controls the right side of the body, and vice versa.

In the brain, the cerebrum has two cerebral hemispheres connected by the corpus callosum. Sensation, reasoning, learning and memory, and language and speech take place in the cerebrum. Each cerebral hemisphere contains a frontal, parietal, occipital, and temporal lobe. The cerebral cortex is a thin layer of gray matter covering the cerebrum. The primary motor area in the frontal lobe sends out motor commands to lower brain centers, which pass them on to motor neurons. The primary somatosensory area in the parietal lobe receives sensory information from lower brain centers in communication with sensory neurons. Association areas are located in all the lobes. A visual association area occurs in the occipital lobe; an auditory area occurs in the temporal lobe; and so forth for the other senses. The prefrontal area of the frontal lobe is especially necessary to higher mental functions.

The brain has a number of other portions. The hypothalamus controls homeostasis, and the thalamus specializes in sending sensory input on to the cerebrum. The cerebellum primarily coordinates skeletal muscle contractions. The medulla oblongata and the pons have centers for vital functions such as breathing and the heartbeat.

13.3 The Limbic System and Higher Mental Functions

The limbic system connects portions of the cerebral cortex with the hypothalamus, the thalamus, and basal nuclei. In the limbic system, the hippocampus acts as a conduit for sending information to long-term memory and retrieving it once again. The amygdala adds emotional overtones to memories. On the cellular level, long-term potentiation seems to be required for long-term memory. Unfortunately, long-term potentiation can go awry when neurons become overexcited and die.

Language and speech are dependent upon Broca's area (a motor speech area) and Wernicke's area (a sensory speech area). These two areas communicate with one another, and interestingly, both are located only in the left hemisphere.

13.4 The Peripheral Nervous System

The peripheral nervous system contains only nerves and ganglia. The cerebrum is always involved in voluntary actions, but reflexes are automatic, and some do not require involvement of the brain. In the somatic system, for example, a stimulus causes sensory receptors to generate nerve impulses that are then conducted by sensory fibers to interneurons in the spinal cord. Interneurons signal motor neurons, which conduct nerve impulses to a skeletal muscle that contracts, producing the response to the stimulus.

The autonomic (involuntary) system controls the smooth muscle of the internal organs and glands. The sympathetic division is associated with responses that occur during times of stress, and the parasympathetic system is associated with responses that occur during times of relaxation.

13.5 Drug Abuse

Although neurological drugs are quite varied, each type has been found to either promote or prevent the action of a particular neurotransmitter.

13.6 Homeostasis

Homeostasis is ultimately under the control of the nervous system. The hypothalamus and the medulla oblongata have various centers that control the functioning of internal organs and the internal environment. These parts of the brain send commands to internal organs by way of the autonomic system. The hypothalamus works closely with endocrine glands to control blood composition, and therefore tissue fluid composition.

Studying the Concepts

1. What are the three types of neurons, and what is their relationship to the CNS? With reference to a motor neuron, describe the structure and function of the three parts of a neuron.   246
2. What is the sodium-potassium pump? What is the resting potential, and how is it brought about?   248–49
3. Describe the two parts of an action potential and the changes

that can be associated with each part.   248–49
4. What is a neurotransmitter, where is it stored, how does it function, and how is it destroyed? Name two well-known neurotransmitters.   250–51
5. The central nervous system contains what structures? Describe the structure and function of the spinal cord.   252–53

6. Name the major parts of the brain, and give a function for each.   254–57
7. Name the lobes of the cerebral hemispheres, and describe the function of motor, somatosensory, and association areas.   255–56
8. What is the reticular formation?   257
9. What is the limbic system, and how is it involved in higher mental functions?   257–59
10. Language and speech require what portions of the brain? What is the left brain/right brain hypothesis?   259
11. The peripheral nervous system contains what three types of nerves? What is meant by a mixed nerve?   260
12. What is the somatic system? Trace the path of a reflex arc.   261
13. What is the autonomic system, and what are its two major divisions? Give several similarities and differences between these divisions.   262–63
14. Describe the physiological effects and mode of action of alcohol, nicotine, cocaine, heroin, and marijuana.   264–65
15. What anatomical abnormalities occur in patients with Alzheimer disease and Parkinson disease?   266

Understanding Key Terms

acetylcholine (ACh) 251
acetylcholinesterase (AChE) 251
action potential 248
Alzheimer disease (AD) 266
amygdala 258
association area 256
autonomic system 263
axon 246
axon bulb 251
basal nuclei 256
brain 254
brain stem 256
cell body 246
central nervous system (CNS) 246
cerebellum 256
cerebral cortex 255
cerebral hemisphere 255
cerebrospinal fluid 252
cerebrum 255
cranial nerve 260
dendrite 246
diencephalon 256
dorsal-root ganglion 260
episodic memory 258
ganglion 260
gray matter 252
hippocampus 258
hypothalamus 256
integration 251
interneuron 246
learning 258
limbic system 257
long-term memory 258
long-term potentiation (LTP) 258

medulla oblongata 256
memory 258
meninges (sing., meninx) 252
midbrain 257
motor neuron 246
myelin sheath 247
nerve 260
nerve impulse 248
neuroglia 246
neuron 246
neurotransmitter 251
node of Ranvier 247
norepinephrine (NE) 251
nuclei 257
oscilloscope 248
parasympathetic division 263
Parkinson disease 266
peripheral nervous system (PNS) 246
pons 257
prefrontal area 256
primary motor area 255
primary somatosensory area 255
reflex 261
refractory period 248
resting potential 248
reticular formation 257
Schwann cell 247
semantic memory 258
sensory neuron 246
short-term memory 258
skill memory 258
sodium-potassium pump 248
somatic system 261
spinal cord 252

spinal nerve 260
sympathetic division 263
synapse 251
synaptic cleft 251
thalamus 256

threshold 248
tract 252
ventricle 252
white matter 252

Testing Your Knowledge of the Concepts

In questions 1–4, match the region of the brain to the functions.
a. medulla oblongata
b. thalamus
c. hypothalamus
d. cerebrum

_____ 1. Interprets sensory input and initiates voluntary muscular movements.

_____ 2. Contains centers for regulating the heart rate, breathing, and blood pressure.

_____ 3. Receives sensory information and channels it to the cerebrum.

_____ 4. Works with the endocrine system to control the salt and water balance of the blood.

In questions 5–7, indicate whether the statement is true (T) or false (F).

_____ 5. There is a preponderance of K^+ outside a resting axon and a preponderance of Na^+ inside the axon.

_____ 6. Transmission of signals at a synapse is carried out by neurotransmitter molecules.

_____ 7. The spinal cord is a part of the peripheral nervous system.

In questions 8 and 9, fill in the blanks.

8. The space between the axon of one neuron and the dendrite of another is called a _____.

9. In the peripheral nervous system, a mixed nerve contains _____ nerve fibers and _____ nerve fibers.

10. Label this diagram.

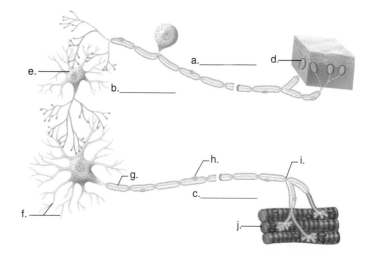

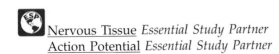

e-Learning Connection

13.1 Nervous Tissue

Nervous Tissue *Essential Study Partner*
Action Potential *Essential Study Partner*

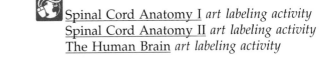

Action Potential *animation activity*
Synapse Structure and Function *art labeling activity*
Development of Membrane Potential *animation activity*
Sodium-Potassium Pump *art quiz*
Myelin Sheath Formation *art quiz*

13.2 The Central Nervous System

Central Nervous System *Essential Study Partner*

Male and Female Brains *reading*

Spinal Cord Anatomy I *art labeling activity*
Spinal Cord Anatomy II *art labeling activity*
The Human Brain *art labeling activity*

13.3 The Limbic System and Higher Mental Functions

Limbic System *art quiz*

13.4 The Peripheral Nervous System

Peripheral Nervous System *Essential Study Partner*

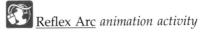

Reflex Arc *animation activity*

13.5 Drug Abuse

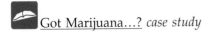

Got Marijuana...? *case study*

13.6 Homeostasis

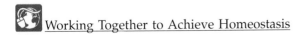

Working Together to Achieve Homeostasis

Chapter Summary

Key Term Flashcards *vocabulary quiz*
Chapter Quiz *objective quiz covering all chapter concepts*

Chapter 14

Senses

Chapter Concepts

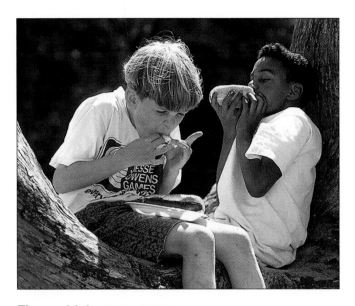

Figure 14.1 **Eating habits.**
Do we like certain foods because we associate their taste and smell with fun?

After a bad day, what's your favorite thing to eat? Chocolate? Mom's homemade pasta? So-called comfort food soothes our spirits and—if only for a minute—makes the world seem okay again.

That's because we learn to link certain tastes and smells with emotion (Fig. 14.1). Sensory cells, like those found in the nose, send messages to the parts of our brain that control emotion and memory. So we remember those freshly baked cookies that once brightened a depressing day. We may then reach for a cookie when we feel down.

Taste and smell are not only essential for experiencing pleasure, they also help ensure our very survival. In nature, animals learn to avoid tempting substances that lead to illness. Likewise, humans shy away from foods linked to negative experiences. A person who comes down with food poisoning after eating at a Japanese restaurant, for example, is unlikely to crave the taste of sushi in the near future. Smell alone can sometimes be protective. Many lives have been saved by a sensitive nose picking up a whiff of smoke or poisonous vapor or detecting spoilage in food.

The senses work together. Once an animal has experienced a noxious substance, it may recognize it by sight alone in the future. Eyes watch out for imminent dangers, and ears hear warning sounds, enabling us to drive in traffic. These same organs allow us to communicate with others and to enrich our lives, as when we see a play or go to the ballet.

Sensory organs at the periphery of the body are the "windows of the brain" because they keep the brain aware of what is going on in the external world. When stimulated, sensory receptors generate nerve impulses that travel to the central nervous system (CNS). Nerve impulses arriving at the cerebral cortex of the brain result in sensation. We see, hear, taste, and smell with our brain, not with our sensory organs.

14.1 Sensory Receptors and Sensations

Sensory receptors are specialized to detect certain types of stimuli (sing., **stimulus**) (Table 14.1). **Exteroceptors** are sensory receptors that detect stimuli from outside the body. Stimuli that result in taste, smell, vision, hearing, and equilibrium all originate outside the body. **Interoceptors** receive stimuli from inside the body. We have had an opportunity in other chapters to mention pressoreceptors that respond to changes in blood pressure, osmoreceptors that detect changes in blood volume, and chemoreceptors that monitor the pH of the blood.

Types of Sensory Receptors

Sensory receptors in humans can be classified into just four categories: chemoreceptors, photoreceptors, mechanoreceptors, and thermoreceptors.

Chemoreceptors respond to chemical substances in the immediate vicinity. As Table 14.1 indicates, taste and smell are dependent on this type of sensory receptor, but certain chemoreceptors in various other organs are sensitive to internal conditions. Chemoreceptors that monitor blood pH are located in the carotid arteries and aorta. If the pH lowers, the breathing rate increases. As more carbon dioxide is expired, the blood pH rises.

Pain receptors (nociceptors) are a type of chemoreceptor. They are naked dendrites that respond to chemicals released by damaged tissues. Pain receptors are protective because they alert us to possible danger. For example, without the pain of appendicitis, we might never seek the medical help needed to avoid a ruptured appendix.

Photoreceptors respond to light energy. Our eyes contain photoreceptors that are sensitive to light rays and thereby provide us with a sense of vision. Stimulation of the photoreceptors known as rod cells results in black and white vision, while stimulation of the photoreceptors known as cone cells results in color vision.

Mechanoreceptors are stimulated by mechanical forces, which most often result in pressure of some sort. When we hear, air-borne sound waves are converted to water-borne pressure waves that can be detected by mechanoreceptors in the inner ear. Similarly, mechanoreceptors are responding to water-borne pressure waves when we detect changes in gravity and motion, helping us keep our balance. These receptors are in the vestibule and semicircular canals of the inner ear, respectively. The sense of touch is dependent on pressure receptors that are sensitive to either strong or slight pressures. Pressoreceptors located in certain arteries detect changes in blood pressure, and stretch receptors in the lungs detect the degree of lung inflation. Proprioceptors, which respond to the stretching of muscle fibers, tendons, joints, and ligaments, make us aware of the position of our limbs.

Thermoreceptors located in the hypothalamus and skin are stimulated by changes in temperature. Those that respond when temperatures rise are called warmth receptors, and those that respond when temperatures lower are called cold receptors.

The sensory receptors are categorized as chemoreceptors, photoreceptors, mechanoreceptors, and thermoreceptors.

Table 14.1	Exteroceptors			
Sensory Receptor	**Stimulus**	**Category**	**Sense**	**Sensory Organ**
Taste cells	Chemicals	Chemoreceptor	Taste	Taste buds
Olfactory cells	Chemicals	Chemoreceptor	Smell	Olfactory epithelium
Rod cells and cone cells in retina	Light rays	Photoreceptor	Vision	Eye
Hair cells in spiral organ	Sound waves	Mechanoreceptor	Hearing	Ear
Hair cells in semi-circular canals	Motion	Mechanoreceptor	Rotational equilibrium	Ear
Hair cells in vestibule	Gravity	Mechanoreceptor	Gravitational equilibrium	Ear

How Sensation Occurs

Sensory receptors respond to environmental stimuli by generating nerve impulses. **Sensation** occurs when nerve impulses arrive at the cerebral cortex of the brain. **Perception** occurs when the cerebral cortex interprets the meaning of sensations.

As we discussed in the previous chapter, sensory receptors are the first element in a reflex arc. We are only aware of a reflex action when sensory information reaches the brain. At that time, the brain integrates this information with other information received from other sensory receptors. After all, if you burn yourself and quickly remove your hand from a hot stove, the brain not only receives information from your skin, it also receive information from your eyes, your nose, and all sorts of sensory receptors.

Some sensory receptors are free nerve endings or encapsulated nerve endings, while others are specialized cells closely associated with neurons. The plasma membrane of a sensory receptor contains receptor proteins that react to the stimulus. For example, the receptor proteins in the plasma membrane of chemoreceptors bind to certain molecules. When this happens, ion channels open and ions flow across the plasma membrane. If the stimulus is sufficient, nerve impulses begin and are carried by a sensory nerve fiber within the PNS to the CNS (Fig. 14.2). The stronger the stimulus, the greater the frequency of nerve impulses. Nerve impulses that reach the spinal cord first are conveyed to the brain by ascending tracts. If nerve impulses finally reach the cerebral cortex, sensation and perception will occur.

Although sensory receptors simply initiate nerve impulses, we have different senses. The brain, as mentioned, is responsible for sensation and perception. Nerve impulses that begin in the optic nerve eventually reach the visual areas of the cerebral cortex and, thereafter, we see objects. Nerve impulses that begin in the auditory nerve eventually reach the auditory areas of the cerebral cortex and, thereafter, we hear sounds. If it were possible to switch these nerves, then stimulation of the eyes would result in hearing! On the other hand, when a blow to the eye stimulates photoreceptors, we "see stars" because nerve impulses from the eyes can only result in sight.

Before sensory receptors initiate nerve impulses, they carry out **integration,** the summing up of signals. One type of integration is called **sensory adaptation,** a decrease in response to a stimulus. We have all had the experience of smelling an odor when we first enter a room and then later not being aware of it at all. Some authorities believe that when sensory adaptation occurs, sensory receptors have stopped sending impulses to the brain. Others believe that the reticular activating system (RAS) has filtered out the ongoing stimuli. You will recall that sensory information is conveyed from the brain stem through the thalamus to the cerebral cortex by the RAS. The thalamus acts as a gatekeeper and only passes on information of immediate importance. Just as we gradually become unaware of particular

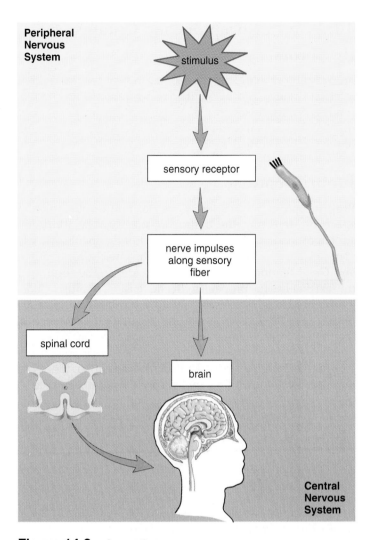

Figure 14.2   Sensation.
The stimulus is received by a sensory receptor, which generates nerve impulses (action potentials). Nerve impulses are conducted to the CNS by sensory nerve fibers within the PNS, and only those impulses that reach the cerebral cortex result in sensation and perception.

environmental stimuli, we can suddenly become aware of stimuli that may have been present for some time. This can be attributed to the workings of the RAS, which has synapses with all the great ascending sensory tracts.

The functioning of our sensory receptors makes a significant contribution to homeostasis. Without sensory input, we would not receive information about our internal and external environment. This information leads to appropriate reflex and voluntary actions to keep the internal environment constant. 🜲

Sensation occurs when nerve impulses reach the cerebral cortex of the brain. Perception, which also occurs in the cerebral cortex, is an interpretation of the meaning of sensations.

14.2 Proprioceptors and Cutaneous Receptors

Proprioceptors in the muscles, joints and tendons, and other internal organs, and also cutaneous receptors in the skin, send nerve impulses to the spinal cord. From there, they travel up the spinal cord in tracts to the somatosensory areas of the cerebral cortex.

Proprioceptors

Proprioceptors help us know the position of our limbs in space by detecting the degree of muscle relaxation, the stretch of tendons, and the movement of ligaments. Muscle spindles act to increase the degree of muscle contraction, and Golgi tendon organs act to decrease it. The result is a muscle that has the proper length and tension, or muscle tone.

Figure 14.3 illustrates the activity of a muscle spindle. In a muscle spindle, sensory nerve endings are wrapped around thin muscle cells within a connective tissue sheath. When the muscle relaxes and undue stretching of the muscle spindle occurs, nerve impulses are generated. The rapidity of the nerve impulses generated by the muscle spindle is proportional to the stretching of a muscle. A reflex action then occurs, which results in contraction of muscle fibers adjoining the muscle spindle. The knee-jerk reflex, which involves muscle spindles, offers an opportunity for physicians to test a reflex action.

The information sent by muscle spindles to the CNS is used to maintain the body's equilibrium and posture despite the force of gravity always acting upon the skeleton and muscles.

Proprioceptors are involved in reflex actions that maintain muscle tone and thereby the body's equilibrium and posture.

Cutaneous Receptors

The skin is composed of two layers: the epidermis and the dermis. In Figure 14.4, the artist has dramatically indicated these two layers by separating the epidermis from the dermis in one location. The epidermis is stratified squamous epithelium in which cells become keratinized as they rise to the surface where they are sloughed off. The dermis is a thick connective tissue layer. The dermis contains sensory receptors for touch, pressure, pain, and temperature. It is a mosaic of these tiny receptors, as you can determine by slowly passing a metal probe over your skin. At certain points, there is a feeling of pressure, and at others, there is a feeling of hot or cold (depending on the probe's temperature).

Figure 14.3 Muscle spindle.
① When a muscle is stretched, a muscle spindle sends sensory nerve impulses to the spinal cord.
② Motor nerve impulses from the spinal cord result in muscle fiber contraction so that muscle tone is maintained.

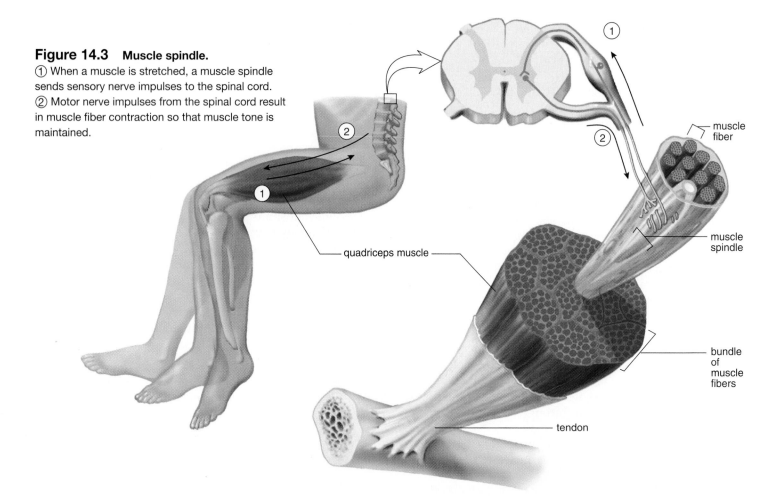

quadriceps muscle

muscle fiber

muscle spindle

bundle of muscle fibers

tendon

Three types of receptors are sensitive to fine touch. The Meissner corpuscles are concentrated in the fingertips, the palms, the lips, the tongue, the nipples, the penis, and the clitoris. Merkel disks are found where the epidermis meets the dermis. A free nerve ending, called a root hair plexus, winds around the base of a hair follicle and fires if the hair is touched.

There are also three different types of pressure receptors: Pacinian corpuscles, Ruffini endings, and Krause end bulbs. Pacinian corpuscles are onion-shaped sensory receptors that lie deep inside the dermis. Ruffini endings and Krause end bulbs are encapsulated by sheaths of connective tissue and contain lacy networks of nerve fibers.

Temperature receptors are simply free nerve endings in the epidermis. Some free nerve endings are responsive to cold; others are responsive to warmth. Cold receptors are far more numerous than warmth receptors, but there are no known structural differences between the two.

Sensory receptors in the human skin are sensitive to touch, pressure, pain, and temperature (warmth and cold).

Pain Receptors

Like the skin, many internal organs have pain receptors, also called nociceptors. Pain receptors are sensitive to extremes in temperature or pressure and to chemicals released by damaged tissues. When inflammation occurs, cells release chemicals that stimulate pain receptors. Aspirin and ibuprofen reduce pain by inhibiting the synthesis of one class of these chemicals.

Sometimes, stimulation of internal pain receptors is felt as pain from the skin as well as internal organs. This is called **referred pain.** Some internal organs have a referred pain relationship with areas located in the skin of the back, groin, and abdomen; pain from the heart is felt in the left shoulder and arm. This most likely happens when nerve impulses from the pain receptors of internal organs travel to the spinal cord and synapse with neurons also receiving impulses from the skin.

Pain receptors, also called nociceptors, are present in the skin and internal organs.

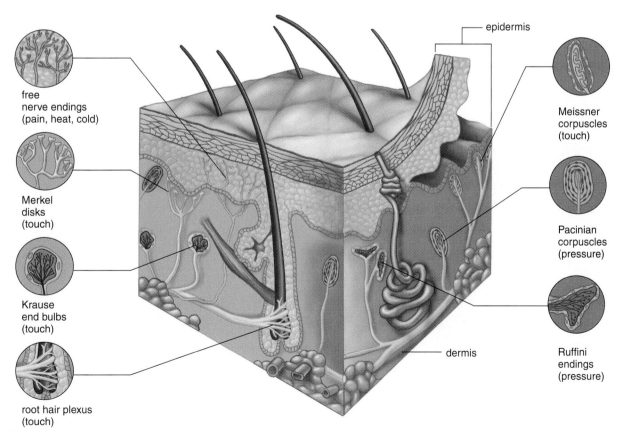

Figure 14.4 Sensory receptors in human skin.
The classical view is that each sensory receptor has the main function shown here. However, investigators report that matters are not so clear-cut. For example, microscopic examination of the skin of the ear shows only free nerve endings (pain receptors), and yet the skin of the ear is sensitive to all sensations. Therefore, it appears that the receptors of the skin are somewhat, but not completely, specialized.

14.3 Chemical Senses

Chemoreceptors in the carotid arteries and in the aorta are primarily sensitive to the pH of the blood. These bodies communicate via sensory nerve fibers with the respiratory center located in the medulla oblongata. When the pH drops, they signal this center, and immediately thereafter the breathing rate increases. The expiration of CO_2 raises the pH of the blood.

Taste and smell are called the chemical senses because their receptors are sensitive to molecules in the food we eat and the air we breathe.

Sense of Taste

The receptors for taste are found in **taste buds** located primarily on the tongue (Fig. 14.5). Many lie along the walls of the papillae, the small elevations on the tongue that are visible to the naked eye. Isolated ones are also present on the hard palate, the pharynx, and the epiglottis.

Taste buds are embedded in tongue epithelium and open at a taste pore. They have supporting cells and a number of elongated taste cells that end in microvilli. The microvilli bear receptor proteins for certain molecules. When these molecules bind to receptor proteins, nerve impulses are generated in associated sensory nerve fibers. These nerve impulses go to the brain, including cortical areas, which interpret them as tastes.

There are four primary types of taste (sweet, sour, salty, bitter), and taste buds for each are concentrated on the tongue in particular regions (Fig. 14.5a). Sweet receptors are most plentiful near the tip of the tongue. Sour receptors occur primarily along the margins of the tongue. Salty receptors are most common on the tip and upper front portion of the tongue. Bitter receptors are located toward the back of the tongue. Actually, the response of taste buds can result in a range of sweet, sour, salty, and bitter tastes. The brain appears to survey the overall pattern of incoming sensory impulses and to take a "weighted average" of their taste messages as the perceived taste.

Taste buds contain taste cells. The microvilli of taste cells have receptor proteins for molecules that cause the brain to distinguish between sweet, sour, salty, and bitter tastes.

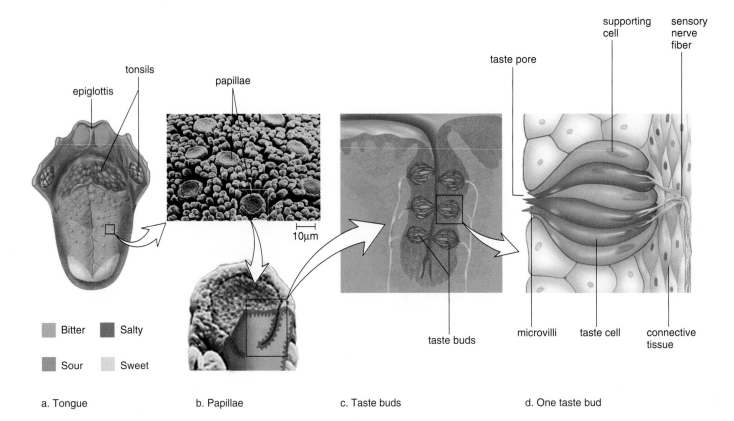

Bitter   Salty

Sour   Sweet

a. Tongue   b. Papillae   c. Taste buds   d. One taste bud

Figure 14.5   Taste buds.
a. Papillae on the tongue contain taste buds that are sensitive to sweet, sour, salty, and bitter tastes in the regions indicated. **b.** Enlargement of papillae. **c.** Taste buds occur along the walls of the papillae. **d.** Taste cells end in microvilli that bear receptor proteins for certain molecules. When molecules bind to the receptor proteins, nerve impulses are generated that go to the brain where the sensation of taste occurs.

Sense of Smell

Our sense of smell is dependent on **olfactory cells** located within olfactory epithelium high in the roof of the nasal cavity (Fig. 14.6). Olfactory cells are modified neurons. Each cell ends in a tuft of about five olfactory cilia, which bear receptor proteins for odor molecules. Each olfactory cell has only one out of 1,000 different types of receptor proteins. Nerve fibers from like olfactory cells lead to the same neuron in the olfactory bulb, an extension of the brain. An odor contains many odor molecules, which activate a characteristic combination of receptor proteins. A rose might stimulate olfactory cells, designated by purple and green, while a daffodil might stimulate a different combination. An odor's signature in the olfactory bulb is determined by which neurons are stimulated. When the neurons communicate this information via the olfactory tract to the olfactory areas of the cerebral cortex, we know we have smelled a rose or a daffodil.

Have you ever noticed that a certain aroma brings to mind a vivid memory of a person or place? A person's perfume may remind you of someone else, or the smell of boxwood may remind you of your grandfather's farm. Look again at Figure 13.9 and notice that the olfactory bulbs have direct connections with the limbic system and its centers for emotions and memory. One investigator showed that when subjects smelled an orange while viewing a painting, they not only remembered the painting, but they had many deep feelings about it.

Actually, the sense of taste and the sense of smell work together to create a combined effect when interpreted by the cerebral cortex. For example, when you have a cold, you think food has lost its taste, but most likely you have lost the ability to sense its smell. This method works in reverse also. When you smell something, some of the molecules move from the nose down into the mouth region and stimulate the taste buds there. Therefore, part of what we refer to as smell may in fact be taste.

Olfactory epithelium contains olfactory cells. The cilia of olfactory cells have receptor proteins for odor molecules that cause the brain to distinguish odors.

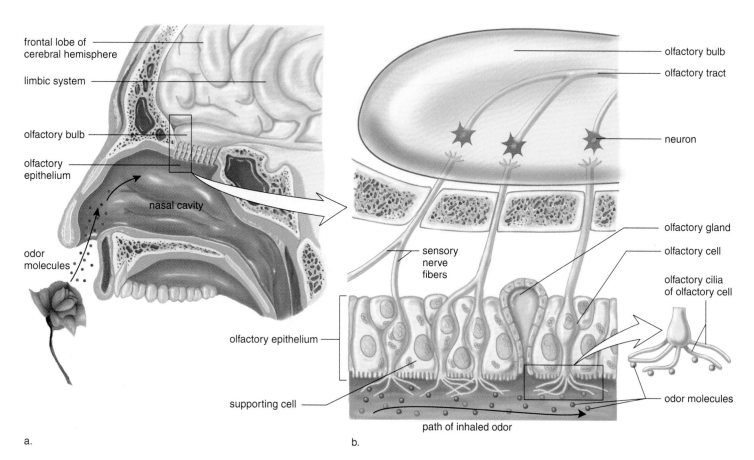

a. b.

Figure 14.6 Olfactory cell location and anatomy.

a. The olfactory epithelium in humans is located high in the nasal cavity. **b.** Olfactory cells end in cilia that bear receptor proteins for specific odor molecules. The cilia of each olfactory cell can bind to only one type of odor molecule (signified here by color). If a rose causes olfactory cells sensitive to "purple" and "green" odor molecules to be stimulated, then neurons designated by purple and green in the olfactory bulb are activated. The primary olfactory area of the cerebral cortex interprets the pattern of neurons stimulated as the scent of a rose.

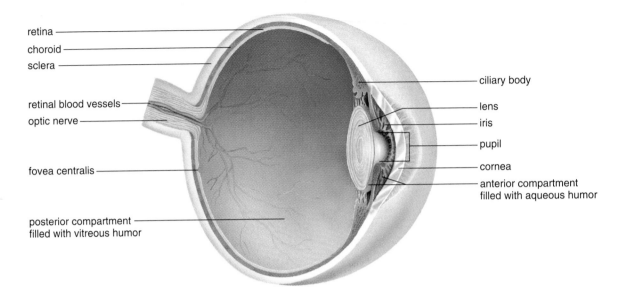

Figure 14.7   Anatomy of the human eye.
Notice that the sclera, the outer layer of the eye, becomes the cornea and that the choroid, the middle layer, is continuous with the ciliary body and the iris. The retina, the inner layer, contains the photoreceptors for vision; the fovea centralis is the region where vision is most acute.

14.4 Sense of Vision

Vision requires the work of the eyes and the brain. As we shall see, much processing of stimuli occurs in the eyes before nerve impulses are sent to the brain. Still, researchers estimate that at least a third of the cerebral cortex takes part in processing visual information.

Anatomy of the Eye

The eyeball, which is an elongated sphere about 2.5 cm in diameter, has three layers, or coats: the sclera, the choroid, and the retina (Fig. 14.7 and Table 14.2). The outer layer, the **sclera,** is white and fibrous except for the transparent **cornea** which is made of transparent collagen fibers. The cornea is the window of the eye.

The middle, thin, darkly pigmented layer, the **choroid,** is vascular and absorbs stray light rays that photoreceptors have not absorbed. Toward the front, the choroid becomes the donut-shaped **iris.** The iris regulates the size of the **pupil,** a hole in the center of the iris through which light enters the eyeball. The color of the iris (and therefore the color of your eyes) correlates with its pigmentation. Heavily pigmented eyes are brown while lightly pigmented eyes are green or blue. Behind the iris, the choroid thickens and forms the circular ciliary body. The **ciliary body** contains the ciliary muscle, which controls the shape of the lens for near and far vision.

The **lens,** attached to the ciliary body by ligaments, divides the eye into two compartments; the one in front of

the lens is the anterior compartment, and the one behind the lens is the posterior compartment. The anterior compartment is filled with a clear, watery fluid called the **aqueous humor.** A small amount of aqueous humor is continually produced each day. Normally, it leaves the anterior compartment by way of tiny ducts. When a person has **glaucoma,** these drainage ducts are blocked, and aqueous humor builds up. If glaucoma is not treated, the resulting

Table 14.2	Functions of the Parts of the Eye
Part	**Function**
Sclera	Protects and supports eyeball
Cornea	Refracts light rays
Choroid	Absorbs stray light
Retina	Contains sensory receptors for sight
Rods	Make black-and-white vision possible
Cones	Make color vision possible
Fovea centralis	Makes acute vision possible
Lens	Refracts and focuses light rays
Ciliary body	Holds lens in place, accommodation
Iris	Regulates light entrance
Pupil	Admits light
Humors	Transmit light rays and support eyeball
Optic nerve	Transmits impulse to brain

pressure compresses the arteries that serve the nerve fibers of the retina, where photoreceptors are located. The nerve fibers begin to die due to lack of nutrients, and the person becomes partially blind. Eventually, total blindness can result.

The third layer of the eye, the **retina,** is located in the posterior compartment, which is filled with a clear gelatinous material called the **vitreous humor.** The retina contains photoreceptors called rod cells and cone cells. The rods are very sensitive to light, but they do not see color; therefore, at night or in a darkened room we see only shades of gray. The cones, which require bright light, are sensitive to different wavelengths of light, and therefore we have the ability to distinguish colors. The retina has a very special region called the **fovea centralis** where cone cells are densely packed. Light is normally focused on the fovea when we look directly at an object. This is helpful because vision is most acute in the fovea centralis. Sensory fibers from the retina form the **optic nerve,** which takes nerve impulses to the brain.

The eye has three layers: the outer sclera, the middle choroid, and the inner retina. Only the retina contains photoreceptors for light energy.

Focusing

When we look at an object, light rays pass through the pupil and are **focused** on the retina (Fig. 14.8*a*). The image produced is much smaller than the object because light rays are bent (refracted) when they are brought into focus. Focusing starts with the cornea and continues as the rays pass through the lens and the humors. Notice that the image on the retina is inverted (upside down) and reversed from left to right.

The lens provides additional focusing power as **visual accommodation** occurs for close vision. The shape of the lens is controlled by the ciliary muscle within the ciliary body. When we view a distant object, the ciliary muscle is relaxed, causing the suspensory ligaments attached to the ciliary body to be taut; therefore, the lens remains relatively flat (Fig. 14.8*b*). When we view a near object, the ciliary muscle contracts, releasing the tension on the suspensory ligaments, and the lens rounds up due to its natural elasticity (Fig. 14.8*c*). Because close work requires contraction of the ciliary muscle, it very often causes muscle fatigue known as eyestrain.

Usually after the age of 40, the lens loses some of its elasticity and is unable to accommodate. Bifocal lenses may then be necessary for those who already have corrective lenses.

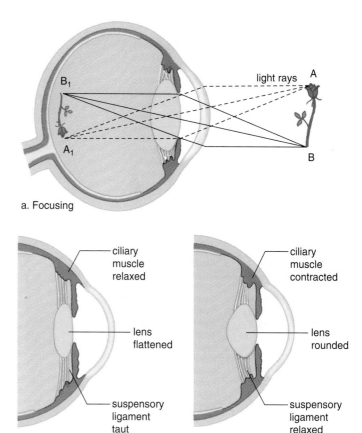

a. Focusing

b. Focusing on distant object   c. Focusing on near object

Figure 14.8 Focusing.
a. Light rays from each point on an object are bent by the cornea and the lens in such a way that an inverted and reversed image of the object forms on the retina. **b.** When focusing on a distant object, the lens is flat because the ciliary muscle is relaxed and the suspensory ligament is taut. **c.** When focusing on a near object, the lens accommodates; it becomes rounded because the ciliary muscle contracts, causing the suspensory ligament to relax.

Also with aging, or possibly exposure to the sun (see the Health Focus on page 288), the lens is subject to **cataracts.** The lens becomes opaque and therefore incapable of transmitting rays of light. Today, the lens is usually surgically replaced with an artificial lens. In the future, it may be possible to restore the original configuration of the proteins making up the lens.

The lens, assisted by the cornea and the humors, focuses images on the retina.

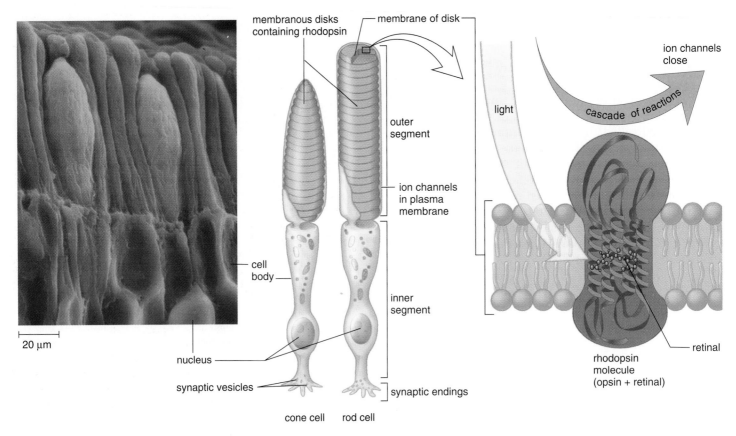

Figure 14.9   Photoreceptors in the eye.
The outer segment of rods and cones contains stacks of membranous disks, which contain visual pigments. In rods, the membrane of each disk contains rhodopsin, a complex molecule containing the protein opsin and the pigment retinal. When rhodopsin absorbs light energy, it splits, releasing opsin, which sets in motion a cascade of reactions that ends when ion channels in the plasma membrane close.

Photoreceptors

Vision begins once light has been focused on the photoreceptors in the retina. Figure 14.9 illustrates the structure of the photoreceptors called **rod cells** and **cone cells.** Both rods and cones have an outer segment joined to an inner segment by a stalk. The outer segment contains many membranous disks, with many pigment molecules embedded in the membrane of these disks. Synaptic vesicles are located at the synaptic endings of the inner segment

The visual pigment in rods is a deep purple pigment called rhodopsin. **Rhodopsin** is a complex molecule made up of the protein opsin and a light-absorbing molecule called **retinal,** which is a derivative of vitamin A. When a rod absorbs light, rhodopsin splits into opsin and retinal, leading to a cascade of reactions and the closure of ion channels in the rod cell's plasma membrane. This stops the release of inhibitory transmitter molecules from the rod's synaptic vesicles and starts the signals that result in nerve impulses going to the brain. Rods are very sensitive to light and therefore are suited to night vision. (Since carrots are

rich in vitamin A, it is true that eating carrots can improve your night vision.) Rod cells are plentiful throughout the entire retina; therefore, they also provide us with peripheral vision and perception of motion.

The cones, on the other hand, are located primarily in the fovea and are activated by bright light. They allow us to detect the fine detail and the color of an object. **Color vision** depends on three different kinds of cones, which contain pigments called the B (blue), G (green), and R (red) pigments. Each pigment is made up of retinal and opsin, but there is a slight difference in the opsin structure of each, which accounts for their individual absorption patterns. Various combinations of cones are believed to be stimulated by in-between shades of color.

The receptors for sight are the rods and the cones. The rods permit vision in dim light at night, and the cones permit vision in bright light needed for color vision.

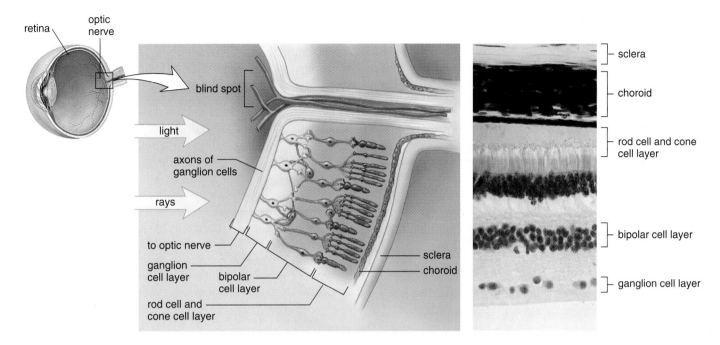

Figure 14.10 **Structure and function of the retina.**
The retina is the inner layer of the eye. Rod cells and cone cells located at the back of the retina synapse with bipolar cells, which synapse with ganglion cells. Integration of signals occurs at these synapses; therefore, much processing occurs in bipolar and ganglion cells. Further, notice that many rod cells share one bipolar cell, but cone cells do not. Certain cone cells synapse with only one ganglion cell. Cone cells, in general, distinguish more detail than do rod cells.

Integration of Visual Signals in the Retina

The retina has three layers of neurons (Fig. 14.10). The layer closest to the choroid contains the rod cells and cone cells; the middle layer contains bipolar cells; and the innermost layer contains ganglion cells, whose sensory fibers become the optic nerve. Only the rod cells and the cone cells are sensitive to light, and therefore light must penetrate to the back of the retina before they are stimulated.

The rod cells and the cone cells synapse with the bipolar cells, which in turn synapse with ganglion cells that initiate nerve impulses. Notice in Figure 14.10 that there are many more rod cells and cone cells than ganglion cells. In fact, the retina has as many as 150 million rod cells and 6 million cone cells but only one million ganglion cells. The sensitivity of cones versus rods is mirrored by how directly they connect to ganglion cells. As many as 100 rods may synapse with the same ganglion cell. No wonder stimulation of rods results in vision that is blurred and indistinct. In contrast, some cone cells in the fovea centralis synapse with only one ganglion cell. This explains why cones, especially in the fovea, provide us with a sharper, more detailed image of an object.

As signals pass to bipolar cells and ganglion cells, integration occurs. Each ganglion cell receives signals from rod cells covering about one square millimeter of retina (about the size of a thumbtack hole). This region is the ganglion cell's receptive field. Some time ago, scientists discovered that a ganglion cell is stimulated when light hits the center of its receptive field and is inhibited when light hits the edges of its receptive field. If all the rod cells in the receptive field are stimulated, the ganglion cell responds in a neutral way—it reacts only weakly or perhaps not at all. This supports the hypothesis that considerable processing occurs in the retina before nerve impulses are sent to the brain. Additional integration occurs in the visual areas of the cerebral cortex.

Synaptic integration and processing begin in the retina before nerve impulses are sent to the brain.

Blind Spot

Figure 14.10 provides an opportunity to point out that there are no rods and cones where the optic nerve exits the retina. Therefore, no vision is possible in this area. You can prove this to yourself by putting a dot to the right of center on a piece of paper. Use your right hand to move the paper slowly toward your right eye while you look straight ahead. The dot will disappear at one point—this is your **"blind spot."**

Integration of Visual Signals in the Brain

Sensory fibers from ganglion cells assemble to form the optic nerves. At the X-shaped optic chiasma, fibers from the right half of each retina converge and continue on together in the right optic tract, and fibers from the left half of each retina converge and continue on together in the left optic tract.

Notice in Figure 14.11, the image is split because the left optic tract carries information about the right portion of the **visual field,** and the right optic tract carries information about the left portion of the visual field.

The optic tracts sweep around the hypothalamus, and most fibers synapse with neurons in nuclei (masses of neuron cell bodies) of the thalamus. Axons from the thalamic nuclei form optic radiations that take nerve impulses to the primary visual areas of the cerebral cortex. Since each primary visual area receives information regarding only half the visual field, these areas must eventually share information to form a unified image. Also, the inverted and reversed image (see also Figure 14.8) must be righted in the brain for us to correctly perceive the visual field.

The most surprising finding has been that the brain has a further way of taking the field apart. Each primary visual area of the cerebral cortex acts like a post office, parceling out information regarding color, form, motion, and possibly other attributes to different portions of the adjoining visual association area. Therefore, the brain has taken the field apart even though we see a unified visual field. The cerebral cortex is believed to rebuild the visual field and give us an understanding of it at the same time.

The visual pathway begins in the retina and passes through the thalamus before reaching the cerebral cortex. The pathway and the visual cortex take the visual field apart, but the visual association areas rebuild it so that we correctly perceive the entire field.

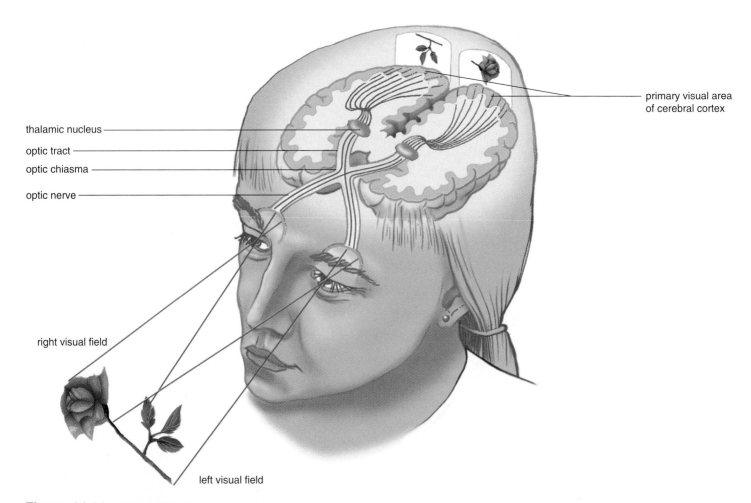

thalamic nucleus

optic tract

optic chiasma

optic nerve

primary visual area of cerebral cortex

right visual field

left visual field

Figure 14.11   Optic chiasma.
Both eyes "see" the entire visual field. Because of the optic chiasma, data from the right half of each retina go to the right visual areas of the cerebral cortex, and data from the left half of the retina go to the left visual areas of the cerebral cortex. These data are combined to allow us to see the entire visual field. Note that the visual pathway to the brain includes the thalamus, which has the ability to filter sensory stimuli.

Abnormalities of the Eye

Color blindness and misshaped eyeballs are two common abnormalities of the eye. More serious abnormalities are discussed in the Health Focus on page 288.

Color Blindness

Complete color blindness is extremely rare. In most instances, a particular type of cone is lacking or deficient in number. The most common mutation is a lack of red or green cones. This abnormality affects 5–8% of the male population. If the eye lacks red cones, the green colors are accentuated, and vice versa.

Distance Vision

The majority of people can see what is designated as a size 20 letter 20 feet away, and so are said to have 20/20 vision. Persons who can see close objects but cannot see the letters from this distance are said to be nearsighted. Nearsighted people can see close objects better than they can see objects at a distance. These individuals have an elongated eyeball, and when they attempt to look at a distant object, the image is brought to focus in front of the retina (Fig. 14.12). They can see close objects because they can adjust the lens to allow the image to focus on the retina, but to see distant objects, these people must wear concave lenses, which diverge the light rays so that the image can be focused on the retina.

Rather than wear glasses or contact lenses, many nearsighted people are now choosing to undergo laser surgery. First, specialists determine how much the cornea needs to be flattened to achieve visual acuity. Controlled by a computer, the laser then removes this amount of the cornea. Most patients achieve at least 20/40 vision, but a few complain of glare and varying visual acuity.

Persons who can easily see the optometrist's chart but cannot see close objects well are farsighted; these individuals can see distant objects better than they can see close objects. They have a shortened eyeball, and when they try to see close objects, the image is focused behind the retina. When the object is distant, the lens can compen-

sate for the short eyeball, but when the object is close, these persons must wear a convex lens to increase the bending of light rays so that the image can be focused on the retina.

When the cornea or lens is uneven, the image is fuzzy. The light rays cannot be evenly focused on the retina. This condition, called **astigmatism,** can be corrected by an unevenly ground lens to compensate for the uneven cornea.

The shape of the eyeball determines the need for corrective lenses.

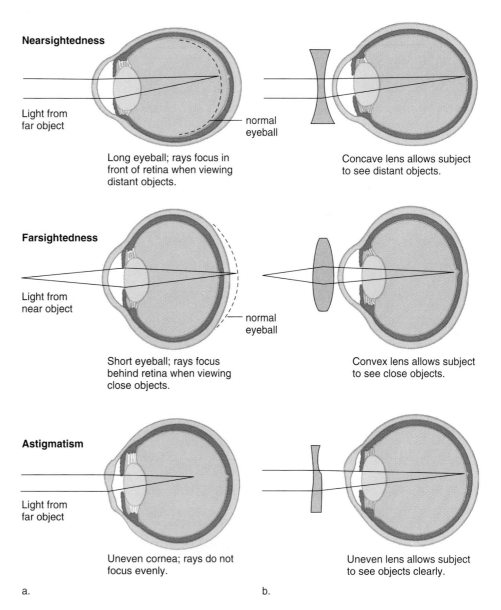

Nearsightedness

Light from far object

normal eyeball

Long eyeball; rays focus in front of retina when viewing distant objects.

Concave lens allows subject to see distant objects.

Farsightedness

Light from near object

normal eyeball

Short eyeball; rays focus behind retina when viewing close objects.

Convex lens allows subject to see close objects.

Astigmatism

Light from far object

Uneven cornea; rays do not focus evenly.

a.

Uneven lens allows subject to see objects clearly.

b.

Figure 14.12 **Common abnormalities of the eye, with possible corrective lenses.**
a. The cornea and the lens function in bringing light rays (lines) to focus, but sometimes they are unable to compensate for the shape of the eyeball or for an irregular curvature of the cornea. **b.** In these instances, corrective lenses can allow the individual to see normally.

14.5 Sense of Hearing

The ear has two sensory functions: hearing and balance (equilibrium). The sensory receptors for both of these are located in the inner ear, and each consists of **hair cells** with stereocilia (long microvilli) that are sensitive to mechanical stimulation. They are mechanoreceptors.

Anatomy of the Ear

Figure 14.13 shows that the ear has three divisions: outer, middle, and inner. The **outer ear** consists of the pinna (external flap) and the auditory canal. The opening of the auditory canal is lined with fine hairs and sweat glands. Modified sweat glands are located in the upper wall of the canal; they secrete earwax, a substance that helps guard the ear against the entrance of foreign materials, such as air pollutants.

The **middle ear** begins at the **tympanic membrane** (eardrum) and ends at a bony wall containing two small openings covered by membranes. These openings are called the **oval window** and the **round window.** Three small bones are found between the tympanic membrane and the oval window. Collectively called the **ossicles,** individually they are the **malleus** (hammer), the **incus** (anvil), and the **stapes** (stirrup) because their shapes resemble these objects. The malleus adheres to the tympanic membrane, and the stapes touches the oval window. An **auditory tube** (eustachian tube), which extends from each middle ear to the nasopharynx, permits equalization of air pressure. Chewing gum, yawning, and swallowing in elevators and airplanes help move air through the auditory tubes upon ascent and descent. As this occurs, we often hear the ears "pop."

Whereas the outer ear and the middle ear contain air, the inner ear is filled with fluid. The **inner ear,** anatomically speaking, has three areas: the **semicircular canals** and the **vestibule** are both concerned with equilibrium; the **cochlea** is concerned with hearing. The cochlea resembles the shell of a snail because it spirals.

Process of Hearing

The process of hearing begins when sound waves enter the auditory canal (Fig. 14.14). Just as ripples travel across the surface of a pond, sound waves travel by the successive vibrations of molecules. Ordinarily, sound waves do not carry much energy, but when a large number of waves strike the tympanic membrane, it moves back and forth (vibrates) ever so slightly. The malleus then takes the pressure from the inner surface of the tympanic membrane and passes it by means of the incus to the stapes in such a way that the pressure is multiplied about 20 times as it moves. The stapes strikes the membrane of the oval window, causing it to vibrate, and in this way, the pressure is passed to the fluid within the cochlea.

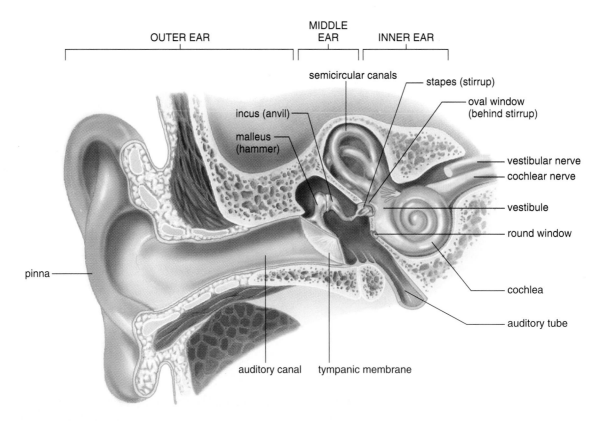

Figure 14.13   Anatomy of the human ear.
In the middle ear, the malleus (hammer), the incus (anvil), and the stapes (stirrup) amplify sound waves. The inner ear contains the mechanoreceptors for equilibrium in the semicircular canals and the vestibule and the mechanoreceptors for hearing in the cochlea.

If the cochlea is unwound and examined in cross section (Fig. 14.14), you can see that it has three canals: the vestibular canal, the **cochlear canal,** and the tympanic canal. The vestibular canal connects with the tympanic canal, which leads to the round window membrane. Along the length of the basilar membrane, which forms the lower wall of the cochlear canal, are little hair cells whose stereocilia are embedded within a gelatinous material called the **tectorial membrane.** The hair cells of the cochlear canal, called the **spiral organ** (organ of Corti), synapse with nerve fibers of the **cochlear nerve** (auditory nerve).

When the stapes strikes the membrane of the oval window, pressure waves move from the vestibular canal to the tympanic canal and across the basilar membrane, and the round window bulges. As the basilar membrane moves up and down, the stereocilia of the hair cells embedded in the tectorial membrane bend. Then nerve impulses begin in the cochlear nerve and travel to the brain stem. When they reach the auditory areas of the cerebral cortex, they are interpreted as a sound.

Each part of the spiral organ is sensitive to different wave frequencies, or pitch. Near the tip, the spiral organ responds to low pitches, such as a tuba, and near the base, it responds to higher pitches, such as a bell or a whistle. The nerve fibers from each region along the length of the spiral organ lead to slightly different areas in the brain. The pitch sensation we experience depends upon which region of the basilar membrane vibrates and which area of the brain is stimulated.

Volume is a function of the amplitude of sound waves. Loud noises cause the fluid of the cochlea to vibrate to a greater degree, and this, in turn, causes the basilar membrane to move up and down to a greater extent. The resulting increased stimulation is interpreted by the brain as volume. It is believed that the tone of a sound is an interpretation of the brain based on the distribution of hair cells stimulated.

The mechanoreceptors for sound are hair cells on the basilar membrane (the spiral organ). When the basilar membrane vibrates, the stereocilia of the hair cells bend, and nerve impulses are transmitted to the brain.

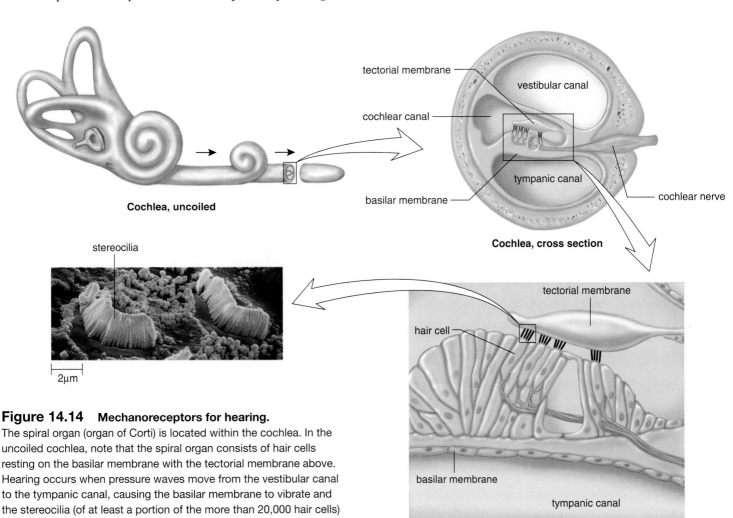

Figure 14.14 **Mechanoreceptors for hearing.**
The spiral organ (organ of Corti) is located within the cochlea. In the uncoiled cochlea, note that the spiral organ consists of hair cells resting on the basilar membrane with the tectorial membrane above. Hearing occurs when pressure waves move from the vestibular canal to the tympanic canal, causing the basilar membrane to vibrate and the stereocilia (of at least a portion of the more than 20,000 hair cells) to bend within the tectorial membrane. Nerve impulses traveling in the cochlear nerve result in hearing.

Cochlea, uncoiled

stereocilia

2μm

tectorial membrane

vestibular canal

cochlear canal

tympanic canal

basilar membrane

cochlear nerve

Cochlea, cross section

tectorial membrane

hair cell

basilar membrane

tympanic canal

Spiral Organ

Visual Focus

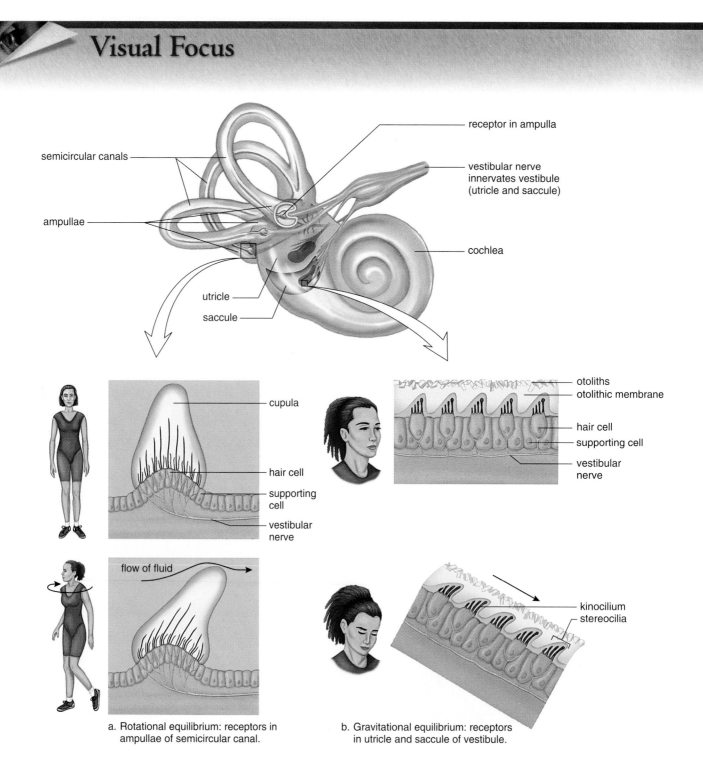

a. Rotational equilibrium: receptors in ampullae of semicircular canal.

b. Gravitational equilibrium: receptors in utricle and saccule of vestibule.

Figure 14.15   Mechanoreceptors for equilibrium.

a. Rotational equilibrium. The ampullae of the semicircular canals contain hair cells with stereocilia embedded in a cupula. When the head rotates, the cupula is displaced, bending the stereocilia. Thereafter, nerve impulses travel in the vestibular nerve to the brain. **b.** Gravitational equilibrium. The utricle and the saccule contain hair cells with stereocilia embedded in an otolithic membrane. When the head bends, otoliths are displaced, causing the membrane to sag and the stereocilia to bend. The rapidity of nerve impulses in the vestibular nerve tells the brain how much the head has moved.

14.6 Sense of Equilibrium

Mechanoreceptors in the semicircular canals detect rotational and/or angular movement of the head **(rotational equilibrium)**, while mechanoreceptors in the utricle and saccule detect movement of the head in the vertical or horizontal planes **(gravitational equilibrium)** (Fig. 14.15).

While these mechanoreceptors help achieve equilibrium, other structures in the body are also involved. For example, we already mentioned that proprioceptors are necessary to maintaining our equilibrium. Vision, if available, is also extremely helpful.

Rotational Equilibrium

Rotational equilibrium involves the semicircular canals, which are arranged so that there is one in each dimension of space. The base of each of the three canals, called the **ampulla,** is slightly enlarged. Little hair cells, whose stereocilia are embedded within a gelatinous material called a cupula, are found within the ampullae. Because there are three semicircular canals, each ampulla responds to head rotation in a different plane of space. As fluid within a semicircular canal flows over and displaces a cupula, the stereocilia of the hair cells bend, and the pattern of impulses carried by the vestibular nerve to the brain changes. Continuous movement of fluid in the semicircular canals causes one form of motion sickness.

Vertigo is dizziness and a sensation of rotation. It is possible to simulate a feeling of vertigo by spinning rapidly and stopping suddenly. When the eyes are rapidly jerked back to a midline position, the person feels like the room is spinning. This shows that the eyes are also involved in our sense of balance.

Gravitational Equilibrium

Gravitational equilibrium depends on the **utricle** and **saccule,** two membranous sacs located in the vestibule. Both of these sacs contain little hair cells, whose stereocilia are embedded within a gelatinous material called an otolithic membrane. Calcium carbonate ($CaCO_3$) granules, or **otoliths,** rest on this membrane. The utricle is especially sensitive to horizontal movements and the bending of the head, while the saccule responds best to vertical (up-down) movements. When the body is still, the otoliths in the utricle and the saccule rest on the otolithic membrane above the hair cells. When the head bends or the body moves in the horizontal and vertical planes, the otoliths are displaced and the otolithic membrane sags, bending the stereocilia of the hair cells beneath. If the stereocilia move toward the kinocilium, the largest stereocilium, nerve impulses in the vestibular nerve increase. If the stereocilia move away from the kinocilium, nerve impulses in the vestibular nerve decrease. These data tell the brain the direction of the movement of the head.

Table 14.3 reviews the functions of the parts of the ear for easy reference.

Movement of a cupula within the semicircular canals contributes to the sense of rotational equilibrium. Movement of the otolithic membrane within the utricle and the saccule accounts for gravitational equilibrium.

Table 14.3	Functions of the Parts of the Ear		
Part	Medium	Function	Mechanoreceptor
OUTER EAR Pinna	Air	Collects sound waves	—
Auditory canal		Filters air	—
MIDDLE EAR Tympanic membrane and ossicles	Air	Amplify sound waves	—
Auditory tube		Equalizes air pressure	—
INNER EAR Semicircular canals	Fluid	Rotational equilibrium	Stereocilia embedded in cupula
Vestibule (contains utricle and saccule)		Gravitational equilibrium	Stereocilia embedded in otolithic membrane
Cochlea (spiral organ)		Hearing	Stereocilia embedded in tectorial membrane

Health Focus

Protecting Vision and Hearing

Preventing a Loss of Vision

The eye is subject to both injuries and disorders. Although flying objects sometimes penetrate the cornea and damage the iris, lens, or retina, careless use of contact lenses is the most common cause of injuries to the eye. However, injuries cause only 4% of all cases of blindness.

The most frequent causes of blindness are retinal disorders, glaucoma, and cataracts, in that order. Retinal disorders are varied. In diabetic retinopathy, which blinds many people between the ages of 20 and 74, capillaries to the retina burst, and blood spills into the vitreous fluid. Careful regulation of blood glucose levels in these patients may be protective. In macular degeneration, the cones are destroyed because thickened choroid vessels no longer function as they should. Glaucoma occurs when the drainage system of the eyes fails, so that fluid builds up and destroys the nerve fibers responsible for peripheral vision. Eye doctors always check for glaucoma, but it is advisable to be aware of the disorder in case it comes on quickly. Those who have experienced acute glaucoma report that the eyeball feels as heavy as a stone. In cataracts, cloudy spots on the lens of the eye eventually pervade the whole lens. The milky, yellow-white lens scatters incoming light and blocks vision.

Regular visits to an eye-care specialist, especially by the elderly, are a necessity in order to catch conditions such as glaucoma early enough to allow effective treatment. It has not been proven that consuming particular foods or vitamin supplements will reduce the risk of cataracts. Even so, it's possible that estrogen replacement therapy may be somewhat protective in postmenopausal women.

Accumulating evidence suggests that both macular degeneration and cataracts, which tend to occur in the elderly, are caused by long-term exposure to the ultraviolet rays of the sun. It is recommended, therefore, that everyone, especially those who live in sunny climates or work outdoors, wear sunglasses that absorb ultraviolet light. Large lenses worn close to the eyes offer further protection. The Sunglass Association of America has devised the following system for categorizing sunglasses:

- Cosmetic lenses absorb at least 70% of UV-B, 20% of UV-A, and 60% of visible light. Such lenses are worn for comfort rather than protection.
- General-purpose lenses absorb at least 95% of UV-B, 60% of UV-A, and 60–92% of visible light. They are good for outdoor activities in temperate regions.
- Special-purpose lenses block at least 99% of UV-B, 60% of UV-A, and 20–97% of visible light. They are good for bright sun combined with sand, snow, or water.

Health-care providers have found an increased incidence of cataracts in heavy cigarette smokers. The risk of cataracts doubles in men who smoke 20 cigarettes or more a day, and in women who smoke 35 cigarettes or more a day. A possible reason is that smoking reduces the delivery of blood, and therefore nutrients, to the lens.

Preventing a Loss of Hearing

Especially when we are young, the middle ear is subject to infections that can lead to hearing impairments if not treated promptly by a physician. The mobility of ossicles decreases with age, and in otosclerosis, new filamentous bone grows over the stirrup, impeding its movement. Surgical treatment is the only remedy for this type of conduction deafness. However, age-associated nerve deafness due to stereocilia damage from exposure to loud noises is preventable. Hospitals are now aware that even the ears of the newborn need to be protected from noise, and are taking steps to make sure neonatal intensive care units and nurseries are as quiet as possible.

In today's society, exposure to the types of noises listed in Table 14A is common. Noise is measured in decibels, and any noise above a level of 80 decibels could result in damage to the hair cells of the organ of Corti. Eventually, the stereocilia and then the hair cells disappear completely (Fig. 14A). If listening to city traffic for extended periods can damage hearing, it stands to reason that frequent attendance at rock concerts, constantly playing a stereo loudly, or using earphones at high volume is also damaging to hearing. The first hint of danger could be temporary hearing loss, a "full" feeling in the ears, muffled hearing, or tinnitus (e.g., ringing in the ears). If you have any of these symptoms, modify your listening habits immediately to prevent further damage. If exposure to noise is unavoidable, specially designed noise-reduction earmuffs are available, and it is also possible to purchase earplugs made from a compressible, spongelike material at the drugstore or sporting-goods store. These earplugs are not the same as those worn for swimming, and they should not be used interchangeably.

Aside from loud music, noisy indoor or outdoor equipment, such as a rug-cleaning machine or a chain saw, is also troublesome. Even motorcycles and recreational vehicles such as snowmobiles and motocross bikes can contribute to a gradual loss of hearing. Exposure to intense sounds of short duration, such as a burst of gunfire, can result in an immediate hearing loss. Hunters may have a significant hearing reduction in the ear opposite the shoulder where the gun is carried. The butt of the rifle offers some protection to the ear nearest the gun when it is shot.

Finally, people need to be aware that some medicines are ototoxic. Anticancer drugs, most notably cisplatin, and certain antibiotics (e.g., streptomycin, kanamycin, and gentamicin) make ears especially susceptible to a hearing loss. Anyone taking such medications needs to be especially careful to protect the ears from loud noises.

Table 14A   Noises That Affect Hearing

Type of Noise	Sound Level (decibels)	Effect
"Boom car," jet engine, shotgun, rock concert	Over 125	Beyond threshold of pain; potential for hearing loss high
Discotheque, "boom box," thunderclap	Over 120	Hearing loss likely
Chain saw, pneumatic drill, jackhammer, symphony orchestra, snowmobile, garbage truck, cement mixer	100–200	Regular exposure of more than one minute risks permanent hearing loss
Farm tractor, newspaper press, subway, motorcycle	90–100	Fifteen minutes of unprotected exposure potentially harmful
Lawn mower, food blender	85–90	Continuous daily exposure for more than eight hours can cause hearing damage
Diesel truck, average city traffic noise	80–85	Annoying; constant exposure may cause hearing damage

Source: National Institute on Deafness and Other Communication Disorders, January 1990, National Institute of Health.

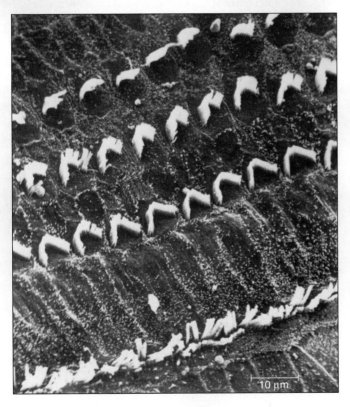

a.

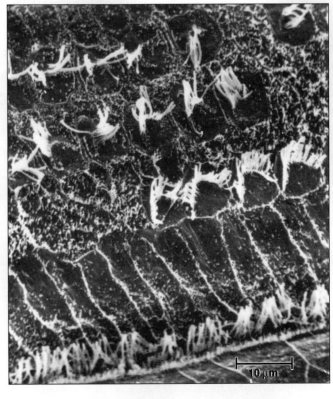

b.

Figure 14A   **The higher the decibel reading, the more likely that a noise will damage hearing.**
a. Normal hair cells in the spiral organ of a guinea pig. **b.** Damaged cells. This damage occurred after 24-hour exposure to a noise level like that at a heavy-metal rock concert (see Table 14A). Hearing is permanently impaired because lost cells will not be replaced, and damaged cells may also die.

Summary

14.1 Sensory Receptors and Sensations
Each type of sensory receptor detects a particular kind of stimulus. When stimulation occurs, sensory receptors initiate nerve impulses that are transmitted to the spinal cord and/or brain. Sensation occurs when nerve impulses reach the cerebral cortex. Perception is an interpretation of the meaning of sensations.

14.2 Proprioceptors and Cutaneous Receptors
Proprioception is illustrated by the action of muscle spindles that are stimulated when muscle fibers stretch. A reflex action, which is illustrated by the knee-reflex, causes the muscle fibers to contract. Proprioception helps maintain equilibrium and posture.

The skin contains sensory receptors, called cutaneous receptors, for touch, pressure, pain, and temperature (warmth and cold). The pain of internal organs is sometimes felt in the skin and is called referred pain.

14.3 Chemical Senses
Taste and smell are due to chemoreceptors that are stimulated by molecules in the environment. The taste buds contain taste cells that communicate with sensory nerve fibers, while the chemoreceptors for smell are neurons.

After molecules bind to plasma membrane receptor proteins on the microvilli of taste cells and the cilia of olfactory cells, nerve impulses eventually reach the cerebral cortex, which determines the taste and odor according to the pattern of chemoreceptors stimulated.

14.4 Sense of Vision
Vision is dependent on the eye, the optic nerves, and the visual areas of the cerebral cortex. The eye has three layers. The outer layer, the sclera, can be seen as the white of the eye; it also becomes the transparent bulge in the front of the eye called the cornea. The middle pigmented layer, called the choroid, absorbs stray light rays. The rod cells (sensory receptors for dim light) and the cone cells (sensory receptors for bright light and color) are located in the retina, the inner layer of the eyeball. The cornea, the humors, and especially the lens bring the light rays to focus on the retina. To see a close object, accommodation occurs as the lens rounds up.

When light strikes rhodopsin within the membranous disks of rod cells, rhodopsin splits into opsin and retinal. A cascade of reactions leads to the closing of ion channels in a rod cell's plasma membrane. Inhibitory transmitter molecules are no longer released, and nerve impulses are carried in the optic nerve to the brain.

Integration occurs in the retina, which is composed of three layers of cells: the rod and cone layer, the bipolar cell layer, and the ganglion cell layer. Integration also occurs in the brain, especially because the visual field is first taken apart by the optic chiasma, and the primary visual area in the cerebral cortex parcels out signals for color, form, and motion to the visual association area.

14.5 Sense of Hearing
Hearing is dependent on the ear, the cochlear nerve, and the auditory areas of the cerebral cortex.

The ear is divided into three parts: outer, middle, and inner. The outer ear consists of the pinna and the auditory canal, which direct sound waves to the middle ear. The middle ear begins with the tympanic membrane and contains the ossicles (malleus, incus, and stapes). The malleus is attached to the tympanic membrane, and the stapes is attached to the oval window, which is covered by a membrane. The inner ear contains the cochlea and the semicircular canals, plus the utricle and the saccule.

Hearing begins when the outer and middle portions of the ear convey and amplify the sound waves that strike the oval window. Its vibrations set up pressure waves within the cochlea, which contains the spiral organ, consisting of hair cells whose stereocilia are embedded within the tectorial membrane. When the stereocilia of the hair cells bend, nerve impulses begin in the cochlear nerve and are carried to the brain.

14.6 Sense of Equilibrium
The ear also contains mechanoreceptors for our sense of equilibrium. Rotational equilibrium is dependent on the stimulation of hair cells within the ampullae of the semicircular canals. Gravitational equilibrium relies on the stimulation of hair cells within the utricle and the saccule.

Studying the Concepts

1. What are the four categories of sensory receptors in the human body? 272
2. Explain sensation, from the reception of stimuli to the passage of nerve impulses to the brain. 273
3. Explain how muscle spindles are involved in proprioception. What are the cutaneous senses? 274–75
4. What causes pain, and what is referred pain? 275
5. Describe the structure of a taste bud, and tell how a taste cell functions. 276
6. Describe the structure and function of the olfactory epithelium. How does the sense of smell come about? 277
7. Describe the anatomy of the eye, and explain focusing and accommodation. 278–79
8. Describe the structure and function of rod cells and cone cells. 280
9. Explain the process of integration in the retina and the brain. 281–82
10. Relate the need for corrective lenses to three possible eyeball shapes. 283
11. Describe the anatomy of the ear and how we hear. 284–85
12. Describe the roles of the semicircular canals, the utricle, and the saccule in equilibrium. 286–87

Testing Your Knowledge of the Concepts

In questions 1-4, match the sense to the descriptions.

a. vision
b. smell
c. hearing
d. cutaneous

_____ 1. A chemical sense that utilizes modified neurons that send nerve impulses directly to the brain.

_____ 2. Much integration occurs in the PNS before nerve impulses are sent to the brain.

_____ 3. Five types of senses associated with the skin.

_____ 4. Sensory receptors are hair cells that communicate with sensory fibers.

In questions 5–7, indicate whether the statement is true (T) or false (F).

_____ 5. The spiral organ is located in the semicircular canals.

_____ 6. The choroid, the middle layer of the eye, becomes the cornea of the eyeball.

_____ 7. The cone cells function in bright light and are sensitive to color.

For questions 8 and 9, fill in the blanks.

8. The _____ and the _____ of the inner ear contain mechanoreceptors for our sense of equilibrium.

9. Because nerve fibers cross at the optic chiasma, the right cerebral hemisphere receives data about the _____ side of the visual field.

10. Label this diagram of an eye.

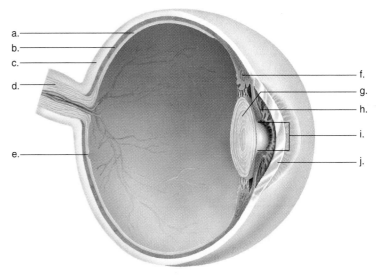

11. Match the terms to these definitions:

a. _____ One of the small bones of the middle ear-malleus, incus, stapes.

b. _____ Type of sensory receptor that responds to chemical stimulation; for example, the receptors for taste and smell.

c. _____ Visual pigment found in rods whose activation by light energy leads to vision.

d. _____ An awareness of a stimulus, occurs only in cerebral cortex.

e. _____ A mechanoreceptor that gives rise to nerve impulses in the cochlear nerve when its stereocilia are bent or tilted.

Understanding Key Terms

ampulla 287
aqueous humor 278
astigmatism 283
auditory tube 284
blind spot 281
cataract 279
chemoreceptor 272
choroid 278
ciliary body 278
cochlea 284
cochlear canal 285
cochlear nerve 285
color vision 280
cone cell 280
cornea 278
exteroceptor 272
focused 279

fovea centralis 279
glaucoma 278
gravitational equilibrium 286
hair cell 284
incus 284
inner ear 284
integration 273
interoceptor 272
iris 278
lens 278
malleus 284
mechanoreceptor 272
middle ear 284
olfactory cell 277
optic nerve 279
ossicle 284
otolith 287

outer ear 284
oval window 284
pain receptor 272
perception 273
photoreceptor 272
proprioceptor 274
pupil 278
referred pain 275
retina 279
retinal 280
rhodopsin 280
rod cell 280
rotational equilibrium 286
round window 284
saccule 287
sclera 278
semicircular canal 284

sensation 273
sensory adaptation 273
sensory receptor 272
spiral organ 285
stapes 284
stimulus 272
taste bud 276
tectorial membrane 285
thermoreceptor 272
tympanic membrane 284
utricle 287
vertigo 287
vestibule 284
visual accommodation 279
visual field 282
vitreous humor 279

e-Learning Connection   **www.mhhe.com/biosci/genbio/maderhuman7/**

14.1 Sensory Receptors and Sensations

 Receptors and Sensations *Essential Study Partner*

 Mixed-Up Senses—Synthesia *reading*

14.3 Chemical Senses

 Sense of Taste *Essential Study Partner*
Sense of Smell *Essential Study Partner*

 Taste *animation activity*
Olfaction *animation activity*

 Smell and Taste Disorders *reading*

14.4 Sense of Vision

 Anatomy of the Human Eye I *art labeling activity*
Anatomy of the Human Eye II *art labeling activity*

14.5 Sense of Hearing

 Sense of Hearing *Essential Study Partner*

 Hearing *animation activity*
Anatomy of the Human Ear *art labeling activity*

14.6 Sense of Equilibrium

 Equilibrium *Essential Study Partner*

 Sense of Balance *animation activity*
Sense of Rotational Acceleration *animation activity*

Chapter Summary

 Key Term Flashcards *vocabulary quiz*
Chapter Quiz *objective quiz covering all chapter concepts*

Chapter *15*

Endocrine System

Chapter Concepts

Figure 15.1   Puberty.
Puberty is a period of rapid physical change when male or female sexual maturation first begins to appear.

Harry stood before the mirror, making sure his appearance would be just right. Tonight was the big junior high dance, and he was taking Mary on his first date. He noticed that he was getting a slight growth of hair on his upper lip and chin, and he couldn't decide whether to shave or let it grow. But the facial hair didn't concern him nearly as much as his voice—the change of pitch, which he could not control, was quite embarrassing. Although some of the kids laughed when this happened, he hoped Mary would not be one of them. Harry even wondered why he cared what Mary thought; just last year he wouldn't even have wanted to go to the dance, never mind go on a date with a girl.

Harry was experiencing normal signs of puberty brought on by an increase in sex hormones. During puberty, sexual organs mature and the secondary sex characteristics appear. As the larynx grows larger, the voice changes more dramatically in boys than in girls. Girls usually undergo a growth spurt before boys, and therefore girls are often taller than boys during early adolescence. Underarm hair appears in both sexes, usually before pubic hair does, but only boys are expected to experience facial hair. Increased activity of oil glands in the skin can cause facial blemishes, or acne. Breasts begin to develop in girls, and menstruation begins. Boys have their first ejaculation.

Sex hormones are produced by the gonads, which are a part of the endocrine system. Along with the nervous system, the endocrine system coordinates the various activities of body parts, including the sexual changes that occur during puberty. The nervous system quickly transmits electrical signals along nerve fibers. The endocrine glands act more slowly because they produce **hormones,** chemical signals which are usually transported in the blood to target organs.

15.1 Endocrine Glands

Endocrine glands are ductless; they secrete their hormones directly into the bloodstream for distribution throughout the body. They can be contrasted with exocrine glands, which have ducts and secrete their products into these ducts for transport to body cavities. For example, the salivary glands send saliva into the mouth by way of the salivary ducts. Figure 15.2 depicts the locations of the endocrine glands in the body, and Table 15.1 lists the hormones they release. Hormones can be categorized as either peptides or steroids as discussed in more depth on page 309.

Like the nervous system, the endocrine system is especially involved in homeostasis—the relative stability of the internal environment. Notice in Table 15.1 that several hormones directly affect the blood glucose, calcium, and sodium levels. Other hormones are involved in the maturation and function of the reproductive organs. In fact, many people are most familiar with the effect of hormones on sexual functions.

The secretion of hormones involved in maintaining homeostasis is usually controlled in two ways: by negative feedback and/or by antagonistic hormonal actions. When controlled by negative feedback, an endocrine gland can be sensitive to either the condition it is regulating or to the blood level of the hormone it is producing. When the blood glucose level rises, the pancreas produces insulin, which causes the liver to store glucose. The stimulus for the production of insulin is thereby inhibited, and the pancreas stops producing insulin. The effect of insulin, however, can also be offset by the production of glucagon by the pancreas. This is regulation by the action of an antagonistic hormone.

Another example of antagonistic hormonal actions occurs when the thyroid lowers the blood calcium level, but the parathyroids raise the blood calcium level. Or consider that insulin lowers the blood glucose level while glucagon raises it. In subsequent sections of this chapter, we will point out other instances in which hormones work opposite to one another and thereby bring about the regulation of a substance in the blood.

The effect of hormones is usually controlled in two ways: (1) negative feedback opposes their release, and (2) antagonistic hormones oppose each other's actions. The end result is maintenance of a bodily substance or function within normal limits.

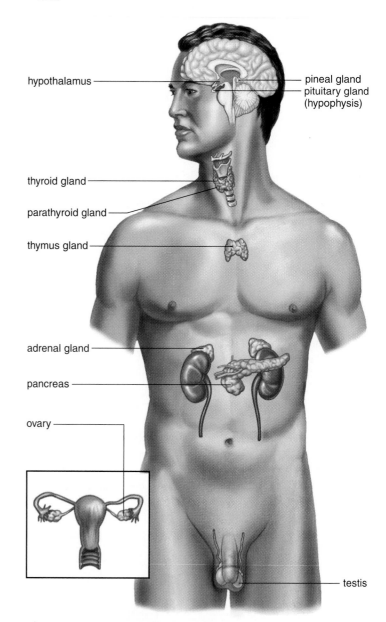

hypothalamus
pineal gland
pituitary gland (hypophysis)
thyroid gland
parathyroid gland
thymus gland
adrenal gland
pancreas
ovary
testis

Figure 15.2   The endocrine system.
Anatomical location of major endocrine glands in the body. The hypothalamus and pituitary gland are located in the brain, the thyroid and parathyroids are located in the neck, while the adrenal glands and pancreas are located in the pelvic cavity. The gonads include the ovaries, located in the pelvic cavity, and the testes, located outside this cavity in the scrotum. Also shown are the pineal gland, located in the brain, and the thymus, which lies ventral to the thorax.

Table 15.1	**Principal Endocrine Glands and Hormones**			
Endocrine Gland	**Hormone Released**	**Chemical Class***	**Target Tissues/Organs**	**Chief Function(s) of Hormone**
Hypothalamus	Hypothalamic-releasing and -inhibiting hormones	Peptide	Anterior pituitary	Regulate anterior pituitary hormones
Posterior pituitary	Antidiuretic (ADH)	Peptide	Kidneys	Stimulates water reabsorption by kidneys
	Oxytocin	Peptide	Uterus, mammary glands	Stimulates uterine muscle contraction; release of milk by mammary glands
Anterior pituitary	Thyroid-stimulating (TSH)	Glycoprotein	Thyroid	Stimulates thyroid
	Adrenocorticotropic (ACTH)	Peptide	Adrenal cortex	Stimulates adrenal cortex
	Gonadotropic (FSH, LH)	Glycoprotein	Gonads	Egg and sperm production; sex hormone production
	Prolactin (PRL)	Protein	Mammary glands	Milk production
	Growth (GH)	Protein	Soft tissues, bones	Cell division, protein synthesis, and bone growth
	Melanocyte-stimulating (MSH)	Peptide	Melanocytes in skin	Unknown function in humans; regulates skin color in lower vertebrates
Thyroid	Thyroxine (T_4) and triiodothyronine (T_3)	Iodinated amino acid	All tissues	Increases metabolic rate; regulates growth and development
	Calcitonin	Peptide	Bones, kidneys, intestine	Lowers blood calcium level
Parathyroids	Parathyroid (PTH)	Peptide	Bones, kidneys, intestine	Raises blood calcium level
Adrenal cortex	Glucocorticoids (cortisol)	Steroid	All tissues	Raise blood glucose level; stimulate breakdown of protein
	Mineralocorticoids (aldosterone)	Steroid	Kidneys	Reabsorb sodium and excrete potassium
	Sex hormones	Steroid	Gonads, skin, muscles, bones	Stimulate reproductive organs and bring about sex characteristics
Adrenal medulla	Epinephrine and norepinephrine	Modified amino acid	Cardiac and other muscles	Emergency situations; raise blood glucose level
Pancreas	Insulin	Protein	Liver, muscles, adipose tissue	Lowers blood glucose level; promotes formation of glycogen
	Glucagon	Protein	Liver, muscles, adipose tissue	Raises blood glucose level
Gonads				
Testes	Androgens (testosterone)	Steroid	Gonads, skin, muscles, bones	Stimulate male secondary sex characteristics
Ovaries	Estrogens and progesterone	Steroid	Gonads, skin, muscles, bones	Stimulate female sex characteristics
Thymus	Thymosins	Peptide	T lymphocytes	Production and maturation of T lymphocytes
Pineal gland	Melatonin	Modified amino acid	Brain	Circadian and circannual rhythms; possibly involved in maturation of sex organs

*See page 309.

15.2 Hypothalamus and Pituitary Gland ⚖

The **hypothalamus** regulates the internal environment through the autonomic system. For example, it helps control heartbeat, body temperature, and water balance. The hypothalamus also controls the glandular secretions of the **pituitary gland.** The pituitary, a small gland about 1 cm in diameter, is connected to the hypothalamus by a stalklike structure. The pituitary has two portions: the posterior pituitary and the anterior pituitary.

Posterior Pituitary ⚖

Neurons in the hypothalamus called neurosecretory cells produce the hormones **antidiuretic hormone (ADH)** and oxytocin (Fig. 15.3, *left*). These hormones pass through axons into the **posterior pituitary** where they are stored in axon endings. Certain neurons in the hypothalamus are sensitive to the water-salt balance of the blood. When these cells determine that the blood is too concentrated, ADH is released from the posterior pituitary. Upon reaching the kidneys, ADH causes water to be reabsorbed. As the blood becomes dilute, ADH is no longer released. This is an example of control by negative feedback because the effect of the hormone (to dilute blood) acts to shut down the release of the hormone. Negative feedback maintains stable conditions and homeostasis.

Inability to produce ADH causes diabetes insipidus (watery urine), in which a person produces copious amounts of urine with a resultant loss of ions from the blood. The condition can be corrected by the administration of ADH.

Oxytocin, the other hormone made in the hypothalamus, causes uterine contraction during childbirth and milk letdown when a baby is nursing. The more the uterus contracts during labor, more nerve impulses reach the hypothalamus, and oxytocin is released. Similarly, the more a baby suckles, more oxytocin is released. In both instances, the release of oxytocin from the posterior pituitary is controlled by **positive feedback**—the stimulus continues to bring about an effect that ever increases in intensity. Positive feedback is not a way to maintain stable conditions and homeostasis.

The posterior pituitary releases ADH and oxytocin, both of which are produced by the hypothalamus.

Anterior Pituitary ⚖

A portal system, consisting of two capillary systems connected by a vein, lies between the hypothalamus and the anterior pituitary (Fig. 15.3, *right*). The hypothalamus controls the anterior pituitary by producing **hypothalamic-releasing**

hormones and **hypothalamic-inhibiting hormones.** For example, there is a thyroid-releasing hormone (TRH) and a thyroid-inhibiting hormone (TIH). TRH stimulates the anterior pituitary to secrete thyroid-stimulating hormone, and TIH inhibits the pituitary from secreting thyroid-stimulating hormone.

Three of the six hormones produced by the **anterior pituitary** (hypophysis) have an effect on other glands: **thyroid-stimulating hormone (TSH)** stimulates the thyroid to produce the thyroid hormones; **adrenocorticotropic hormone (ACTH)** stimulates the adrenal cortex to produce cortisol; and **gonadotropic hormones** (FSH and LH) stimulate the gonads—the testes in males and the ovaries in females—to produce gametes and sex hormones. In each instance, the blood level of the last hormone in the sequence exerts negative feedback control over the secretion of the first two hormones:

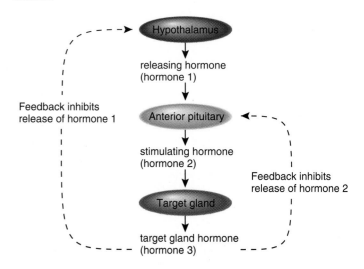

The other three hormones produced by the anterior pituitary do not affect other endocrine glands. **Prolactin (PRL)** is produced in quantity only after childbirth. It causes the mammary glands in the breasts to develop and produce milk. It also plays a role in carbohydrate and fat metabolism.

Melanocyte-stimulating hormone (MSH) causes skin-color changes in many fishes, amphibians, and reptiles that have melanophores, special skin cells that produce color variations. The concentration of this hormone in humans is very low.

Growth hormone (GH), or somatotropic hormone, promotes skeletal and muscular growth. It stimulates the rate at which amino acids enter cells and protein synthesis occurs. It also promotes fat metabolism as opposed to glucose metabolism.

The hypothalamus, the anterior pituitary, and other glands controlled by the anterior pituitary are all involved in self-regulating negative feedback mechanisms that maintain stable conditions.

Visual Focus

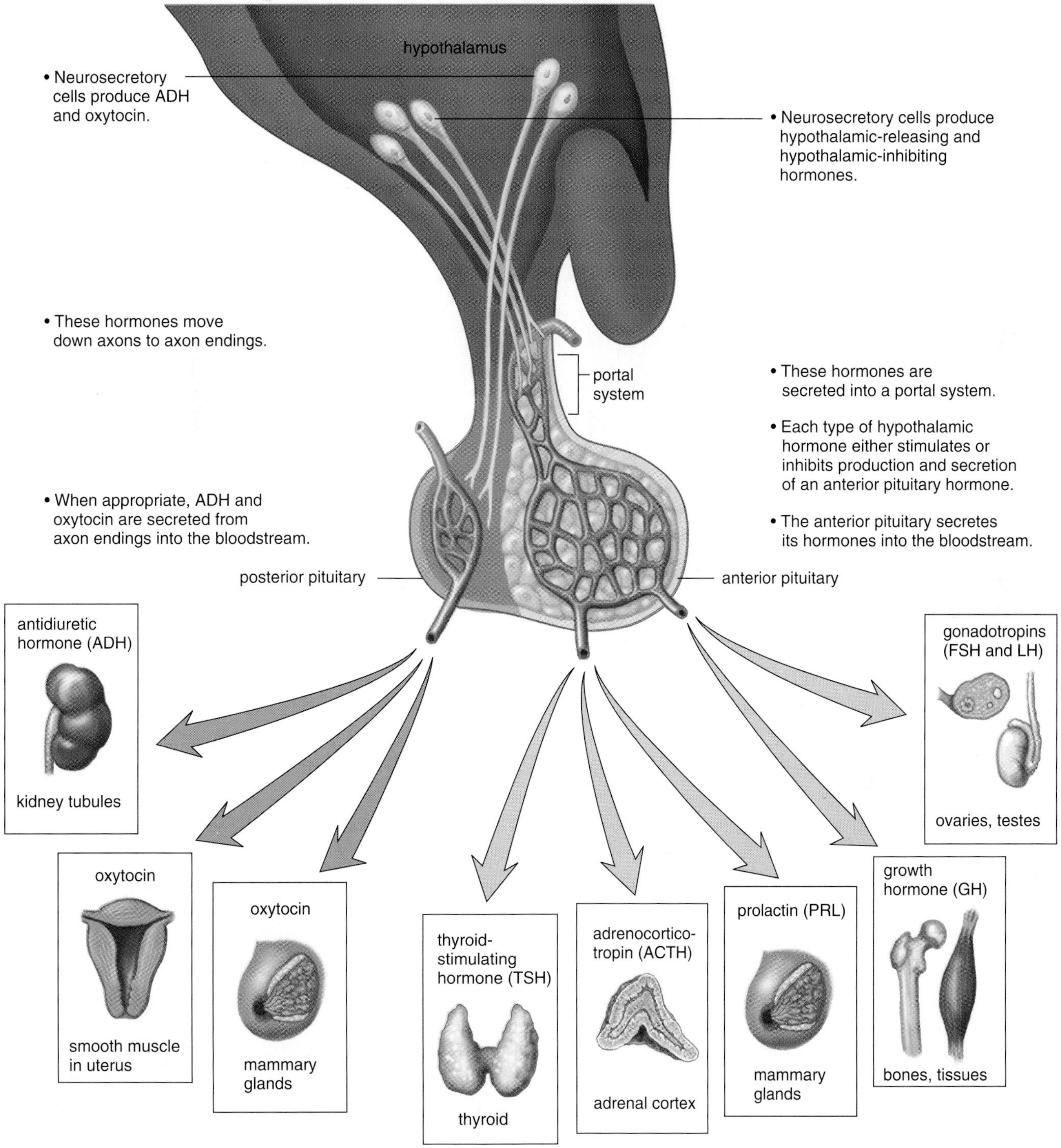

hypothalamus

• Neurosecretory cells produce ADH and oxytocin.

• Neurosecretory cells produce hypothalamic-releasing and hypothalamic-inhibiting hormones.

• These hormones move down axons to axon endings.

portal system

• These hormones are secreted into a portal system.

• Each type of hypothalamic hormone either stimulates or inhibits production and secretion of an anterior pituitary hormone.

• When appropriate, ADH and oxytocin are secreted from axon endings into the bloodstream.

• The anterior pituitary secretes its hormones into the bloodstream.

posterior pituitary

anterior pituitary

antidiuretic hormone (ADH)

kidney tubules

gonadotropins (FSH and LH)

ovaries, testes

oxytocin

smooth muscle in uterus

oxytocin

mammary glands

thyroid-stimulating hormone (TSH)

thyroid

adrenocorticotropin (ACTH)

adrenal cortex

prolactin (PRL)

mammary glands

growth hormone (GH)

bones, tissues

Figure 15.3   Hypothalamus and the pituitary.

(Left) The hypothalamus produces two hormones, ADH and oxytocin, which are stored and secreted by the posterior pituitary. *(Right)* The hypothalamus controls the secretions of the anterior pituitary, and the anterior pituitary controls the secretions of the thyroid, adrenal cortex, and gonads, which are also endocrine glands.

Effects of Growth Hormone

GH is produced by the anterior pituitary. The quantity is greatest during childhood and adolescence, when most body growth is occurring. If too little GH is produced during childhood, the individual has **pituitary dwarfism,** characterized by perfect proportions but small stature. If too much GH is secreted, a person can become a giant (Fig. 15.4). Giants usually have poor health, primarily because GH has a secondary effect on the blood sugar level, promoting an illness called diabetes mellitus.

On occasion, there is overproduction of growth hormone in the adult, and a condition called **acromegaly** results. Since long bone growth is no longer possible in adults, only the feet, hands, and face (particularly the chin, nose, and eyebrow ridges) can respond, and these portions of the body become overly large (Fig. 15.5).

The amount of growth hormone during childhood affects the height of an individual.

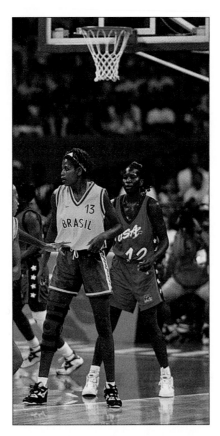 

Figure 15.4   Effect of growth hormone.
The amount of growth hormone produced by the anterior pituitary during childhood affects the height of an individual. Plentiful growth hormone produces very tall basketball players and even giants. Little growth hormone results in limited stature and even pituitary dwarfism.

Age 9   Age 16   Age 33   Age 52

Figure 15.5   Acromegaly.
Acromegaly is caused by overproduction of GH in the adult. It is characterized by enlargement of the bones in the face, the fingers, and the toes of an adult.

15.3 Thyroid and Parathyroid Glands

The **thyroid gland** is a large gland located in the neck, where it is attached to the trachea just below the larynx (see Fig. 15.2). The parathyroid glands are embedded in the posterior surface of the thyroid gland.

Thyroid Gland

The thyroid gland is composed of a large number of follicles, each a small spherical structure made of thyroid cells filled with triiodothyronine (T_3), which contains three iodine atoms, and **thyroxine (T_4),** which contains four iodine atoms.

Effects of Thyroid Hormones

To produce thyroxine and triiodothyronine, the thyroid gland actively acquires iodine. The concentration of iodine in the thyroid gland can increase to as much as 25 times that of blood. If iodine is lacking in the diet, the thyroid gland is unable to produce the thyroid hormones. In response to constant stimulation by the anterior pituitary, the thyroid enlarges, resulting in a **simple goiter** (Fig. 15.6). Some years ago, it was discovered that the use of iodized salt allows the thyroid to produce the thyroid hormones, and therefore helps prevent simple goiter.

Thyroid hormones increase the metabolic rate. They do not have one target organ; instead, they stimulate all cells of the body to metabolize at a faster rate. More glucose is broken down and more energy is utilized.

If the thyroid fails to develop properly, a condition called **cretinism** results (Fig. 15.7). Individuals with this condition are short and stocky and have had extreme hypothyroidism (undersecretion of thyroid hormone) since infancy or childhood. Thyroid hormone therapy can initiate growth, but unless treatment is begun within the first two months, mental retardation results. The occurrence of hypothyroidism in adults produces the condition known as **myxedema,** which is characterized by lethargy, weight gain, loss of hair, slower pulse rate, lowered body temperature, and thickness and puffiness of the skin. The administration of adequate doses of thyroid hormones restores normal function and appearance.

In the case of hyperthyroidism (oversecretion of thyroid hormone), or Graves disease, the thyroid gland is overactive, and a goiter forms. The eyes protrude because of edema in eye socket tissues and swelling of the muscles that move the eyes. This type of goiter is called **exophthalmic goiter.** The patient usually becomes hyperactive, nervous, irritable, and suffers from insomnia. Removal or destruction of a portion of the thyroid by means of radioactive iodine is sometimes effective in curing the condition. Hyperthyroidism can also be caused by a thyroid tumor, which is usually detected as a lump during physical examination. Again, the treatment is surgery in combination with administration of radioactive iodine. The prognosis for most patients is excellent.

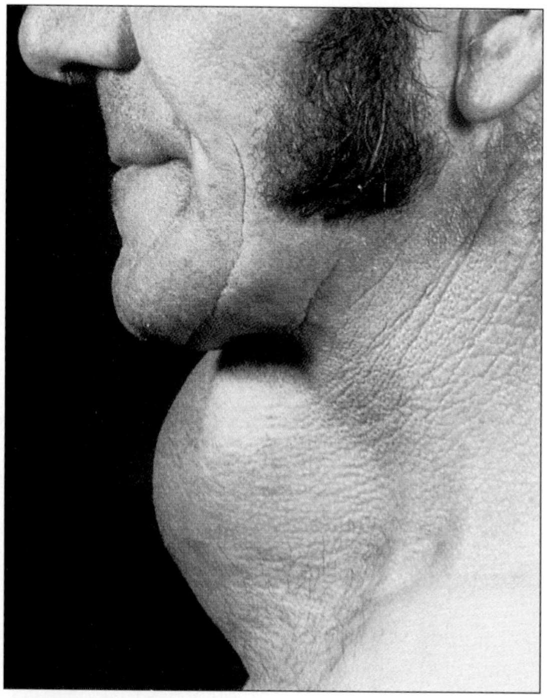

Figure 15.6 **Simple goiter.**
An enlarged thyroid gland often is caused by a lack of iodine in the diet. Without iodine, the thyroid is unable to produce thyroxine, and continued anterior pituitary stimulation causes the gland to enlarge.

Figure 15.7 **Cretinism.**
Individuals who have hypothyroidism since infancy or childhood do not grow and develop as others do. Unless medical treatment is begun, the body is short and stocky; mental retardation is also likely.

Calcitonin

Calcium (Ca^{2+}) plays a significant role in both nervous conduction and muscle contraction. It is also necessary to blood clotting. The blood calcium level is regulated in part by **calcitonin,** a hormone secreted by the thyroid gland when the blood calcium level rises (Fig. 15.8). The primary effect of calcitonin is to bring about the deposit of calcium in the bones. It does this by temporarily reducing the activity and number of osteoclasts. When the blood calcium lowers to normal, the release of calcitonin by the thyroid is inhibited, but a low level stimulates the release of **parathyroid hormone (PTH)** by the parathyroid glands.

Parathyroid Glands

Many years ago, the four parathyroid glands were sometimes mistakenly removed during thyroid surgery because they are so small. Parathyroid hormone (PTH), the hormone produced by the **parathyroid glands,** causes the blood phosphate (HPO_4^{2-}) level to decrease and the blood calcium level to increase.

A low blood calcium level stimulates the release of PTH. PTH promotes the activity of osteoclasts and the release of calcium from the bones. PTH also promotes the reabsorption of calcium by the kidneys, where it activates vitamin D. Vitamin D, in turn, stimulates the absorption of calcium from the intestine. These effects bring the blood calcium level back to the normal range so that the parathyroid glands no longer secrete PTH.

When insufficient parathyroid hormone production leads to a dramatic drop in the blood calcium level, tetany results. In **tetany,** the body shakes from continuous muscle contraction. The effect is brought about by increased excitability of the nerves, which initiate nerve impulses spontaneously and without rest.

The antagonistic actions of calcitonin, from the thyroid gland, and parathyroid hormone, from the parathyroid glands, maintain the blood calcium level within normal limits.

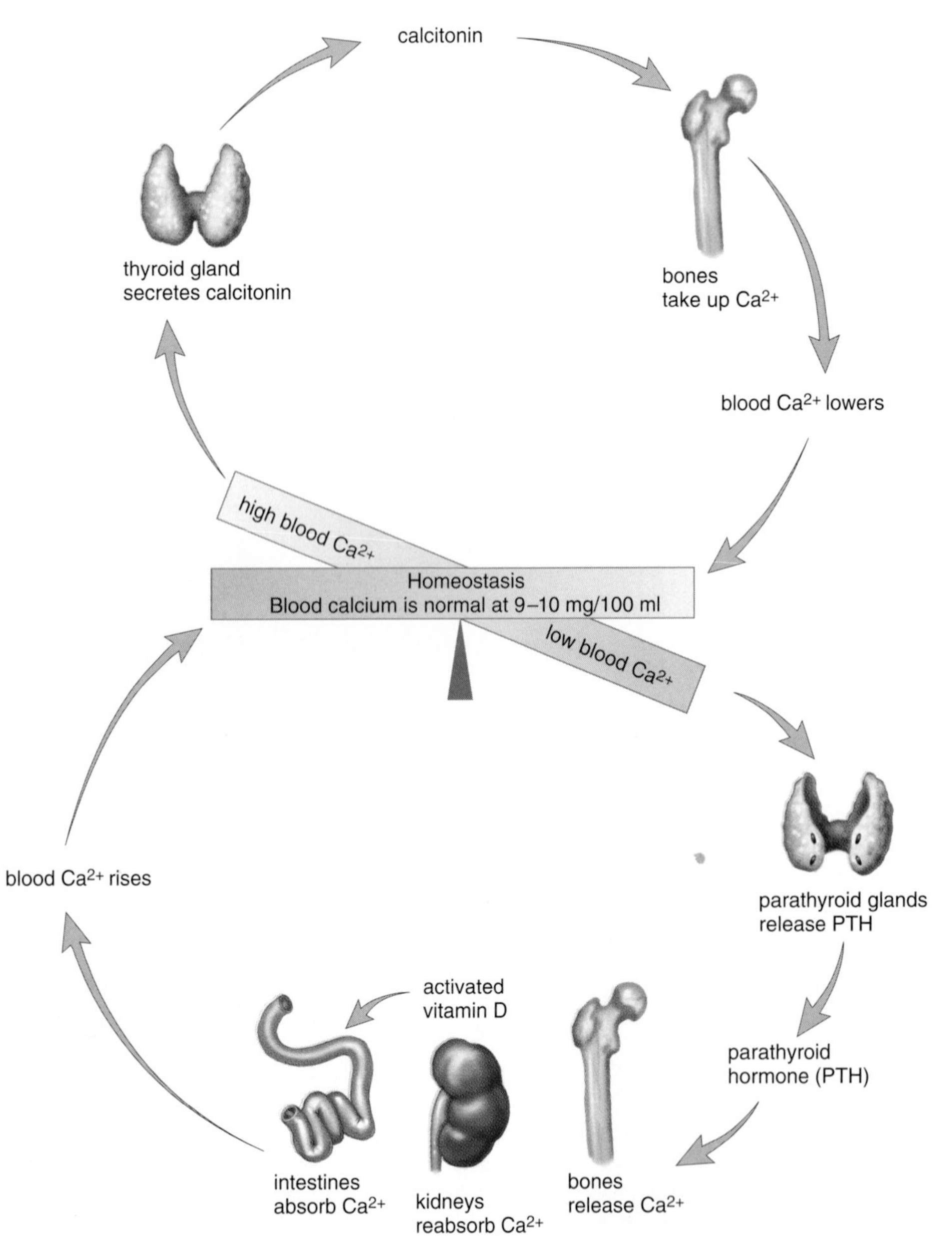

calcitonin

thyroid gland
secretes calcitonin

bones
take up Ca^{2+}

blood Ca^{2+} lowers

high blood Ca^{2+}

Homeostasis
Blood calcium is normal at 9–10 mg/100 ml

low blood Ca^{2+}

blood Ca^{2+} rises

parathyroid glands
release PTH

parathyroid
hormone (PTH)

activated
vitamin D

intestines
absorb Ca^{2+} kidneys
reabsorb Ca^{2+} bones
release Ca^{2+}

Figure 15.8   Regulation of blood calcium level.
(Top) When the blood calcium (Ca^{2+}) level is high, the thyroid gland secretes calcitonin. Calcitonin promotes the uptake of Ca^{2+} by the bones, and therefore the blood Ca^{2+} level returns to normal. *(Bottom)* When the blood Ca^{2+} level is low, the parathyroid glands release parathyroid hormone (PTH). PTH causes the bones to release Ca^{2+}, the kidneys to reabsorb Ca^{2+} and activate vitamin D, and thereafter the intestines absorb Ca^{2+}. Therefore, the blood Ca^{2+} level returns to normal.

15.4 Adrenal Glands

We have two **adrenal glands** that sit atop the kidneys (see Fig. 15.2). Each adrenal gland consists of an inner portion called the **adrenal medulla** and an outer portion called the **adrenal cortex.** These portions, like the anterior pituitary and the posterior pituitary, have no physiological connection with one another.

The hypothalamus exerts control over the activity of both portions of the adrenal glands. It initiates nerve impulses that travel by way of the brain stem, spinal cord, and sympathetic nerve fibers to the adrenal medulla, which then secretes its hormones. The hypothalamus, by means of ACTH-releasing hormone, controls the anterior pituitary's secretion of ACTH, which in turn, stimulates the adrenal cortex. Stress of all types, including both emotional and physical trauma, prompts the hypothalamus to stimulate the adrenal glands.

Epinephrine (adrenaline) and **norepinephrine** (noradrenaline) produced by the adrenal medulla rapidly bring about all the bodily changes that occur when an individual reacts to an emergency situation. The effects of these hormones are short term (Fig. 15.9). In contrast, the hormones produced by the adrenal cortex provide a long-term response to stress. The two major types of hormones produced by the adrenal cortex are the mineralocorticoids and the glucocorticoids. The **mineralocorticoids** regulate salt and water balance, leading to increases in blood volume and blood pressure. The **glucocorticoids** regulate carbohydrate, protein, and fat metabolism, leading to an increase in blood glucose level. Cortisone, the medication that is often administered for inflammation of joints, is a glucocorticoid.

The adrenal cortex also secretes a small amount of male sex hormones and a small amount of female sex hormones in both sexes—that is, in the male, both male and female sex hormones are produced by the adrenal cortex, and in the female, both male and female sex hormones are also produced by the adrenal cortex.

The adrenal medulla is under nervous control, and the adrenal cortex is under the control of ACTH, an anterior pituitary hormone. The adrenal hormones help us respond to stress.

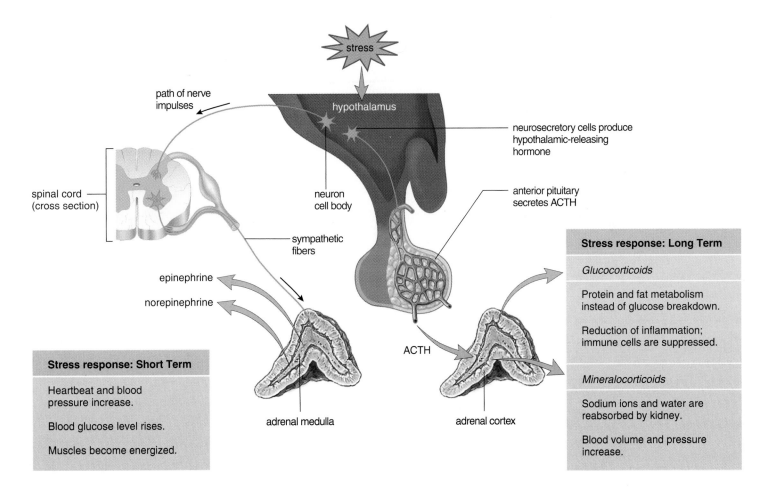

Figure 15.9 Adrenal glands.
Both the adrenal medulla and the adrenal cortex are under the control of the hypothalamus when they help us respond to stress. *(Left)* The adrenal medulla provides a rapid, but short-term, stress response. *(Right)* The adrenal cortex provides a slower, but long-term, stress response.

Glucocorticoids

Cortisol is a biologically significant glucocorticoid produced by the adrenal cortex. Cortisol raises the blood glucose level in at least two ways. It promotes the breakdown of muscle proteins to amino acids, which are taken up by the liver from the bloodstream. The liver breaks down these excess amino acids to glucose, which enters the blood. Cortisol also promotes the metabolism of fatty acids rather than carbohydrates, and this spares glucose.

Cortisol also counteracts the inflammatory response that leads to the pain and the swelling of joints in arthritis and bursitis. The administration of cortisol aids these conditions because it reduces inflammation. Very high levels of glucocorticoids in the blood can suppress the body's defense system, including the inflammatory response that occurs at infection sites. Cortisone and other glucocorticoids can relieve swelling and pain from inflammation, but by suppressing pain and immunity, they can also make a person highly susceptible to injury and infection.

Mineralocorticoids

Aldosterone is the most important of the mineralocorticoids. The primary target organ of aldosterone is the kidney, where it promotes renal absorption of sodium (Na^+) and renal excretion of potassium (K^+).

The secretion of mineralocorticoids is not under the control of the anterior pituitary. When the blood sodium level and therefore blood pressure are low, the kidneys secrete **renin** (Fig. 15.10). Renin is an enzyme that converts the plasma protein angiotensinogen to angiotensin I, which is changed to an-giotensin II by a converting enzyme found in lung capillaries. Angiotensin II stimulates the adrenal cortex to release aldosterone. The effect of this system, called the renin-angiotensin-aldosterone system, is to raise blood pressure in two ways. Angiotensin II constricts the arterioles, and aldosterone causes the kidneys to reabsorb sodium. When the blood sodium level rises, water is reabsorbed, in part because the hypothalamus secretes ADH (see page 296). Then blood pressure increases to normal.

There is an antagonistic hormone to aldosterone, as you might suspect. When the atria of the heart are stretched due to increased blood volume, cardiac cells release a hormone called **atrial natriuretic hormone (ANH),** which inhibits the secretion of aldosterone from the adrenal cortex. The effect of this hormone is, therefore, to cause the excretion of sodium—that is, *natriuresis*. When sodium is excreted, so is water, and therefore blood pressure lowers to normal.

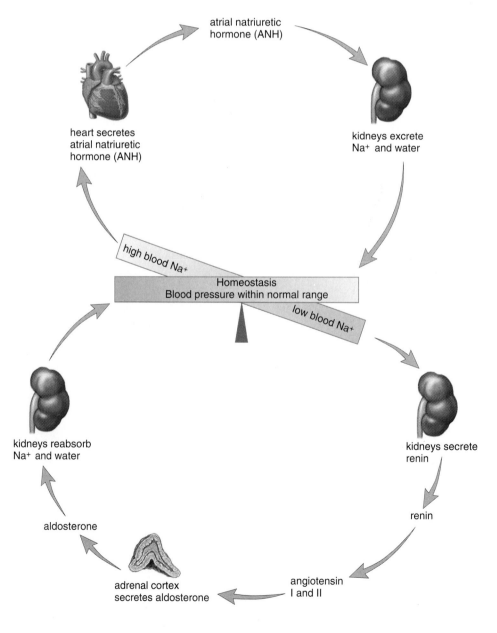

Figure 15.10 **Regulation of blood pressure and volume.**
(Bottom) When the blood sodium (Na^+) level is low, a low blood pressure causes the kidneys to secrete renin. Renin leads to the secretion of aldosterone from the adrenal cortex. Aldosterone causes the kidneys to reabsorb Na^+, and water follows, so that blood volume and pressure return to normal. *(Top)* When the blood Na^+ is high, a high blood volume causes the heart to secrete atrial natriuretic hormone (ANH). ANH causes the kidneys to excrete Na^+, and water follows. The blood volume and pressure return to normal.

Malfunction of the Adrenal Cortex

When there is a low level of adrenal cortex hormones due to hyposecretion, a person develops **Addison disease.** The presence of ACTH, which is in excess but ineffective, causes a bronzing of the skin because ACTH like MSH can lead to a buildup of melanin (Fig. 15.11). Without cortisol, glucose cannot be replenished when a stressful situation arises. Even a mild infection can lead to death. The lack of aldosterone results in a loss of sodium and water and the development of low blood pressure and possibly severe dehydration. Left untreated, Addison disease can be fatal.

When there is a high level of adrenal cortex hormones due to hypersecretion, a person develops **Cushing syndrome** (Fig. 15.12). The excess of cortisol results in a tendency toward diabetes mellitus as muscle protein is metabolized and subcutaneous fat is deposited in the midsection. The trunk is obese while the arms and legs remain a normal size. An excess of aldosterone and reabsorption of sodium and water by the kidneys leads to a basic blood pH and hypertension. The face is moon shaped due to edema. Masculinization may occur in women because of excess adrenal male sex hormones.

The adrenal cortex hormones are essential to homeostasis. Addison disease is due to adrenal cortex hyposecretion, and Cushing syndrome is due to adrenal cortex hypersecretion.

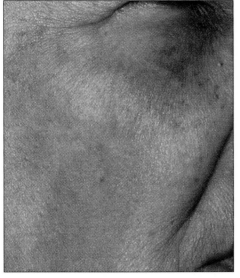

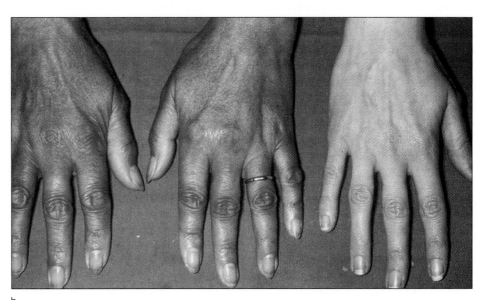

a.

b.

Figure 15.11   Addison disease.
Addison disease is characterized by a peculiar bronzing of the skin, particularly noticeable in these light-skinned individuals. Note the color of (**a**) the face and (**b**) the hands compared to the hand of an individual without the disease.

Figure 15.12   Cushing syndrome.
Cushing syndrome results from hypersecretion of hormones due to an adrenal cortex tumor.
a. Patient first diagnosed with Cushing syndrome.
b. Four months later, after therapy.

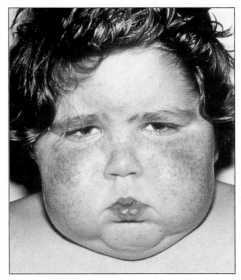

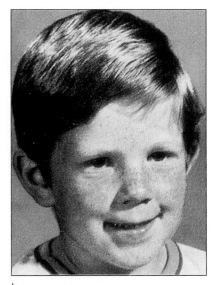

a.

b.

15.5   Pancreas

The **pancreas** is a long organ that lies transversely in the abdomen between the kidneys and near the duodenum of the small intestine. It is composed of two types of tissue. Exocrine tissue produces and secretes digestive juices that go by way of ducts to the small intestine. Endocrine tissue, called the **pancreatic islets** (islets of Langerhans), produces and secretes the hormones **insulin** and **glucagon** directly into the blood (Fig. 15.13).

Insulin is secreted when there is a high blood glucose level, which usually occurs just after eating. Insulin stimulates the uptake of glucose by cells, especially liver cells, muscle cells, and adipose tissue cells. In liver and muscle cells, glucose is then stored as glycogen. In muscle cells, the breakdown of glucose supplies energy for protein metabolism, and in fat cells the breakdown of glucose supplies

glycerol for the formation of fat. In these various ways, insulin lowers the blood glucose level.

Glucagon is secreted from the pancreas, usually between meals, when there is a low blood glucose level. The major target tissues of glucagon are the liver and adipose tissue. Glucagon stimulates the liver to break down glycogen to glucose and to use fat and protein in preference to glucose as energy sources. Adipose tissue cells break down fat to glycerol and fatty acids. The liver takes these up and uses them as substrates for glucose formation. In these various ways, glucagon raises the blood glucose level.

The two antagonistic hormones insulin and glucagon, both produced by the pancreas, maintain the normal level of glucose in the blood.

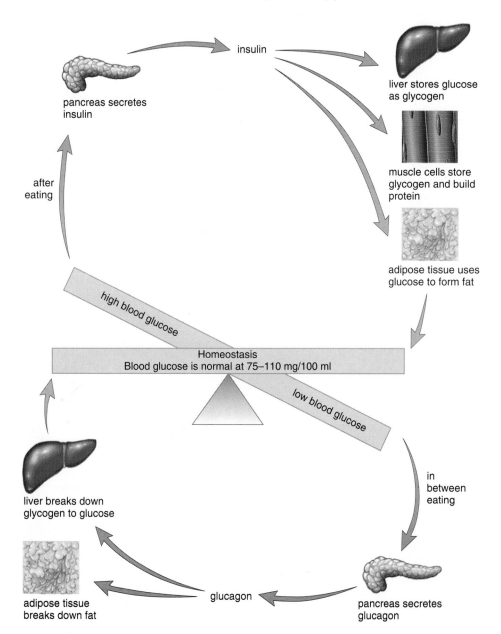

pancreas secretes insulin

insulin

liver stores glucose as glycogen

muscle cells store glycogen and build protein

adipose tissue uses glucose to form fat

after eating

high blood glucose

Homeostasis
Blood glucose is normal at 75–110 mg/100 ml

low blood glucose

liver breaks down glycogen to glucose

adipose tissue breaks down fat

glucagon

in between eating

pancreas secretes glucagon

Figure 15.13   Regulation of blood glucose level.

(Top) When the blood glucose level is high, the pancreas secretes insulin. Insulin promotes the storage of glucose as glycogen and the synthesis of proteins and fats (as opposed to their use as energy sources). Therefore, insulin lowers the blood glucose level. (Bottom) When the blood glucose level is low, the pancreas secretes glucagon. Glucagon acts opposite to insulin; therefore, glucagon raises the blood glucose level to normal.

Diabetes Mellitus

Diabetes mellitus is a fairly common hormonal disease in which liver cells, and indeed all body cells, are unable to take up and/or metabolize glucose. Therefore, cellular famine exists in the midst of plenty and the person becomes extremely hungry. As the blood glucose level rises, glucose, along with water, is excreted in the urine. The loss of water in this way causes the diabetic to be extremely thirsty. Since glucose is not being metabolized, the body turns to the breakdown of protein and fat for energy. The metabolism of fat leads to the buildup of ketones in the blood and acidosis (acid blood), which can eventually cause coma and death. The symptoms of hyperglycemia (high blood sugar) develop slowly and there is time to get adequate medical care.

In addition to testing for glucose in the urine, the glucose tolerance test is often used to assist the diagnosis of diabetes mellitus. After the patient is given 100 g of glucose, the blood glucose concentration is measured at intervals. In a diabetic, the blood glucose level rises greatly and remains elevated for several hours. In a nondiabetic, the blood glucose level rises somewhat and then returns to normal in about one and one-half hours. In the meantime, glucose appears in the urine (Fig. 15.14).

There are two types of diabetes mellitus. In *type I (insulin-dependent) diabetes,* the pancreas is not producing insulin. The condition is believed to be brought on by exposure to an environmental agent, most likely a virus, whose presence causes cytotoxic T cells to destroy the pancreatic islets. As a result, the individual must have daily insulin injections. These injections control the diabetic symptoms but still can cause inconveniences, since either an overdose of insulin or missing a meal can bring on the symptoms of hypoglycemia (low blood sugar). These symptoms include perspiration, pale skin, shallow breathing, and anxiety. Because the brain requires a constant supply of sugar, unconsciousness can result. The cure is quite simple: immediate ingestion of a sugar cube or fruit juice can very quickly counteract hypoglycemia.

It's possible to transplant a working pancreas into patients with type I diabetes. To do away with the necessity of taking immunosuppressive drugs, fetal pancreatic islet cells have been injected into patients. Another experimental procedure is to place pancreatic islet cells in a capsule that allows insulin to get out but prevents antibodies and T lymphocytes from getting in. This artificial organ is implanted in the abdominal cavity.

Of the 16 million people who now have diabetes in the United States, most have *type II (noninsulin-dependent) diabetes.* This type of diabetes mellitus usually occurs in people of any age who are obese and inactive. The pancreas produces insulin, but the liver and muscle cells do not respond to it in the usual manner. They may increasingly lack the receptor proteins that bind to insulin. If type II diabetes is untreated, the results can be as serious as those of type I diabetes. It is possible to prevent or at least control type II diabetes by adhering to a low-fat, low-sugar diet and exercising regularly. If this fails, oral drugs that stimulate the pancreas to secrete more insulin and enhance the metabolism of glucose in the liver and muscle cells are available.

The symptoms of type I diabetes are compelling, and therefore most people seek help right away. The symptoms of type II diabetes are more likely to be overlooked. It's projected that there may be as many as 7 million Americans who have type II diabetes and are not aware of it. Yet the results of type II diabetes are as serious as those for type I diabetes.

Long-term complications of both types of diabetes are blindness, kidney disease, and circulatory disorders including atherosclerosis, heart disease, stroke, and reduced circulation. The latter can lead to gangrene in the arms and legs. Pregnancy carries an increased risk of diabetic coma, and the child of a diabetic is somewhat more likely to be stillborn or to die shortly after birth. The complications of diabetes are not expected to appear if the blood glucose level is carefully regulated and kept within normal limits.

Diabetes mellitus is caused by the lack of insulin or the insensitivity of cells to insulin, a hormone that lowers the blood glucose level, particularly by causing the liver to store glucose as glycogen.

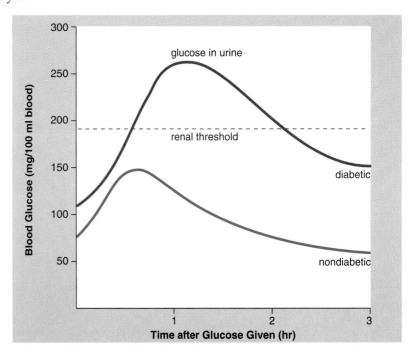

Figure 15.14   Glucose tolerance test.
Following the administration of 100 g of glucose, the blood glucose level rises dramatically in the diabetic but not in the nondiabetic. Glucose appears in the urine when its level exceeds 190 mg/100 ml.

15.6 Other Endocrine Glands

The **gonads** are the testes in males and the ovaries in females. The gonads are endocrine glands. There are also lesser known glands and some tissues that produce hormones.

Testes and Ovaries

The **testes** are located in the scrotum, and the **ovaries** are located in the pelvic cavity. The testes produce **androgens** (e.g., **testosterone**), which are the male sex hormones, and the ovaries produce estrogens and progesterone, the female sex hormones. The hypothalamus and the pituitary gland control the hormonal secretions of these organs in the same manner previously described for the thyroid gland.

Greatly increased testosterone secretion at the time of puberty stimulates the growth of the penis and the testes. Testosterone also brings about and maintains the male secondary sex characteristics that develop at the time of puberty. Testosterone causes growth of a beard, axillary (underarm) hair, and pubic hair. It prompts the larynx and the vocal cords to enlarge, causing the voice to change. It is partially responsible for the muscular strength of males,

and this is the reason some athletes take supplemental amounts of **anabolic steroids,** which are either testosterone or related chemicals. The contraindications of taking anabolic steroids are listed in Figure 15.15. Testosterone also stimulates oil and sweat glands in the skin; therefore, it is largely responsible for acne and body odor. Another side effect of testosterone is baldness. Genes for baldness are probably inherited by both sexes, but baldness is seen more often in males because of the presence of testosterone.

The female sex hormones, **estrogens** and **progesterone,** have many effects on the body. In particular, estrogens secreted at the time of puberty stimulate the growth of the uterus and the vagina. Estrogen is necessary for egg maturation and is largely responsible for the secondary sex characteristics in females, including female body hair and fat distribution. In general, females have a more rounded appearance than males because of a greater accumulation of fat beneath the skin. Also, the pelvic girdle is wider in females than in males, resulting in a larger pelvic cavity. Both estrogen and progesterone are required for breast development and regulation of the uterine cycle, which includes monthly menstruation (discharge of blood and mucosal tissues from the uterus).

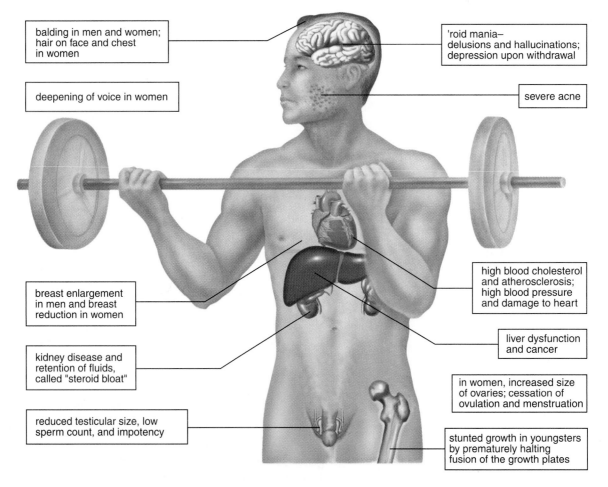

Figure 15.15   The effects of anabolic steroid use.

Thymus Gland

The **thymus** is a lobular gland that lies just beneath the sternum (see Fig. 15.2). This organ reaches its largest size and is most active during childhood. With aging, the organ gets smaller and becomes fatty. Lymphocytes that originate in the bone marrow and then pass through the thymus are transformed into T lymphocytes. The lobules of the thymus are lined by epithelial cells that secrete hormones called thymosins. These hormones aid in the differentiation of lymphocytes packed inside the lobules. Although the hormones secreted by the thymus ordinarily work in the thymus, there is hope that these hormones could be injected into AIDS or cancer patients where they would enhance T lymphocyte function.

Pineal Gland

The **pineal gland,** which is located in the brain (see Fig. 15.2), produces the hormone called **melatonin,** primarily at night. Melatonin is involved in our daily sleep-wake cycle; normally we grow sleepy at night when melatonin levels increase and awaken once daylight returns and melatonin levels are low. Daily 24-hour cycles such as this are called **circadian rhythms** and, as discussed in the Health Focus on page 308, circadian rhythms are controlled by a biological clock located in the hypothalamus.

Based on animal research, it appears that melatonin also regulates sexual development. It is of interest that children whose pineal gland has been destroyed due to a brain tumor experience early puberty.

The gonads, thymus, and pineal gland are also endocrine organs. The gonads secrete the sex hormones, the thymus secretes thymosins, and the pineal gland secretes melatonin.

Hormones from Other Tissues

Some organs that are usually not considered endocrine glands do indeed secrete hormones. We have already mentioned that the heart produces atrial natriuretic hormone. And you will recall that the stomach and the small intestine produce peptide hormones that regulate digestive secretions. A number of other types of tissues produce hormones.

Leptin

Leptin is a protein hormone produced by adipose tissue. Leptin acts on the hypothalamus, where it signals satiety—that the individual had enough to eat. Strange to say, the blood of obese individuals may be rich in leptin. It is possible that the leptin they produce is ineffective because of a genetic mutation, or else their hypothalamic cells lack a suitable number of receptors for leptin.

Growth Factors

A number of different types of organs and cells produce peptide **growth factors,** which stimulate cell division and mitosis. They are like hormones in that they act on cell types with specific receptors to receive them. Some, like lymphokines, are released into the blood; others diffuse to nearby cells. Growth factors of particular interest are:

Granulocyte and macrophage colony-stimulating factor (GM-CSF) is secreted by many different tissues. GM-CSF causes bone marrow stem cells to form either granulocyte or macrophage cells, depending on whether the concentration is low or high.

Platelet-derived growth factor is released from platelets and from many other cell types. It helps in wound healing and causes an increase in the number of fibroblasts, smooth muscle cells, and certain cells of the nervous system.

Epidermal growth factor and nerve growth factor stimulate the cells indicated by their names, as well as many others. These growth factors are also important in wound healing.

Tumor angiogenesis factor stimulates the formation of capillary networks and is released by tumor cells. One treatment for cancer is to prevent the activity of this growth factor.

Prostaglandins

Prostaglandins are potent chemical signals produced within cells from arachidonate, a fatty acid. Prostaglandins are not distributed in the blood; instead they act locally, quite close to where they were produced. In the uterus, prostaglandins cause muscles to contract; therefore, they are implicated in the pain and discomfort of menstruation in some women. Also, prostaglandins mediate the effects of pyrogens, chemicals that are believed to reset the temperature regulatory center in the brain. Aspirin reduces body temperature and controls pain because of its effect on prostaglandins.

Certain prostaglandins reduce gastric secretion and have been used to treat ulcers; others lower blood pressure and have been used to treat hypertension; and yet others inhibit platelet aggregation and have been used to prevent thrombosis. However, different prostaglandins have contrary effects, and it has been very difficult to successfully standardize their use. Therefore, prostaglandin therapy is still considered experimental.

Many tissues aside from the traditional endocrine glands produce hormones. Some of these enter the bloodstream, and some act only locally.

Melatonin

The hormone melatonin has been sold as a nutritional supplement for about five years. The popular press promotes its use in pill form for sleep, aging, cancer treatment, sexuality, and more. At best, melatonin may have some benefits in certain sleep disorders. But most physicians will not yet recommend it for that use because so little is known about its dosage requirements and possible side effects.

Melatonin is a hormone produced by the pineal gland in greatest quantity at night and smallest quantity during the day. Notice in Figure 15A that melatonin's production cycle accompanies our natural sleep-wake cycle. Rhythms with a period of about 24 hours are called circadian ("about a day") rhythms. All circadian rhythms seem to be controlled by an internal biological clock because they are free-running—that is, they have a regular cycle even in the absence of environmental cues. In scientific experiments, humans have lived in underground bunkers where they never see the light of day. In a few people, the sleep-wake cycle drifts badly, but in most, the daily activity schedule is just about 25 hours. How do we normally manage to stay on a 24-hour schedule? An individual's internal biological clock is reset each day by the environmental day-night cycle. Characteristically, biological clocks that control circadian rhythms are reset by environmental cues, or else they drift out of phase with the environment.

Recent research suggests that our biological clock lies in a cluster of neurons within the hypothalamus called the suprachiasmatic nucleus (SCN). The SCN undergoes spontaneous cyclical changes in activity, and therefore it can act as a pacemaker for circadian rhythms like the rise and fall of body temperature and our sleep-wake cycle. Neural connections between the retina and the SCN indicate that reception of light by the eyes most likely resets the SCN and keeps our biological rhythms on a 24-hour cycle. Some people suffer from seasonal affective disorder, or SAD. As the days get darker and darker during the fall and winter, they become depressed, sometimes severely. They find it difficult to keep up because their biological clock has fallen behind without early morning light to reset it. If so, a half-hour dose of simulated daylight from a portable light box first thing in the morning makes them feel operational again.

The SCN also controls the secretion of melatonin by the pineal gland, and in turn melatonin may quiet the operation of the neurons in the SCN. So, taking melatonin at twilight may help bring on sleep and help reset your biological clock for the next day.

Research is still going forward to see if melatonin will be effective for circadian rhythm disorders such as SAD, jet lag, sleep phase problems, recurrent insomnia in the totally blind, and some other less common disorders. So-called jet lag occurs when you travel across several time zones and your biological clock is out of phase with local time. You feel wide awake when it is time to sleep because your body is receiving internal signals that it is morning. Likewise, acute periods of sleepiness and fatigue occur during the day, because of internal signals indicating it is nighttime. Jet-lag symptoms gradually disappear as your biological clock adjusts to the environmental signals of the local time zone. Many young people have a sleep phase problem because their circadian cycle lasts 25 to 26 hours. As they lengthen their day, they get out of sync with normal times for sleep and activity. The totally blind have the same problem because they have no chance of using the environmental day-night cycle to reset their internal clock.

In clinical trials, it was found that melatonin could shift circadian rhythms and reset our biological clock. The experimenters found that melatonin given in the afternoon shifts rhythms earlier, while melatonin given in the morning shifts rhythms later. For most people, the process was gradual: the average rate of change was about an hour a day. Before you try melatonin, however, you might want to consider that melatonin is known to affect reproductive behavior in other mammals. It's a matter of deciding if any potential side effect from melatonin use is worth the possible benefits.

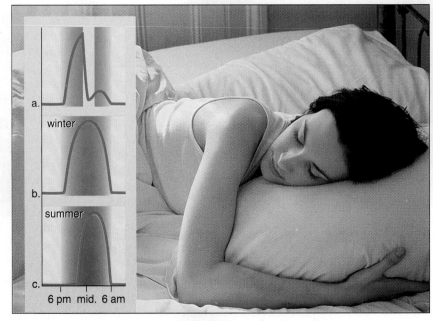

Figure 15A   Melatonin production.
Melatonin production is greatest at night when we are sleeping. Light suppresses melatonin production (**a**) so its duration is longer in the winter (**b**) than in the summer (**c**).

15.7 Chemical Signals

Both the nervous and endocrine systems utilize chemical signals. Chemical signals affect only those cells that have appropriate receptor proteins. You know already that neurotransmitter substances diffuse across a synaptic cleft and bind to receptor proteins in the receiving neuron. Similarly, a hormone binds to receptor proteins that combine with it in a lock-and-key manner. After binding occurs, the hormone affects the metabolism of the cell.

Hormones fall into two basic chemical classes: (1) a **steroid hormone** always has the same complex of four-carbon rings, but each one has different side chains; (2) a **peptide hormone** includes those hormones that are peptides, proteins, glycoproteins, or modified amino acids.

Steroid hormones are lipids, and therefore they cross cellular membranes (Fig. 15.16a). Only after they are inside the cell do steroid hormones, such as estrogen and progesterone, bind to receptor proteins. The hormone-receptor complex then binds to DNA, activating particular genes. Activation leads to production of a cellular enzyme in multiple quantities.

Most peptide hormones cannot pass through cellular membranes, and they bind to a receptor protein in the plasma membrane (Fig. 15.16b). After epinephrine binds to a receptor protein, a relay system leads to the conversion of ATP to **cyclic AMP** (cyclic adenosine monophosphate). Cyclic AMP (cAMP) contains only one phosphate group attached to the adenosine portion of the molecule at two spots. Thus, the molecule is cyclic. The peptide hormone is called the first messenger, and cAMP is called the second messenger. Calcium is also a common second messenger, and this helps explain why calcium regulation in the body is so important.

The second messenger sets in motion an enzyme cascade. In muscle cells, the enzyme cascade leads to the breakdown of glycogen to glucose (Fig. 15.16b). An enzyme cascade is so called because each enzyme in turn activates another. The binding of a single peptide hormone molecule can result in as much as a thousandfold response.

Hormones are chemical signals that influence the metabolism of the cell either indirectly by regulating the production of a particular protein (steroid hormone) or directly by activating an enzyme cascade (peptide hormone).

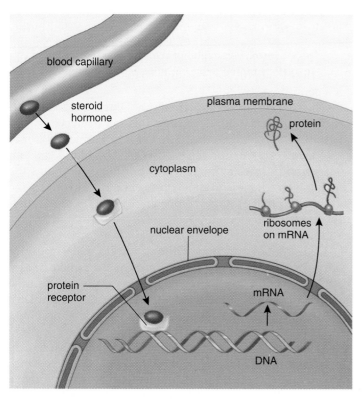

a. Action of steroid hormone

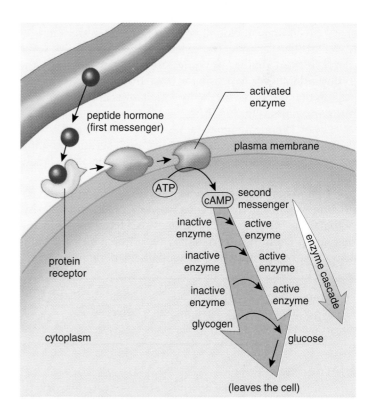

b. Action of peptide hormone

Figure 15.16 Cellular activity of hormones.
a. After passing through the plasma membrane and nuclear envelope, a steroid hormone binds to a receptor protein inside the nucleus. The hormone-receptor complex then binds to DNA, and this leads to activation of certain genes and protein synthesis. **b.** Peptide hormones, called first messengers, bind to a specific receptor protein in the plasma membrane. A protein relay in the membrane ends when an enzyme converts ATP to cAMP, the second messenger, which activates an enzyme cascade.

Figure 15.17
Chemical signals.
a. Endocrine hormones and neurosecretions typically are carried in the bloodstream and act at a distance within the body of a single organism. **b.** Some chemical signals have local effects only; they pass between cells that are adjacent to one another. Neurotransmitters belong to this category, as do local hormones such as prostaglandins.

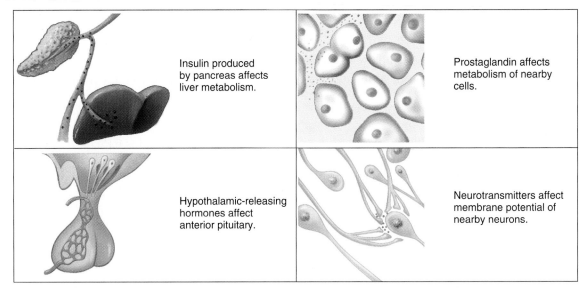

Insulin produced by pancreas affects liver metabolism.

Prostaglandin affects metabolism of nearby cells.

Hypothalamic-releasing hormones affect anterior pituitary.

Neurotransmitters affect membrane potential of nearby neurons.

a. Signal acts at a distance between body parts.

b. Signal acts locally between adjacent cells.

Hormonal Versus Neural Signals

Most hormones are chemical signals that act from a distance. Pituitary hormones influence the function of numerous organs throughout the body, and insulin produced by the pancreas affects liver metabolism. The nervous system also at times utilizes chemical signals that are produced some distance from the organ being affected. For example, the hypothalamic-releasing hormones travel in the portal system that lies between the hypothalamus and the anterior pituitary gland (Fig. 15.17a).

Other chemical signals act locally—that is, cell to cell. Prostaglandins are local hormones, and certainly neurotransmitter substances released by one neuron affect a neuron nearby (Fig. 15.17b).

There is an overlap between the functioning of the endocrine system and the nervous system. Consider that the hypothalamus produces the hormones ADH and oxytocin and that norepinephrine is secreted by the adrenal medulla but is also a neurotransmitter in the sympathetic nervous system.

Some chemical signals work from a distance, and other chemical signals act locally between adjacent cells.

Homeostasis ⚖

The endocrine system and the nervous system work together to achieve and maintain the relative stability of the internal environment. Consider that the hypothalamus produces ADH, a hormone that helps maintain the osmolarity of the blood. The osmolarity of the blood is also affected by aldosterone, which is produced by the adrenal cortex, and by antidiuretic hormone, which is secreted by the heart.

Through its effect on the autonomic system, the hypothalamus controls the secretion of epinephrine and norepinephrine by sympathetic nerve endings and by the adrenal medulla. Epinephrine and norepinephrine allow the body to respond to emergency situations in a fight or flight manner. The stress response is sustained by cortisol secreted by the adrenal cortex. Cortisol permits the body to ignore any injuries that may have occurred during a confrontation and keeps the blood glucose level high so that energy is available for a continued struggle.

The concentration of calcium (Ca^{2+}) in the blood is critical because of the importance of this ion to nervous conduction and muscle contraction. As you know, the bones serve as a reservoir for calcium. When the blood calcium concentration lowers, parathyroid hormone promotes the breakdown of bone and the reabsorption of calcium by the kidneys and the intestines. Opposing the action of parathyroid hormone, calcitonin secreted by the thyroid brings about the deposit of calcium in the bones.

Cells cannot function without a continual supply of ATP. Just after eating, insulin encourages the uptake of glucose by cells and the storage of glucose as glycogen in the liver and muscles. In between eating, glucagon stimulates the liver to break down glycogen to glucose so that the blood level stays constant.

Once we see how many glands, organs, and tissues produce chemical signals, we begin to realize their importance in keeping the body working in an efficient and productive manner, as illustrated on the next page.

Hormones are intimately involved in regulating the internal environment.

Human Systems Work Together

Integumentary System

Androgens activate sebaceous glands and help regulate hair growth.

Skin provides sensory input that results in the activation of certain endocrine glands.

Skeletal System

Growth hormone regulates bone development; parathyroid hormone and calcitonin regulate Ca^{2+} content.

Bones provide protection for glands; store Ca^{2+} used as second messenger.

Muscular System

Androgens promote growth of skeletal muscle; epinephrine stimulates heart and constricts blood vessels.

Muscles help protect glands.

Nervous System

Sex hormones affect development of brain.

Hypothalamus is part of endocrine system; nerves innervate glands of secretion.

Cardiovascular System

Epinephrine increases blood pressure; ADH, aldosterone, and atrial natriuretic hormone help regulate blood volume; growth factors control blood cell formation.

Blood vessels transport hormones from glands; blood services glands; heart produces atrial natriuretic hormones.

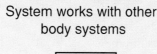

How the Endocrine System works with other body systems

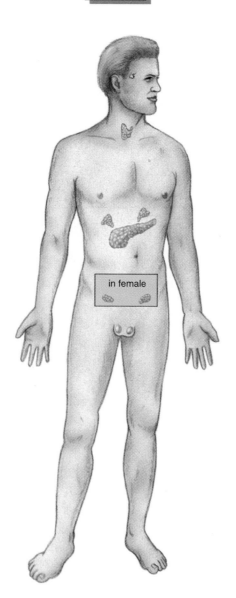

in female

Lymphatic System/Immunity

Thymus is necessary for maturity of T lymphocytes.

Lymphatic vessels pick up excess tissue fluid; immune system protects against infections.

Respiratory System

Epinephrine promotes ventilation by dilating bronchioles; growth factors control production of red blood cells that carry oxygen.

Gas exchange in lungs provides oxygen and rids body of carbon dioxide.

Digestive System

Hormones help control secretion of digestive glands and accessory organs; insulin and glucagon regulate glucose storage in liver.

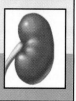

Stomach and small intestine produce hormones.

Urinary System

ADH, aldosterone, and atrial natriuretic hormone regulate reabsorption of water and Na^{+} by kidneys.

Kidneys keep blood values within normal limits so that transport of hormones continues.

Reproductive System

Hypothalamic, pituitary, and sex hormones control sex characteristics and regulate reproductive processes.

Gonads produce sex hormones.

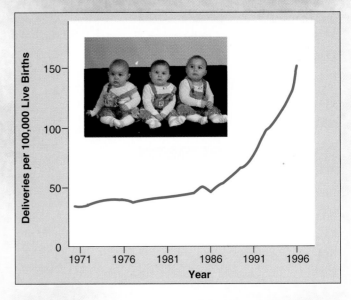

Bioethical Focus

Fertility Drugs

When Patti Frustaci gave birth to septuplets in California on May 21, 1985, the father was at first thrilled. But one septuplet was stillborn, another three died within 19 days, and the three surviving children now suffer from a myriad of health problems, including lung and heart damage. Higher-order multiple births (triplets or more) in the United States increased 19% between 1980 and 1994 (Fig. 15B). During these years, it became customary to use fertility drugs (gonadotropic hormones) to stimulate the ovaries.

The risks for premature delivery, low birth weight, and developmental abnormalities rise sharply for higher-order multiple births. And the physical and emotional burden placed on the parents is extraordinary. They face endless everyday chores and find it difficult to maintain normal social relationships, if only because they get insufficient sleep. Finances are strained in order to provide for the children's needs, including housing and child-care assistance. About one-third report that they received no help from relatives, friends, or neighbors in the first year after the birth. Trips to the hospital for accidental injury are more frequent because parents with only two arms and legs cannot keep so many children safe at one time.

Many clinicians are now urging that all possible steps be taken to ensure that the risk of higher-order multiple births be reduced. However, none of the ethical choices to bring this about are attractive. If fertility drugs are outlawed, some couples might be denied the possibility of ever having a child. A higher-order multiple pregnancy can be terminated, or selective reduction can be done. During selective reduction, one or more of the fetuses is killed by an injection of potassium chloride. Selective reduction could very well result in psychological and social complications for the mother and surviving children. The parents could opt to utilize in vitro fertilization (in which the eggs are fertilized in the lab), with the intent that only one or two zygotes will be placed in the woman's womb. But then, as

Figure 15B Higher-order multiple births from 1971 to 1996.

is the case with selective reduction, any leftover zygotes may never have an opportunity to continue development.

Decide Your Opinion

1. Do you approve of the use of fertility drugs despite the risk of higher-order multiple births? Why or why not?
2. Not many clinics will carry out selective reduction. Should a woman be forced to carry all fetuses to term even if it increases the likelihood that some children will be born with physical abnormalities?
3. Should society decide the fate of zygotes left over from in vitro fertilization, or should that be left up to the couple?

Looking at Both Sides   www.mhhe.com/biosci/genbio/maderhuman7/

Every bioethical issue has at least two sides. Even if you already have an opinion, it is important to explore the opposite opinion before finalizing your position. The Online Learning Center at www.mhhe.com/biosci/genbio/maderhuman7/ will help you fine-tune your initial opinion, explore both sides, and finalize your position. Either you will acquire new arguments for your original opinion or you may even change your opinion. Be sure to complete these activities in sequence:

Taking Sides   Decide your initial opinion by answering a series of questions. Then see if your opinion changes after completing the next two activities.

Further Debate   Read opposing articles that give you further information on this particular bioethical issue.

Explain Your Position   Answer another series of questions and then defend your original or changed opinion. You can e-mail your position to your instructor if he or she wishes.

Summarizing the Concepts

15.1 Endocrine Glands
Endocrine glands secrete hormones into the bloodstream, and from there they are distributed to target organs or tissues. The major endocrine glands are listed in Table 15.1.

15.2 Hypothalamus and Pituitary Gland
Neurosecretory cells in the hypothalamus produce antidiuretic hormone (ADH) and oxytocin, which are stored in axon endings in the posterior pituitary until they are released.

The hypothalamus produces hypothalamic-releasing and hypothalamic-inhibiting hormones, which pass to the anterior pituitary by way of a portal system. The anterior pituitary produces at least six types of hormones, and some of these stimulate other hormonal glands to secrete hormones.

15.3 Thyroid and Parathyroid Glands
The thyroid gland requires iodine to produce thyroxine and triiodothyronine, which increase the metabolic rate. If iodine is available in limited quantities, a simple goiter develops; if the thyroid is overactive, an exophthalmic goiter develops. The thyroid gland also produces calcitonin, which helps lower the blood calcium level. The parathyroid glands secrete parathyroid hormone, which raises the blood calcium and decreases the blood phosphate levels.

15.4 Adrenal Glands
The adrenal glands respond to stress: immediately, the adrenal medulla secretes epinephrine and norepinephrine, which bring about responses we associate with emergency situations. On a long-term basis, the adrenal cortex produces the glucocorticoids (e.g., cortisol) and the mineralocorticoids (e.g., aldosterone). Cortisol stimulates hydrolysis of proteins to amino acids that are converted to glucose; in this way, it raises the blood glucose level.

Aldosterone causes the kidneys to reabsorb sodium ions (Na^+) and to excrete potassium ions (K^+). Addison disease develops when the adrenal cortex is underactive, and Cushing syndrome develops when the adrenal cortex is overactive.

15.5 Pancreas
The pancreatic islets secrete insulin, which lowers the blood glucose level, and glucagon, which has the opposite effect. The most common illness caused by hormonal imbalance is diabetes mellitus, which is due to the failure of the pancreas to produce insulin or the failure of the cells to take it up.

15.6 Other Endocrine Glands
The gonads produce the sex hormones; the pineal gland produces melatonin, which may be involved in circadian rhythms and the development of the reproductive organs; and the thymus secretes thymosins, which stimulate T lymphocyte production and maturation.

Tissues also produce hormones. Adipose tissue produces leptin, which acts on the hypothalamus, and various tissues produce growth factors. Prostaglandins are produced and act locally.

15.7 Chemical Signals
Hormones are either steroids or peptides. Steroid hormones combine with a receptor in the cell, and the complex attaches to and activates DNA. Reception of a peptide hormone at the plasma membrane activates an enzyme cascade.

In the human body, some chemical signals, such as traditional endocrine hormones and secretions of neurosecretory cells, act at a distance. Others, such as prostaglandins, growth factors, and neurotransmitters, act locally. The endocrine system works with the other organ systems to maintain homeostasis in the ways described in the illustration on page 311.

Studying the Concepts

1. Describe a mechanism by which the production of a hormone is regulated and another by which the effect of a hormone is controlled. 294
2. Explain the relationship of the hypothalamus to the posterior pituitary gland and to the anterior pituitary gland. List the hormones secreted by the posterior and anterior pituitary glands. 296
3. Give an example of the negative feedback relationship among the hypothalamus, the anterior pituitary, and other endocrine glands. 296
4. Discuss the effect of growth hormone on the body and the result of having too much or too little growth hormone when a young person is growing. What is the result if the anterior pituitary produces growth hormone in an adult? 298
5. What types of goiters are associated with a malfunctioning thyroid? Explain each type. 299
6. How do the thyroid and the parathyroid work together to control the blood calcium level? 300
7. How do the adrenal glands respond to stress? What hormones are secreted by the adrenal medulla, and what effects do these hormones have? 301
8. Name the most significant glucocorticoid and mineralocorticoid, and discuss their functions. Explain the symptoms of Addison disease and Cushing syndrome. 302–3
9. Draw a diagram to explain how insulin and glucagon maintain the blood glucose level. Use your diagram to explain the major symptoms of type I diabetes mellitus. 304–5
10. Name the other endocrine glands discussed in this chapter, and discuss the functions of the hormones they secrete. 306–7
11. What are leptin, growth factors, and prostaglandins? How do these substances act? 307
12. Explain how steroid hormones and peptide hormones affect the metabolism of the cell. 309
13. Contrast hormonal and neural signals and show that there is an overlap between the mode of operation of the nervous system and that of the endocrine system. 310

Testing Your Knowledge of the Concepts

In questions 1–6, match the endocrine gland to its hormone(s).
a. anterior pituitary
b. posterior pituitary
c. thyroid
d. adrenal medulla
e. adrenal cortex
f. pancreas

_____ 1. insulin
_____ 2. growth hormone
_____ 3. cortisol
_____ 4. oxytocin
_____ 5. norepinephrine
_____ 6. calcitonin

In questions 7–10, indicate whether the statement is true (T) or false (F).

_____ 7. The hypothalamus produces the hormones secreted by the anterior pituitary.

_____ 8. Growth hormone is sometimes produced in adults.

_____ 9. Parathyroid hormone acts on the bones, intestines, and kidneys to cause the secretion of sodium.

_____ 10. The release of insulin just after eating keeps the blood glucose level within normal range.

In questions 11–13, fill in the blanks.

11. Aldosterone and its antagonistic hormone _____ help keep the osmolarity of the blood within its normal range.

12. The secretion of many hormones is regulated by a _____ feedback system.

13. The male sex hormone _____ and the female sex hormones _____ and _____ maintain the secondary sex characteristics.

14. Fill in this diagram to explain the negative feedback relationship between the hypothalamus, the anterior pituitary, and the target gland.

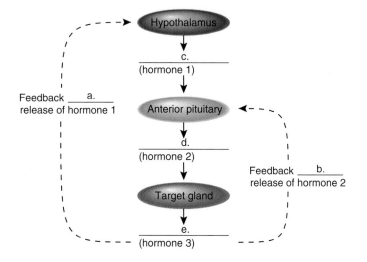

15. Match the terms to these definitions:

a. _____ Organ that is in the neck and secretes several important hormones, including thyroxine and calcitonin.

b. _____ Condition characterized by high blood glucose level and the appearance of glucose in the urine.

c. _____ Hormone secreted by the anterior pituitary that stimulates the adrenal cortex.

d. _____ Type of hormone that causes the activation of an enzyme cascade in cells.

e. _____ Hormone released by the posterior pituitary that causes contraction of the uterus and milk letdown.

Understanding Key Terms

acromegaly 298
Addison disease 303
adrenal cortex 301
adrenal gland 301
adrenal medulla 301
adrenocorticotropic
   hormone (ACTH) 296
aldosterone 302
anabolic steroid 306
androgen 306
anterior pituitary 296
antidiuretic hormone
   (ADH) 296
atrial natriuretic hormone
   (ANH) 302
calcitonin 300
circadian rhythm 307
cortisol 302
cretinism 299
Cushing syndrome 303
cyclic AMP 309
diabetes mellitus 305
endocrine gland 294
epinephrine 301
estrogen 306
exophthalmic goiter 299
glucagon 304
glucocorticoid 301
gonad 306
gonadotropic hormone 296
growth factor 307
growth hormone (GH) 296
hormone 294
hypothalamic-inhibiting
   hormone 296
hypothalamic-releasing
   hormone 296

hypothalamus 296
insulin 304
leptin 307
melanocyte-stimulating
   hormone (MSH) 296
melatonin 307
mineralocorticoids 301
myxedema 299
norepinephrine 301
ovary 306
oxytocin 296
pancreas 304
pancreatic islets (islets of
   Langerhans) 304
parathyroid gland 300
parathyroid
   hormone (PTH) 300
peptide hormone 309
pineal gland 307
pituitary dwarfism 298
pituitary gland 296
positive feedback 296
posterior pituitary 296
progesterone 306
prolactin (PRL) 296
prostaglandin (PG) 307
renin 302
simple goiter 299
steroid hormone 309
testes 306
testosterone 306
tetany 300
thymus 307
thyroid gland 299
thyroid-stimulating
   hormone (TSH) 296
thyroxine (T$_4$) 299

e-Learning Connection

15.1 Endocrine Glands

 <u>Introduction</u> *Essential Study Partner*
<u>Hormone Action</u> *Essential Study Partner*
<u>Endocrine Glands</u> *Essential Study Partner*

 <u>The Endocrine System</u> *art labeling activity*
<u>Human Endocrine System</u> *art quiz*

15.2 Hypothalamus and Pituitary Gland

 <u>The Hypothalamus</u> *Essential Study Partner*

 <u>Endocrine System Regulation</u> *animation activity*
<u>Anterior Pituitary Control by Hypothalamus</u> *art quiz*

15.3 Thyroid and Parathyroid Glands

 <u>Parathyroid Hormone</u> *animation activity*

15.4 Adrenal Glands

 <u>Disorders of the Adrenal Cortex</u> *reading*

15.5 Pancreas

 <u>Glucose Regulation</u> *animation activity*

15.6 Other Endocrine Glands

 <u>Endocrine System</u> *case study*

15.7 Chemical Signals

 <u>Peptide Hormone Action</u> *animation activity*
<u>Cyclic AMP</u> *art quiz*
<u>Working Together to Achieve Homeostasis</u>

Chapter Summary

 <u>Additional Activities</u> *Essential Study Partner*

<u>Key Term Flashcards</u> *vocabulary quiz*
<u>Chapter Quiz</u> *objective quiz covering all chapter concepts*

Further Readings

Azari, N. P., and Seitz, R. J. September/October 2000. Brain plasticity and recovery from stroke. *American Scientist* 88(5):426. Imaging by PET and fMRI allows observation of neural activity in the living brain.

Barinaga, M. June 23, 2000. A critical issue for the brain. *Science* 288(5474):2116. Research into the young brain to determine if critical periods determine what we learn and when.

Barinaga, M. March 24, 2000. Family of bitter taste receptors found. *Science* 287(5461):2133. A huge family of receptors, each of which seems to respond to different bitter-tasting compounds, has been identified.

Barkley, R. A. September 1998. Attention-deficit hyperactivity disorder. *Scientific American* 279(3):66. ADHD may result from neurological abnormalities with a genetic basis.

Cynader, M. March 17, 2000. Strengthening visual connections. *Science* 287(5460):1943. A good description is given of research to discover connections in the cerebral cortex for vision.

Debinski, W. May/June 1998. Anti-brain tumor cytotoxins. *Science & Medicine* 5(3):36. Delivery of bacterial toxins specifically to tumor cells is a new therapeutic strategy for the treatment of brain tumors.

Ezzell, C. March 2000. Brain terrain. *Scientific American* 282(3):22. Mapping the functions of various areas of the human brain is difficult and controversial.

Gadsby, P. July 2000. Tourist in a taste lab. *Discover* 21(7):70. Some tongues are far more sensitive than others in taste detection.

Gazzaniga, M. S. July 1998. The split brain revisited. *Scientific American* 279(1):50. Recent research on split brains has led to new insights into brain organization and consciousness.

Gould, J. L., and Gould, C. G. 1999. *The animal mind.* New York: Scientific American Library. A readable introduction to the study of cognition in animals.

Halstead, L. S. April 1998. Post-polio syndrome. *Scientific American* 278(4):42. Recovered polio victims are experiencing fatigue, pain, and weakness, resulting from degeneration of motor neurons.

Jordan, V. C. October 1998. Designer estrogens. *Scientific American* 279(4):60. Selective estrogen receptor modulators may protect against breast and endometrial cancers, osteoporosis, and heart disease.

Kempermann, G., and Gage, F. May 1999. New nerve cells for the adult brain. *Scientific American* 280(5):48. The human brain has been found to produce new nerve cells in adulthood.

Koch, C., and Laurent, G. April 2, 1999. Complexity and the nervous system. *Science* 284(5411):96. Advances in the neurosciences have revealed the staggering complexity of even "simple" nervous systems.

Kunzig, R. February 1999. What's a pinna for? *Discover* 20(2):24. Article discusses what the folds of the outer ear are for.

LeVay, S. March 2000. Brain invaders. *Scientific American* 282(3):27. A new auditory prosthesis implanted directly into the brain stem may restore hearing.

Mader, S. S. 2001. *Understanding human anatomy and physiology.* 4th ed. Dubuque, Iowa: The McGraw-Hill Companies, Inc. A text that emphasizes the basics for beginning allied health students.

Marcus, D. M., and Camp, M. W. May/June 1998. Age-related macular degeneration. *Science & Medicine* 5(3):10. New therapies are needed for this common cause of vision loss.

Mattson, M. P. March/April 1998. Experimental models of Alzheimer's disease. *Science & Medicine* 5(2):16. In this disease, mutations accelerate changes that occur during normal aging.

McDonald, J. W., et al. September 1999. Repairing the damaged spinal cord. *Scientific American* 281(3):64. New treatments aim to minimize or reverse damage to the spinal cord; some restoration of the injured spinal cord seems feasible.

Nolte, J. 1999. *The human brain.* 4th ed. St. Louis: Mosby, Inc. Beginners are guided through the basic aspects of brain structure and function.

Powledge, T. M. July 1999. Addiction and the brain. *BioScience* 49(7):513. Addiction is seen as a brain disease triggered by drugs that change the biochemistry and anatomy of neurons.

Sack, R. L. September/October 1998. Melatonin. *Science & Medicine* 5(5):8. Certain mood and sleep disorders can be managed with melatonin treatments.

Sapolsky, R. March 1999. Stress and your shrinking brain. *Discover* 20(3):116. Excess stress hormones may damage the brain, particularly the hippocampus.

Saunders, F. January 1999. Iris ID. *Discover* 20(1):34. An identification system has been developed in which the iris of the eye is scanned and its unique patterns are recognized.

Smith, E., and Jonides, J. March 12, 1999. Storage and executive processes in the frontal lobes. *Science* 283(5408):1657. The human frontal cortex helps mediate working memory, which is used for temporary storage of information.

Springer, S., and Deutsch, G. 1998. *Left brain, right brain.* 5th ed. New York: W. H. Freeman & Co. A readable overview of brain asymmetry and its implications; introduces current brain-behavior research.

Squire, L. R., and Kandel, E. R. 2000. *Memory: From mind to molecules.* New York: Scientific American Library. For general readership, this book offers a comprehensive overview of memory from molecules and cells to brain systems and cognition.

Staff writer. February 11, 2000. A new clue to how alcohol damages brains. *Science* 287(5455):947. Article presents new information about alcohol and brain damage.

Staff writer. July 23, 1999. Mapping smells in the brain. *Science* 285(5427):508. Article discusses how a particular smell calls up complex memories.

Staff writers. October 22, 1999. Olfaction. *Science* 286(5440):703. This issue contains a series of articles on the sense of smell.

Tessier-Lavigne, M., and Goodman, C. S. February 4, 2000. Regeneration in the Nogo Zone. *Science* 287(5454):813. Article discusses the role of Schwann cells and oligodendrocyte cells in nerve regeneration.

Vogel, G. June 2, 2000. Brain cells reveal surprising versatility. *Science* 288(5471):1559. Stem cells from the brains of adult mice can become functional blood cells and, when injected into embryos, can develop into other types of body tissues.

Whitacre, C., et al. February 26, 1999. A gender gap in autoimmunity. *Science* 283(5406):1277. The predominance of autoimmune disease among women suggests that sex hormones may regulate susceptibility.

Youdim, M. B., and Riederer, P. January 1997. Understanding Parkinson's disease. *Scientific American* 276(1):52. The tremors and immobility of Parkinson disease can be traced to damage in a part of the brain that regulates movement.

PART

V

Reproduction in Humans

H uman beings are either male or female. The reproductive organs of each sex produce the sex cells that join prior to the development of a new individual. The embryo develops into a fetus within the body of the female, and birth occurs when there is a reasonable chance for independent existence.

We are in the midst of a sexual revolution. We have the freedom to engage in varied sexual practices and to reproduce by alternative methods of conception, such as in vitro fertilization. With freedom comes a responsibility to be familiar with the biology of reproduction and health-related issues, such as sexually transmitted diseases, not only for ourselves but for our potential offspring.

Chapter 16

Reproductive System

Chapter Concepts

16.1 Male Reproductive System
- The male reproductive system is designed for continuous sperm production and delivery within a fluid medium.   318

16.2 Female Reproductive System
- The female reproductive system is designed for the monthly production of an egg and preparation of the uterus to house the developing fetus.   322

16.3 Female Hormone Levels
- Hormones control the monthly reproductive cycle in females and play a significant role in maintaining pregnancy, should it occur.   325

16.4 Control of Reproduction
- Birth control measures vary in effectiveness from those that are very effective to those that are minimally effective.   329
- There are assisted reproductive technologies today, including in vitro fertilization followed by transfer of an embryo to the uterus.   332

16.5 Homeostasis
- The reproductive system works with the other systems of the body to maintain homeostasis.   333

It had seemed simple enough. Leigh Anne and Joe graduated from college, launched their careers, and got married. Some years later, they bought a house. Then, they decided to have a baby.

That's when things got complicated.

For some reason, Leigh Anne just didn't get pregnant. After two years of trying to conceive, the couple headed to a well-known fertility specialist. After some tests, the doctor explained a variety of fertility treatments and drugs designed to help couples conceive.

Leigh Anne and Joe weighed their options and decided to try in vitro fertilization. During a series of visits to the clinic, a doctor removed eggs from Leigh Anne and combined them with sperm from Joe. Nurtured in the lab, the combination formed fertilized eggs, which the doctor then placed in Leigh Anne's uterus.

Fortunately, the procedure worked. Leigh Anne's pregnancy was normal and healthy. Today, the couple's three-year-old races around the house, plays leapfrog over the family dog, and eats everything in sight. At some point, Leigh Anne and Joe might try in vitro fertilization again. For now, though, they'd just like their toddler to try a nap.

16.1 Male Reproductive System

The male reproductive system includes the organs depicted in Figure 16.1. The male gonads are paired testes (sing., **testis**), which are suspended within the scrotal sacs of the **scrotum.**

Sperm produced by the testes mature within the **epididymis** (pl., epididymides), which is a tightly coiled duct lying just outside each testis. Maturation seems to be required for sperm to swim to the egg. When sperm leave an epididymis, they enter a **vas deferens** (pl., vasa deferentia) where they may also be stored for a time. Each vas deferens passes into the abdominal cavity where it curves around the bladder and empties into an ejaculatory duct. The ejaculatory ducts enter the **urethra.**

At the time of ejaculation, sperm leave the penis in a fluid called seminal fluid **(semen).** The seminal vesicles, the prostate gland, and the bulbourethral glands (Cowper glands) add secretions to seminal fluid. The pair of **seminal vesicles** lie at the base of the bladder, and each has a duct that joins with a vas deferens. The **prostate gland** is a single doughnut-shaped gland that surrounds the upper portion of the urethra just below the bladder. In older men, the

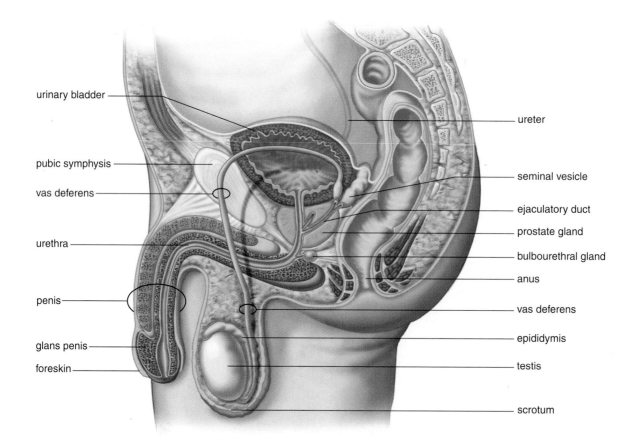

Figure 16.1   The male reproductive system.
The testes produce sperm. The seminal vesicles, the prostate gland, and the bulbourethral glands provide a fluid medium for the sperm. Circumcision is the removal of the foreskin. Notice that the penis in this drawing is not circumcised because the foreskin is present.

prostate can enlarge and squeeze off the urethra, making urination painful and difficult. The condition can be treated medically. **Bulbourethral glands** are pea-sized organs that lie posterior to the prostate on either side of the urethra.

Each component of seminal fluid seems to have a particular function. Sperm are more viable in a basic solution, and seminal fluid, which is milky in appearance, has a slightly basic pH (about 7.5). Swimming sperm require energy, and seminal fluid contains the sugar fructose, which presumably serves as an energy source. Seminal fluid also contains prostaglandins, chemicals that cause the uterus to contract. Some investigators believe that uterine contractions help propel the sperm toward the egg.

Orgasm in Males

The **penis** (Fig. 16.2) is the male organ of sexual intercourse. The penis has a long shaft and an enlarged tip called the glans penis. The glans penis is normally covered by a layer of skin called the foreskin. Circumcision is the surgical removal of the foreskin, usually soon after birth.

Spongy, erectile tissue containing distensible blood spaces extends through the shaft of the penis. During sexual arousal, autonomic nerve impulses lead to the production of cGMP (cyclic guanosine monophosphate) in smooth muscle cells, and the erectile tissue fills with blood. The veins that take blood away from the penis are compressed, and the penis becomes erect. **Erectile dysfunction** (formerly called impotency) exists when the erectile tissue doesn't expand enough to compress the veins. The drug Viagra inhibits an enzyme that breaks down cGMP, ensuring that a full erection will take place. However, vision problems may occur when taking Viagra because the same enzyme occurs in the retina.

As sexual stimulation intensifies, sperm enter the urethra from each vas deferens, and the glands contribute secretions to the seminal fluid. Once seminal fluid is in the urethra, rhythmic muscle contractions cause it to be expelled from the penis in spurts. During ejaculation, a sphincter closes off the bladder so that no urine enters the urethra. (Notice that the urethra carries either urine or semen at different times.)

The contractions that expel seminal fluid from the penis are a part of male orgasm, the physiological and psychological sensations that occur at the climax of sexual stimulation. The psychological sensation of pleasure is centered in the brain, but the physiological reactions involve the genital (reproductive) organs and associated muscles, as well as the entire body. Marked muscular tension is followed by contraction and relaxation.

Following ejaculation and/or loss of sexual arousal, the penis returns to its normal flaccid state. After ejaculation, a male typically experiences a period of time, called the refractory period, during which stimulation does not bring about an erection. The length of the refractory period increases with age.

There may be in excess of 400 million sperm in the 3.5 ml of semen expelled during ejaculation. The sperm count can be much lower than this, however, and fertilization of the egg by a sperm still can take place.

Sperm are produced in the testes, mature in the epididymis, and pass from the vas deferens to the urethra. After glands add fluid to sperm, semen is ejaculated from the penis at the time of male orgasm.

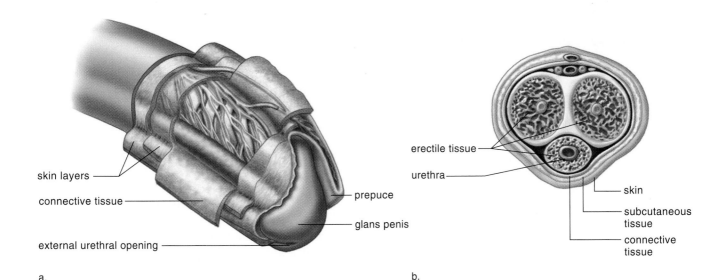

a.

b.

Figure 16.2   Penis anatomy.
a. Beneath the skin and the connective tissue lies the urethra, surrounded by erectile tissue. This tissue expands to form the glans penis, which in uncircumcised males is partially covered by the foreskin. **b.** Two other columns of erectile tissue in the penis are located dorsally.

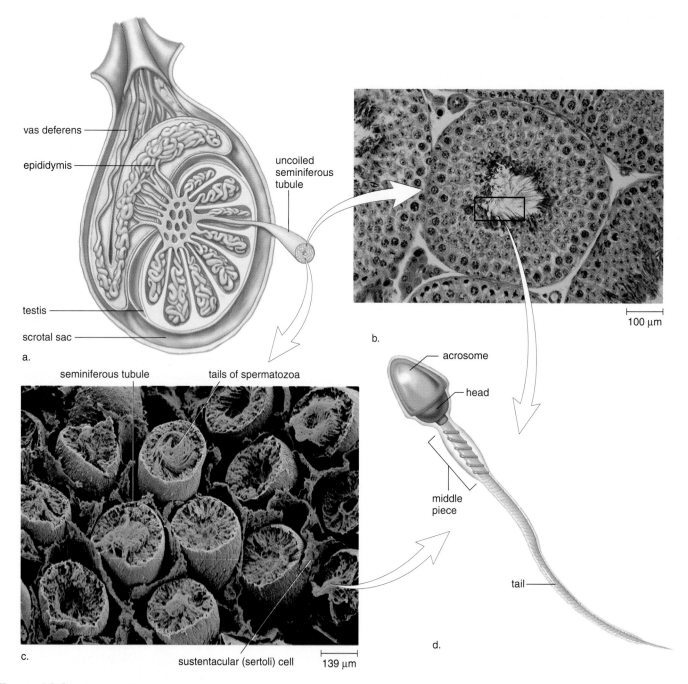

Figure 16.3 **Testis and sperm.**
a. The lobules of a testis contain seminiferous tubules. **b.** Light micrograph of a cross section of the seminiferous tubules. **c.** Scanning electron micrograph of a cross section of the seminiferous tubules, where spermatogenesis occurs. **d.** A sperm has a head, a middle piece, and a tail. The nucleus is the head, capped by the enzyme-containing acrosome.

Male Gonads, the Testes

The testes lie outside the abdominal cavity of the male within the scrotum. The testes begin their development inside the abdominal cavity but descend into the scrotal sacs during the last two months of fetal development. If, by chance, the testes do not descend and the male is not treated or operated on to place the testes in the scrotum, sterility—the inability to produce offspring—usually follows. This is because the internal temperature of the body is too high to produce viable sperm. The scrotum helps regulate the temperature of the testes by holding them closer or further away from the body.

A longitudinal section of a testis shows that it is composed of compartments called lobules, each of which contains one to three tightly coiled **seminiferous tubules** (Fig. 16.3a). Altogether, these tubules have a combined length of approximately 250 meters. A microscopic cross section of a

Table 16.1	Male Reproductive Organs
Organ	**Function**
Testes	Produce sperm and sex hormones
Epididymides	Where sperm mature and some sperm are stored
Vasa deferentia	Conduct and store sperm
Seminal vesicles	Contribute nutrients and fluid to semen
Prostate gland	Contributes basic fluid to semen
Urethra	Conducts sperm
Bulbourethral glands	Contribute mucoid fluid to semen
Penis	Organ of sexual intercourse

seminiferous tubule shows that it is packed with cells undergoing **spermatogenesis** (Fig. 16.3b), the production of sperm. Also present are **sustentacular** (Sertoli) **cells,** which support, nourish, and regulate the spermatogenic cells (Fig. 16.3c).

Mature **sperm,** or spermatozoa, have three distinct parts: a head, a middle piece, and a tail (Fig. 16.3d). Mitochondria in the middle piece provide energy for the movement of the tail, which has the structure of a flagellum. The head contains a nucleus covered by a cap called the **acrosome,** which stores enzymes needed to penetrate the egg. The ejaculated semen of a normal human male contains several hundred million sperm, but only one sperm normally enters an egg. Sperm usually do not live more than 48 hours in the female genital tract.

Hormonal Regulation in Males

The hypothalamus has ultimate control of the testes' sexual function because it secretes a hormone called **gonadotropin-releasing hormone,** or **GnRH,** that stimulates the anterior pituitary to secrete the gonadotropic hormones. There are two gonadotropic hormones—**follicle-stimulating hormone (FSH)** and **luteinizing hormone (LH)**—in both males and females. In males, FSH promotes the production of sperm in the seminiferous tubules, which also release the hormone inhibin.

LH in males is sometimes given the name **interstitial cell-stimulating hormone (ICSH)** because it controls the production of testosterone by the **interstitial cells,** which are found in the spaces between the seminiferous tubules. All these hormones are involved in a negative feedback relationship that maintains the fairly constant production of sperm and testosterone (Fig. 16.4). ♊

Testosterone, the main sex hormone in males, is essential for the normal development and functioning of the organs listed in Table 16.1. Testosterone also brings about and maintains the male secondary sex characteristics that develop at the time of puberty. Males are generally taller than females and have broader shoulders and longer legs relative to trunk length. The deeper voices of males compared to

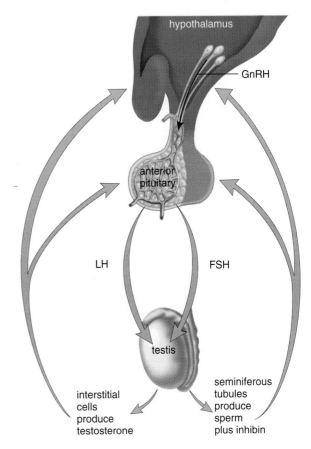

Figure 16.4   Hormonal control of testes.
GnRH (gonadotropin-releasing hormone) stimulates the anterior pituitary to secrete the gonadotropic hormones: FSH stimulates the production of sperm, and LH stimulates the production of testosterone. Testosterone and inhibin exert negative feedback control over the hypothalamus and the anterior pituitary, and this regulates the level of testosterone in the blood.

females are due to a larger larynx with longer vocal cords. Since the so-called Adam's apple is a part of the larynx, it is usually more prominent in males than in females. Testosterone causes males to develop noticeable hair on the face, chest, and occasionally other regions of the body such as the back. Testosterone also leads to the receding hairline and pattern baldness that occur in males.

Testosterone is responsible for the greater muscular development in males. Knowing this, males and females sometimes take anabolic steroids, which are either testosterone or related steroid hormones resembling testosterone. Health problems involving the kidneys, the circulatory system, and hormonal imbalances can arise from such use. The testes shrink in size, and feminization of other male traits occurs.

The gonads in males are the testes, which produce sperm as well as testosterone, the most significant male sex hormone.

16.2 Female Reproductive System

The female reproductive system includes the organs depicted in Figure 16.5 and listed in Table 16.2. The female gonads are paired **ovaries** that lie in shallow depressions, one on each side of the upper pelvic cavity. **Oogenesis** is the production of an **egg,** the female gonad. The ovaries alternate in producing one egg a month. **Ovulation** is the process by which an egg bursts from an ovary and usually enters an oviduct.

The Genital Tract

The **oviducts,** also called uterine or fallopian tubes, extend from the uterus to the ovaries; however, the oviducts are not attached to the ovaries. Instead, they have fingerlike projections called fimbriae (sing., **fimbria**) that sweep over the ovaries. When an egg bursts from an ovary during ovulation, it usually is swept into an oviduct by the combined action of the fimbriae and the beating of cilia that line the oviducts.

Once in the oviduct, the egg is propelled slowly by ciliary movement and tubular muscle contraction toward the uterus. Because an egg only lives approximately 6 to 24 hours, fertilization and **zygote** formation occur while the egg is still in an oviduct. The developing embryo normally arrives at the uterus after several days and then embeds, or implants, itself in the uterine lining, which has been prepared to receive it.

The **uterus** is a thick-walled, muscular organ about the size and shape of an inverted pear. Normally, it lies above and is tipped over the urinary bladder. The oviducts join the uterus at its upper end, while at its lower end the **cervix** enters the vagina nearly at a right angle.

Cancer of the cervix is a common form of cancer in women. Early detection is possible by means of a **Pap test,** which requires the removal of a few cells from the region

Table 16.2	Female Reproductive Organs
Organ	**Function**
Ovaries	Produce egg and sex hormones
Oviducts (fallopian tubes)	Conduct egg; location of fertilization
Uterus (womb)	Houses developing fetus
Cervix	Contains opening to uterus
Vagina	Receives penis during sexual intercourse; serves as birth canal and as an exit for menstrual flow

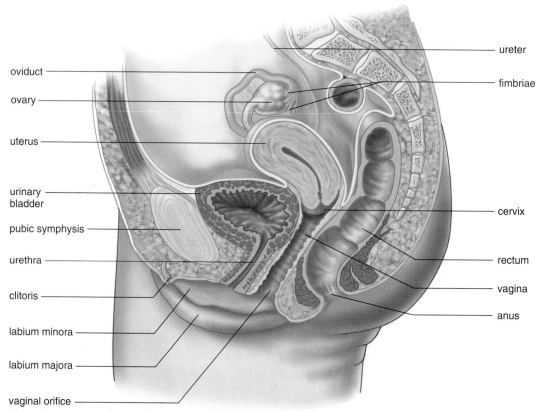

oviduct
ovary
uterus
urinary bladder
pubic symphysis
urethra
clitoris
labium minora
labium majora
vaginal orifice
ureter
fimbriae
cervix
rectum
vagina
anus

Figure 16.5   The female reproductive system.
The ovaries release one egg a month; fertilization occurs in the oviduct, and development occurs in the uterus. The vagina is the birth canal and the organ of sexual intercourse.

of the cervix for microscopic examination. If the cells are cancerous, a hysterectomy may be recommended. A hysterectomy is the removal of the uterus, including the cervix. Removal of the ovaries in addition to the uterus is technically termed an ovariohysterectomy. Because the vagina remains, the woman still can engage in sexual intercourse.

Development of the embryo normally takes place in the uterus. This organ, sometimes called the womb, is approximately 5 cm wide in its usual state but is capable of stretching to over 30 cm wide to accommodate a growing baby. The lining of the uterus, called the **endometrium,** participates in the formation of the placenta (see p. 328), which supplies nutrients needed for embryonic and fetal development. The endometrium has two layers—a basal layer and an inner, functional layer. In the nonpregnant female, the functional layer of the endometrium varies in thickness according to a monthly reproductive cycle called the uterine cycle.

A small opening in the cervix leads to the vaginal canal. The **vagina** is a tube that lies at a 45° angle with the small of the back. The mucosal lining of the vagina lies in folds and can extend. This is especially important when the vagina serves as the birth canal and it facilitates sexual intercourse, when the vagina receives the penis. The vagina acts as an exit for menstrual flow.

Once each month, an egg produced by an ovary enters an oviduct. If fertilization occurs, the developing embryo is propelled by cilia to the uterus, where it implants itself in the endometrium.

External Genitals

The external genital organs of the female are known collectively as the **vulva** (Fig. 16.6). The vulva includes two large, hair-covered folds of skin called the labia majora. The labia majora extend backward from the mons pubis, a fatty prominence underlying the pubic hair. The labia minora are two small folds lying just inside the labia majora. They extend forward from the vaginal opening to encircle and form a foreskin for the glans clitoris. The glans clitoris is the organ of sexual arousal in females and, like the penis, contains a shaft of erectile tissue that becomes engorged with blood during sexual stimulation.

The cleft between the labia minora contains the openings of the urethra and the vagina. The vagina may be partially closed by a ring of tissue called the hymen. The hymen is ordinarily ruptured by sexual intercourse or by other types of physical activities. If remnants of the hymen persist after sexual intercourse, they can be surgically removed.

Notice that the urinary and reproductive systems in the female are entirely separate. For example, the urethra carries only urine, and the vagina serves only as the birth canal and the organ for sexual intercourse.

Orgasm in Females

Upon sexual stimulation, the labia minora, the vaginal wall, and the clitoris become engorged with blood. The breasts also swell and the nipples become erect. The labia majora enlarge, redden, and spread away from the vaginal opening.

The vagina expands and elongates. Blood vessels in the vaginal wall release small droplets of fluid that seep into the vagina and lubricate it. Mucus-secreting glands beneath the labia minora on either side of the vagina also provide lubrication for entry of the penis into the vagina. Although the vagina is the organ of sexual intercourse in females, the clitoris plays a significant role in the female sexual response. The extremely sensitive clitoris can swell to two or three times its usual size. The thrusting of the penis and the pressure of the pubic symphyses of the partners acts to stimulate the clitoris.

Orgasm occurs at the height of the sexual response. Blood pressure and pulse rate rise, breathing quickens, and the walls of the uterus and oviducts contract rhythmically. A sensation of intense pleasure is followed by relaxation when organs return to their normal size. There is no refractory period, and multiple orgasms can occur during a single sexual experience.

The vagina and the external genitals, especially the clitoris, play an active role in the sexual response of females, which culminates in uterine and oviduct contractions.

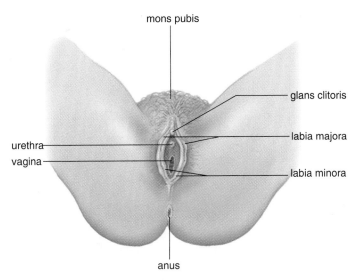

Figure 16.6   External genitals of the female.
At birth, the opening of the vagina is partially blocked by a membrane called the hymen. Physical activities and sexual intercourse disrupt the hymen.

Visual Focus

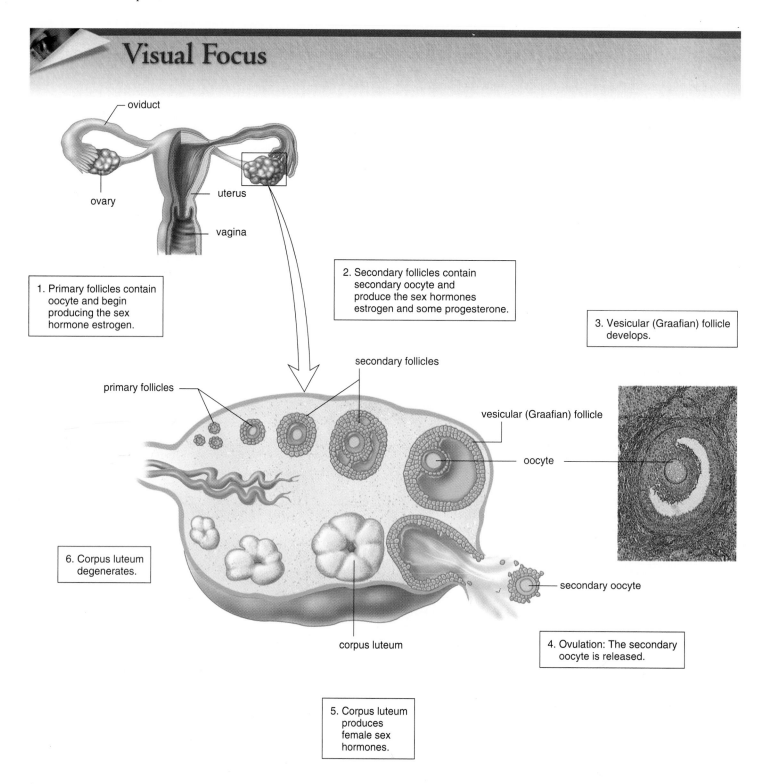

oviduct

ovary

uterus

vagina

1. Primary follicles contain oocyte and begin producing the sex hormone estrogen.

2. Secondary follicles contain secondary oocyte and produce the sex hormones estrogen and some progesterone.

3. Vesicular (Graafian) follicle develops.

primary follicles

secondary follicles

vesicular (Graafian) follicle

oocyte

6. Corpus luteum degenerates.

secondary oocyte

corpus luteum

4. Ovulation: The secondary oocyte is released.

5. Corpus luteum produces female sex hormones.

Figure 16.7   Anatomy of ovary and follicle.
As a follicle matures, the oocyte enlarges and is surrounded by layers of follicular cells and fluid. Eventually, ovulation occurs, the mature follicle ruptures, and the secondary oocyte is released. A single follicle actually goes through all stages in one place within the ovary.

16.3 Female Hormone Levels

Hormone levels cycle in the female on a monthly basis, and the ovarian cycle drives the uterine cycle as discussed in this section.

The Ovarian Cycle

A longitudinal section through an ovary shows that it is made up of an outer cortex and an inner medulla (Fig. 16.7). There are many **follicles** in the cortex, and each one contains an immature egg, called an oocyte. A female is born with as many as 2 million follicles, but the number is reduced to 300,000–400,000 by the time of puberty. Only a small number of follicles (about 400) ever mature because a female usually produces only one egg per month during her reproductive years. Since oocytes are present at birth, they age as the woman ages. This may be one reason why older women are more likely to produce children with genetic defects.

As the follicle undergoes maturation, it develops from a primary follicle to a secondary follicle to a vesicular (Graafian) follicle (Fig. 16.7). A primary follicle consists of a primary oocyte surrounded by a simple squamous epithelium. The secondary follicle has a stratified cuboidal epithelium and begins to develop pools of follicular fluid surrounding the oocyte. In a vesicular follicle, a fluid-filled cavity increases to the point that the follicle wall balloons out on the surface of the ovary.

As the vesicular follicle develops, the primary oocyte divides, producing a secondary oocyte. The vesicular follicle bursts, releasing the secondary oocyte (often called an egg for convenience) surrounded by a clear membrane. This is referred to as ovulation. Once a follicle has lost its egg, it develops into a **corpus luteum,** a glandlike structure. If pregnancy does not occur, the corpus luteum begins to degenerate after about 10 days.

These events, called the **ovarian cycle,** are under the control of the gonadotropic hormones, follicle-stimulating hormone (FSH), and luteinizing hormone (LH) (Fig. 16.8). The gonadotropic hormones are not present in constant amounts but instead are secreted at different rates during the cycle. During the first half, or follicular phase, of the ovarian cycle, FSH promotes the development of a follicle in the ovary, which secretes estrogen and some progesterone. As the estrogen level in the blood rises, it exerts feedback control over the anterior pituitary secretion of FSH so that the follicular phase comes to an end.

Presumably, the high level of estrogen in the blood also causes a sudden secretion of a large amount of GnRH from the hypothalamus. This leads to a surge of LH production by the anterior pituitary and to ovulation at about the 14th day of a 28-day cycle.

During the second half, or luteal phase, of the ovarian cycle, LH promotes the development of the corpus luteum, which secretes progesterone and some estrogen. Progesterone causes the endometrium to build up. As the blood level of progesterone rises, it exerts feedback control over the anterior pituitary secretion of LH so that the corpus luteum in the ovary begins to degenerate. As the luteal phase comes to an end, menstruation occurs.

One ovarian follicle per month produces a secondary oocyte. Following ovulation, the follicle develops into the corpus luteum.

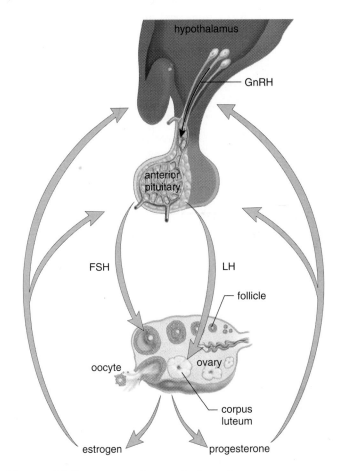

Figure 16.8 Hormonal control of ovaries.
The hypothalamus produces GnRH (gonadotropin-releasing hormone). GnRH stimulates the anterior pituitary to produce FSH (follicle-stimulating hormone) and LH (luteinizing hormone). FSH stimulates the follicle to produce estrogen, and LH stimulates the corpus luteum to produce progesterone. Estrogen and progesterone maintain the sexual organs (e.g., uterus) and the secondary sex characteristics, and exert feedback control over the hypothalamus and the anterior pituitary.

The Uterine Cycle

The female sex hormones, **estrogen** and **progesterone,** have numerous functions. These hormones affect the endometrium of the uterus, causing the uterus to undergo a cyclical series of events known as the **uterine cycle** (Table 16.3 and Fig. 16.9). Twenty-eight-day cycles are divided as follows.

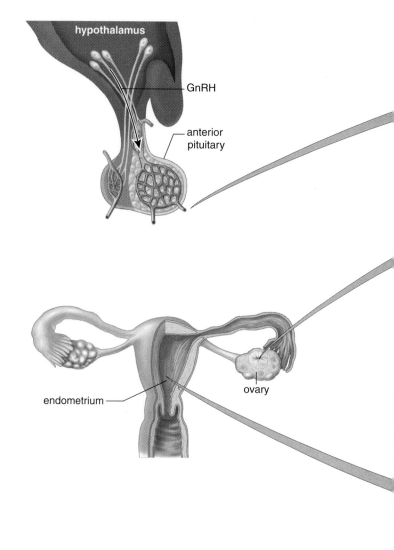

- During *days 1–5,* a low level of female sex hormones in the body causes the endometrium to disintegrate and its blood vessels to rupture. On day one of the cycle, a flow of blood and tissues, known as the menses, passes out of the vagina during **menstruation,** also called the menstrual period.

- During *days 6–13,* increased production of estrogen by a new ovarian follicle in the ovary causes the endometrium to thicken and to become vascular and glandular. This is called the proliferative phase of the uterine cycle.

- Ovulation usually occurs on the 14th day of a 28-day cycle.

- During *days 15–28,* increased production of progesterone by the corpus luteum in the ovary causes the endometrium of the uterus to double or triple in thickness (from 1 mm to 2–3 mm) and the uterine glands to mature, producing a thick mucoid secretion. This is called the secretory phase of the uterine cycle. The endometrium now is prepared to receive the developing embryo. If this does not occur, the corpus luteum in the ovary degenerates, and the low level of sex hormones in the female body results in the endometrium breaking down.

During the uterine cycle, the endometrium of the uterus builds up and then is broken down during menstruation.

Table 16.3	Ovarian and Uterine Cycles		
Ovarian Cycle	**Events**	**Uterine Cycle**	**Events**
Follicular phase—Days 1–13	FSH secretion begins	Menstruation—Days 1–5	Endometrium breaks down
	Follicle maturation occurs	Proliferative phase—Days 6–13	Endometrium rebuilds
	Estrogen secretion is prominent		
Ovulation—Day 14*	LH spike occurs		
Luteal phase—Days 15–28	LH secretion continues	Secretory phase—Days 15–28	Endometrium thickens and glands are secretory
	Corpus luteum form		
	Progesterone secretion is prominent		

*Assuming 28-day cycle.

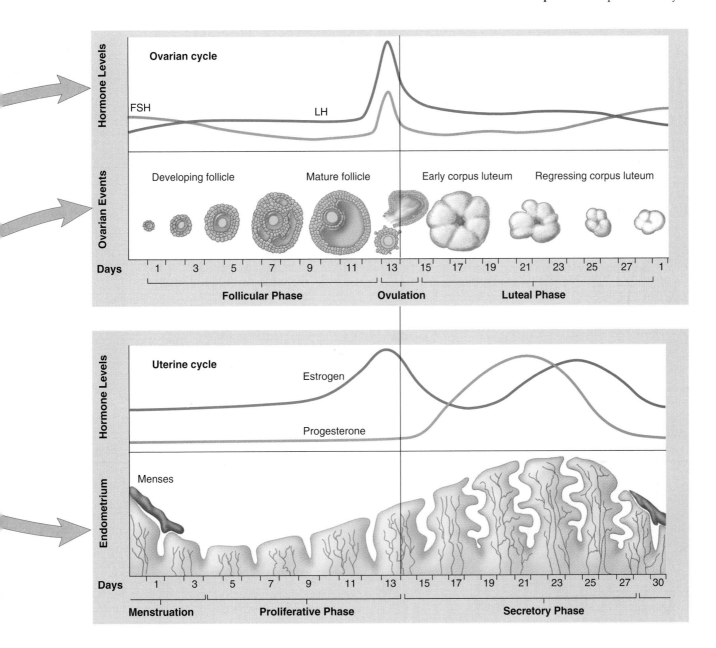

Figure 16.9 **Female hormone levels.**

During the follicular phase of the ovarian cycle, FSH released by the anterior pituitary promotes the maturation of a follicle in the ovary. The ovarian follicle produces increasing levels of estrogen, which causes the endometrium to thicken during the proliferative phase of the uterine cycle. After ovulation and during the luteal phase of the ovarian cycle, LH promotes the development of the corpus luteum. This structure produces increasing levels of progesterone, which causes the endometrial lining to become secretory. Menses due to the breakdown of the endometrium begins when progesterone production declines to a low level.

Fertilization and Pregnancy

If fertilization does occur, an embryo begins development even as it travels down the oviduct to the uterus. The endometrium is now prepared to receive the developing embryo, which becomes embedded in the lining several days following fertilization (Fig. 16.10). The **placenta** originates from both maternal and fetal tissues. It is the region of exchange of molecules between fetal and maternal blood, although there is rarely any mixing of the two. At first, the placenta produces **human chorionic gonadotropin (HCG),** which maintains the corpus luteum in the ovary until the placenta begins its own production of progesterone and estrogen. Progesterone and estrogen have two effects: they shut down the anterior pituitary so that no new follicle in the ovaries matures, and they maintain the endometrium so that the corpus luteum in the ovary is no longer needed. Usually, there is no menstruation during pregnancy.

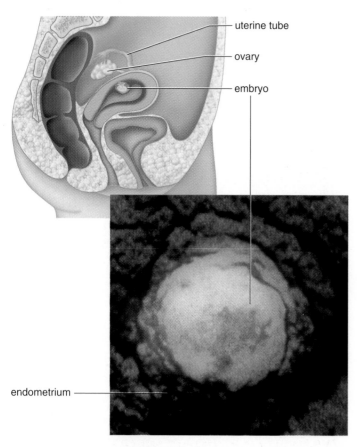

Figure 16.10   Implantation.
A scanning electron micrograph showing an embryo implanted in the endometrium on day 12 following fertilization.

Estrogen and Progesterone

Estrogen and progesterone affect not only the uterus but other parts of the body as well. Estrogen is largely responsible for the secondary sex characteristics in females, including body hair and fat distribution. In general, females have a more rounded appearance than males because of a greater accumulation of fat beneath the skin. Like males, females develop axillary and pubic hair during puberty. In females, the upper border of pubic hair is horizontal, but in males, it tapers toward the navel. Both estrogen and progesterone are also required for breast development. Other hormones are involved in milk production following a pregnancy and milk letdown when a baby begins to nurse.

The pelvic girdle is wider and deeper in females, so the pelvic cavity usually has a larger relative size compared to males. This means that females have wider hips than males and that the thighs converge at a greater angle toward the knees. Because the female pelvis tilts forward, females tend to have more of a lower back curve than males, an abdominal bulge, and protruding buttocks.

Menopause

Menopause, the period in a woman's life during which the ovarian and uterine cycles cease, is likely to occur between ages 45 and 55. The ovaries are no longer responsive to the gonadotropic hormones produced by the anterior pituitary, and the ovaries no longer secrete estrogen or progesterone. At the onset of menopause, the uterine cycle becomes irregular, but as long as menstruation occurs, it is still possible for a woman to conceive. Therefore, a woman is usually not considered to have completed menopause until there has been no menstruation for a year.

The hormonal changes during menopause often produce physical symptoms, such as "hot flashes" (caused by circulatory irregularities), dizziness, headaches, insomnia, sleepiness, and depression. These symptoms may be mild or even absent. If they are severe, medical attention should be sought. Women sometimes report an increased sex drive following menopause. It has been suggested that this may be due to androgen production by the adrenal cortex.

Estrogen and progesterone produced by the ovaries are the female sex hormones. They foster the development of the reproductive organs; maintain the uterine cycle, and the secondary sex characteristics in females.

16.4 Control of Reproduction

Several means are available to dampen or enhance our reproductive potential. **Contraceptives** are medications and devices that reduce the chance of pregnancy.

Birth Control Methods

The most reliable method of birth control is abstinence—that is, not engaging in sexual intercourse. This form of birth control has the added advantage of preventing transmission of a sexually transmitted disease. Table 16.4 lists other means of birth control used in the United States, and rates their effectiveness. For example, with the least effective method given in the table, we expect that within a year, 70 out of 100, or 70%, of sexually active women will not get pregnant, while 30 women will get pregnant.

Figure 16.11 features some of the most effective and commonly used means of birth control. Oral contraception (birth control pills) often involves taking a combination of estrogen and progesterone on a daily basis. The estrogen and progesterone in the birth control pill effectively shut down the pituitary production of both FSH and LH so that no follicle in the ovary begins to develop in the ovary; and since ovulation does not occur, pregnancy cannot take place. There are possible side effects, so women taking birth control pills should see a physician regularly.

An **intrauterine device (IUD)** is a small piece of molded plastic that is inserted into the uterus by a physician. IUDs are believed to alter the environment of the uterus and oviducts so that fertilization probably will not occur—but if fertilization should occur, implantation cannot take place. The type of IUD featured in Figure 16.11 has copper wire wrapped around the plastic.

The **diaphragm** is a soft latex cup with a flexible rim that lodges behind the pubic bone and fits over the cervix. Each woman must be properly fitted by a physician, and the diaphragm can be inserted into the vagina no more than two

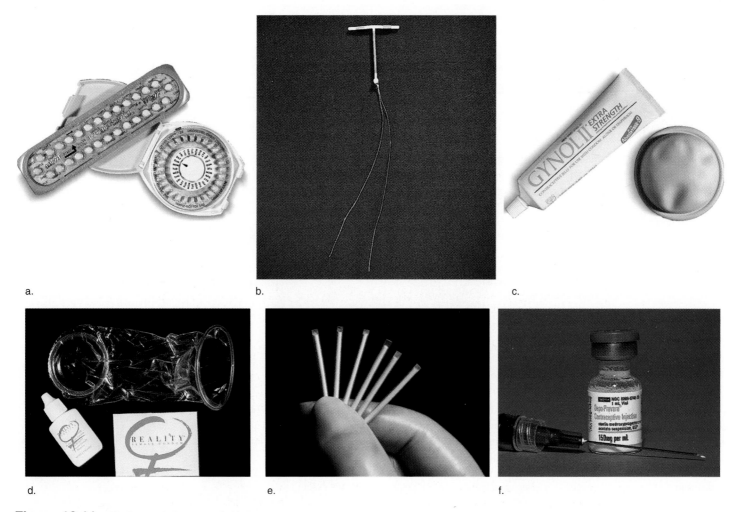

a.   b.   c.

d.   e.   f.

Figure 16.11 **Various birth control devices.**
a. Oral contraception (birth control pills). **b.** Intrauterine device. **c.** Spermicidal jelly and diaphragm. **d.** Female condom. **e.** Contraceptive implants.
f. Depo-Provera injection.

hours before sexual relations. Also, it must be used with spermicidal jelly or cream and should be left in place at least six hours after sexual relations. The cervical cap is a mini-diaphragm.

A male **condom** is most often a latex sheath that fits over the erect penis. The ejaculate is trapped inside the sheath, and thus does not enter the vagina. When used in conjunction with a spermicide, the protection is better than with the condom alone. The latex condom is generally recognized as protecting against sexually transmitted diseases.

Contraceptive implants utilize a synthetic progesterone to prevent ovulation by disrupting the ovarian cycle. Six match-sized, time-release capsules are surgically inserted under the skin of a woman's upper arm. The effectiveness of such an implant may last five years. Depo-Provera injections, which change the endometrium, utilize a synthetic progesterone that must be administered every three months. Changes occur in the endometrium that make pregnancy less likely to occur.

There has been a revival of interest in barrier methods of birth control, including the male condom, because these methods offer some protection against sexually transmitted diseases. A female condom, now available, consists of a large polyurethane tube with a flexible ring that fits onto the cervix. The open end of the tube has a ring that covers the external genitals.

Investigators have long searched for a "male pill." In a recent clinical trial, sperm production was suppressed in a small number of men who received a daily oral dose of progesterone. Skin implants of testosterone were required to maintain secondary sex characteristics. Larger studies are being planned.

Contraceptive vaccines are now being developed. For example, a vaccine developed to immunize women against HCG, the hormone so necessary to maintaining the **implantation** of the embryo, was successful in a limited clinical trial. Since HCG is not normally present in the body, no untoward autoimmune reaction is expected, but the immunization does wear off with time. Others believe that it would also be possible to develop a safe antisperm vaccine that could be used in women.

Table 16.4	Common Birth Control Methods			
Name	**Procedure**	**Methodology**	**Effectiveness**	**Risk**
Abstinence	Refrain from sexual intercourse	No sperm in vagina	100%	None
Vasectomy	Vasa deferentia cut and tied	No sperm in seminal fluid	Almost 100%	Irreversible sterility
Tubal ligation	Oviducts cut and tied	No eggs in oviduct	Almost 100%	Irreversible sterility
Oral contraception	Hormone medication taken daily	Anterior pituitary does not release FSH and LH	Almost 100%	Thromboembolism, especially in smokers
Depo-Provera injection	Four injections of progesterone-like steroid given per year	Anterior pituitary does not release FSH and LH	About 99%	Breast cancer? Osteoporosis?
Contraceptive implants	Tubes of progestin (form of progesterone) implanted under skin	Anterior pituitary does not release FSH and LH	More than 90%	Presently none known
Intrauterine device (IUD)	Plastic coil inserted into uterus by physician	Prevents implantation	More than 90%	Infection (pelvic inflammatory disease, PID)
Diaphragm	Latex cup inserted into vagina to cover cervix before intercourse	Blocks entrance of sperm to uterus	With jelly, about 90%	Presently none known
Cervical cap	Latex cap held by suction over cervix	Delivers spermicide near cervix	Almost 85%	Cancer of cervix
Male condom	Latex sheath fitted over erect penis	Traps sperm and prevents STDs	About 85%	Presently none known
Female condom	Polyurethane liner fitted inside vagina	Blocks entrance of sperm to uterus and prevents STDs	About 85%	Presently none known
Coitus interruptus	Penis withdrawn before ejaculation	Prevents sperm from entering vagina	About 75%	Presently none known
Jellies, creams, foams	These spermicidal products inserted before intercourse	Kills a large number of sperm	About 75%	Presently none known
Natural family planning	Day of ovulation determined by record keeping; various methods of testing	Intercourse avoided on certain days of the month	About 70%	Presently none known
Douche	Vagina cleansed after intercourse	Washes out sperm	Less than 70%	Presently none known

Endometriosis

In the female reproductive tract, there is a small space between an ovary and the oviduct, the tube that leads to the uterus. The lining of the uterus, called the endometrium, loses its outer layer during menstruation, and sometimes a portion of the menstrual discharge is carried backwards up the oviduct and into the abdominal cavity instead of being discharged through the opening in the cervix (Fig. 16A). There, endometrial tissue can become attached to and implanted in various organs such as the ovaries; the wall of the vagina, bowel, or bladder; or even the nerves that serve the lower back or legs. This painful condition is called endometriosis, and it affects 1–3% of women of reproductive age.

Women with uterine cycles of less than 27 days and with a menstrual flow lasting longer than one week have an increased chance of endometriosis. Women who have taken the birth control pill for a long time or have had several pregnancies have a decreased chance of endometriosis.

When a woman has endometriosis, the displaced endometrial tissue reacts as if it were still in the uterus—thickening, becoming secretory, and then breaking down. The discomfort of menstruation is then felt in other organs of the abdominal cavity, resulting in pain. An area of endometriosis can degenerate and become a scar. Scars that hold two organs together, called adhesions, can distort organs and lead to infertility.

Only direct observation of the abdominal organs can confirm endometriosis. First, a half-inch incision is made near the navel, and carbon dioxide gas is injected into the abdominal cavity to separate the organs. Then, a laparoscope (optical telescope) is inserted into the abdomen, allowing the physician to see the organs. Patches of endometriosis show up as purple, blue, or red spots, and there may be dark brown cysts filled with blood. When these rupture, there is a great deal of pain. In the severest cases, there is scarring and the formation of adhesions and abnormal masses around the pelvic organs. A second incision, usually at the pubic hairline, allows the insertion of other instruments that can be used to remove endometrial implants.

Further treatment can take one of two courses. Hormone therapy is effective, especially the use of GnRH antagonists, which shut down the anterior pituitary and therefore ovarian production of estrogen. Or the ovaries can be removed, stopping the uterine cycle. In either case, the woman suffers symptoms of menopause that can be relieved by taking estrogen in doses that do not reactivate the uterine cycle. Some women report relief from pelvic pain after acupuncture, Yoga, and meditative techniques that promote relaxation.

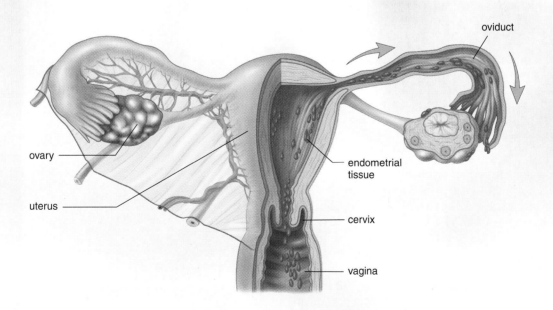

Figure 16A Endometriosis.
It has been suggested that endometriosis is caused by a backward menstrual flow, as represented by the arrows in this drawing. This backward flow allows endometrial cells to enter the abdominal cavity, where they take up residence and respond to monthly cyclic changes in hormonal levels. The result is extreme discomfort.

Morning-after Pills

The expression "morning-after pill" refers to a medication that will prevent pregnancy after unprotected intercourse. The expression is a misnomer in that medication can begin one to several days after unprotected intercourse.

One type is a kit called Preven, made up of four synthetic progesterone pills; two are taken up to 72 hours after unprotected intercourse, and two more are taken 12 hours later. The medication upsets the normal uterine cycle, making it difficult for an embryo to implant itself in the endometrium. In a recent study, it was estimated that the medication was 85% effective in preventing unintended pregnancies.

Mifepristone, better known as RU-486, is a pill that is presently used to cause the loss of an implanted embryo by blocking the progesterone receptor proteins of endometrial cells. Without functioning receptors for progesterone, the endometrium sloughs off, carrying the embryo with it. When taken in conjunction with a prostaglandin to induce uterine contractions, RU-486 is 95% effective. It is possible that some day this medication will also be a "morning-after pill," taken when menstruation is late without evidence that pregnancy has occurred.

The birth control methods and devices now available vary in effectiveness. New methods of birth control are expected to be developed.

Infertility

Infertility is the failure of a couple to achieve pregnancy after one year of regular, unprotected intercourse. The American Medical Association estimates that 15% of all couples are infertile. The cause of infertility can be attributed to the male (40%), the female (40%), or both (20%).

Causes of Infertility

The most common causes of infertility in females are blocked oviducts and endometriosis. **Endometriosis** is the presence of uterine tissue outside the uterus, particularly in the oviducts and on the abdominal organs. As discussed in the Health Focus on page 331, endometriosis can contribute to infertility. Endometriosis occurs when the menstrual discharge flows up into the oviducts and out into the abdominal cavity. This backward flow allows living uterine cells to establish themselves in the abdominal cavity where they go through the usual uterine cycle, causing pain and structural abnormalities that make it more difficult for a woman to conceive.

Sometimes the causes of infertility can be corrected by medical intervention so that couples can have children (Fig. 16.12). If no obstruction is apparent and body weight is normal, it is possible to give females fertility drugs, which are gonadotropic hormones that stimulate the ovaries and

bring about ovulation. Such hormone treatments may cause multiple ovulations and higher-order multiple births (see the Bioethical Focus in Chapter 15).

The most frequent cause of infertility in males is low sperm count and/or a large proportion of abnormal sperm. Disease, radiation, chemical mutagens, high testes temperature, and the use of psychoactive drugs can contribute to this condition. A healthy lifestyle can sometimes lead to an improved sperm count, but thus far no hormonal treatment has proven to be especially successful. For those who have had a vasectomy (a portion of the vasa deferentia removed) reversal surgery is available, but the pregnancy success rate is only about 40% unless the vasectomy occurred less than three years earlier.

When reproduction does not occur in the usual manner, many couples adopt a child. Others sometimes first try one of the assisted reproductive technologies discussed in the following paragraphs. If all the alternative methods discussed were employed simultaneously, it would be possible for a baby to have five parents: (1) sperm donor, (2) egg donor, (3) surrogate mother, and (4) and (5) contracting mother and father.

Assisted Reproductive Technologies

Assisted reproductive technologies (ART) consist of techniques used to increase the chances of pregnancy. Often, sperm and/or eggs are retrieved from the testes and ovaries, and fertilization takes place in a clinical or laboratory setting.

Artificial Insemination by Donor (AID) During artificial insemination, sperm are placed in the vagina by a physician. Sometimes a woman is artificially inseminated by her partner's sperm. This is especially helpful if the partner has a low sperm count—the sperm can be collected over a period of time and concentrated so that the sperm count is sufficient to result in fertilization. Often, however, a woman is inseminated by sperm acquired from a donor who is a complete stranger to her. At times, a combination of partner and donor sperm is used.

A variation of AID is intrauterine insemination (IUI). IUI involves the use of fertility drugs to stimulate the ovaries, followed by placement of the donor's sperm in the uterus rather than in the vagina.

If desired by the parents, sperm can be sorted into those that are believed to be X-bearing or Y-bearing to increase the chances of having a child of the desired sex.

In Vitro Fertilization (IVF) During IVF, conception occurs in laboratory glassware. The newer ultrasound machines can spot follicles in the ovaries that hold immature eggs; therefore, the latest method is to forego the administration of fertility drugs and retrieve immature eggs by using a needle. The immature eggs are then brought to maturity in glassware before concentrated sperm are then added. After about two to four days, the embryos are ready to be transferred to the uterus of the woman, who is now in the secretory phase

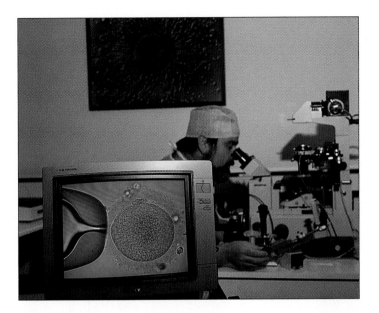

Figure 16.12   In vitro fertilization.
A researcher is using a microscope connected to a television screen (at left) to carry out in vitro fertilization. A pipette (at left of egg) holds the egg steady while a needle (not visible) introduces the sperm into the egg. In this way, in vitro fertilization ensures that fertilization will take place.

of her uterine cycle. If desired, the embryos can be tested for a genetic disease, and only those found to be free of the disease will be used. If implantation is successful, development is normal and continues to term.

Gamete Intrafallopian Transfer (GIFT)   The term **gamete** refers to a sex cell, either a sperm or an egg. Gamete intrafallopian transfer was devised as a means to overcome the low success rate (15–20%) of in vitro fertilization. The method is exactly the same as in vitro fertilization, except the eggs and the sperm are placed in the oviducts immediately after they have been brought together. GIFT has an advantage in that it is a one-step procedure for the woman—the eggs are removed and reintroduced all in the same time period. A variation on this procedure is to fertilize the eggs in the laboratory and then place the zygotes in the oviducts.

Surrogate Mothers   In some instances, women are paid to have babies. These women are called surrogate mothers. The sperm and even the egg can be contributed by the contracting parents.

Intracytoplasmic Sperm Injection (ICSI)   In this highly sophisticated procedure, one single sperm is injected into an egg. It is used effectively when a man has severe infertility problems.

When corrective procedures fail to reverse infertility, it is possible to consider assisted reproductive technologies.

16.5 Homeostasis

Regulation of sex hormone blood level is an example of homeostatic control. Figure 16.4 shows how the blood level of testosterone is maintained, and Figure 16.8 shows how the blood levels of estrogen and progesterone are maintained within normal limits. Negative feedback results in a self-regulatory mechanism that maintains the appropriate level of these hormones in the blood.

The illustration on the next page shows how the reproductive system works with the other systems of the body to maintain homeostasis. Usually we stress that the function of sex hormones is to foster the maturation of the reproductive organs and to maintain the secondary sex characteristics. These functions of sex hormones have nothing to do with homeostasis. Why? Because homeostasis pertains to the constancy of the internal environment of cells. Other activities of the sex hormones do affect the internal environment. For example, estrogen promotes fat deposition which serves as a source of energy for cells and which helps the body maintain its normal temperature because of its insulating effect.

In recent years it's been discovered that the sex hormones have still other activities that affect homeostasis even more directly. Estrogen stimulates the liver and the bones. Estrogen induces the liver to produce many types of proteins that transport substances in the blood. These include proteins that bind iron and copper and lipoproteins that transport cholesterol. Iron and copper are enzyme cofactors necessary to cellular metabolism. While we associated cholesterol with cardiovascular diseases, in fact, it is a substance that contributes to the functioning of the plasma membrane. Estrogen induces synthesis of bone matrix proteins and counteracts the loss of bone mass. At menopause, when the rate of estrogen secretion is drastically reduced, osteoporosis (decrease in bone density) may develop.

Similarly, besides the action of androgens (e.g. testosterone) on the sexual organs and function of males, androgens play a metabolic role in cells. They stimulate synthesis of structural proteins in skeletal muscles and bone, and affect the activity of various enzymes in the liver and kidneys. In the kidney, androgens stimulate synthesis of erythropoietin, the protein that signals the bone marrow to increase production of red blood cells. We are just now beginning to discover the role that estrogen and androgens play in the metabolism of cells and therefore their role in homeostasis in general.

The sex hormones greatly affect our appearance because they maintain the secondary sex characteristics. Some of these characteristics, such as fat deposition in females, help maintain homeostasis. In addition, it is now known that both estrogen and androgens have general metabolic effects that pertain to homeostasis.

Human Systems Work Together

Integumentary System

Androgens activate oil glands; sex hormones stimulate fat deposition, affect hair distribution in males and females.

Skin receptors respond to touch; modified sweat glands produce milk; skin stretches to accommodate growing fetus.

Skeletal System

Sex hormones influence bone growth and density in males and females.

Bones provide support and protection of reproductive organs.

Muscular System

Androgens promote growth of skeletal muscle.

Muscle contraction occurs during orgasm and moves gametes; abdominal and uterine muscle contractions occur during childbirth.

Nervous System

Sex hormones masculinize or feminize the brain, exert feedback control over the hypothalamus, and influence sexual behavior.

Brain controls onset of puberty; nerves are involved in erection of penis and clitoris, movement of gametes along ducts, and contraction of uterus.

Endocrine System

Gonads produce the sex hormones.

Hypothalamic, pituitary, and sex hormones control sex characteristics and regulate reproductive processes.

How the Reproductive System works with other body systems

Cardiovascular System

Sex hormones influence cardiovascular health; sexual activities stimulate **cardiovascular** system.

Blood vessels transport sex hormones; vasodilation causes genitals to become erect; blood services the reproductive organs.

Lymphatic System/Immunity

Sex hormones influence immune functioning; acidity of vagina helps prevent pathogen invasion of body; milk passes antibodies to newborn.

Immune system does not attack sperm or fetus, even though they are foreign to the body.

Respiratory System

Sexual activity increases breathing; pregnancy causes breathing rate and vital capacity to increase.

Gas exchange increases during sexual activity.

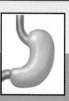

Digestive System

Pregnancy crowds digestive organs and promotes heartburn and constipation.

Digestive tract provides nutrients for growth and repair of organs and for development of fetus.

Urinary System

Penis in males contains the urethra and performs urination; prostate enlargement hinders urination.

Semen is discharged through the urethra in males; kidneys excrete wastes and maintain electrolyte levels for mother and child.

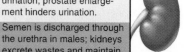

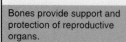

Bioethical Focus

Assisted Reproductive Technologies

The dizzying array of assisted reproductive technologies has progressed from simple in vitro fertilization to the ability to freeze eggs or sperm or even embryos for future use. Older women who never had the opportunity to freeze their eggs can still have children if they use donated eggs—perhaps today harvested from a fetus.

Legal complications abound, ranging from which mother has first claim to the child—the surrogate mother, the woman who donated the egg, or the primary caregiver—to which partner has first claim to frozen embryos following a divorce. Legal issues about who has the right to use what techniques have rarely been discussed, much less decided upon. Some clinics will help anyone, male or female, no questions asked, as long as they have the ability to pay. And most clinics are heading toward doing any type of procedure, including guaranteeing the sex of the child and making sure the child will be free from a particular genetic disorder. It would not be surprising if, in the future, zygotes could be engineered to have any particular trait desired by the parents.

Even today eugenic (good gene) goals are evidenced by the fact that reproductive clinics advertise for egg and sperm donors, primarily in elite college newspapers. The question becomes, "Is it too late for us as a society to make ethical decisions about reproductive issues?" Should we come to a consensus about what techniques should be allowed and who should be able to use them? We all want to avoid, if possible, what happened to Jonathan Alan Austin. Jonathan, who was born to a surrogate mother, later died from injuries inflicted by his father. Perhaps if a background check were legally required, surrogate mothers would only make themselves available to individuals or couples who are known to have certain psychological characteristics.

Figure 16B Couple and children.
Are assisted reproductive technologies a boon to society? To individuals who wish to have children? Or a detriment to both?

Decide Your Opinion

1. As a society, we have never been in favor of regulating reproduction. Should we regulate assisted reproduction if we do not regulate unassisted reproduction? Why or why not?

2. Should the state be the guardian of frozen embryos and make sure they all get a chance for life? Why or why not?

3. Is it appropriate for physicians and parents to select which embryos will be implanted in the uterus? On the basis of sex? On the basis of genetic inheritance? Why or why not?

Looking at Both Sides   www.mhhe.com/biosci/genbio/maderhuman7/

Every bioethical issue has at least two sides. Even if you already have an opinion, it is important to explore the opposite opinion before finalizing your position. The Online Learning Center at www.mhhe.com/biosci/genbio/maderhuman7/ will help you fine-tune your initial opinion, explore both sides, and finalize your position. Either you will acquire new arguments for your original opinion or you may even change your opinion. Be sure to complete these activities in sequence:

Taking Sides   Decide your initial opinion by answering a series of questions. Then see if your opinion changes after completing the next two activities.

Further Debate   Read opposing articles that give you further information on this particular bioethical issue.

Explain Your Position   Answer another series of questions and then defend your original or changed opinion. You can e-mail your position to your instructor if he or she wishes.

Summarizing the Concepts

16.1 Male Reproductive System

In males, spermatogenesis, occurring in seminiferous tubules of the testes, produces sperm that mature and are stored in the epididymides and may also be stored in the vasa deferentia before entering the urethra, along with secretions produced by the seminal vesicles, prostate gland, and bulbourethral glands. Sperm and these secretions are called semen.

The external genitals of males are the penis, the organ of sexual intercourse, and the scrotum, which contains the testes. Orgasm in males is a physical and emotional climax during sexual intercourse that results in ejaculation of semen from the penis.

Hormonal regulation, involving secretions from the hypothalamus, the anterior pituitary, and the testes, maintains testosterone, produced by the interstitial cells of the testes, at a fairly constant level. FSH from the anterior pituitary promotes spermatogenesis in the seminiferous tubules, and LH promotes testosterone production by the interstitial cells.

16.2 Female Reproductive System

In females, oogenesis occurring within the ovaries typically produces one mature follicle each month. This follicle balloons out of the ovary and bursts, releasing an egg, which enters an oviduct. The oviducts lead to the uterus where implantation and development occur. The external genital area includes the vaginal opening, the clitoris, the labia minora, and the labia majora.

The vagina is the organ of sexual intercourse and the birth canal in females. The vagina and the external genitals, especially the clitoris, play an active role in orgasm, which culminates in uterine and oviduct contractions.

16.3 Female Hormone Levels

In the nonpregnant female, the ovarian and uterine cycles are under hormonal control of the hypothalamus, anterior pituitary, and the female sex hormones estrogen and progesterone. During the first half of the ovarian cycle, FSH from the anterior pituitary causes maturation of a follicle that secretes estrogen and some progesterone. After ovulation and during the second half of the cycle, LH from the anterior pituitary converts the follicle into the corpus luteum, which secretes progesterone and some estrogen. Estrogen and progesterone regulate the uterine cycle. Estrogen causes the endometrium to rebuild. Ovulation usually occurs on day 14 of a 28-day cycle. As progesterone is produced by the corpus luteum, the endometrium thickens and becomes secretory. Then, a low level of hormones causes the endometrium to break down, as menstruation occurs.

If fertilization and implantation occur, the corpus luteum in the ovary is maintained because of HCG production by the placenta. Progesterone production does not cease, and the embryo implants itself in the thickened endometrium. Menstruation does not occur during pregnancy because of HCG production.

16.4 Control of Reproduction

Numerous birth control methods and devices such as the birth control pill, diaphragm, and condom are available for those who wish to prevent pregnancy. Effectiveness varies, and research is being conducted to find new and possibly better methods. A morning-after pill is now on the market, and a male pill may soon be available. Some couples are infertile, and if so, they may make use of assisted reproductive technologies in order to have a child. Artificial insemination and in vitro fertilization have been followed by more sophisticated techniques such as intracytoplasmic sperm injection.

16.5 Homeostasis

The reproductive system works with the other systems of the body in the ways described in the illustration on page 334.

Studying the Concepts

1. Outline the path of sperm. What glands contribute fluids to semen? 318–19
2. Discuss the anatomy and physiology of the testes. Describe the structure of sperm. 320–21
3. Name the endocrine glands involved in maintaining the sex characteristics of males and the hormones produced by each. 321
4. Describe the organs of the female genital tract. Where do fertilization and implantation occur? Name two functions of the vagina. 322–23
5. Name and describe the external genitals in females. 323
6. Discuss the anatomy and the physiology of the ovaries. Describe the ovarian cycle. 324–25
7. Describe the uterine cycle, and relate it to the ovarian cycle. In what way is menstruation prevented if pregnancy occurs? 326–28
8. Name three functions of the female sex hormones. 328
9. Discuss the various means of birth control and their relative effectiveness in preventing pregnancy. 329–30, 332
10. Describe how in vitro fertilization is carried out. 332–33

Testing Your Knowledge of the Concepts

In questions 1–4, match the organs to the functions below.
a. testes and ovaries
b. clitoris and penis
c. oviduct and vas deferens
d. seminal vesicle and prostate gland

_____ 1. part of the genital tract that conducts gametes

_____ 2. found only in the male

_____ 3. gonads that produce gametes

_____ 4. organs involved in sexual response

In questions 5–7, indicate whether the statement is true (T) or false (F).

_____ 5. In tracing the path of sperm, the bladder would be mentioned after the epididymis.

_____ 6. The testes are endocrine glands.

_____ 7. In the ovarian cycle, estrogen causes the endometrial lining to become secretory.

In questions 8–16, fill in the blanks.

8. An erection is caused by the entrance of _____ into sinuses within the penis.

9. Estrogen is to females as _____ is to males.

10. The prostate gland, the bulbourethal glands, and the _____ all contribute to seminal fluid.

11. The main male sex hormone is _____.

12. In the female reproductive system, the uterus lies between the oviducts and the _____.

13. In the ovarian cycle, once each month a(n) _____ releases an egg. In the uterine cycle, the _____ of the uterus is prepared to receive the embryo.

14. The female sex hormones are _____ and _____.

15. Pregnancy in the female is detected by the presence of _____ in blood or urine.

16. In vitro fertilization occurs in _____.

17. Match the key terms to these definitions:
    a. _____ Organ that leads from the uterus to the vestibule and serves as the birth canal and organ of sexual intercourse.
    b. _____ Gland located around the male urethra below the urinary bladder, adds secretions to semen.
    c. _____ External genitals that surround the opening of the vagina.
    d. _____ Coiled tubule next to the testes where sperm mature and may be stored for a short time.
    e. _____ Organ that produces sex cells; ovaries in the female and testes in the male.

18. Label this diagram of the male reproductive system, and trace the path of sperm.

Understanding Key Terms

acrosome 321
assisted reproductive technologies (ART) 332
birth control pill 329
bulbourethral gland 319
cervix 322
condom 330
contraceptive 329
corpus luteum 325
diaphragm 329
egg 322
endometriosis 332
endometrium 323
epididymis 318
erectile dysfunction 319
estrogen 326
fimbria 322
follicle 325
follicle-stimulating hormone (FSH) 321
gamete 333
gonadotropin-releasing hormone (GnRH) 321
human chorionic gonadotropin (HCG) 328
implantation 330
infertility 332
interstitial cell 321
interstitial cell-stimulating hormone (ICSH) 321
intrauterine device (IUD) 329

luteinizing hormone (LH) 321
menopause 328
menstruation 326
oogenesis 322
ovarian cycle 325
ovary 322
oviduct 322
ovulation 322
Pap test 323
penis 319
placenta 328
progesterone 326
prostate gland 318
scrotum 318
semen 318
seminal vesicle 318
seminiferous tubule 320
sperm 321
spermatogenesis 321
sustentacular (Sertoli) cell 321
testis 318
testosterone 321
urethra 318
uterine cycle 326
uterus 322
vagina 323
vas deferens 318
vulva 323
zygote 322

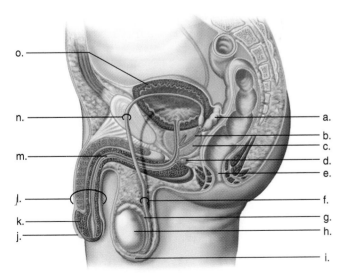

16.1 Male Reproductive System

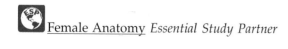Male Anatomy *Essential Study Partner*

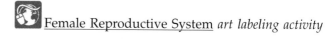Penile Erection *animation activity*
Circumcision *animation activity*
Spermatogenesis *animation activity*
Male Reproductive System *art labeling activity*
Penis Anatomy I *art labeling activity*
Penis Anatomy II *art labeling activity*

16.2 Female Reproductive System

Female Anatomy *Essential Study Partner*

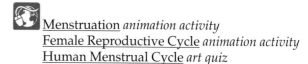Female Reproductive System *art labeling activity*

16.3 Female Hormone Levels

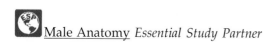Female Physiology *Essential Study Partner*

Menstruation *animation activity*
Female Reproductive Cycle *animation activity*
Human Menstrual Cycle *art quiz*

16.4 Control of Reproduction

Birth Control *Essential Study Partner*

Endometriosis *art labeling activity*
Looking at Both Sides *critical thinking activity*

16.5 Homeostasis

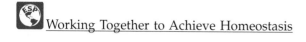Working Together to Achieve Homeostasis

Chapter Summary

Key Term Flashcards *vocabulary quiz*
Chapter Quiz *objective quiz covering all chapter concepts*

Chapter 17

Sexually Transmitted Diseases

Chapter Concepts

Figure 17.1   Sexual relationships.
Everyone who has sexual relationships should be aware of the possibility of acquiring a sexually transmitted disease and take all necessary precautions.

By the time Jennifer W. realizes what has happened, it's too late. She is sitting in a doctor's office, quietly explaining that every time she goes to the bathroom, she has this terrible burning pain. She doesn't want to tell her friends. She can't tell her parents.

At least she is telling the doctor.

The source of Jennifer's agony is genital herpes, a very common—and very embarrassing—condition. **Sexually transmitted diseases** are contagious diseases caused by pathogens that are passed from one human to another by sexual contact. Genital herpes is a sexually transmitted disease (STD) caused by a virus. As in Jennifer's case, most people catch the disease by having sex with someone who's infected but has no symptoms (Fig. 17.1). That's because herpes viruses—as well as AIDS and a host of other STDs—can lie latent in the body, hiding from sight. Even if you can't see the disease, you should try to prevent passage by taking any one of the following precautions: (1) practicing abstinence; (2) having a monogamous (always the same partner) sexual relationship with someone who does not have an STD, and in the case of AIDS and hepatitis B, is not an intravenous drug user; or (3) always using a female or male latex condom in the proper manner with a water-based vaginal spermicide containing nonoxynol-9. You should also avoid oral/genital contact—just touching the genitals can transfer an STD in some cases.

After taking a prescription drug, Jennifer recovers from her present symptoms of a herpes infection. But she'll never be free of the virus, a possible recurrence of symptoms, or the knowledge that her pain could have been avoided. Perhaps if Jennifer had taken Human Biology she would have known about STDs and how to protect herself.

STDs are now a major worldwide health problem, especially because many young people are sexually active and the age of sexual intercourse has been declining. This chapter discusses pathogens such as viruses, bacteria, and fungi and the STDs they cause. In addition, more information about human immunodeficiency viral (HIV) infection and AIDS can be found in the supplement that follows this chapter.

17.1 Viral Infectious Diseases

Pathogens are viruses, bacteria, and other organisms that cause diseases. Viruses cause numerous diseases in humans (Table 17.1), including AIDS, herpes, genital warts, and hepatitis B, four sexually transmitted diseases of great concern today.

Viruses are incapable of independent reproduction and reproduce only inside a living **host** cell. For this reason, viruses are called obligate intracellular parasites. To maintain viruses in the laboratory, they are injected into laboratory-bred animals, live chick embryos, or animal cells maintained in tissue culture. Viruses infect all sorts of cells—from bacterial cells to human cells—but they are very specific. For example, viruses called bacteriophages infect only bacteria, the tobacco mosaic virus infects only plants, and the rabies virus infects only mammals. Human viruses even specialize in a particular tissue. Human immunodeficiency virus (HIV) enters certain blood cells, the polio virus reproduces in spinal nerve cells, and the hepatitis viruses infect only liver cells.

Viruses are noncellular, and they have a unique construction. These tiny particles always have at least two parts: an inner core of nucleic acid, which constitutes their genetic material, and an outer capsid composed of protein subunits.

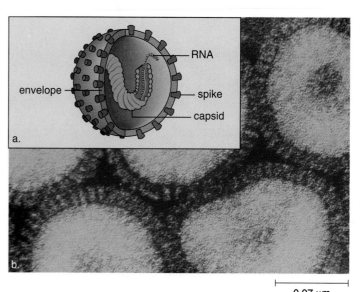

Figure 17.2 Influenza virus.
This RNA virus causes influenza in humans. **a.** Diagram of virus. **b.** Electron micrograph of actual virus.

For some viruses, the nucleic acid is DNA, the same as our genetic material. For other viruses, the genetic material is in the form of RNA. The capsid of a virus surrounds and protects the genetic material (Fig. 17.2).

In addition to these components, viruses that infect animal cells are often surrounded by an outer envelope derived from the host-cell plasma membrane. The envelope contains proteins that allow the virus to adhere to the plasma membrane. After the virus enters a cell, it takes over the cell's machinery in order to reproduce.

Figure 17.3 illustrates the events in the reproductive cycle of a typical DNA animal virus:

1. *Binding of the virus to the plasma membrane.* Viruses have envelope proteins that allow them to bind only to specific types of host cells. These cells have receptor proteins to which a virus can adhere.

2. *Penetration of the virus into the cell.* After a virus has bound to the plasma membrane of a cell, the viral particle consisting of the capsid and nucleic acid is brought into the host cell. Uncoating—the removal of the capsid—follows. Now the genetic material is released and can express itself.

3. *Replication of the viral genetic material.* When a virus takes over the machinery of a cell, it uses that cell's enzymes to help it reproduce. One important aspect of reproduction for a DNA animal virus is the replication of its DNA. During replication, the host cell's enzymes make many copies of viral DNA.

4. *Production of viral proteins.* The capsid proteins, as well as envelope proteins in the viral envelope, are synthesized by host cell ribosomes according to viral DNA instructions. The function of any genetic material is to specify the production of particular proteins. In order to bring this about, DNA must utilize RNA as an intermediary.

Table 17.1	Infectious Diseases Caused by Viruses
Site of Infection	**Diseases**
Not Sexually Transmitted	
Respiratory tract	Common colds, flu,* viral pneumonia, hantavirus pulmonary syndrome
Skin	Measles,* German measles,* chicken pox*, shingles, warts
Nervous system	Encephalitis, polio,* rabies*
Liver	Yellow fever,* hepatitis A, C, and D
Cardiovascular	Ebola
Other systems	Mumps,* cancer
Sexually Transmitted	
Immune system	AIDS
Reproductive system	Genital warts, genital herpes
Liver	Hepatitis B*

*Vaccines are available. Yellow fever, rabies, and flu vaccines are given only if the situation requires them. Smallpox vaccinations are no longer required.

5. *Assembly of new viruses.* Viral proteins and DNA replicates are assembled to form new viral particles. Reproduction of the virus has now taken place.

6. *Budding of new viruses from the host cell.* During budding, the virus gets its envelope, which consists of host plasma membrane components and envelope proteins that were coded for by viral DNA. The process of budding does not necessarily kill the host cell.

Some viruses that enter human cells—for example, the papillomaviruses, the herpesviruses, the hepatitis viruses, and the adenoviruses—can undergo a period of **latency** during which they are hidden in the cell and do not reproduce. Certain environmental factors, such as ultraviolet radiation, can induce the virus to reproduce and bud from the cell. Latent viruses are of special concern because their presence in the host cell can alter the cell and make it become cancerous.

Viruses are transferred from one host animal to another by a variety of mechanisms, including food and water, human or animal bites, contact, and aerosols (droplets in the air). Sexually transmitted viral diseases require intimate contact between persons for their successful transmission. Viral diseases are controlled by preventing transmission, by administering vaccines, and only recently by administering antiviral drugs as discussed in the Health Focus on page 344.

Knowing how a particular pathogen is transmitted can help prevent its spread. Covering the mouth and nose when coughing or sneezing helps prevent the spread of a cold, and use of a latex condom during intercourse helps prevent the transmission of sexually transmitted diseases. Vaccines, which are antibodies administered to stimulate immunity to a pathogen so that it cannot later cause disease (see page 156), are available for some viral diseases such as polio, measles, mumps, and hepatitis B, a sexually transmitted disease (see Table 17.1). Antibiotics, which are designed to interfere with bacterial metabolism, have no effect on viral illnesses. Instead, drugs must be designed that interfere with either entry of the virus into the host cell, reproduction of the virus, or exit of the virus from the cell.

A number of viruses cause diseases in humans. Some of these are significant sexually transmitted diseases (STDs).

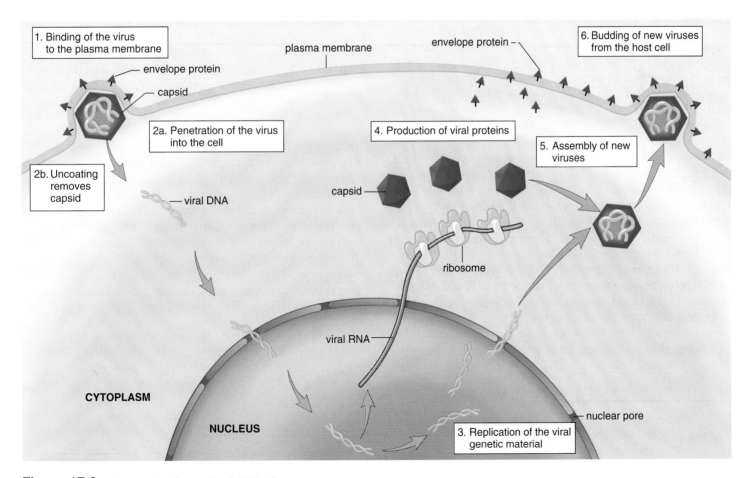

Figure 17.3   Life cycle of an animal DNA virus.
Viruses reproduce within a host cell. DNA animal viruses typically utilize this series of events to accomplish reproduction.

HIV Infections

The **AIDS (acquired immunodeficiency syndrome)** supplement, which begins on page 355, discusses HIV infections at greater length than this brief summary. Two basic types of human immunodeficiency viruses (HIV) are known to infect humans. Of these, HIV-2, which is found mainly in Africa, is less virulent than HIV-1, which is now pandemic. HIV-1 occurs as many subtypes; in the United States, HIV-1B is prevalent. HIV-1C is now rampaging through sub-Saharan Africa, where a large percentage of teenagers and young adults may die from AIDS in the next few years. HIV-1C could very well be the cause of a new AIDS epidemic in the United States unless all persons follow the guidelines for preventing transmission outlined in the introduction (see page 339).

AIDS is the last stage of an HIV infection. It's estimated that 34.3 million people worldwide are now infected with the HIV virus and that 19 million persons have died from AIDS. In the United States, AIDS was first seen among the homosexual community, and they still are the most exposed group. However, new HIV infections are more likely to occur in minority women. From this you can conclude that HIV is now being transmitted between heterosexuals, and more frequently between African Americans and Hispanics than Caucasians.

The primary host for HIV is an immune cell such as a T lymphocyte or macrophage. Since these are the very cells the body uses to fight infections, the immune system becomes severely impaired in persons with AIDS. During the first stage of an HIV infection, symptoms are few, but the individual is highly contagious. Several months to several years after infection, the helper T lymphocyte count falls, and infections, such as other sexually transmitted diseases, begin to appear. In the last stage of infection, called AIDS, the T lymphocyte count is less than 200 per mm^3, and at least one **opportunistic infection** is present. Such diseases only have the *opportunity* to occur because the immune system is severely weakened. Persons with AIDs typically die from an opportunistic disease, such as *P. carinii* pneumonia.

There is no cure for AIDS, but a treatment called highly active antiretroviral therapy (HAART) is usually able to stop HIV reproduction to the extent that the virus becomes undetectable in the blood. The sooner drug therapy begins after infection, the better the chances the immune system will not be destroyed by HIV. The medication must be continued indefinitely because as soon as HAART is discontinued, the virus rebounds.

More people, many of whom are heterosexuals, are infected with the HIV virus each year. All possible steps should be taken to avoid transmission because treatment is costly and there is no cure.

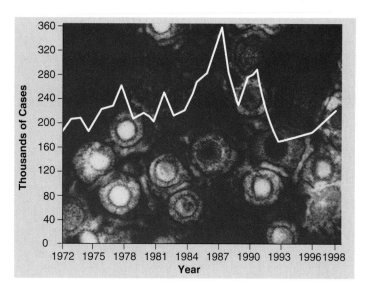

Figure 17.4   Genital warts.
A graph depicting the incidence of new cases of genital warts reported in the United States from 1972 to 1998 is superimposed on a photomicrograph of human papillomaviruses.

Genital Warts

Human papillomaviruses (HPVs) cause warts, including common warts, plantar warts, and also genital warts, which are sexually transmitted. Over one million persons become infected each year with a form of HPV that causes **genital warts**, but only a portion seek medical help (Fig. 17.4).

Transmission and Symptoms

Quite often, carriers of genital warts do not detect any sign of warts, although flat lesions may be present. When present, the warts commonly are seen on the penis and foreskin of men and near the vaginal opening in women. A newborn can become infected by passage through the birth canal.

Genital warts are associated with cancer of the cervix, as well as tumors of the vulva, the vagina, the anus, the penis, and the mouth. Some researchers believe that HPVs are involved in 90–95% of all cases of cancer of the cervix. Teenagers with multiple sex partners seem to be particularly susceptible to HPV infections. More cases of cancer of the cervix are being seen among this age group.

Treatment

Presently, there is no cure for an HPV infection, but it can be treated effectively by surgery, freezing, application of an acid, or laser burning, depending on severity. If the warts are removed, they may recur. Also, even after treatment, the virus can be transmitted. Therefore, abstinence or use of a condom with a vaginal spermicide containing nonoxynol-9 is necessary to prevent the spread of genital warts.

Genital warts is a prevalent STD associated with cervical cancer in women.

Herpes Infections

The herpesviruses cause various illnesses. Chicken pox and shingles are caused by a type of herpesvirus called the varicella-zoster virus, and mononucleosis is caused by another type of herpesvirus called the Epstein-Barr virus. These conditions do not appear to be spread by sexual contact. The herpes simplex viruses that do cause sexually transmitted disease are large DNA viruses that infect the mucosal linings of the body such as the mouth and vagina. There are two types of herpes simplex viruses: type 1 usually causes cold sores and fever blisters, while type 2 more often causes genital herpes. Crossover infections do occur, however. That is, type 1 has been known to cause a genital infection, while type 2 has been known to cause cold sores and fever blisters.

Cold Sores and Fever Blisters

Cold sores are usually caused by herpes simplex virus type 1. Infection with this virus usually occurs during childhood with symptoms of a mild upper respiratory illness. In some cases, however, painful blisters may appear in the throat, on or below the tongue, and on other mouth parts, including the lips. The sores heal, but the virus remains latent in ganglia outside the central nervous system and spinal cord. Thereafter, a recurrence of symptoms takes the form of a cold sore around the lips. A high fever, stress, colds, menstruation, or exposure to sunlight seem to trigger a reactivation of the virus. It is important to realize that herpes lesions are infectious for at least three to four days until the sores begin to heal. Contact with the sores or any contaminated object can cause the virus to be transmitted. Oral/genital contact can lead to a genital herpes infection.

Genital Herpes

Genital herpes, usually caused by the herpes simplex virus type 2, now infects about 45 million Americans; therefore, millions of persons could be having recurring symptoms at the same time. As Figure 17.5 shows, the incidence of reported new cases of genital herpes has increased.

Persons usually get infected with herpes simplex virus type 2 when they are adults. In some people, there are no symptoms. Or there may be a tingling or itching sensation before blisters appear on the genitals (within 2–20 days). Once the blisters rupture, they leave painful ulcers that may take as long as three weeks or as little as five days to heal. The blisters may be accompanied by fever, pain on urination, swollen lymph nodes in the groin, and in women, a copious vaginal discharge.

After the ulcers heal, the disease is only latent. Blisters can recur, although usually at less frequent intervals and with milder symptoms. Again, fever, stress, sunlight, and menstruation are associated with a recurrence of symptoms. When there are no symptoms, the virus primarily resides in the ganglia of sensory nerves associated with the affected skin. Although herpes simplex virus type 2 was formerly

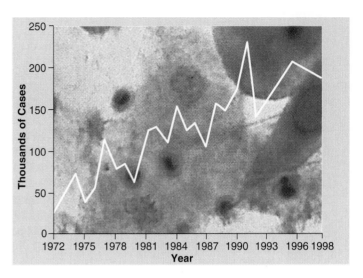

Figure 17.5 **Genital herpes.**
A graph depicting the incidence of new reported cases of genital herpes in the United States from 1972 to 1998 is superimposed on a photomicrograph of cells infected with the herpesvirus.

thought to be a cause of cervical cancer, this is no longer believed to be the case.

Infection of the newborn can occur if the child comes in contact with a lesion in the birth canal. At worst, the infant can be gravely ill, become blind, have neurological disorders including brain damage, or die. Birth by cesarean section prevents these occurrences; therefore, all pregnant women infected with the virus should be sure to tell their healthcare provider.

Transmission

Genital herpes lesions shed infective viruses. Live viruses have occasionally been cultured from the skin of persons with no lesions; therefore, it is believed that the virus can be spread even when there are no visible lesions. Persons who are infected should take extreme care to prevent the possibility of spreading the infection to other parts of their own body, such as the eyes. Certainly, sexual contact should be avoided until all lesions are completely healed, and then the general directions given in the introduction for avoiding STD transfer should be followed (see page 339).

Treatment

Presently, there is no cure for genital herpes. The drugs acyclovir and vidarabine disrupt viral reproduction. The ointment form of acyclovir relieves initial symptoms, and the oral form prevents the recurrence of symptoms as long as it is being taken. Research is being conducted in an attempt to develop a vaccine.

Herpes simplex viruses cause genital herpes, an extremely infectious disease whose symptoms recur and for which there is no cure.

STDs and Medical Treatment

Treatment of STDs is troublesome at best. Presently, there are no vaccines available except for a hepatitis B infection, and to date, not many persons have been inoculated. Unfortunately, there is no treatment once a hepatitis B infection has occurred. The antiviral drugs acyclovir and vidarabine are helpful against genital herpes, but they do not cure herpes. Once an individual stops the therapy, the lesions are apt to recur. The current recommended therapy for AIDS includes an expensive combination of drugs that prevent HIV reproduction and curtail its ability to infect other cells.

Antibiotics are used to treat sexually transmitted diseases caused by bacteria. Antibiotics such as penicillin, streptomycin, and tetracycline interfere with metabolic pathways unique to bacteria, and therefore, they are not expected to harm host cells. Still, there are problems associated with antibiotic therapy. Some individuals are allergic to antibiotics, and the reaction may even be fatal. Antibiotics not only kill off disease-causing bacteria, they also reduce the number of beneficial bacteria in the intestinal tract. The use of antibiotics sometimes prevents natural immunity from occurring, leading to the necessity for recurring antibiotic therapy.

Most important, perhaps, is the growing resistance of certain strains of bacteria to a particular antibiotic. Antibiotics were introduced in the 1940s, and for several decades they worked so well it appeared that infectious diseases had been brought under control. However, we now know that bacterial strains can mutate and become resistant to a particular antibiotic. Worse yet, bacteria can swap bits of DNA, and in this way, resistance

can pass to other strains of bacteria. Penicillin and tetracycline, long used to cure gonorrhea, now have a failure rate of more than 20% against certain strains of gonococcus. To help prevent resistant bacteria, antibiotics should be administered only when absolutely necessary, and the prescribed therapy should be finished. Also, as a society, we have to continue to develop new antibiotics to which resistance has not yet occurred.

An alarming new concern has been the upsurge of tuberculosis in persons with sexually transmitted diseases. Tuberculosis is a chronic lung infection caused by the bacterium *Mycobacterium tuberculosis*. This bacterium is often called the tubercle bacillus (TB) because usually the bacilli are walled off within tubercles that can sometimes be seen in X rays of the lungs. TB can pass from an HIV-infected person to a healthy person more easily than HIV. Sexual contact is not needed—only close personal contact, and therefore, it is possible that TB will now spread from those with STDs to the general populace. Some of the modern strains of TB are resistant to antibiotic therapy; therefore, just as with HIV, research is needed for the development of an effective vaccine.

Prevention is still the best way to manage STDs. Several STDs are primarily transmitted by infected white blood cells that penetrate the single layer of cells lining the inside of the uterine cervix and uterus. The use of a condom along with a spermicide that contains nonoxynol-9 and the avoidance of oral/genital contact are essential unless you have a monogamous relationship (always the same partner) with someone who is free of STDs and is not an intravenous drug user.

Hepatitis Infections

There are several types of **hepatitis.** Hepatitis A, caused by HAV (hepatitis A virus), is usually acquired from sewage-contaminated drinking water. Hepatitis A can also be sexually transmitted through oral/anal contact.

Hepatitis C, caused by HCV, is called the posttransfusion form of hepatitis. This type of hepatitis, which is usually acquired by contact with infected blood, is also of great concern. Infection can lead to chronic hepatitis, liver cancer, and death.

Hepatitis E, caused by HEV, is usually seen in developing countries. Only imported cases—that is, occurring in travelers to the country or in visitors to endemic regions—have been reported in the United States.

Hepatitis B

HBV is a DNA virus that is spread in the same way as HIV, the cause of AIDS—through sharing needles by drug abusers and through sexual contact between heterosexuals or between homosexual men. Therefore, it is common for an AIDS patient to also have an HBV infection. Also, like HIV, HBV can

be passed from mother to child by way of the placenta.

Only about 50% of infected persons have flulike symptoms, including fatigue, fever, headache, nausea, vomiting, muscle aches, and dull pain in the upper right of the abdomen. Jaundice, a yellowish cast to the skin, can also be present. Some persons have an acute infection that lasts only three to four weeks. Others have a chronic form of the disease that leads to liver failure and the need for a liver transplant.

Since there is no treatment for an HBV infection, prevention is imperative. The general directions given on page 339 should be followed, but inoculation with the HBV vaccine is the best protection. The vaccine, which is safe and does not have any major side effects, is now on the list of recommended immunizations for children.

Hepatitis B is an infection that can lead to liver failure. Because it is spread in the same way as AIDS, many persons are infected with both viruses at the same time.

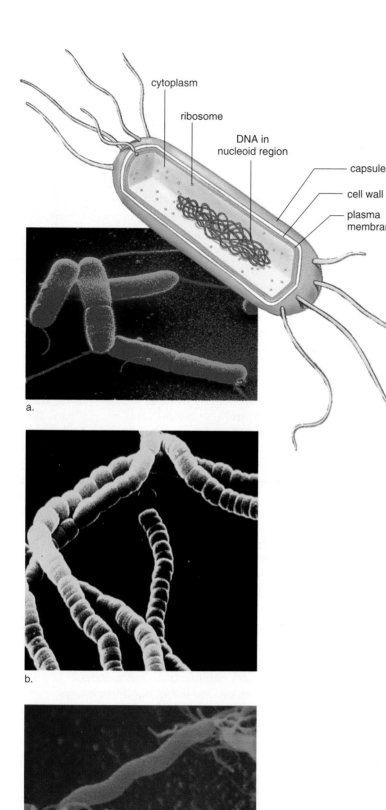

cytoplasm

ribosome

DNA in
nucleoid region

capsule

cell wall

plasma
membrane

a.

b.

c.

Figure 17.6   **Bacteria.**
Bacteria occur as **(a)** a bacillus (rod shape), **(b)** a coccus (round shape),
and **(c)** a spirillum. The structure of a bacterium is also shown in **(a).**

17.2 Bacterial Infectious Diseases

Although **bacteria** are generally larger than viruses, they are still microscopic. As a result, it is not always obvious that they are abundant in the air, water, and soil and on most objects. It has even been suggested that the combined weight of all bacteria would exceed that of any other type of organism on earth. Bacteria have long been used by humans to produce various products commercially. Chemicals, such as ethyl alcohol, acetic acid, butyl alcohol, and acetone, are produced by bacteria. Bacterial action is involved in the production of butter, cheese, sauerkraut, rubber, cotton, silk, coffee, wine, and cocoa. By means of genetic engineering, bacteria are now used to produce human insulin and interferon, as well as other types of proteins. Even certain antibiotics are produced by bacteria.

Bacteria occur in three basic shapes: rod (bacillus); round, or spherical (coccus); and spiral (a curved shape called a spirillum) (Fig. 17.6). Human cells contain a membrane-bounded nucleus and several kinds of membranous organelles; therefore, they are called eukaryotic (true nucleus) cells. Bacterial cells lack a membrane-bounded nucleus and the other membranous organelles typical of human cells; therefore, they are called prokaryotic (before nucleus) cells. A nucleoid region contains their DNA, and a cytoplasm contains the enzymes of their many metabolic pathways. Bacteria are enclosed in a cell wall, and some are also surrounded by a polysaccharide or polypeptide capsule that inhibits destruction by the host. Motile bacteria often have flagella (sing., flagellum).

Most bacteria are free-living organisms that perform many useful services in the environment. They are able to live in just about any habitat and use just about any substance as a food source because of their wide metabolic ability. Most bacteria are decomposers—they break down dead organic matter in the environment by secreting digestive enzymes, and then they absorb the nutrient molecules.

Many bacteria are symbiotic—that is, they live in close association with another species. A number of bacteria are normal inhabitants of our bodies. These bacteria, called the normal microbial flora, cause disease only when the opportunity arises. The bacterium *Escherichia coli,* which lives in our intestines, breaks down remains that we have not digested. The microflora of our intestines, including *E. coli,* provide us with the vitamins thiamine, riboflavin, pyridoxine, B_{12}, and K. The last two vitamins are necessary for the production of blood components. *E. coli* causes cystitis if it gets into the urinary tract. And *Staphylococcus aureus,* which lives on our skin, causes an infection if it gets into a wound.

Bacteria usually reproduce asexually by **binary fission.** First, the single chromosome duplicates, and then the two chromosomes move apart into separate areas. Next, the plasma membrane and cell wall grow inward and partition the cell into two daughter cells, each of which has its own chromosome (Fig. 17.7). Under favorable conditions, growth

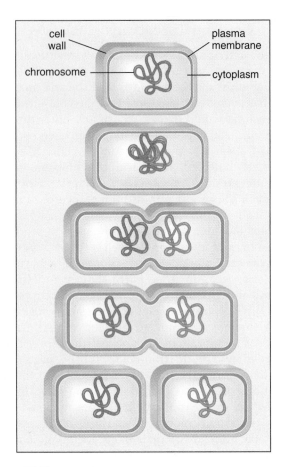

Figure 17.7   Bacterial reproduction.
When bacteria reproduce, the single chromosome is attached to the plasma membrane, where it is replicating. As the plasma membrane and the cell wall lengthen, the two chromosomes separate. Once fission has taken place, each bacterium has its own chromosome.

may be very rapid, with cell division occurring as often as every 20 minutes in *E. coli.* When faced with unfavorable environmental conditions, some bacteria can form **endospores.** During spore formation, the cell shrinks, rounds up within the former plasma membrane, and secretes a new, thicker wall inside the old one. Endospores are amazingly resistant to extreme temperatures, drying out, and harsh chemicals, including acids and bases. When conditions are again suitable for growth, the spore absorbs water, breaks out of the inner shell, and becomes a typical bacterial cell capable of reproducing.

The diseases listed in Table 17.2 are caused by bacteria that are ordinarily pathogenic; that is, they usually cause an illness. Most bacteria are aerobic, and they require a supply of oxygen just as we do, but several human diseases are caused by anaerobic bacteria. Among them are botulism, gas gangrene, and tetanus.

To be a pathogen, a bacterium must
- have an ability to pass from one host to the next
- penetrate into the host's tissues
- withstand the host's defense mechanisms
- induce illness in the host

Table 17.2   Infectious Diseases Caused by Bacteria

Site of Infection	Diseases
Not Sexually Transmitted	
Respiratory tract	Strep throat, pneumonia,* whooping cough,* tuberculosis*
Skin	Staph (pimples and boils)
Nervous system	Tetanus,* botulism, meningitis
Digestive tract	Food poisoning, dysentery, cholera*
Other systems	Gas gangrene* (wound infections), diphtheria,* typhoid fever*
Sexually Transmitted	
Reproductive system	Chlamydia, gonorrhea, syphilis (can spread to other systems)

*Vaccines are available. Tuberculosis vaccine is not used in this country. Typhoid fever, cholera, and gas gangrene vaccines are given if the situation requires it. Others are routinely given.

Bacterial diseases are controlled by preventing transmission, by administering vaccines, and by antibiotic therapy. **Antibiotics** are medications that kill bacteria by interfering with one of their unique metabolic pathways. Antibiotic therapy is highly successful if it is carried out in the recommended manner. Otherwise, as discussed in the Health Focus on page 344, resistant strains can develop. Unfortunately, there are no vaccines for STDs caused by bacteria; therefore, preventing transmission becomes all the more important. Abstinence or monogamous relations (always the same partner) with someone who is free of an STD will prevent transmission. Otherwise, the use of a condom with a vaginal spermicide containing nonoxynol-9 and avoidance of oral/genital contact are recommended.

Bacteria lack a membrane-bounded nucleus and other membranous organelles found in eukaryotic cells, but they are capable of an independent existence. Most are free-living, but a few cause human diseases that can often be cured by antibiotic therapy.

Chlamydia

Chlamydia is named for the tiny bacterium that causes it (*Chlamydia trachomatis*). For years, chlamydiae were considered to be more closely related to viruses than to bacteria, but today it is known that these organisms are cellular. Even so, they are obligate parasites due to their inability to produce ATP molecules. After chlamydia enters a cell by endocytosis, the life cycle occurs inside the endocytic vacuole, which eventually bursts and liberates many new infective chlamydiae.

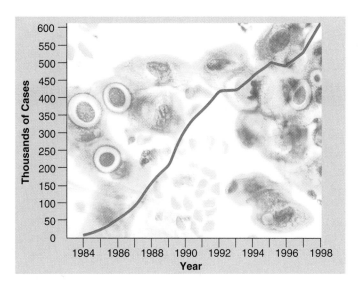

Figure 17.8   **Chlamydial infection.**
A graph depicting the incidence of reported cases of chlamydia in the United States from 1984 to 1998 is superimposed on a photomicrograph of a cell containing different stages of the organism.

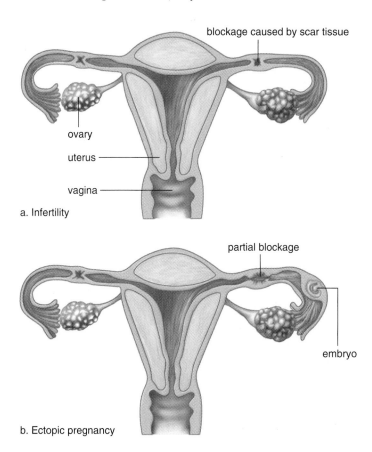

a. Infertility

b. Ectopic pregnancy

Figure 17.9   **Pelvic inflammatory disease.**
Following a chlamydial infection, the oviducts can be **(a)** completely blocked by scar tissue so that infertility results or **(b)** partially blocked so that fertilization occurs but the embryo is unable to pass to the uterus. The growing embryo can cause the oviduct to burst.

New chlamydial infections are more numerous than any other sexually transmitted disease. From 1984 through 1998, reported incidences of chlamydia increased dramatically, from less than 50,000 cases per year to about 600,000 cases (Fig. 17.8). Some estimate that the actual incidence could be as high as 6 million new cases per year. For every reported case in men, more than five cases are detected in women. This is mainly due to increased detection of asymptomatic infections through screening. The low rates in men suggest that many of the sex partners of women with chlamydia are not diagnosed or reported.

Symptoms

Chlamydial infections of the lower reproductive tract usually are mild or asymptomatic, especially in women. About 8–21 days after infection, men may experience a mild burning sensation on urination and a mucoid discharge. Women may have a vaginal discharge along with the symptoms of a urinary tract infection. Unfortunately, a physician mistakenly may diagnose these symptoms as a gonorrheal or urinary infection and prescribe the wrong type of antibiotic, or the person may never seek medical help. In these instances, there is a particular risk of the infection spreading from the cervix to the oviducts so that **pelvic inflammatory disease (PID)** results. This very painful condition can result in a blockage of the oviducts, with the possibility of sterility or ectopic pregnancy (Fig. 17.9).

Some believe that chlamydial infections increase the possibility of premature and stillborn births. If a newborn comes in contact with chlamydia during delivery, pneumonia or inflammation of the eyes can result. Erythromycin eyedrops at birth prevent this occurrence. If a chlamydial infection is detected in a pregnant woman, treatment with erythromycin should be used during pregnancy.

Diagnosis and Treatment

New and faster laboratory tests are now available for detecting a chlamydial infection. Their expense sometimes prevents public clinics from using them, however. Criteria that could help physicians decide which women should be tested include: no more than 24 years old; having had a new sex partner within the preceding two months; having a cervical discharge; bleeding during parts of the vaginal exam; and using a nonbarrier method of contraception. A chlamydial infection is cured with the antibiotics tetracycline, doxycycline, and azithromycin; therefore, treatment should begin immediately.

PID and sterility are possible effects of a chlamydial infection in women. This condition may accompany a gonorrheal infection, discussed next.

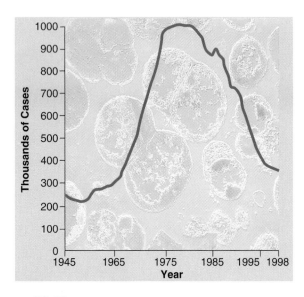

Figure 17.10 Gonorrhea.
A graph depicting the incidence of new cases of gonorrhea in the United States from 1945 to 1998 is superimposed on a photomicrograph of a urethral discharge from an infected male. Gonorrheal bacteria (*Neisseria gonorrhoeae*) occur in pairs; for this reason, they are called diplococci.

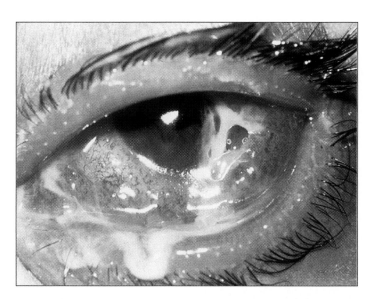

Figure 17.11 Secondary sites for a gonorrheal infection.
Gonorrheal infection of the eye is possible whenever the bacterium comes in contact with the eyes. This can happen when a newborn passes through the birth canal. Manual transfer from the genitals to the eyes is also possible.

Gonorrhea

Gonorrhea is caused by the bacterium *Neisseria gonorrhoeae,* which is a diplococcus, meaning that generally there are two spherical cells in close proximity. The incidence of gonorrhea has been declining since an all-time high in 1978 (Fig. 17.10). However, it can be noted that gonorrhea rates among African Americans is 30 times greater than the rate among Caucasians. Also, women using the birth control pill have a greater risk of contracting gonorrhea because hormonal contraceptives cause the genital tract to be more receptive to pathogens.

Symptoms

The diagnosis of gonorrhea in men is not difficult as long as they display typical symptoms (as many as 20% of men may be asymptomatic). The patient complains of pain on urination and has a milky urethral discharge three to five days after contact with the pathogen. In women, the bacteria may first settle within the vagina or near the cervix, from which they may spread to the oviducts. Unfortunately, the majority of women are asymptomatic until they develop severe pains in the abdominal region due to PID (pelvic inflammatory disease) (see Fig. 17.9). PID from gonorrhea is especially apt to occur in women using an IUD (intrauterine device) as a birth control measure.

PID due to a chlamydial or gonorrheal infection affects many thousands of women each year in the United States. PID-induced scarring of uterine tubes will cause infertility in about 20%, ectopic pregnancy in 9%, and chronic pelvic pain in 18%.

Gonorrhea proctitis is an infection of the anus, with symptoms including anal pain, and blood or pus in the feces. Oral/genital contact can cause infection of the mouth, throat, and the tonsils. Gonorrhea can spread to internal parts of the body, causing heart damage or arthritis. If, by chance, the person touches infected genitals and then his or her eyes, a severe eye infection can result (Fig. 17.11).

Eye infection leading to blindness can occur as a baby passes through the birth canal. Because of this, all newborn infants receive erythromycin eyedrops as a protective measure.

Transmission and Treatment

The chances of getting a gonorrheal infection from an infected partner are good. Women have a 50–60% risk, while men have a 20% risk of contracting the disease after even a single exposure to an infected partner. Therefore, the preventive measures listed in the introduction on page 339 should be followed.

Blood tests for gonorrhea are being developed, but in the meantime, it is necessary to diagnose the condition by microscopically examining the discharge of men or by growing a culture of the bacterium from either the male or the female to positively identify the organism. Because there is no blood test, it is very difficult to recognize asymptomatic carriers, who are capable of passing on the condition without realizing it. If the infection is diagnosed, gonorrhea can usually be cured using the antibiotics penicillin or tetracycline; however, resistance to antibiotic therapy is becoming more common. In 1998, resistance was noted in as many as 30% of strains tested.

Gonorrheal infections are presently on the decline; however, resistance to antibiotic therapy is becoming increasingly prevalent.

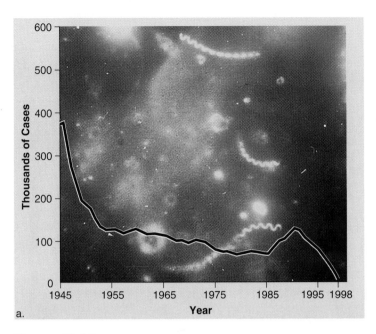

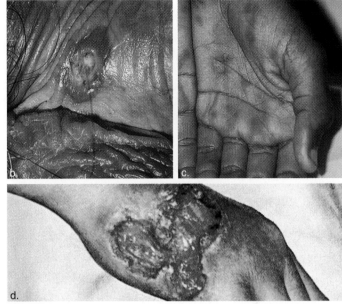

Figure 17.12   Syphilis.
a. A graph depicting the incidence of total cases of syphilis in the United States from 1945 to 1998. **b.** The first stage of syphilis is a chancre where the bacterium enters the body. **c.** The second stage is a body rash that occurs even on the palms of the hands and soles of the feet. **d.** In the tertiary stage, gummas may appear on skin or internal organs.

Syphilis

Syphilis is caused by a bacterium called *Treponema pallidum*, an actively motile, corkscrewlike organism that is classified as a spirochete. The number of new cases of syphilis in 1998 were the fewest reported in the United States since 1945 (Fig. 17.12*a*).

Syphilis has three stages, which can be separated by latent periods during which the bacteria are not multiplying (Fig. 17.12*b–d*). During the primary stage, a hard chancre (ulcerated sore with hard edges) indicates the site of infection. The chancre can go unnoticed, especially since it usually heals spontaneously, leaving little scarring. During the secondary stage, proof that bacteria have invaded and spread throughout the body is evident when the individual breaks out in a rash. Curiously, the rash does not itch and is seen even on the palms of the hands and the soles of the feet. There can be hair loss and infectious gray patches on the mucous membranes, including the mouth. These symptoms disappear of their own accord.

Not all cases of secondary syphilis go on to the tertiary stage. Some spontaneously resolve the infection, and some do not progress beyond the secondary stage. During a tertiary stage, which lasts until the patient dies, syphilis may affect the cardiovascular system, and weakened arterial walls (aneurysms) are seen, particularly in the aorta. In other instances, the disease may affect the nervous system. The patient may become mentally impaired, blind, walk with a shuffle, or show signs of insanity. Gummas, large destructive ulcers, may develop on the skin or within the internal organs.

Congenital syphilis is caused by syphilitic bacteria crossing the placenta. The child is stillborn, or born blind, with many other possible anatomical malformations.

Diagnosis and Treatment

Diagnosis of syphilis can be made by blood tests or by microscopic examination of fluids from lesions. One blood test is based on the presence of reagin, an antibody that appears during the course of the disease. Currently, the most used test is Rapid Plasma Reagin (RPR), which is used for screening large numbers of test serums. Because the blood tests can give a false positive, they are followed up by microscopic detection. Dark-field microscopic examination of fluids from lesions can detect the living organism. Alternately, test serum is added to a freeze-dried *T. pallidum* on a slide. Any antibodies present that react to *T. pallidum* are detected by fluorescent-labeled antibodies to human gamma globulin. The organism will now fluoresce when examined under a fluorescence microscope.

Syphilis is a devastating disease. Control of syphilis depends on prompt and adequate treatment of all new cases; therefore, it is crucial for all sexual contacts to be traced so that they can be treated. The cure for all stages of syphilis is some form of penicillin. To prevent transmission, the general instructions given in the introduction (see page 339) should be faithfully followed.

Syphilis is a devastating sexually transmitted disease, which is curable by antibiotic therapy. The chance of resistance is always a threat.

Bioethical Focus

HIV Vaccine Testing in Africa

The United Nations estimates that 16,000 people become newly infected with the human immunodeficiency virus (HIV) each day, or 5.8 million per year. Ninety percent of these infections occur in sub-Saharan Africa, where infected persons do not have access to antiviral therapy. In Uganda, for example, there is only one physician per 100,000 people, and only $6 is spent annually on health care per person. In the United States, $12,000–$15,000 is usually spent on treating an HIV-infected person per year.

The only methodology presently available to prevent the spread of HIV in a less-developed country* is counseling against behaviors that increase the risk of infection. Clearly, an effective vaccine would be most beneficial to these countries. Several HIV vaccines are in various stages of development, and all need to be clinically tested in order to see if they are effective. It seems reasonable to carry out such trials in less-developed countries, but there are many ethical questions.

A possible way to carry out the trial is this: vaccinate the uninfected sexual partners of HIV-infected individuals. After all, if the uninfected partner remains free of the disease, then the vaccine is effective. But is it ethical to allow a partner identified as having an HIV infection to remain untreated for the sake of the trial?

And should there be a placebo group—a group that does not get the vaccine? After all, if a greater number of persons in the placebo group become infected than those in the vaccine group, then the vaccine is effective. But if members of the placebo group become infected, shouldn't they be given effective treatment? For that matter, even participants in the vaccine group might become infected. Shouldn't any participant in the trial be given proper treatment if they become infected? Who would pay for such treatment when the trial could involve thousands of persons?

*Country in which the population is expanding rapidly and the majority of people live in poverty.

Figure 17A   AIDS in Africa.
Performing clinical trials and treating HIV infections in Africa raises many ethical questions.

Decide Your Opinion

1. Should HIV vaccine trials be done in developing countries, which stand to gain the most from an effective vaccine? Why or why not?
2. Should the trial be carried out using the same standards as in developed countries? Why or why not?
3. Who should pay for the trial—the drug company, the participants, or the country of the participants?

Looking at Both Sides   www.mhhe.com/biosci/genbio/maderhuman7/

Every bioethical issue has at least two sides. Even if you already have an opinion, it is important to explore the opposite opinion before finalizing your position. The Online Learning Center at www.mhhe.com/biosci/genbio/maderhuman7/ will help you fine-tune your initial opinion, explore both sides, and finalize your position. Either you will acquire new arguments for your original opinion or you may even change your opinion. Be sure to complete these activities in sequence:

Taking Sides   Decide your initial opinion by answering a series of questions. Then see if your opinion changes after completing the next two activities.

Further Debate   Read opposing articles that give you further information on this particular bioethical issue.

Explain Your Position   Answer another series of questions and then defend your original or changed opinion. You can e-mail your position to your instructor if he or she wishes.

17.3 Other Infectious Diseases

Bacteria are organisms classified in the kingdom Monera. Three other kingdoms also contain organisms that cause sexually transmitted diseases. They are kingdoms Protista, Fungi, and Animalia. Protists are unicellular, but fungi are usually multicellular and animals, including humans, are always multicellular. All the organisms in these kingdoms are eukaryotic. Their cells contain a nucleus, chromosomes, mitochondria, and other organelles.

Infectious Diseases Caused by Protozoa

Protozoa, in the kingdom Protista, are unicellular organisms that are said to be animal-like because their single cell functions as an animal cell does. Protozoa are often found in aquatic environments, such as freshwater ponds, and the ocean simply teems with them. Protozoa can reproduce by simple cell division but they usually have some means of sexual exchange. All protozoa require an outside source of nutrients, and only the parasitic ones take their nourishment from a host.

Being animal-like, you would think that protozoa would have some means of locomotion, and many do. However, the protozoan that causes malaria does not locomote— during its asexual phase, it simply produces nonmotile spores inside red blood cells. Amoebic dysentery is due to an infection by a protozoan that moves by simply extending its cytoplasm in one direction just like a white blood cell. Giardiasis, an infection of the digestive tract, and trichomoniasis, an infection of the vagina, are caused by zooflagellates. Zooflagellates have flagella which allow them to move about.

Trichomoniasis is a sexually transmitted disease caused by the zooflagellate *Trichomonas vaginalis* (Fig. 17.13a). This form of vaginitis is characterized by a frothy white or yellow, foul-smelling vaginal discharge accompanied by itching. Trichomoniasis is most often acquired through sexual intercourse, and an asymptomatic partner is usually the reservoir of infection.

 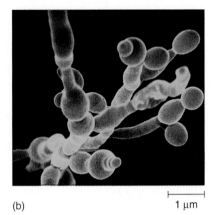

(a)   8 μm   (b)   1 μm

Figure 17.13 **Organisms that cause vaginitis.**
a. *Trichomonas vaginalis*, a protozoan. **b.** *Candida albicans*, a yeast.

Infectious Diseases Caused by Fungi

Two types of fungi are well known: the unicellular yeasts and the multicellular molds. Most fungi are nonpathogenic decomposers, such as the mold that appears on bread if kept in the refrigerator for too long. A few types of fungi are parasitic. Infectious fungi typically live on the skin, the nails, and in the vagina.

Almost everyone is familiar with the unicellular yeasts. Certain types are used to make bread rise and to produce wine, beer, and whiskey because of their ability to ferment sugar. *Candida albicans* (Fig. 17.13b) is a yeast that can cause diaper rash, thrush (a mouth infection), and a form of vaginitis in women. *Candida albicans* is usually found in the vagina; its growth simply increases beyond normal under certain circumstances. Antibiotic therapy and taking the birth control pill can disrupt the normal balance of flora that live on mucous membranes, and the result can be candidiasis, a yeast infection.

The body of a multicellular fungus is composed of fine filaments called hyphae. A multicellular fungus is the cause of the skin infections called athlete's foot and ringworm.

Table 17.3	**Infectious Diseases Caused by Protozoa, Fungi, and Animals**		
Site of Infection	**Protozoa**	**Fungi**	**Animals**
Not Sexually Transmitted			
Skin or mucous membranes	—	Athlete's foot, ringworm, vaginitis, thrush	Scabies, lice
Nervous system	Sleeping sickness	Cryptococcosis	—
Digestive tract	Amoebic dysentery, giardiasis	Ergot poisoning	Tapeworms, trichinosis
Cardiovascular or lymphatic	Malaria	—	Schistosomiasis, elephantiasis
Sexually Transmitted			
Reproductive system	Trichomoniasis	—	—
Skin	—		Pubic lice

Infectious Diseases Caused by Animals

Although we tend to think of organisms in the kingdom Animalia as being macroscopic, actually some are quite small, even microscopic. An animal is a motile, multicellular organism with specialized tissues. Since animals require an outside source of nutrients, some are parasitic. Various types of worms, such as pinworms, tapeworms, and hookworms, are well-known parasites that inhabit the bodies of other animals. Lice are insects that can infect the hair of humans. The head louse is well known for infecting the hair of schoolchildren; the crab louse is sexually transmitted.

Pubic Lice (Crabs)

The parasitic crab louse *Phthirus pubis* (an insect) takes its name from its resemblance to a small crab (Fig. 17.14). Crabs, as it is commonly called, can be contracted by direct contact with an infested person or by contact with his or her clothing or bedding. Females lay their eggs around the base of the hair, and these eggs hatch within a few days to produce a larger number of animals that suck blood from their host and cause severe itching, particularly at night. The pubic hair, underarm hair, and even the eyebrows can be infected.

In contrast to most types of sexually transmitted diseases, self-diagnosis and self-treatment that does not require shaving is possible. Usually, an infected person finds and

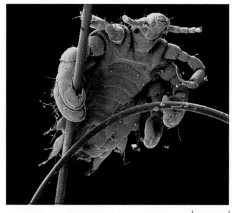

Figure 17.14
Sexually transmitted animal.
This parasitic crab louse, *Phthirus pubis*, infects the pubic hair of humans.

├──────┤
250 µm

identifies adult lice or their numerous eggs on or near pubic hairs. Medications such as lindane are applied to the infected area; all sexual partners need to be treated also. Undergarments, sheets, and night clothing should be washed by machine in hot water.

Aside from viruses and bacteria, a protozoan and a yeast can cause vaginitis, and an animal is the cause of a condition called pubic lice.

Summarizing the Concepts

17.1 Viral Infectious Diseases

Viruses are tiny particles that always have an outer capsid of protein and an inner core of nucleic acid that may be DNA or RNA. Viruses reproduce within living cells. After an animal virus enters a cell, uncoating occurs; viral DNA is replicated and capsid proteins are made. Following assembly, viruses bud from the cell.

Viruses are the cause of many infectious diseases in humans from common colds to cancer (Table 17.1). HIV is the cause of AIDS, a disease that is now pandemic. HIV-1C is now rampaging through Africa, and may eventually cause a new epidemic in the United States where HIV-1B is the cause of most infections. An HIV infection has three stages; the last stage is called AIDS. By this time, the immune system is devastated, and the individual dies of an opportunistic disease.

Genital warts are caused by human papillomavirus (HPV), which is a cuboidal DNA virus that reproduces in the nuclei of skin cells. Genital warts is a disease characterized by warts on the penis and foreskin in men and near the vaginal opening in women. Genital herpes is caused by herpes simplex virus: type 1 usually causes cold sores and fever blisters, while type 2 often causes genital herpes. Genital herpes is a disease that causes painful blisters on the genitals. Hepatitis B, which is spread in the same manner as AIDS, can lead to liver failure.

Medications have been developed to control AIDS and genital herpes, but there is no cure for these conditions, and no vaccines are available. Of all the viral STDs, only a vaccine for hepatitis B exists at the present time.

17.2 Bacterial Infectious Diseases

Bacteria are cells that lack a nucleus and the membranous organelles typical of animal cells. Bacteria reproduce independently by binary fis-

sion and may form endospores. Antibiotics are available to cure bacterial STDs unless a resistant strain is involved.

Bacteria are the cause of many infectious diseases in humans from strep throat to syphilis (Table 17.2). Chlamydia is caused by a tiny bacterium of the same name. These bacteria develop inside phagocytic vacuoles that eventually burst and liberate infective chlamydiae. Chlamydia can be asymptomatic but can produce symptoms of a urinary tract infection. Both chlamydia and gonorrhea can result in PID, leading to sterility and ectopic pregnancy. Gonorrhea is caused by the bacterium *Neisseria gonorrhoeae*, a diplococcus. Gonorrhea may not cause symptoms, particularly in women, but in men there may be painful urination and a thick, milky discharge. Syphilis is caused by a bacterium called *Treponema pallidum*, an actively motile, corkscrewlike organism. Syphilis is a systemic disease that should be cured in its early stages before deterioration of the nervous system and cardiovascular system possibly takes place.

Abstinence, a monogamous relationship, or use of a condom with a vaginal spermicide that contains nonoxynol-9 can help prevent the transmission of viral and bacterial STDs.

17.3 Other Infectious Diseases

The cells of protozoa, fungi, and other animals resemble our own. Protozoa and yeasts are unicellular but other fungi and all animals are multicellular. These organisms reproduce independently. Protozoa cause several types of infections in humans; trichomoniasis, a form of vaginitis in women, is sexually transmitted. Athlete's foot is a well known infection caused by a fungus and so is candidiasis, also a type of vaginitis. Public lice is a sexually transmitted disease caused by an animal.

Studying the Concepts

1. Describe the structure and life cycle of a DNA virus. 340–41
2. Describe the cause and symptoms of an HIV infection. 342
3. Among which groups of society is AIDS now increasing most rapidly? How might transmission be prevented? 342
4. Give the cause and symptoms of genital warts, genital herpes, and a hepatitis B infection. 342–44
5. Discuss the treatment for STDs caused by viruses. How might these viruses be prevented from spreading? 344
6. State the three shapes of bacteria, and describe the structure of a prokaryotic cell. 345
7. Describe the symptoms and results of a chlamydial infection and gonorrhea in men and in women. What is PID, and how does it affect reproduction? 346–48
8. Describe the three stages of syphilis. 349
9. How does the newborn acquire an infection of genital warts, herpes, chlamydia, gonorrhea, or syphilis? What effects do these infections have on infants? 342, 343, 347, 348, 349
10. Describe the symptoms of vaginitis and pubic lice. 351–52

Testing Your Knowledge of the Concepts

In questions 1–4, match the disease to the descriptions.

a. HIV
b. genital warts
c. genital herpes
d. syphilis
e. gonorrhea

_____ 1. Often appears with chlamydia, and both conditions can cause PID.

_____ 2. Primary host is a T lymphocyte or a macrophage.

_____ 3. Painful ulcers can lead to passage from mother to child.

_____ 4. Has a tertiary stage in which cardiovascular and nervous system involvement appears.

In questions 5–7, indicate whether the statement is true (T) or false (F).

_____ 5. Most new cases of HIV infection now occur in minority women.

_____ 6. There is a vaccine available for hepatitis B, a disease that can lead to liver failure.

_____ 7. Penicillin therapy will usually cure both gonorrhea and syphilis.

In questions 8–17, fill in the blanks.

8. Bacteria reproduce asexually by _____.

9. All viruses have an inner core of _____ and a capsid of _____.

10. Herpes simplex virus type 1 causes _____, and type 2 causes _____.

11. Bacterial cells have one of three shapes: _____, _____, and _____.

12. Women are often asymptomatic for a chlamydial infection until they develop _____.

13. The use of a _____ can help reduce the risk of acquiring an STD.

14. In the tertiary stage of syphilis, there may be large sores called _____.

15. These three sexually transmitted diseases are usually curable by antibiotic therapy: _____, _____, and _____.

16. Women who take the birth control pill are more likely to acquire vaginitis due to a _____ infection.

17. Viruses can only reproduce in a _____ host cell.

18. Compose a sentence for each box in the following drawing to explain the life cycle of a DNA virus.

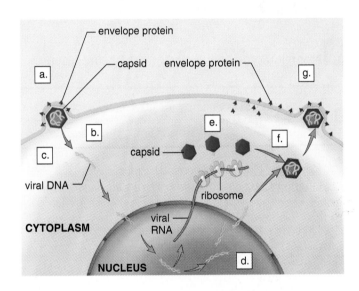

Understanding Key Terms

AIDS (acquired immunodeficiency syndrome) 342
antibiotic 346
bacteria 345
binary fission 345
chlamydia 346
endospore 346
genital herpes 343
genital warts 342
gonorrhea 348

hepatitis 344
host 340
latency 341
opportunistic infection 342
pathogen 340
pelvic inflammatory disease (PID) 347
sexually transmitted disease 339
syphilis 349
virus 340

17.1 Viral Infectious Diseases

Characteristics of Viruses *Essential Study Partner*
Life Cycles *Essential Study Partner*

Looking at Both Sides *critical thinking activity*

17.2 Bacterial Infectious Diseases

Characteristics of Bacteria *Essential Study Partner*
Diversity *Essential Study Partner*

Nonphotosynthetic Bacterium *art labeling activity*

Chapter Summary

Key Term Flashcards *vocabulary quiz*
Chapter Quiz *objective quiz covering all chapter concepts*

AIDS Supplement

Supplement Concepts

Figure S.1 HIV (green) budding from an infected T lymphocyte.

Figure S.1 shows HIV (human immunodeficiency virus) budding from a T lymphocyte, its primary host. No wonder the immune system falters in a person with an HIV infection. The very cells that orchestrate the immune response are under viral attack.

While we in the United States bear a tremendous expense to treat people infected with HIV, this luxury is not available to the many more that are infected in the developing countries. Will there be a vaccine sometime soon? If so, the immune system would be primed to control the infection before it establishes itself. There might be something about the virus—its mode of transmission or the course of the disease—that will make any vaccine ineffective. Then, too, a vaccine for one type of HIV infection may not be effective against another type. HIV is a family of viruses, and each type has different subtypes. Both HIV-1 and HIV-2 infections are found in Africa, while only HIV-1 is prevalent in the United States and most other countries of the world. There are several subtypes of HIV-1, and among these, HIV-1C is extremely virulent and is now wreaking havoc in sub-Saharan Africa. It may hit the United States sometime soon.

The burden of avoiding an HIV infection is on the individual. We all must come to realize the importance of our T lymphocytes and take measures to protect them from possible destruction by all types of HIV. The various ways to prevent infection discussed on page 362 should be followed faithfully by all.

An HIV infection usually leads to AIDS (acquired immunodeficiency syndrome). The full name of AIDS can be explained in this way: *acquired* means that the condition is caught rather than inherited; *immune deficiency* means that the virus attacks the immune system so there is greater susceptibility to certain infections and cancer; and *syndrome* means that some fairly typical opportunistic infections and cancers usually occur in the infected person. These are the conditions that cause AIDS patients to die.

S.I   Origin and Scope of the AIDS Pandemic

It's generally accepted that HIV originated in Africa and then spread to the United States and Europe by way of the Caribbean. HIV has been found in a preserved 1959 blood sample taken from a man who lived in an African country now called the Democratic Republic of the Congo. Even before this discovery, scientists speculated that an immunodeficiency virus may have evolved into HIV during the late 1950s.

Of the two types of HIV, HIV-2 corresponds to a type of immunodeficiency virus found in the green monkey which lives in western Africa. Recently, it was announced that researchers have found a virus identical to HIV-1 in a subgroup of chimpanzees once common in west-central Africa. Perhaps, HIV viruses were originally found only in nonhuman primates. They could have mutated to HIV after humans ate nonhuman primates for meat.

British scientists have been able to show that AIDS came to their country perhaps as early as 1959. They examined the preserved tissues of a Manchester seaman who died that year and concluded that he most likely died of AIDS. Similarly, it's thought that HIV entered the United States on numerous occasions as early as the 1950s. But the first documented case is a 15-year-old male who died in Missouri in 1969 with skin lesions now known to be characteristic of an AIDS-related cancer. Doctors froze some of his tissues because they could not identify the cause of death. Researchers also want to test the preserved tissue samples of a 49-year-old Haitian who died in New York in 1959 of the type of pneumonia now known to be AIDS related.

Throughout the 1960s, it was the custom in the United States to list leukemia as the cause of death in immunodeficient patients. Most likely, some of these people actually died of AIDS. Since HIV is not extremely infectious, it took several decades for the number of AIDS cases to increase to the point that AIDS became recognizable as a specific and separate disease. The name AIDS was coined in 1982, and HIV was found to be the cause of AIDS in 1983–84.

AIDS most likely originated in the 1950s, but it wasn't until 1983 that HIV was recognized as the cause of AIDS.

Prevalence of AIDS

AIDS is said to be pandemic because the disease is prevalent in the entire human population around the globe (Fig. S.2). Today, 34.3 million adults and children are estimated to be living with an HIV infection, and 19 million persons have now died of AIDS worldwide. HIV is transmitted by sexual contact with an infected person, including vaginal or rectal intercourse and oral/genital contact. Also, needle sharing among intravenous drug users is high-risk behavior. A less common mode of transmission (and now rare in countries where blood is screened for HIV) is through transfusions of infected blood or blood-clotting factors. Babies born to HIV-infected women may become infected before or during birth, or through breast-feeding after birth. Table S.1 summarizes the most frequent ways by which HIV is transmitted.

The incidence of HIV infection in the more developed countries of North America and western Europe and elsewhere is modest. The incidence ranges from 0.02% in Japan to 0.6% in the United States and 0.7% in Portugal. Further, the introduction of combination drug therapy has resulted in a drastic decrease in overall AIDS deaths since 1995. In the United States, HIV first spread through the homosexual community, and male-to-male sexual contact still accounts for the largest percentage of new AIDS cases. But the rate of new HIV infections is now rising faster among heterosexuals than homosexuals. Even now, 23% of all people with AIDS in the United States are women and they account for 32% of all newly diagnosed cases of HIV infection. In 1986, the majority of AIDS cases occurred among Caucasians; today this proportion has shifted to minorities. There is one more statistic to call to your attention. Most new HIV infections are occurring among teenagers and young adults; even with drug therapy, these young people will probably not escape eventually coming down with AIDS.

In the largest countries of Latin America and the Caribbean, with an incidence of 2.0% and 0.6% respectively,

Table S.1   Transmission of HIV
Possible Routes
Homosexual and heterosexual contact
Intravenous drug use
Transfusion (unlikely in U.S.)
Crossing placenta during pregnancy; breast-feeding
Increasing the Risk
Promiscuous behavior (large number of partners, sex with a prostitute)
Drug abuse with needle sharing
Presence of another sexually transmitted disease

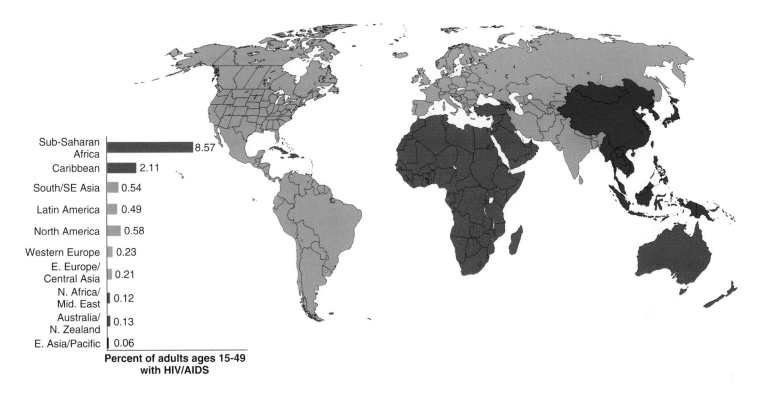

Figure S.2 **HIV prevalence rates in adults at the end of 1999.**

HIV/AIDS occurs in all continents and countries of the globe, and presently 34.3 million adults and children are infected worldwide.

the AIDS epidemic has charted a course similar to that in North America. HIV/AIDS was first seen among homosexuals and intravenous drug users. Now it is increasingly being spread by heterosexual contact.

Haiti, with an overall HIV prevalence in adults of about 5%, is the worst-affected country outside of Africa. In sub-Saharan Africa, 24.5 million people are infected with HIV. This is almost 9% of the total adult population between 15 and 49 years of age. The increasing number of deaths among young adults means that there will be more people in their 60s and 70s than in their 40s and 50s. Many are concerned about how families in Africa will cope when the old have to care for their grandchildren and when these grandchildren have to assume adult responsibilities much sooner than otherwise.

Some countries in Africa are affected more than others. In ten years, HIV prevalence in South Africa grew from 1% of the adult population in 1990 to about 20% today. Similarly, Botswana was essentially free of the disease in 1990. Now 45–50% of young adults aged 20–30 years are infected with HIV. A certain subtype of HIV-1, namely HIV-1C, has brought about this great devastation. HIV-1C has a greater ease of transmission, and also multiplies and mutates faster than all the other subtypes. Most likely, HIV-1C has already spread from Africa to western India, and a hybrid virus con-

taining some HIV-1C genetic material has reached mainland China. It's quite possible that in five to twenty years the more developed countries, including the United States, will experience a new epidemic of AIDS caused by HIV-1C. Therefore, it behooves the more developed countries to do all they can to help African countries aggressively seek a solution to this new HIV epidemic.

AIDS in the United States is presently caused by HIV-1B, and drug therapy has brought the condition under control. But the use of drug therapy has two dangers. People may become lax in their efforts to avoid infection because they know that drug therapy is available. Also, the use of drugs leads to drug-resistant viruses. Even now, some HIV-1B viruses have become drug resistant when patients have failed to adhere to their drug regimens. We cannot escape the conclusion that all persons should do everything they can to avoid becoming infected. Behaviors that help prevent transmission are discussed in the Health Focus on page 362.

New HIV infections are increasing faster among heterosexuals than homosexuals; women in the United States now account for nearly 25% of AIDS cases.

S.2   Phases of an HIV Infection

The following description of the phases of an HIV infection pertains to an HIV-1B infection, the type now prevalent in the United States. The HIV viruses primarily infect helper T lymphocytes that are also called CD4 T lymphocytes, or simply CD4 T cells. These cells display a molecule called CD4 on their surface. Helper T cells, you will recall, ordinarily stimulate B lymphocytes to produce antibodies and cytotoxic T cells to destroy cells that are infected with a virus.

The Centers for Disease Control and Prevention now recognize three categories of an HIV-1B infection. During an acute phase, there are usually no symptoms, yet the person is highly infectious. During the chronic phase, the individual loses weight, suffers from diarrhea, and most likely develops infections like thrush or genital herpes. The final phase of an HIV infectiion is AIDS, when a person comes down with pneumonia, cancer, and other serious conditions.

Category A: Acute Phase

A normal CD4 T cell count is at least 800 cells per cubic millimeter of blood. This first phase of an HIV infection is characterized by a CD4 T cell count of 500 per mm^3 or greater (Fig. S.3). This count is sufficient for the immune system to function normally.

Today investigators are able not only to track the blood level of CD4 T cells, but also to monitor the viral load. The viral load is the number of HIV particles in the blood. At the start of an HIV-1B infection, the virus is replicating ferociously, and the killing of CD4 T cells is evident because the blood level of these cells drops dramatically. For a few weeks, however, people don't usually have any symptoms at all. Then, a few (1–2%) do have mononucleosis-like symptoms that may include fever, chills, aches, swollen lymph nodes, and an itchy rash. These symptoms disappear, and there are no other symptoms for quite some time. The HIV blood test commonly used at clinics is not yet positive because it tests for the presence of antibodies and not for the presence of HIV itself. This means that the person is highly infectious, even though the HIV blood test is negative. For this reason, all persons need to follow the guidelines for preventing the transmission of HIV as outlined in the Health Focus on page 362.

After a period of time, the body responds to the infection by increased activity of immune cells, and the HIV blood test becomes positive. During this phase, the number of CD4 T cells is greater than the viral load (Fig. S.3). But some investigators believe that a great unseen battle is going on. The body is staying ahead of the hordes of viruses entering the blood by producing as many as one to two billion new helper T lymphocytes each day. This is called the "kitchen sink model" for CD4 loss. The sink's faucet (production of new CD4 T cells) and the sink's drain (destruction of CD4 T cells) are wide open. As long as the body's production of new CD4 T cells is able to keep pace with the destruction of these cells by HIV and by cytotoxic T cells, the person has a healthy immune system that can deal with the infection.

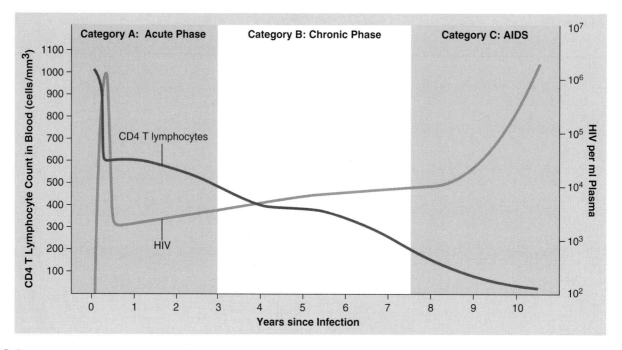

Figure S.3   **Stages of an HIV infection.**
In category A individuals, the number of HIV in plasma rises upon infection and then falls. The number of CD4 T lymphocytes falls, but stays above 400 per mm^3. In category B individuals, the number of HIV in plasma is slowly rising and the number of T lymphocytes is decreasing. In category C individuals, the number of HIV in plasma rises dramatically as the number of T lymphocytes falls below 200 per mm^3.

Category B: Chronic Phase

Several months to several years after infection, an untreated individual will probably progress to category B. During this stage, the CD4 T cell count is 200 to 499 per mm^3 of blood, and most likely, symptoms will begin to appear. Symptoms of category B include swollen lymph nodes in the neck, armpits, or groin that persist for three months or more; severe fatigue not related to exercise or drug use; unexplained persistent or recurrent fevers, often with night sweats; persistent cough not associated with smoking, a cold, or the flu; and persistent diarrhea. Also possible are signs of nervous system impairment, including loss of memory, inability to think clearly, loss of judgment, and/or depression.

When the individual develops non-life-threatening but recurrent infections, it is a signal that full-blown AIDS will occur shortly. One possible infection is thrush, a *Candida albicans* infection that is identified by the presence of white spots and ulcers on the tongue and inside the mouth. The fungus may also spread to the vagina, resulting in a chronic infection there. Another frequent infection is herpes simplex, with painful and persistent sores on the skin surrounding the anus, the genital area, and/or the mouth.

Category C: AIDS

When a person has AIDS, the CD4 T cell count has fallen below 200 per mm^3, and the lymph nodes have degenerated. The patient is extremely thin and weak due to persistent diarrhea and coughing, and will most likely develop one of the opportunistic infections. An **opportunistic infection** is one that has the *opportunity* to occur only because the immune system is severely weakened. Persons with AIDS die from one or more of the following diseases rather than from the HIV infection itself:

- *Pneumocystis carinii* pneumonia. The lungs become useless as they fill with fluid and debris due to an infection with a protozoan.
- *Mycobacterium tuberculosis.* This bacterial infection, usually of the lungs, is seen more often as an infection of lymph nodes and other organs in patients with AIDS.
- Toxoplasmic encephalitis is caused by a protozoan parasite that ordinarily lives in cats and other animals. Many people harbor a latent infection in the brain or muscle, but in AIDS patients, the infection leads to loss of brain cells, seizures, and weakness.
- Kaposi's sarcoma is an unusual cancer of the blood vessels, which gives rise to reddish-purple, coin-sized spots and lesions on the skin.
- Invasive cervical cancer. This cancer of the cervix spreads to nearby tissues. This condition was added to the list when AIDS became more common in women.

Although there are newly developed drugs to deal with opportunistic diseases, most AIDS patients are repeatedly hospitalized due to weight loss, constant fatigue, and multiple infections (Fig. S.4). Death usually follows in 2–4 years.

During the acute phase of an HIV infection, the person is highly infectious although there may be no symptoms; during the chronic phase, there may be swollen lymph nodes and various infections; during the last phase, which is called AIDS, the patient usually succumbs to an opportunistic infection.

a. AIDS patient, Tom Moran July 1987

b. AIDS patient Tom Moran, early January 1988

c. AIDS patient Tom Moran, late January 1988

Figure S.4   The course of an AIDS infection.
These photos show the effect of an HIV infection in one individual who progressed through all the stages of AIDS.

S.3   Treatment for HIV

Until a few years ago, an HIV infection almost invariably led to AIDS and an early death. The medical profession was able to treat the opportunistic infections that stemmed from immune failure, but had no drugs for controlling HIV itself. But since late 1995, scientists have gained a much better understanding of the structure of HIV and its life cycle. Now therapy is available that successfully controls HIV replication and keeps patients in the chronic phase of infection for a variable number of years so that the development of AIDS is postponed.

HIV Structure and Life Cycle

The genetic material for an HIV virus consists of RNA instead of DNA. Inside the HIV capsid are RNA and three enzymes of interest: reverse transcriptase, integrase, and protease. The HIV particle has an envelope acquired when it buds from an infected cell (Fig. S.5).

The events that occur in the reproductive cycle of an HIV virus are essentially the same as those for the DNA virus shown in Figure 17.3, but there is one extra event because HIV is a retrovirus.

1. *Binding of the virus to the plasma membrane.* HIV has an envelope protein known as gp120. This envelope protein allows the virus to bind to a CD4 receptor in the host-cell plasma membrane. Ordinarily, a CD4 receptor is a binding site for various signaling molecules.
2. *Penetration of the virus into the cell.* After binding occurs, the HIV virus fuses with the plasma membrane, and the virus enters the cell. Uncoating removes the capsid, and RNA is released.
3. *Production of viral DNA.* This event in the reproductive cycle is unique to **retroviruses.** The enzyme called reverse transcriptase makes a DNA copy of their RNA genetic material. Usually in cells DNA is transcribed into RNA. Retroviruses can do the opposite only because they have this unique enzyme from which they take their name. (*Retro* in Latin means reverse.)

The viral enzyme integrase now splices viral DNA into a host chromosome. The term **HIV provirus** refers to viral DNA integrated into host DNA. HIV is usually transmitted to another person by means of cells that contain proviruses. Also, proviruses serve as a latent reservoir for HIV during drug treatment. Even if drug therapy results in an undetectable viral load, investigators know that there are still proviruses inside infected lymphocytes.

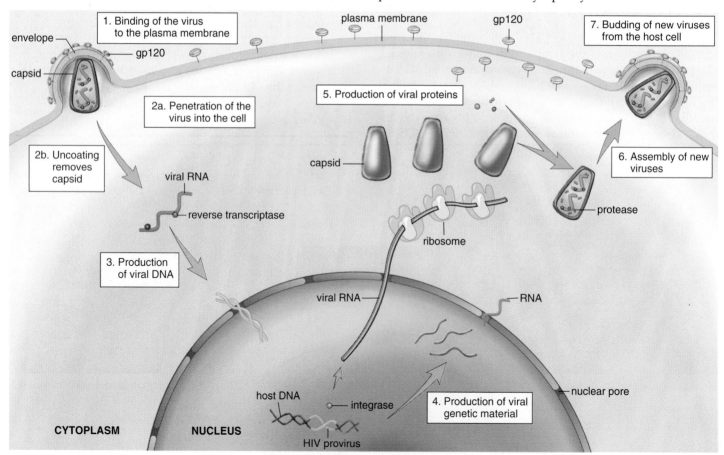

Figure S.5   Reproduction of HIV.
HIV-1 is a retrovirus that utilizes reverse transcription to produce viral DNA. Viral DNA integrates into the cell's chromosomes before it reproduces and buds from the cell.

4. *Production of the viral genetic material.* When the provirus is activated, perhaps by a new and different infection, the normal cell machinery directs the production of more viral RNA. Some of this RNA becomes the genetic material for new virus particles.

5. *Production of viral proteins.* The rest of viral RNA brings about the synthesis of viral proteins (including capsid proteins, viral enzymes, and gp120) at host ribosomes.

6. *Assembly of new viruses.* Capsid proteins, viral enzymes, and RNA are assembled to form new viral particles. The viral enzyme protease cleaves viral proteins so that they are a size suitable for viral assembly. Reproduction of the virus has now taken place.

7. *Budding of new viruses from the host cell.* During budding, the virus gets its envelope and gp120 coded for by the viral genetic material.

The life cycle of an HIV virus includes transmission to a new host. Body secretions, such as semen from an infected male, contain proviruses inside CD4 T lymphocytes. When this semen is discharged into the vagina, rectum, or mouth, infected CD4 T cells migrate through the organ's lining and enter the body. The receptive partner in anal-rectal intercourse appears to be most at risk because the lining of the rectum is a thin, single-cell layer. CD4 macrophages present in tissues are believed to be the first infected when proviruses enter the body. When these macrophages move to the lymph nodes, HIV begins to infect CD4 T cells. HIV can hide out in local lymph nodes for some time, but eventually the lymph nodes degenerate, and large numbers of HIV enter the general bloodstream. Now the viral load begins to increase; when it exceeds the CD4 T cell count, the individual progresses to the final stage of an HIV infection.

HIV is a retrovirus that infects immune cells, such as helper T lymphocytes, carrying a CD4 receptor. HIV is transmitted as a provirus inside infected CD4 cells.

Drug Therapy

There is no cure for AIDS, but a treatment called highly active antiretroviral therapy (HAART) is usually able to stop HIV replication to such an extent that the viral load becomes undetectable. Even so, investigators know that about one million viruses are still present, including those that exist only as inactive proviruses.

HAART utilizes a combination of two types of drugs: reverse transcriptase and protease inhibitors. The well-publicized drug called AZT and several others are reverse transcriptase inhibitors, which prevent viral DNA from being produced. When HIV protease is blocked, the resulting viruses lack the capacity to cause infection. It is important to realize that these drugs are very expensive, that they

cause side effects such as diarrhea, neuropathy (painful or numb feet), hepatitis, and possibly diabetes, and that the regimen of pill taking throughout the day is very demanding. If the drugs are not taken as prescribed or if therapy is stopped, resistance may occur. Researchers are trying to develop new drugs that might be helpful against resistant strains. A new class of drugs called fusion inhibitors seems to block HIV's entry into cells. Even more experimental is the discovery of compounds that sabotage integrase, the enzyme that splices viral RNA into host-cell DNA.

Investigators have found that when HAART is discontinued, the virus rebounds. It may then be possible to help the immune system counter the virus by injecting the patient with a stimulatory cytokine like interferon or interleukin-2.

A pregnant woman who is infected with HIV and takes reverse transcriptase inhibitors during her pregnancy reduces the chances of HIV transmission to her newborn by nearly 66%. If possible, drug therapy should be delayed until the 10th to 12th week of pregnancy to minimize any adverse effects of AZT on fetal development. Treatment with reverse transcriptase inhibitors during only the last few weeks of a pregnancy cuts transmission of HIV to an offspring by half.

Vaccines

There is a general consensus that control of the AIDS epidemic will not occur until a vaccine that prevents an HIV infection is developed. An effective vaccine should bring about a twofold immune response: production of antibodies by B cells and stimulation of cytotoxic T cells.

Traditionally, vaccines are made by weakening a pathogen so that it will not cause disease when it is injected into the body. One group of investigators using this approach announced that they have found a way to expose hidden parts of gp120 so the immune system can better learn to recognize this antigen. It will be several years before this vaccine is ready for clinical trials.

Another group has developed a pill that might keep HIV from ever entering the body by turning on cytotoxic T cells within mucous membranes. This vaccine will soon be tested in Uganda, an African country.

Others have been working on subunit vaccines that utilize just a single HIV protein, such as gp120, as the vaccine. So far this approach has not resulted in sufficient antibodies to keep an infection at bay. After many clinical trials, none too successful, most investigators now agree that a combination of various vaccines may be the best strategy to bring about a response in both B lymphocytes and cytotoxic T cells.

Combination drug therapy has met with encouraging success against an HIV infection. Development of a vaccine is also being pursued.

Preventing Transmission of HIV

SEXUAL ACTIVITIES TRANSMIT HIV.

Abstain from sexual intercourse or develop a long-term monogamous (always the same partner) sexual relationship with a partner who is free of HIV.

Refrain from multiple sex partners or having relations with someone who has multiple sex partners. If you have sex with two other people and each of these has sex with two people and so forth, the number of people who are relating is quite large.

Remember that the prevalence of AIDS is presently higher among homosexuals and bisexuals than among those who are heterosexual.

Be aware that having relations with an intravenous drug user is risky because the behavior of this group risks AIDS. Be aware that anyone who already has another sexually transmitted disease is more susceptible to an HIV infection.

Avoid anal-rectal intercourse (in which the penis is inserted into the rectum) because this behavior increases the risk of infection. The lining of the rectum is thin, and infected CD4 T cells can easily enter the body there. Also, the rectum is supplied with many blood vessels, and insertion of the penis into the rectum is likely to cause tearing and bleeding that facilitate the entrance of HIV. The vaginal lining is thick and difficult to penetrate, but the lining of the uterus is only one cell thick and does allow CD4 T cells to enter (Fig. S.A).

Uncircumcised males are more likely to become infected than circumcised males because vaginal secretions can remain under the foreskin for a long period of time.

PRACTICE SAFE SEX.

Always use a latex condom during sexual intercourse if you do not know for certain that your partner has been free of HIV for the past five years. Be sure to follow the directions supplied by the manufacturer. Use of a water-based spermicide containing nonoxynol-9 in addition to the condom can offer further protection because nonoxynol-9 immobilizes the virus and virus-infected lymphocytes.

Figure S.A **Transmission by way of the uterus.**
HIV is spread by passing virus-infected CD4 cells found in body secretions or in blood from one person to another. Arrows show lymphocytes moving from the uterine cavity through the endometrium to enter the body.

Avoid fellatio (kissing and insertion of the penis into a partner's mouth) *and cunnilingus* (kissing and insertion of the tongue into the vagina) because they may be a means of transmission. The mouth and gums often have cuts and sores that facilitate the entrance of infected CD4 T cells.

Practice penile, vaginal, oral, and hand cleanliness. Be aware that hormonal contraceptives make the female genital tract receptive to the transmission of HIV and other sexually transmitted diseases.

Be cautious about the use of alcohol or any drug that may prevent you from being able to control your behavior.

DRUG USE TRANSMITS HIV.

Stop, if necessary, or do not start the habit of injecting drugs into your veins. Be aware that HIV can be spread by blood-to-blood contact.

Always use a new sterile needle for injection or one that has been cleaned in bleach if you are a drug user and cannot stop your behavior.

Chapter 18

Development
and Aging

Chapter Concepts

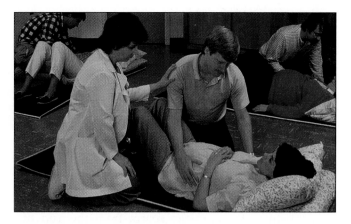

Figure 18.1   Childbirth classes.
At childbirth classes, both expectant parents learn how to facilitate the birthing process.

"At last," thought Mary Jean as she left the doctor's office. "I'm going to have a baby." Bill was sure to be overjoyed, and so would the prospective grandparents. Bill and Mary Jean had been trying to have a baby for three years with no success. Recently, they had been talking to their doctor about artificial insemination, in vitro fertilization, and other methods of getting pregnant. It was a wonderful surprise and a relief to find that the pregnancy had happened naturally.

Now there were many decisions for Mary Jean to make. "Where should I have the baby—at a birthing center or a hospital? Should I breast- or bottle-feed? How much weight can I gain?" She would discuss all these concerns with Bill and have them sign up soon for childbirth classes (Fig. 18.1). She wanted Bill to be with her every step of the way.

Thinking back over her Human Biology course, Mary Jean knew that by now the embryo had implanted itself in the endometrium (lining) of the uterus. At two months, it would have a human appearance and become a fetus floating within a watery sac. The fetus would be completely dependent upon the placenta, a special structure through which nourishment and oxygen are supplied and wastes are removed. Only at eight months or more would her child have an excellent chance of becoming an independently functioning human being. She was glad to have a doctor who would oversee her health while she waited for that all-important day.

This chapter is about the processes and stages that occur as a single cell becomes a complex organism. As discussed in chapter 16, a male continually produces sperm in the testes, and a female releases one egg a month from the ovaries. The release of the egg, called ovulation, usually occurs on about day 14 of an ovarian cycle that lasts 28 days. The hormones produced by the ovary control the uterine cycle, which consists of a proliferative phase and a secretory phase. During the proliferative phase, the endometrium (uterine lining) first breaks down during menstruation and then begins to thicken. During the secretory phase, the endometrium becomes more vascular and glandular. As a result, it contains nutrient substances which can sustain a developing embryo. These two phases prepare the uterus to accept the fertilized egg called a **zygote.**

This time, everything went perfectly for Mary Jean and Bill. The 200 to 600 million sperm released by Bill during sexual intercourse were ejaculated against the opening of the cervix at the far end of the vagina. The semigelatinous seminal fluid protected the sperm from the acid of the vagina for several minutes, and many managed to enter the uterus. Whether or not sperm enter the uterus depends in part on the consistency of the cervical mucus. Three to four days prior to ovulation and on the day of ovulation, the mucus is watery, and the sperm can penetrate it easily. During the other days of the uterine cycle, the mucus is thicker and has a sticky consistency, and the sperm can rarely penetrate it. Some high-speed sperm reach the oviducts in only 30 minutes, but generally the journey from vagina to oviduct takes several hours. Many sperm get lost along the way, and only several hundred sperm ever reach the oviducts.

18.1   Fertilization

Fertilization normally occurs in the upper third of the oviduct, where the sperm encounter an egg if ovulation has occurred. The sperm must swim against the downward current created by the ciliary action of the epithelium lining the oviducts. However, it is believed that uterine and oviduct contractions help transport the sperm and that prostaglandins within seminal fluid promote these contractions.

Fertilization is a series of events that brings the egg nucleus and the sperm nucleus together. A sperm has three distinct parts: a head, a middle piece, and a tail. The head contains a nucleus and is capped by a membrane-bounded acrosome. The tail is a flagellum, which allows the sperm to swim toward the egg, and the middle piece contains energy-producing mitochondria. The plasma membrane of the egg is surrounded by an extracellular matrix termed the zona pellucida. In turn, the zona pellucida is surrounded by a few layers of adhering follicular cells, collectively called the corona radiata. These cells nourished the egg when it was in a follicle of the ovary.

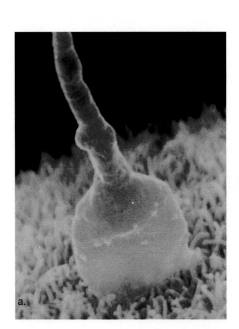

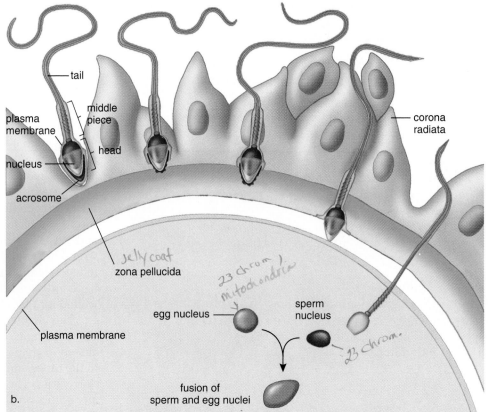

Figure 18.2   Fertilization.

a. During fertilization, a single sperm enters the egg. **b.** The head of a sperm has a membrane-bounded acrosome filled with enzymes. When released, these enzymes digest a pathway for the sperm through the zona pellucida. After it binds to the plasma membrane of the egg, a sperm enters the egg. When the sperm nucleus fuses with the egg nucleus, fertilization is complete.

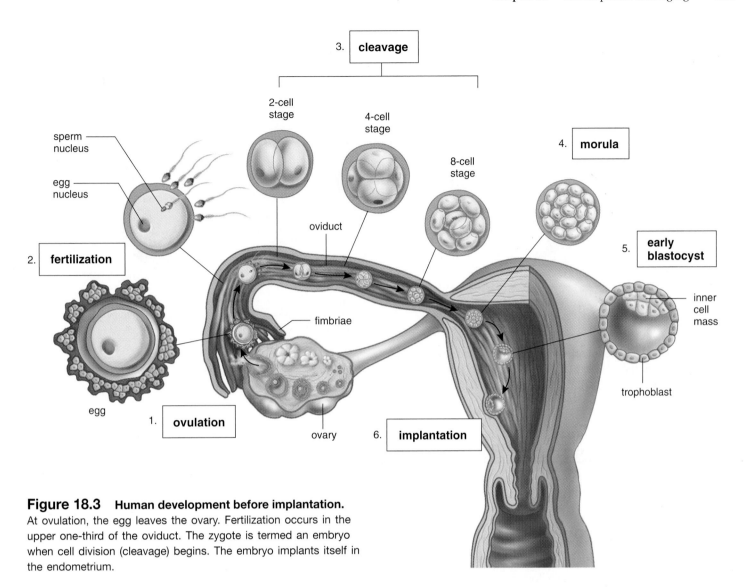

Figure 18.3 **Human development before implantation.**
At ovulation, the egg leaves the ovary. Fertilization occurs in the
upper one-third of the oviduct. The zygote is termed an embryo
when cell division (cleavage) begins. The embryo implants itself in
the endometrium.

During fertilization, (1) several sperm penetrate the co-rona radiata, (2) several sperm attempt to penetrate the zona pellucida, and (3) one sperm enters the egg and their nuclei fuse (Fig. 18.2). The acrosome plays a role in allowing sperm to penetrate the zona pellucida. After a sperm head binds tightly to the zona pellucida, the acrosome releases digestive enzymes that forge a pathway for the sperm through the zona pellucida. When a sperm binds to the egg, their plasma membranes fuse, and this sperm (the head, the middle piece, and usually the tail) enters into the egg. Fusion of the sperm nucleus and the egg nucleus follows.

To ensure proper development, only one sperm should enter an egg. Prevention of polyspermy (entrance of more than one sperm) depends on changes in the egg's plasma membrane and in the zona pellucida. As soon as a sperm touches an egg, the egg's plasma membrane depolarizes (from −65 Mv to 10 Mv), and this prevents the binding of any other sperm. Then the egg releases substances that lead to a lifting of the zona pellucida away from the surface of the egg. Now sperm cannot bind to the zona pellucida either.

When fertilization is complete, the egg is termed a zygote, and when the zygote begins dividing, it is called an **embryo.** The developing embryo travels very slowly down the oviduct to the uterus, where it implants itself in the endometrium (Fig. 18.3). As discussed in chapter 16, a membrane about the embryo begins to produce a hormone called HCG (human chorionic gonadotropin), and this hormone prevents the breakdown of the endometrium so that menstruation does not occur. The presence of HCG in the blood and urine confirms that a woman is pregnant. The signs that often prompt a woman to have a pregnancy test are cessation of menstruation, increased frequency of urination, morning sickness, and increase in the size and fullness of the breasts, as well as darkening of the areola, a pigmented area surrounding the nipple.

Following fertilization of an egg by a sperm, the
zygote begins developing, and the embryo
implants itself in the endometrium.

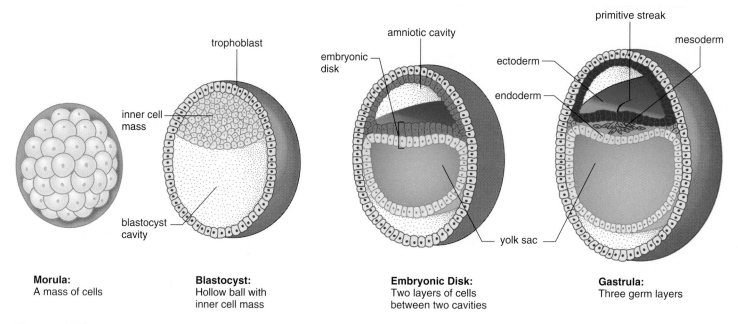

Figure 18.4   Early developmental stages in cross section.
Cleavage results in the inner cell mass. Morphogenesis occurs as cells rearrange themselves, and differentiation is first exemplified by the formation of three different germ layers.

18.2   Development Before Birth

This section considers the major processes and events that take place from the time of fertilization to the time of birth.

Processes of Development

Embryonic development includes these processes:

Cleavage   Immediately after fertilization, the zygote begins to divide so that at first there are 2, then 4, 8, 16, and 32 cells, and so forth. Increase in size does not accompany these divisions.

Morphogenesis   Morphogenesis refers to the shaping of the embryo and is first evident when certain cells are seen to move, or migrate, in relation to other cells. By these movements, the embryo begins to assume various shapes.

Differentiation   Differentiation occurs as cells take on a specific structure and function. For example, nerve cells have long processes that conduct nerve impulses, and muscle cells contain contractile elements.

Growth   During most of embryonic development, cell division is accompanied by an increase in the size of the daughter cells, and growth (in the true sense of the term) takes place.

The following processes can be observed in the early developmental stages, which humans share with all animals (Fig. 18.4).

Morula

Cleavage is a process that occurs during the first stage of development. During cleavage, cell division without growth results in a mass of tiny cells. The cells are uniform in size because the cytoplasm has been equally divided among them. This solid mass of cells is called a morula, which means a bunch of berries.

Blastula

Morphogenesis begins as the cells of the morula form an empty ball of cells called the blastula. All animal blastulas have an empty cavity, but since the human blastula is called a **blastocyst,** the cavity is called the blastocyst cavity.

In humans, an inner cell mass becomes an embryonic disk composed of two layers of cells. The lower layer of cells becomes the yolk sac. A cavity called the amniotic cavity occurs above the embryonic disk.

Gastrula

Gastrulation is a movement of cells that results in a gastrula, an embryo composed of three differentiated tissue layers. These tissue layers, called **ectoderm, mesoderm,** and **endoderm,** are known as the embryonic germ layers because they give rise to all the other tissues and organs of the body (Table 18.1).

During gastrulation, a thickening called the primitive streak marks the midline region of the embryo (Fig. 18.5*a*). Cells are passing through the primitive streak and forming the germ layers endoderm and mesoderm. The outer layer of cells that remains is the ectoderm. Later, the germ layers fold to form a tubular body (Fig. 18.5*b*).

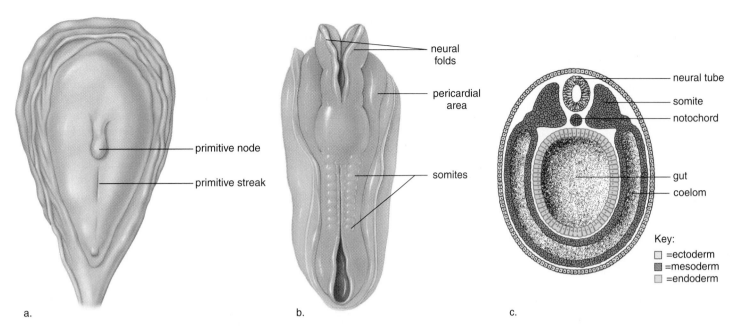

Figure 18.5   **Primitive streak and neurula.**
a. At 16 days, the primitive node marks the extent of the primitive streak where invagination results in a three-layered embryo. **b.** At 21 days, neural folds are seen, and when they meet, a neural tube forms along the midline of the body. The pericardial area contains the primitive heart, and the somites give rise to the muscles and the vertebrae, which replace the notochord. **c.** Generalized cross section of a human neurula. Each of the germ layers can be associated with the later development of particular organs as listed in Table 18.1.

Neurula

Mesodermal cells that turn in at the primitive node become the notochord, a dorsal supporting rod. (In humans, the notochord is later replaced by the vertebral column.) The ner- vous system develops from ectoderm located just above the notochord. A neural plate thickens into neural folds that fuse, forming a neural tube (Fig. 18.5*b*). The neural tube develops into the spinal cord and the brain.

Neurulation involves **induction,** a process by which one tissue influences the development of another tissue. Experi- ments have shown that the nervous system does not form unless there is a notochord present. Today, investigators believe that induction explains the process of differentiation. Induction requires direct contact or the production of a chemical by one tissue that most likely activates certain genes in the cells of the other tissue. These genes then direct how differentiation is to occur.

Midline mesoderm not contributing to the formation of the notochord becomes two longitudinal masses of tis- sue. From these blocklike portions of mesoderm, called somites, the muscles of the body and the vertebrae of the spine develop. The coelom, an embryonic body cavity that forms at this time, is completely lined by mesoderm. In humans, the coelom becomes the thoracic and abdomi- nal cavities. Figure 18.5c gives a generalized cross section of the embryo indicating the location of the three germ layers.

Table 18.1	Embryonic Germ Layers and Organ Development	
Ectoderm	**Mesoderm**	**Endoderm**
Skin epidermis, including hair, nails, and sweat glands	All muscles	Lining of digestive tract, trachea, bronchi, lungs, gallbladder, and urethra
Nervous system, including brain, spinal cord, ganglia, and nerves	Dermis of skin	Liver
Retina, lens, and cornea of eye	All connective tissue, including bone, cartilage, and blood	Pancreas
Inner ear	Blood vessels	Thyroid, parathyroid, and thymus glands
Lining of nose, mouth, and anus	Kidneys	Urinary bladder
Tooth enamel	Reproductive organs	

Development involves certain processes that can be related to particular stages of early development.

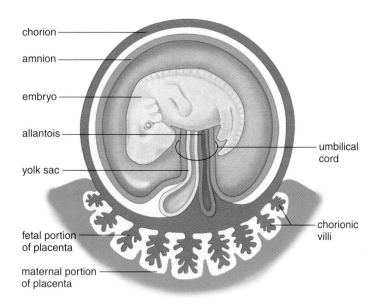

chorion

amnion

embryo

allantois

yolk sac

umbilical cord

fetal portion of placenta

chorionic villi

maternal portion of placenta

Figure 18.6   The extraembryonic membranes.
The chorion and amnion surround the embryo. The two other extraembryonic membranes, the yolk sac and allantois, contribute to the umbilical cord.

Extraembryonic Membranes

One of the major events in early development is the establishment of the extraembryonic membranes (Fig. 18.6). The term **extraembryonic membranes** is apt because these membranes extend out beyond the embryo. One of the membranes, the **amnion,** provides a fluid environment for the developing embryo and fetus. It is a remarkable fact that all animals, even land-dwelling humans, develop in a watery medium. One authority describes the functions of amniotic fluid in this way:

> It prevents the walls of the uterus from cramping the fetus and allows it unhampered growth and movement. It encompasses the fetus with a fluid of constant temperature which is a marvelous insulator against cold and heat. Above all, it acts as an excellent shock absorber.[1]

The **yolk sac** is another extraembryonic membrane. Yolk is a nutrient material utilized by other animal embryos—the yellow of a chick's egg is yolk. However, in humans, the yolk sac contains no yolk and is the first site of red blood cell formation. Part of this membrane becomes incorporated into the umbilical cord. Another extraembryonic membrane, the **allantois,** contributes to the circulatory system: its blood vessels become umbilical blood vessels that transport fetal blood to and from the placenta. The **chorion,** the outer extraembryonic membrane, becomes part of the **placenta** (Fig. 18.7), where the fetal blood exchanges gases, nutrients, and wastes with the maternal blood.

[1]A.F. Guttmacher. *Pregnancy, Birth and Family Planning.* (New York: New American Library, 1974), p. 74.

Fetal Circulation

Fetal circulation involves the placenta, which begins forming once the embryo is implanted fully. The placenta has a fetal side contributed by the chorion and a maternal side consisting of uterine tissues. Notice in Figure 18.7 how projections called **chorionic villi** are immersed in maternal blood. The blood of the mother and the fetus never mix since exchange always takes place across the placenta. Carbon dioxide and other wastes move from the fetal side to the maternal side, and nutrients and oxygen move from the maternal side to the fetal side of the placenta by diffusion. The **umbilical cord,** which stretches between the placenta and the fetus, is the lifeline of the fetus because it contains the umbilical arteries and veins. These vessels transport waste molecules (carbon dioxide and urea) to the placenta for disposal and take oxygen and nutrient molecules from the placenta to the rest of the fetal circulatory system.

By the tenth week, the placenta is formed fully and begins to produce progesterone and estrogen. These hormones have two effects due to their negative feedback effect on the mother's hypothalamus and anterior pituitary. They prevent any new follicles from maturing, and they maintain the endometrium. There is usually no menstruation during pregnancy.

Harmful chemicals in the mother's blood can cross the placenta, and this is of particular concern during the embryonic period, when various structures are first forming. Each organ or part seems to have a sensitive period during which a substance can alter its normal function. The Health Focus on pages 370–71 concerns the origination of birth defects and explains ways to detect genetic defects before birth.

Path of Fetal Blood

Blood within the fetal aorta travels to its various branches, including the iliac arteries, which connect to the *umbilical arteries* leading to the placenta (Fig. 18.7). Exchange between maternal blood and fetal blood takes place here. The *umbilical vein* carries blood rich in nutrients and oxygen away from the placenta to the fetus. The umbilical vein enters the liver and then joins the *venous duct,* which merges with the inferior vena cava, a vessel that returns blood to the heart.

Features of the fetal heart can be related to the fact that the fetus does not use its lungs for gas exchange. The blood entering the right atrium from the inferior vena cava would ordinarily be sent to the lungs. However, in the fetus this blood is O_2-rich and is shunted instead into the left atrium through the *oval opening* (foramen ovale) between the two atria. Also, any blood that does enter the right ventricle and is pumped into the pulmonary trunk is shunted into the aorta by way of the *arterial duct* (ductus arteriosus).

The most common of all cardiac defects in the newborn is the persistence of the oval opening. With the tying of the cord and the expansion of the lungs, blood enters the lungs in quantity. Return of this blood to the left side of the heart usually causes a flap to cover the opening. Incomplete closure occurs in nearly one out of four individuals, but even so, pas-

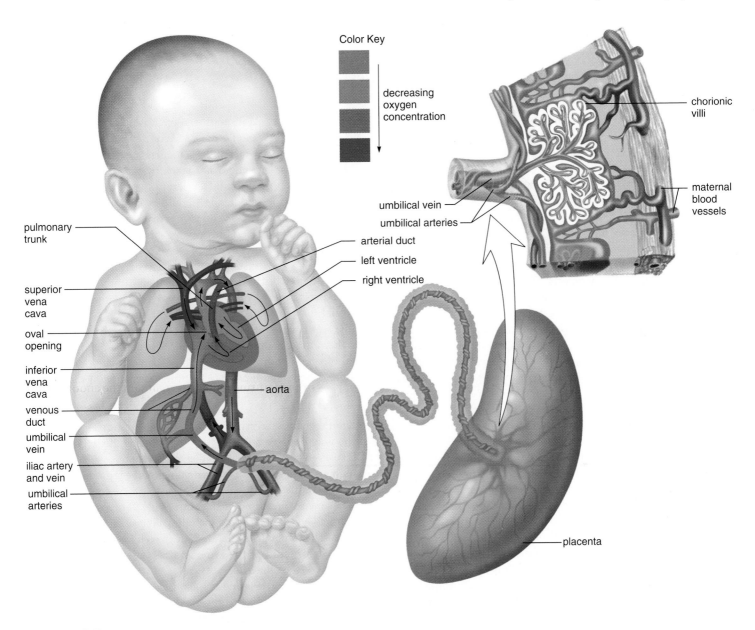

Color Key

decreasing
oxygen
concentration

pulmonary trunk

arterial duct

left ventricle

right ventricle

superior vena cava

oval opening

inferior vena cava

aorta

venous duct

umbilical vein

iliac artery and vein

umbilical arteries

chorionic villi

maternal blood vessels

umbilical vein

umbilical arteries

placenta

Figure 18.7 **Fetal circulation and the placenta.**

The lungs are not functional in the fetus, and the blood passes directly from the right atrium to the left atrium or from the right ventricle to the aorta. The umbilical arteries take fetal blood to the placenta where exchange of molecules between fetal and maternal blood takes place across the walls of the chorionic villi. Oxygen and nutrient molecules diffuse into the fetal blood, and carbon dioxide and urea diffuse from the fetal blood. The umbilical vein returns blood from the placenta to the fetus.

sage of the blood from the right atrium to the left atrium rarely occurs because either the opening is small or it closes when the atria contract. In a small number of cases, the passage of O_2-poor blood from the right side to the left side of the heart is sufficient to cause a "blue baby." Such a condition now can be corrected by open heart surgery.

The arterial duct closes at birth because endothelial cells divide and block off the duct. Remains of the arterial duct and parts of the umbilical arteries and vein later are transformed into connective tissue. The umbilical blood vessels

are severed when the umbilical cord is cut at birth, leaving only the umbilicus (navel).

Fetal circulation shunts blood away from the lungs, toward and away from the placenta within the umbilical blood vessels located within the umbilical cord. Exchange of substances between fetal blood and maternal blood takes place at the placenta, which forms from the chorion and uterine tissue.

Preventing Birth Defects

It is believed that at least 1 in 16 newborns has a birth defect, either minor or serious, and the actual percentage may be even higher. It is estimated that only 20% of all birth defects are due to heredity. Those that are hereditary can sometimes be detected before birth. Amniocentesis allows the fetus to be tested for abnormalities of development; chorionic villi sampling allows the embryo to be tested; and a new method has been developed for screening eggs to be used for in vitro fertilization (Fig. 18A).

It is recommended that all females take everyday precautions to protect any future and/or presently developing embryos and fetuses from defects. Proper nutrition is a must (deficiency in folic acid causes neural tube defects). X-ray diagnostic therapy should be avoided during pregnancy because X rays cause mutations in the developing embryo or fetus. Children born to women who received X-ray treatment are apt to have birth defects and/or to develop leukemia later. Toxic chemicals, such as pesticides and many organic industrial chemicals, which are also mutagenic, can cross the placenta. Cigarette smoke not only contains carbon monoxide but also other fetotoxic chemicals. Babies born to smokers are often underweight and subject to convulsions.

Pregnant Rh⁻ women should receive an Rh immunoglobulin injection to prevent the production of Rh antibodies. These antibodies can cause nervous system and heart defects.

Sometimes, birth defects are caused by pathogens. Females can be immunized before the childbearing years for rubella (German measles), which in particular causes birth defects such as deafness. Unfortunately, immunization for sexually transmitted diseases is not possible. The AIDS virus can cross the placenta and cause mental retardation. Proper medication can greatly reduce the chance of this happening. When a mother has herpes, gonorrhea, or chlamydia, newborns can become infected as they pass through the birth canal. Blindness and other physical and mental defects may develop. Birth by cesarean section could prevent these occurrences.

Pregnant women should not take any type of drug except with a doctor's prescription. Certainly illegal drugs, such as marijuana, cocaine, and heroin, should be completely avoided. "Cocaine babies" now make up 60% of drug-affected babies. Severe fluctuations in blood pressure that are produced by the use of cocaine temporarily deprive the developing brain of oxygen. Cocaine babies have visual problems, lack coordination, and are mentally retarded. The drugs aspirin, caffeine (present in coffee, tea, and cola), and alcohol should be severely limited. It is not unusual for babies of drug addicts and alcoholics to display withdrawal symptoms and to have various abnormalities. Babies born to women who drink while pregnant are apt to have fetal alcohol syndrome (FAS). These babies have decreased weight, height, and head size, with malformation of the head and face. Mental retardation is common in FAS infants.

A woman has to be very careful about taking medications while pregnant. Excessive vitamin A, sometimes used to treat acne, may damage an embryo. When the synthetic hormone DES was given to pregnant women to prevent miscarriage, their daughters showed various abnormalities of the reproductive organs and an increased tendency toward cervical cancer. Other sex hormones, including birth control pills, can possibly cause abnormal fetal development, including abnormalities of the sex organs. The tranquilizer thalidomide is well known for having caused deformities of the arms and legs in children born to women who took the drug.

Now that physicians and laypeople are aware of the various ways birth defects can be prevented, it is hoped that the incidence of birth defects will decrease in the future.

Figure 18A **Three methods for genetic-defect testing before birth.**

a. Amniocentesis is usually performed from the 15th to the 17th week of pregnancy. A long needle is passed through the abdominal wall to withdraw a small amount of amniotic fluid, along with fetal cells. Since there are only a few cells in the amniotic fluid, testing may be delayed as long as four weeks until cell culture produces enough cells for testing purposes. About 40 tests are available for different defects.
b. Chorionic villi sampling is usually performed from the 5th to the 12th week of pregnancy. The doctor inserts a long, thin tube through the vagina into the uterus. With the help of ultrasound, which gives a picture of the uterine contents, the tube is placed between the lining of the uterus and the chorion. Then a sampling of the chorionic villi cells is obtained by suction. Chromosome analysis and biochemical tests for genetic defects can be done immediately on these cells. However, chorionic villi sampling poses a greater threat to the unborn child than does amniocentesis.
c. Screening eggs for genetic defects is a new technique. Preovulatory eggs are removed by aspiration after a laparoscope (optical telescope) is inserted into the abdominal cavity through a small incision in the region of the navel. The first polar body is tested. If the woman is heterozygous *(Aa)* and the defective gene *(a)* is found in the polar body, the egg must have received the normal gene *(A)*. Normal eggs then undergo in vitro fertilization and are placed in the prepared uterus. At present, only one in ten attempts results in a birth, but it is known ahead of time that the child will be normal for the genetic traits tested.

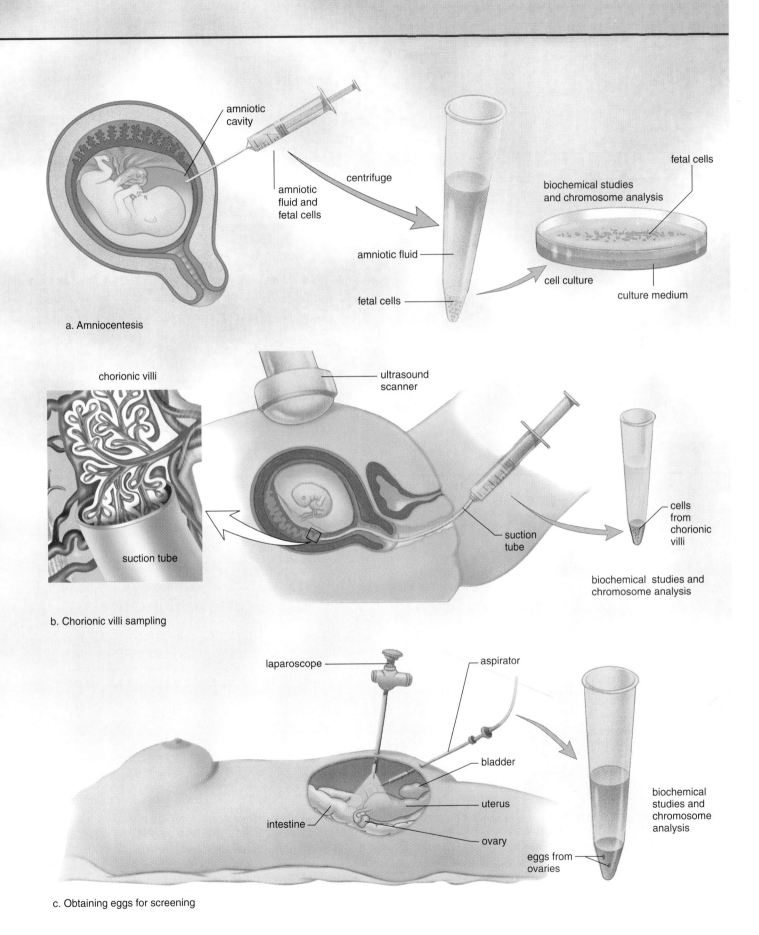

a. Amniocentesis

b. Chorionic villi sampling

c. Obtaining eggs for screening

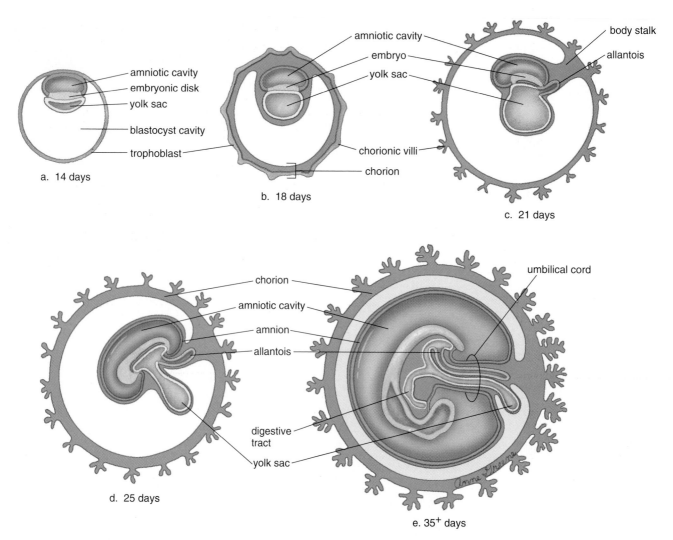

Figure 18.8 **Embryonic development.**
a. At first, no organs are present in the embryo, only tissues. The amniotic cavity is above the embryo, and the yolk sac is below. **b.** The chorion is developing villi, so important to exchange between mother and child. **c.** The allantois and yolk sac are two more extraembryonic membranes. **d.** These extraembryonic membranes are positioned inside the body stalk as it becomes the umbilical cord. **e.** At 35+ days, the embryo has a head region and a tail region. The umbilical cord takes blood vessels between the embryo and the chorion (placenta).

Embryonic Development

Earlier in this section, we discussed the processes of development. Now we consider these processes in the context of embryonic development. **Embryonic development** is the second week through the eighth week, and **fetal development** is the third month through the ninth month of human development.

First Month

Immediately after fertilization, the embryo divides repeatedly as it passes down the oviduct to the uterus. The resulting morula becomes the hollow blastocyst with an inner cell mass to one side. The inner cell mass is the embryo. Researchers call these early embryonic cells **stem cells** because they give rise to all the different types of cells in the human body. Grown in the correct culture medium in the labora-

tory, stem cells can become a liver cell, a neuron, a muscle cell, or any type of cell. Scientists foresee the use of stem cells to cure human conditions like Parkinson disease, Alzheimer disease, diabetes, and more. Work with embryonic stem cells is controversial, however, and much research is needed before this promise can come to fruition.

The blastocyst is bounded by a layer of cells that becomes the chorion. The early appearance of the chorion emphasizes the complete dependence of the developing embryo on this extraembryonic membrane. The blastocyst arrives in the uterus on the fourth or fifth day after fertilization. Then, after two or three days, the blastocyst begins to implant itself in the endometrium (see Fig. 18.3).

By the end of the second week, implantation is complete. The ever-growing number of cells becomes a two-layered embryonic disk. The amniotic cavity is seen above

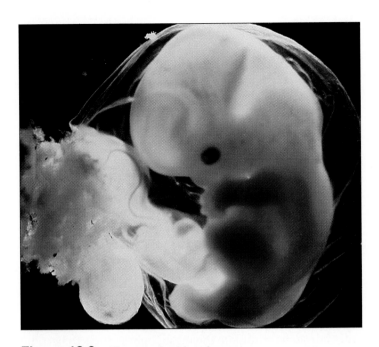

Figure 18.9   Five-week old embryo.
The head is much larger than the rest of the body; the arms resemble paddles and the legs are limb buds. Note the tail.

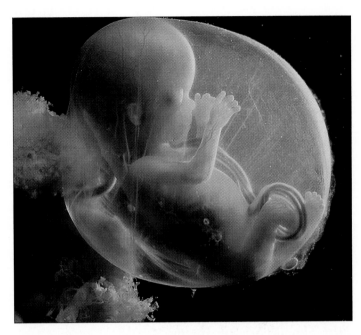

Figure 18.10   Three- to four-month-old fetus.
Facial features, arms and legs including fingers and toes are forming. The umbilical cord is quite visible.

the embryo, and the yolk sac is below (Fig. 18.8*a*). By the end of the third week, the embryo is a gastrula, and another extraembryonic membrane, the allantois, makes its appearance. Later, the allantois and the yolk sac become part of the umbilical cord as it forms (Fig. 18.8*c–e*). Some organs are already developed, including the spinal cord and heart.

By the end of the first month, the placenta is forming. The embryo has a nonhuman appearance largely due to the presence of a tail, but also because the arms and legs, which begin as limb buds, resemble paddles. The head is much larger than the rest of the embryo, and the whole embryo bends under its weight (Fig. 18.9). The eyes, ears, and nose are just appearing. The enlarged heart beats, and the bulging liver takes over the production of blood cells, which will carry nutrients to the developing organs and wastes from the developing organs.

Second Month

At the end of two months, the embryo's tail has disappeared, and the arms and legs are more developed, with fingers and toes apparent. The head is very large, the nose is flat, the eyes are far apart, and the ears are distinctively present. Internally, all major organs have appeared. Embryonic development is now finished.

By the end of the embryonic period, all organ systems have appeared, and there is a mature and functioning placenta. The embryo is only about 38 mm (1½ inches) long.

Fetal Development

During fetal development, the fetus has a human appearance, but refinements are still taking place.

Third and Fourth Months

At the beginning of the third month, head growth begins to slow down as the rest of the body increases in length. Epidermal refinements, such as eyelashes, eyebrows, hair on the head, fingernails, and nipples, appear (Fig. 18.10).

Cartilage is replaced by bone as ossification centers appear in the bones. The skull has six large fontanels (membranous areas or soft spots), which later permit a certain amount of flexibility as the head passes through the birth canal and allow rapid growth of the brain during infancy. The fontanels usually close by 2 years of age.

Sometime during the third month, it is possible to distinguish males from females. As discussed in the next section, once the testes form they produce androgens, the male sex hormones. The androgens stimulate the growth and differentiation of the male genitals. In the absence of androgens, female genitals form. The ovaries do not produce estrogen because it crosses the placenta from the mother's bloodstream. At this time, the testes or ovaries are located within the abdominal cavity. Later, in the last trimester of male fetal development, the testes descend into the scrotal sacs of the scrotum. Sometimes the testes fail to descend, in which case an operation can be performed to place them in their proper location.

By the end of the fourth month, the fetus is less than 150 mm (6 in) in length and weighs a little more than 170 grams (6 oz).

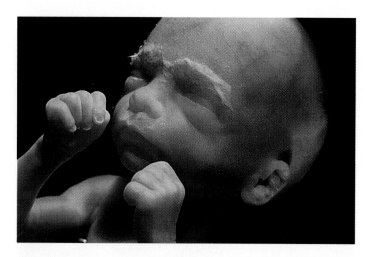

Figure 18.11   Six-month-old fetus.
Fetus looks fully formed but weighs only about three pounds. A greasy, cheeselike substance covers the eyebrows and the rest of the body

Fifth Through Seventh Months

During the fifth through seventh months (Fig. 18.11), the mother begins to feel fetal movement. At first, there is only a fluttering sensation, but as the legs grow and develop, kicks and jabs are felt. The fetal heartbeat is loud enough to be heard when a physician applies a stethoscope to the mother's abdomen. The fetus is in the fetal position with the head bent down and in contact with the flexed knees.

The wrinkled skin is covered by a fine down called **lanugo.** The lanugo is coated with a white, greasy, cheeselike substance called **vernix caseosa,** which is believed to protect the delicate skin from the amniotic fluid. During these months, the eyelids open fully.

At the end of this period, the fetus is almost 300 mm (12 in) long, and the weight has increased to almost 1,350 grams (3 lb). It is possible that if born now, the baby will survive; however, the lungs lack surfactant, which reduces surface tension within the alveoli (air sacs). Babies born without surfactant risk respiratory distress syndrome or collapsed lungs.

Eighth and Ninth Months

As the end of development approaches, the fetus usually rotates so that the head is pointed toward the cervix. However, if the fetus does not turn, a breech birth (rump first) is likely. It is very difficult for the cervix to expand enough to accommodate this form of birth, and asphyxiation of the baby is more likely to occur. Thus, a cesarean section may be prescribed.

At the end of nine months, the fetus is about 530 mm (20½ in) long and weighs about 3,400 grams (7½ lb). Weight gain is due largely to an accumulation of fat beneath the skin. Full-term babies have the best chance of survival.

From the fifth to the ninth month, the fetus continues to grow and to gain weight. Full-term babies have a better chance of survival.

18.3   Development of Male and Female Sex Organs

The sex of an individual is determined at the moment of fertilization. Both males and females have 23 pairs of chromosomes; in males one of these pairs is an X and Y, while females have two X chromosomes. During the first several weeks of development, it is impossible to tell whether the unborn child is a boy or girl by external inspection. Gonads don't start developing until the seventh week of development. The tissue that gives rise to the gonads is called indifferent because it can become testes or ovaries depending on the action of hormones. Genes on the Y chromosome cause testes to develop and produce androgenic hormones, and these determine the course of development.

In Figure 18.12*a,* notice that at six weeks both males and females have the same type of tissues and ducts. During this indifferent stage, an embryo has the potential to develop into a male or a female. If a Y chromosome is present, androgens stimulate the mesonephric ducts to become male genital ducts. The mesonephric ducts enter the urethra, which belongs to both the urinary and reproductive systems in males. Androgens suppress development of paramesonephric ducts in males.

In the absence of a Y chromosome and in the presence of two X chromosomes, ovaries develop instead of testes from the same indifferent tissue. Now the mesonephric ducts regress, and the paramesonephric ducts develop into the uterus and uterine tubes. A developing vagina also extends from the uterus. There is no connection between the urinary and genital systems in females.

At fourteen weeks, both the primitive testes and ovaries are located deep inside the abdominal cavity. An inspection of the interior of the testes would show that sperm are even now starting to develop, and similarly, the ovaries already contain large numbers of tiny follicles, each containing an ovum. Toward the end of development, the testes descend into the scrotal sac; the ovaries remain in the abdominal cavity.

Figure 18.12b shows the development of the external genitals. These tissues are also indifferent at first—they can develop into either male or female genitals. At six weeks, a small bud appears between the legs that can develop into the male penis or the female clitoris, depending on the presence or absence of the Y chromosome and androgenic hormones. At nine weeks, there is a groove called the urogenital groove bordered by two swellings. By fourteen weeks, this groove has disappeared in males, and the scrotum has formed from the original swellings. In females, the groove persists and becomes the vaginal opening. Labia majora and labia minora are present instead of a scrotum.

The Y chromosome triggers male genitals. If the Y is absent, female genitals form.

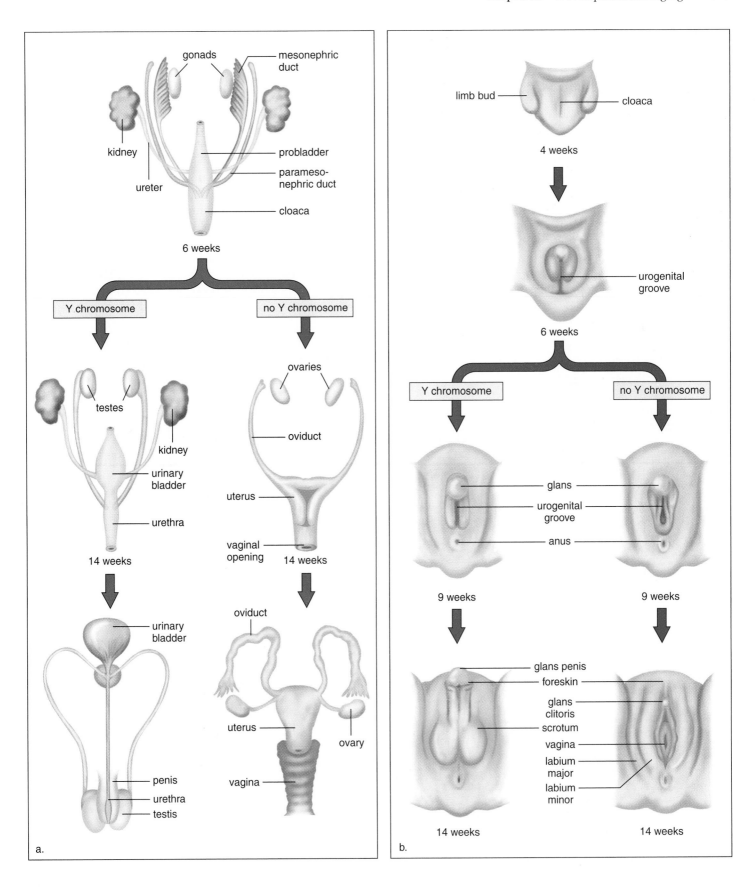

Figure 18.12   Male and female organs.
a. Development of gonads and ducts. **b.** Development of external genitals.

18.4   Birth

The uterus has contractions throughout pregnancy. At first, these are light, lasting about 20–30 seconds and occurring every 15–20 minutes. Near the end of pregnancy, the contractions may become stronger and more frequent so that a woman may think that she is in labor. However, the onset of true labor is marked by uterine contractions that occur regularly every 15–20 minutes and last for 40 seconds or more.

A positive feedback mechanism can explain the onset and continuation of labor. Uterine contractions are induced by a stretching of the cervix, which also brings about the release of oxytocin from the posterior pituitary. Oxytocin stimulates the uterine muscles, both directly and through the action of prostaglandins. Uterine contractions push the fetus downward, and the cervix stretches even more. This cycle keeps repeating itself until birth occurs.

Stage I

Prior to or at the first stage of **parturition,** which is the process of giving birth to an offspring, there can be a "bloody show" caused by expulsion of a mucous plug from the cervical canal. This plug prevents bacteria and sperm from entering the uterus during pregnancy.

At first, the uterine contractions of labor occur in such a way that the cervical canal slowly disappears as the lower part of the uterus is pulled upward toward the baby's head (Fig. 18.13b). This process is called effacement, or "taking up the cervix." With further contractions, the baby's head acts as a wedge to assist cervical dilation. If the amniotic membrane has not already ruptured, it is apt to do so during this stage, releasing the amniotic fluid, which leaks out the vagina (sometimes referred to as "breaking water"). The first stage of parturition ends once the cervix is dilated completely.

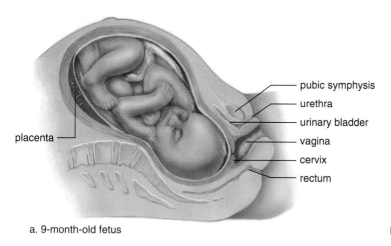

placenta

pubic symphysis
urethra
urinary bladder
vagina
cervix
rectum

a. 9-month-old fetus

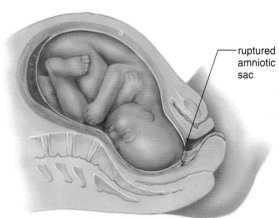

ruptured amniotic sac

b. First stage of birth: cervix dilates

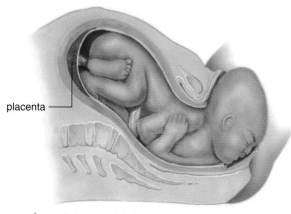

placenta

c. Second stage of birth: baby emerges

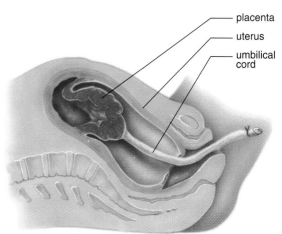

placenta
uterus
umbilical cord

d. Third stage of birth: expelling afterbirth

Figure 18.13   Three stages of parturition (birth).
a. Position of fetus just before birth begins. **b.** Dilation of cervix. **c.** Birth of baby. **d.** Expulsion of afterbirth.

Stage 2

During the second stage of parturition, the uterine contractions occur every 1–2 minutes and last about one minute each. They are accompanied by a desire to push, or bear down. As the baby's head gradually descends into the vagina, the desire to push becomes greater. When the baby's head reaches the exterior, it turns so that the back of the head is uppermost (Fig. 18.13c). Since the vaginal orifice may not expand enough to allow passage of the head, an **episiotomy** is often performed. This incision, which enlarges the opening, is sewn together later. As soon as the head is delivered, the baby's shoulders rotate so that the baby faces either to the right or the left. At this time, the physician may hold the head and guide it downward, while one shoulder and then the other emerges. The rest of the baby follows easily.

Once the baby is breathing normally, the umbilical cord is cut and tied, severing the child from the placenta. The stump of the cord shrivels and leaves a scar, which is the umbilicus.

Stage 3

The placenta, or **afterbirth,** is delivered during the third stage of parturition (Fig. 18.13d). About 15 minutes after delivery of the baby, uterine muscular contractions shrink the uterus and dislodge the placenta. The placenta then is expelled into the vagina. As soon as the placenta and its membranes are delivered, the third stage of parturition is complete.

During the first stage of parturition, the cervix dilates; during the second stage, the child is born; and during the third stage, the afterbirth is expelled.

Female Breast and Lactation

A female breast contains 15 to 25 lobules, each with a milk duct, which begins at the nipple and divides into numerous other ducts that end in blind sacs called alveoli (Fig. 18.14).

During pregnancy, the breasts enlarge as the ducts and alveoli increase in number and size. The same hormones that affect the mother's breasts can also affect the child's. Some newborns, including males, even secrete a small amount of milk for a few days.

Usually, there is no production of milk during pregnancy. The hormone prolactin is needed for lactation to begin, and the production of this hormone is suppressed because of the feedback control that the increased amount of estrogen and progesterone during pregnancy has on the pituitary. Once the baby is delivered, however, the pituitary begins secreting prolactin. It takes a couple of days for milk production to begin, and in the meantime, the breasts produce **colostrum,** a thin, yellow, milky fluid rich in protein, including antibodies.

The continued production of milk requires a suckling child. When a breast is suckled, the nerve endings in the areola are stimulated, and a nerve impulse travels along neural pathways from the nipples to the hypothalamus, which directs the pituitary gland to release the hormone oxytocin. When this hormone arrives at the breast, it causes contraction of the lobules so that milk flows into the ducts (called milk letdown), where it may be drawn out of the nipple by the suckling child.

Whether to breast-feed or not is a private decision based in part on a woman's particular circumstances. However, it is well known that breast milk contains antibodies produced by the mother that can help a baby survive. Babies have immature immune systems, less stomach acid to destroy foreign antigens, and also unsanitary habits. Breast-fed babies are less likely to develop stomach and intestinal illnesses, including diarrhea, during the first thirteen weeks of life. Breast-feeding also has physiological benefits for the mother. Suckling causes uterine contractions that can help the uterus return to its normal size, and breast-feeding uses up calories and can help a woman return to her normal weight.

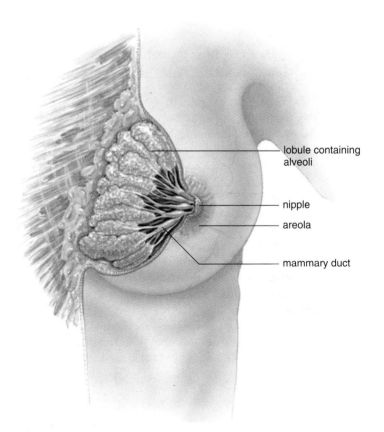

Figure 18.14   Female breast anatomy.
The female breast contains lobules consisting of ducts and alveoli. The alveoli are lined by milk-producing cells in the lactating (milk-producing) breast.

lobule containing alveoli

nipple

areola

mammary duct

18.5 Development After Birth

Development does not cease once birth has occurred but continues throughout the stages of life: infancy, childhood, adolescence, and adulthood. **Aging** encompasses these progressive changes that contribute to an increased risk of infirmity, disease, and death (Fig. 18.15).

Today, there is great interest in **gerontology,** the study of aging, because there are now more older individuals in our society than ever before, and the number is expected to rise dramatically. In the next half-century, the number of people over age 75 will rise from the present 8 million to 14.5 million, and the number over age 80 will rise from 5 million to 12 million. The human life span is judged to be a maximum of 120–125 years. The present goal of gerontology is not necessarily to increase the life span, but to increase the health span, the number of years that an individual enjoys the full functions of all body parts and processes.

Theories of Aging

There are many theories about what causes aging. Three of these are considered here.

Genetic in Origin

Several lines of evidence indicate that aging has a genetic basis: (1) The number of times a cell divides is species-specific. The maximum number of times human cells divide is around 50. Perhaps as we grow older, more and more cells are unable to divide, and instead, they undergo degenerative changes and die. (2) Some cell lines may become nonfunctional long before the maximum number of divisions has occurred. Whenever DNA replicates, mutations can occur, and this can lead to the production of nonfunctional proteins. Eventually, the number of inadequately functioning cells can build up, which contributes to the aging process. (3) The children of long-lived parents tend to live longer than those of short-lived parents. Recent work suggests that when an animal produces fewer free radicals, it lives longer. Free radicals are unstable molecules that carry an extra electron. In order to stabilize themselves, free radicals donate an electron to another molecule like DNA or proteins (e.g., enzymes) or lipids found in plasma membranes. Eventually these molecules are unable to function, and the cell is destroyed. There are genes that code for antioxidant enzymes that detoxify free radicals. This research suggests that animals with particular forms of these genes—and therefore more efficient antioxidant enzymes—live longer.

Whole-Body Process

A decline in the hormonal system can affect many different organs of the body. For example, type II diabetes is common in older individuals. The pancreas makes insulin, but the cells lack the receptors that enable them to respond. Menopause in women occurs for a similar reason. There is plenty of follicle-

Figure 18.15 Aging.
Aging is a slow process during which the body undergoes changes that eventually bring about death, even if no marked disease or disorder is present. Medical science is trying to extend the human life span and the health span, the length of time the body functions normally.

stimulating hormone in the bloodstream, but the ovaries do not respond. Perhaps aging results from the loss of hormonal activities and a decline in the functions they control.

The immune system, too, no longer performs as it once did, and this can affect the body as a whole. The thymus gland gradually decreases in size, and eventually most of it is replaced by fat and connective tissue. The incidence of cancer increases among the elderly, which may signify that the immune system is no longer functioning as it should. This idea is also substantiated by the increased incidence of autoimmune diseases in older individuals.

It is possible, though, that aging is not due to the failure of a particular system that can affect the body as a whole, but to a specific type of tissue change that affects all organs and even the genes. It has been noticed for some time that proteins—such as the collagen fibers present in many support tissues—become increasingly cross-linked as people age. Undoubtedly, this cross-linking contributes to the stiffening and loss of elasticity characteristic of aging tendons and ligaments. It may also account for the inability of such organs as the blood vessels, the heart, and the lungs to function as they once did. Some researchers have now found that glucose has the tendency to attach to any type of protein, which is the first step in a cross-linking process. They are presently experimenting with drugs that can prevent cross-linking.

Extrinsic Factors

The current data about the effects of aging are often based on comparisons of the elderly to younger age groups. But perhaps today's elderly were not as aware when they were younger of the importance of, for example, diet and exercise to general health. It is possible, then, that much of what we attribute to aging is instead due to years of poor health habits.

Consider, for example, osteoporosis. This condition is associated with a progressive decline in bone density in both males and females so that fractures are more likely to occur after only minimal trauma. Osteoporosis is common in the elderly—by age 65, one-third of women will have vertebral fractures, and by age 81, one-third of women and one-sixth of men will have suffered a hip fracture. While there is no denying that a decline in bone mass occurs as a result of aging, certain extrinsic factors are also important. The occurrence of osteoporosis itself is associated with cigarette smoking, heavy alcohol intake, and inadequate calcium intake. Not only is it possible to eliminate these negative factors by personal choice, but it is also possible to add a positive factor. A moderate exercise program has been found to slow down the progressive loss of bone mass.

Even more important, a sensible exercise program and a proper diet that includes at least five servings of fruits and vegetables a day will most likely help eliminate cardiovascular disease. Experts no longer believe that the cardiovascular system necessarily suffers a large decrease in functioning ability with age. Persons 65 years of age and older can have well-functioning hearts and open coronary arteries if their health habits are good and they continue to exercise regularly.

Effect of Age on Body Systems

Data about how aging affects body systems are necessarily based on past events. It's possible that in the future age will not have these effects or at least not to same degree as those described here.

Skin

As aging occurs, skin becomes thinner and less elastic because the number of elastic fibers decreases and the collagen fibers undergo cross-linking, as discussed previously. Also, there is less adipose tissue in the subcutaneous layer; therefore, older people are more likely to feel cold. The loss of thickness partially accounts for sagging and wrinkling of the skin.

Homeostatic adjustment to heat is also limited because there are fewer sweat glands for sweating to occur. There are fewer hair follicles, so the hair on the scalp and the extremities thins out. The number of oil (sebaceous) glands is reduced, and the skin tends to crack. Older people also experience a decrease in the number of melanocytes, making hair gray and skin pale. In contrast, some of the remaining pigment cells are larger, and pigmented blotches appear on the skin.

Processing and Transporting

Cardiovascular disorders are the leading cause of death today. The heart shrinks because there is a reduction in cardiac muscle cell size. This leads to loss of cardiac muscle strength and reduced cardiac output. Still, it is observed that the heart, in the absence of disease, is able to meet the demands of increased activity. It can increase its rate to double or triple the amount of blood pumped each minute even though the maximum possible output declines.

Because the middle layer of arteries contains elastic fibers, which most likely are subject to cross-linking, the arteries become more rigid with time, and their size is further reduced by plaque, a buildup of fatty material. Therefore, blood pressure readings gradually rise. Such changes are common in individuals living in Western industrialized countries but not in agricultural societies. A low cholesterol and saturated fatty acid diet has been suggested as a way to control degenerative changes in the cardiovascular system.

There is reduced blood flow to the liver, and this organ does not metabolize drugs as efficiently as before. This means that as a person gets older, less medication is needed to maintain the same level in the bloodstream.

Cardiovascular problems are often accompanied by respiratory disorders, and vice versa. Growing inelasticity of lung tissue means that ventilation is reduced. Because we rarely use the entire vital capacity, these effects are not noticed unless there is increased demand for oxygen.

There is also reduced blood supply to the kidneys. The kidneys become smaller and less efficient at filtering wastes. Salt and water balance are difficult to maintain, and the elderly dehydrate faster than young people. Difficulties involving urination include incontinence (lack of bladder control) and the inability to urinate. In men, the prostate gland may enlarge and reduce the diameter of the urethra, making urination so difficult that surgery is often needed.

The loss of teeth, which is frequently seen in elderly people, is more apt to be the result of long-term neglect than aging. The digestive tract loses tone, and secretion of saliva and gastric juice is reduced, but there is no indication of reduced absorption. Therefore, an adequate diet, rather than vitamin and mineral supplements, is recommended. Elderly people commonly complain of constipation, increased gas, and heartburn; gastritis, ulcers, and cancer can also occur.

Integration and Coordination

It is often mentioned that while most tissues of the body regularly replace their cells, some at a faster rate than others, the brain and the muscles ordinarily do not. However, contrary to previous opinion, recent studies show that few neural cells of the cerebral cortex are lost during the normal aging process. This means that cognitive skills remain unchanged even though there is characteristically a loss in short-term memory. Although the elderly learn more slowly than the young, they can acquire and remember new material. It is noted that when more time is given for the subject to respond, age differences in learning decrease.

Neurons are extremely sensitive to oxygen deficiency, and if neuron death does occur, it may be due not to aging itself but to reduced blood flow in narrowed blood vessels. Specific disorders, such as depression, Parkinson disease, and Alzheimer disease, are sometimes seen, but they are not common. Reaction time, however, does slow, and more stimulation is needed for hearing, taste, and smell receptors to function as before. After age 50, there is a gradual reduction in the ability to hear tones at higher frequencies, and this can make it difficult to identify individual voices and to understand conversation in a group. The lens of the eye does not accommodate as well and also may develop a cataract. Glaucoma, the buildup of pressure due to increased fluid, is more likely to develop because of a reduction in the size of the anterior cavity of the eye.

Loss of skeletal muscle mass is not uncommon, but it can be controlled by a regular exercise program. There is a reduced capacity to do heavy labor, but routine physical work should be no problem. A decrease in the strength of the respiratory muscles and inflexibility of the rib cage contribute to the inability of the lungs to expand as before, and reduced muscularity of the urinary bladder contributes to an inability to empty the bladder completely, and therefore to the occurrence of urinary infections.

As noted before, aging is accompanied by a decline in bone density. Osteoporosis, characterized by a loss of calcium and mineral from bone, is not uncommon, but there is evidence that proper health habits can prevent its occurrence. Arthritis, which causes pain upon movement of a joint, is also seen.

Weight gain occurs because the basal metabolism decreases and inactivity increases. Muscle mass is replaced by stored fat and retained water.

The Reproductive System

Females undergo menopause, and thereafter the level of female sex hormones in the blood falls markedly. The uterus and the cervix are reduced in size, and there is a thinning of the walls of the oviducts and the vagina. The external genitals become less pronounced. In males, the level of androgens falls gradually over the age span of 50–90, but sperm production continues until death.

Figure 18.16 **Remaining active.**
The aim of gerontology is to allow the elderly to enjoy living.

It is of interest that as a group, females live longer than males. Although their health habits may be better, it is also possible that the female sex hormone estrogen offers women some protection against cardiovascular disorders when they are younger. Males suffer a marked increase in heart disease in their forties, but an increase is not noted in females until after menopause, when women lead men in the incidence of stroke. Men are still more likely than women to have a heart attack, however.

Aging Well

We have listed many adverse effects of aging, but it is important to emphasize that while such effects are seen, they are not a necessary occurrence (Fig. 18.16). We must discover any extrinsic factors that precipitate these adverse effects and guard against them. Just as it is wise to make the proper preparations to remain financially independent when older, it is also wise to realize that biologically successful old age begins with the health habits developed when we are younger.

Maternal Health Habits

The fetus is subject to harm by maternal use of medicines and drugs of abuse, including nicotine and alcohol. Also, various sexually transmitted diseases, notably HIV infection, can be passed on to the fetus by way of the placenta. Women need to know how to protect their unborn children from harm. Indeed their behavior should be protective if they are sexually active, even if they are using a recognized form of birth control. Harm can occur before a woman realizes she is pregnant!

Because maternal health habits can affect a child before it is born, there has been a growing acceptance of prosecuting women when a newborn has a condition, such as fetal alcohol syndrome, that could only have been caused by the drinking habits of the mother. Employers have also become aware that they might be subject to prosecution if the workplace exposes pregnant employees to toxins. To protect themselves, Johnson Controls, a U.S. battery manufacturer, developed a fetal protection policy. No woman who could bear a child was offered a job that might expose her to toxins that could negatively affect the development of her baby. To get such a job, a woman had to show that she had been sterilized or was otherwise incapable of having children. In 1991, the U.S. Supreme Court declared this policy unconstitutional on the basis of sexual discrimination. The decision was hailed as a victory for women, but was it? The decision was written in such a way that women alone, and not an employer, are responsible for any harm done to the fetus by workplace toxins.

Some have noted that prosecuting women for causing prenatal harm can itself have a detrimental effect. The women may tend to avoid prenatal treatment, thereby increasing the risk to their children. Or they may opt for an abortion in order to avoid the possibility of prosecution. Women feel they are in a no-win situation. If they have a child that has been harmed due to their behavior, they are bad mothers; or if they abort, they are also bad mothers.

Figure 18B   Health habits.
Should a pregnant woman be liable for poor health habits that could harm her child?

Decide Your Opinion

1. Do you believe a woman should be prosecuted if her child is born with a preventable condition? Why or why not?
2. Is the woman or the physician responsible when a woman of childbearing age takes a prescribed medication that does harm to the unborn? Is the employer or the woman responsible when a workplace toxin does harm to an unborn child?
3. Should sexually active women who can bear a child be expected to avoid substances or situations that could possibly harm an unborn child, even if they are using birth control? Why or why not?

Looking at Both Sides   www.mhhe.com/biosci/genbio/maderhuman7/

Every bioethical issue has at least two sides. Even if you already have an opinion, it is important to explore the opposite opinion before finalizing your position. The Online Learning Center at www.mhhe.com/biosci/genbio/maderhuman7/ will help you fine-tune your initial opinion, explore both sides, and finalize your position. Either you will acquire new arguments for your original opinion or you may even change your opinion. Be sure to complete these activities in sequence:

Taking Sides   Decide your initial opinion by answering a series of questions. Then see if your opinion changes after completing the next two activities.

Further Debate   Read opposing articles that give you further information on this particular bioethical issue.

Explain Your Position   Answer another series of questions and then defend your original or changed opinion. You can e-mail your position to your instructor if he or she wishes.

Summarizing the Concepts

18.1 Fertilization
Only one sperm actually enters the egg, and this sperm's nucleus fuses with the egg nucleus. When the zygote begins to divide, it is an embryo, which travels down the oviduct and embeds itself in the endometrium. Cells surrounding the embryo produce HCG, and the presence of this hormone indicates that the female is pregnant.

18.2 Development Before Birth
The processes of development (cleavage, morphogenesis, differentiation, and growth) begin during the early developmental stages, which consist of the morula, blastocyst (in humans), gastrula, and neurula. The extraembryonic membranes, including the placenta, are special features of human development. Fetal lungs do not operate, and fetal circulation takes blood to the placenta, where exchange takes place.

Human development consists of embryonic (first two months) and fetal (third through ninth month) development. During the embryonic period, the extraembryonic membranes appear and serve important functions: the embryo acquires organ systems. During the fetal period, there is a refinement of these systems.

During the third and fourth months, it is obvious that the skeleton is becoming ossified. The sex of the individual is now distinguishable. From the fifth to the ninth months, the fetus continues to grow and to gain weight. Babies born after six or seven months may survive, but full-term babies have a better chance of survival.

18.3 Development of Male and Female Sex Organs
Male and female organs develop from the same indifferent tissue, depending on whether the chromosomes are XY or XX. External examination does not reveal the sex of the fetus until after the third month.

18.4 Birth
Parturition has three phases. During the first stage, the cervix dilates to allow passage of the baby's head and body. The amniotic membrane usually bursts sometime during this stage. During the second stage, the baby is born, and the umbilical cord is cut. During the third stage, the placenta is delivered.

Milk is not produced during pregnancy because of hormonal suppression, but once the child is born, milk production begins. Prolactin promotes the production of milk, and oxytocin allows milk letdown.

18.5 Development After Birth
Development after birth consists of infancy, childhood, adolescence, and adulthood. Young adults are at their prime, and then the aging process begins. Aging encompasses progressive changes from about age 20 on that contribute to an increased risk of infirmity, disease, and death. Perhaps aging is genetic in origin, perhaps it is due to a change that affects the whole body, or perhaps it is due to extrinsic factors.

Studying the Concepts

1. Describe the process of fertilization and the events immediately following it. 364–65
2. What is the basis of the pregnancy test? 365
3. What are the processes of development? Relate these processes to the early developmental stages. 366–67
4. Name the four extraembryonic membranes, and give a function for each. 368
5. Describe the structure and function of the placenta. 368
6. During which period of pregnancy does a woman have to be the most careful about the intake of medications and other drugs? 368
7. Describe the vascular components of the umbilical cord and how they relate to fetal circulation. 368
8. Specifically, what events normally occur during the first, second, third, and fourth weeks of development? 372–73 What events normally happen during the second through the ninth months? 373–74
9. Describe the development of the male and female sex organs. 374–75
10. What are the three stages of birth? Describe the events of each stage. 376–77
11. Describe the suckling reflex. 377
12. Discuss three theories of aging. What are the major changes in body systems that have been observed as adults age? 378–80

Testing Your Knowledge of the Concepts

In questions 1–4, match the stage of development to the descriptions.
a. morula   c. gastrula
b. blastula   d. neurula

_____ 1. Invagination of cells along the primitive streak occurs.

_____ 2. Notochord induces formation of neural tube.

_____ 3. Morphogenesis begins as a cavity forms.

_____ 4. Cleavage has resulted in a ball of cells.

In questions 5–7, indicate whether the statement is true (T) or false (F).

_____ 5. All major organs form during embryonic development.

_____ 6. The umbilical vein carries blood rich in nutrients and oxygen to the fetus.

_____ 7. In most deliveries, the head appears before the rest of the body.

In questions 8–15, fill in the blanks.

8. The _____ membranes include the chorion, the _____, the yolk sac, and the allantois.

9. Once the embryo arrives at the uterus, it begins to _____ itself in the endometrium.

10. Fertilization occurs when the _____ nucleus fuses with the _____.

11. The zygote divides as it passes down a uterine tube. This process is called _____.

12. When cells take on a specific structure and function, _____ has occurred.

13. Fetal development begins at the end of the _____ month.

14. During development, the nutrient needs of the developing embryo (fetus) are served by the _____.

15. The hormone _____ is required for milk letdown during the suckling reflex.

16. Label this diagram.

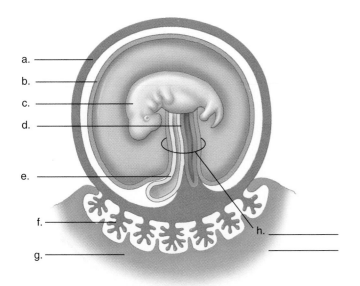

a. _____
b. _____
c. _____
d. _____
e. _____
f. _____
g. _____
h. _____

Understanding Key Terms

afterbirth   377
aging   378
allantois   368
amnion   368
blastocyst   366
chorion   368
chorionic villi   368
colostrum   377
ectoderm   366
embryo   365
embryonic development   372
endoderm   366
episiotomy   377
extraembryonic membrane   368

fertilization   364
fetal development   372
gerontology   378
induction   367
lanugo   374
mesoderm   366
parturition   376
placenta   368
stem cell   372
umbilical cord   368
vernix caseosa   374
yolk sac   368
zygote   364

Further Readings

Cox, F. D. 2000.*The AIDS booklet.* 6th ed. Dubuque, Iowa: Brown & Benchmark Publishers. This easy-to-read, informative booklet covers the transmission, prevention, and treatment of AIDS.

Crooks, R., and Baur, K. 1999. *Our sexuality.* 7th ed. Redwood City, Calif.: Benjamin/Cummings Publishing. Introduction to the biological, psychosocial, behavioral, and cultural aspects of sexuality.

Editors. Summer 2000. The quest to beat aging. *Scientific American Presents* 11(2). This special issue features articles on aging, stem cell research, Alzheimer disease, cancer, heart disease, and quality of life issues.

Hanke, T., and McMichael, A. J. September 2000. Design and construction of an experimental HIV-1 vaccine. *Nature Medicine* 6(9):951. A new candidate HIV vaccine has been designed for clinical trials in Kenya.

Ingber, D. E. January 1998. The architecture of life. *Scientific American* 278(1):48. Simple mechanical rules may govern development, tissue organization, and cellular movement.

Mader, S. S. 1990. *Human reproductive biology.* 2d ed. Dubuque, Iowa: Wm. C. Brown Publishers. An introductory text covering human reproduction in a clear, easily understood manner.

Markowitz, M. H. June 1998. A new dawn in AIDS treatments. *Discover* 19(6):S-6. A new combination therapy greatly reduces viral replication.

Newman, J. December 1995. How breast milk protects newborns. *Scientific American* 273(6):76. Human milk contains special antibodies that boost the newborn's immune system.

Packer, C. July/August 1998. Why menopause? *Natural History* 107(6):24. Article addresses reasons why menopause occurs so early in life, compared to other aging processes.

Pennisi, E. July 17, 1998. Genome reveals wiles and weak points of syphilis. *Science* 281(5375):324. Gene sequencing of the spirochete that causes syphilis may finally make it possible to culture spirochetes for study.

Prescott, L. M., et al. 1999. *Microbiology.* 4th ed. Dubuque, Iowa: Wm. C. Brown Publishers. This introductory text covers all major areas of microbiology.

Rose, M. R. December 1999. Can human aging be postponed? *Scientific American* 281(6):68. Future anti-aging therapies will need to counter many destructive biochemical processes to maintain youthfulness.

Schwartlander, B., et al. July 7, 2000. AIDS in a new millenium. *Science* 289(5476):64. This report gives statistics and summarizes current knowledge of the spread and impact of AIDS.

Smith, R. The timing of birth. March 1999. *Scientific American* 280(3):68. A hormone in the human placenta influences the timing of delivery; this finding could result in ways to predict and prevent premature labor.

Staff writer. October 22, 1999. Do mitochondrial mutations dim the fire of life? *Science* 286(5440):664. Changes in the genomes of mitochondria could contribute to aging.

Wallace, D. C. August 1997. Mitochondrial DNA in aging and disease. *Scientific American* 277(2):40. Genes in mitochondria have been linked to certain diseases, and could be important in age-related disorders.

Weiss, R. November 1997. Aging—New answers to old questions. *National Geographic* 192(5):2. The mechanics of human aging are studied.

e-Learning Connection   www.mhhe.com/biosci/genbio/maderhuman7/

18.1 Fertilization

 Fertilization *Essential Study Partner*

18.2 Development Before Birth

 Preembryonic Development *Essential Study Partner*
Embryonic and Fetal Development *Essential Study Partner*

 Human Extraembryonic Membranes *art labeling activity*
Extraembryonic Membranes *art quiz*
Embryonic Stem Cells—The Future of Organ Transplants *case study*

18.4 Birth

 Parturition *Essential Study Partner*

 Anatomy of the Breast *art labeling activity*

Chapter Summary

 Additional Activities *Essential Study Partner*

 Key Term Flashcards *vocabulary quiz*
Chapter Quiz *objective quiz covering all chapter concepts*

VI

Human Genetics

H uman beings practice sexual reproduction, which requires gamete production. Gametes carry half the total number of chromosomes as well as variable combinations of chromosomes and genes. It is sometimes possible to determine the chances of an offspring receiving a particular parental gene and, therefore, inheriting a genetic disorder.

Genes, now known to be constructed of DNA, control not only the metabolism of the cell but also, ultimately, the characteristics of the individual. The step-by-step procedure by which DNA specifies protein synthesis has been discovered. Biotechnology is a new and burgeoning field that permits DNA to be extracted from one organism and inserted into a different organism for a purpose useful to human beings.

Cancer is a cellular disease brought on by DNA mutations that transform a normal cell into a cancer cell. Cancer-causing genes are derived from normal genes, which keep cell division under control. Because environmental factors play a major role in causing or promoting cancer, it may be possible to reduce its incidence. Also, methods to detect and treat cancer are improving daily.

Chapter 19

Chromosomal Inheritance

Chapter Concepts

19.1 Human Life Cycle
- The human life cycle involves two types of cell divisions: mitosis and meiosis. 386

19.2 Mitosis
- Mitosis, nuclear division in which the chromosomal number remains constant, is involved in growth and repair. 387
- Cytokinesis divides the cytoplasm and organelles between the daughter cells. 389

19.3 Meiosis
- Meiosis, nuclear division in which the chromosomal number is reduced by one-half, is involved in gamete production. 390

19.4 Chromosomal Inheritance
- Normally, humans inherit 22 pairs of autosomes and one pair of sex chromosomes for a total of 46 chromosomes. 395
- Normally, males have the sex chromosomes XY and females have the sex chromosomes XX. 395
- Abnormalities arise when humans inherit abnormal chromosomes or an abnormal number of chromosomes. 396–98

Mary was a very attractive young woman who liked to play ice hockey. Her high school gym teacher took an interest in her ability and gave her extra coaching. She hoped that one day Mary would play on an Olympic team.

But something was wrong. Mary was 16 and still not menstruating. Her parents decided to have her undergo a series of medical tests. Much to the surprise of everyone, Mary had an X and Y chromosome in the nucleus of her cells. She was a chromosomal male.

The doctor explained to Mary and her parents that Mary had testicular feminization syndrome. A syndrome is a group of medical characteristics that appear together. An individual like Mary has internal testes that produce testosterone, but her cells won't respond to it. Her external genitals are like those of a female, and she has well-developed breasts. However, she will never be able to have children.

Mary will be able to go on and play hockey in the Olympics, but she will always have to carry a letter explaining her condition. Otherwise, she will be disqualified because of her sex chromosomes.

This chapter describes cell division and various syndromes that occur when people inherit an abnormal number of chromosomes. Chromosomal abnormalities are apt to be due to abnormal meiosis, the type of cell division needed for gamete production.

19.1 Human Life Cycle

The human life cycle involves growth and sexual reproduction (Fig. 19.1). During growth, a type of nuclear division called mitosis ensures that each and every cell has a complete number of chromosomes. Sexual reproduction requires the production of sex cells, which have half the number of chromosomes. A type of nuclear division called meiosis reduces the chromosomal number by one-half.

Meiosis occurs in the sex organs, also called the gonads. In males, the testes produce sperm; in females, the ovaries produce cells that become eggs. The sperm and the egg are the sex cells, or **gametes.** Gametes contain the **haploid (n)** number of chromosomes; the haploid number of chromosomes in humans is 23.

A new individual comes into existence when a haploid sperm fertilizes a haploid egg. Each parent contributes one chromosome of each type to a zygote, which then has the **diploid (2n)** number of chromosomes. As the individual develops, mitosis occurs, and each **somatic** (body) **cell** has the diploid number of chromosomes. In humans, the diploid number is 46, and there are 23 pairs of chromosomes.

The life cycle of humans requires two types of nuclear division: mitosis and meiosis.

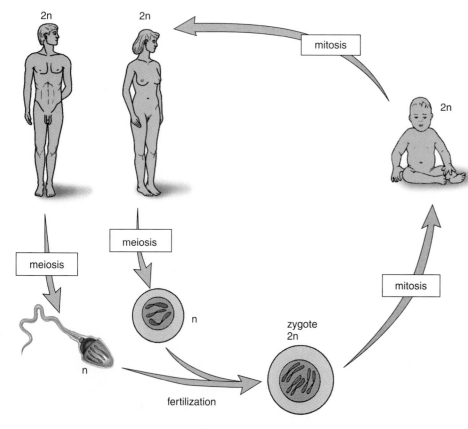

Figure 19.1   Life cycle of humans.
Meiosis in males is a part of sperm production, and meiosis in females is a part of egg production. When a haploid sperm fertilizes a haploid egg, the zygote is diploid. The zygote undergoes mitosis as it develops into a newborn child. Mitosis continues after birth until the individual reaches maturity, and then the life cycle begins again.

19.2 Mitosis

Mitosis is nuclear division that produces *two daughter cells, each with the same number and kinds of chromosomes as the parental cell, the cell that divides.* Therefore, following mitosis, the parental cell and the daughter cells are genetically identical. Mitosis occurs as part of the cell cycle.

Cell Cycle

The **cell cycle** consists of interphase, mitosis, and cytokinesis which is division of the cytoplasm and organelles. The cell divides, and then it enters interphase before dividing again. Therefore, **interphase** is the interval of time between cell divisions. The length of time required for the entire cell cycle varies according to the type of cell, but 18–24 hours is typical. Mitosis and cytokinesis last from less than an hour to slightly more than 2 hours; for the rest of the time, the cell is in interphase.

It used to be said that interphase was a resting stage, but we now know that this is not the case. The organelles are metabolically active and are carrying on their normal functions. If the cell is going to divide, DNA replication occurs. During replication, DNA is copied, and each chromosome becomes duplicated. A duplicated chromosome is composed of two sister chromatids. The sister chromatids are genetically identical, meaning that they contain the same genes. Also, organelles, including the centrioles, duplicate. A nondividing cell has one pair of centrioles, but in a cell that is going to divide, this pair duplicates, and there are two pairs of centrioles outside the nucleus.

The cell cycle includes interphase, mitosis, and cytokinesis. During interphase, DNA replication results in each chromosome having two sister chromatids. The centrioles and other organelles also duplicate.

Overview of Mitosis: 2n → 2n

When mitosis is going to occur, chromatin in the nucleus becomes condensed, and the chromosomes become visible. Before mitosis begins, the parental cell is 2n, and the sister chromatids are held together in a region called the **centromere.** At the completion of mitosis, each chromosome consists of a single **chromatid.**

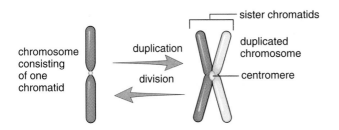

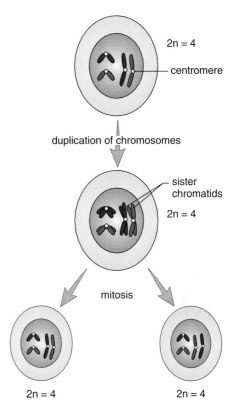

Figure 19.2 Overview of mitosis.
The blue chromosomes were inherited from one parent, and the red chromosomes were inherited from the other parent.

Figure 19.2 gives an overview of mitosis; for simplicity, only four chromosomes are depicted. (In determining the number of chromosomes, it is necessary to count only the number of independent centromeres.) During mitosis, the centromeres divide, the sister chromatids separate, and one of each kind of chromosome goes into each daughter cell. Therefore, each daughter cell gets a complete set of chromosomes and is 2n. (Following separation, each chromatid is called a chromosome.) Since each daughter cell receives the same number and kinds of chromosomes as the parental cell, each is genetically identical to the other and to the parental cell.

Mitosis occurs in humans when tissues grow or when repair occurs. Following fertilization, the zygote begins to divide mitotically, and mitosis continues during development and the life span of the individual. Also, when a cut heals or a broken bone mends, mitosis has occurred. In the adult, some tissues divide more readily than other tissues. But apparently most tissues contain stem cells, which can continually divide. Stem cells in the red bone marrow divide to produce millions of blood cells every day.

Following mitosis, each of two daughter cells has the same number and kinds of chromosomes as the parental cell.

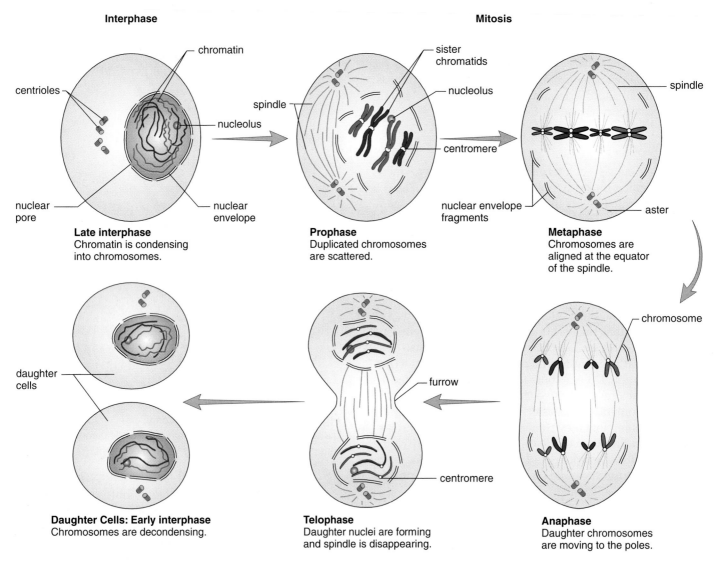

Figure 19.3   Interphase and mitosis.
The blue chromosomes were inherited from one parent, and the red chromosomes were inherited from the other parent.

Stages of Mitosis

As an aid in describing the events of mitosis, the process is divided into four phases: prophase, metaphase, anaphase, and telophase (Fig. 19.3). Although the stages of mitosis are depicted as if they were separate, they are actually continuous, and one stage flows from the other with no noticeable interruption.

Prophase

The events of **prophase** indicate that nuclear division is about to occur. The two pairs of centrioles outside the nucleus begin moving away from each other toward opposite ends of the nucleus. Spindle fibers appear between the separating centriole pairs, the nuclear envelope begins to fragment, and the nucleolus begins to disappear.

The chromosomes are now visible. Each is composed of two sister chromatids held together at a centromere. Spindle fibers attach to the centromeres as the chromosomes continue to shorten and to thicken. During prophase, chromosomes are randomly placed in the nucleus (Fig. 19.4).

Structure of the Spindle   At the end of prophase, a cell has a fully formed spindle. A **spindle** has poles, asters, and fibers. The **asters** are arrays of short microtubules that radiate from the poles, and the fibers are bundles of microtubules that stretch between the poles. Microtubule organizing centers (MTOC) are associated with the centrioles at the poles. These centers organize microtubules when the cell is not dividing; it is likely that they also organize the spindle. Centrioles may assist in this function, but their location at the poles of a spindle could be simply to ensure that each daughter cell receives a pair of centrioles.

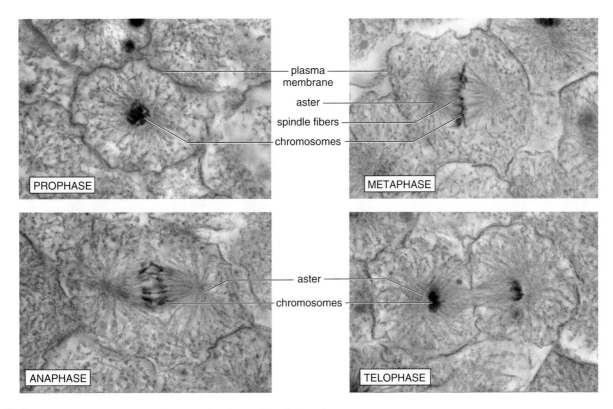

Figure 19.4   Micrographs of mitosis occurring in a whitefish embryo.

Metaphase

During **metaphase,** the nuclear envelope is fragmented, and the spindle occupies the region formerly occupied by the nucleus. The chromosomes are now at the equator (center) of the spindle. Metaphase is characterized by a fully formed spindle, and the chromosomes, each with two sister chromatids, are aligned at the equator (Fig. 19.4).

Anaphase

At the start of **anaphase,** the sister chromatids separate. *Once separated, the chromatids are called chromosomes.* Separation of the sister chromatids ensures that each cell receives a copy of each type of chromosome and thereby has a full complement of genes. During anaphase, the daughter chromosomes move to the poles of the spindle. Anaphase is characterized by the diploid number of chromosomes moving toward each pole.

Function of the Spindle   The spindle brings about chromosomal movement. Two types of spindle fibers are involved in the movement of chromosomes during anaphase. One type extends from the poles to the equator of the spindle; there they overlap. As mitosis proceeds, these fibers increase in length, and this helps push the chromosomes apart. The chromosomes themselves are attached to other spindle fibers that simply extend from their centromeres to the poles. These fibers get shorter and shorter as the chromosomes move toward the poles. Therefore, they pull the chromosomes apart.

Spindle fibers, as stated earlier, are composed of microtubules. Microtubules can assemble and disassemble by the addition or subtraction of tubulin (protein) subunits. This is what enables spindle fibers to lengthen and shorten and what ultimately causes the movement of the chromosomes.

Telophase

Telophase begins when the chromosomes arrive at the poles. During telophase, the chromosomes become indistinct chromatin again. The spindle disappears as nucleoli appear, and nuclear envelope components reassemble in each cell. Telophase is characterized by the presence of two daughter nuclei.

Cytokinesis

Cytokinesis is division of the cytoplasm and organelles. In animal cells, a slight indentation called a **cleavage furrow** passes around the circumference of the cell. Actin filaments form a contractile ring, and as the ring gets smaller and smaller, the cleavage furrow pinches the cell in half. As a result, each cell becomes enclosed by its own plasma membrane.

Following mitosis, each daughter cell is 2n. When the sister chromatids separate during anaphase, each newly forming cell receives the same number and kinds of chromosomes as the parental cell.

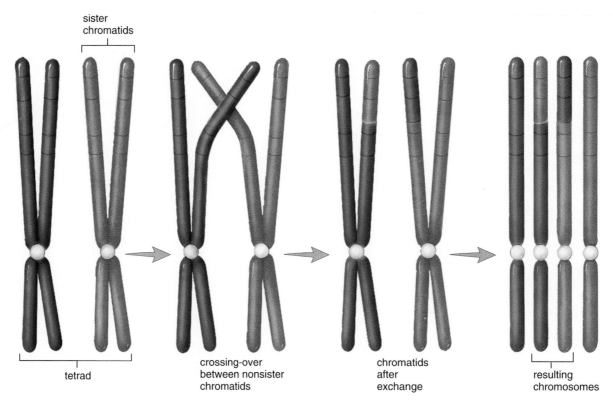

sister
chromatids

tetrad

crossing-over
between nonsister
chromatids

chromatids
after
exchange

resulting
chromosomes

Figure 19.5   Crossing-over.

When homologous chromosomes are in synapsis, the nonsister chromatids exchange genetic material. The illustration shows only one crossover per chromosome pair, but the average is slightly more than two per chromosome pair in humans. Following crossing-over, the sister chromatids may no longer be identical and instead may have different combinations of genes.

19.3 Meiosis

Meiosis, which requires two nuclear divisions, results in *four daughter cells, each having one of each kind of chromosome and therefore half the number of chromosomes as the parental cell.* The parental cell has the 2n number of chromosomes, while the daughter cells have the n number of chromosomes. Therefore, meiosis is often called reduction division. The daughter cells that result from meiosis go on to become the gametes.

Overview of Meiosis: 2n → n

Meiosis results in four daughter cells because it consists of two divisions, called meiosis I and meiosis II. Before meiosis I begins, each chromosome has duplicated and is composed of two sister chromatids. The parental cell is 2n. When a cell is 2n, the chromosomes occur in pairs. For example, the 46 chromosomes of humans occur in 23 pairs. These pairs are called **homologous chromosomes.**

During meiosis I, the homologous chromosomes of each pair come together and line up side-by-side due to a means of attraction still unknown. This so-called **synapsis** results in a **tetrad,** an association of four chromatids that stay in close proximity until they separate. During synap-

sis, nonsister chromatids may exchange genetic material. The exchange of genetic material between chromatids is called **crossing-over.** Crossing-over is significant because it recombines the genes of the parental cell and increases the variability of the gametes and therefore the offspring (Fig. 19.5).

Following synapsis during meiosis I, the homologous chromosomes of each pair separate. This separation means that one chromosome from each homologous pair will be found in each daughter cell. There are no restrictions as to which chromosome goes to each daughter cell, and therefore, all possible combinations of chromosomes may occur within the gametes.

Following meiosis I, the daughter cells have half the number of chromosomes, and the chromosomes are still duplicated.

During meiosis I, homologous chromosomes separate, and the daughter cells receive one of each pair. The daughter cells are not genetically identical. The chromosomes are still composed of two chromatids.

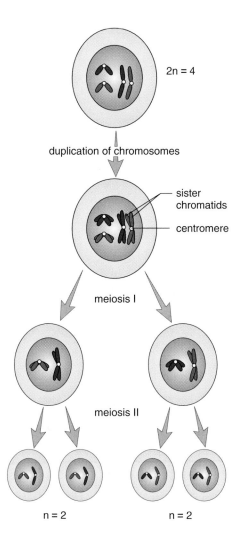

2n = 4

duplication of chromosomes

sister
chromatids

centromere

meiosis I

meiosis II

n = 2   n = 2

Figure 19.6   Overview of meiosis.
Following duplication of chromosomes, the parental cell undergoes two divisions, meiosis I and meiosis II. During meiosis I, homologous chromosomes separate, and during meiosis II, chromatids separate. The final daughter cells are haploid. (The blue chromosomes were inherited from one parent, and the red chromosomes were inherited from the other parent.)

When meiosis II begins, the chromosomes are still duplicated. Therefore, no duplication of chromosomes is needed between meiosis I and meiosis II. The chromosomes are **dyads** because each one is composed of two sister chromatids. During meiosis II, the sister chromatids separate in each of the cells from meiosis I. Each of the resulting four daughter cells has the haploid number of chromosomes.

Following meiosis, the daughter cells are not genetically identical to the parental cell. Also, notice in Figure 19.6 that the daughter cells on the right do not have the same chromosomes as the daughter cells on the left. Why not? Because the homologous chromosomes of each pair separated during meiosis I. What other chromosome combinations are possible in the daughter cells in addition to those depicted? It's possible that the gametes could contain chromosomes from only one parent (either the father or the mother) instead of both parents as depicted.

During meiosis II, the sister chromatids separate, and the resulting four daughter cells are each haploid.

The Importance of Meiosis

Because of meiosis, the chromosomal number stays constant in each generation of humans. In humans, meiosis occurs in the testes and ovaries during the production of the gametes. When a haploid sperm fertilizes a haploid egg, the new individual has the diploid number of chromosomes. There are three ways the new individual is assured a different combination of genes than either parent has:

1. Crossing-over recombines the genes on the sister chromatids of homologous pairs of chromosomes.
2. Following meiosis, gametes have all possible combinations of chromosomes.
3. At fertilization, recombination of chromosomes occurs because the sperm and egg carry varied combinations of chromosomes.

Meiosis and fertilization ensure that the chromosomal number stays constant in each generation and that the new individual has a different combination of chromosomes and genes than either parent.

Stages of Meiosis

The same four stages of mitosis—prophase, metaphase, anaphase, and telophase—occur during both meiosis I and meiosis II.

The First Division

The stages of meiosis I are diagrammed in Figure 19.7a. During prophase I, the spindle appears while the nuclear envelope fragments and the nucleolus disappears. The homologous chromosomes, each having two sister chromatids, undergo synapsis, forming tetrads. Crossing-over occurs now, but for simplicity, this event has been omitted from Figure 19.7. In metaphase I, tetrads line up at the equator of the spindle. During anaphase I, homologous chromosomes of each pair separate and move to opposite poles of the spindle. During telophase I, nucleoli appear, and nuclear envelopes form as the spindle disappears.

During cytokinesis, the plasma membrane furrows to give two cells. Each daughter cell contains only one chromosome from each homologous pair. The chromosomes are dyads, and each has two sister chromatids. No replication of DNA occurs during a period of time called interkinesis.

The Second Division

The stages of meiosis II are diagrammed in Figure 19.7b. At the beginning of prophase II, a spindle appears while the nuclear envelope disassembles and the nucleolus disappears. Dyads (one dyad from each pair of homologous chromosomes) are present, and each attaches to the spindle independently. During metaphase II, the dyads are lined up at the equator. At the start of anaphase II, the centromeres split. The sister chromatids of each dyad separate and move toward the poles. Each pole receives the same number of chromosomes. In telophase II, the spindle disappears as nuclear envelopes form.

During cytokinesis, the plasma membrane furrows to give two complete cells, each of which has the haploid, or n, number of chromosomes. Since each cell from meiosis I undergoes meiosis II, there are four daughter cells altogether.

Meiosis involves two cell divisions. During meiosis I, tetrads form and crossing-over occurs. Homologous chromosomes separate, and each daughter cell receives one of each kind of chromosome. During meiosis II, the sister chromatids separate, and there are four daughter cells, each with the haploid number of chromosomes.

Meiosis I

Prophase I

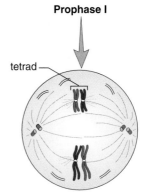

tetrad

Metaphase I

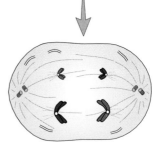

Anaphase I

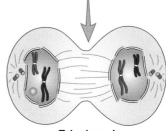

Telophase I

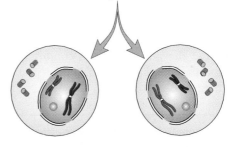

a.   **Daughter Cells: Late interphase**

Meiosis II

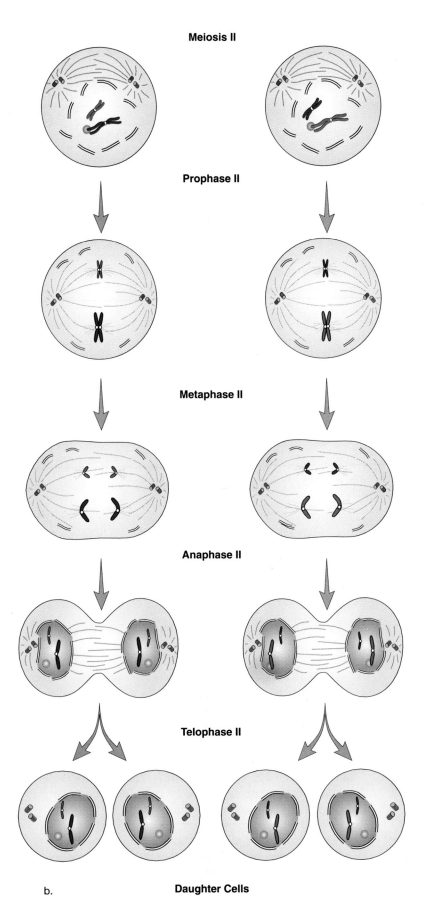

Prophase II

Metaphase II

Anaphase II

Telophase II

b.   **Daughter Cells**

Figure 19.7   Meiosis I and meiosis II.
a. During meiosis I, homologous chromosomes undergo synapsis and then separate so that each daughter cell has only one chromosome from each original homologous pair. For simplicity's sake, the results of crossing-over have not been depicted. Notice that each daughter cell is haploid and each chromosome still has two chromatids. **b.** During meiosis II, sister chromatids separate. Each daughter cell is haploid, and each chromosome consists of one chromatid. (The blue chromosomes were inherited from one parent, and the red chromosomes were inherited from the other parent.)

Spermatogenesis

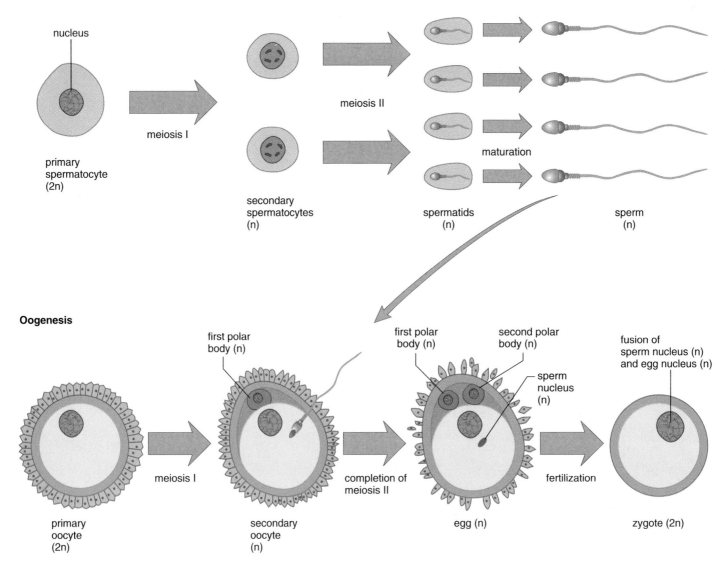

Oogenesis

Figure 19.8   Spermatogenesis and oogenesis.
Spermatogenesis produces four viable sperm, whereas oogenesis produces one egg and at least two polar bodies. Notice that oogenesis does not go to completion unless the secondary oocyte is fertilized. In humans, both sperm and egg have 23 chromosomes each; therefore, following fertilization, the zygote has 46 chromosomes.

Spermatogenesis and Oogenesis

Meiosis is a part of **spermatogenesis,** production of sperm, and **oogenesis,** production of eggs. Spermatogenesis and oogenesis occur in the sex organs—the testes in males and the ovaries in females. The gametes appear differently in the two sexes (Fig. 19.8), and meiosis is different, too. The process of meiosis in males always results in four cells that become sperm. Meiosis in females produces only one egg. Meiosis I results in one large cell called a secondary oocyte and one polar body. After meiosis II, there is one egg and two (or possibly three) polar bodies. **Polar bodies** are products of oogenesis that contain chromosomes but little cytoplasm. The plentiful cytoplasm of the egg serves as a source of nutrients and cellular organelles for the developing embryo.

Spermatogenesis, once started, continues to completion, and mature sperm result. In contrast, oogenesis does not necessarily go to completion. Only if a sperm fertilizes the secondary oocyte does it undergo meiosis II and become an egg. Regardless of this complication, however, both the sperm and the egg contribute the haploid number of chromosomes to the zygote (fertilized egg). In humans, each gamete contributes 23 chromosomes.

Spermatogenesis, which occurs in the testes of males, produces sperm. Oogenesis, which occurs in the ovaries of females, produces eggs. Meiosis is a part of spermatogenesis and oogenesis; therefore, both sperm and egg are haploid.

19.4 Chromosomal Inheritance

In a nondividing cell, the nucleus contains indistinct and diffuse chromatin, but in a dividing cell, chromatin becomes the short and thick chromosomes. A cell may be photographed just prior to division so that a picture of the chromosomes is obtained. The picture may be entered into a computer and the chromosomes electronically arranged by pairs. The resulting display of pairs of chromosomes is called a **karyotype** (Fig. 19.9).

Individual chromosomes are recognized by their size, location of the centromere (a constriction), and characteristic banding due to staining. Although both males and females have 23 pairs of chromosomes, one of these pairs is of unequal length in males. The larger chromosome of this pair is the **X chromosome,** and the smaller is the **Y chromosome.** Females have two X chromosomes in their karyotype; males

have one X and one Y. The X chromosome and Y chromosome are called the **sex chromosomes** because they contain the genes that determine sex.

The other chromosomes, known as **autosomes,** include all of the pairs of chromosomes except the X and Y chromosomes. Each pair of autosomes in the human karyotype is numbered.

Nondisjunction

An abnormal chromosomal makeup in an individual can be due to nondisjunction. **Nondisjunction** is the failure of homologous chromosomes or sister chromatids to separate during the formation of gametes. Nondisjunction occurs during meiosis I when both members of a homologous pair of chromosomes go into the same daughter cell, or during meiosis II when daughter chromosomes fail to separate and both go into the same gamete.

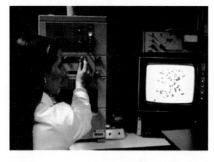

plasma

white blood cells

red blood cells

treatment

1. Blood is centrifuged to separate out blood cells.

2. Only white blood cells are transferred and treated to stop cell division.

3. Sample is fixed, stained, and spread on a microscope slide.

4. Slide is examined microscopically, and the chromosomes are photographed. Computer arranges the chromosomes into pairs.

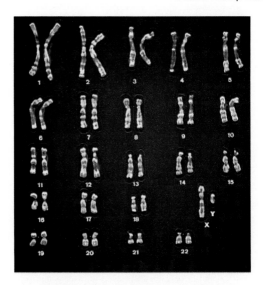

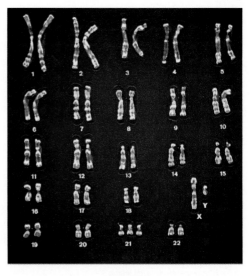

5. Karyotype: Chromosomes are paired by size, centromere location, and banding patterns. This is a normal karyotype.

6. Down syndrome karyotype, with an extra chromosome 21.

Figure 19.9   **Human karyotype preparation.**

As illustrated here, the stain used can result in chromosomes with a banded appearance. The bands help researchers identify and analyze the chromosomes.

Figure 19.10*a* shows nondisjunction during meiosis I. The daughter cells are abnormal because the daughter cell on the left received both members of the homologous pair and the daughter cell on the right received neither. Following meiosis II, all four daughter cells are abnormal. In Figure 19.10*b*, nondisjunction occurs during meiosis II. Meiosis I occurred normally, and each daughter cell has one of the members of the homologous pair. During meiosis II, the daughter chromosomes fail to separate, and they go into the same daughter cell. Figure 19.10 assumes that any one of the daughter cells will develop into an egg. Nondisjunction can also occur during spermatogenesis.

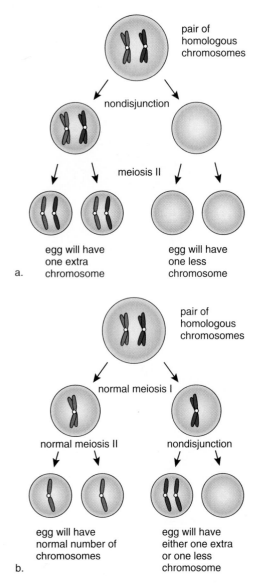

a. egg will have one extra chromosome / egg will have one less chromosome

b. egg will have normal number of chromosomes / egg will have either one extra or one less chromosome

Figure 19.10 **Nondisjunction of chromosomes during meiosis.**

a. Nondisjunction can occur during meiosis I if homologous chromosomes fail to separate and **(b)** during meiosis II if the sister chromatids fail to separate completely. In either case, certain abnormal gametes carry an extra chromosome (n + 1) or lack a chromosome (n − 1).

Down Syndrome

Down syndrome (Fig. 19.11) is easily recognized by these characteristics: short stature, an eyelid fold, stubby fingers, a wide gap between the first and second toes, a large, fissured tongue, a round head, a palm crease (the so-called simian line), and unfortunately, mental retardation, which can sometimes be severe.

Down syndrome is also called trisomy 21 because the individual usually has three copies of chromosome 21. In most instances, the egg had two copies instead of one of this chromosome. (In 23% of the cases studied, however, the sperm had the extra chromosome 21.) The chance of a woman having a Down syndrome child increases rapidly with age, starting at about age 40. The frequency of Down syndrome is 1 in 800 births for mothers under 40 years of age and 1 in 80 for mothers over 40 years of age. Most Down syndrome babies are born to women younger than age forty, however, because this is the age group having the most babies. Amniocentesis (removing fluid and cells from the amniotic sac surrounding the fetus) followed by karyotyping can detect a Down syndrome child.

It is known that the genes that cause Down syndrome are located on the bottom third of chromosome 21, and extensive investigative work has been directed toward discovering the specific genes responsible for the characteristics of the syndrome. One day it might be possible to control the expression of these genes even before birth, so that the symptoms of Down syndrome do not appear.

Figure 19.11 **Down syndrome.**

Down syndrome occurs when the egg has an extra chromosome 21 due to nondisjunction in either meiosis I or meiosis II. Characteristics include a wide, rounded face and narrow, slanting eyelids. Mental retardation to varying degrees is usually present.

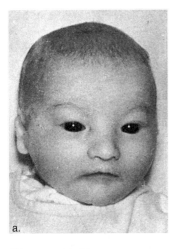

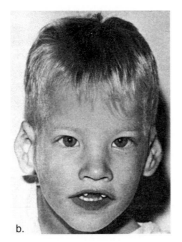

Figure 19.12   **Cri du chat syndrome.**
a. An infant and (**b**) an older child with this syndrome.

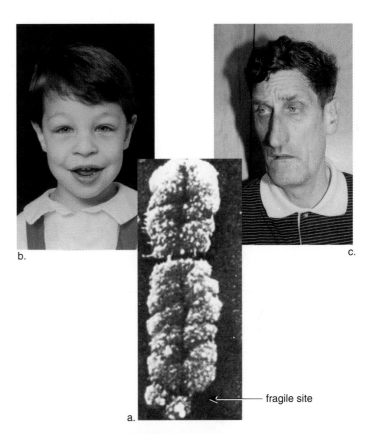

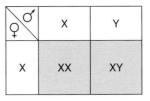

fragile site

Figure 19.13   **Fragile X syndrome.**
a. An arrow points out the fragile site of this type X chromosome.
b. A young person with the syndrome appears normal, but (**c**) with age, the elongated face has a prominent jaw, and the ears noticeably protrude.

Cri du Chat Syndrome

A chromosomal deletion is responsible for cri du chat (cat's cry) syndrome, which has a frequency of one in 50,000 live births (Fig. 19.12). An infant with this syndrome has a moon face, a small head, and a cry that sounds like the meow of a cat because of a malformed larynx. An older child has an eyelid fold and misshapen ears that are placed low on the head. Severe mental retardation becomes evident as the child matures. A karyotype shows that a portion of one chromosome 5 is missing (deleted), while the other chromosome 5 is normal, as are all the other chromosomes.

Fragile X Syndrome

Males outnumber females by about 25% in institutions for the mentally retarded. In some of these males, the X chromosome is nearly broken, leaving the tip hanging by a flimsy thread. These males are said to have fragile X syndrome (Fig. 19.13a).

Fragile X syndrome occurs in one in 1,000 male births and one in 2,500 female births. As children, fragile X syndrome individuals appear to be normal except they may be hyperactive or autistic. Their speech is delayed in development and is often repetitive in nature. As adults, they are short in stature with a long face. The jaw is prominent, and there are big, usually protruding ears (Fig. 19.4b, c). Males also have large testicles. Stubby hands, lax joints, and a heart defect may also be seen. The symptoms, including mental retardation, are not as severe in females.

The inheritance of an abnormal number of chromosomes or a defective chromosome can result in a syndrome with particular symptoms.

Sex Chromosomal Inheritance

As stated, the sex chromosomes in humans are called X and Y. Because women are XX, an egg always bears an X, but since males are XY, a sperm can bear an X or a Y. Therefore, the sex of the newborn child is determined by the father. If a Y-bearing sperm fertilizes the egg, then the XY combination results in a male. On the other hand, if an X-bearing sperm fertilizes the egg, the XX combination results in a female. All factors being equal, there is a 50% chance of having a girl or a boy.

♀ \ ♂	X	Y
X	XX	XY

However, for reasons that are not clear, more males than females are conceived, but from then on, the death rate among males is higher than for females. By age 85, there are twice as many females as males.

Too Many/Too Few Sex Chromosomes

Turner syndrome occurs in one in 6,000 births. The individual is XO, with one sex chromosome, an X; the O signifies the absence of a second sex chromosome. These females are short, with a broad chest and a webbed neck. The ovaries, oviducts, and uterus are very small and nonfunctional. Turner females do not undergo puberty or menstruate, and there is a lack of breast development (Fig. 19.14a). They are usually of normal intelligence and can lead fairly normal lives, but they are infertile even if they receive hormone supplements.

Klinefelter syndrome occurs in one in 1,500 births. These males, who have two or more X chromosomes in addition to a Y chromosome, are sterile. The testes and prostate gland are underdeveloped, and there is no facial hair. Also, there may be some breast development (Fig. 19.14b). Affected individuals have large hands and feet and very long arms and legs. They are usually slow to learn but not mentally retarded unless they inherit more than two X chromosomes.

The poly-X syndrome occurs in one in 1,500 births. A poly-X female has three or more X chromosomes. It might be supposed that the XXX female is especially feminine, but this is not the case. Although in some cases there is a tendency toward learning disabilities, most poly-X females have no apparent physical abnormalities except that they may have menstrual irregularities, including early onset of menopause.

Jacob syndrome occurs in one in 1,000 births. These XYY males are usually taller than average, suffer from persistent acne, and tend to have speech and reading problems. At one time, it was suggested that these men were likely to be criminally aggressive, but it has since been shown that the incidence of such behavior among them may be no greater than among XY males.

Individuals sometimes are born with the sex chromosomes XO (Turner syndrome), XXY (Klinefelter syndrome), XXX (poly-X syndrome), and XYY (Jacob syndrome). No matter how many X chromosomes there are, an individual with a Y chromosome develops into a male.

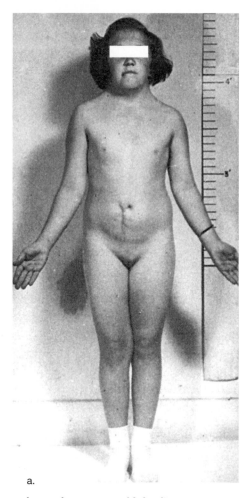

a.

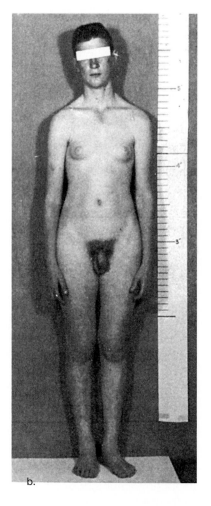

b.

Figure 19.14 **Abnormal sex chromosomal inheritance.**
a. Female with Turner (XO) syndrome, which includes a web neck, short stature, and immature sexual features. **b.** A male with Klinefelter (XXY) syndrome, which is marked by small testes and breast development in some cases.

Bioethical Focus

Cloning in Humans

The term cloning means making exact multiple copies of DNA or a cell or an organism. Cloning has been around for some time. Identical twins are clones of a single zygote. When a single bacterium reproduces asexually on a petri dish, a colony of cells results. Each member of the colony is a clone of the original cell. Through biotechnology, bacteria now produce cloned copies of human DNA.

Now, for the first time in our history, it is possible to produce a clone of a mammal. No sperm and egg are required to produce a zygote. The nucleus of an adult cell is placed in an enucleated egg that undergoes development to become an exact copy of the organism that donated the nucleus. Some people fear that billionaires and celebrities will hasten to make multiple copies of themselves. Others feel that this is unlikely. Rather, they see possibilities for a different type of cloning.

Suppose it were possible to put a nucleus from a cell of a burn victim into an embryonic stem cell that is cajoled to become skin cells. These cells could be used to provide grafts of new skin. Would this be a proper use of cloning in humans?

Or suppose parents want to produce a child free of a genetic disease. Scientists produce a zygote through in vitro fertilization, and then they clone the zygote to produce any number of cells. Genetic engineering to correct the defect doesn't work on all the cells—only a few. They implant just those few in the uterus, where development continues to term. Would this be a proper use of cloning in humans?

What if in the future scientists are able to produce children with increased intelligence or athletic prowess using this same technique? Would this be an acceptable use of cloning in humans?

Figure 19A   Identical twins.
Cloning is a natural phenomenon in the biological world. Identical twins are clones of the same zygote.

Decide Your Opinion

1. Presently, research in the cloning of humans is banned in the United States. Should it be? Why or why not?
2. Under what circumstances might cloning in humans be acceptable? Explain.
3. Is cloning to produce improved breeds of farm animals acceptable? Why or why not?

Looking at Both Sides   www.mhhe.com/biosci/genbio/maderhuman7/

Every bioethical issue has at least two sides. Even if you already have an opinion, it is important to explore the opposite opinion before finalizing your position. The Online Learning Center at www.mhhe.com/biosci/genbio/maderhuman7/ will help you fine-tune your initial opinion, explore both sides, and finalize your position. Either you will acquire new arguments for your original opinion or you may even change your opinion. Be sure to complete these activities in sequence:

Taking Sides   Decide your initial opinion by answering a series of questions. Then see if your opinion changes after completing the next two activities.

Further Debate   Read opposing articles that give you further information on this particular bioethical issue.

Explain Your Position   Answer another series of questions and then defend your original or changed opinion. You can e-mail your position to your instructor if he or she wishes.

Summarizing the Concepts

19.1 Human Life Cycle
The life cycle of higher organisms requires two types of cell divisions: mitosis and meiosis. Mitosis is responsible for growth and repair, while meiosis is required for gamete production.

19.2 Mitosis
Mitosis assures that all cells in the body have the diploid number and the same kinds of chromosomes. The cell cycle includes interphase, mitosis, and cytokinesis. During interphase, DNA replication causes each chromosome to have sister chromatids.

Mitosis has the following phases: prophase—at first the chromosomes have no particular arrangement, but later the chromosomes are attached to spindle fibers; metaphase, when the chromosomes are aligned at the equator; anaphase, when the sister chromatids separate, becoming daughter chromosomes that move toward the poles; and telophase, when new nuclear envelopes form around the daughter chromosomes. Cytokinesis occurs by furrowing.

19.3 Meiosis
Meiosis involves two cell divisions. During meiosis I, the homologous chromosomes (following crossing-over between nonsister chromatids) separate, and during meiosis II, the sister chromatids separate. The result is four cells with the haploid number of chromosomes, each consisting of one chromatid. Meiosis is a part of gamete formation in humans. Spermatogenesis in males usually produces four viable sperm, while oogenesis in females produces one egg and at least two polar bodies. Oogenesis does not go on to completion unless a sperm fertilizes the developing egg.

Among sexually reproducing organisms, such as humans, each individual is genetically different because (1) crossing-over recombines the genes on the sister chromatids of homologous pairs of chromosomes; (2) following meiosis, gametes have all possible combinations of chromosomes; and (3) upon fertilization, recombination of chromosomes occurs. Tables 19.1 and 19.2 contrast meiosis with mitosis.

19.4 Chromosomal Inheritance
A human karyotype ordinarily shows 22 homologous pairs of autosomes and one pair of sex chromosomes. The sex pair is an X and a Y chromosome in males or two X chromosomes in females. Nondisjunction during meiosis leads to gametes with an abnormal number of chromosomes and several possible syndromes.

Abnormalities in chromosome makeup sometimes occur. The major inherited autosomal abnormality is Down syndrome, in which the individual inherits three copies of chromosome 21. Cri du chat syndrome occurs when part of chromosome 5 is deleted. Examples of abnormal sex chromosomal inheritance are a fragile X chromosome and abnormal chromosomal numbers: Turner syndrome (XO), Klinefelter syndrome (XXY), poly-X syndrome (XXX and higher), and Jacob syndrome (XYY).

Table 19.1 **Meiosis I Versus Mitosis**	
Meiosis I	**Mitosis**
Prophase I Pairing of chromosomes	Prophase No pairing of chromosomes
Metaphase I Homologous pairs of duplicated chromosomes (tetrads) at equator	Metaphase Duplicated chromosomes (dyads) at equator
Anaphase I Members of homologous pairs of chromosomes separate.	Anaphase Sister chromatids separate, becoming daughter chromosomes that move to the poles.
Telophase I Daughter cells are haploid.	Telophase Daughter cells are diploid.

Table 19.2 **Meiosis II Versus Mitosis**	
Meiosis II	**Mitosis**
Prophase II No pairing of chromosomes	Prophase No pairing of chromosomes
Metaphase II Haploid number of dyads at equator	Metaphase Diploid number of dyads at equator
Anaphase II Sister chromatids separate, becoming daughter chromosomes that move to the poles.	Anaphase Sister chromatids separate, becoming daughter chromosomes that move to the poles.
Telophase II Four haploid daughter cells that are not genetically identical	Telophase Two diploid daughter cells that are genetically identical

Studying the Concepts

1. Draw a diagram describing the human life cycle. What is mitosis? What is meiosis? 386
2. Describe the stages of mitosis, including in your description the terms centrioles, nucleolus, spindle, and furrowing. 387–89
3. How do the terms diploid (2n) and haploid (n) pertain to meiosis? 391
4. Describe the stages of meiosis I, including the terms tetrad and dyad in your description. 390–92
5. Compare the stages of meiosis II to a mitotic division. 392–93, 400
6. What is the importance of mitosis and meiosis in the life cycle of humans? 386, 391
7. How does spermatogenesis in males compare to oogenesis in females? 394
8. Describe the normal karyotype of a human being. What is the difference between a male and a female karyotype? 395
9. What is nondisjunction, and how does it occur? 395–96
10. Describe Down syndrome, the most common autosomal abnormality in humans. 396
11. What is cri du chat syndrome? 397
12. List four sex chromosomal abnormalities that have to do with an abnormal number of sex chromosomes. 398

Testing Your Knowledge of the Concepts

In questions 1–5, match the syndrome to its chromosomal makeup.
  a. Down syndrome
  b. Turner syndrome
  c. Klinefelter syndrome
  d. cri du chat syndrome
  e. Jacob syndrome

_____ 1. deletion in chromosome 5

_____ 2. extra chromosome 21

_____ 3. XO

_____ 4. XXY

_____ 5. XYY

In questions 6–11, indicate whether the statement is true (T) or false (F).

_____ 6. Normally, a person has 46 chromosomes in his or her karyotype.

_____ 7. Nondisjunction can occur during meiosis I or meiosis II.

_____ 8. During meiosis I, chromatids separate, and during meiosis II the members of a homologous pair separate.

_____ 9. The chromosomes are aligned at the equator of the spindle during anaphase.

_____ 10. Half of your chromosomes were inherited from your father, and half were inherited from your mother.

_____ 11. Homologous chromosomes pair during prophase of mitosis.

In questions 12–16, fill in the blanks.

12. A karyotype shows an individual's _____.

13. The sex chromosomes of a male are called _____ and _____.

14. If the parental cell has 24 chromosomes, the daughter cells following mitosis will have _____ chromosomes.

15. Meiosis in males is a part of _____, and meiosis in females is a part of _____.

16. Oogenesis will not go to completion unless _____ occurs.

17. Match the terms to these definitions.

  a. _____ One of the two identical parts of a chromosome following replication of DNA.
  b. _____ Repeating sequence of events that includes interphase, mitosis, and cytokinesis.
  c. _____ Structure consisting of asters, poles, and fibers to which the chromosomes are attached during cell division.
  d. _____ Nonfunctioning daughter cell that has little cytoplasm and is formed during oogenesis.
  e. _____ Half the diploid number; the number of chromosomes in the gametes.

18. Which of these drawings represents metaphase I?

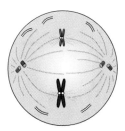 

19. How do you know your choice represents metaphase I?

Understanding Key Terms

anaphase 389	meiosis 390
aster 388	metaphase 389
autosome 395	mitosis 387
cell cycle 387	nondisjunction 395
centromere 387	oogenesis 394
chromatid 387	polar body 394
cleavage furrow 389	prophase 388
crossing-over 390	sex chromosome 395
cytokinesis 389	somatic cell 386
diploid (2n) 386	spermatogenesis 394
dyad 391	spindle 388
gamete 386	synapsis 390
haploid (n) 386	syndrome 386
homologous chromosome 390	telophase 389
interphase 387	tetrad 390
karyotype 395	X chromosome 395
	Y chromosome 395

e-Learning Connection

www.mhhe.com/biosci/genbio/maderhuman7/

19.2   Mitosis

 <u>Mitosis</u> *animation activity*

 <u>Mitosis Overview I</u> *art labeling activity*
<u>Mitosis Overview II</u> *art labeling activity*

19.3   Meiosis

 <u>Meiosis</u> *Essential Study Partner*

 <u>Meiosis</u> *animation activity*
<u>Meiosis I</u> *art labeling activity*

19.4   Chromosomal Inheritance

 <u>Abnormal Chromosomes</u> *Essential Study Partner*

 <u>Exploring Meiosis—Down Syndrome</u> *simulation*
<u>Looking at Both Sides</u> *critical thinking activity*

Chapter Summary

 <u>Additional Activities</u> *Essential Study Partner*

 <u>Key Term Flashcards</u> *vocabulary quiz*
<u>Chapter Quiz</u> *objective quiz covering all chapter concepts*

Chapter 20

Genes and Medical Genetics

Chapter Concepts

Figure 20.1   Inheritance.
We know that physical traits pass from parent to child. What about behavioral traits? Do the likes and dislikes of parents influence their children?

Intelligent. Homosexual. Loving. Aggressive. Overweight. Do any of these characteristics describe you? If so, you can probably place some of the responsibility on your genes.

Biologists have long linked physical **traits,** such as facial characteristics, to the genes. Now, after deciphering the human body's vast network of genes, they have extended the influence of genes to even more traits. It's possible the findings could lead to new treatments for various ills. For example, some biologists are studying newfound genes involved in obesity in order to create drugs that fight fat. Others are engineering lab mice so that they lack specific genes for aggression, and then watching the animals to see how they act.

Even though dozens of new, behavior-related genes are sure to surface in the future, we may never know just how genetics and environment combine to make us who we are. Social scientists argue that behavioral traits may be controlled to a degree by genes, but that child rearing, peer groups, and other social conditions also shape the personality. Most scientists agree that behavioral traits are due to a combination of genetics and environmental influences.

Granted that genes control our physical features and, at least to a degree, our behavioral features—but just what are genes? This chapter shows that it is sometimes helpful to think of genes as units of chromosomes, particularly when we want to calculate the chances of passing on a genetic disorder. But we have already mentioned that genes are actually DNA molecules that specify protein synthesis. We are now beginning to decipher how this function can directly affect the structure of the cell and the organism.

20.1 Genotype and Phenotype

Genotype refers to the genes of the individual. Alternate forms of a gene having the same position (locus) on a pair of chromosomes and affecting the same trait are called **alleles.** It is customary to designate an allele by a letter, which rep-

resents the specific characteristic it controls; a **dominant allele** is assigned an uppercase (capital) letter, while a **recessive allele** is given the same letter but in lowercase. In humans, for example, unattached (free) earlobes are dominant over attached earlobes, so a suitable key would be E for unattached earlobes and e for attached earlobes.

Since autosomal alleles occur in pairs, the individual normally has two alleles for a trait. Just as one of each pair of chromosomes is inherited from each parent, so too is one of each pair of alleles inherited from each parent. Figure 20.2 shows three possible fertilizations and the resulting genetic makeup of the zygote and, therefore, the individual. In the first instance, the chromosomes of both the sperm and the egg carry an E. Consequently, the zygote and subsequent individual have the alleles EE, which may be called a **homozygous dominant** genotype. A person with genotype EE obviously has unattached earlobes. The physical appearance of the individual, in this case unattached earlobes, is called the **phenotype.**

In the second fertilization, the zygote received two recessive alleles (ee), and the genotype is called **homozygous recessive.** An individual with this genotype has the recessive phenotype, which is attached earlobes. In the third fertilization, the resulting individual has the alleles Ee, which is called a **heterozygous** genotype. A heterozygote shows the dominant characteristic; therefore, the phenotype of this individual is unattached earlobes.

These examples show that a dominant allele contributed from only one parent can bring about a particular dominant phenotype. A recessive allele must be received from both parents to bring about the recessive phenotype.

The genotype, whether homozygous dominant (*EE*), homozygous recessive (*ee*), or heterozygous (*Ee*), tells what alleles a person carries. The phenotype—for example, attached or unattached earlobes—tells what the person looks like.

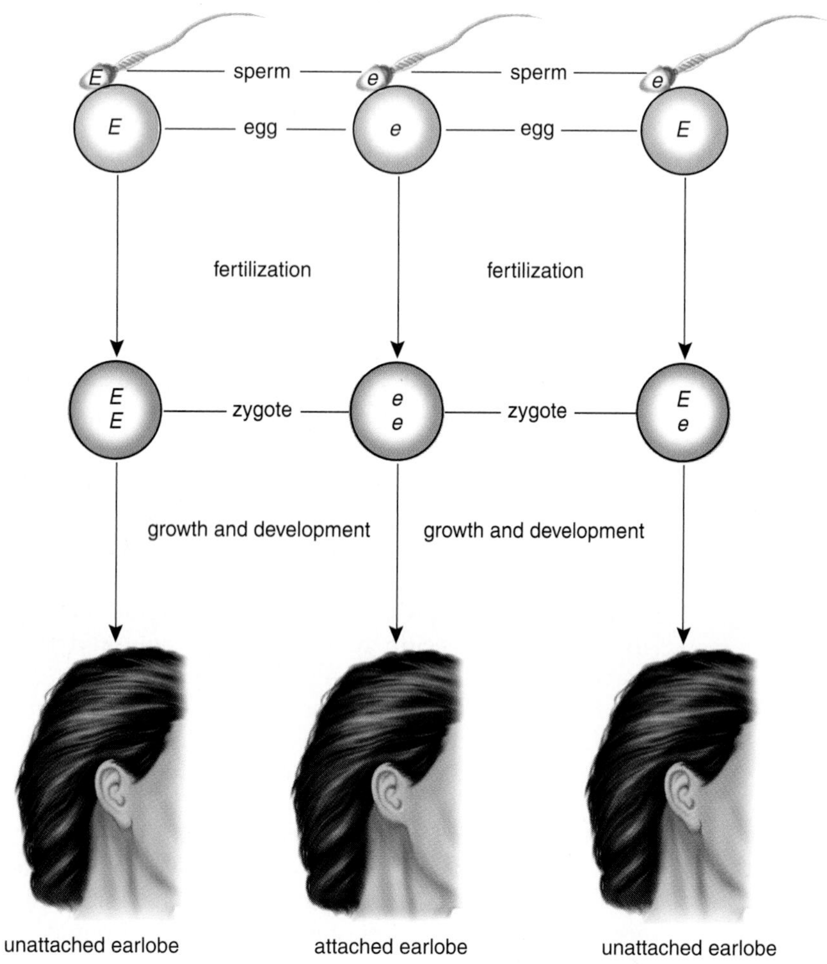

unattached earlobe   attached earlobe   unattached earlobe

Figure 20.2   Genetic inheritance.
Individuals inherit a minimum of two alleles for every characteristic of their anatomy and physiology. The inheritance of a single dominant allele (*E*) causes an individual to have unattached earlobes; two recessive alleles (*ee*) cause an individual to have attached earlobes. Notice that each individual receives one allele from the father (by way of a sperm) and one allele from the mother (by way of an egg).

20.2   Dominant/Recessive Traits

The alleles designated by *E* and *e* are on a certain part of the chromosomes. An individual has two alleles for each trait because a chromosome pair carries alleles for the same traits. How many alleles for each trait will be in the gametes? One, because chromosome pairs separate during meiosis I.

Forming the Gametes

During gametogenesis, the chromosome number is reduced. Whereas the individual has 46 chromosomes, a gamete has only 23 chromosomes. (If this did not happen, each new generation of individuals would have twice as many chromosomes as their parents.) Reduction of the chromosome number occurs when the pairs of chromosomes separate as meiosis occurs. Since the alleles are on the chromosomes, they also separate during meiosis, and therefore, the gametes carry only one allele for each trait. If an individual carried the alleles *EE*, all the gametes would carry an *E* since that is the only choice. Similarly, if an individual carried the alleles *ee*, all the gametes would carry an *e*. What if an individual were *Ee*? Figure 20.3 shows that half of the gametes would carry an *E* and half would carry an *e*. Figure 20.4 shows the genotypes and phenotypes for certain other traits in humans, and you can practice deciding what alleles the gametes would carry for these genotypes.

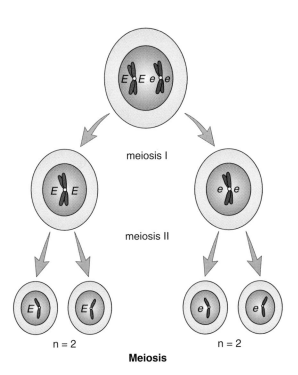

Figure 20.3   Gametogenesis.
Because the pairs of chromosomes separate during meiosis, which occurs during gametogenesis, the gametes have only one allele for each trait.

a.   Widow's peak: *WW* or *Ww*

b.   Straight hairline: *ww*

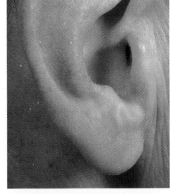

c.   Unattached earlobes: *EE* or *Ee*

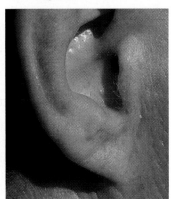

d.   Attached earlobes: *ee*

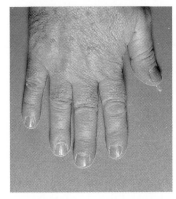

e.   Short fingers: *SS* or *Ss*

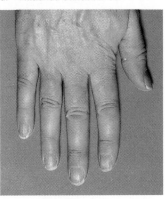

f.   Long fingers: *ss*

g.   Freckles: *FF* or *Ff*

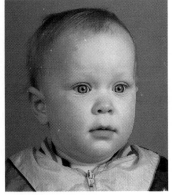

h.   No freckles: *ff*

Figure 20.4   Common inherited characteristics in human beings.
The notations indicate which characteristics are dominant and which are recessive.

Figuring the Odds

Many times parents would like to know the chances of having a child with a certain genotype and, therefore, a certain phenotype. If one of the parents is homozygous dominant (*EE*), the chances of their having a child with unattached earlobes is 100% because this parent has only a dominant allele (*E*) to pass on to the offspring. On the other hand, if both parents are homozygous recessive (*ee*), there is a 100% chance that each of their children will have attached earlobes. However, if both parents are heterozygous, then what are the chances that their child will have unattached or attached earlobes? To solve a problem of this type, it is customary first to indicate the genotype of the parents and their possible gametes.

Genotypes:	*Ee*	*Ee*
Gametes:	*E* and *e*	*E* and *e*

Second, a **Punnett square** is used to determine the phenotypic ratio among the offspring when all possible sperm are given an equal chance to fertilize all possible eggs (Fig. 20.5). The possible sperm are lined up along one side of the square, and the possible eggs are lined up along the other side of the square (or vice versa). The ratio among the offspring in this case is 3:1 (three children with unattached earlobes to one with attached earlobes). This means that there is a 3/4 chance (75%) for each child to have unattached earlobes and a 1/4 chance (25%) for each child to have attached earlobes.

Another cross of particular interest is that between a heterozygous individual (*Ee*) and a pure recessive (*ee*). In this case, the Punnett square shows that the ratio among the offspring is 1:1, and the chance of having the dominant or recessive phenotype is 1/2, or 50% (Fig. 20.6).

Comparing the two crosses (Fig. 20.5 and Fig. 20.6), each child has a 75% chance of having the dominant phenotype if the two parents are heterozygous and a 50% chance if one parent is heterozygous and the other is recessive. Each child has a 25% chance of having the recessive phenotype if the parents are heterozygous and a 50% chance if one parent is heterozygous and the other is homozygous recessive.

If both parents are heterozygous, each child has a 25% chance of exhibiting the recessive phenotype. If one parent is heterozygous and the other is homozygous recessive, each child has a 50% chance of exhibiting the recessive phenotype.

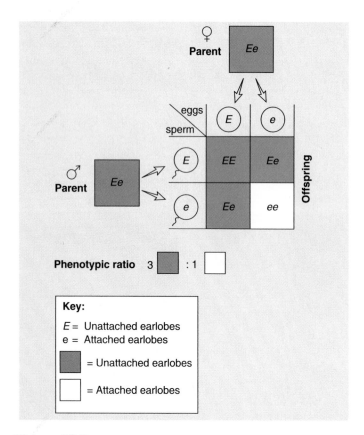

Figure 20.5   Heterozygous-by-heterozygous cross.
When the parents are heterozygous, each child has a 75% chance of having the dominant phenotype and a 25% chance of having the recessive phenotype.

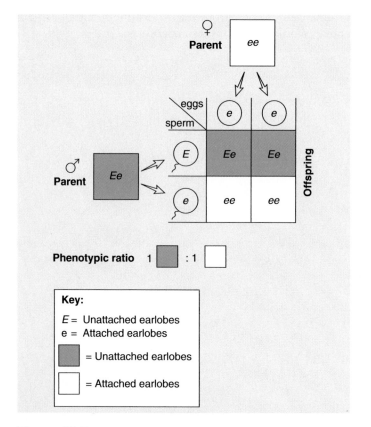

Figure 20.6   Heterozygous-by-homozygous recessive cross.
When one parent is heterozygous and the other homozygous recessive, each child has a 50% chance of having the dominant phenotype and a 50% chance of having the recessive phenotype.

Recessive Disorders

Of the many autosomal (non-sex-linked) recessive disorders, we will discuss only three.

Tay-Sachs Disease

Tay-Sachs disease is a well-known genetic disease that usually occurs among Jewish people in the United States, most of whom are of central and eastern European descent. At first it is not apparent that a baby has Tay-Sachs disease. However, development begins to slow down between four months and eight months of age, and neurological impairment and psychomotor difficulties then become apparent. The child gradually becomes blind and helpless, develops uncontrollable seizures, and eventually becomes paralyzed. There is no treatment or cure for Tay-Sachs disease, and most affected individuals die by the age of three or four.

Tay-Sachs disease results from a lack of the enzyme hexosaminidase A (Hex A) and the subsequent storage of its substrate, a fatty substance known as glycosphingolipid, in lysosomes. Although more and more lysosomes build up in many body cells, the primary sites of storage are the cells of the brain, which account for the onset and the progressive deterioration of psychomotor functions.

Persons heterozygous for Tay-Sachs have about half the level of Hex A activity found in normal individuals, but no problems result because enzymes are useable. Prenatal diagnosis of the disease is possible following either amniocentesis or chorionic villi sampling.

Cystic Fibrosis

Cystic fibrosis is the most common lethal genetic disease among Caucasians in the United States. About one in 20 Caucasians is a carrier, and about one in 2,500 newborns has the disorder. The mucus in the bronchial tubes and pancreatic ducts is particularly thick and viscous, interfering with the function of the lungs and pancreas. To ease breathing, the thick mucus in the lungs has to be manually loosened periodically (Fig. 20.7), but still the lungs become infected frequently. The clogged pancreatic ducts prevent digestive enzymes from reaching the small intestine, and patients usually take digestive enzymes mixed with applesauce before every meal. Life expectancy has now increased to as much as 28 years of age.

Research has demonstrated that chloride ions (Cl⁻) fail to pass through plasma membrane channel proteins in cystic fibrosis patients. Ordinarily, after chloride ions have passed through the membrane, water follows. It is thought that lack of water is the cause of the abnormally thick mucus in the bronchial tubes and pancreatic ducts. The cystic fibrosis allele, which is located on chromosome 7, has been isolated, and attempts have been made to insert it into nasal epithelium, so far with little success. Genetic testing for the allele in adult carriers and in fetuses is possible; if present, couples have to decide whether to risk having a child with the condition or whether abortion is an option.

Figure 20.7 **Cystic fibrosis.**
The mucus in the lungs of a child with cystic fibrosis should be periodically loosened by clapping the back.

Phenylketonuria (PKU)

Phenylketonuria (PKU) occurs once in 5,000 newborns, so it is not as frequent as the disorders previously discussed. However, it is the most commonly inherited metabolic disorder that affects nervous system development. First cousins who marry are more apt to have a PKU child.

Affected individuals lack an enzyme that is needed for the normal metabolism of the amino acid phenylalanine, and an abnormal breakdown product, phenylketone, accumulates in the urine. The PKU allele is located on chromosome 12, and a prenatal DNA test can determine the presence of this allele. Newborns are routinely tested in the hospital for elevated levels of phenylalanine in the blood. If elevated levels are detected, newborns are placed on a diet low in phenylalanine, which must be continued until the brain is fully developed, around age seven, or else severe mental retardation develops. Some doctors recommend that the diet continue for life, but in any case, a pregnant woman with phenylketonuria must be on the diet in order to protect her unborn child from harm.

Practice Problems 1*

1. What is the genotype of the child in Figure 20.7? What were the genotypes of her parents if neither one has cystic fibrosis? (Use this key: C = normal; c = cystic fibrosis.)

2. What are the chances that these parents will have a child with cystic fibrosis? (Hint: See Figure 20.5 and substitute the key given in question 1.)

3. A mother and daughter have PKU. A son does not have PKU. What is the genotype of the father?

*Answers to Practice Problems appear in Appendix A.

Dominant Disorders

Of the many autosomal dominant disorders, we will discuss only neurofibromatosis and Huntington disease.

Neurofibromatosis

Neurofibromatosis, sometimes called von Recklinghausen disease, is one of the most common genetic disorders. It affects roughly one in 3,000 people, including an estimated 100,000 in the United States. It is seen equally in every racial and ethnic group throughout the world.

At birth or later, the affected individual may have six or more large, tan spots (known as cafe-au-lait) on the skin. Such spots may increase in size and number and may get darker. Small benign tumors (lumps) called neurofibromas may occur under the skin or in various organs. Neurofibromas are made up of nerve cells and other cell types.

The expression of neurofibromatosis varies. In most cases, the symptoms are mild, and patients live a normal life. In some cases, however, the effects are severe. Skeletal deformities, including a large head, are seen, and eye and ear tumors can lead to blindness and hearing loss. Many children with neurofibromatosis have learning disabilities and are hyperactive.

In 1990, researchers isolated the allele for neurofibromatosis, which was known to be on chromosome 17. By analyzing the DNA (deoxyribonucleic acid), they determined that the allele was huge and actually included three even smaller alleles. This was only the second time that nested alleles have been found in humans. The allele for neurofibromatosis is active in controlling cell division. When it mutates, a benign tumor develops.

Huntington Disease

One in 20,000 persons in the United States has Huntington disease, a neurological disorder that leads to progressive degeneration of brain cells, which in turn causes severe muscle spasms and personality disorders (Fig. 20.8). Most people appear normal until they are of middle age and have already had children who might also be stricken. Occasionally, the first signs of the disease are seen in these children when they are teenagers or even younger. There is no effective treatment, and death comes ten to fifteen years after the onset of symptoms.

Several years ago, researchers found that the allele for Huntington disease was located on chromosome 4. A test was developed for the presence of the allele, but few people want to know if they have inherited the allele because as yet there is no treatment for Huntington disease. It appears that persons most at risk have inherited the disorder from their fathers. The latter observation is consistent with a new hypothesis called **genomic imprinting.** The genes are imprinted differently during formation of sperm and egg, and therefore, the sex of the parent passing on the disorder becomes important.

Figure 20.8 **Huntington disease.**
Persons with this condition gradually lose psychomotor control of the body. At first there are only minor disturbances, but the symptoms become worse over time.

Practice Problems 2*

1. What is the genotype of the woman in Figure 20.8 if she is heterozygous? What would be the genotype of a husband who is homozygous recessive? (Use this key: H = Huntington; h = normal.)

2. If the couple in question 1 reproduces, what are the chances that their children will develop Huntington disease? Will be free of Huntington disease? (Hint: See Figure 20.6 and use the key given in question 1.)

3. Which of these children could pass on Huntington disease? Why?

4. A child has neurofibromatosis. The mother appears normal. What is the genotype of the father?

*Answers to Practice Problems appear in Appendix A.

Recessive genetic disorders can be passed on by parents who appear to be normal. Dominant genetic disorders are passed on by a parent who has or will develop the disorder.

Pedigree Charts

When a genetic disorder is autosomal dominant, an individual with the alleles *AA* or *Aa* will have the disorder. When a genetic disorder is autosomal recessive, only individuals with the alleles *aa* will have the disorder. As mentioned in the Health Focus on page 410, genetic counselors often construct pedigree charts to determine whether a condition is dominant or recessive. A pedigree chart shows the pattern of inheritance for a particular condition. Consider these two possible patterns of inheritance:

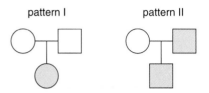

In both patterns, males are designated by squares and females by circles. Shaded circles and squares are affected individuals. A line between a square and a circle represents a union. A vertical line going downward leads, in these patterns, to a single child. (If there are more children, they are placed off a horizontal line.) Which pattern of inheritance do you suppose represents an autosomal dominant characteristic, and which represents an autosomal recessive characteristic?

In pattern I, the child is affected, but neither parent is; this can happen if the condition is recessive and the parents are *Aa*. Notice that the parents are **carriers** because they appear to be normal but are capable of having a child with a genetic disorder. Figure 20.9 shows a typical pedigree chart for a recessive genetic disorder. Other ways to recognize an autosomal recessive pattern of inheritance are also listed in the figure. Notice that when both parents are affected, all the children are affected. Why? Because the parents can pass on only recessive alleles for this condition.

It is important to realize that "chance has no memory"; therefore, each child born to heterozygous parents has a 25% chance of having the disorder. In other words, it is possible that if a heterozygous couple has four children, each child might have the condition.

In pattern II, the child is affected, as is one of the parents. When a disorder is dominant, an affected child usually has at least one affected parent. Of the two patterns, this one shows a dominant pattern of inheritance. Figure 20.10 shows a typical pedigree chart for a dominant disorder. Other ways to recognize an autosomal dominant pattern of inheritance are also listed. Notice that two affected parents can have a child that is unaffected. Why? Because the child inherited a recessive allele from each parent.

Dominant and recessive alleles have different patterns of inheritance.

---

[1]See Appendix A for answers.

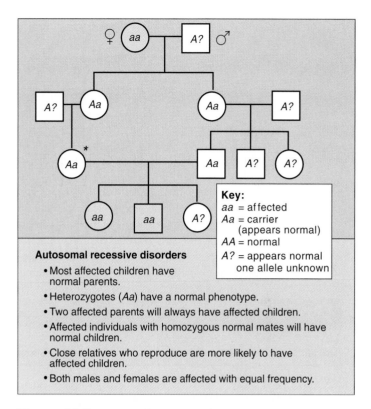

Autosomal recessive disorders
- Most affected children have normal parents.
- Heterozygotes (*Aa*) have a normal phenotype.
- Two affected parents will always have affected children.
- Affected individuals with homozygous normal mates will have normal children.
- Close relatives who reproduce are more likely to have affected children.
- Both males and females are affected with equal frequency.

Key:
aa = affected
Aa = carrier (appears normal)
AA = normal
A? = appears normal one allele unknown

Figure 20.9 **Autosomal recessive pedigree chart.**
The list gives ways to recognize an autosomal recessive disorder. How would you know the individual at the asterisk is heterozygous?[1]

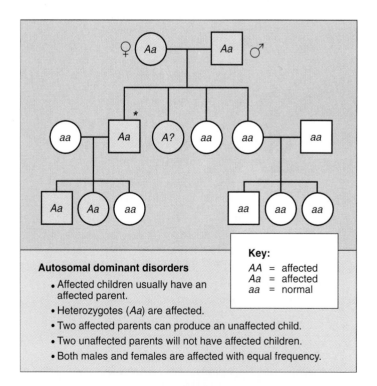

Autosomal dominant disorders
- Affected children usually have an affected parent.
- Heterozygotes (*Aa*) are affected.
- Two affected parents can produce an unaffected child.
- Two unaffected parents will not have affected children.
- Both males and females are affected with equal frequency.

Key:
AA = affected
Aa = affected
aa = normal

Figure 20.10 **Autosomal dominant pedigree chart.**
The list gives ways to recognize an autosomal dominant disorder. How would you know the individual at the asterisk is heterozygous?[1]

Genetic Counseling

Now that potential parents are becoming aware that many illnesses are caused by faulty genes, more couples are seeking genetic counseling. The counselor studies the backgrounds of the couple and tries to determine if any immediate ancestor may have had a genetic disorder. A pedigree chart may be constructed. Then the counselor studies the couple. As much as possible, laboratory tests are performed on all persons involved.

Tests are now available for a large number of genetic diseases. For example, chromosomal tests are available for cystic fibrosis, neurofibromatosis, and Huntington disease. Blood tests can identify carriers of thalassemia and sickle-cell disease. By measuring enzyme levels in blood, tears, or skin cells, carriers of enzyme defects can also be identified for certain inborn metabolic errors, such as Tay-Sachs disease. From this information, the counselor can sometimes predict the chances of having a child with the disorder.

Whenever the woman is pregnant, chorionic villi sampling can be done early, and amniocentesis can be done later in the pregnancy. These procedures, which were illustrated in chapter 18, allow the testing of embryonic and fetal cells, respectively, to determine if the unborn child has a genetic disorder. If so, treatment may be available even before birth, or parents may decide whether or not to end the pregnancy (Table 20A).

Table 20A	Test and Treatment for Some Human Genetic Disorders			
Name	**Description**	**Chromosome**	**Incidence Among Newborns in the U.S.**	**Status**
Autosomal Recessive Disorders				
Cystic fibrosis	Mucus in the lungs and digestive tract is thick and viscous, making breathing and digestion difficult	7	One in 2,500 Caucasians	Allele located; chromosome test now available;* treatment being investigated
Tay-Sachs disease	Neurological impairment and psychomotor difficulties develop early, followed by blindness and uncontrollable seizures before death occurs, usually before age 5	15	One in 3,600 Jews of eastern European descent	Biochemical test now available*
Phenylketonuria	Inability to metabolize phenylalanine, and if a special diet is not begun, mental retardation develops	12	One in 5,000 Caucasians	Biochemical test now available; treatment available
Autosomal Dominant Disorders				
Neurofibromatosis	Benign tumors occur under the skin or deeper	17	One in 3,000	Allele located; chromosome test now available*
Huntington disease	Minor disturbances in balance and coordination develop in middle age and progress toward severe neurological disturbances leading to death	4	One in 20,000	Allele located; chromosome test now available*
Incomplete Dominance				
Sickle-cell disease	Poor circulation, anemia, internal hemorrhaging, due to sickle-shaped red blood cells	11	One in 100 African Americans	Chromosome test now available*
X-Linked Recessive				
Hemophilia A	Propensity for bleeding, often internally, due to the lack of a blood clotting factor	X	One in 15,000 male births	Treatment available
Duchenne muscular dystrophy	Muscle weakness develops early and progressively intensifies until death occurs, usually before age 20	X	One in 5,000 male births	Allele located; biochemical tests of muscle tissue available; treatment being investigated

*Prenatal testing is done.

20.3   Beyond Simple Inheritance Patterns

Certain traits, such as those just studied, follow the rules of simple dominant or recessive inheritance. But there are other more complicated patterns of inheritance.

Polygenic Inheritance

Polygenic inheritance occurs when one trait is governed by two or more sets of alleles, and the individual has a copy of all allelic pairs, possibly located on many different pairs of chromosomes. Each dominant allele has a quantitative effect on the phenotype, and these effects are additive. The result is a continuous variation of phenotypes, resulting in a distribution of these phenotypes that resembles a bell-

shaped curve. The more genes involved, the more continuous the variation and the distribution of the phenotypes. Also, environmental effects cause many intervening phenotypes; in the case of height, differences in nutrition assure a bell-shaped curve (Fig. 20.11).

Skin Color

Just how many pairs of alleles control skin color is not known, but a range in colors can be explained on the basis of two pairs. When a very dark person reproduces with a very light person, the children have medium-brown skin; when two people with medium-brown skin reproduce with one another, the children may range in skin color from very dark to very light. This can be explained by assuming that skin color is controlled by two pairs of alleles and that *each capital letter contributes pigment to the skin:*

Genotypes	Phenotypes
AABB	Very dark
AABb or *AaBB*	Dark
AaBb or *AAbb* or *aaBB*	Medium brown
Aabb or *aaBb*	Light
aabb	Very light

Notice again that there is a range in phenotypes and that there are several possible phenotypes in between the two extremes. Therefore, the distribution of these phenotypes is expected to follow a bell-shaped curve—few people have the extreme phenotypes, and most people have the phenotype that lies in the middle between the extremes.

Polygenic Disorders

Many human disorders, such as cleft lip and/or palate, clubfoot, congenital dislocations of the hip, hypertension, diabetes, schizophrenia, and even allergies and cancers, are most likely controlled by polygenes and are subject to environmental influences. Therefore, many investigators are in the process of considering the nature versus nurture question; that is, what percentage of the trait is controlled by genes and what percentage is controlled by the environment? Thus far, it has not been possible to come to precise, generally accepted percentages for any particular trait.

In recent years, reports have surfaced that all sorts of behavioral traits, such as alcoholism, phobias, and even suicide, can be associated with particular genes. No doubt behavioral traits are to a degree controlled by genes, but again, it is impossible at this time to determine to what degree. And very few scientists would support the idea that these behavioral traits are predetermined by our genes.

a.

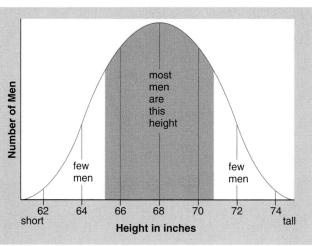
b.

Figure 20.11   Polygenic inheritance.
When you record the heights of a large group of young men (**a**), the values follow a bell-shaped curve (**b**). Such a continuous distribution is due to control of a trait by several sets of alleles. Environmental effects are also involved.

Many human traits most likely controlled by polygenes are subject to environmental influences. The frequency of the phenotypes of such traits follows a bell-shaped curve.

Multiple Allelic Traits

When a trait is controlled by **multiple alleles,** the gene exists in several allelic forms. But each person usually has only two of the possible alleles. For example, a person's blood type is determined by multiple alleles.

ABO Blood Types

Three alleles for the same gene control the inheritance of ABO blood types. These alleles determine the presence or absence of antigenic glycoproteins on the red blood cells:

A = A antigen on red blood cells
B = B antigen on red blood cells
O = Neither A nor B antigen on red blood cells

Alleles A and B are dominant over O, and are fully expressed in the presence of the other. This is called codominance, as mentioned on page 413.

The possible genotypes and phenotypes for blood type are as follows:

Genotypes	Phenotypes
AA, AO	Type A
BB, BO	Type B
AB	Type AB
OO	Type O

Figure 20.12 shows that matings between certain genotypes can have surprising results.

Blood typing can sometimes aid in paternity suits. However, a blood test of a supposed father can only suggest that he might be the father, not that he definitely is the father. For example, it is possible, but not definite, that a man with type A blood (having genotype AO) is the father of a child with type O blood. On the other hand, a blood test sometimes can definitely prove that a man is not the father. For example, a man with type AB blood cannot possibly be the father of a child with type O blood. Therefore, blood tests can be used in legal cases only to exclude a man from possible paternity.

As a point of interest, the Rh factor is inherited separately from A, B, AB, or O blood types. When you are Rh positive, there is a particular antigen on the red blood cells, and when you are Rh negative, it is absent. It can be assumed that the inheritance of this antigen is controlled by a single allelic pair in which simple dominance prevails: the Rh-positive allele is dominant over the Rh-negative allele.

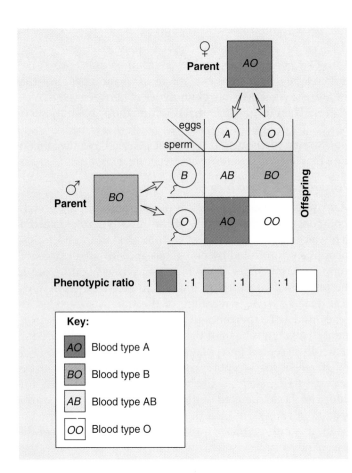

Figure 20.12   Inheritance of blood type.
A mating between blood type A and blood type B can result in any one of the four blood types. Why? Because the parents are AO and BO. If both parents are blood type AB, the children can have what blood type?

Practice Problems 3✱

1. What is the darkest child that could result from a mating between a light individual and a very light individual?

2. What is the lightest child that could result from a mating between two medium-brown individuals?

3. From the following blood types, determine which baby belongs to which parents:

   Baby 1, type O   Mrs. Doe, type A
   Baby 2, type B   Mr. Doe, type A
                               Mrs. Jones, type A
                               Mr. Jones, type AB

*Answers to Practice Problems appear in Appendix A.

Inheritance by multiple alleles occurs when a gene exists in more than two allelic forms. However, each individual usually inherits only two alleles for these genes.

Incompletely Dominant Traits

The field of human genetics also has examples of codominance and incomplete dominance. **Codominance** occurs when alleles are equally expressed in a heterozygote. We have already mentioned that the multiple alleles controlling blood type are codominant. An individual with the genotype *AB* has type AB blood. Also, skin color is controlled by polygenes in which all dominants (capital letters) add equally to the phenotype. For this reason, it is possible to observe a range of skin colors from very dark to very light.

Incomplete dominance is exhibited when the heterozygote has an intermediate phenotype between that of either homozygote. For example, incomplete dominance pertains to the inheritance of curly versus straight hair in Caucasians. When a curly-haired Caucasian reproduces with a straight-haired Caucasian, their children have wavy hair. When two wavy-haired persons reproduce, the expected phenotypic ratio among the offspring is 1:2:1—that is, one curly-haired child to two with wavy hair to one with straight hair (Fig. 20.13).

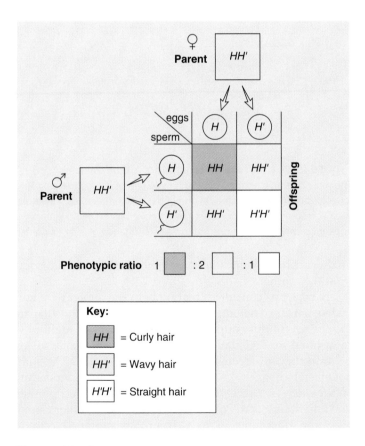

Figure 20.13   Incomplete dominance.
Among Caucasians, neither straight nor curly hair is dominant. When two wavy-haired individuals reproduce, each offspring has a 25% chance of having either straight or curly hair and a 50% chance of having wavy hair, the intermediate phenotype.

Sickle-Cell Disease

Sickle-cell disease is an example of a human disorder that is controlled by incompletely dominant alleles. In persons with sickle-cell disease, the red blood cells aren't biconcave disks like normal red blood cells; they are irregular. In fact, many are sickle-shaped. The defect is caused by an abnormal hemoglobin (Hb^S). Normal hemoglobin (Hb^A) differs from Hb^S by one amino acid in the protein globin. This one change causes Hb^S to be less soluble than Hb^A.

Individuals with the Hb^AHb^A genotype are normal, those with the Hb^SHb^S genotype have sickle-cell disease, and those with the Hb^AHb^S genotype have the sickle-cell trait. Two individuals with sickle-cell trait can produce children with all three phenotypes in the same manner as shown in Figure 20.13 for an incompletely dominant trait.

Because sickle-shaped cells can't pass along narrow capillary passageways like disk-shaped cells, they clog the vessels and break down. This is why persons with sickle-cell disease suffer from poor circulation, anemia, and poor resistance to infection. Internal hemorrhaging leads to further complications, such as jaundice, episodic pain in the abdomen and joints, and damage to internal organs. Persons with sickle-cell trait do not usually have any sickle-shaped cells unless they experience dehydration or mild oxygen deprivation. Still, at present, most investigators believe that no restrictions on physical activity are needed for persons with the sickle-cell trait.

Among regions of malaria-infested Africa, infants with sickle-cell disease die, but infants with sickle-cell trait have a better chance of survival than the normal homozygote. The malaria parasite normally reproduces inside red blood cells. But a red blood cell of a sickle-cell trait infant becomes sickle-shaped if it becomes infected with a malaria-causing parasite. As the cell becomes sickle-shaped and loses potassium, the parasite dies. The protection afforded by the sickle-cell trait keeps the allele prevalent in populations exposed to malaria. As many as 60% of the population in malaria-infested regions of Africa have the allele. In the United States, about 10% of the African American population carries the allele.

Both standard and innovative therapies are being explored. For example, a bone marrow transplant can sometimes be successful in curing sickle-cell disease. On the other hand, persons with sickle-cell disease produce normal fetal hemoglobin during development, and drugs that turn on the genes for fetal hemoglobin in adults are being developed. Mice have been genetically engineered to produce sickled red blood cells in order to test new anti-sickling drugs and various genetic therapies.

There are three genotypes for sickle-cell disease. When heterozygotes reproduce, a child can have any one of the three phenotypes.

20.4 Sex-Linked Traits

The sex chromosomes contain genes just as the autosomal chromosomes do. Some of these genes determine whether the individual is a male or a female. Investigators have found that the Y chromosome has an *SRY* gene (sex-determining region of the Y chromosome). When this gene is lacking from the Y chromosome, the individual is a female even though the chromosomal inheritance is XY.

Traits controlled by alleles on the **sex chromosomes** are said to be **sex-linked;** an allele that is only on the X chromosome is **X-linked,** and an allele that is only on the Y chromosome is **Y-linked.** Most sex-linked alleles are on the X chromosome, and the Y chromosome is blank for these. Very few alleles have been found on the Y chromosome, as you might predict, since it is much smaller than the X chromosome.

The X chromosomes also carry many genes unrelated to the sex of the individual, and we will look at a few of these in depth. It would be logical to suppose that a sex-linked trait is passed from father to son or from mother to daughter, but this is not the case. A male always receives a sex-linked condition from his mother, from whom he inherited an X chromosome. The Y chromosome from the father does not carry an allele for the trait. Usually the trait is recessive; therefore, a female must receive two alleles, one from each parent, before she has the condition.

X-Linked Alleles

When considering X-linked traits, the allele on the X chromosome is shown as a letter attached to the X chromosome. For example, this is the key for red-green color blindness:

X^B = normal vision
X^b = color blindness

The possible genotypes and phenotypes in both males and females are:

Genotypes	Phenotypes
$X^B X^B$	Female who has normal color vision
$X^B X^b$	Carrier female who has normal color vision
$X^b X^b$	Female who is color blind
$X^B Y$	Male who has normal vision
$X^b Y$	Male who is color blind

Recall that carriers are individuals who appear normal but can pass on an allele for a genetic disorder. Note here that the second genotype is a carrier female because although a female with this genotype appears normal, she is capable of passing on an allele for color blindness. Color-blind females are rare because they must receive the allele from both parents; color-blind males are more common since they need only one recessive allele to be color blind. The allele for color blindness has to be inherited from their mother because it is

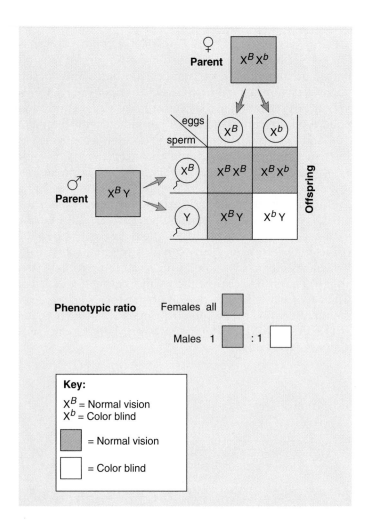

Figure 20.14　Cross involving an X-linked allele.
The male parent is normal, but the female parent is a carrier; an allele for color blindness is located on one of her X chromosomes. Therefore, each son stands a 50% chance of being color blind. The daughters will appear normal, but each one stands a 50% chance of being a carrier.

on the X chromosome; males only inherit the Y chromosome from their father.

Now, let us consider a mating between a man with normal vision and a heterozygous woman (Fig. 20.14). What are the chances of this couple having a color-blind daughter? A color-blind son? All daughters will have normal color vision because they all receive an X^B from their father. The sons, however, have a 50% chance of being color blind, depending on whether they receive an X^B or an X^b from their mother. The inheritance of a Y chromosome from their father cannot offset the inheritance of an X^b from their mother. Notice in Figure 20.14 that the phenotypic results for sex-linked traits are given separately for males and females.

The X chromosome carries alleles that are not on the Y chromosome. Therefore, a recessive allele on the X chromosome is expressed in males.

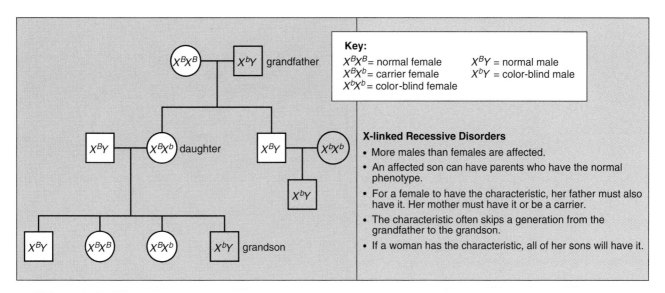

Figure 20.15 **X-linked recessive pedigree chart.**
The list gives ways of recognizing an X-linked recessive disorder.

X-Linked Disorders

Figure 20.15 gives a pedigree chart for an X-linked recessive condition. It also lists ways to recognize this pattern of inheritance. X-linked conditions can be dominant or recessive, but most of the X-linked conditions we know about are recessive. More males than females have the trait because recessive alleles on the X chromosome are always expressed in males since the Y chromosome does not have a corresponding allele. If a male has an X-linked recessive condition, his daughters are often carriers; therefore, the condition passes from grandfather to grandson. Females who have the condition inherited the allele from both their mother and their father, and all the sons of such a female will have the condition. Three well-known X-linked recessive disorders are color blindness, muscular dystrophy, and hemophilia.

Color Blindness

In humans, there are three different classes of cone cells, the receptors for color vision in the retina of the eyes. Only one pigment protein is present in each type of cone cell; there are blue-sensitive, red-sensitive, and green-sensitive cone cells. The allele for the blue-sensitive protein is autosomal, but the alleles for the red- and green-sensitive proteins are on the X chromosome. About 8% of Caucasian men have red-green color blindness. Most of these see brighter greens as tans, olive greens as browns, and reds as reddish-browns. A few cannot tell reds from greens at all. They see only yellows, blues, blacks, whites, and grays. Opticians use special charts to detect those who are color blind.

Muscular Dystrophy

Muscular dystrophy, as the name implies, is characterized by a wasting away of the muscles. The most common form,

Duchenne muscular dystrophy, is X-linked and occurs in about one out of every 3,600 male births. Symptoms, such as waddling gait, toe walking, frequent falls, and difficulty in rising, may appear as soon as the child starts to walk. Muscle weakness intensifies until the individual is confined to a wheelchair. Death usually occurs by age 20; therefore, affected males are rarely fathers. The recessive allele remains in the population through passage from carrier mother to carrier daughter.

Recently, the allele for Duchenne muscular dystrophy was isolated, and it was discovered that the absence of a protein now called dystrophin is the cause of the disorder. Much investigative work determined that dystrophin is involved in the release of calcium from the sarcoplasmic reticulum in muscle fibers. The lack of dystrophin causes calcium to leak into the cell, which promotes the action of an enzyme that dissolves muscle fibers. When the body attempts to repair the tissue, fibrous tissue forms, and this cuts off the blood supply so that more and more cells die.

A test is now available to detect carriers of Duchenne muscular dystrophy. Also, various treatments are being attempted. Immature muscle cells can be injected into muscles, and for every 100,000 cells injected, dystrophin production occurs in 30–40% of muscle fibers. The allele for dystrophin has been inserted into the thigh muscle cells of mice, and about 1% of these cells then produced dystrophin.

Hemophilia

About one in 10,000 males is a hemophiliac. There are two common types of hemophilia: hemophilia A is due to the absence or minimal presence of a clotting factor known as factor IX, and hemophilia B is due to the absence of clotting factor VIII. Hemophilia is called the bleeder's disease

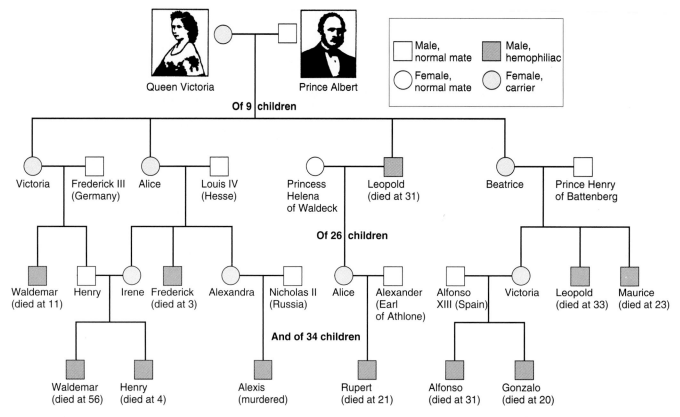

Figure 20.16   A simplified pedigree showing the X-linked inheritance of hemophilia in European royal families.
Because Queen Victoria was a carrier, each of her sons had a 50% chance of having the disease, and each of her daughters had a 50% chance of being a carrier. This pedigree shows only the affected individuals. Many others are unaffected, such as the members of the present British royal family.

because the affected person's blood does not clot or clots very slowly. Although hemophiliacs bleed externally after an injury, they also bleed internally, particularly around joints. Hemorrhages can be stopped with transfusions of fresh blood (or plasma) or concentrates of the clotting protein. Also, factor VIII is now available as a biotechnology product.

At the turn of the century, hemophilia was prevalent among the royal families of Europe, and all of the affected males could trace their ancestry to Queen Victoria of England. Figure 20.16 shows that of Queen Victoria's 26 grandchildren, four grandsons had hemophilia and four granddaughters were carriers. Because none of Queen Victoria's forebears or relatives were affected, it seems that the faulty allele she carried arose by mutation either in Victoria or in one of her parents. Her carrier daughters Alice and Beatrice introduced the allele into the ruling houses of Russia and Spain, respectively. Alexis, the last heir to the Russian throne before the Russian Revolution, was a hemophiliac. There are no hemophiliacs in the present British royal family because Victoria's eldest son, King Edward VII, did not receive the allele and therefore could not pass it on to any of his descendants.

Practice Problems 4*

1. Both the mother and the father of a male hemophiliac appear to be normal. From whom did the son inherit the allele for hemophilia? What are the genotypes of the mother, the father, and the son?

2. A woman is color blind. What are the chances that her sons will be color blind? If she is married to a man with normal vision, what are the chances that her daughters will be color blind? Will be carriers?

3. Both the husband and wife have normal vision. The wife gives birth to a color-blind daughter. What can you deduce about the girl's parentage?

*Answers to Practice Problems appear in Appendix A.

Certain traits that have nothing to do with the gender of the individual are controlled by genes on the X chromosomes. Males have only one X chromosome, and therefore X-linked recessive alleles are expressed.

Bioethical Focus

Genetic Profiling

Your genetic profile is expected to tell physicians which diseases you are likely to develop. Knowledge of your genes might indicate your susceptibility to various types of cancer, for example. This information could be used to develop a prevention program, including the avoidance of environmental influences associated with the disease. No doubt, you would be less inclined to smoke if you knew your genes make it almost inevitable that smoking will give you lung cancer.

People worry, however, that their genetic profile could be used against them. Perhaps employers will not hire, or insurance companies will not insure, those who have a propensity for a particular diseases. About 25 states have passed laws prohibiting genetic discrimination by health insurers, and 11 have passed laws prohibiting genetic discrimination by employers. Is such legislation enough to allay our fears of discrimination?

On the other hand, employers may fear that one day they will be required to provide an environment specific to every employee's need to prevent future illness. Would you approve of this, or should individuals be required to leave an area or job that exposes them to an environmental influence that could be detrimental to their health?

People's medical records are usually considered private. But if scientists could match genetic profiles to environmental conditions that bring on illnesses, they could come up with better prevention guidelines for the next generation. Should genetic profiles and health records become public information under these circumstances? It would particularly help in the study of complex diseases like cardiovascular disorders, noninsulin-dependent diabetes, and juvenile rheumatoid arthritis.

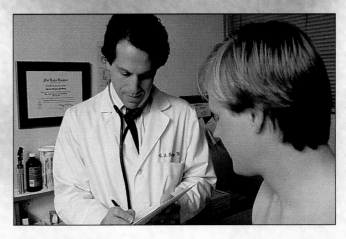

Figure 20A   Genetic profiling.
Will genetic profiling become a standard part of routine medical examinations in the future?

Decide Your Opinion

1. Should people be encouraged or even required to have their genes analyzed so that they can develop programs to possibly prevent future illness?
2. Should employers be encouraged or required to provide an environment suitable to a person's genetic profile? Or should the individual avoid a work environment that could bring on an illness?
3. How can we balance individual rights with the public health benefit of matching genetic profiles to detrimental environments?

Looking at Both Sides   www.mhhe.com/biosci/genbio/maderhuman7/

Every bioethical issue has at least two sides. Even if you already have an opinion, it is important to explore the opposite opinion before finalizing your position. The Online Learning Center at www.mhhe.com/biosci/genbio/maderhuman7/ will help you fine-tune your initial opinion, explore both sides, and finalize your position. Either you will acquire new arguments for your original opinion or you may even change your opinion. Be sure to complete these activities in sequence:

Taking Sides   Decide your initial opinion by answering a series of questions. Then see if your opinion changes after completing the next two activities.

Further Debate   Read opposing articles that give you further information on this particular bioethical issue.

Explain Your Position   Answer another series of questions and then defend your original or changed opinion. You can e-mail your position to your instructor if he or she wishes.

Sex-Influenced Traits

Not all traits we associate with the gender of an individual are sex-linked traits. Some are simply **sex-influenced traits;** that is, the phenotype is determined by autosomal genes that are expressed differently in males and females.

It is possible that the sex hormones determine whether these genes are expressed or not. Pattern baldness (Fig. 20.17) is thought to be influenced by the male sex hormone testosterone because males who take the hormone to increase masculinity begin to lose their hair. A more detailed explanation has been suggested by some investigators. It has been reasoned that due to the effect of hormones, men require only one allele for baldness in order for the condition to appear, whereas women require two alleles. In other words, the allele for baldness acts as a dominant in men but as a recessive in women. This means that men who have a bald father and a mother with a normal hairline have a 50% chance at best and a 100% chance at worst of going bald. Women who have a bald father and a mother with a normal hairline have no chance at best and a 50% chance at worst of going bald.

Another sex-influenced trait of interest is the length of the index finger. In females, an index finger longer than the fourth finger (ring finger) seems to be dominant. In males, an index finger longer than the fourth finger seems to be recessive.

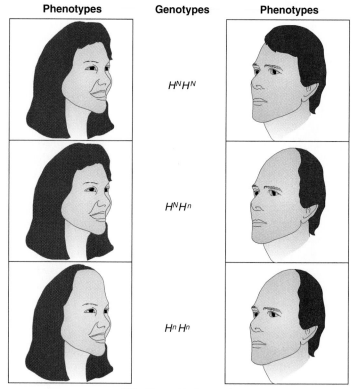

Phenotypes	Genotypes	Phenotypes

$H^N H^N$

$H^N H^n$

$H^n H^n$

H^N = Normal hair growth
H^n = Pattern baldness

Figure 20.17 Pattern baldness, a sex-influenced characteristic.
Due to hormonal influences, the presence of only one allele for baldness causes the condition in men, whereas the condition does not occur in women unless they possess both alleles for baldness.

Summarizing the Concepts

20.1 Genotype and Phenotype
It is customary to use letters to represent the genotypes of individuals. Homozygous dominant (two capital letters) and heterozygous (a capital letter and a lowercase letter) exhibit the dominant phenotype. Homozygous recessive (two lowercase letters) exhibits the recessive phenotype.

20.2 Dominant/Recessive Traits
Whereas the individual has two alleles (copies of a gene) for each trait, the gametes have only one allele for each trait. A Punnett square can help determine the phenotype ratio among offspring because all possible sperm types of a particular male are given an equal chance to fertilize all possible egg types of a particular female. When a heterozygous individual reproduces with another heterozygote, there is a 75% chance the child will have the dominant phenotype and a 25% chance the child will have the recessive phenotype. When a heterozygous individual reproduces with a pure recessive, the offspring has a 50% chance of having either phenotype.

A pedigree chart shows the phenotype of family members for a particular condition for several generations. The pattern of inheritance often reveals whether the condition is dominant, recessive, or sex-linked recessive.

20.3 Beyond Simple Inheritance Patterns
Polygenic traits are controlled by more than one pair of alleles, and the individual inherits each allelic pair. Skin color illustrates polygenic inheritance. When a trait is controlled by multiple alleles, each person has one pair of all possible alleles. ABO blood types illustrate inheritance by multiple alleles. Determination of inheritance is sometimes complicated by codominance (e.g., blood type) and by incomplete dominance (e.g., hair curl among Caucasians).

20.4 Sex-Linked Traits
Sex-linked genes are located on the sex chromosomes, and most of these alleles are X-linked because they are on the X chromosome. Since the Y chromosome is blank for X-linked alleles, males are more apt to exhibit the recessive phenotype, which they inherited from their mother. If a female does have the phenotype, then her father must also have it, and her mother must be a carrier or have the trait. Usually, X-linked traits skip a generation and go from maternal grandfather to grandson by way of a carrier daughter.

Sex-influenced traits are controlled by an allele that acts as if it were dominant in one sex but recessive in the other. Most likely, the activity of such genes is influenced by the sex hormones.

Studying the Concepts

1. What is the difference between the genotype and the phenotype of an individual? For which type phenotype— dominant or recessive—are there two possible genotypes? 404
2. Why do the gametes have only one allele for each trait? 405
3. What is the chance of producing a child with the dominant phenotype from each of these crosses? 406

$AA \times AA$

$Aa \times AA$

$Aa \times Aa$

$aa \times aa$

4. From which of the crosses in question 3 can there be an offspring with the recessive phenotype? Explain. 406
5. Give examples of genetic disorders caused by inheritance of two recessive alleles and disorders caused by inheritance of a single dominant allele. 407–8
6. List ways to recognize an autosomal recessive genetic disorder and an autosomal dominant genetic disorder when examining a pedigree chart. 409
7. Give examples of these patterns of inheritance: polygenic inheritance, multiple alleles, and incomplete dominance. 411–13
8. Give examples of genetic disorders caused by the patterns of inheritance in question 7. 411–13
9. If a trait is on an autosome, how would you designate a homozygous dominant female? If a trait is on an X chromosome, how would you designate a homozygous dominant female? 414
10. List ways to recognize an X-linked recessive disorder when examining a pedigree chart. Why do males exhibit such disorders more often than females? 415
11. Give examples of genetic disorders caused by X-linked recessive alleles. 415–16

Testing Your Knowledge of the Concepts

In questions 1–4, match the answers to the genetics problem.
a. 0%
b. 25%
c. 50%
d. 100%

_____ 1. A woman heterozygous for polydactyly (6 fingers and toes) reproduces with a man without the condition. Polydactyly is dominant; what are the chances a child will have the condition?

_____ 2. Parents who do not have Tay-Sachs disease produce a child who has Tay-Sachs disease (recessive). What are the chances the next child will have Tay-Sachs?

_____ 3. One parent has sickle-cell disease (incompletely dominant), and the other is perfectly normal. What are the chances a child will have sickle-cell trait?

_____ 4. Black hair is dominant over blond hair. What are the chances a homozygous black-haired woman who reproduces with a blond-haired man will have brown-haired children?

5. Identify each of these genetic disorders.
   a. Mucus in lungs and digestive tract is thick and viscous.
   b. Neurological impairment and psychomotor difficulties develop early.
   c. Benign tumors occur under the skin or deeper.
   d. Minor disturbances in balance and coordination develop in middle age and progress toward severe mental disturbances.

In questions 6–8, indicate whether the statement is true (T) or false (F).

_____ 6. In a pedigree chart, it is observed that the trait skips a generation and passes from grandfather to grandson. The trait must be sex-linked dominant.

_____ 7. For a girl to have hemophilia (X-linked recessive), both of her parents must have it.

_____ 8. The genotype of a female who has a color-blind father but a homozygous normal mother is $X^B X^b$.

In questions 9 and 10, fill in the blanks.

9. Mary has a widow's peak, and John has a straight hairline. This would be a description of their _____.

10. The genotype of a color-blind man with attached earlobes (recessive) is _____.

11. Determine if the characteristic possessed by the shaded males and females in the pedigree chart below is autosomal dominant, autosomal recessive, or X-linked recessive.

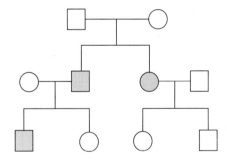

Understanding Key Terms

allele 404
carrier 409
codominance 413
dominant allele 404
genomic imprinting 408
genotype 404
heterozygous 404
homozygous dominant 404
homozygous recessive 404
incomplete dominance 413

multiple alleles 412
phenotype 404
polygenic inheritance 411
Punnett square 406
recessive allele 404
sex chromosome 414
sex-influenced trait 418
sex-linked 414
trait 403
X-linked 414

e-Learning Connection   **www.mhhe.com/biosci/genbio/maderhuman7/**

20.1 Genotype and Phenotype

Introduction *Essential Study Partner*

20.2 Dominant/Recessive Traits

Monohybrid Cross *Essential Study Partner*

20.3 Beyond Simple Inheritance Patterns

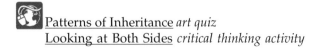
Correlation Between Sickle Cell Allele and Malaria *art quiz*

20.4 Sex-Linked Traits

Sex Chromosomes *Essential Study Partner*

Patterns of Inheritance *art quiz*
Looking at Both Sides *critical thinking activity*

Chapter Summary

Heredity in Families *simulation*

Key Term Flashcards *vocabulary quiz*
Chapter Quiz *objective quiz covering all chapter concepts*

Chapter 21

DNA and Biotechnology

Chapter Concepts

Figure 21.1   Cindy Cutshall.
Cindy was sick and couldn't play baseball until she underwent gene therapy. She doesn't like to talk about the procedure. She says, "It's too weird."

Cindy Cutshall was born with a rare immune disorder called SCID (severe combined immunodeficiency). Her T and B lymphocytes were ineffective because they couldn't produce a critical enzyme known as ADA (adenosine deaminase). She was in and out of the hospital with infections, and her parents feared she would die young.

And so, in 1993, the Cutshalls agreed to let Cindy undergo gene therapy, sponsored by the National Institutes of Health (NIH), the top biomedical funding agency in the United States. Gene therapy is based on the idea that it is possible to supply a patient with a missing or defective gene. NIH researchers took white blood stem cells from Cindy, added the gene that specifies ADA, and injected the improved cells back into her. Stem cells for white blood cells were chosen because they reside in the bone marrow where they continually produce more white blood cells. Combined with drug treatment, the therapy apparently worked. Today, Cindy is well, and her white blood cells contain functioning ADA genes.

Like Cindy, over 600 patients have now received some form of gene therapy in treatment for AIDS, cancer, cystic fibrosis, and other diseases. In many cases, the patient's cells have not received an adequate number of working genes to make a difference. But once efficient methods of delivery have been developed, gene therapy may one day improve the lives of millions.

The **gene** that Cindy received was a specific piece of **DNA (deoxyribonucleic acid)** taken from a human chromosome. This piece of DNA is responsible for the production of ADA, the enzyme that Cindy needed in her cells. In this chapter, we will learn that any particular gene is just one section of one of the chromosomes, and that a gene codes for a specific sequence of amino acids in a protein. Gene therapy is only one aspect of the burgeoning field of biotechnology.

21.1   DNA and RNA Structure and Function

DNA, the genetic material, is found principally in the chromosomes, which are located in the nucleus of a cell. (Small amounts of DNA are also found in mitochondria.)

Figure 21.2 shows the levels of organization of a chromosome. When a cell is about to divide, the chromosomes are very compact structures in which DNA is coiled and condensed. In

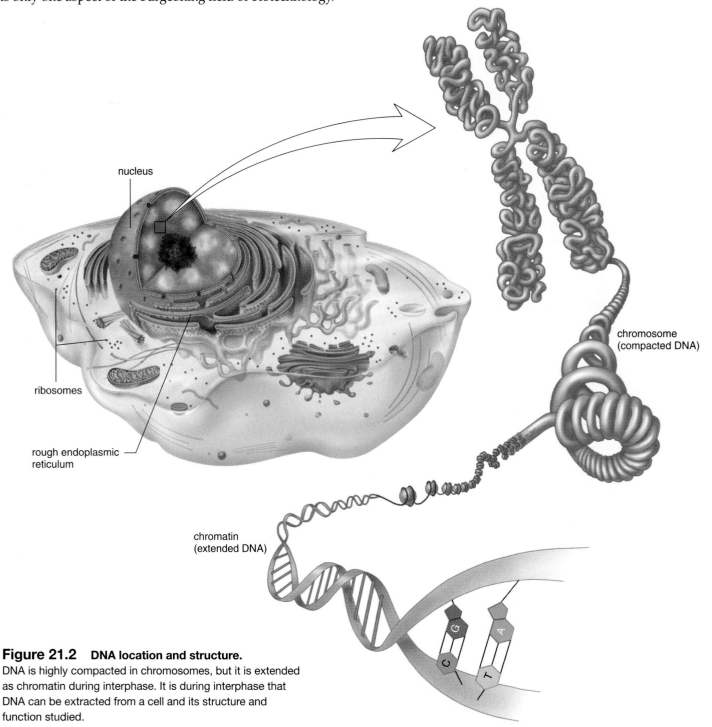

nucleus

ribosomes

rough endoplasmic reticulum

chromosome (compacted DNA)

chromatin (extended DNA)

Figure 21.2   DNA location and structure.
DNA is highly compacted in chromosomes, but it is extended as chromatin during interphase. It is during interphase that DNA can be extracted from a cell and its structure and function studied.

between cell divisions, chromosomes uncoil and decondense. At that time, the chromosomes exist as a mass of very long fine threads known as chromatin. Chromosomes, but not chromatin, can be seen with the light microscope. Within chromatin, the DNA molecule is a double-stranded helix, and the strands are held together by hydrogen bonding between the bases as discussed in chapter 2. The structure of DNA permits replication of the molecule in between cell divisions.

DNA Structure and Replication

DNA is a type of nucleic acid, and like all nucleic acids, it is formed by the sequential joining of molecules called nucleotides. Nucleotides, in turn, are composed of three smaller molecules—a phosphate, a sugar, and a nitrogen containing base. The sugar in DNA is deoxyribose, which accounts for the name deoxyribonucleic acid. The four nucleotides that make up the DNA molecule have the following bases: **adenine (A), thymine (T), cytosine (C),** and **guanine (G)** (Fig. 21.3).

When nucleotides join, the sugar and the phosphate molecules become the backbone of a strand, and the bases project to the side. In DNA, there are two strands of nucleotides; consequently, DNA is double stranded. Weak hydrogen bonds between the bases hold the strands together. Each base is bonded to another particular base, called **complementary base pairing**—adenine (A) is always paired with thymine (T), and cytosine (C) is always paired with guanine (G), and

vice versa. The dotted lines in Figure 21.4a represent the hydrogen bonds between the bases. The structure of DNA is said to resemble a ladder; the sugar-phosphate backbone makes up the sides of the ladder, and the paired bases are the rungs. The ladder structure of DNA twists to form a spiral staircase called a double helix (Fig. 21.4b).

DNA is a double helix with a sugar-phosphate backbone on the outside and paired bases on the inside. Complementary base pairing occurs: adenine (A) pairs with thymine (T), and cytosine (C) pairs with guanine (G).

Replication of DNA

Between cell divisions, chromosomes must duplicate before cell division can occur again. **Replication** of DNA occurs as a part of chromosome duplication. Replication is facilitated by the structure of DNA.

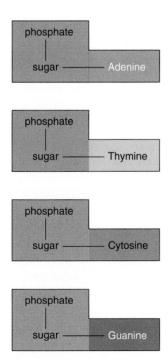

Figure 21.3   DNA nucleotides.
DNA contains four different kinds of nucleotides, molecules that in turn contain a phosphate, a sugar, and a base. The base of a DNA nucleotide can be either adenine (A), thymine (T), cytosine (C), or guanine (G).

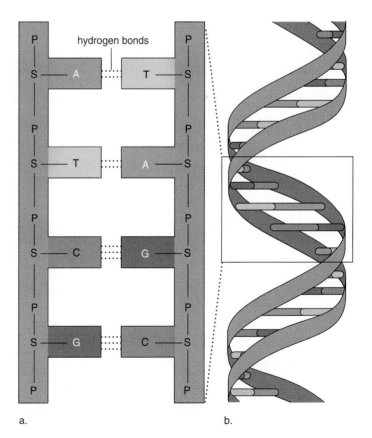

a.   b.

Figure 21.4   DNA is double stranded.
a. When the DNA nucleotides join, they form two strands so that the structure of DNA resembles a ladder. The phosphate (P) and sugar (S) molecules make up the sides of the ladder, and the bases make up the rungs of the ladder. Each base is weakly bonded (dotted lines) to its complementary base (T is bonded to A and vice versa; G is bonded to C and vice versa). **b.** The DNA ladder twists to form a double helix. Each chromatid of a duplicated chromosome contains one double helix.

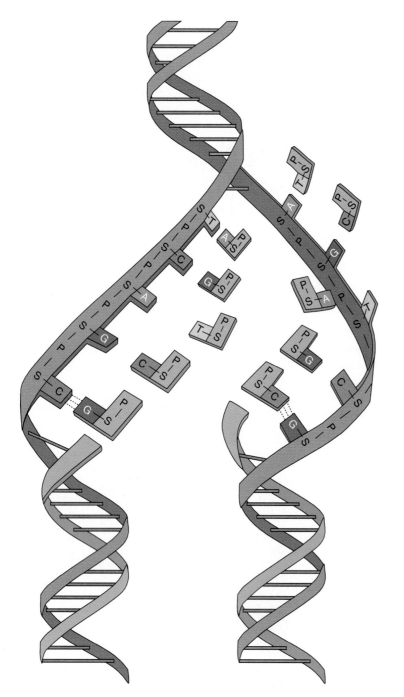

Region of parental DNA helix. (Both backbones are shown in dark color.)

Region of replication (simplified). Parental DNA helix is unwound and unzipped. New nucleotides are pairing with those in parental strands.

Region of completed replication. Each double helix is composed of an old parental strand (dark) and a new daughter strand (light).

Figure 21.5   **DNA replication.**
When replication is finished, two complete DNA double-helix molecules are present. Note that DNA replication is semiconservative because each new double helix is composed of an old parental strand and a new daughter strand.

Replication requires these steps:
1. The hydrogen bonds between the two strands of DNA break as enzymes unwind and "unzip" the molecule.
2. New nucleotides, always present in the nucleus, fit into place beside each old (parental) strand by the process of complementary base pairing.
3. These new nucleotides become joined by an enzyme called **DNA polymerase.**
4. When the process is finished, two complete DNA molecules are present, identical to each other and to the original molecule.

Each new double helix is composed of an old (parental) and a new (daughter) strand (Fig. 21.5). Because each strand of DNA serves as a **template,** or mold, for the production of a complementary strand, DNA replication is called semiconservative. Rarely, a replication error occurs—the sequence of the complementary bases is not correct. But if an error does occur, the cell has repair enzymes that usually fix it. A replication error that persists is a **mutation,** a permanent change in a gene that can cause a change in the phenotype.

DNA replication results in two double helixes. DNA unwinds and unzips, and new (daughter) strands form, each complementary to an old (parental) strand.

The Structure and Function of RNA

RNA (ribonucleic acid) is made up of nucleotides containing the sugar ribose. This sugar accounts for the scientific name of this polynucleotide. The four nucleotides that make up the RNA molecule have the following bases: adenine (A), **uracil (U),** cytosine (C), and guanine (G) (Fig. 21.6a). Notice that in RNA, the base uracil replaces the base thymine.

RNA, unlike DNA, is single stranded (Fig. 21.6b), but the single RNA strand sometimes doubles back on itself. Similarities and differences between these two nucleic acid molecules are listed in Table 21.1.

In general, RNA is a helper to DNA, allowing protein synthesis to occur according to the genetic information that DNA provides. There are three types of RNA, each with a specific function in protein synthesis.

Ribosomal RNA

Ribosomal RNA (rRNA) is produced in the nucleolus of a nucleus where a portion of DNA serves as a template for its formation. Ribosomal RNA joins with proteins made in the cytoplasm to form the subunits of ribosomes. The subunits leave the nucleus and come together in the cytoplasm when protein synthesis is about to begin. Proteins are synthesized at the ribosomes, which in low-power electron micrographs look like granules arranged along the endoplasmic reticulum, a system of tubules and saccules within the cytoplasm. Some ribosomes appear free in the cytoplasm or in clusters called polyribosomes.

Messenger RNA

Messenger RNA (mRNA) is produced in the nucleus where DNA serves as a template for its formation. This type of RNA carries genetic information from DNA to the ribosomes in the cytoplasm where protein synthesis occurs. Messenger RNA is a linear molecule.

Transfer RNA

Transfer RNA (tRNA) is produced in the nucleus, and a portion of DNA also serves as a template for its production. Appropriate to its name, tRNA transfers amino acids to the ribosomes, where the amino acids are joined, forming a protein. There are 20 different types of amino acids in proteins; therefore, there has to be at least this number of tRNAs functioning in the cell. Each type of tRNA carries only one type of amino acid.

RNA is a polynucleotide that functions during protein synthesis in various ways. There are three types of RNA, each with a specific role.

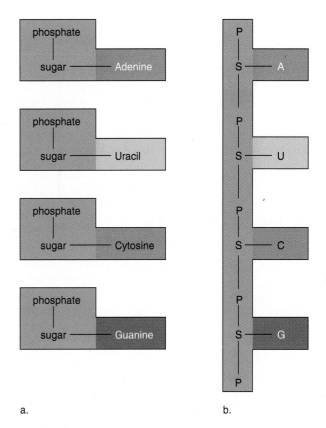

a.

b.

Figure 21.6 **RNA structure.**
a. The four nucleotides in RNA each have a phosphate (P) molecule; the sugar (S) is ribose; and the base may be either adenine (A), uracil (U), cytosine (C), or guanine (G). **b.** RNA is single stranded. The sugar and phosphate molecules join to form a single backbone, and the bases project to the side.

Table 21.1	DNA-RNA Similarities and Differences
DNA-RNA Similarities	
Both are nucleic acids	
Both are composed of nucleotides	
Both have a sugar-phosphate backbone	
Both have four different type bases	
DNA-RNA Differences	
DNA	RNA
Found in nucleus	Found in nucleus and cytoplasm
The genetic material	Helper to DNA
Sugar is deoxyribose	Sugar is ribose
Bases are A, T, C, G	Bases are A, U, C, G
Double stranded	Single stranded
Is transcribed (to give mRNA)	Is translated (to give proteins)

21.2   Gene Expression

Genetic information is inherited from one's parents in the form of DNA. As we shall see, DNA provides the cell with a blueprint for the synthesis of proteins in a cell. DNA resides in the nucleus, and protein synthesis occurs in the cytoplasm. Messenger RNA (mRNA) carries a copy of DNA's blueprint into the cytoplasm, and the other RNA molecules we just discussed are involved in bringing about protein synthesis.

Before discussing the mechanics of protein synthesis, let's review the structure of proteins.

Structure and Function of Proteins

Proteins are composed of individual units called amino acids. Twenty different amino acids are commonly found in proteins, which are synthesized at the ribosomes in the cytoplasm of cells. Proteins differ because the number and order of their amino acids differ. Notice in Figure 21.7 that the sequence of amino acids in a portion of one protein can differ completely from the sequence of amino acids in a portion of another protein. The unique sequence of amino acids in a protein leads to its particular shape, and the shape of a protein helps determine its function.

Proteins are found in all parts of the body; some are structural proteins and some are enzymes. The protein hemoglobin is responsible for the red color of red blood cells. Albumin and globulins (antibodies) are well-known plasma proteins. Muscle cells contain the proteins actin and myosin, which give them substance and ability to contract.

Enzymes are organic catalysts that speed reactions in cells. The reactions in cells form metabolic or chemical pathways. A pathway can be represented as follows:

$$E_A \quad E_B \quad E_C \quad E_D$$
$$A \rightarrow B \rightarrow C \rightarrow D \rightarrow E$$

In this pathway, the letters are molecules, and the notations over the arrows are enzymes: molecule A becomes molecule B, and enzyme E_A speeds the reaction; molecule B becomes molecule C, and enzyme E_B speeds the reaction, and so forth. Notice that each reaction in the pathway has its own enzyme: enzyme E_A can only convert A to B, enzyme E_B can only convert B to C, and so forth. For this reason, enzymes are said to be *specific*.

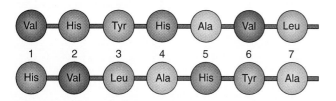

Figure 21.7   Structure of proteins.
Proteins differ by the sequence of their amino acids; the top row shows one possible sequence, and the bottom row shows another possible sequence in a portion of a protein.

The DNA Code

The discovery that metabolic disorders run in certain families suggested to early investigators that genes control the the metabolism of a cell. The mental retardation seen in persons with phenylketonuria (PKU) is caused by an inability to convert phenylalanine to tyrosine. The lack of pigment in albinos is caused by an inability to convert tyrosine to melanin. These and other human conditions are caused by the deficiency of particular enzymes.

Laboratory experiments performed over many years finally showed that a gene is a segment of DNA that codes for a protein. If a person inherits a faulty gene, the sequence of bases in DNA is abnormal, and a particular enzyme is not constructed normally. We now know that DNA contains a triplet code—every three bases (a *triplet*) represents (*codes for*) for one amino acid (Table 21.2).

The genetic code is essentially universal. The same codons stand for the same amino acids in most organisms, from bacteria to humans. This illustrates the remarkable biochemical unity of living things and suggests that all living things have a common evolutionary ancestor.

Transcription

Gene expression requires two steps, called transcription and translation. It is helpful to remember that in other applications the word transcription means making an exact copy of written information, while translation means to put this information into another language.

Table 21.2	**Some DNA Codes and RNA Codons**		
DNA Code	**mRNA Codon**	**tRNA Anticodon**	**Amino Acid**
TTT	AAA	UUU	Lysine
TGG	ACC	UGG	Threonine
CCG	GGC	CCG	Glycine
CAT	GUA	CAU	Valine
CTC	GAG	CUC	Glutamate
GAG	CUC	GAG	Leucine
AGA	UCU	AGA	Serine
ACT	UGA	ACU	Stop
GCT	CGA	GCU	Arginine
TAA	AUU	UAA	Isoleucine
GGA	CCU	GGA	Proline
CGA	GCU	CGA	Alanine
TTG	AAC	UUG	Asparagine

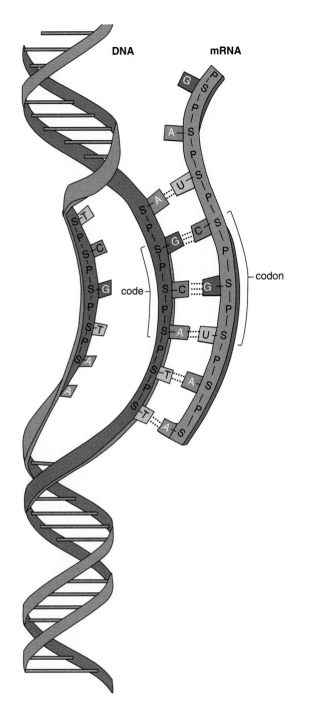

During **transcription** of DNA, a strand of mRNA forms that is complementary to a portion of DNA. Preparatory to transcription, a segment of the DNA helix unwinds and unzips. Then mRNA is produced as complementary RNA nucleotides pair with the nucleotides of one DNA strand: G (in RNA) pairs with C (in DNA), U pairs with A, and A pairs with T. An enzyme called **RNA polymerase** joins the nucleotides together, and the RNA that results has a sequence of bases complementary to those of a gene. While DNA contains a **triplet code** in which every three bases stand for one amino acid, mRNA contains **codons,** each of which is made up of three bases that also stand for the same amino acid (see Table 21.2).

In Figure 21.8, mRNA has formed and is ready to be processed as discussed next.

Processing of mRNA

Most genes in humans are interrupted by segments of DNA that are not part of the gene. These portions are called introns because they are intragene segments. The other portions of the gene are called exons because they are ultimately expressed. Only exons result in a protein product.

When DNA is transcribed, the mRNA contains bases that are complementary to both exons and introns, but before the mRNA exits the nucleus, it is *processed*. During processing, the introns are removed, and the exons are joined to form an mRNA molecule consisting of continuous exons. There has been much speculation about the role of introns. It is possible that they allow crossing-over within a gene during meiosis. It is also possible that introns divide a gene into domains that can be joined in different combinations to yield mRNAs that result in different protein products (Fig. 21.9).

In human cells, processing occurs in the nucleus. After the mRNA strand is processed, it passes from the cell nucleus into the cytoplasm. There it becomes associated with the ribosomes.

Figure 21.8   Transcription.
During transcription, mRNA is formed when nucleotides complementary to the sequence of bases in a portion of DNA (i.e., a gene) join. Note that DNA contains a triplet code (sequences of three bases) and that mRNA contains codons (sequences of three complementary bases). *Top:* mRNA transcript is ready to move into the cytoplasm. *Middle:* transcription has occurred, and mRNA nucleotides have joined together. *Bottom:* rest of DNA molecule.

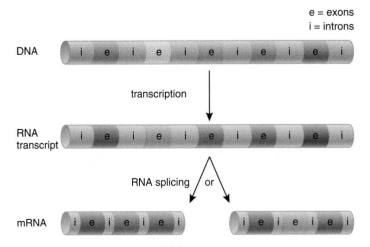

Figure 21.9   Function of introns.
Introns allow alternative splicing and therefore the production of different versions of mature mRNA from the same gene.

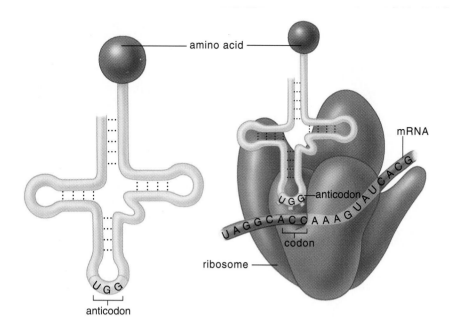

a. tRNA–amino acid   b. tRNA–amino acid at ribosome

Figure 21.10   Anticodon-codon base pairing.
a. tRNA molecules have an amino acid attached to one end and an anticodon at the other end. **b.** The anticodon of a tRNA molecule is complementary to a codon. The pairing between codon and anticodon at a ribosome ensures that the sequence of amino acids in a polypeptide is that directed originally by DNA. If the codon is ACC, the anticodon is UGG, and the amino acid is threonine.

Translation

Translation is the synthesis of a polypeptide under the direction of an mRNA molecule. (Some proteins, like insulin, consist of one polypeptide, and others, like hemoglobin, require more than one polypeptide.) During translation, transfer RNA (tRNA) molecules bring amino acids to the ribosomes. Usually, there is more than one tRNA molecule for each of the 20 amino acids found in proteins. The amino acid binds to one end of the molecule (Fig. 21.10a). Therefore, the entire complex is designated as tRNA–amino acid.

At the other end of each tRNA, there is a specific **anticodon,** a group of three bases that is complementary to an mRNA codon. The tRNA molecules come to the ribosome (Fig. 21.10b) where each anticodon pairs with a codon in the order directed by the sequence of the mRNA codons. In this way, the order of codons in mRNA brings about a particular order of amino acids in a protein.

If the codon sequence is ACC, GUA, and AAA, what will be the sequence of amino acids in a portion of the polypeptide? Inspection of Table 21.2 (p. 426) allows us to determine this:

Codon	Anticodon	Amino Acid
ACC	UGG	Threonine
GUA	CAU	Valine
AAA	UUU	Lysine

Polypeptide synthesis requires three steps: initiation, elongation, and termination.

1. During *initiation,* mRNA binds to the smaller of the two ribosomal subunits; then the larger subunit joins the smaller one.
2. During *elongation,* the polypeptide lengthens one amino acid at a time (Fig. 21.11). A ribosome is large enough to accommodate two tRNA molecules: the incoming tRNA molecule and the outgoing tRNA molecule. The incoming tRNA–amino acid complex receives the peptide from the outgoing tRNA. The ribosome then moves laterally so that the next mRNA codon is available to receive an incoming tRNA–amino acid complex. In this manner, the peptide grows, and the linear structure of a polypeptide comes about. (The particular shape of a polypeptide comes about later.)
3. Then *termination* of synthesis occurs at a codon that means stop and does not code for an amino acid. The ribosome dissociates into its two subunits and falls off the mRNA molecule.

As soon as the initial portion of mRNA has been translated by one ribosome, and the ribosome has begun to move down the mRNA, another ribosome attaches to the mRNA. Therefore, several ribosomes, collectively called a **polyribosome,** can move along one mRNA at a time. And several polypeptides of the same type can be synthesized using one mRNA molecule (Fig. 21.12).

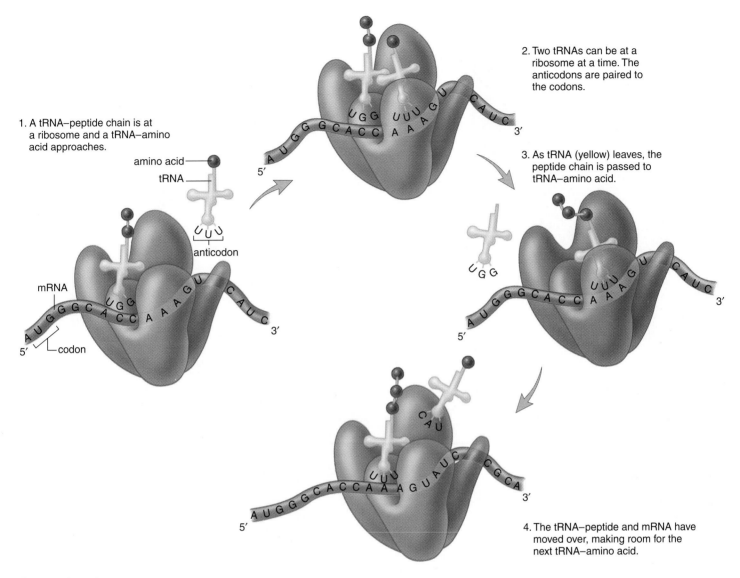

1. A tRNA–peptide chain is at a ribosome and a tRNA–amino acid approaches.

amino acid
tRNA
mRNA
codon
anticodon

2. Two tRNAs can be at a ribosome at a time. The anticodons are paired to the codons.

3. As tRNA (yellow) leaves, the peptide chain is passed to tRNA–amino acid.

4. The tRNA–peptide and mRNA have moved over, making room for the next tRNA–amino acid.

Figure 21.11   Translation.
Transfer RNA (tRNA)–amino acid molecules arrive at the ribosome, and the sequence of messenger RNA (mRNA) codons dictates the order in which amino acids become incorporated into a polypeptide.

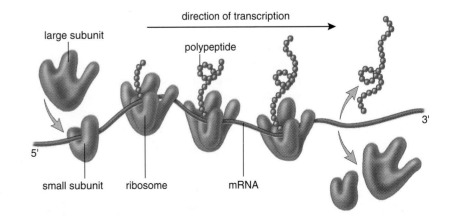

direction of transcription

large subunit

polypeptide

small subunit   ribosome   mRNA

Figure 21.12   Polyribosome structure.
Several ribosomes, collectively called a polyribosome, move along a messenger RNA (mRNA) molecule at one time. They function independently of one another; therefore, several polypeptides can be made at the same time.

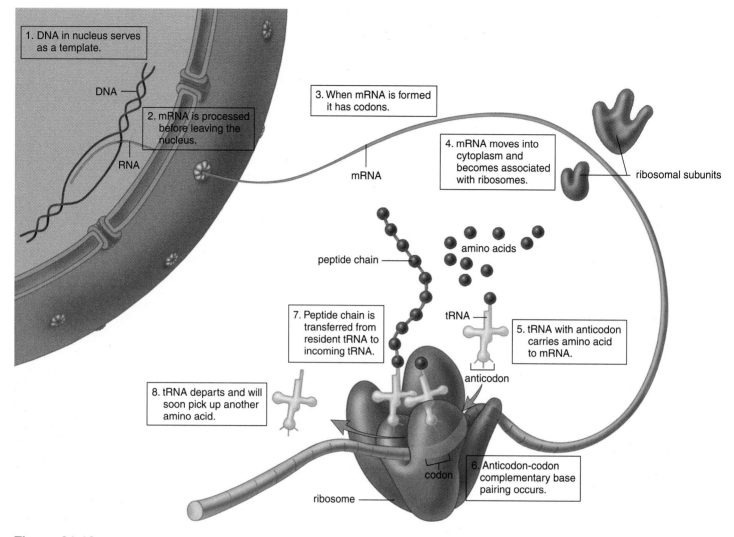

Figure 21.13   Summary of gene expression.

Review of Gene Expression

The following list, along with Table 21.3 and Figure 21.13, provides a brief summary of the events involved in gene expression that results in a protein product.

1. DNA in the nucleus contains a triplet code. Each group of three bases stands for a specific amino acid.
2. During transcription, a segment of a DNA strand serves as a template for the formation of mRNA. The bases in mRNA are complementary to those in DNA; every three bases is a codon for a certain amino acid.
3. Messenger RNA (mRNA) is processed before it leaves the nucleus, during which time the introns are removed.
4. Messenger RNA (mRNA) carries a sequence of codons to the ribosomes, which are composed of rRNA and proteins.
5. Transfer RNA (tRNA) molecules, each of which is bonded to a particular amino acid, have anticodons that pair complementarily to the codons in mRNA.

Table 21.3	Participants in Gene Expression	
Molecule	**Special Significance**	**Definition**
DNA	Triplet code	Sequence of bases in threes
mRNA	Codon	Complementary sequence of bases in threes
tRNA	Anticodon	Sequence of three bases complementary to codon
Amino acids	Building blocks	Transported to ribosomes by tRNAs
Protein	Enzymes and structural proteins	Amino acids joined in a predetermined order

6. During translation, tRNA molecules and their attached amino acids arrive at the ribosomes, and the linear sequence of codons of the mRNA determines the order in which the amino acids become incorporated into a protein.

The Regulation of Gene Expression

All cells receive a copy of all genes; however, cells differ as to which genes are being actively expressed. Muscle cells, for example, have a different set of genes that are turned on in the nucleus and different proteins that are active in the cytoplasm than do nerve cells. In eukaryotic cells, a variety of mechanisms regulates gene expression, from transcription to protein activity. These mechanisms can be grouped under four primary levels of control, two that pertain to the nucleus and two that pertain to the cytoplasm.

1. *Transcriptional control:* In the nucleus, a number of mechanisms regulate which genes are transcribed and/or the rate at which transcription of the genes occurs. These include the organization of chromatin and the use of transcription factors that initiate transcription, the first step in gene expression.
2. *Posttranscriptional control:* Posttranscriptional control occurs in the nucleus after DNA is transcribed and mRNA is formed. How mRNA is processed before it leaves the nucleus and also how fast mature mRNA leaves the nucleus can affect the amount of gene expression.
3. *Translational control:* Translational control occurs in the cytoplasm after mRNA leaves the nucleus and before there is a protein product. The life expectancy of mRNA molecules (how long they exist in the cytoplasm) can vary, as can their ability to bind ribosomes. It is also possible that some mRNAs may need additional changes before they are translated at all.
4. *Posttranslational control:* Posttranslational control, which also occurs in the cytoplasm, occurs after protein synthesis. The polypeptide product may have to undergo additional changes before it is biologically functional. Also, a functional enzyme is subject to feedback control—the binding of an enzyme's product can change its shape so that it is no longer able to carry out its reaction.

Activated Chromatin

For a gene to be transcribed in human cells, the chromosome in that region must first decondense. The chromosomes within the developing egg cells of many vertebrates are called lampbrush chromosomes because they have many loops that appear to be bristles (Fig. 21.14). Here mRNA is being synthesized in great quantity; then protein synthesis can be carried out, despite rapid cell division during development.

Transcription Factors

In human cells, **transcription factors** are DNA-binding proteins. Every cell contains many different types of transcription factors, and a specific combination is believed to regulate the activity of any particular gene. After the right combination of transcription factors binds to DNA, an RNA polymerase attaches to DNA and begins the process of transcription.

As cells mature, they become specialized. Specialization is determined by which genes are active, and therefore, perhaps, by which transcription factors are present in that cell. Signals received from inside and outside the cell could turn on or off genes that code for certain transcription factors. For example, the gene for fetal hemoglobin ordinarily gets turned off as a newborn matures—one possible treatment for sickle-cell disease is to turn this gene on again.

Regulation of gene expression occurs at four levels. The first level, transcriptional control, involves organization of chromatin and the use of transcription factors that bind to DNA.

Figure 21.14   Lampbrush chromosomes.
These chromosomes, seen in amphibian egg cells, give evidence that when mRNA is being synthesized, chromosomes decondense. Each chromosome has many loops extending from its axis (white on the micrograph). Many mRNA transcripts are being made off these DNA loops (red).

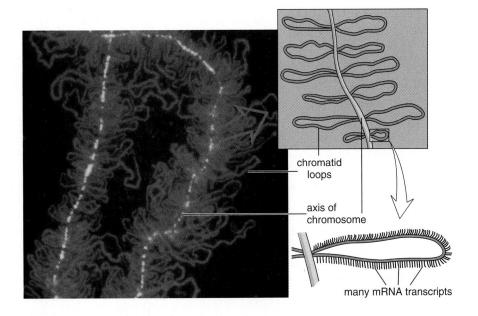

chromatid loops

axis of chromosome

many mRNA transcripts

21.3   Biotechnology

Genetic engineering is the use of technology to alter the genomes of viruses, bacteria, and other cells for medical or industrial purposes. **Biotechnology** includes genetic engineering and other techniques that make use of natural biological systems to produce a product or to achieve an end desired by human beings. Genetic engineering has the capability of altering the genotypes of unicellular organisms as well as the genotypes of plants and animals, including ourselves.

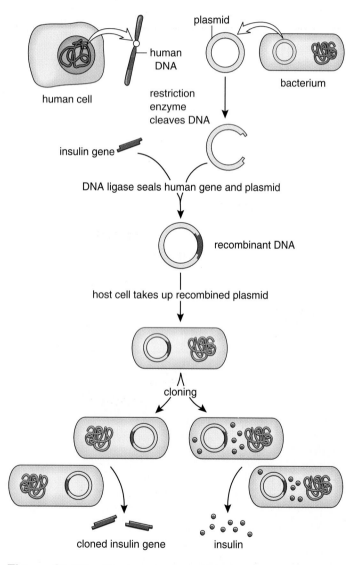

Figure 21.15   Cloning of a human gene.
Human DNA and plasmid DNA are cleaved by a specific type of restriction enzyme and spliced together by the enzyme DNA ligase. Gene cloning is achieved when a host cell takes up the recombinant plasmid, and as the cell reproduces, the plasmid is replicated. Multiple copies of the gene are now available to an investigator. If the insulin gene functions normally as expected, insulin may also be retrieved.

The Cloning of a Gene

The **cloning** of a gene produces many identical copies. Recombinant DNA technology is used when a very large quantity of the gene is required. The use of the polymerase chain reaction (PCR) creates fewer copies within a laboratory test tube.

Recombinant DNA Technology

Recombinant DNA (rDNA) contains DNA from two or more different sources (Fig. 21.15). To make rDNA, a technician often begins by selecting a **vector,** the means by which rDNA is introduced into a host cell. One common type of vector is a plasmid. **Plasmids** are small accessory rings of DNA from bacteria. The ring is not part of the bacterial chromosome and replicates on its own. Plasmids were discovered by investigators studying the reproduction of the intestinal bacterium *Escherichia coli* (*E. coli*).

Two enzymes are needed to introduce foreign DNA into vector DNA (Fig. 21.15). The first enzyme, called a **restriction enzyme,** cleaves the vector's DNA, and the second, called **DNA ligase,** seals foreign DNA into the opening created by the restriction enzyme.

Hundreds of restriction enzymes occur naturally in bacteria, where they stop viral reproduction by cutting up viral DNA. They are called restriction enzymes because they *restrict* the growth of viruses, but they also act as *molecular scissors* because each one cuts DNA at a specific cleavage site. For example, the restriction enzyme called *Eco*RI always cuts double-stranded DNA in this manner when it has this sequence of bases:

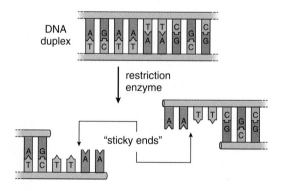

Notice there is now a gap into which a piece of foreign DNA can be placed if it ends in bases complementary to those exposed by the restriction enzyme. To assure this, it is only necessary to cleave the foreign DNA with the same type of restriction enzyme. The single-stranded, but complementary, ends of the two DNA molecules are called "sticky ends" because they can bind a piece of foreign DNA by complementary base pairing. Sticky ends facilitate the insertion of foreign DNA into vector DNA.

The second enzyme needed for preparation of rDNA, DNA ligase, is a cellular enzyme that seals any breaks in a

DNA molecule. Genetic engineers use this enzyme to seal the foreign piece of DNA into the vector. DNA splicing is now complete; an rDNA molecule has been prepared.

Bacterial cells take up recombinant plasmids, especially if they are treated to make them more permeable. Thereafter, if the inserted foreign gene is replicated and actively expressed, the investigator can recover either the cloned gene or a protein product (Fig. 21.15).

In order for gene expression to occur in a bacterium, the gene has to be accompanied by regulatory regions unique to bacteria. Also, the gene should not contain introns because bacteria don't have introns. However, it's possible to make a human gene that lacks introns. The enzyme called reverse transcriptase can be used to make a DNA copy of mRNA. This DNA molecule, called **complementary DNA** (cDNA), does not contain introns. Alternatively, it is possible to manufacture small genes in the laboratory. A machine called a DNA synthesizer joins together the correct sequence of nucleotides, and this resulting gene also lacks introns.

The Polymerase Chain Reaction

The **polymerase chain reaction (PCR)** can create millions of copies of a single gene, or any specific piece of DNA, in a test tube. PCR is very specific—the targeted DNA sequence can be less than one part in a million of the total DNA sample! This means that a single gene among all the human genes can be amplified (copied) using PCR.

PCR requires the use of DNA polymerase, the enzyme that carries out DNA replication, and a supply of nucleotides for the new DNA strands. PCR is a chain reaction because the targeted DNA is repeatedly replicated as long as the process continues. The colors in Figure 21.16 distinguish old from new DNA. Notice that with each replication cycle the amount of DNA doubles.

PCR has been in use for several years, and now almost every laboratory has automated PCR machines to carry out the procedure. Automation became possible after a temperature-insensitive (thermostable) DNA polymerase was extracted from the bacterium *Thermus aquaticus,* which lives in hot springs. The enzyme can withstand the high temperature used to separate double-stranded DNA; therefore, replication need not be interrupted by the need to add more enzyme.

Analyzing DNA Segments   Following PCR, DNA can be subjected to **DNA fingerprinting.** When the DNA of a person is treated with restriction enzymes, the result is a unique collection of different-sized fragments. During a process called gel electrophoresis, the fragments can be separated according to their charge/size ratios, and the result is a pattern of distinctive bands. If two DNA patterns match, there is a high probability that the DNA came from the same person. Because of PCR, DNA from a single sperm is enough to identify a suspected rapist. DNA was used to successfully identify a teenage murder victim from eight-year-old

remains. Skeletal DNA was compared to the DNA of the victim's parents. (A DNA profile resembles that of one's parents because it is inherited.) It has even been possible to sequence DNA taken from a 76,000-year-old mummified human brain.

DNA fingerprinting following PCR has other applications. When the DNA matches that of a virus or mutated gene, we know that a viral infection, genetic disorder, or cancer is present. Mitochondrial DNA sequences in modern living populations have been used to determine the evolutionary history of human populations.

Recombinant DNA technology and the polymerase chain reaction are two ways to clone a gene.

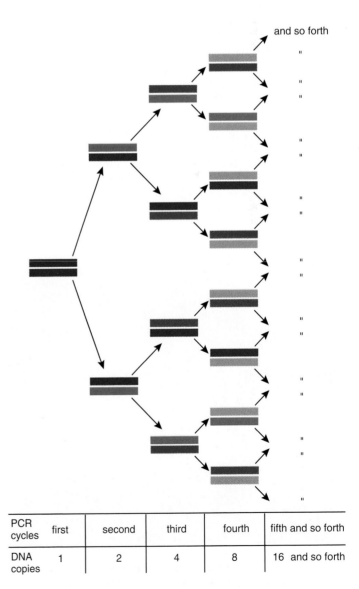

PCR cycles	first	second	third	fourth	fifth and so forth
DNA copies	1	2	4	8	16 and so forth

Figure 21.16   Polymerase chain reaction (PCR).
PCR allows the production of many identical copies of DNA in a laboratory test tube.

Biotechnology Products

Today, bacteria, plants, and animals are genetically engineered to produce biotechnology products (Fig. 21.17). Organisms that have had a foreign gene inserted into them are called **transgenic organisms.**

From Bacteria

Recombinant DNA technology is used to produce bacteria that reproduce in large vats called bioreactors. When the foreign gene is replicated and actively expressed, a large amount of protein product can be obtained. Biotechnology products produced by bacteria, such as insulin, human growth hormone, t-PA (tissue plasminogen activator), and hepatitis B vaccine, are now on the market.

Transgenic bacteria have been produced to promote the health of plants. For example, bacteria that normally live on plants and encourage the formation of ice crystals have been changed from frost-plus to frost-minus bacteria. Also, a bacterium that normally colonizes the roots of corn plants has now been endowed with genes (from another bacterium) that code for an insect toxin. The toxin protects the roots from insects.

Bacteria can be selected for their ability to degrade a particular substance, and this ability can then be enhanced by genetic engineering. For instance, naturally occurring bacteria that eat oil can be genetically engineered to do an even better job of cleaning up beaches after oil spills. Industry has found that bacteria can be used as biofilters to prevent airborne chemical pollutants from being vented into the air. They can also remove sulfur from coal before it is burned and help clean up toxic waste dumps. One such strain was given genes that allowed it to clean up levels of toxins that would have killed other strains. Further, these bacteria were given "suicide" genes that caused them to self-destruct when the job had been accomplished.

Organic chemicals are often synthesized by having catalysts act on precursor molecules or by using bacteria to carry out the synthesis. Today, it is possible to go one step further and to manipulate the genes that code for these enzymes. For instance, biochemists discovered a strain of bacteria that is especially good at producing phenylalanine, an organic chemical needed to make aspartame, the dipeptide sweetener better known as NutraSweet. They isolated, altered, and formed a vector for the appropriate genes so that various bacteria could be genetically engineered to produce phenylalanine.

Many major mining companies already use bacteria to obtain various metals. Genetic engineering can enhance the ability of bacteria to extract copper, uranium, and gold from low-grade sources. Some mining companies are testing genetically engineered organisms that have improved bioleaching capabilities.

From Plants

Techniques have been developed to introduce foreign genes into immature plant embryos or into plant cells that have had the cell wall removed and are called protoplasts. It is possible to treat protoplasts with an electric current while they are suspended in a liquid containing foreign DNA. The electric current makes tiny, self-sealing holes in the plasma membrane through which genetic material can enter. Protoplasts go on to develop into mature plants.

Foreign genes transferred to cotton, corn, and potato strains have made these plants resistant to pests because their cells now produce an insect toxin. Similarly, soybeans have been made resistant to a common herbicide. Some corn and cotton plants are both pest and herbicide resistant. These and other genetically engineered crops are now sold commercially.

Plants are also being engineered to produce human proteins, such as hormones, clotting factors, and antibodies, in their seeds. One type of antibody made by corn can deliver radioisotopes to tumor cells, and another made by soybeans can be used to treat genital herpes.

A weed called mouse-eared cress has been engineered to produce a biodegradable plastic (polyhydroxybutyrate, or PHB) in cell granules.

Figure 21.17 **Biotechnology products.**
Products like clotting factor VIII, which is administered to hemophiliacs, can be made by transgenic bacteria, plants, or animals. After being processed and packaged, it is sold as a commercial product.

Genetically engineered bacteria produce products in factories or they have environmental functions. Genetically engineered crops have better yields, and some plants produce products that assist in the treatment of human disease.

Health Focus

Organs for Transplant

Although it is now possible to transplant various organs, there are not enough human donors to go round. Thousands of patients die each year while waiting for an organ. It's no wonder, then, that scientists are suggesting that we should get organs from a source other than another human. **Xenotransplantation** is the use of animal organs instead of human organs in transplant patients. You might think that apes, such as the chimpanzee or the baboon, might be a scientifically suitable species for this purpose. But apes are slow breeders and probably cannot be counted on to supply all the organs needed. Also, many people might object to using apes for this purpose. In contrast, animal husbandry has long included the raising of pigs as a meat source, and pigs are prolific. A female pig can become pregnant at six months and can have two litters a year, each averaging about ten offspring.

Ordinarily, humans would violently reject transplanted pig organs. Genetic engineering, however, can make these organs less antigenic. Scientists have produced a strain of pigs whose organs would most likely, even today, survive for a few months in humans. They could be used to keep a patient alive until a human organ was available. The ultimate goal is to make pig organs as widely accepted by humans as type O blood. A person with type O blood is called a universal donor because the red blood cells carry no A or B antigens.

As xenotransplantation draws near, other concerns have been raised. Some experts fear that animals might be infected with viruses, akin to Ebola virus or the virus that causes "mad cow" disease. After infecting a transplant patient, these viruses might spread into the general populace and begin an epidemic. Scientists believe that HIV was spread to humans from monkeys when humans ate monkey meat. Those in favor of using pigs for xenotransplantation point out that pigs have been around humans for centuries without infecting them with any serious diseases.

Just a few years ago, scientists believed that transplant organs had to come from humans or other animals. Now, however, tissue engineering is demonstrating that it is possible to make some bioartificial organs—hybrids created from a combination of living cells and biodegradable polymers. Presently, only lab-grown hybrid tissues are on the market. A product composed of skin cells growing on a polymer is used to temporarily cover the wounds of burn patients. Similarly, damaged cartilage can be replaced with a hybrid tissue produced after chondrocytes are harvested from a patient. Another connective tissue product made from fibroblasts and collagen is available to help heal deep wounds without scarring. Soon to come are a host of other products including replacement corneas, heart valves, bladder valves, and breast tissue.

Tissue engineers have also created implants—cells producing a useful product encapsulated within a narrow plastic tube or a capsule the size of a dime or quarter (Fig. 21A). The pores of the container are large enough to allow the product to diffuse out but are too small for immune cells to enter and destroy the cells. An implant whose cells secrete natural pain killers will survive for months in the spinal cord and it can be easily withdrawn when desired. A "bridge to a liver transplant" is a bedside vascular apparatus. The patient's blood passes through porous tubes surrounded by pig liver cells. These cells convert toxins in the blood to nonpoisonous substances.

The goal of tissue engineering is to produce fully functioning organs for transplant. After nine years, a Harvard Medical School team headed by Anthony Atala has produced a working urinary bladder in the laboratory. After testing the bladders in laboratory animals, the Harvard group is ready to test them in humans whose own bladders have been damaged by accident or disease, or will not function properly due to a congenital birth defect. Another group of scientists has been able to grow arterial blood vessels in the laboratory. Still to come, tissue engineers are hopeful they can one day produce larger internal organs like the liver and kidney.

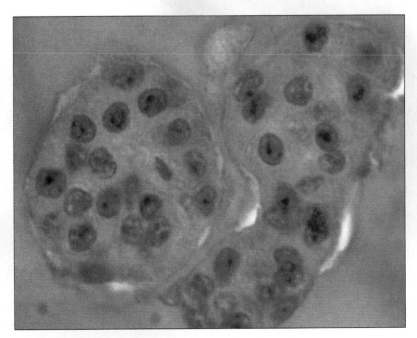

Figure 21A Microreactors.
These pancreatic cells from pigs are in a plastic container whose pores are too small for immune cells to enter and destroy them.

From Animals

Techniques have been developed to insert genes into the eggs of animals. It is possible to microinject foreign genes into eggs by hand, but another method uses vortex mixing. The eggs are placed in an agitator with DNA and silicon-carbide needles, and the needles make tiny holes through which the DNA can enter. When these eggs are fertilized, the resulting offspring are transgenic animals. Using this technique, many types of animal eggs have taken up the gene for bovine growth hormone (bGH). The procedure has been used to produce larger fishes, cows, pigs, rabbits, and sheep.

Gene pharming, the use of transgenic farm animals to produce pharmaceuticals, is being pursued by a number of firms. Genes that code for therapeutic and diagnostic proteins are incorporated into the animal's DNA, and the proteins appear in the animal's milk. There are plans to produce drugs for the treatment of cystic fibrosis, cancer, blood diseases, and other disorders. Figure 21.18 outlines the procedure for producing transgenic mammals: DNA containing the gene of interest is injected into donor eggs. Following in vitro fertilization, the zygotes are placed in host females where they develop. After female offspring mature, the product is secreted in their milk.

Cloning of Transgenic Animals   Imagine that an animal has been genetically altered to produce a biotechnology product. What would be the best possible method of getting identical copies of the animal? Asexual reproduction through cloning the animal would be the preferred procedure. Cloning is a form of asexual reproduction because it requires only the genes of that one animal. For many years, it was believed that adult vertebrate animals could not be cloned. Although each cell contains a copy of all the genes, certain genes are turned off in mature, specialized cells. Cloning of an adult vertebrate requires that all genes of an adult cell be turned on again if development is to proceed normally. It had long been thought this would be impossible.

In 1997, Scottish scientists announced that they had achieved this feat and produced a cloned sheep called Dolly. Since then, calves and goats have also been cloned. Figure 21.18 also shows that after enucleated eggs have been injected with 2n nuclei of adult cells, they can be coaxed to begin development. The offspring have the genotype and phenotype of the adult that donated the nuclei; therefore, the adult has been cloned. In a procedure that produced cloned mice, the 2n nuclei were taken from corona radiata cells. Corona radiata cells are those that cling to an egg after ovulation occurs. Now that scientists have a way to clone mammals, this procedure will undoubtedly be used routinely.

Gene pharming, the use of genetically engineered animals to produce pharmaceuticals in milk, is now a reality. Procedures have also been developed to allow the cloning of these animals.

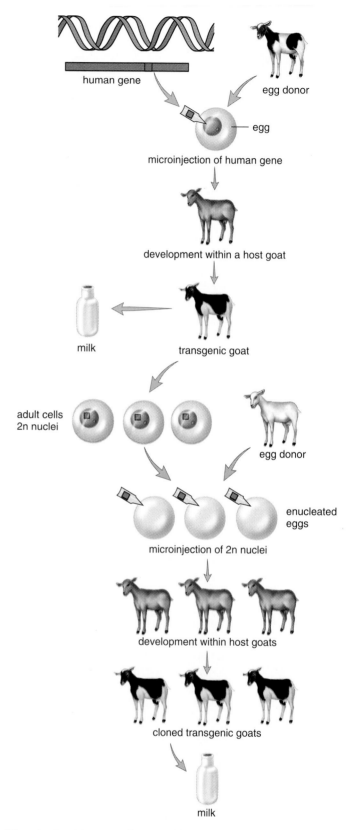

Figure 21.18   Transgenic animals.
A genetically engineered egg develops in a host to give a transgenic goat that produces a biotechnology product in its milk. Nuclei from the transgenic goat are transferred into donor eggs which develop into cloned transgenic goats.

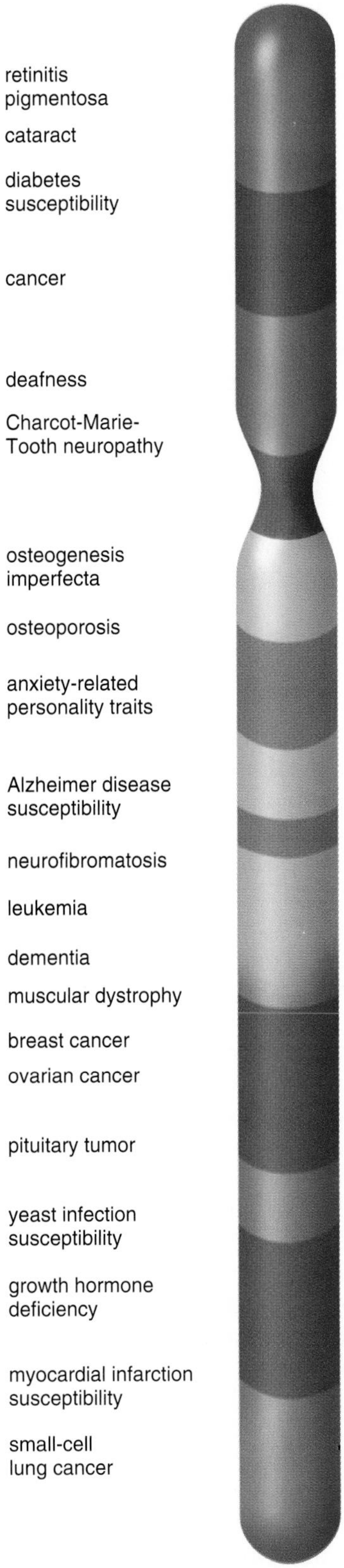

retinitis
pigmentosa

cataract

diabetes
susceptibility

cancer

deafness

Charcot-Marie-
Tooth neuropathy

osteogenesis
imperfecta

osteoporosis

anxiety-related
personality traits

Alzheimer disease
susceptibility

neurofibromatosis

leukemia

dementia

muscular dystrophy

breast cancer

ovarian cancer

pituitary tumor

yeast infection
susceptibility

growth hormone
deficiency

myocardial infarction
susceptibility

small-cell
lung cancer

Figure 21.19 Genetic map of chromosome 17.

This map shows the sequence of mutant genes which cause the diseases noted.

The Human Genome

The DNA in the human nucleus has three billion base pairs, and researchers now know the sequence of these base pairs, one after the other, along the length of our chromosomes. It took some fifteen years for researchers to complete this monumental task. At first, the project was slow-going, but the use of ever-more powerful computers allowed researchers to speed up enormously in recent years. Two rival groups have been at work on the project. The International Human Genome Sequencing Consortium, which consists of laboratories in many and various countries, was supported by public funds, notably the United States Government. On the other hand, Celera Genomics, a private company supported by a pharmaceutical firm, has been sequencing the genome for only a few years. These competing groups used slightly different techniques but came to the same conclusions. The rivalry between them put the Human Genome Project on a fast track that resulted in complete sequencing earlier than originally projected.

A lot of our DNA consists of nucleotide repeats that do not code for a protein. So far, researchers have found only 30,000 genes that do code for proteins. This number seems terribly low because a roundworm has 20,000 genes. We are certainly more complex than roundworms and, by rights, should have many more genes. How many more is being hotly debated at this time. Those who believe that they have found most of our genes, speculate that each of our genes could code for about three proteins simply by using different combinations of exons (see Fig. 21.9).

We do know that there is little difference between the sequence of our bases and other organisms whose DNA sequences are also known. We even share a large number of genes with bacteria! It's possible that we will discover that our uniqueness is due to the regulation of our genes. Even though we now know the sequence of bases in the human genome, much work still needs to be done to make sense out of what we have discovered.

Researchers are hopeful that sequencing the human genome will help them discover the root causes of many illnesses. Most testing for genetic diseases is based on genetic maps of the chromosomes. A genetic map tells the location of mutant genes along a chromosome. Figure 21.19 shows the loci of significant mutant genes on chromosome 17, for example. Completing the chromosomal genetic map should accelerate now because researchers need only know a short sequence of bases in a gene of interest in order for the computer to search the genome for a match. Then, the computer will tell the researcher where this gene is located.

In just two years, scientists have located millions of single nucleotide polymorphisms (SNPs, pronounced "snips"). These are spots where the DNA differs by only one nucleotide between individuals. SNPs may allow researchers to determine why one person in the same environment succumbs to a disease and another does not. The hope is to discover genetic differences that lead to skin cancer, diabetes, inflammatory bowel disease, breast cancer, heart disease, and mental illnesses, among others. One day we may be able to tailor treatments to the individual. For example, only certain hypertension patients benefit from a low-salt diet, and it would be useful to know which patients these are. Myriad Genetics, a genome company, has developed a test for a mutant angiotensinogen gene because they want to see if patients with this mutation are the ones who benefit from a low-salt diet. Several other mutant genes have been correlated with specific drug treatments (Table 21.4). One day the medicine you take might carry a label stating that is effective only in persons with genotype #101!

Table 21.4	Customizing Drug Treatments	
Mutant Gene For	**Disease**	**Treatment**
Apolipoprotein E	Alzheimer	Experimental Glaxo Wellcome drug
Cytochrome P-450	Cancer	Amonafide
Chloride gate	Cystic fibrosis	Pulmozyme
Dopamine receptor D4	Schizophrenia	Clozapine
Angiotensinogen	Hypertension	Low-salt diet

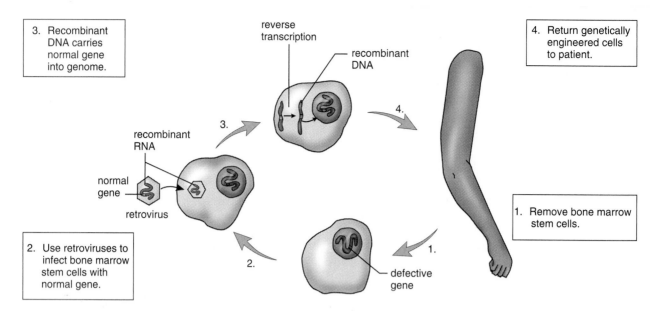

Figure 21.20   Ex vivo gene therapy in humans.
Bone marrow stem cells are withdrawn from the body, an RNA retrovirus is used to insert a normal gene into them, and they are returned to the body.

Gene Therapy

Gene therapy is the insertion of genetic material into human cells for the treatment of a disorder. It includes procedures that give a patient healthy genes to make up for faulty genes, and also includes the use of genes to treat various other human illnesses such as cardiovascular disease and cancer. Currently there are approximately 1,000 patients enrolled in nearly 200 approved gene therapy trials in the United States. Gene therapy includes both ex vivo (outside the body) and in vivo (inside the body) methods.

Genetic Disorders

As an example of an ex vivo method of gene therapy, consider Figure 21.20, which describes in more detail the methodology for the treatment of children like Cindy Cutshall (see page 421) who have SCID (severe combined immunodeficiency). As previously described, these children lack the enzyme ADA (adenosine deaminase) that is involved in the maturation of T and B cells. In order to carry out gene therapy, bone marrow stem cells are removed from the blood and infected with an RNA retrovirus that carries a normal gene for the enzyme. Then the cells are returned to the patient. Bone marrow stem cells are preferred for this procedure because they divide to produce more cells with the same genes. With an ex vivo method of gene therapy, it is possible to test and make sure gene transfer has occurred before the cells are returned to the patient. Patients who have undergone this procedure do have a significant improvement in their immune function that is associated with a sustained rise in the level of ADA enzyme activity in the blood. Whereas previously patients were also given ADA enzyme as a safety measure, gene therapy alone was

recently used to cure two infants in France. Before being treated, the infants had to be hospitalized and kept in a sterile bubble. Now they are home with no need for any special protection.

Among the many gene therapy trials, one is for the treatment of familial hypercholesterolemia, a condition that develops when liver cells lack a receptor for removing cholesterol from the blood. The high levels of blood cholesterol make the patient subject to fatal heart attacks at a young age. In a newly developed procedure, a small portion of the liver is surgically excised and infected with a retrovirus containing a normal gene for the receptor before being returned to the patient. Several patients have experienced a lowering of serum cholesterol levels following this procedure.

Cystic fibrosis patients lack a gene that codes for the transmembrane carrier of the chloride ion. Patients often die due to numerous infections of the respiratory tract. Despite many setbacks, a number of gene therapy studies are in process. In these trials, the gene needed to cure cystic fibrosis is sprayed into the nose or delivered to the lower respiratory tract by adenoviruses or by the use of liposomes, microscopic vesicles that spontaneously form when lipoproteins are put into a solution. With all three types of vehicles, too few respiratory tract epithelial cells take up the gene. Investigators are trying to improve uptake, and also hypothesizing that a combination of all three vectors might be more successful. There is also the possibility of starting treatment before the respiratory tract has been altered by the effects of cystic fibrosis.

In a recent and surprising move, some researchers are investigating the possibility of directly correcting the DNA base sequence of patients with a genetic disorder. The exact procedure has not yet been published.

Bioethical Focus

Transgenic Plants

Transgenic plants can possibly allow crop yields to keep up with the ever-increasing worldwide demand for food. And some of these plants have the added benefit of requiring less fertilizer and/or pesticides, which are harmful to human health and the environment.

But some scientists believe that transgenic plants pose their own threat to the environment, and many activists believe that transgenic plants are themselves dangerous to our health. Studies have shown that wind-carried pollen can cause transgenic plants to hybridize with nearby weedy relatives. Although it has not happened yet, some fear that characteristics acquired in this way might cause weeds to become uncontrollable pests. Or any toxin transgenic plants produce could possibly hurt other organisms. The pollen from a plant engineered to produce Bt, a pesticide, was coated on milkweed plants, and monarch caterpillars who ate the plants died. Also, although transgenic crops have not caused any illnesses in humans, some scientists concede the possibility that people could be allergic to the transgene's protein product. After unapproved genetically modified corn was detected in Taco Bell taco shells, a massive recall pulled about 2.8 million boxes of the product from grocery stores.

Already, transgenic plants must be approved by the Food and Drug Administration before they are considered safe for human consumption, and they must meet certain Environmental Protection Administration standards. Are you in favor of strengthening safety standards for transgenic crops even if it means food production may suffer? Should all biotech foods be labeled as such so the buyer can choose whether or not to eat them? At the very least, should "organic" mean that the food was produced by natural rather than by transgenic plants or animals?

Figure 21B Regulation of transgenic plants.
Activists protest the present regulation of transgenic plants, which they believe to be too permissive.

Decide Your Opinion

1. Do you believe that research and development of transgenic plants and animals should be halted? Why or why not?
2. Do you instead approve of regulating the use of transgenic plants? Why or why not?
3. Should those who feel that transgenic plants are safe still support legislation to regulate them? Why or why not?

Looking at Both Sides www.mhhe.com/biosci/genbio/maderhuman7/

Every bioethical issue has at least two sides. Even if you already have an opinion, it is important to explore the opposite opinion before finalizing your position. The Online Learning Center at www.mhhe.com/biosci/genbio/maderhuman7/ will help you fine-tune your initial opinion, explore both sides, and finalize your position. You may acquire new arguments for your original opinion, or you may even change your opinion. Be sure to complete these activities in sequence:

Taking Sides   Decide your initial opinion by answering a series of questions. Then see if your opinion changes after completing the next two activities.

Further Debate   Read opposing articles that give you further information on this particular bioethical issue.

Explain Your Position   Answer another series of questions and then defend your original or changed opinion. You can e-mail your position to your instructor if he or she wishes.

Other Illnesses

During a coronary bypass operation, the surgeon grafts a blood vessel between the aorta and the coronary vessels, bypassing areas of blockage. During angioplasty, a balloon catheter is sometimes used to open up a closed artery. Unfortunately, repeat bypass operations or angioplasty may be required because the new or repaired blood vessels become obstructed again with new plaque deposits. To counter this inevitable outcome, doctors have now come up with an alternative procedure. It has been known for some time that VEGF (vascular endothelial growth factor) can cause the growth of new blood vessels. The gene that codes for this growth factor can be injected alone or within a virus into the heart. Over one thousand patients have undergone this procedure, and they report much less chest pain and the ability to run longer on a treadmill. This therapy can accompany surgeries or can be used alone.

Perhaps it will also be possible to use in vivo gene therapy to cure hemophilia, diabetes, Parkinson disease, or AIDS. To treat hemophilia, patients could get regular doses of cells bioengineered to contain normal clotting-factor genes. Or such cells could be placed in organoids, artificial organs that can be implanted in the abdominal cavity. To cure Parkinson disease, dopamine-producing cells could be grafted directly into the brain.

Gene therapy is increasingly being tried in clinical trials on cancer patients. Since chemotherapy often kills off healthy cells as well as cancer cells, researchers have given genes to cancer patients that either make healthy cells more tolerant of chemotherapy or make tumors more vulnerable to it. In one clinical trial, bone marrow stem cells from about 30 women with late-stage ovarian cancer were infected with a virus carrying a gene to make them more tolerant of chemotherapy. Once the bone marrow stem cells were protected, it was possible to increase the level of chemotherapy to kill the cancer cells.

Cancer cells often lack a *p53* gene. This gene helps regulate the cell cycle and also brings about apoptosis in cells with damaged DNA. As discussed in the next chapter, there is much interest in using gene therapy to introduce a working *p53* gene into cancer cells and in that way kill them off.

Knowing the sequence of base pairs in the human genome will be useful to gene therapists. Researchers are envisioning all sorts of ways to cure human genetic disorders, as well as many other types of illnesses.

Summarizing the Concepts

21.1 DNA and RNA Structure and Function

DNA is a double helix composed of two nucleic acid strands that are held together by weak hydrogen bonds between the bases: A is bonded to T, and C is bonded to G. During replication, the DNA strands unzip, and then a new complementary strand forms opposite each old strand. This results in two identical DNA molecules.

RNA is a single-stranded nucleic acid in which U (uracil) occurs instead of T (thymine). Both DNA and RNA are involved in gene expression.

21.2 Gene Expression

Proteins differ from one another by the sequence of their amino acids. DNA has a code that specifies this sequence. Gene expression requires transcription and translation. During transcription, the DNA code (triplet of three bases) is passed to an mRNA that then contains codons. Introns are removed from mRNA during mRNA processing. During translation, tRNA molecules bind to their amino acids, and then their anticodons pair with mRNA codons. In the end, each protein has a sequence of amino acids according to the blueprint provided by the sequence of nucleotides in DNA.

Control of gene expression can occur at four levels in a human cell: at the time of transcription; after transcription and during mRNA processing; during translation; and after translation—before, at, or after protein synthesis. It is known that chromatin has to be extended for transcription to occur and that transcription factors control the activity of genes in human cells.

21.3 Biotechnology

Two methods are currently available for making copies of DNA. Recombinant DNA contains DNA from two different sources. Human DNA can be inserted into a plasmid, which is taken up by bacteria. When the plasmid replicates, the gene is cloned and its protein is produced. The polymerase chain reaction (PCR) uses the enzyme DNA polymerase to make multiple copies of target DNA. Following PCR, it is possible to determine if the DNA of an infectious organism or any particular sequence of DNA is present. DNA fingerprinting is done to see if DNA from one source matches DNA from another source.

Transgenic bacteria are now routinely made. Crops resistant to pests and herbicides are commercially available. Transgenic animals have been given various genes, in particular the one for bovine growth hormone (bGH). Animals have also been genetically engineered to secrete a protein product in their milk. Cloning of animals is now possible.

Although researchers involved in the Human Genome Project now know the sequence of the bases on each chromosome, they do not yet know the sequence of all the genes on each chromosome. When researchers have this information, they will be able to determine how an individual's genes differ from the norm. This information is expected to lead to improved pharmaceuticals and enhance the possibility of gene therapy for more and various ills.

Human gene therapy is undergoing clinical trials. Ex vivo therapy involves withdrawing cells from the patient, inserting a functioning gene, usually via a retrovirus, and then returning the treated cells to the patient. Many investigators are trying to develop in vivo therapy in which viruses, laboratory-grown cells, or synthetic carriers could be used to carry healthy genes directly into the patient.

Studying the Concepts

1. Describe the structure of DNA and how this structure contributes to the ease of DNA replication. 423–24
2. Describe the structure of RNA and compare it to the structure of DNA. 425
3. Name and discuss the roles of three different types of RNA. 425
4. Describe the structure and function of a protein and the manner in which DNA codes for a particular protein. 426
5. Describe the process of transcription and the three steps of translation. If the code is TTT, CAT, TGG, CCG, what are the codons, and what is the sequence of amino acids? 426–29
6. What are the four levels of genetic control in human cells? Describe two means by which transcription is regulated. 431
7. Describe precisely how you might clone a gene using recombinant DNA technology. When might you use a PCR instrument? 432–33
8. Naturally occurring bacteria have been genetically engineered to perform what services? 434
9. What types of genetically engineered plants are now available and/or are expected in the near future? 434
10. What types of animals have been genetically engineered and for what purposes? Discuss the advantages of cloning transgenic animals. 436
11. What are the two primary goals of the Human Genome Project? What are the possible benefits of the project for human society? 437
12. How is gene therapy in humans currently being done? 438

In questions 8–12, fill in the blanks.

8. The current ex vivo gene therapy clinical trials use a(n) _____ as a vector to insert healthy genes into the patient's cells.

9. The polymerase chain reaction makes many _____ of a short DNA segment.

10. The DNA code is a _____ code, meaning that every three bases stands for a(n) _____.

11. The sequence of mRNA codons dictates the sequence of amino acids in a protein. This step in protein synthesis is called _____.

12. Bacteria, plants, and animals that have been genetically engineered are called _____ organisms.

13. Following is a segment of a DNA molecule. (Remember that only the template strand is transcribed.) What are (a) the RNA codons and (b) the tRNA anticodons?

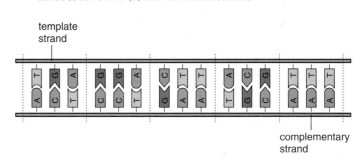

template
strand

complementary
strand

Testing Your Knowledge of the Concepts

In questions 1–4, match the nucleotide to its function.
a. DNA
b. mRNA
c. tRNA
d. rRNA

_____ 1. Joins with proteins to form subunits of a ribosome.

_____ 2. Contains codons that determine the sequence of amino acids in a polypeptide.

_____ 3. Contains a code and serves as a template for the production of RNA.

_____ 4. Brings amino acids to the ribosomes during the process of transcription.

In questions 5–7, indicate whether the statement is true (T) or false (F).

_____ 5. The two types of enzymes needed to make a recombinant DNA plasmid are a restriction enzyme and DNA ligase.

_____ 6. Replication of DNA is semiconservative, meaning that DNA contains a code and mRNA contains codons.

_____ 7. If the DNA code is ATC, GGG, CGC, then mRNA is TAG, CCC, GCG.

Understanding Key Terms

adenine (A) 423
anticodon 428
biotechnology 432
cloning 432
codon 427
complementary base pairing 423
complementary DNA 433
cytosine (C) 423
DNA (deoxyribonucleic acid) 422
DNA fingerprinting 433
DNA ligase 432
DNA polymerase 424
gene 422
gene therapy 438
genetic engineering 432
guanine (G) 423
messenger RNA (mRNA) 425
mutation 424
plasmid 432

polymerase chain reaction (PCR) 433
polyribosome 428
recombinant DNA (rDNA) 432
replication 423
restriction enzyme 432
ribosomal RNA (rRNA) 425
RNA (ribonucleic acid) 425
RNA polymerase 427
template 424
thymine (T) 423
transcription 427
transcription factor 431
transfer RNA (tRNA) 425
transgenic organism 434
translation 428
triplet code 427
uracil (U) 425
vector 432
xenotransplantation 435

e-Learning Connection   www.mhhe.com/biosci/genbio/maderhuman7/

21.1   DNA and RNA Structure and Function

DNA Structure *Essential Study Partner*
DNA Replication *Essential Study Partner*
Concept Review *Essential Study Partner*

DNA Structure *animation activity*
Base Pairing *art quiz*

21.2   Gene Expression

Gene Activity *Essential Study Partner*
Transcription *Essential Study Partner*
Translation *Essential Study Partner*

Transcription *animation activity*
Translation *animation activity*
Polyribosome *animation activity*
Translation *art quiz*

21.3   Biotechnology

Polymerase Chain Reaction *animation activity*
Looking at Both Sides *critical thinking activity*

DNA in Criminal Investigations *reading*
Tissue Engineering *reading*

Chapter Summary

Key Term Flashcards *vocabulary quiz*
Chapter Quiz *objective quiz covering all chapter concepts*

Chapter 22

Cancer

Chapter Concepts

22.1 Cancer Cells
- Cancer cells have characteristics that can be associated with their ability to grow uncontrollably. 444

22.2 Origin of Cancer
- Mutations of proto-oncogenes and tumor-suppressor genes are linked to the development of cancer. 446

22.3 Causes of Cancer
- A number of agents are known to bring about the mutations that cause cancer, and a mutated gene can be inherited. 448
- Environmental factors play a role in the development of cancer. 448

22.4 Diagnosis and Treatment
- There are few routine screening tests for specific cancers, and cancer often goes undetected until it is advanced. New ways to diagnose cancer are under investigation. 450
- Surgery, radiation, and chemotherapy are presently the standard ways of treating cancers. 452
- Molecular knowledge of cancer has produced many specific types of drugs and therapies that are now in clinical trials. 454
- It is possible to take protective steps to reduce the risk of cancer. 456

Figure 22.1 Sunbathing.
Sunbathing can lead to skin cancer. Melanoma is a particularly malignant skin cancer.

If there's one thing 25-year-old Emily loves, it's lying in the sun. "Baking on the beach" is what she calls it (Fig. 22.1). A little baby oil, a radio, some sunglasses. She does it for hours. Unfortunately, while Emily rubs in the oil, she doesn't notice the small mole on the back of her right arm. Dark and curvy, the little mole looks like a splotch of spilled ink. Doctors call it melanoma.

If Emily is lucky, she or her physician will see the mole before the cancer cells underneath break away and spread. If no one notices the mole, the cells will likely travel through Emily's circulatory system, making their way to her lungs, liver, brain, and ovaries. When that happens, it will be too late.

About 32,000 cases of melanoma are diagnosed each year. One in five persons diagnosed dies within five years. For most people, it's easy to avoid the disease. Most important, don't "bake on the beach" and don't visit tanning machines. Protective clothing is a must, and wearing sunscreen may also help.

The melanoma that developed on Emily's right arm went through certain phases. During initiation, a single cell underwent a mutation that caused it to begin to divide repeatedly. Then promotion occurred as other factors encouraged a tumor to develop, and the tumor cells continued to divide. As they divided, other mutations occurred. Finally, progression took place as more mutations gave one cell a selective advantage over the other cells. This process can be repeated several times, until there is a cell that has the ability to invade surrounding tissues. This is called metastasis.

22.1   Cancer Cells

Cancer is a genetic disease requiring a series of mutations, each propelling cells toward the development of a **tumor,** an accumulation of cancer cells that no longer function properly. **Carcinogenesis,** the development of cancer, is a gradual, stepwise process, and it may be decades before a person notices any sign or symptom of a tumor. Although tumor cells share common characteristics, each type of cancer has its own particular sequence of mutations that has resulted in uncontrolled growth.

Characteristics of Cancer Cells

Cancer cells share characteristics that distinguish them from normal cells (Table 22.1).

Cancer Cells Lack Differentiation

Most cells are specialized; they have a specific form and function that suits them to the role they play in the body. Cancer cells are nonspecialized and do not contribute to the functioning of a body part. A cancer cell does not look like a differentiated epithelial, muscular, nervous, or connective tissue cell; instead, it looks distinctly abnormal. Normal cells can undergo the cell cycle for about 50 times, and then they die. Cancer cells can enter the cell cycle repeatedly, and in this way, they are potentially immortal.

Cancer Cells Have Abnormal Nuclei

The nuclei of cancer cells are enlarged, and there may be an abnormal number of chromosomes. The chromosomes have mutated; for example, some parts may be duplicated and some may be deleted. In addition, *gene amplification* (extra copies of specific genes) is seen much more frequently than in normal cells. Ordinarily, cells with damaged DNA undergo **apoptosis,** a series of enzymatic reactions that lead to the death of the cell. Cancer cells fail to undergo apoptosis.

Cancer Cells Form Tumors

Normal cells anchor themselves to a substratum and/or adhere to their neighbors. They exhibit *contact inhibition*— that is, when they come in contact with a neighbor, they stop dividing. In culture, normal cells form a single layer that covers the bottom of a petri dish. Cancer cells have lost all restraint; they pile on top of one another to grow in multiple layers.

Normal cells do not grow and divide unless they are stimulated to do so by a growth factor. Cancer cells have a reduced need for a growth factor, such as epidermal growth factor, in order to grow and divide. Conversely, cancer cells no longer respond to inhibitory growth factors such as transforming growth factor beta (TGF-β) from their neighbors. Their growth, termed *neoplasia,* contains cells that are disorganized, a condition termed *anaplasia.* During carcinogenesis, the most aggressive cell becomes the dominant cell of the tumor (Fig. 22.2).

Table 22.1	Characteristics of Normal Cells Versus Cancer Cells	
Characteristic	**Normal Cells**	**Cancer Cells**
Differentiation	Yes	No
Nuclei	Normal	Abnormal
Tumor formation	No	Yes
Contact inhibition	Yes	No
Growth factors	Required	Not required
Angiogenesis	No	Yes
Metastasis	No	Yes

Cancer Cells Induce Angiogenesis

Angiogenesis is the formation of new blood vessels. To grow larger than, say, a pea, a tumor must have a well-developed capillary network to bring it nutrients and oxygen. Tumors secrete angiogenic proteins that stimulate nearby blood vessels to branch and send capillaries to the tumor. This blood flow not only nourishes the tumor, it also enables it to grow larger and to metastasize. Some modes of cancer treatment are aimed at preventing angiogenesis from occurring.

Cancer Cells Metastasize

A *benign tumor* is a disorganized, usually encapsulated, mass that does not invade adjacent tissue. *Cancer in situ* is a tumor located in its place of origin, before there has been any invasion of normal tissue. *Malignancy* is present when **metastasis** establishes new tumors distant from the primary tumor. To metastasize, cancer cells must make their way across a basement membrane and into a blood vessel or lymphatic vessel. Cancer cells produce proteinase enzymes that degrade the membrane and allow them to invade underlying tissues. Cancer cells tend to be motile because they have a disorganized internal cytoskeleton and lack intact actin filament bundles. After traveling through the blood or lymph, cancer cells start tumors elsewhere in the body.

The patient's prognosis (probable outcome) is dependent on the degree to which the cancer has progressed: (1) whether the tumor has invaded surrounding tissues; (2) if so, whether lymph nodes are involved; and (3) whether there are metastatic tumors in distant parts of the body. With each progressive step, the prognosis becomes less favorable.

Cancer cells are nonspecialized, have abnormal chromosomes, and divide uncontrollably. Cancer in situ is a tumor located in its place of origin. Malignant tumors establish new tumors distant from the primary tumor.

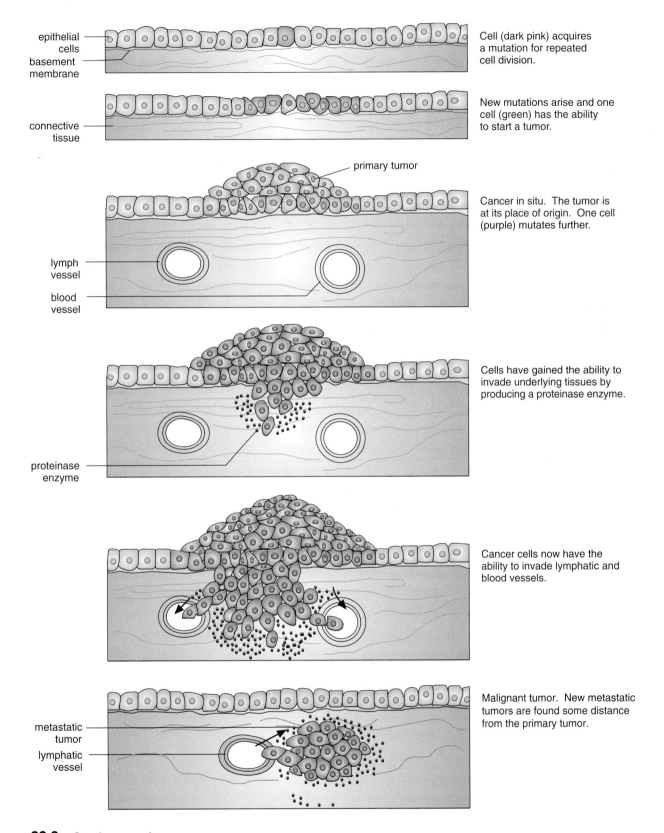

epithelial cells
basement membrane

Cell (dark pink) acquires a mutation for repeated cell division.

connective tissue

New mutations arise and one cell (green) has the ability to start a tumor.

primary tumor

Cancer in situ. The tumor is at its place of origin. One cell (purple) mutates further.

lymph vessel

blood vessel

Cells have gained the ability to invade underlying tissues by producing a proteinase enzyme.

proteinase enzyme

Cancer cells now have the ability to invade lymphatic and blood vessels.

metastatic tumor
lymphatic vessel

Malignant tumor. New metastatic tumors are found some distance from the primary tumor.

Figure 22.2   Carcinogenesis.
The development of cancer requires a series of mutations leading first to a localized tumor and then to metastatic tumors. With each successive step toward cancer, the most genetically altered and aggressive cell becomes the dominant type of the tumor.

22.2   Origin of Cancer

Mutations in at least four classes of genes are associated with the development of cancer. (1) The nucleus has a DNA repair system that corrects mutations occurring when DNA replicates. There are enzymes that read a newly forming DNA strand, making sure it has the same sequence of bases as the template strand. Any mistakes are then corrected by DNA repair enzymes. Mutations that arise in the genes encoding these various enzymes can contribute to carcinogenesis. (2) When a gene is being actively transcribed, the chromosome must decondense in a particular area. Mutations in genes that code for proteins involved in regulating the structure of chromatin help send cells down the road toward cancer. (3) **Proto-oncogenes** code for proteins that stimulate the cell cycle, and **tumor-suppressor genes** code for proteins that inhibit the cell cycle. When cancer develops, normal regulation of the cell cycle no longer occurs due to mutations in these two types of genes. (4) DNA segments called **telomeres** occur at the ends of chromosomes and protect them from damage. Telomeres usually get shorter every time chromosomes replicate prior to cell division. Once the cell has divided about 50 times, the shortened telomeres signal the cell to enter senescence and stop dividing. In cancer cells, an enzyme called telomerase is active and keeps telomeres at a constant length so that cancer cells can keep on dividing over and over again.

Regulation of the Cell Cycle

The cell cycle is a series of stages that occur in this sequence: G_1 stage—organelles begin to double in number; S stage—replication of DNA, which results in the duplication of chromosomes, occurs; G_2 stage—synthesis of proteins occurs, preparing the cell for mitosis; M stage—mitosis occurs. The passage of a cell from G_1 to the S stage is tightly regulated, as is the passage of a cell from G_2 to mitosis.

Proto-oncogenes are a part of a *stimulatory pathway* that extends from the plasma membrane to the nucleus. In the stimulatory pathway, a growth factor released by a neighboring cell is received by a receptor protein, and this sets in motion a whole series of enzymatic reactions that ends when signaling proteins that can trigger cell division enter the nucleus and turn on oncogenes (Fig. 22.3). Tumor-suppressor genes are a part of an *inhibitory pathway* that extends from the plasma membrane to the nucleus. In the inhibitory pathway, a growth-inhibitory factor released by a neighboring cell is received by a receptor protein, and this sets in motion a whole series of enzymatic reactions that ends when signaling proteins that inhibit cell division enter the nucleus. The balance between stimulatory signals and inhibitory signals determines whether proto-oncogenes are active or tumor-suppressor genes are active.

Oncogenes

Proto-oncogenes are so called because a mutation can cause them to become **oncogenes** (cancer-causing genes). An oncogene may code for a faulty receptor in the stimulatory pathway. A faulty receptor may be able to start the stimulatory process even when no growth factor is present! Or an oncogene may produce either an abnormal protein product or abnormally high levels of a normal product that stimulates the cell cycle to begin or to go to completion. In either case, uncontrolled growth threatens.

Researchers have identified perhaps one hundred oncogenes that can cause increased growth and lead to tumors. The oncogenes most frequently involved in human cancers belong to the *ras* gene family. An alteration of only a single nucleotide pair is sufficient to convert a normally functioning *ras* proto-oncogene to an oncogene. The *ras*K oncogene is found in about 25% of lung cancers, 50% of colon cancers, and 90% of pancreatic cancers. The *ras*N oncogene is associated with **leukemias** (cancer of blood-forming cells) and lymphomas (cancers of lymphoid tissue), and both *ras* oncogenes are frequently found in thyroid cancers.

Tumor-Suppressor Genes

When a tumor-suppressor gene undergoes a mutation, inhibitory proteins fail to be active, and the regulatory balance shifts in favor of cell cycle stimulation. Researchers have identified about a half-dozen tumor-suppressor genes. The *RB* tumor-suppressor gene was discovered when the inherited condition retinoblastoma was being studied. If a child receives only one normal *RB* gene and that gene mutates, eye tumors develop in the retina by the age of three. The *RB* gene has now been found to malfunction in cancers of the breast, prostate, and bladder, among others. Loss of the *RB* gene through chromosome deletion is particularly frequent in a type of lung cancer called small-cell lung carcinoma. How the RB protein fits into the inhibitory pathway is known. When a particular growth-inhibitory factor attaches to a receptor, the RB protein is activated. An active RB protein turns off the expression of a proto-oncogene, whose product initiates cell division.

Another major tumor-suppressor gene is called *p53*, a gene that is more frequently mutated in human cancers than any other known gene. It has been found that the p53 protein acts as a transcription factor and as such is involved in turning on the expression of genes whose products are cell cycle inhibitors. *p53* can also stimulate apoptosis, programmed cell death.

Each cell contains regulatory pathways involving proto-oncogenes and tumor-suppressor genes that code for components active in the pathways.

Visual Focus

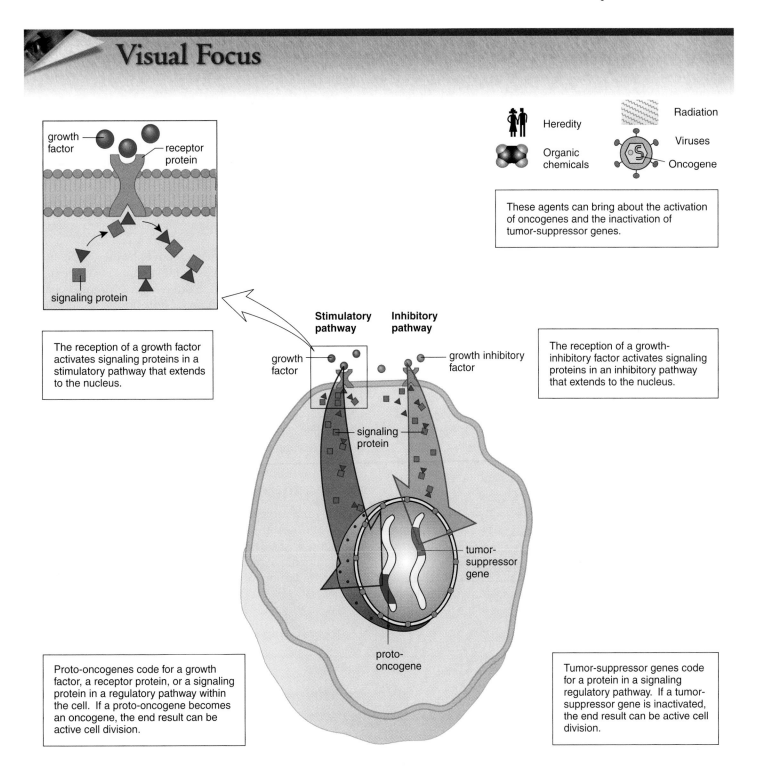

Stimulatory pathway

Inhibitory pathway

growth factor

receptor protein

signaling protein

Heredity

Organic chemicals

Radiation

Viruses

Oncogene

These agents can bring about the activation of oncogenes and the inactivation of tumor-suppressor genes.

The reception of a growth factor activates signaling proteins in a stimulatory pathway that extends to the nucleus.

growth factor

growth inhibitory factor

The reception of a growth-inhibitory factor activates signaling proteins in an inhibitory pathway that extends to the nucleus.

signaling protein

tumor-suppressor gene

proto-oncogene

Proto-oncogenes code for a growth factor, a receptor protein, or a signaling protein in a regulatory pathway within the cell. If a proto-oncogene becomes an oncogene, the end result can be active cell division.

Tumor-suppressor genes code for a protein in a signaling regulatory pathway. If a tumor-suppressor gene is inactivated, the end result can be active cell division.

Figure 22.3 Origin of cancer.
Two types of regulatory pathways extend from the plasma membrane to the nucleus. In the stimulatory pathway, receptor proteins receive growth-stimulatory factors. Then, proteins within the cytoplasm and proto-oncogenes within the nucleus stimulate the cell cycle. In the inhibitory pathway, receptor proteins receive growth-inhibitory factors. Then, proteins within the cytoplasm and tumor-suppressor genes within the nucleus inhibit the cell cycle from occurring. Whether cell division occurs or not depends on the balance of stimulatory and inhibitory signals received. Hereditary and environmental factors cause mutations of proto-oncogenes and tumor-suppressor genes. These mutations can cause uncontrolled growth and a tumor.

Apoptosis

Apoptosis, defined as programmed cell death, is a set sequence of cellular changes involving shattering the nucleus, chopping up the chromosomes, digesting the cytoskeleton, and packaging the cellular remains into membrane-enclosed vesicles (apoptotic bodies) that can be engulfed by macrophages. The appearance of membrane blisters is a tip-off that a cell is undergoing apoptosis (Fig. 22.4).

A remarkable finding of the past few years is that cells routinely harbor the enzymes, now called caspases, that bring about apoptosis. There are two sets of caspases. The first set are the "initiators" that receive the message to activate the second set, "executioners," which then activate the enzymes that dismantle the cell. The enzymes are ordinarily held in check by inhibitors, but they can be unleashed during development and in adults. For example, during human development, exterior signals cause certain cells to die, allowing paddles to be converted into fingers and toes.

In adults, cells that have DNA damage beyond repair usually go ahead and kill themselves. This is regarded as a safeguard to prevent the development of tumors in the body. If DNA is damaged in any way, the p53 protein inhibits the cell cycle, and this gives cellular enzymes the opportunity to repair the damage. But if DNA damage should persist, the *p53* gene goes on to bring about apoptosis of the cell. No wonder many types of tumors contain cells that lack an active *p53* gene. Some cancer cells have a *p53* gene but make a large amount of bcl-2, a protein that binds to and inactivates the p53 protein.

Apoptosis of cells that contain oncogenes and mutated tumor-suppressor genes is a reasonable way for the body to prevent the development of cancer.

22.3   Causes of Cancer

Increasingly, it seems that cancer is caused by heredity plus environmental effects.

Heredity

Already we know that particular types of cancer seem to run in families. For instance, the risk of developing breast, lung, and colon cancers increases two-to-threefold when first-degree relatives have had these cancers. We now know that the mutated tumor-suppressor gene *BRCA1* (breast cancer gene #1) runs in certain families and can be used to predict which members will develop breast cancer.

Certain childhood cancers seem to be due to the inheritance of a dominant gene. Retinoblastoma is an eye tumor that usually develops by age three; Wilms tumor is characterized by numerous tumors in both kidneys. In adults, several family syndromes (e.g., Li-Fraumeni cancer family syndrome, Lynch cancer family syndrome, and Warthin cancer family syndrome) are known. Those who inherit a dominant allele develop tumors in various parts of the body.

Now that the human genome has been sequenced, it may soon be possible to determine which genetic profiles make a person susceptible to cancer when exposed to particular environmental carcinogens.

Environmental Carcinogens

A **mutagen** is an agent that increases the chances of a mutation, while a **carcinogen** is an environmental agent that can contribute to the development of cancer. Carcinogens are often mutagenic. Among the best-known mutagenic carcinogens are radiation, organic chemicals, and viruses.

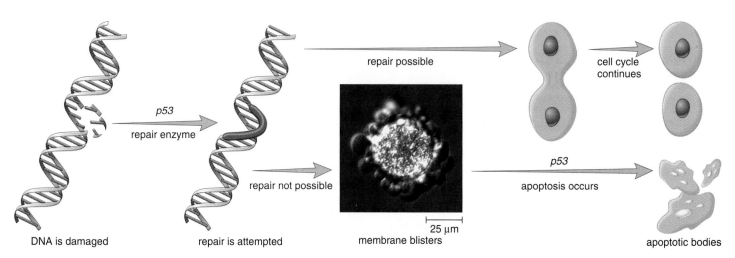

repair possible

cell cycle continues

p53

repair enzyme

repair not possible

25 μm

membrane blisters

p53

apoptosis occurs

DNA is damaged

repair is attempted

apoptotic bodies

Figure 22.4   Functions of *p53*.

If DNA is damaged by a mutagen, *p53* is instrumental in stopping the cell cycle and activating repair enzymes. If repair is impossible, the p53 protein promotes apoptosis.

Radiation

Ultraviolet radiation in sunlight and tanning lamps is most likely responsible for the dramatic increases seen in skin cancer the past several years. Today there are at least six cases of skin cancer for every one case of lung cancer. Nonmelanoma skin cancers are usually curable through surgery, but melanoma skin cancer tends to metastasize and is responsible for 1–2% of total cancer deaths in the United States.

Another natural source of radiation is radon gas. In the very rare house with an extremely high level of exposure, the risk of developing lung cancer is thought to be equivalent to smoking a pack of cigarettes a day. Therefore, the combination of radon gas and smoking cigarettes can be particularly dangerous. Radon levels can be lowered by improving the ventilation of a building.

Most of us have heard about the damaging effects of the nuclear bomb explosions or accidental emissions from nuclear power plants. For example, more cancer deaths are expected in the vicinity of the Chernobyl Power Station (in the former U.S.S.R.), which suffered a terrible accident in 1986. Usually, however, diagnostic X rays account for most of our exposure to artificial sources of radiation. The benefits of these procedures can far outweigh the possible risk, but it is still wise to avoid any X-ray procedures that are not medically warranted.

Despite much publicity, scientists have not been able to show a clear relationship between cancer and radiation from electric power lines, household appliances, and cellular telephones.

Organic Chemicals

Tobacco smoke, foods, certain hormones, and pollutants all contain organic chemicals that can be carcinogenic.

Tobacco Smoke   Tobacco smoke contains a number of organic chemicals that are known carcinogens, and it is estimated that smoking is also implicated in the development of cancers of the mouth, larynx, bladder, kidney, and pancreas. The greater the number of cigarettes smoked per day, the earlier the habit starts, and the higher the tar content, the more likely it is that cancer will develop. When smoking is combined with drinking alcohol, the risk of these cancers increases even more.

Passive smoking, or inhalation of someone else's tobacco smoke, is also dangerous and probably causes a few thousand deaths each year.

Foods and Hormones   Animal testing of certain food additives, such as red dye #2, has shown that some additives are cancer-causing in very high doses. In humans, salty foods and fats appear to make a significant contribution to the development of cancer.

Increasingly, studies suggest that hormone replacement therapy can contribute to the development of uterine and breast cancer in women who have taken these hormones for five to ten years.

Figure 22.5   Industrial chemicals.
Organic chemicals in air and particularly in drinking water are associated with a higher incidence of cancer.

Pollutants   Industrial chemicals, such as benzene and carbon tetrachloride, and industrial materials, such as vinyl chloride and asbestos fibers, are also associated with the development of cancer (Fig. 22.5). Pesticides and herbicides are dangerous not only to pets and plants but also to our own health because they contain organic chemicals that can cause mutations. It is now generally accepted that dioxin, a contaminant of herbicides, can cause cancers of lymphoid tissues, muscular tissues, and certain connective tissues.

Viruses

At least three DNA viruses—hepatitis B virus, Epstein-Barr virus, and human papillomavirus—have been linked to human cancers.

In China, almost all persons have been infected with the hepatitis B virus, and this correlates with the high incidence of liver cancer in that country. For a long time, circumstances suggested that cervical cancer was a sexually transmitted disease, and now human papillomaviruses are routinely isolated from cervical cancers. Burkitt lymphoma occurs frequently in Africa, where virtually all children are infected with the Epstein-Barr virus. In China, the Epstein-Barr virus is isolated in nearly all nasopharyngeal cancer specimens.

RNA-containing retroviruses, in particular, are known to cause cancers in animals. In humans, the retrovirus HTLV-1 (human T-cell lymphotropic virus, type 1) has been shown to cause adult T-cell leukemia. This disease occurs frequently in parts of Japan, the Caribbean, and Africa, particularly in regions where people are known to be infected with the virus.

Development of cancer is determined by a person's genetic profile plus exposure to environmental carcinogens.

22.4 Diagnosis and Treatment

Due to improved means of detection and treatment, there were 1,700 fewer deaths due to cancer in 1999 in the United States than in 1998. This seems like a small number until you realize that the population grew by 2.5 million and the average age continued to increase. The present cancer death rate is one in four persons. Researchers are optimistic that the rate of decline will accelerate and that each passing year will have far better statistics than the year before.

Diagnosis of Cancer

Diagnosis of cancer before metastasis is difficult, although treatment at this stage is usually more successful. The American Cancer Society, Inc., publicizes seven warning signals that spell out the word CAUTION and that everyone should be aware of:

C   hange in bowel or bladder habits;
A   sore that does not heal;
U   nusual bleeding or discharge;
T   hickening or lump in breast or elsewhere;
I   ndigestion or difficulty in swallowing;
O   bvious change in wart or mole;
N   agging cough or hoarseness.

Unfortunately, some of these symptoms are not obvious until cancer has progressed to one of its later stages.

Routine Screening Tests

An aim of medicine is to develop tests for cancer that are relatively easy to do, cost little, and are fairly accurate. So far, only the Pap test for cervical cancer fulfills these three requirements. A physician merely takes a sample of cells from the cervix, which are then examined microscopically for signs of abnormality. Regular Pap tests are credited with preventing over 90% of deaths from cervical cancer.

Breast cancer is not as easily detected, but three procedures are recommended. First, every woman should do a monthly breast self-examination (see the Health Focus on page 451). Second, during an annual physical examination, which is recommended especially for women above age 40, a physician does this same procedure. While helpful, this type of examination may not detect lumps before metastasis has already taken place. That is the goal of the third recommended procedure, *mammography,* which is an X-ray study of the breast (Fig. 22.6). However, mammograms do not show all cancers, and new tumors may develop in the interval between mammograms. The objective is that a mammogram will reveal a lump that is too small to be felt and at a time when the cancer is still highly curable.

Screening for colon cancer is also dependent upon three types of testing. A digital rectal examination performed by a physician is actually of limited value because only a portion of the rectum can be reached by finger. With flexible

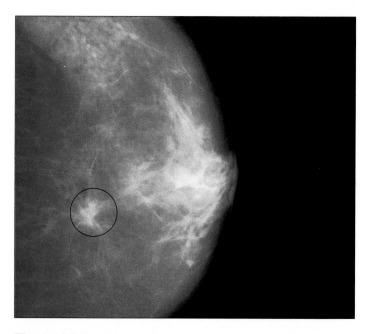

Figure 22.6   Mammogram.
An X-ray image of the breast can find tumors too small to be palpable.

sigmoidoscopy, the second procedure, a much larger portion of the colon can be examined by using a thin, pliable, lighted tube. Finally, a stool blood test (fecal occult blood test) consists of examining a stool sample to detect any hidden blood. The sample is smeared on a slide, and a chemical is added that changes color in the presence of hemoglobin. This procedure is based on the supposition that a cancerous polyp bleeds, although some polyps do not bleed, and bleeding is not always due to a polyp. Therefore, the percentage of false negatives and false positives is high. All positive tests are followed up by a colonoscopy, an examination of the entire colon, or by X ray after a barium enema. If the colonoscope detects polyps, they can be destroyed by laser therapy.

Other tests routinely used are blood tests to detect leukemia and urinalysis for the diagnosis of bladder cancer. Newer tests under consideration are tumor marker tests and tests for oncogenes.

Tumor Marker Tests

Tumor marker tests are blood tests for tumor antigens/antibodies. They are possible because tumors release substances that provoke an antibody response in the body. For example, if an individual has already had colon cancer, it is possible to use the presence of an antigen called CEA (for carcinoembryonic antigen) to detect any relapses. When the CEA level rises, additional tumor growth has occurred.

There are also tumor marker tests that can be used as an adjunct procedure to detect cancer in the first place. They are not reliable enough to count on solely, but in conjunction with physical examination and ultrasound (see following), they are considered useful. For example, there is a prostate-

Shower Check for Cancer

The American Cancer Society urges women to do a breast self-exam and men to do a testicle self-exam every month. Breast cancer and testicular cancer are far more curable if found early, and we must all take on the responsibility of checking for one or the other.

Breast Self-Exam for Women

1. Check your breasts for any lumps, knots, or changes about one week after your period.
2. Place your right hand behind your head. Move your *left* hand over your *right* breast in a circle. Press firmly with the pads of your fingers (Fig. 22A). Also check the armpit.
3. Now place your left hand behind your head and check your *left* breast with your *right* hand in the same manner as before. Also check the armpit.
4. Check your breasts while standing in front of a mirror right after you do your shower check. First, put your hands on your hips and then raise your arms above your head (Fig. 22B). Look for any changes in the way your breasts look: dimpling of the skin, changes in the nipple, or redness or swelling.

5. If you find any changes during your shower or mirror check, see your doctor right away.

You should know that the best check for breast cancer is a mammogram. When your doctor checks your breasts, ask about getting a mammogram.

Testicle Self-Exam for Men

1. Check your testicles once a month.
2. Roll each testicle between your thumb and finger as shown in Figure 22C. Feel for hard lumps or bumps.
3. If you notice a change or have aches or lumps, tell your doctor right away so he or she can recommend proper treatment.

Cancer of the testicles can be cured if you find it early. You should also know that prostate cancer is the most common cancer in men. Men over age 50 should have an annual health checkup that includes a prostate examination.

Information provided by the American Cancer Society. Used by permission.

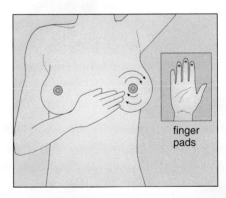

Figure 22A Shower check for breast cancer.

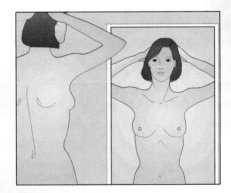

Figure 22B Mirror check for breast cancer.

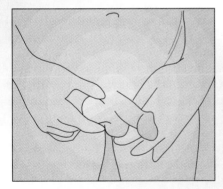

Figure 22C Shower check for testicular cancer.

specific antigen (PSA) test for prostate cancer, a CA-125 test for ovarian cancer, and an alpha-fetoprotein (AFP) test for liver tumors.

Genetic Tests

Tests for genetic mutations in proto-oncogenes and tumor-suppressor genes are making it possible to detect the likelihood of cancer before the development of a tumor. Tests are available that signal the possibility of colon, bladder, breast, thyroid cancers, and melanoma.

A *ras* oncogene can be detected in stool and urine sam-

ples. If the test for *ras* oncogene in the stool is positive, the physician suspects colon cancer, and if the test for *ras* oncogene in urine is positive, the physician suspects the presence of bladder cancer.

A genetic test is also available for the presence of *BRCA1* (breast cancer gene #1). A woman who has inherited this gene can choose to either have prophylactic surgery or to be frequently examined for signs of breast cancer.

Physicians now believe that a mutated *RET* gene means that thyroid cancer is present or may occur in the future, and a mutated *p16* gene appears to be associated with melanoma. Genetic testing can also be used to determine if cancer cells still remain after a tumor has been removed. In one study, 50% of the patients still had tumor cells in apparently "clean" margins or in lymph nodes believed to be free of them.

Microsatellites are small regions of DNA that always have di-, tri-, or tetranucleotide repeats. Physicians can use microsatellites to detect chromosomal deletions that accompany bladder cancer. They compare the number of nucleotide DNA repeats in a lymphocyte microsatellite to the number in a microsatellite of a cell found in urine. When the number of repeats is less in the cell from urine, a bladder tumor is suspected. Clinical trials involving this procedure have been so successful that a larger trial involving many institutions is now under way.

Telomerase, you will recall, is the enzyme that keeps telomeres a constant length in cells. The gene that codes for telomerase is turned off in normal cells but is active in cancer cells. Therefore if the test for the presence of telomerase is positive, the cell is cancerous.

Confirming the Diagnosis

There are ways to confirm a diagnosis of cancer without major surgery. Needle biopsies allow removal of a few cells for examination, and sophisticated techniques such as laparoscopy permit viewing of body parts. Computerized axial tomography (or CT scan) uses computer analysis of scanning X-ray images to create cross-sectional pictures that portray a tumor's size and location. Magnetic resonance imaging (MRI) is another type of imaging technique that depends on computer analysis. MRI is particularly useful for analysis of tumors in tissues surrounded by bone, such as tumors of the brain or spinal cord. A radioactive scan obtained after a radioactive isotope is administered can reveal any abnormal isotope accumulation due to a tumor. During ultrasound, echoes of high-frequency sound waves directed at a part of the body are used to reveal the size, shape, and location of tissue masses. Ultrasound can confirm tumors of the stomach, prostate, pancreas, kidney, uterus, and ovary.

There are standard procedures to detect specific cancers—for example, the Pap smear for cervical cancer, a mammogram for breast cancer, and the stool blood test for colon cancer. Tumor marker tests and genetic tests are new ways to test for cancer. Biopsy and imaging are used to confirm the diagnosis of cancer.

Treatment of Cancer

Certain standard methods have been used to treat cancer for quite some time. Other methods of therapy are in clinical trials and, if they prove to be successful, will become more generally available in the future.

Traditional Therapies

Surgery, radiation, and chemotherapy are the standard methods of cancer therapy (Fig. 22.7). Surgery alone is sufficient for cancer in situ. But because there is always the danger that some cancer cells were left behind, surgery is often preceded by and/or followed by radiation therapy.

Radiation   Radiation is mutagenic, and dividing cells such as cancer cells are more susceptible to its effects than other cells. The theory is that radiation will cause cancer cells to mutate and undergo apoptosis. Powerful X rays or gamma rays can be administered by using an externally applied beam or, in some instances, by implanting tiny radioactive sources into the patient's body. Cancer of the cervix and larynx, early stages of prostate cancer, and Hodgkin disease are often treated with radiation therapy alone.

Although X rays and gamma rays are the mainstays of radiation therapy, protons and neutrons also work well. Proton beams can be aimed at the tumor like a rifle bullet hitting the bull's-eye of a target.

Chemotherapy   Chemotherapy is a way to catch cancer cells that have spread throughout the body. Most chemotherapeutic drugs kill cells by damaging their DNA or interfering with DNA synthesis. The hope is that all cancer cells will be killed while leaving untouched enough normal cells to allow the body to keep functioning. Whenever possible, chemotherapy is specifically designed for the particular cancer. In Allen's lymphoma/leukemia, for example, it is known that a small portion of chromosome 9 is missing, and therefore DNA metabolism differs in the cancerous cells compared to normal cells. Specific chemotherapy for this cancer uses a drug designed to exploit this metabolic difference and destroy the cancerous cells.

One drug, taxol, extracted from the bark of the Pacific yew tree, was found to be particularly effective against advanced ovarian cancers as well as breast, head, and neck tumors. Taxol interferes with microtubules needed for cell division. Now chemists have synthesized a family of related drugs, called taxoids, which may be more powerful with less side effects than taxol itself.

Certain types of cancer, such as leukemias, lymphomas, and testicular cancer, are now successfully treated by combination chemotherapy alone. There is an 80% survival rate of children with childhood leukemia. Hodgkin disease, a lymphoma, once killed two out of three patients. Now, combination therapy using four different drugs can

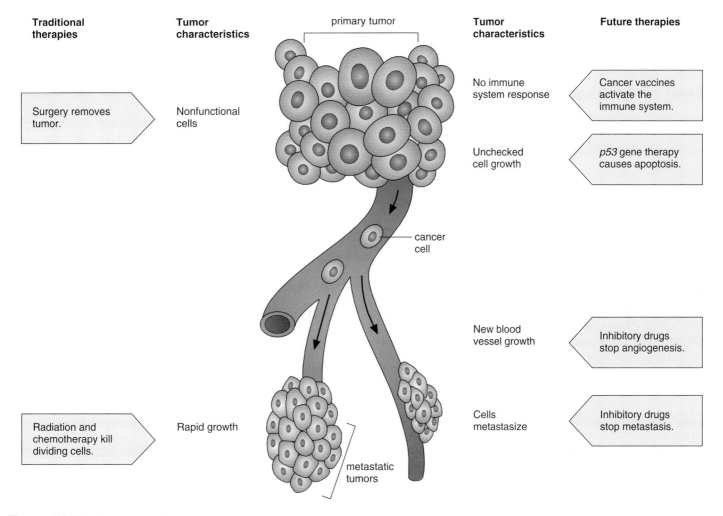

Figure 22.7   Treatment of cancer.
(Left) Traditional therapies are aimed at these stated tumor characteristics. *(Right)* Future therapies are aimed at these stated tumor characteristics.

wipe out the disease in a matter of months in three out of four patients, even when the cancer is not diagnosed immediately. In other cancers—most notably breast and colon cancer—chemotherapy can reduce the chance of recurrence after surgery has removed all detectable traces of the disease.

Chemotherapy sometimes fails because cancer cells become resistant to one or several chemotherapeutic drugs, a phenomenon called multidrug resistance. This occurs because all the drugs are capable of interacting with a plasma membrane carrier that pumps them out of the cell. Researchers are testing drugs known to poison the pump in an effort to restore the efficacy of the drugs.

Some physicians have found that they get better results when chemotherapy is used first, followed by surgery or radiation. This is called induced chemotherapy. Another possibility is to use combinations of drugs with nonoverlapping patterns of toxicity because cancer cells can't become resistant to many different types at once.

Bone Marrow Transplants   The red bone marrow contains large populations of dividing cells; therefore, red bone marrow is particularly prone to destruction by chemotherapeutic drugs. In bone marrow autotransplantation, a patient's stem cells are harvested and stored before chemotherapy begins. Quite high doses of radiation or chemotherapeutic drugs are then given within a relatively short period of time. This prevents multidrug resistance from occurring, and the treatment is more likely to catch each and every cancer cell. Then the stored stem cells, which are needed to produce blood cells, are returned to the patient by injection. They automatically make their way to bony cavities where they initiate blood cell formation.

Future Therapies

In the years ahead, doctors are likely to tailor therapy to suit each tumor's specific genetic mutations. The treatments used will probably be a combination of these approaches in order to combat cancer from many directions.

Cancer Vaccine Therapy   When cancer develops, the immune system has failed to dispose of cancer cells even though they bear antigens that make them different from the body's normal cells. A cancer vaccine stimulates the immune system to react to cancer cells' antigens. The first generation of cancer vaccines simply used treated tumor cells mixed with cytokines in an attempt to awaken the body's immune system. Increasingly, investigators are now using immune cells, genetically engineered to bear the tumor's antigens (Fig. 22.8). When these cells are returned to the body, they produce cytokines and present the antigen to cytotoxic T cells, which then go forth and destroy tumor cells in the body. A new vaccine of this type called Melacine mobilizes the immune system against melanoma. This vaccine is heading toward possible FDA approval because it has fewer serious side effects than chemotherapy.

Monoclonal Antibody Therapy   Monoclonal antibodies are antibodies of the same type because they are produced by the same plasma cell. Some investigators are experimenting with monoclonal antibodies designed to zero in on the receptor proteins of cancer cells. To increase the killing power of monoclonal antibodies, they are linked to radioactive isotopes or chemotherapeutic drugs. Herceptin is a monoclonal antibody that binds to a growth factor receptor found on the surface of about 30% of breast cancer cells. Because this monoclonal antibody can be combined with a chemother-apy agent, it delivers a double punch. In a recent trial, tumors shrank in 49% of the women receiving treatment.

p53 Gene Therapy   With greater understanding of genes and carcinogenesis, it is not unreasonable to think that gene therapy can help cure human cancers.

Recently, a retrovirus carrying a normal *p53* gene was injected directly into tumor cells of patients with lung cancer. The tumors shrank in three patients and stopped growing in the other three. Researchers believe that *p53* expression is only needed for 24 hours to trigger apoptosis, programmed cell death. And the *p53* gene seems to trigger cell death only in cancer cells—that is, elevating the *p53* level in a normal cell doesn't do any harm, possibly because apoptosis requires extensive DNA damage.

Some investigators prefer working with adenoviruses rather than retroviruses. Ordinarily when adenoviruses infect a cell, they first produce a protein that inactivates *p53*. In a cleverly designed procedure, investigators genetically engineered an adenovirus that lacks the gene for this protein. Now, the adenovirus can infect and kill only cells that lack a *p53* gene. Which cells are those? Tumor cells, of course. Another plus to this procedure is that the injected adenovirus spreads through the cancer, killing tumor cells as it goes. This genetically engineered virus is now in clinical trials.

Inhibitory Drug Therapy   The aim of inhibitory drugs is to prevent the spread of cancer or even prevent the disease from occurring in the first place. If angiogenesis and metastasis can be controlled, cancer might become a disease that you live with rather than die from.

Drugs that inhibit angiogenesis are a proposed therapy that is now well under way. Antiangiogenic drugs confine and reduce tumors by breaking up the network of new cap-

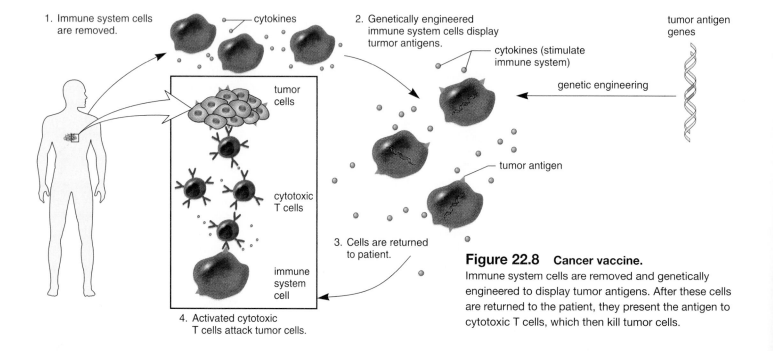

1. Immune system cells are removed.

cytokines

2. Genetically engineered immune system cells display turmor antigens.

tumor antigen genes

cytokines (stimulate immune system)

genetic engineering

tumor cells

cytotoxic T cells

tumor antigen

immune system cell

3. Cells are returned to patient.

4. Activated cytotoxic T cells attack tumor cells.

Figure 22.8   Cancer vaccine.
Immune system cells are removed and genetically engineered to display tumor antigens. After these cells are returned to the patient, they present the antigen to cytotoxic T cells, which then kill tumor cells.

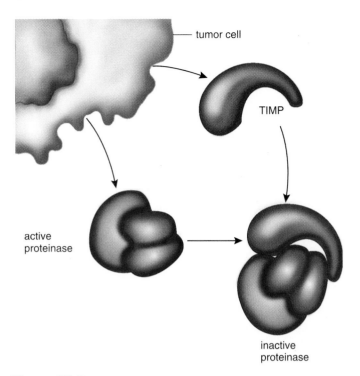

Figure 22.9   **Inhibitory drug therapy.**
Cancer cells produce proteinase enzymes that allow them to metastasize and TIMP, a proteinase inhibitor. Perhaps it would be possible to isolate, produce, and administer the TIMP to cancer patients. In this way, metastasis would be prevented.

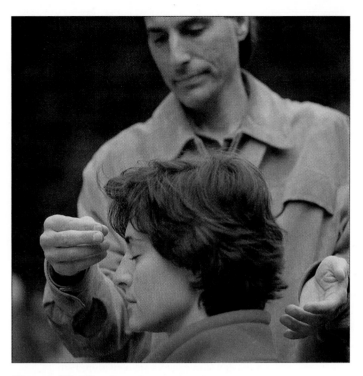

Figure 22.10   **Complementary therapy.**
Some cancer patients turn to complementary therapy for assistance. One form of complementary therapy is the transfer of energy from one person to another in order to restore the recipient to good health.

illaries in the vicinity of a tumor. A number of antiangiogenic compounds are currently being tested in clinical trials. Two highly effective drugs, called angiostatin and endostatin, have been shown to inhibit angiogenesis in laboratory animals and are expected to do the same in humans.

Drugs that inhibit metastasis are a distinct possibility. When cancer cells derived from epithelial cells metastasize, they produce proteinase enzymes that allow them to make their way through the basement membrane and the wall of capillaries. Surprisingly, cancer cells themselves produce an inhibitor of proteinase enzymes. Proteinase action occurs only if the number of enzyme molecules is greater than the number of TIMP (tissue inhibitor metalloproteinase) molecules. This means that TIMPs or drugs that act like them may offer an approach to prevent metastasis (Fig. 22.9).

Other types of inhibitory drugs are also known. In promyelocytic leukemia, for example, cells have too many retinoic acid (a cousin of vitamin A) receptor proteins. Strangely, if retinoic acid is administered, differentiation occurs and cell division ceases. The ultimate in inhibitory drugs is one that prevents carcinogenesis. The hormones estrogen and progesterone sometimes promote breast cancer. Tamoxifen is a drug that binds to estrogen receptor proteins and thereby prevents estrogen from binding. For 25 years, tamoxifen has been used to limit breast cancer recurrence in women already treated for the disease. In a recent trial to see if tamoxifen could prevent breast cancer from ever occurring,

twice as many women receiving a placebo developed breast cancer compared to the group that received tamoxifen. Unfortunately, however, tamoxifen seemed to increase the risk of uterine cancer. Still, researchers and others are cheered by the fact that breast cancer prevention appears to be a possibility. Another drug, called raloxifene, is now being tested to see if it can prevent breast cancer without increasing the risk of uterine cancer.

Complementary Therapy   Many patients are interested in therapies outside the mainstream of conventional medicine. The list of possible alternative therapies includes tai chi, homeopathy, biofeedback, acupuncture, and exotic foods. Although there have been no studies to show which of these might be beneficial, many patients still want to avail themselves of therapies that might better be called complementary therapies. Due to the insistence of patients, institutions are beginning to make these therapies available (Fig. 22.10). The goal is to develop a way of working with cancer patients that integrates both conventional and alternative therapies.

Surgery, followed by radiation and/or chemotherapy, is the standard method of treating cancer. Many new therapies are undergoing clinical trials. The hope is that combination therapy can allow people to live with cancer rather than to die from cancer.

Prevention of Cancer

There is clear evidence that the risk of certain types of cancer can be reduced by adopting protective behaviors and the right diet.

Protective Behaviors

These behaviors help prevent cancer:

Don't smoke   Cigarette smoking accounts for about 30% of all cancer deaths. Smoking is responsible for 90% of lung cancer cases among men and 79% among women—about 87% altogether. Those who smoke two or more packs of cigarettes a day have lung cancer mortality rates 15 to 25 times greater than nonsmokers. Smokeless tobacco (chewing tobacco or snuff) increases the risk of cancers of the mouth, larynx, throat, and esophagus.

Don't sunbathe   Almost all cases of basal-cell and squamous-cell skin cancers are considered sun related. Further, sun exposure is a major factor in the development of melanoma, and the incidence of this cancer increases for those living near the equator.

Avoid alcohol   Cancers of the mouth, throat, esophagus, larynx, and liver occur more frequently among heavy drinkers, especially when accompanied by tobacco use (cigarettes or chewing tobacco).

Avoid radiation   Excessive exposure to ionizing radiation can increase cancer risk. Even though most medical and dental X rays are adjusted to deliver the lowest dose possible, unnecessary X rays should be avoided. Excessive radon exposure in homes increases the risk of lung cancer, especially in cigarette smokers. It is best to test your home and take the proper remedial actions.

Be tested for cancer   Do the shower check for breast cancer or testicular cancer. Have other exams done regularly by a physician.

Be aware of occupational hazards   Exposure to several different industrial agents (nickel, chromate, asbestos, vinyl chloride, etc.) and/or radiation increases the risk of various cancers. Risk from asbestos is greatly increased when combined with cigarette smoking.

Be aware of hormone therapy   Estrogen therapy to control menopausal symptoms increases the risk of endometrial cancer. Including progesterone in estrogen replacement therapy helps minimize this risk. Some studies suggest that estrogen alone is less likely to cause breast cancer than is the combined therapy.

The Right Diet

Statistical studies have suggested that persons who follow certain dietary guidelines are less likely to have cancer. The following dietary guidelines greatly reduce your risk of developing cancer:

Avoid obesity   The risk of cancer (especially colon, breast, and uterine cancers) is 55% greater among obese women and 33% greater among obese men, compared to people of normal weight.

Lower total fat intake   A high-fat intake has been linked to development of colon, prostate, and possibly breast cancers.

Eat plenty of high-fiber foods   These include whole-grain cereals, fruits, and vegetables. Studies have indicated that a high-fiber diet protects against colon cancer, a frequent cause of cancer deaths. It is worth noting that foods high in fiber also tend to be low in fat!

Increase consumption of foods that are rich in vitamins A and C   Beta-carotene, a precursor of vitamin A, is found in dark green, leafy vegetables; carrots; and various fruits. Vitamin C is present in citrus fruits. These vitamins are called antioxidants because in cells they prevent the formation of free radicals (organic ions that have an unpaired electron) that can possibly damage DNA. Vitamin C also prevents the conversion of nitrates and nitrites into carcinogenic nitrosamines in the digestive tract.

Reduce consumption of salt-cured, smoked, or nitrite-cured foods   Salt-cured or pickled foods may increase the risk of stomach and esophageal cancers. Smoked foods, like ham and sausage, contain chemical carcinogens similar to those in tobacco smoke. Nitrites are sometimes added to processed meats (e.g., hot dogs and cold cuts) and other foods to protect them from spoilage; as mentioned previously, nitrites are converted to nitrosamines in the digestive tract.

Include vegetables from the cabbage family in the diet   The cabbage family includes cabbage, broccoli, brussels sprouts, kohlrabi, and cauliflower. These vegetables may reduce the risk of gastrointestinal and respiratory tract cancers.

Be moderate in the consumption of alcohol   People who drink and smoke are at an unusually high risk for cancers of the mouth, larynx, and esophagus.

Bioethical Focus

Tobacco and Alcohol Use

The adjacent Health Focus promotes a healthy lifestyle as a way to prevent the occurrence of cancer. Tobacco and alcohol use in particular increase our chances of developing cancer. Smoking one to two packs of cigarettes a day increases the chances of lung cancer some twenty times, cancer of the larynx eight times, cancer of the mouth and pharynx four times, cancer of the esophagus three times, and cancer of the bladder two times. Users of smokeless tobacco (snuff and chewing tobacco) are at greater risk for oral cancer. A nonsmoker married to a smoker still has a 25% greater risk of lung cancer because of passive smoking. Alcohol plays a role in the development of cancers of the oral cavity, pharynx, esophagus, larynx, liver, and possibly breast. The combination of alcohol and tobacco is particularly dangerous. Together, they account for 75–85% of all cancers of the mouth, pharynx, and esophagus in the U.S.

Do you think society has an obligation to educate young people about the detrimental health effects of using tobacco and alcohol? If yes, do you think public schools should assume this responsibility? Why or why not? What else could be done to educate the public? Should young people be punished if they are caught using tobacco and alcohol? By whom? Or should they be allowed to choose their own lifestyle despite the possibility of future illness?

When people are incapacitated and no longer able to work, they often become a burden on society. The government pays their medical bills and may even be called upon to provide their daily needs. Should the taxes of those of us who have been protective of our health go to pay these expenses?

Figure 22D Tobacco smoke and alcohol.
Smoking cigarettes, especially when combined with drinking alcohol, is associated with cancer of the lungs, larynx, mouth and pharynx, bladder, and many other organs.

Decide Your Opinion

1. Should society require that all people be educated about the detrimental effects of tobacco and alcohol use?
2. Even so, should people be allowed to smoke and drink alcohol if it is their choice? Why or why not?
3. Should companies be required to employ people who use alcohol and tobacco? Should society be required to pay the medical bills of those who are ill due to tobacco and alcohol use? Explain your answers.

Looking at Both Sides   www.mhhe.com/biosci/genbio/maderhuman7/

Every bioethical issue has at least two sides. Even if you already have your opinion, it is important to explore the opposite opinion before finalizing your position. The Online Learning Center at www.mhhe.com/biosci/genbio/maderhuman7/ will help you fine-tune your initial opinion, explore both sides, and finalize your position. You may acquire new arguments for your original opinion, or you may even change your opinion. Be sure to complete these activities in sequence:

Taking Sides   Decide your initial opinion by answering a series of questions. Then see if your opinion changes after completing the next two activities.

Further Debate   Read opposing articles that give you further information on this particular bioethical issue.

Explain Your Position   Answer another series of questions and then defend your original or changed opinion. You can e-mail your position to your instructor if he or she wishes.

Summarizing the Concepts

22.1 Cancer Cells

Cancer cells have characteristics that distinguish them from normal cells. They are nondifferentiated and enter the cell cycle repeatedly. They have abnormal nuclei with many chromosomal irregularities. They form tumors because they do not exhibit contact inhibition. They induce angiogenesis and cause nearby blood vessels to form a capillary network that services the tumor. They metastasize and form tumors in other parts of the body.

22.2 Origin of Cancer

Each cell contains two types of regulatory pathways. The receptor proteins for growth factors, signaling proteins, and proto-oncogenes in the stimulatory network promote the cell cycle and cause the cell to divide. The receptor proteins for growth-inhibitory factors, the signaling proteins, and tumor-suppressor genes in the inhibitory pathway repress the cell cycle and cause the cell to stop dividing. Whether the cell divides or not depends on which pathway is most active at the time.

When proto-oncogenes mutate, becoming oncogenes, and tumor-suppressor genes mutate, the regulatory pathways no longer function as they should, and uncontrolled growth results. Usually, cells that bear damaged DNA undergo apoptosis. Apoptosis fails to take place in cancer cells.

22.3 Causes of Cancer

Certain environmental factors are carcinogens. They bring about mutations that lead to cancer. Ultraviolet radiation is a well-known car-

cinogen for melanoma. Carcinogenic organic chemicals include those in tobacco smoke, foods, and pollutants. Tobacco smoke and diet are believed to account for 60% of all cancer deaths. Industrial chemicals, including pesticides and herbicides, are carcinogenic. Certain viruses, such as hepatitis B, human papillomaviruses, and Epstein-Barr virus, cause specific cancers. Cancers that run in families are most likely due to the inheritance of mutated genes that contribute to carcinogenesis.

22.4 Diagnosis and Treatment

There are procedures to detect specific cancers—for example, the Pap test for cervical cancer, a mammogram for breast cancer, and the stool blood test for colon cancer. But new ways, such as tumor marker tests and tests for oncogenes and mutated tumor-suppressor genes, are being developed. Biopsy and imaging are used to confirm the diagnosis of cancer.

Surgery, followed by radiation and/or chemotherapy, is the traditional method of treating cancer. Chemotherapy involving bone marrow transplants has now become fairly routine. Future methods of therapy include cancer vaccines, *p53* gene therapy, inhibitory drugs for angiogenesis and metastasis, and complementary therapies. Cancer vaccines stimulate the immune system to find and destroy cancer cells. One type of *p53* gene therapy assures that cancer cells undergo apoptosis. Inhibitory drugs are being developed to prevent angiogenesis and metastasis. There are also drugs that inhibit cell growth, and others that prevent carcinogenesis in the first place. Many patients are increasingly interested in complementary therapy of various types.

Studying the Concepts

1. Why is cancer called a genetic disease? 444
2. List and discuss five characteristics of cancer cells that distinguish them from normal cells. 444
3. What are oncogenes and tumor-suppressor genes? What role do they play in the regulatory pathways that control cell division and involve receptor proteins and signaling proteins? 446–47
4. What is apoptosis, and what role does it play in carcinogenesis? 448
5. Name three types of carcinogens, and give examples of each type. What role does heredity play in the development of cancer? 448–49
6. What are the standard ways to detect cervical cancer, breast cancer, and colon cancer? 450–51
7. Describe and give examples of tumor marker tests and genetic tests for oncogenes and mutated tumor-suppressor genes. 450–52
8. What are the traditional therapies for treatment for cancer? Explain why a bone marrow transplant is sometimes used in conjunction with chemotherapy. 452–53
9. What is the most recent way to prepare a cancer vaccine? 454

10. Explain the manner in which monoclonal antibodies are used to fight cancer. 454
11. Describe two investigative therapies utilizing the *p53* gene. Why would you expect them to be successful? 454
12. Explain the rationale for inhibitory drugs for angiogenesis and metastasis. 454–55
13. Why were investigators cheered by the results of a clinical trial in which women took tamoxifen? 455
14. Give some examples of complementary therapies. 455

Understanding Key Terms

angiogenesis 444	mutagen 448
apoptosis 444	oncogene 446
cancer 444	proto-oncogene 446
carcinogen 448	telomere 446
carcinogenesis 444	tumor 444
leukemia 446	tumor marker tests 450
metastasis 444	tumor-suppressor gene 446

Testing Your Knowledge of the Concepts

In questions 1–4, match the therapy to the description.
a. cancer vaccine therapy
b. *p53* gene therapy
c. inhibitory drug therapy
d. radiation

_____ 1. Inject a virus that carries a gene for apoptosis.

_____ 2. Administer angiostatin to prevent angiogenesis.

_____ 3. Activate cytotoxic T cells to attack tumor cells.

_____ 4. Use proton beams that can be aimed directly at a tumor.

In questions 5–7, indicate whether the statement is true (T) or false (F).

_____ 5. Without angiogenesis, cancer cells are not able to survive.

_____ 6. The newer cancer vaccines utilize immune cells that have been genetically engineered to display a tumor's antigens.

_____ 7. Since standard chemotherapy works so well, there is no need to develop new methods of therapy for cancer.

In questions 8–13, fill in the blanks.

8. The _____ virus is a carcinogen for cancer of the cervix.

9. Autotransplants of bone marrow stem cells permit a much higher dosage of _____ than otherwise.

10. Cancer cells _____, meaning that they travel to distant body parts and start new tumors.

11. The mutation of _____ and _____ genes leads to uncontrolled cell growth and cancer.

12. Cancer cells are sensitive to radiation therapy and chemotherapy because they are constantly _____.

13. Monoclonal _____ carrying a chemotherapeutic drug are able to find and destroy cancer cells.

14. Label this diagram:

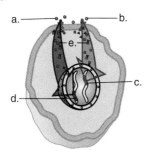

Further Readings

Berns, M. W. April 1998. Laser scissors and tweezers. *Scientific American* 278(62):4. New laser techniques allow manipulation of chromosomes and other structures inside cells.

Carr, A. M. March 10, 2000. Piecing together the *p53* puzzle. *Science* 287(5459):1765. Article discusses *p53* gene in detail.

Ferber, D. November 26, 1999. GM (genetically modified) crops in the cross hairs. *Science* 286(5445):1662. The risks of GM crops to health and the environment are examined.

Guerinot, M. L. January 14, 2000. The green revolution strikes gold. *Science* 287(5451):241. A genetically modified new grain, β-carotene-enriched "golden rice," will be available to developing nations of the world.

Karow, J. The "other" genomes. July 2000. *Scientific American* 283(1):53. Simple organisms are being harnessed to find new drugs for humans.

Kufe, D. W. March 2000. Smallpox, polio and now a cancer vaccine? *Nature Medicine* 6(3):252. Fusions of dendritic cells and renal carcinoma cells have been used as a vaccine in patients with metastatic renal cancer.

Lander, E. S., and Weinberg, R. A. March 10, 2000. Journey to the center of biology. *Science* 287(5459):1777. Article chronicles the discovery of the mechanisms of heredity, from Mendel to genomic science and technology; also discusses the future of genetics.

McKinnell, R. G., and DiBerardino, M. A. November 1999. The biology of cloning: History and rationale. *BioScience* 49(11):875. Cloning of various organisms including Dolly, amphibians, insects, fish, and mammals is discussed.

Nagel, R. L. September 2000. Molecule, heal thyself. *Natural History* 109(7):30. Genetics is rapidly transforming how we think about disease—probably the most important insight of the past decade is that all diseases are always both genetic and environmental.

Pennisi, E. May 12, 2000. Chromosome 21 done, phase two begun. *Science* 288(5468):939. With the complete sequencing of chromosome 21 in hand, the task of figuring out just what goes wrong in Down syndrome patients will be far easier.

Science 291(5507) February 16, 2001. This issue contains a series of articles on the human genome and how genomic data may be used.

Service, R. F. April 7, 2000. Hazel trees offer new source of cancer drug. *Science* 288(5463):27. Researchers announced that they have isolated taxol from hazelnut trees and fungi, which could lead to an abundant new source of the drug.

Shreeve, J. October 1999. Secrets of the gene. *National Geographic* 196(4):43. Article covers many genetic advances, such as gene therapy, forensics, and evolution.

Van Noorden, C. J. F., et al. March/April 1998. Metastasis. *American Scientist* 86(2):130. The mechanisms by which cancer cells metastasize are discussed.

Weiner, D. B., and Kennedy, R. C. July 1999. Genetic vaccines. *Scientific American* 281(1):50. Bits of DNA or RNA, if introduced into cells, can stimulate powerful immune responses against viruses, bacteria, and some cancers.

Wikonkal, N. M., and Brash, D. E. September/October 1998. Squamous cell carcinoma. *Science & Medicine* 5(5):18. Mutations of tumor-suppressor gene *p53* are commonly found in squamous cell carcinomas.

Zumstein, L. A., and Lundblad, V. October 1999. Telomeres: Has cancer's Achilles heel been exposed? *Nature Medicine* 5(10):1129. Telomerase is expressed in the majority of human cancers, but new data suggest that inhibition of telomere replication can lessen cancer cell growth.

e-Learning Connection **www.mhhe.com/biosci/genbio/maderhuman7/**

22.1 Cancer Cells

Gene Regulation *Essential Study Partner*

22.2 Origin of Cancer

Cell Cycle *art quiz*

22.3 Causes of Cancer

Potential Carcinogen—Pollution *art quiz*
Potential Carcinogen—Diet *art quiz*
Potential Carcinogen—Smoking *art quiz*

Skin Cancer *reading*

22.4 Diagnosis and Treatment

Looking at Both Sides *critical thinking activity*

Chapter Summary

Key Term Flashcards *vocabulary quiz*
Chapter Quiz *objective quiz covering all chapter concepts*

PART

VII

Human Evolution and Ecology

Like all living things, human beings can trace their ancestry to the first cell(s) to have evolved about 3.5 billion years ago. A chemical evolution produced the first cell(s), and then biological evolution began. The evidence for biological evolution is strong. Living things share biochemical and anatomical characteristics, and the fossil record tells the history of life. Researchers have found that human evolution is complex but still follows the same patterns of evolution as other groups.

Living things live in ecosystems, such as a forest or a pond, where they interact with each other and with the physical environment. Almost all ecosystems today have been impacted by human activities. We find it hard to predict the result of our many activities, and only now do we realize that the biodiversity of the biosphere is threatened. Conservation biology is a new field that arose in response to this crisis. Its goal is to protect biodiversity by preserving ecosystems for the benefit of all living things, including human beings.

Chapter 23

Human Evolution

Chapter Concepts

23.1 Origin of Life
- A chemical evolution produced the first cell(s).   462

23.2 Biological Evolution
- All living things are descended from the first cell(s), and therefore they share a cellular structure and a common chemistry.   463
- Darwin gathered evidence for common descent and proposed a mechanism for natural selection.   464
- The process of natural selection accounts for the great diversity of living things.   465

23.3 Humans Are Primates
- Humans (*Homo sapiens*) are in the order Primates, which are mammals adapted to living in trees.   466
- Primate characteristics include an enlarged brain, opposable thumb, stereoscopic vision, and an emphasis on learned behavior.   466

23.4 Evolution of Australopithecines
- About 3 MYA,[1] several species of australopithecines (the first hominids) evolved in Africa. The australopithecines had a small brain, but they walked erect.   468

23.5 Evolution of Humans
- About 2 MYA, *Homo habilis*, the first hominid to make tools, was most likely a hunter, and may have been able to speak.   469
- About 1.9 to 1.6 MYA, *Homo erectus* migrated out of Africa and was a big-game hunter that possessed fire.   470
- The out-of-Africa hypothesis says that modern humans evolved in Africa, and after migrating to Europe and Asia (about 100,000 BP[2]), they replaced any *Homo* species there, including the Neanderthals.   471
- Cro-Magnon is the name given to modern humans who made sophisticated tools and definitely had culture.   472

[1]MYA = millions of years ago
[2]BP= before present

Wambui was busy tidying the inside of her hut in the grasslands of western Kenya. She looked down at her son sleeping on a mat. She hoped that this child would not die of malaria as had her daughter the year before. She thought that her sister, Saki, who lived in the same compound, was extremely lucky because none of her children had died of malaria. But Saki was a widow—her husband had died of sickle-cell disease. Now her sons had no father to guide them when they tended animals and worked in the fields.

Wambui didn't realize that her children were susceptible to malaria because they were homozygous for normal hemoglobin ($Hb^A Hb^A$; see page 413). The malaria parasite is able to live and multiply in normal red blood cells. Saki's children were heterozygous ($Hb^A Hb^S$) and had sickle-cell trait because her husband had passed them the Hb^S allele. The malaria parasite can't live in the red blood cells of sickle-cell trait individuals because the parasite makes red cells become sickle-shaped and lose potassium. Sometimes Saki's children tired easily, but they did not contract malaria.

Biologists are interested in why sickle-cell disease persists in Africa when it is clearly a debilitating disease. The answer is that the heterozygote ($Hb^A Hb^S$) has an evolutionary advantage—namely, protection from malaria. The survival of the heterozygote in malaria-infested Africa assures the continuance of individuals with sickle-cell disease who die from circulatory problems caused by red blood cells that are always sickle-shaped.

23.1   Origin of Life

A fundamental principle of biology states that all living things are made of cells and that every cell comes from a pre-existing cell. But if this is so, how did the first cell come about? Since it was the very first living thing, it had to come from nonliving chemicals. Could there have been a slow increase in the complexity of chemicals—could a **chemical evolution** have produced the first cell(s) on the primitive earth?

The Primitive Earth

We know that the solar system was in place about 4.6 billion years ago (BYA) and that the gravitational field of the earth allowed it to have an atmosphere. The earth's primitive atmosphere was not the same as today's atmosphere. Most likely, the primitive atmosphere was formed by gases escaping from volcanoes. If so, the primitive atmosphere would have consisted mostly of water vapor (H_2O), nitrogen (N_2), and carbon dioxide (CO_2), with only small amounts of hydrogen (H_2) and carbon monoxide (CO). The primitive atmosphere had little, if any, free oxygen.

At first, the earth and its atmosphere were extremely hot, and water, existing only as a gas, formed dense, thick clouds. Then, as the earth cooled, water vapor condensed to liquid water, and rain began to fall. It rained in such enormous quantities over hundreds of millions of years that the oceans of the world were produced.

Small Organic Molecules

The rain washed the other gases, such as nitrogen and carbon dioxide, into the oceans (Fig. 23.1a). The primitive earth had many sources of energy, including volcanoes, meteorites, radioactive isotopes, lightning, and ultraviolet radiation. In the presence of so much available energy, the primitive gases may have reacted with one another and produced small organic compounds, such as nucleotides and amino acids (Fig. 23.1b). In 1953, Stanley Miller performed an experiment. He placed a mixture of primitive gases in a closed system, heated the mixture, and circulated it past an electric spark. After cooling, Miller discovered a variety of small organic molecules in the resulting liquid. Other investigators have achieved the same results using various mixtures of gases.

Macromolecules

The newly formed small organic molecules likely joined to produce organic macromolecules (Fig. 23.1c). There are two hypotheses of special interest concerning this stage in the origin of life. One is the **RNA-first hypothesis,** which suggests that only the macromolecule RNA (ribonucleic acid) was needed at this time to progress toward formation of the first cell(s). This hypothesis was formulated after the discovery that RNA can sometimes be both a substrate and an enzyme. Such RNA molecules are called ribozymes. Perhaps RNA could have carried out the processes of life commonly associated today with DNA (deoxyribonucleic acid) and proteins. Those who support this hypothesis say that it was an "RNA world" some 3.5 billion years ago.

Another hypothesis is termed the **protein-first hypothesis.** Sidney Fox has shown that amino acids join together when exposed to dry heat. He suggests that amino acids collected in shallow puddles along the rocky shore, and the heat of the sun caused them to form proteinoids, small polypeptides that have some catalytic properties. When proteinoids are returned to water, they form microspheres, structures composed only of protein that have many of the properties of a cell.

The Protocell

A cell has a lipid-protein membrane. Fox has shown that if lipids are made available to microspheres, the two tend to become associated, producing a lipid-protein membrane. A **protocell,** which could carry on metabolism but could not reproduce, would have come into existence in this manner.

The protocell would have been able to use the still abundant small organic molecules in the ocean as food. Therefore, the protocell is believed to have been a **heterotroph,** an organism that takes in preformed food. Further, the protocell would have been a fermenter because there was no free oxygen.

The True Cell

A true cell can reproduce, and in today's cells DNA replicates before cell division occurs. Enzymatic proteins carry out the replication process.

How did the first cell acquire both DNA and enzymatic proteins? Those who support the RNA-first hypothesis propose a series of steps. According to this hypothesis, the first cell had RNA genes, which like messenger RNA, could have specified protein synthesis. Some of the proteins formed would have been enzymes. Perhaps one of these enzymes, like reverse transcriptase found in retroviruses, could use RNA as a template to form DNA. Replication of DNA would then proceed normally.

Those who support the protein-first hypothesis suggest that some of the proteins in the protocell would have evolved the enzymatic ability to synthesize DNA from nucleotides in the ocean. Then DNA would have gone on to specify protein synthesis, and in this way the cell acquired all its enzymes, even the ones that replicate DNA.

A chemical evolution produced the first cell(s), and then biological evolution began.

23.2 Biological Evolution

The first true cells were the simplest of life forms; therefore, they must have been **prokaryotic cells** (monerans) which lack a nucleus. Later, **eukaryotic cells** (protists) which have nuclei evolved. Then, multicellularity and the other kingdoms (fungi, plants, and animals) evolved (see Figure 1.2). Obviously, all these types of organisms—even prokaryotic cells—are alive today. Each type of organism has its own evolutionary history that is traceable back to the first cell(s).

Biological evolution is a change in life forms that has taken place in the past and will take place in the future. Biological evolution has two important aspects: descent from a common ancestor and adaptation to the environment. Descent from the original cell(s) explains why all living things have a common chemistry and a cellular structure. An **adaptation** is a characteristic that makes an organism able to survive and reproduce in its environment. Adaptations to different environments explain the diversity of life—why there are so many different types of living things.

Biological evolution explains both the unity (sameness) and diversity of life.

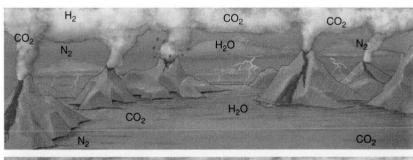
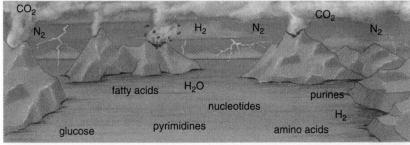
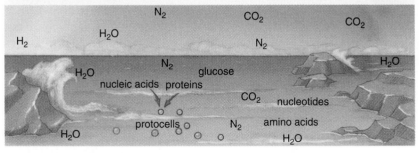

a. The primitive atmosphere contained gases, including H_2O, CO_2, and N_2, that escaped from volcanoes. As the water vapor cooled, some gases were washed into the oceans by rain.

b. The availability of energy from volcanic eruption and lightning allowed gases to form small organic molecules, such as nucleotides and amino acids.

c. Small organic molecules could have joined to form proteins and nucleic acids, which became incorporated into membrane-bounded spheres. The spheres became the first cells, called protocells. Later protocells became true cells that could reproduce.

Figure 23.1 Chemical evolution.
A chemical evolution is believed to have produced the protocell, which became a true cell once it had genes composed of DNA and could reproduce.

Common Descent

Charles Darwin was an English naturalist who first formulated the theory of evolution that has since been supported by so much independent data. At the age of 22, Darwin sailed around the world as the naturalist on board the HMS *Beagle.* Between 1831 and 1836, the ship sailed in the tropics of the Southern Hemisphere, where life forms are more abundant and varied than in Darwin's native England.

Even though it was not his original intent, Darwin began to realize and to gather evidence that life forms change over time and from place to place. The types of evidence that convinced Darwin that common descent occurs were fossil, biogeographical, anatomical, and biochemical.

Fossil Evidence

Fossils are the remains and traces of past life or any other direct evidence of past life. Traces include trails, footprints, burrows, worm casts, or even preserved droppings. Most fossils consist only of hard parts, such as shells, bones, or teeth, because these parts are usually not consumed or destroyed.

The **fossil record** is the history of life recorded by remains from the past. The fossil record supports common descent because similarity of form allows fossils to be linked over time despite observed changes. Darwin found the remains of a sloth and an armadillo, and the similarity of the fossils to the living animals made him realize that the fossils

Figure 23.2   Fossils.

The modern horse, *Equus,* and an extinct ancestor, *Mesohippus,* from the fossil record.

were related to the modern animals. Figure 23.2 shows the similarities between the modern horse and one of its ancestors from the fossil record.

Biogeographical Evidence

Biogeography is the study of the distribution of plants and animals throughout the world. Darwin noted that South America had no rabbits, even though the environment was quite suitable to them. He concluded there are no rabbits in South America because rabbits originated somewhere else and had no means to reach South America.

Instead of rabbits, the Patagonian hare exists in the grasslands of South America. This hare has long legs and ears but the face of a guinea pig. Both rabbits and Patagonian hares eat grass, hide in bushes, and move rapidly using long hind legs. Darwin began to think that the Patagonian hare was related to a guinea pig through common descent and that rabbits and Patagonian hares share the same characteristics because they are both adapted to the same type of environment.

Anatomical Evidence

Darwin was able to show that a common descent hypothesis offers a plausible explanation for anatomical similarities among living organisms. Vertebrate forelimbs are used for flight (birds and bats), orientation during swimming (whales and seals), running (horses), climbing (arboreal lizards), or swinging from tree branches (monkeys). Despite dissimilar functions, all vertebrate forelimbs contain the same sets of bones organized in similar ways. The most plausible explanation for this unity is that the basic forelimb plan belonged to a common ancestor and the plan was modified in the various groups as each continued along its own evolutionary pathway. The shared characteristics of vertebrates also explain why their embryological development is so similar.

Vertebrate forelimbs are **homologous structures.** Homologous structures indicate that organisms are related to one another through common descent. **Analogous structures,** such as the wings of birds and insects, have the same function but do not have a common ancestry.

Biochemical Evidence

Darwin was not aware as we are today that almost all living organisms use the same basic biochemical molecules, including DNA and ATP. Also, organisms use the DNA triplet code and the same 20 amino acids in their proteins. There are no functional reasons why these chemicals need to be so similar. But their similarity can be explained by hypothesizing descent from a common ancestor.

Natural Selection

When Darwin returned home, he spent the next 20 years gathering data to support the principle of biological evolution. His most significant contribution to this principle was a process for **natural selection,** by which a species becomes adapted to its environment. On his trip aboard the *Beagle*, Darwin visited the Galápagos Islands. There he saw a number of finches that resembled one another but had different ways of life. Some were seed-eating ground finches, some cactus-eating ground finches, and some insect-eating tree finches. A warbler-type finch had a beak that could take honey from a flower. A woodpecker- type finch lacked the long tongue of a woodpecker but could use a cactus spine or twig to ferret out insects from the bark of a tree. Darwin thought the finches were all descended from a mainland ancestor whose offspring had spread out among the islands and become adapted to different environments. Darwin believed that organisms do not strive to adapt themselves to the environment; instead, the environment selects for survival and reproduction those organisms that happen to have a characteristic that gives them an advantage in a particular environment.

In order to emphasize the nonteleological nature of the natural selection process, it is often contrasted with a process espoused by Jean-Baptiste Lamarck, another nineteenth-century naturalist. Lamarck's explanation for the long neck of the giraffe was based on the assumption that the ancestors of the modern giraffe were trying to reach into the trees to browse on high-growing vegetation (Fig. 23.3). Continual stretching of the neck caused it to become longer, and this acquired characteristic was passed on to the next generation. Lamarck's process is **teleological** because, according to him, the desired outcome is known ahead of time. This type of explanation has not stood the test of time. Natural selection, on the other hand, has been fully substantiated by later investigators.

Here are the critical elements of the process:

- Variations. Individual members of a species vary in physical characteristics. Physical variations can be passed from generation to generation. (Darwin was never aware of genes, but we know today that the inheritance of the genotype determines the phenotype.)
- Struggle for existence. Even though each species could eventually produce many descendants, the number in each generation usually stays about the same. Why? Because resources are limited, and only certain ones survive and reproduce.
- Survival of the fittest. Some members of a species are able to capture more resources in a particular environment because they have a characteristic that gives them an advantage over other members of the species. The environment "selects" these better-adapted members to have offspring and therefore to pass on this characteristic.
- Adaptation. Each subsequent generation includes more individuals that are adapted in the same way to the environment.

Can natural selection account for the great diversity of life? Yes, if we are aware that environments differ widely and life has been evolving for a very long time.

Early giraffes probably had short necks that they stretched to reach food.

Early giraffes probably had necks of various lengths.

Their offspring had longer necks that they stretched to reach food.

Natural selection due to competition led to survival of the longer-necked giraffes and their offspring.

Eventually, the continued stretching of the neck resulted in today's giraffe.

Eventually, only long-necked giraffes survived the competition.

a. Lamarck's theory

b. Darwin's theory

Figure 23.3   Mechanism of evolution.
This diagram contrasts Jean-Baptiste Lamarck's process of acquired characteristics with Charles Darwin's process of natural selection. Only natural selection is supported by data.

Biological evolution explains both the unity and diversity of life. Living things share characteristics because they are all descended from the first cell(s), and they are diverse because they are adapted to different environments.

23.3   Humans Are Primates

For further study of human evolution, we can turn to the classification of humans. Biologists classify organisms according to their evolutionary relatedness. The organisms in a particular classification category share a set of common anatomical and molecular characteristics. The classification categories are kingdom, phylum, class, order, family, genus, and species. The **binomial name** of an organism gives its genus and species. Organisms in the same kingdom have general characteristics in common; those in the same genus have quite specific characteristics in common. Table 23.1 lists some of the characteristics that help classify humans into major categories. The dates in the first column of the table tell us when these groups of animals first appear in the fossil record.

Characteristics of Primates

Humans belong to a group of mammals called primates. **Primates,** in contrast to other types of mammals, are adapted to an arboreal life—that is, for living in trees. Primate limbs are mobile, as are the hands, because the thumb (and in nonhuman primates, the big toe as well) is opposable—meaning that the thumb can touch each of the other fingers. Therefore, a primate can easily reach out and bring food such as fruit to the mouth. When locomoting, tree limbs can be grasped and released freely because nails have replaced claws.

In primates, the snout is shortened considerably, allowing the eyes to move to the front of the head. The stereoscopic vision (or depth perception) that results permits primates to make accurate judgments about the distance and position of adjoining tree limbs.

Primates have a well-developed brain; the visual portion of the brain is proportionally large, as are the centers responsible for hearing and touch. One birth at a time is the norm in primates since it is difficult to care for several offspring while moving from limb to limb. The juvenile period of dependency is extended, and there is an emphasis on learned behavior and complex social interactions.

These characteristics especially distinguish primates from other mammals:

- Opposable thumb (and in some cases, big toe)
- Well developed brain
- Nails (not claws)
- Single birth
- Extended period of parental care
- Emphasis on learned behavior

Relationships Among Primates

Once biologists have studied the characteristics of a group of organisms, they can construct an **evolutionary tree** that shows their past history. The evolutionary tree in Figure 23.4 shows that all primates share one common ancestor and that the other types of primates diverged from the human line of descent over time. For example, prosimians, represented by tarsiers, were the first type of primate to diverge, and African apes were the last group to diverge from our line of descent. Notice that the tree indicates that humans are most closely related to African apes. One of the most unfortunate misconceptions concerning human evolution is the belief that Darwin and others suggested that humans evolved from apes. On the contrary, humans and apes are believed to have shared a common apelike ancestor. Today's apes are our distant cousins, and we couldn't have evolved from our cousins because we are contemporaries—living on earth at the same time. Presently, it is believed that this common an-

Table 23.1	Evolution and Classification of Humans	
MYA*	Classification Category	Characteristics
600	Kingdom Animalia	Usually motile, multicellular organisms, without cell walls or chlorophyll; usually have an internal cavity for digestion of nutrients
540	Phylum Chordata	Organisms that at one time in their life history have a dorsal hollow nerve cord, a notochord, and pharyngeal pouches
120	Class Mammalia	Warm-blooded vertebrates possessing mammary glands; body more or less covered with hair; well-developed brain
60	Order Primates	Good brain development; opposable thumb and sometimes big toe; lacking claws, scales, horns, and hoofs
6	Family Hominidae	Limb anatomy suitable for upright stance and bipedal locomotion
3	Genus *Homo*	Maximum brain development, especially in regard to particular portions; hand anatomy suitable to the making and use of tools
0.1	Species *Homo sapiens***	Body proportions of modern humans; speech centers of brain well developed

*MYA = millions of years ago.
**To specify an organism, you must use the full binomial name, such as *Homo sapiens*.

cestor existed about 6 MYA.

Biologists have not been able to agree on which extinct form known only by the fossil record is our common ancestor with apes. Molecular data have therefore been used to determine the date of the split between hominids and apes.* When two lines of descent, called a **lineage,** first diverge from a common ancestor, the genes and proteins of the two lineages are nearly identical. But as time goes by, each lineage accumulates genetic changes, which lead to RNA and protein changes. Many genetic changes are neutral (not tied to adaptation) and accumulate at a fairly constant rate; such changes can be used as a kind of **molecular clock** to indicate the relatedness of two groups and when they diverged from one another. Molecular data suggest that among African apes, we are most closely related to chimpanzees.

Until recently, many scientists thought that hominids began to walk upright on two feet (called **bipedalism**) because of a dramatic change in climate that caused forests to be replaced by grassland. Now, some biologists suggest that

*Humans and austalopithecines (see page 468) are hominids in the family Hominidae.

the first hominid began to develop bipedalism even while it lived in trees. Why? Because they cannot find evidence of a dramatic shift in vegetation about 6 MYA. The first hominid's environment is now thought to have included some forest, some woodland, and some grassland. While still living in trees, the first hominids may have walked upright on large branches as they collected fruit from overhead. Then, when they began to forage on the ground among bushes, it would have been easier to shuffle along on their hind limbs. Bipedalism would also have prevented them from getting heatstroke because an upright stance exposes more of the body to breezes. Bipedalism may have been an advantage in still another way. Males may have acquired food far afield, and if they could carry it back to females, they would have been more assured of sex.

The evolution of bipedalism is believed to be the distinctive feature that separates hominids from apes. Scientists are still looking for reasons bipedalism evolved.

Figure 23.4   Primate evolutionary tree.

The ancestor to all primates climbed into one of the first fruit-bearing forests about 66 MYA. The descendants of this ancestor adapted to the new way of life and developed traits such as a shortened snout and nails instead of claws. The time when the other primates diverged from the human line of descent is largely known from the fossil record. A common ancestor was living at each point of divergence; for example, there was a common ancestor for monkeys, apes, and humans about 33 MYA, one for all apes and humans about 15 MYA; and one for just African apes and humans about 6 MYA.

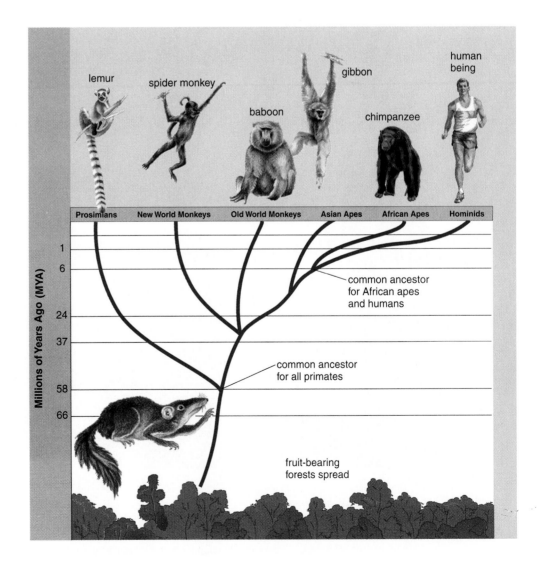

23.4   Evolution of Australopithecines

The hominid line of descent begins with the **australopithecines,** a group of individuals that evolved and diversified in Africa. Some australopithecines were slight of frame and termed gracile (slender) types. Some were robust (powerful) and tended to have strong upper bodies and especially massive jaws, with chewing muscles anchored to a prominent bony crest along the top of the skull. The gracile types most likely fed on soft fruits and leaves, while the robust types had a more fibrous diet that may have included hard nuts.

Southern Africa

The first australopithecine to be discovered was unearthed in southern Africa by Raymond Dart in the 1920s. This hominid, named *Australopithecus africanus*, is a gracile type dated about 2.8 MYA. *A. robustus*, dated from 2 to 1.5 MYA, is a robust type from southern Africa. Both *A. africanus* and *A. robustus* had a brain size of about 500 cc; their skull differences are essentially due to dental and facial adaptations to different diets.

These hominids are believed to have walked upright. Nevertheless, the proportions of the limbs are apelike: the forelimbs are longer than the hind limbs. Some argue that *A. africanus,* with its relatively large brain, is the best ancestral candidate for early *Homo,* whose limb proportions are similar to those of this fossil.

Eastern Africa

More than 20 years ago, a team led by Donald Johanson unearthed nearly 250 fossils of a hominid called *A. afarensis.* A now-famous female skeleton dated at 3.18 MYA is known worldwide by its field name, Lucy. (The name derives from the Beatles song "Lucy in the Sky with Diamonds.") Although her brain was quite small (400 cc), the shapes and relative proportions of her limbs indicate that Lucy stood upright and walked bipedally (Fig. 23.5*a*). Even better evidence of bipedal locomotion comes from a trail of footprints in Laetoli dated about 3.7 MYA. The larger prints are double, as though a smaller-sized being was stepping in the footfalls of another—and there are additional small prints off to the side, within hand-holding distance (Fig. 23.5*b*).

Since the australopithecines were apelike above the waist (small brain) and humanlike below the waist (walked erect), it seems that human characteristics did not evolve all at one time. The term **mosaic evolution** is applied when different body parts change at different rates and therefore at different times.

A. afarensis, a gracile type, is believed to be ancestral to the robust types found in eastern Africa: *A. aethiopicus* and *A. boisei. A. boisei* had a powerful upper body and the largest molars of any hominid. These robust types died out, and therefore, it is possible that *A. afarensis* is ancestral to both *A. africanus* and early *Homo.*

Australopithecines, which arose in Africa, were the first hominids. Their remains show that bipedalism was the first humanlike feature to evolve. It is unknown at this time which australopithecine is ancestral to early *Homo.*

a.

b.

Figure 23.5   *Australopithecus afarensis.*

a. A reconstruction of Lucy on display at the St. Louis Zoo. **b.** These fossilized footprints occur in ash from a volcanic eruption some 3.7 million years ago. The larger footprints are double, and a third, smaller individual was walking to the side. (A female holding the hand of a youngster may have been walking in the footprints of a male.) The footprints suggest that *A. afarensis* walked bipedally.

23.5   Evolution of Humans

Fossils are assigned to the genus *Homo* if (1) the brain size is 600 cc or greater, (2) the jaw and teeth resemble those of humans, and (3) tool use is evident (Fig. 23.6).

Early *Homo*

Homo habilis, dated between 2.0 and 1.9 MYA, may be ancestral to modern humans. Some of these fossils have a brain size as large as 775 cc, which is about 45% larger than that of *A. afarensis*. The cheek teeth are smaller than even those of the gracile australopithecines. Therefore, it is likely that these early *Homo*s were omnivores who ate meat in addition to plant material. Bones at their campsites bear cut marks, indicating that they used tools to strip them of meat.

The stone tools made by *H. habilis*, whose name means "handy man," are rather crude. It's possible that these are the cores from which they took flakes sharp enough to scrape away hide, cut tendons, and easily remove meat from bones.

The skulls of early *Homo*s suggest that the portions of the brain associated with speech areas were enlarged. We can speculate that the ability to speak may have led to hunting cooperatively. Other members of the group may have remained plant gatherers, and if so, both hunters and gatherers most likely ate together and shared their food. In this way, society and culture could have begun. **Culture,** which encompasses human behavior and products (such as technology and the arts), is dependent upon the capacity to speak and transmit knowledge. We can further speculate that the advantages of a culture to *H. habilis* may have hastened the extinction of the australopithecines.

H. habilis warrants classification as *Homo* because of brain size, dentition, and tool use. These early *Homo*s may have had the rudiments of a culture.

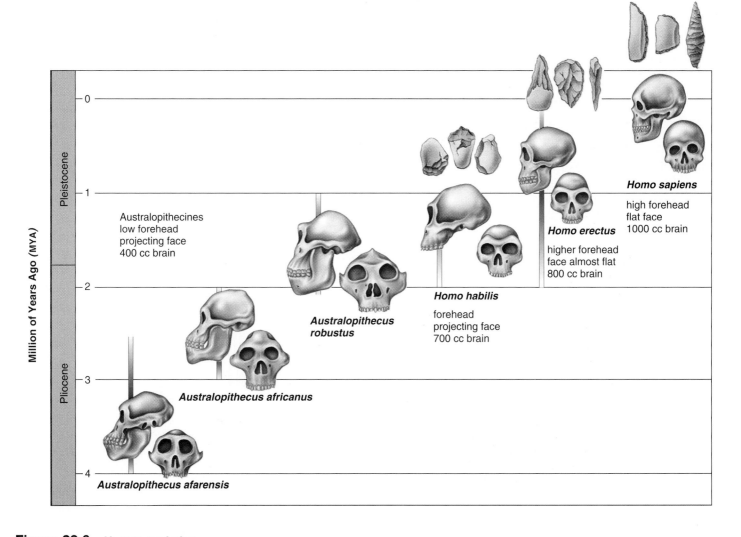

Figure 23.6   Human evolution.
The length of time each species existed is indicated by the vertical gray lines. Notice that there have been times when two or more hominids existed at the same time. Therefore, human evolution resembles a "bush" rather than a single branch.

Homo erectus

Homo erectus and like fossils are found in Africa, Asia, and Europe and dated between 1.9 and 0.3 MYA. A Dutch anatomist named Eugene Dubois was the first to unearth *H. erectus* bones in Java in 1891, and since that time many other fossils have been found in the same area. Although all fossils assigned the name *H. erectus* are similar in appearance, there is enough discrepancy to suggest that several different species have been included in this group. In particular, some experts believe that the African and Asian forms are two different species.

Compared to *H. habilis*, *H. erectus* had a larger brain (about 1,000 cc) and a flatter face. The nose projected, however. This type of nose is adaptive for a hot, dry climate because it permits water to be removed before air leaves the body. The recovery of an almost complete skeleton of a 10-year-old boy indicates that *H. erectus* was much taller than the hominids discussed thus far (Fig. 23.7). Males were 1.8 meters tall (about 6 feet), and females were 1.55 meters (approaching 5 feet). Indeed, these hominids were erect and most likely had a striding gait like ours. The robust and most likely heavily muscled skeleton still retained some australopithecine features. Even so, the size of the birth canal indicates that infants were born in an immature state that required an extended period of care.

It is believed that *H. erectus* first appeared in Africa and then migrated into Asia and Europe. At one time, the migration was thought to have occurred about 1 MYA, but recently *H. erectus* fossil remains in Java and the Republic of Georgia have been dated at 1.9 and 1.6 MYA, respectively. These remains push the eveolution of H. erectus in Africa to an earlier date than has yet been determined. In any case, such an extensive population movement is a first in the history of humankind and a tribute to the intellectual and physical skills of the species.

H. erectus was the first hominid to use fire, and it also fashioned more advanced tools than early *Homo*s. These hominids used heavy, teardrop-shaped axes and cleavers as well as flakes, which were probably used for cutting and scraping. Some believe that *H. erectus* was a systematic hunter and brought kills to the same site over and over again. In one location, researchers have found over 40,000 bones and 2,647 stones. These sites could have been "home bases" where social interaction occurred and a prolonged childhood allowed time for learning. Perhaps a language evolved and a culture more like our own developed.

H. erectus, which evolved from *H. habilis*, had a striding gate, made well-fashioned tools (perhaps for hunting), and could control fire. This hominid migrated into Europe and Asia from Africa between 2 and 1 MYA.

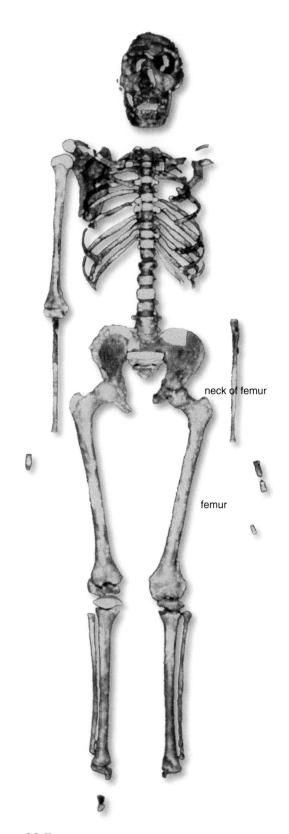

neck of femur

femur

Figure 23.7   *Homo erectus.*
This skeleton of a 10-year-old boy who lived 1.6 MYA in eastern Africa shows femurs that are angled, with necks much longer than in the femurs of modern humans.

Evolution of Modern Humans

Most researchers believe that *Homo sapiens* (modern humans) evolved from *H. erectus*, but they differ as to the details. Perhaps *Homo sapiens* evolved from *H. erectus* separately in Asia, Africa, and Europe. The hypothesis that *Homo sapiens* evolved in several different locations is called the **multiregional continuity hypothesis** (Fig. 23.8*a*). This hypothesis proposes that evolution to modern humans was essentially similar in several different places. If so, each region should show a continuity of its own anatomical characteristics from the time when *H. erectus* first arrived in Europe and Asia.

Opponents argue that it seems highly unlikely that evolution would have produced essentially the same result in these different places. They suggest, instead, the **out-of-Africa hypothesis,** which proposes that *H. sapiens* evolved from *H. erectus* only in Africa and thereafter *H. sapiens* migrated to Europe and Asia about 100,000 years BP (before present) (Fig. 23.8*b*). If so, there would be no continuity of characteristics between fossils dated 200,000 BP and 100,000 BP in Europe and Asia.

According to which hypothesis would modern humans be most genetically alike? With the multiregional hypothesis, human populations have been evolving separately for a long time, and therefore genetic differences are expected. With the out-of-Africa hypothesis, we are all descended from a few individuals from about 100,000 years BP. Therefore, the out-of-Africa hypothesis suggests that we are more genetically similar.

A few years ago, a study attempted to show that all the people of Europe (and the world for that matter) have essentially the same mitochondrial DNA. Called the "mitochondrial Eve" hypothesis by the press (note that this is a misnomer because no single ancestor is proposed), the statistics that calculated the date of the African migration were found to be flawed. Still, the raw data—which indicate a close genetic relationship among all Europeans—support the out-of-Africa hypothesis.

These opposing hypotheses have sparked many other innovative studies to test them. The final conclusions are still being determined.

Investigators are currently testing two hypotheses: that modern humans (1) evolved separately in Asia, Africa, and Europe, or (2) evolved in Africa and then migrated to Asia and Europe.

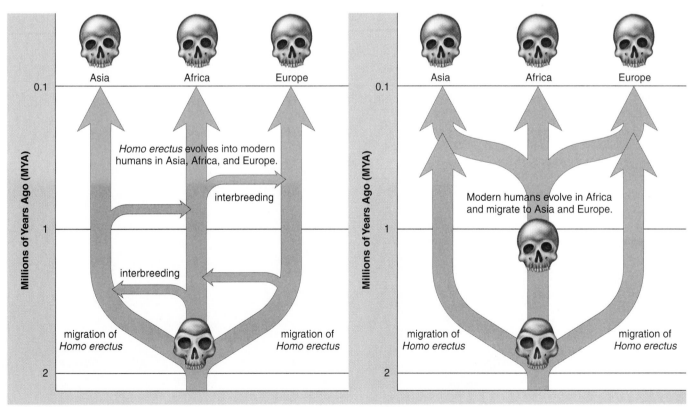

a. Multiregional continuity

b. Out of Africa

Figure 23.8   Evolution of modern humans.
a. The multiregional continuity hypothesis proposes that *Homo sapiens* evolved separately in at least three different places: Asia, Africa, and Europe. Therefore, continuity of genotypes and phenotypes is expected in these regions. **b.** The out-of-Africa hypothesis proposes that *Homo sapiens* evolved only in Africa; then this species migrated and supplanted populations of *Homo* in Asia and Europe about 100,000 years ago.

Neanderthals

Neanderthals *(H. neanderthalensis)* take their name from Germany's Neander Valley, where one of the first Neanderthal skeletons, dated some 200,000 years ago, was discovered. The Neanderthals had massive brow ridges, and the nose, the jaws, and the teeth protruded far forward. The forehead was low and sloping, and the lower jaw lacked a chin. New fossils show that the pubic bone was long compared to ours.

According to the out-of-Africa hypothesis, (see page 471) Neanderthals were supplanted by modern humans. Surprisingly, however, the Neanderthal brain was, on the average, slightly larger than that of *Homo sapiens* (1,400 cc, compared to 1,360 cc in most modern humans). The Neanderthals were heavily muscled, especially in the shoulders and neck (Fig. 23.9). The bones of the limbs were shorter and thicker than those of modern humans. It is hypothesized that a larger brain than that of modern humans was required to control the extra musculature. The Neanderthals lived in Europe and Asia during the last Ice Age, and their sturdy build could have helped conserve heat.

The Neanderthals give evidence of being culturally advanced. Most lived in caves, but those living in the open may have built houses. They manufactured a variety of stone tools, including spear points, which could have been used for hunting, and scrapers and knives, which would have helped in food preparation. They most likely successfully hunted bears, woolly mammoths, rhinoceroses, reindeer, and other contemporary animals. They used and could control fire, which probably helped in cooking frozen meat and in keeping warm. They even buried their dead with flowers and tools and may have had a religion.

Cro-Magnons

Cro-Magnons are the oldest fossils to be designated *Homo sapiens*. In keeping with the out-of-Africa hypothesis, the Cro-Magnons, who are named after a fossil location in France were the modern humans who entered Asia and Europe from Africa 100,000 years ago or even earlier. Cro-Magnons had a thoroughly modern appearance (Fig. 23.10). They made advanced stone tools, including compound tools, as when stone flakes were fitted to a wooden handle. They may have been the first to throw spears, enabling them to kill animals from a distance, and to make knifelike blades. They were such accomplished hunters that some researchers believe they were responsible for the extinction of many larger mammals, such as the giant sloth, the mammoth, the saber-toothed tiger, and the giant ox, during the late Pleistocene epoch.

Cro-Magnons hunted cooperatively, and perhaps they were the first to have a language. They are believed to have lived in small groups, with the men hunting by day while the women remained at home with the children. It's quite possible that this hunting way of life among prehistoric people influences our behavior even today. The Cro-Magnon culture included art. They sculpted small figurines out of reindeer bones and antlers. They also painted beautiful drawings of animals on cave walls in Spain and France (Fig. 23.10).

Most likely, modern humans migrated out of Africa and supplanted other *Homo* species in Asia and Europe. Today there is only one *Homo* species in existence; namely, *Homo sapiens*.

Figure 23.9   Neanderthals.
This drawing shows that the nose and the mouth of the Neanderthals protruded from their faces, and their muscles were massive. They made stone tools and were most likely excellent hunters.

Figure 23.10   Cro-Magnons.
Cro-Magnon people are the first to be designated *Homo sapiens*. Their toolmaking ability and other cultural attributes, such as their artistic talents, are legendary.

Bioethical Focus

The Theory of Evolution

The term "theory" in science is reserved for those ideas that scientists have found to be all-encompassing because they are based on data collected in a number of different fields. Evolution is a scientific theory. So is the cell theory, which says that all organisms are composed of cells, and so is the atomic theory, which says that all matter is composed of atoms. No one argues that schools should teach alternatives to the cell theory or the atomic theory. Yet confusion reigns over the use of the expression "the theory of evolution."

No wonder most scientists in our country are dismayed when state legislatures or school boards rule that teachers must put forward a variety of "theories" on the origin of life, including one that runs contrary to the mass of data that supports the theory of evolution. An organization in California called the Institute for Creation Research advocates that students be taught an "intelligent-design theory," which says that DNA could never have arisen without the involvement of an "intelligent agent," and that gaps in the fossil record mean that species arose fully developed with no antecedents.

Since our country forbids the mingling of church and state—no purely religious ideas can be taught in the schools—the advocates for an intelligent-design theory are careful not to mention the Bible or any strictly religious ideas (i.e., God created the world in seven days). Still, teachers who have a solid scientific background do not feel comfortable teaching an intelligent-design theory because it does not meet the test of a scientific theory. Science is based on hypotheses that have been tested by observation and/or experimentation. A scientific theory has stood the test of time—that is, no hypotheses have been supported by observation and/or experimentation that run contrary to the theory. On the contrary, the theory of evolution is supported by data collected in such wide-ranging fields as development, anatomy, geology, and biochemistry.

The polls consistently show that nearly half of all Americans

Figure 23A *Australopithecus africanus.*
An evolution of humans is supported by fossil remains.

prefer to believe the Old Testament account of creation. That, of course, is their right, but should schools be required to teach an intelligent-design theory that traces its roots back to the Old Testament and is not supported by observation and experimentation?

Decide Your Opinion

1. Should teachers be required to teach an intelligent-design theory of the origin of life in schools? Why or why not?
2. Should schools rightly teach that science is based on data collected by the testing of hypotheses by observation and experimentation? Why or why not?
3. Should schools be required to show that the intelligent-design theory does not meet the test of being scientific? Why or why not?

Looking at Both Sides   www.mhhe.com/biosci/genbio/maderhuman7/

Every bioethical issue has at least two sides. Even if you already know your opinion, it is important to explore the opposite opinion before finalizing your position. The Online Learning Center at www.mhhe.com/biosci/genbio/maderhuman7/ will help you fine tune your initial opinion, explore both sides, and finalize your position. You may acquire new arguments for your original opinion, or you may even change your opinion. Be sure to complete these activities in sequence:

Taking Sides   Decide your initial opinion by answering a series of questions. Then see if your opinion changes after completing the next two activities.

Further Debate   Read opposing articles that give you further information on this particular bioethical issue.

Explain Your Position   Answer another series of questions and then defend your original or changed opinion. You can e-mail your position to your instructor if he or she wishes.

We Are One Species

Human beings are diverse, but even so, we are all classified as *Homo sapiens*. The biological definition of species is a group of organisms able to interbreed and bear fertile offspring. Any two types of humans are able to reproduce with one another, signifying that all humans belong to the same species. While it may appear that there are various "races," molecular data show that the DNA base sequence varies as much between individuals of the same ethnicity as between individuals of different ethnicities.

It is generally accepted that the human phenotype is adapted to the climate of a region. Although it might seem as if dark skin is a protection against the hot rays of the sun, it has been suggested that it is actually a protection against ultraviolet ray absorption. Dark-skinned persons living in southern regions and light-skinned persons living in northern regions absorb the same amount of radiation. (Some absorption is required for vitamin D production.) Other features that correlate with skin color, such as hair type and eye color, may simply be side effects of genes that control skin color.

Differences in body shape represent adaptations to temperature. A squat body with short limbs and nose retains more heat than an elongated body with long limbs and nose. Also, almond-shaped eyes, a flat nose and forehead, and broad cheeks are believed to be adaptations to the last Ice Age.

While it always has seemed to some that physical differences warrant assigning humans to different "races," this contention is not borne out by the molecular data mentioned in this chapter.

Summarizing the Concepts

23.1 Origin of Life
A chemical evolution produced the first cell. In the presence of an outside energy source, such as ultraviolet radiation, the primitive atmospheric gases reacted with one another to produce small organic molecules.

Next, macromolecules evolved and interacted. The RNA-first hypothesis is supported by the discovery of RNA enzymes called ribozymes. The protein-first hypothesis is supported by the observation that amino acids polymerize abiotically when exposed to dry heat. The protocell must have been a heterotrophic fermenter living on the preformed organic molecules in the ocean. Eventually, the DNA → RNA → protein self-replicating system evolved, and a true cell that could reproduce came into being.

23.2 Biological Evolution
Descent from a common ancestor explains the unity of living things—for example, why all living things have a cellular structure and a common chemistry. Adaptation to different environments explains the great diversity of living things.

Darwin discovered much evidence for common descent. The fossil record gives us the history of life in general and allows us to trace the descent of a particular group. Biogeography shows that the distribution of organisms on earth is explainable by assuming organisms evolved in one locale. The common anatomies and development of a group of organisms adapted to different environments are explainable by descent from a common ancestor. All organisms have similar biochemical molecules, and this supports the concept of common descent.

Darwin developed a mechanism for adaptation known as natural selection. Members of a population exhibit inherited variations and compete with one another for limited resources. The members with variations that help them survive and reproduce have more offspring, and in this way the adaptive characteristics become widespread in the next generation. The process of natural selection is nonteleological.

23.3 Humans Are Primates
The classification of humans can be used to trace their ancestry. Humans are primates, mammals adapted to living in trees. An evolutionary diagram of primates based on anatomical, molecular, and fossil evidence shows that we share a common ancestor with African apes. This common ancestor lived about 6 MYA. Researchers are trying to find environmental reasons why humans came down out of trees and walked erect.

23.4 Evolution of Australopithecines
The first hominid (humans are in this family) was an australopithecine that lived about 3 MYA. Australopithecines could walk erect, but they had a small brain. This testifies to a mosaic evolution for humans—that is, not all advanced features evolved at the same time. It is uncertain which australopithecine is ancestral to early *Homo*.

23.5 Evolution of Humans
H. habilis made tools, but *H. erectus* was the first fossil to have a brain size of more than 1,000 cc. *H. erectus* migrated from Africa into Europe and Asia. They used fire and may have been big-game hunters.

Whereas the multiregional continuity hypothesis suggests that modern humans evolved separately in Europe, Africa, and Asia, the out-of-Africa hypothesis says that *H. sapiens* evolved in Africa but then migrated to Asia and Europe. The Neanderthals were already living in Europe and Asia before modern humans arrived. The Neanderthals did not have the physical traits of modern humans, but they did have culture. Cro-Magnon is a name often given to modern humans. Their tools were sophisticated, and they definitely had a culture, as shown by the paintings on the walls of caves.

Studying the Concepts

1. List and discuss the steps by which a chemical evolution could have produced a protocell. 462
2. What is a true cell? How might DNA replication have come about? 463
3. Biological evolution explains what two general observations about living things? 463
4. Show that the fossil record, biogeography, comparative anatomy, and biochemistry all give evidence of evolution. 464
5. Explain Darwin's mechanism of natural selection. How does natural selection result in adaptation to the environment? 464–65
6. Name the major classification categories. Classify humans, and state some of the characteristics for each category. 466
7. In general, primates are adapted to what type of life? List and discuss several primate characteristics. 466
8. Describe an evolutionary tree. What can we learn from the primate evolutionary tree? 466–67
9. Describe the characteristics of australopithecines, early *Homo*, *Homo erectus*, Neanderthals, and Cro-Magnon. 468–72
10. Give evidence that human evolution resembles a bush rather than a single branch. 469
11. Which humans walked erect? Used tools? Used fire? Drew pictures? 468–72
12. Contrast the multi-regional continuity hypothesis with the out-of-Africa hypothesis. 471

Testing Your Knowledge of the Concepts

In questions 1–4, match the evolutionary evidence to the description.
a. biogeography
b. fossil record
c. comparative biochemistry
d. comparative anatomy

_____ 1. Species change over time.

_____ 2. Forms of life are variously distributed.

_____ 3. A group of related species have homologous structures.

_____ 4. The same types of molecules are found in all living things.

In questions 5–7, indicate whether the statement is true (T) or false (F).

_____ 5. The result of natural selection is adaptation to the environment.

_____ 6. An unusually dry wind is believed to have produced the first cell(s).

_____ 7. *Homo habilis* is named for his ability to make stone tools.

In questions 8–15, fill in the blanks.

8. The protocell could carry on metabolism, but it could not _____.

9. Along with monkeys and apes, humans are classified as _____.

10. _____ were the first hominids to evolve.

11. The out-of-Africa hypothesis proposes that modern humans evolved in _____ only.

12. The australopithecines could probably walk _____, but they had a _____ brain.

13. The only fossil to be called *Homo sapiens is* _____.

14. Modern humans evolved _____ (choose billions, millions, thousands) of years ago.

15. To describe the out-of-Africa hypothesis, in which boxes of the following diagram would you place a skull of *Homo sapiens*? _____

16. A skull of *Homo erectus*? _____

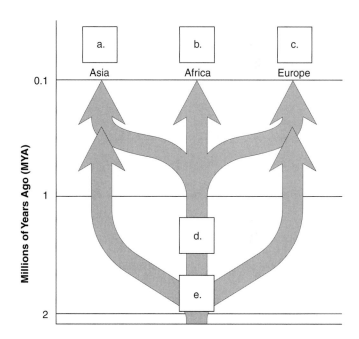

Understanding Key Terms

adaptation 463
analogous structure 464
australopithecine 468
binomial name 466
biogeography 464
biological evolution 463
bipedalism 467
chemical evolution 462
Cro-Magnon 472
culture 469
eukaryotic cell 463
evolutionary tree 466
fossil 464
fossil record 464
heterotroph 462
hominid 467
Homo erectus 470

Homo habilis 469
Homo sapiens 471
homologous structure 464
lineage 467
molecular clock 467
mosaic evolution 468
multiregional continuity hypothesis 471
natural selection 464
Neanderthal 472
out-of-Africa hypothesis 471
primate 466
prokaryotic cell 463
protein-first hypothesis 462
protocell 462
RNA-first hypothesis 462
teleological 465

23.1 Origin of Life

Origin of Life *Essential Study Partner*

Looking at Both Sides *critical thinking activity*

23.2 Biological Evolution

Before Darwin *Essential Study Partner*
Voyage of the Beagle *Essential Study Partner*
Natural Selection *Essential Study Partner*
Evidence for Evolution *Essential Study Partner*
Concept Quiz *Essential Study Partner*

23.3 Humans are Primates

Primates *Essential Study Partner*

23.5 Evolution of Humans

Hominid History *Essential Study Partner*
Concept Quiz *Essential Study Partner*

Chapter Summary

Key Term Flashcards *vocabulary quiz*
Chapter Quiz *objective quiz covering all chapter concepts*

Chapter 24

Ecosystems and Human Interferences

Chapter Concepts

24.1 The Nature of Ecosystems

24.2 Energy Flow and Chemical Cycling

24.3 Global Biogeochemical Cycles

Figure 24.1 **Piñon trees.**
Piñon trees are a population of organisms in an ecosystem that also includes deer mice, which feed on piñon nuts.

When people in the Southwest first began dying of an unknown cause in 1993, no one knew to blame the warm, wet weather or the fat little deer mice scampering about. And absolutely no one would have pointed a finger at a new strain of hantavirus. These unlikely factors combined to launch one of the decade's most deadly emerging diseases: hantavirus pulmonary syndrome.

First identified in Korea some 40 years ago, hantaviruses can be carried by common deer mice. Normally, that's not a problem. But in 1993, deer mice in New Mexico and nearby states experienced a population boom. The boom stemmed from unusually heavy spring rains, which nourished trees carrying piñon nuts, a favorite food of the deer mouse (Fig. 24.1). This slight change in the weather was enough to spur a tenfold increase in the deer mouse population. Suddenly, the mice were everywhere—inside garages, in backyards, even on Indian reservations in the region. With the mice came exposure to hantaviruses carried in their feces and urine.

Natural **ecosystems** contain populations of organisms—deer mice, piñon trees, and yes, humans—that interact among themselves and the physical environment. The saying goes that in an ecosystem everything is connected to everything else. Thus, in this example, warm, wet weather led to plentiful piñon nuts, which led to a deer mouse explosion, and finally, to illness in humans. This chapter examines the interworkings of ecosystems and how they have been impacted by human beings. While an example like this might make it seem that we should further restrict natural ecosystems, this is not the case. Rather, the point of this chapter is that they need to be preserved, albeit managed, if the human species is to continue.

Thankfully, the illness outbreak caused by the hantavirus ended. Rodent control efforts, disease surveillance programs, and new medications should prevent similar outbreaks in the future.

24.1   The Nature of Ecosystems

When the earth was formed, the outer crust was covered by ocean and barren land. Over time, aquatic organisms filled the seas, and terrestrial organisms colonized the land so that eventually there were many complex communities of living things. A **community** is made up of all the **populations** in a particular area, such as a forest or pond. When we study a community, we are considering only the populations of organisms that make up that community, but when we study an ecosystem, we are concerned with the community and its physical environment. Table 24.1 defines these important terms in the study of ecology and shows how they relate to the biosphere as a whole.

Succession

To the untrained human eye, it seems as if communities stay pretty much the same from year to year. But actually, one of the most outstanding characteristics of natural communities is their changing nature. The dynamic nature of communities is dramatically illustrated by the phenomenon of succession. **Succession** is a sequential change in the relative dominance of species within a community. You may have observed that an abandoned field slowly changes to be more like the surrounding natural area. Although the final result will be determined by the species that migrate there, it is possible to suggest the interim stages of succession.

Table 24.1	Ecological Terms
Term	**Definition**
Ecology	Study of the interactions of organisms with each other and with the physical environment
Population	All the members of the same species that inhabit a particular area
Community	All the populations that are found in a particular area
Ecosystem	A community and its physical environment, including both nonliving (abiotic) and living (biotic) components
Biosphere	All the communities on earth whose members exist in air and water and on land

Primary succession begins on bare rock. At first the rock is subjected to weathering by wind and rain, followed by the invasion of lichens and mosses. They cause the buildup of soil, permitting low-lying grasses and then shrubs to take over. Depending on the area, pine trees and then broadleaf trees eventually take root (Fig. 24.2). Secondary succession begins in an area disturbed but no longer used by humans. This area too will also go through a herbaceous stage before shrubs and then trees possibly appear. When ecologists first observed succession, they called the final stage of succession the climax community. They felt that each particular area had its own climax community. For example, in the United States, a deciduous forest is typical of the Northeast, a prairie is natural to the Midwest, and a semidesert covers the Southwest (Fig. 24.3). Today, we realize that many factors influence what particular species are found within a community, and it is not predetermined from the start.

The dynamic nature of communities is shown by their changing nature when succession occurs. Climax communities are threatened by disturbances.

a.   b.   c.   d.

Figure 24.2   Example of primary succession.
a. Mosses and lichens can grow on bare rock. **b.** Grasses take hold. **c.** Shrubs spread throughout the area. **d.** Trees become established.

a. Southwest desert

c. Coastal Pacific Northwest rain forest

b. California chaparral

d. Midwest prairie

Figure 24.3 Biological communities of the United States.
a. A desert gets limited rainfall a year. Because daytime temperatures are high, most desert animals are active at night when it is cooler. **b.** The California chaparral recovers quickly after a fire. **c.** A coastal rain forest of the Pacific Northwest receives a plentiful amount of rain. **d.** Midwest prairies contain grasses and rich soil, built up by the remains of grasses.

Biotic Components of an Ecosystem

An ecosystem possesses both nonliving (abiotic) and living (biotic) components. The abiotic components include resources, such as sunlight and inorganic nutrients, and conditions, such as type of soil, water availability, prevailing temperature, and amount of wind.

The biotic components of an ecosystem are the various populations of organisms. Each population in an ecosystem has a habitat and a niche. The **habitat** of an organism is its place of residence—that is, where it can be found, such as under a log or at the bottom of a pond. The **niche** of an organism is its profession or total role in the community. A description of an organism's niche includes its interactions with the physical environment and with the other organisms in the community. For example, the trees in Figure 24.4*a* take solar energy, carbon dioxide, and water from the abiotic environment but have many interactions with other organisms, such as woodpeckers (Fig. 24.4*b*). Woodpeckers feed on parasitic grubs from a tree that also provides a habitat for the woodpecker's young. All of these behaviors are aspects of the woodpecker's niche, as is apparent from reviewing the list given in the figure.

The populations in an ecosystem are often categorized as producers or consumers. **Producers** produce organic nutrients and are autotrophic organisms. **Autotrophic organisms** are able to carry on photosynthesis and make organic nutrients for themselves (and indirectly for the other populations as well). In terrestrial ecosystems, the producers are predominantly green plants, while in freshwater and marine ecosystems, the dominant producers are various species of algae.

Consumers consume organic nutrients and are heterotrophic organisms. **Heterotrophic organisms** feed on producers directly or on organisms that have fed on producers. It is possible to distinguish four types of consumers, depending on their food source:

- **Herbivores** feed directly on green plants; they are termed primary consumers.
- **Carnivores** feed only on other animals and are thus secondary or tertiary consumers.
- **Omnivores** feed on both plants and animals.
- **Decomposers** (bacteria and fungi) feed on and thereby break down **detritus**, the remains of plants and animals following their death.

Therefore, a caterpillar feeding on a leaf is a herbivore; a green heron feeding on a fish is a carnivore; and a human being eating both leafy green vegetables and beef is an omnivore. The bacteria and fungi of decay are important detritus feeders, but so are other soil organisms, such as earthworms and various small arthropods. The importance of the latter can be demonstrated by placing leaf litter in bags with mesh too fine to allow soil animals to enter; the leaf litter does not decompose well, even though bacteria and fungi are present. Small soil organisms precondition the detritus so that bacteria and fungi can break it down to inorganic matter that producers can use again.

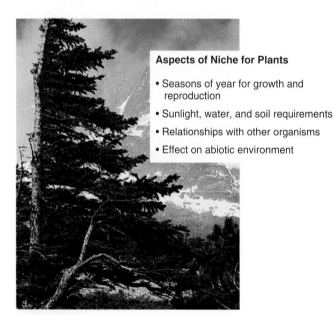

Aspects of Niche for Plants

- Seasons of year for growth and reproduction
- Sunlight, water, and soil requirements
- Relationships with other organisms
- Effect on abiotic environment

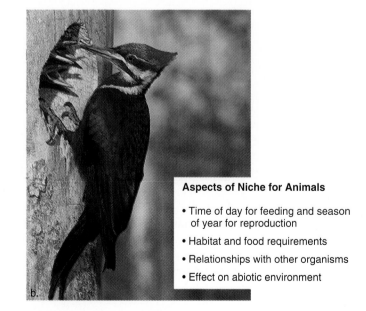

Aspects of Niche for Animals

- Time of day for feeding and season of year for reproduction
- Habitat and food requirements
- Relationships with other organisms
- Effect on abiotic environment

Figure 24.4   Biotic components of an ecosystem.
The biotic components of an ecosystem are influenced by the abiotic components as in (**a**), where the force of the wind has affected tree growth. Each biotic component has a niche whose aspects are listed for plants (**a**) and animals (**b**).

24.2 Energy Flow and Chemical Cycling

When we diagram all the biotic components of an ecosystem, as in Figure 24.5, it is possible to illustrate that every ecosystem is characterized by two fundamental phenomena: energy flow and chemical cycling. Energy flow begins when producers absorb solar energy, and chemical cycling begins when producers take in inorganic nutrients from the physical environment. Thereafter, producers make organic nutrients (food) for themselves and indirectly for the other populations of the ecosystem. Energy flow occurs because all the energy content of organic nutrients is eventually converted to heat, which dissipates in the environment. Therefore, most ecosystems cannot exist without a continual supply of solar energy. Chemicals cycle when inorganic nutrients are returned to the producers from the atmosphere or soil, as appropriate.

Only a portion of the organic nutrients made by autotrophs is passed on to heterotrophs because plants use organic molecules to fuel their own cellular respiration. Similarly, only a small percentage of nutrients taken in by heterotrophs is available to higher-level consumers. Figure 24.6 shows why. A certain amount of the food eaten by a herbivore is never digested and is eliminated as feces. Metabolic wastes are excreted as urine. Of the assimilated energy, a large portion is utilized during cellular respiration and thereafter becomes heat. Only the remaining food, which is converted into increased body weight (or additional offspring), becomes available to carnivores.

The elimination of feces and urine by a heterotroph, and indeed the death of all organisms, does not mean that substances are lost to an ecosystem. They represent the nutrients made available to decomposers. Since decomposers can be food for other heterotrophs of an ecosystem, the situation can get a bit complicated. Still, we can conceive that all the solar energy that enters an ecosystem eventually becomes heat. And this is consistent with the observation that ecosystems are dependent on a continual supply of solar energy.

The laws of thermodynamics support the concept that energy flows through an ecosystem. The first law states that energy cannot be created (or destroyed). This explains why ecosystems are dependent on a continual outside source of energy, usually solar energy, which is used by photosynthesizers to produce organic nutrients. The second law states that, with every transformation, some energy is degraded into a less available form such as heat. Because plants carry on cellular respiration, for example, only about 55% of the original energy absorbed by plants is available to an ecosystem.

Energy flows through an ecosystem, while chemicals cycle within and between ecosystems.

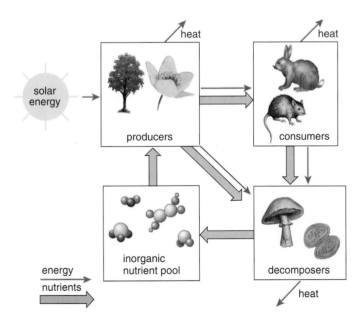

Figure 24.5 Nature of an ecosystem.
Chemicals cycle, but energy flows through an ecosystem. As energy transformations repeatedly occur, all the energy derived from the sun eventually dissipates as heat.

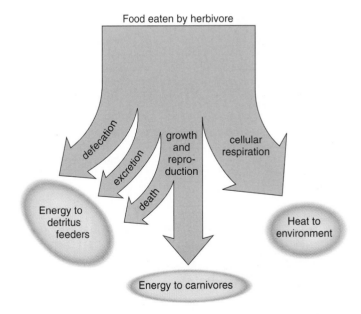

Figure 24.6 Energy balances.
Only about 10% of the food energy taken in by a herbivore is passed on to carnivores. A large portion goes to detritus feeders in the ways indicated, and another large portion is used for cellular respiration.

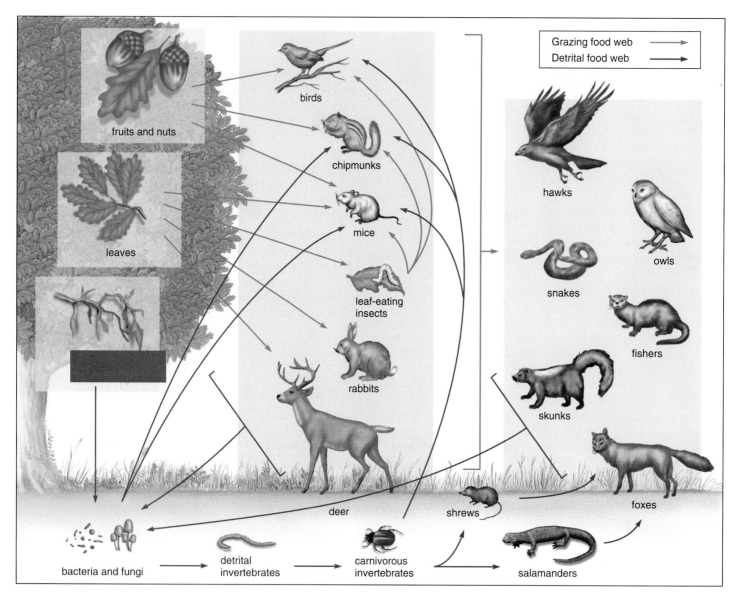

Figure 24.7   Forest food webs.
Two linked food webs are shown for a forest ecosystem: a grazing food web and a detrital food web.

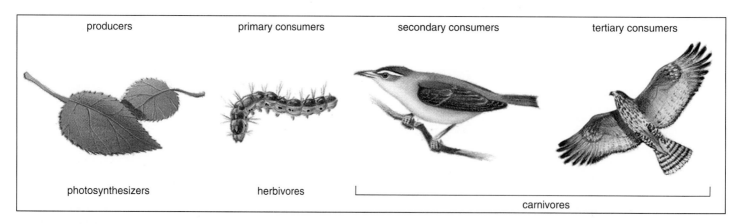

Figure 24.8   Food chain.
Trace a similar grazing food chain in the grazing food web depicted in Figure 24.7.

Food Webs and Trophic Levels

The principles we have been discussing can now be applied to an actual ecosystem—a forest. The feeding relationships in an ecosystem are interconnected as shown in Figure 24.7. Therefore, they create a **food web;** the upper part of Figure 24.7 is a **grazing food web** because it begins with trees, such as the oak trees depicted. A **detrital food web** begins with detritus, partially decayed matter in the soil. Insects in the form of caterpillars feed on leaves, while mice, rabbits, and deer feed on leaf tissue at or near the ground. Birds, chipmunks, and mice feed on fruits and nuts, but they are in fact omnivores because they also feed on caterpillars. These herbivores and omnivores all provide food for a number of different carnivores.

In the detrital food web, detritus, along with the bacteria and fungi of decay, is food for larger decomposers. Because some of these, like shrews and salamanders, become food for aboveground carnivores, the detrital and the grazing food webs are joined.

We naturally tend to think that aboveground plants like trees are the largest storage form of organic matter and energy, but this is not necessarily the case. In this particular forest, the organic matter lying on the forest floor and mixed into the soil contains much more energy than does the leaf matter of living trees. The soil contains over twice as much energy as the leaves of the trees. Therefore, more energy in a forest may be funneling through the detrital food web than through the grazing food web.

Trophic Levels

You can see that Figure 24.7 would allow us to link organisms one to another in a straight line manner, according to who eats whom. Such diagrams are called **food chains** (Fig. 24.8). For example, in the grazing food web we can find this **grazing food chain:**

leaves → caterpillars → tree birds → hawks

And in the detrital food web we could find this **detrital food chain:**

dead organic matter → soil microbes → earthworms, etc.

A **trophic level** is composed of all the organisms that feed at a particular link in a food chain. In the grazing food web in Figure 24.7, going from left to right, the trees are primary producers (first trophic level), the first series of animals are primary consumers (second trophic level), and the next group of animals are secondary consumers (third trophic level).

Ecological Pyramids

Ecologists portray the energy relationships between trophic levels in the form of **ecological pyramids,** diagrams whose building blocks designate the various trophic levels (Fig. 24.9). (We need to keep in mind that sometimes organisms don't fit into only one trophic level. For example, chipmunks feed on fruits and nuts, but they also feed on leaf-eating insects.)

A pyramid of numbers simply tells how many organisms there are at each trophic level. It's easy to see that a pyramid of numbers could be completely misleading. For example, in Figure 24.7 you would expect each tree to contain numerous caterpillars; therefore, there would be more herbivores than autotrophs! The problem, of course, has to do with size. Autotrophs can be tiny, like microscopic algae, or they can be big, like beech trees; similarly, herbivores can be as small as caterpillars or as large as elephants.

Pyramids of biomass eliminate size as a factor since **biomass** is the number of organisms multiplied by their weight. You would certainly expect the biomass of the producers to be greater than the biomass of the herbivores, and that of the herbivores to be greater than that of the carnivores. In aquatic ecosystems such as lakes and open seas, where algae are the only producers, the herbivores may have a greater biomass than the producers when you take their measurements. Why? The reason is that over time, the algae reproduce rapidly, but they are also consumed at a high rate. Pyramids like this one, that have more herbivores than producers, are called inverted pyramids:

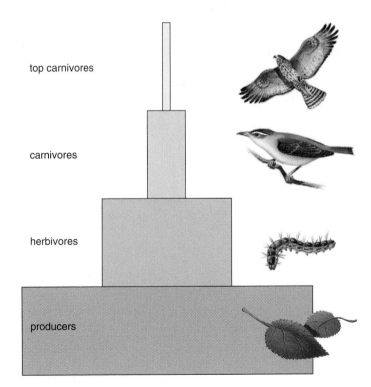

Figure 24.9 Ecological pyramid.
An ecological pyramid shows the relationship between either the number of organisms, the biomass, or the amount of energy theoretically available at each trophic level.

zooplankton

phytoplankton

relative
dry weight

There are ecological pyramids of energy also, and they generally have the appearance of Figure 24.9. Ecologists are now beginning to rethink the usefulness of utilizing pyramids to describe energy relationships. One problem is what to do with the decomposers, which are rarely included in pyramids, even though a large portion of energy becomes detritus in many ecosystems.

There is a rule of 10% with regard to biomass (or energy) pyramids. It says that, in general, the amount of biomass (or energy) from one level to the next is reduced by a magnitude of 10. Thus, if an average of 1,000 kg of plant material is consumed by herbivores, about 100 kg is converted to herbivore tissue, 10 kg to first-level carnivores, and 1 kg to second-level carnivores. The rule of 10% suggests that few carnivores can be supported in a food web. This is consistent with the observation that each food chain has from three to four links, rarely five.

24.3   Global Biogeochemical Cycles

All organisms require a variety of organic and inorganic nutrients. Carbon dioxide and water are necessary for photosynthesis. Nitrogen is a component of all the structural and functional proteins and nucleic acids that sustain living tissues. Phosphorus is essential for ATP and nucleotide production. In contrast to energy, inorganic nutrients are used over and over again by autotrophs.

Since the pathways by which chemicals circulate through ecosystems involve both living (biospheric) and nonliving (geological) components, they are known as **biogeochemical cycles.** For each element, chemical cycling may involve (1) a reservoir—a source normally unavailable to producers, such as fossilized remains, rocks, and deep-sea sediments; (2) an exchange pool—a source from which organisms do generally take chemicals, such as the atmosphere or soil; and (3) the biotic community—through which chemicals move along food chains, perhaps never entering a pool (Fig. 24.10).

There are two general categories of biogeochemical cycles. In a gaseous cycle, exemplified by the carbon and nitrogen cycles, the element returns to and is withdrawn from the atmosphere as a gas. In the sedimentary cycle, exemplified by the phosphorus cycle, the element is absorbed from the sediment by plant roots, passed to heterotrophs, and eventually returned to the soil by decomposers, usually in the same general area.

The diagrams on the next few pages make it clear that nutrients can flow between terrestrial and aquatic ecosystems. In the nitrogen and phosphorus cycles, these nutrients run off from a terrestrial to an aquatic ecosystem and in that way enrich aquatic ecosystems. Decaying organic matter in aquatic ecosystems can be a source of nutrients for intertidal inhabitants such as fiddler crabs. Sea birds feed on fish but deposit guano (droppings) on land, and in that way phosphorus from the water is deposited on land. It would seem that anything put into the environment in one ecosystem could find its way to another ecosystem. Scientists find the soot from urban areas and pesticides from agricultural fields in the snow and animals of the Arctic.

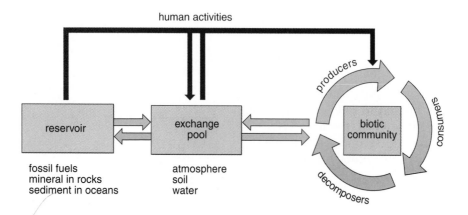

Figure 24.10   Model for chemical cycling.
Chemical nutrients cycle between these components of ecosystems: Reservoirs such as fossil fuels, minerals in rocks, and sediments in oceans are normally relatively unavailable sources, but pools such as those in the atmosphere, soil, and water are available sources of chemicals for the biotic community. When human activities remove chemicals from reservoirs and pools and make them available to the biotic community, the result can be pollution.

The Water Cycle

The **water (hydrologic) cycle** is described in Figure 24.11. Fresh water is distilled from salt water. The sun's rays cause fresh water to evaporate from seawater, and the salts are left behind. Vaporized fresh water rises into the atmosphere, cools, and falls as rain over the oceans and the land.

Water evaporates from land and from plants (evaporation from plants is called transpiration). It also evaporates from bodies of fresh water, but since land lies above sea level, gravity eventually returns all fresh water to the sea. In the meantime, water is contained within standing waters (lakes and ponds), flowing water (streams and rivers), and groundwater.

Some of the water from **precipitation** (e.g., rain, snow, sleet, hail, and fog) sinks or percolates into the ground and saturates the earth to a certain level. The top of the saturation zone is called the groundwater table, or simply, the water table. Sometimes groundwater is also located in **aquifers,** rock layers that contain water and release it in appreciable quantities to wells or springs. Aquifers are recharged when rainfall and melted snow percolate into the soil. In some parts of the country, especially the arid West and southern Florida, withdrawals from aquifers exceed any possibility of recharge. This is called "groundwater mining." In these locations, the groundwater is dropping, and residents may run out of groundwater, at least for irrigation purposes, within a few short years. Fresh water, which makes up only about 3% of the world's supply of water, is called a renewable resource because a new supply is always being produced. But it is possible to run out of fresh water when the available supply is not adequate or is polluted so that it is not usable.

In the water cycle, fresh water evaporates from the bodies of water. Precipitation on land enters the ground, surface waters, or aquifers. Water ultimately returns to the ocean—even the quantity that remains in aquifers for some time.

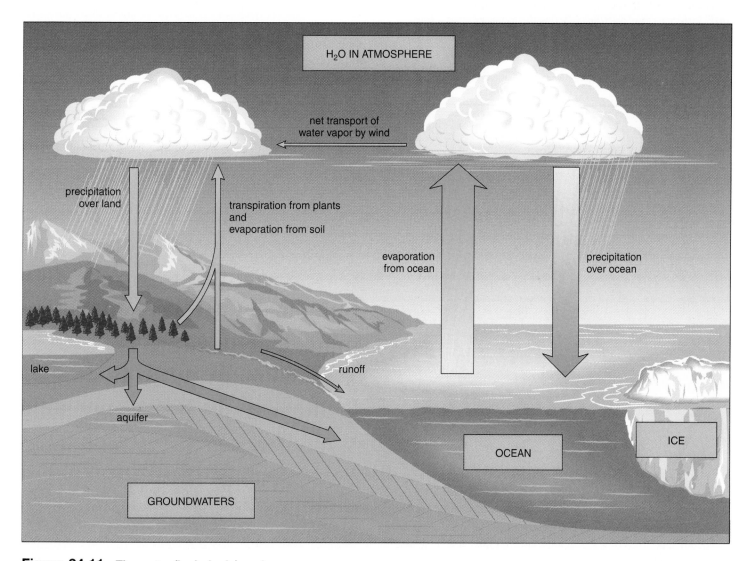

Figure 24.11 The water (hydrologic) cycle.

The Carbon Cycle

In the **carbon cycle**, organisms exchange carbon dioxide with the atmosphere (Fig. 24.12). On land, plants take up carbon dioxide from the air, and through photosynthesis they incorporate carbon into organic nutrients that are used for other living things. When organisms (e.g., plants, animals, and decomposers) respire, a portion of this carbon is returned to the atmosphere as carbon dioxide. In aquatic ecosystems, the exchange of carbon dioxide with the atmosphere is indirect. Carbon dioxide from the air combines with water to produce bicarbonate ion (HCO_3^-), a source of carbon for protists, which also produce organic nutrients through photosynthesis. And when aquatic organisms respire, the carbon dioxide they give off becomes bicarbonate ion.

Living and dead organisms are reservoirs for carbon. If decomposition of dead remains fails to occur, they are subject to physical processes that transform them into coal, oil, and natural gas. We call these reservoirs for carbon the **fossil fuels.** Most of the fossil fuels were formed during the Carboniferous period, 286 to 360 million years ago, when an exceptionally large amount of organic matter was buried before decomposing. Another reservoir for carbon is calcium carbonate shells from marine organisms which accumulate in ocean bottom sediments.

Carbon Dioxide and Global Warming

A **transfer rate** is defined as the amount of a nutrient that moves from one component of the environment to another within a specified period of time. The width of the arrows in Figure 24.12 indicates the transfer rate of carbon dioxide (CO_2). The transfer rates due to photosynthesis and respiration, which includes decay, are just about even. However, more carbon dioxide is now being deposited in the atmosphere than is being removed. In 1850, atmospheric carbon dioxide was about 280 parts per million (ppm), and today it is about 350 ppm. This increase is largely due to the burning of wood and fossil fuels and the destruction of forests to make way for farmland and pasture.

Other gases are also being emitted due to human activities, including nitrous oxide (N_2O, from fertilizers and animal wastes) and methane (CH_4, from bacterial decomposition, particularly in the guts of animals, in sediments, and in flooded rice paddies). These gases are known as **greenhouse gases,** because just like the panes of a greenhouse, they allow solar radiation to pass through but hinder the escape of infrared rays (heat) back into space. Thus, these emissions are contributing significantly to an overall rise in the earth's ambient temperature, a phenomenon called **global warming,** and their effect is known as the **greenhouse effect.**

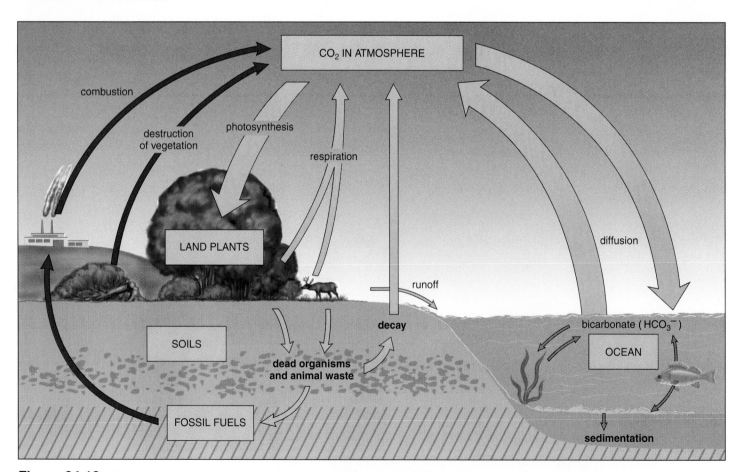

Figure 24.12   The carbon cycle. (Purple arrows represent human activities; gray arrows represent natural events.)

Figure 24.13 shows the earth's radiation balances. One thing to be learned from this diagram is that water vapor is a greenhouse gas, so clouds (which are composed of water vapor) also reradiate heat back to earth. If the earth's temperature rises, more water will evaporate, forming more clouds and setting up a positive feedback effect that could increase global warming still more.

Today, data collected around the world show a steady rise in the concentration of greenhouse gases. For example, methane is increasing by about 1% a year. Such data are used to generate computer models that predict the environmental temperature may become warmer than ever before experienced by living things. The global climate has already warmed about 0.6°C since the industrial revolution. Computer models are unable to consider all possible variables, but the earth's temperature may rise 1.5–4.5°C by 2060 if greenhouse emissions continue at the current rates.

Global warming will bring about other effects, which computer models attempt to forecast. It is predicted that as the oceans warm, temperatures in the polar regions will rise to a greater degree than in other regions. If so, glaciers will melt, and sea levels will rise, not only due to this melting but also because water expands as it warms. Water evaporation will increase, and most likely there will be increased rainfall along the coasts and dryer conditions inland. The occurrence of droughts will reduce agricultural yields and also cause trees to die off. Expansion of forests into Arctic areas might not offset the loss of forests in the temperate zones. Coastal agricultural lands, such as the deltas of Bangladesh and China, will be inundated, and billions of dollars will have to be spent to keep coastal cities like New York, Boston, Miami, and Galveston from disappearing into the sea. Species extinction is also likely, as will be discussed in chapter 25.

The atmosphere is an exchange pool for carbon dioxide. Fossil fuel combustion in particular has increased the amount of carbon dioxide in the atmosphere. Global warming is predicted because carbon dioxide and other gases impede the escape of infrared radiation from the surface of the earth.

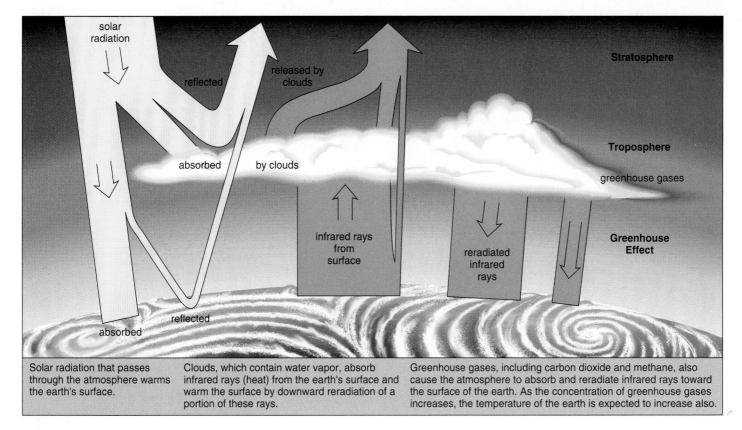

| Solar radiation that passes through the atmosphere warms the earth's surface. | Clouds, which contain water vapor, absorb infrared rays (heat) from the earth's surface and warm the surface by downward reradiation of a portion of these rays. | Greenhouse gases, including carbon dioxide and methane, also cause the atmosphere to absorb and reradiate infrared rays toward the surface of the earth. As the concentration of greenhouse gases increases, the temperature of the earth is expected to increase also. |

Figure 24.13   Earth's radiation balances.
The contribution of greenhouse gases (*far right*) to the earth's surface causes global warming.

The Nitrogen Cycle

Nitrogen is an abundant element in the atmosphere. Nitrogen makes up about 78% of the atmosphere by volume, yet nitrogen deficiency sometimes limits plant growth. Plants cannot incorporate nitrogen gas into organic compounds, and therefore they depend on various types of bacteria to make nitrogen available to them in the **nitrogen cycle** (Fig. 24.14).

Nitrogen fixation occurs when nitrogen (N_2) is converted to a form that plants can use. Some nitrogen-fixing bacteria live in nodules on the roots of legumes (plants of the pea family, such as peas, beans, and alfalfa). They make nitrogen-containing organic compounds available to a host plant. Free-living bacteria in bodies of water and in the soil are able to fix nitrogen gas as ammonium (NH_4^+). Plants can use NH_4^+ and nitrate (NO_3^-) from the soil. After NO_3^- is taken up, it is enzymatically reduced to NH_4^+, which is used to produce amino acids and nucleic acids.

Nitrification is the production of nitrates. Nitrogen gas (N_2) is converted to nitrate (NO_3^-) in the atmosphere when cosmic radiation, meteor trails, and lightning provide the high energy needed for nitrogen to react with oxygen.

Ammonium (NH_4^+) in the soil is converted to nitrate by chemoautotrophic soil bacteria in a two-step process. First, nitrite-producing bacteria convert ammonium to nitrite (NO_2^-), and then nitrate-producing bacteria convert nitrite to nitrate. Notice the subcycle in the nitrogen cycle that involves dead organisms and animal wastes, ammonium, nitrites, nitrates, and plants. This subcycle does not necessarily depend on nitrogen gas at all (Fig. 24.14).

Denitrification is the conversion of nitrate to nitrous oxide and nitrogen gas. There are denitrifying bacteria in both aquatic and terrestrial ecosystems. Denitrification balances nitrogen fixation, but not completely.

Nitrogen and Air Pollution

Human activities significantly alter transfer rates in the nitrogen cycle. Because we produce fertilizers, thereby converting N_2 to NO_3^-, and burn fossil fuels, the atmosphere contains three times more nitrogen oxides (NO_x) than it would otherwise. Fossil fuel combustion also pumps much sulfur dioxide (SO_2) into the atmosphere (Fig. 24.15). Both nitrogen oxides and sulfur dioxide are converted to acids when they combine with water vapor in the atmosphere.

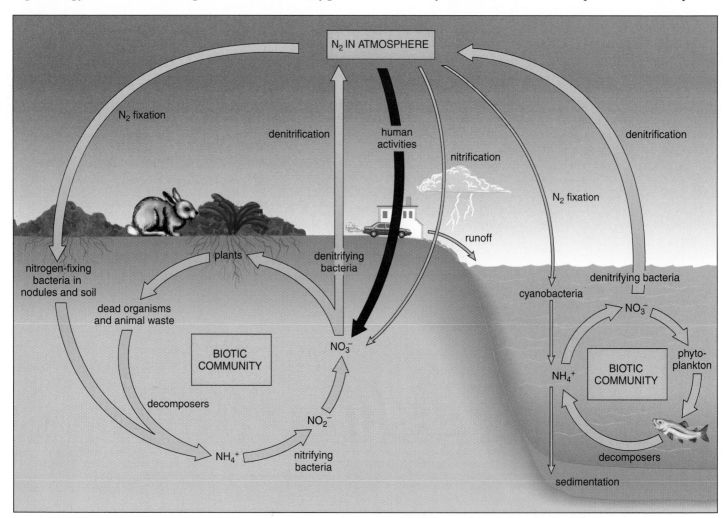

Figure 24.14   The nitrogen cycle.

These acids return to earth as either wet deposition (acid rain or snow) or dry deposition (sulfate and nitrate salts).

Increased deposition of acids has drastically affected forests and lakes in northern Europe, Canada, and the northeastern United States because their soils are naturally acidic and their surface waters are only mildly alkaline (basic) to begin with. The forests in these areas are dying (Fig. 24.15), and their waters cannot support normal fish populations. **Acid deposition** reduces agricultural yields and corrodes marble, metal, and stonework, an effect that is noticeable in cities.

Nitrogen oxides (NO$_x$) and hydrocarbons (HC) react with one another in the presence of sunlight to produce **photochemical smog,** which contains ozone (O$_3$) and **PAN (peroxyacetylnitrate).** Hydrocarbons come from fossil fuel combustion, but additional amounts come from various other sources, including paint solvents and pesticides. Breathing ozone affects the respiratory and nervous systems, resulting in respiratory distress, headache, and exhaustion. These symptoms are particularly apt to appear in young people. Ozone is especially damaging to plants, resulting in leaf mottling and reduced growth.

Normally, warm air near the earth is able to escape into the atmosphere, taking pollutants with it. However, during a **thermal inversion,** pollutants are trapped near the earth beneath a layer of warm stagnant air. Because the air does not circulate, pollutants can build up to dangerous levels. Some areas surrounded by hills are particularly susceptible to the effects of a temperature inversion because the air tends to stagnate, and there is little turbulent mixing (Fig. 24.16).

Fertilizer use also results in the release of nitrous oxide (N$_2$O), a greenhouse gas and a contributor to ozone shield depletion, a topic to be discussed in the Health Focus on page 492.

Atmospheric N$_2$, a reservoir and exchange pool for nitrogen, must be fixed by bacteria in order to make nitrogen available to plants. Environmental problems are associated with the release of nitrous oxide (N$_2$O) and nitrogen oxides (NO$_x$) due to the action of bacteria on fertilizers and fossil fuel combustion, respectively.

Figure 24.15 Acid deposition.
a. Many forests in higher elevations of northeastern North America and Europe are dying due to acid deposition. **b.** Air pollution due to emissions from factories and fossil fuel burning is the major cause of acid deposition, which contains nitric acid (H$_2$NO$_3$) and sulfuric acid (H$_2$SO$_4$).

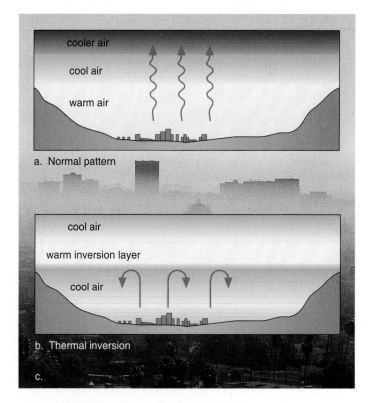

Figure 24.16 Thermal inversion.
a. Normally, pollutants escape into the atmosphere when warm air rises. **b.** During a thermal inversion, a layer of warm air (warm inversion layer) overlies and traps pollutants in cool air below. **c.** Los Angeles is particularly susceptible to thermal inversions, and this accounts for why this city is the "air pollution capital" of the United States.

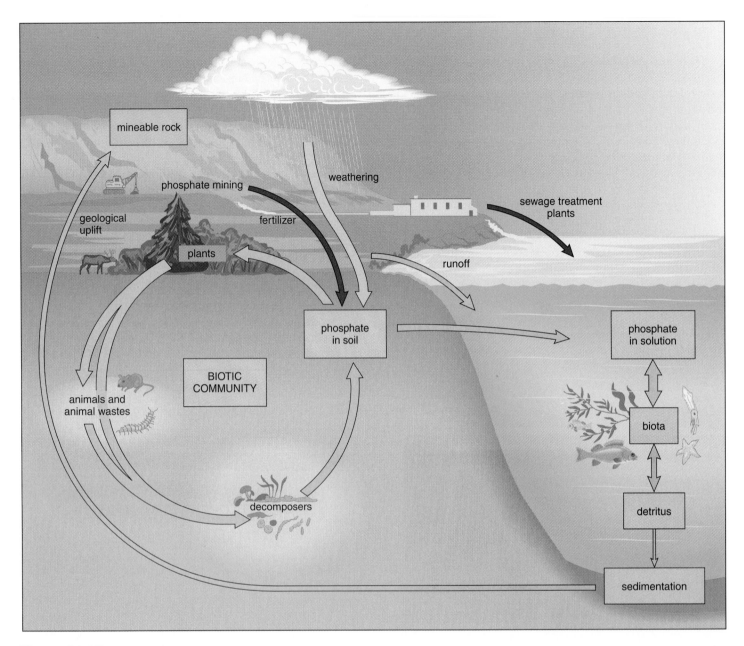

Figure 24.17   **The phosphorus cycle.**

The Phosphorus Cycle

On land, the weathering of rocks makes phosphate ions (PO_4^{3-} and HPO_4^{2-}) available to plants, which take up phosphate from the soil (Fig. 24.17). Some of this phosphate runs off into aquatic ecosystems where algae take up phosphate from the water before it becomes trapped in sediments. Phosphate in sediments only becomes available when a geological upheaval exposes sedimentary rocks to weathering once more. Phosphorus does not enter the atmosphere; therefore, the **phosphorus cycle** is called a sedimentary cycle.

The phosphate taken up by producers is incorporated into a variety of organic molecules, including phospholipids and ATP or the nucleotides that become a part of DNA and

RNA. Animals eat producers and incorporate some of the phosphate into teeth, bones, and shells that do not decompose for very long periods. However, death and decay of all organisms and also decomposition of animal wastes do make phosphate ions available to producers once again. Because available phosphate is generally taken up very quickly, it is a limiting inorganic nutrient in ecosystems. A limiting nutrient is one that regulates the growth of organisms because it is in shorter supply than other nutrients in the environment.

Phosphorus and Water Pollution

Human beings boost the supply of phosphate by mining phosphate ores for fertilizer and detergent production.

Runoff of phosphate and nitrogen due to fertilizer use, animal wastes from livestock feedlots, and discharge from sewage treatment plants results in **eutrophication** (overenrichment) of waterways. Eutrophication can lead to an algal bloom, apparent when green scum floats on the water. When the algae die off, decomposers use up all available oxygen during cellular respiration. The result is a massive fish kill.

Figure 24.18 lists the various sources of water pollution. Point sources are sources of pollution that are specific, and nonpoint sources are those caused by runoff from the land. Industrial wastes can include heavy metals and organochlorides, such as those in some pesticides. These materials are not degraded readily under natural conditions or in conventional sewage treatment plants. They enter bodies of water and are subject to **biological magnification** because they remain in the body and are not excreted. Therefore, they become more concentrated as they pass along a food chain. Biological magnification occurs more readily in aquatic food chains, which have more links than terrestrial food chains. Humans are the final consumers in food chains, and in some areas, human milk contains detectable amounts of DDT and PCBs, which are organochlorides.

Coastal regions are the immediate receptors for local pollutants and the final receptors for pollutants carried by rivers that empty at a coast. Waste dumping occurs at sea, but ocean currents sometimes transport both trash and pollutants back to shore. Offshore mining and shipping add pollutants to the oceans. Some 5 million metric tons of oil a year—or more than one gram per 100 square meters of the oceans' surfaces—end up in the oceans. Large oil spills kill plankton, fish fry, and shellfishes, as well as birds and marine mammals. The largest tanker spill in U.S. territorial waters occurred on March 24, 1989, when the tanker *Exxon Valdez* struck a reef in Alaska's Prince William Sound and leaked 44 million liters of crude oil.

In the last 50 years, we have polluted the seas and exploited their resources to the point that many species are at the brink of extinction. Fisheries once rich and diverse, such as George's Bank off the coast of New England, are in severe decline. Haddock was once the most abundant species in this fishery, but now it accounts for less than 2% of the total catch. Cod and bluefin tuna have suffered a 90% reduction in population size. In warm, tropical regions, many areas of coral reefs are now overgrown with algae because the fish that normally keep the algae under control have been killed off.

Sedimentary rock is a reservoir for phosphorus; for the most part, producers are dependent on decomposers to make phosphate available to them. Fertilizer production and other human activities add phosphate to aquatic ecosystems, contributing to water pollution.

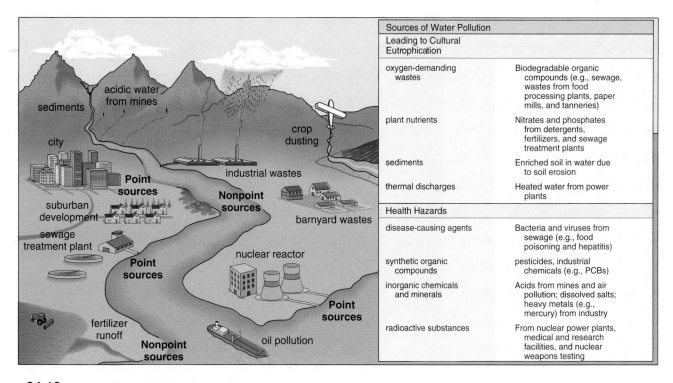

Figure 24.18   Sources of surface water pollution.
Many bodies of water are dying due to the introduction of pollutants from point sources, which are easily identifiable, and nonpoint sources, which cannot be specifically identified.

Health Focus

Ozone Shield Depletion

The earth's atmosphere is divided into layers. The troposphere envelops us as we go about our day-to-day lives. When ozone (O_3) is present in the troposphere (called ground-level ozone), it is considered a pollutant because it adversely affects a plant's ability to grow and our ability to breathe oxygen (O_2). In the stratosphere, some 50 kilometers above the earth, ozone forms the **ozone shield,** a layer of ozone that absorbs much of the ultraviolet (UV) rays of the sun so that fewer rays strike the earth.

UV radiation causes mutations that can lead to skin cancer and can make the lens of the eye develop cataracts. It also is believed to adversely affect the immune system and our ability to resist infectious diseases. Crop and tree growth is impaired, and UV radiation also kills off small plants (phytoplankton) and tiny shrimplike animals (krill) that sustain oceanic life. Without an adequate ozone shield, our health and food sources are threatened.

Ozone shield depletion in recent years is, therefore, of serious concern. It became apparent in the 1980s that depletion of ozone had occurred worldwide and that there was a severe depletion (40–50% of the ozone) above the Antarctic every spring. A vortex of cold wind (a whirlpool in the atmosphere) circles the pole during the winter months, creating ice crystals in which chemical reactions occur that break down ozone. Severe depletions of the ozone layer are commonly called "ozone holes." Detection devices now tell us that the ozone hole above the Antarctic is about the size of the United States and growing (Fig. 24A). Of even greater concern, an ozone hole has now appeared above the Arctic as well, and ozone holes could also occur within northern and southern latitudes, where many people live. Whether or not these holes develop depends on prevailing winds, weather conditions, and the type of particles in the atmosphere. A United Nations Environment Program report predicts a 26% rise in cataracts and nonmelanoma skin cancers for every 10% drop in the ozone level. A 26% increase translates into 1.75 million additional cases of cataracts and 300,000 more skin cancers every year, worldwide.

The cause of ozone depletion can be traced to the release of chlorine atoms (Cl) into the stratosphere. Chlorine atoms combine with ozone and strip away the oxygen atoms, one by one. One atom of chlorine can destroy up to 100,000 molecules of ozone before settling to the earth's surface as chloride years later. These chlorine atoms come from the breakdown of **chlorofluorocarbons** (CFCs), chemicals much in use by humans. The best known CFC is Freon, a coolant found in refrigerators and air conditioners. CFCs are also used as cleaning agents and foaming agents during the production of styrofoam used in coffee cups, egg artons, insulation, and paddings. Formerly, CFCs were used as propellants in spray cans, but this application is now banned in the United States and several European countries.

Most of the countries of the world have stopped using CFCs. The United States halted production in 1995. Computer projections suggest that an 85% reduction in CFC emissions is needed to stabilize CFC levels in the atmosphere. Otherwise, they keep on increasing. There are many available CFC substitutes that will not release chlorine atoms (nor bromine atoms) to harm the ozone shield.

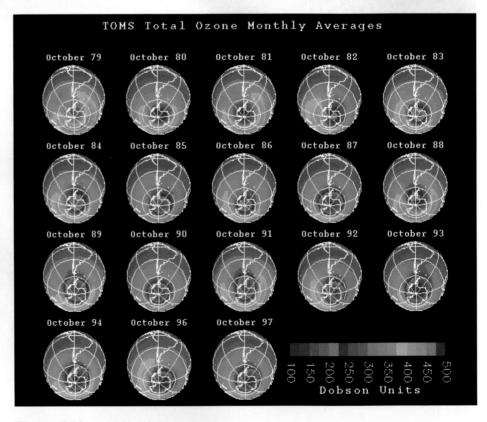

Figure 24A Ozone shield depletion.

These satellite observations show that the amount of ozone over the South Pole between October 1979 and October 1997 fell by more than 50%. Green represents an average amount of ozone, blue less, and purple still less. Yellow, orange, and red represent above-average amounts of ozone.

Preserving Ecosystems Abroad

Peter Jutro, a scientist working for the U.S. Environmental Protection Agency, wants to study native traditions for clues on how to preserve ecosystems. He has run into opposition from the indigenous groups because they mistrust conservationists. Take, as an example, the fact that the Kuna people in Panama refused to renew the lease for a Smithsonian Institution study of reef ecology for the past 21 years. Much of the trouble seems to have come from the failure of scientists to explain their program to local communities. In a meeting of the Kuna congress, the scientists were accused by the Kuna of "stealing their knowledge, stealing their reefs, stealing their sand." Local people find it hard to see a difference between a scientific study and commercial ventures that exploit their areas for minerals, timber, and other resources.

Laura Snook, a forester from Duke University, researches the growing habits of mahogany trees in Mexico. She came to the conclusion that because the trees grow slowly, logging shouldn't be done too fast. Most local foresters resented her findings, but she was able to establish a good relationship with two women foresters who were open to her ideas. Now local people are changing the way they replant mahogany trees so that the resource will be there for some time to come. The point is that conservationists working with people from different cultural backgrounds probably need to communicate their goals more clearly and involve local people in project planning. Perhaps they should also learn to accept slower timescales and styles of decision making different from our own.

Figure 24B Preserving ecosystems.
These ecologists are measuring tropical trees. Preserving ecosystems abroad or at home requires understanding local concerns.

Decide Your Opinion

1. Should U.S. scientists study ecosystems in such far-flung places as Panama, Alaska, Mexico, and the Amazon? Why or why not?
2. To what degree should conservationists be politically correct when negotiating with local groups? Explain.
3. Suggest other ways conservationists could gain the cooperation of local groups.

Looking at Both Sides   www.mhhe.com/biosci/genbio/maderhuman7/

Every bioethical issue has at least two sides. Even if you already have an opinion, it is important to explore the opposite opinion before finalizing your position. The Online Learning Center at www.mhhe.com/biosci/genbio/maderhuman7/ will help you fine-tune your initial opinion, explore both sides, and finalize your position. You may acquire new arguments for your original opinion, or you may even change your opinion. Be sure to complete these activities in sequence:

Taking Sides   Decide your initial opinion by answering a series of questions. Then see if your opinion changes after completing the next two activities.

Further Debate   Read opposing articles that give you further information on this particular bioethical issue.

Explain Your Position   Answer another series of questions and then defend your original or changed opinion. You can e-mail your position to your instructor if he or she wishes.

Summarizing the Concepts

24.1 The Nature of Ecosystems

The process of succession from either bare rock or disturbed land results in a climax community. An ecosystem is a community of organisms plus the physical environment. Each population in an ecosystem has a habitat and a niche. Some populations are producers and some are consumers. Producers are autotrophs that produce their own organic food. Consumers are heterotrophs that take in organic food. Consumers may be herbivores, carnivores, omnivores, or decomposers.

24.2 Energy Flow and Chemical Cycling

Energy flows through an ecosystem. Producers transform solar energy into food for themselves and all consumers. As herbivores feed on plants (or algae), and carnivores feed on herbivores, some energy is converted to heat. Feces, urine, and dead bodies become food for decomposers. Eventually, all the solar energy that enters an ecosystem is converted to heat, and thus ecosystems require a continual supply of solar energy.

Inorganic nutrients are not lost from the biosphere as is energy. They recycle within and between ecosystems. Decomposers return some proportion of inorganic nutrients to autotrophs, and other portions are imported or exported between ecosystems in global cycles.

Ecosystems contain food webs, and a diagram of a food web shows how the various organisms are connected by eating relationships. In a grazing food web, food chains begin with a producer. In a detrital food web, food chains begin with detritus. The two food webs are joined when the same consumer is a link in both a grazing and detrital food chain. A trophic level is all the organisms that feed at a particular link in a food chain. Ecological pyramids show trophic levels stacked one on top of the other like building blocks. Generally they show that biomass and energy content decrease from one trophic level to the next. Most pyramids pertain to grazing food webs and largely ignore the detrital food web portion of an ecosystem.

24.3 Global Biogeochemical Cycles

Biogeochemical cycles contain reservoirs, which are components of ecosystems, such as fossil fuels, sediments, and rocks, that contain elements available on a limited basis to living things. Pools are components of ecosystems, such as the atmosphere, soil, and water, which are ready sources of nutrients for living things.

In the water cycle, evaporation over the ocean is not compensated for by rainfall. Evaporation from terrestrial ecosystems includes transpiration from plants. Rainfall over land results in bodies of fresh water plus groundwater, including aquifers. Eventually, all water returns to the oceans.

In the carbon cycle, organisms add as much carbon dioxide to the atmosphere as they remove. Shells in ocean sediments, organic compounds in living and dead organisms, and fossil fuels are reservoirs for carbon. Human activities such as burning fossil fuels and trees are adding carbon dioxide to the atmosphere. Like the panes of a greenhouse, carbon dioxide and other gases allow the sun's rays to pass through but impede the release of infrared wavelengths. It is predicted that a buildup of these "greenhouse gases" will lead to a global warming. The effects of global warming could be a rise in sea level and a change in climate patterns, with disastrous effects.

In the nitrogen cycle, the biotic community, which includes several types of bacteria, keeps recycling nitrogen back to the producers. Certain bacteria in water, soil, and root nodules, can fix atmospheric nitrogen. Other bacteria return nitrogen to the atmosphere. Human activities convert atmospheric nitrogen to fertilizer, which is broken down by soil bacteria; humans also burn fossil fuels. In this way, a large quantity of nitrogen oxide (NO_x) and sulfur dioxide (SO_2) is added to atmosphere where it reacts with water vapor to form acids that contribute to acid deposition. Acid deposition can kill lakes and forests and corrode marble, metal, and stonework. Nitrogen oxides and hydrocarbons (HC) react to form smog, which contains ozone and PAN (peroxyacetylnitrate). These oxidants are harmful to animal and plant life.

In the phosphorus cycle, the biotic community recycles phosphorus back to the producers, and only limited quantities are made available by the weathering of rocks. Phosphates are mined for fertilizer production; when phosphates and nitrates enter lakes and ponds, overenrichment occurs. Many kinds of wastes enter the rivers and then flow to the oceans, which have now become degraded from added pollutants.

Studying the Concepts

1. What is succession, and how does it result in a climax community? 478
2. Distinguish between the abiotic and biotic components of an ecosystem. What are the aspects of niche for a plant? An animal? 480
3. Distinguish between autotrophs and heterotrophs, and describe four different types of heterotrophs found in natural ecosystems. Explain the terms producer and consumer. 480
4. Tell why energy must flow but chemicals can cycle in an ecosystem. 481
5. Describe two types of food webs and two types of food chains typically found in terrestrial ecosystems. Which of these typically moves more energy through an ecosystem? 482–83
6. What is a trophic level? An ecological pyramid? 483–84
7. Give examples of reservoirs and exchange pools in biogeochemical cycles. 484
8. Draw one diagram to illustrate the water cycle and another to represent the carbon cycle. 485–86
9. How and why is the global climate expected to change, and what are the predicted consequences of this change? 486–87
10. Draw a diagram of the nitrogen cycle. What types of bacteria are involved in this cycle? 488
11. What causes acid deposition, and what are its effects? How does photochemical smog develop, and what is a thermal inversion? 488–89
12. Draw a diagram of the phosphorus cycle. 490
13. What are several ways in which fresh water and marine water can be polluted? What is biological magnification? 490–91

Testing Your Knowledge of the Concepts

In questions 1–4, match the population to the description.

a. producer
b. consumer
c. decomposer
d. herbivore

_____ 1. Heterotroph that feeds on plant material.

_____ 2. Autotroph that manufactures organic nutrients.

_____ 3. Any type of heterotroph that feeds on plant material or on other animals.

_____ 4. Heterotroph that breaks down detritus as a source of nutrients.

In questions 5–8, indicate whether the statement is true (T) or false (F).

_____ 5. If global warming occurs, it is predicted that rising waters will threaten many coastal cities.

_____ 6. Carbon dioxide from the burning of fossil fuels is one of the major greenhouse gases.

_____ 7. Acid deposition affects just those areas in which there are factories and power plants.

_____ 8. Thermal inversions make the tropics colder and the Arctic warmer.

In questions 9 and 10, fill in the blanks.

9. During the process of denitrification, nitrate is converted to _____.

10. In the carbon cycle, when living organisms, _____, carbon dioxide (CO_2) is returned to the exchange pool.

11. Label the following diagram of an ecosystem:

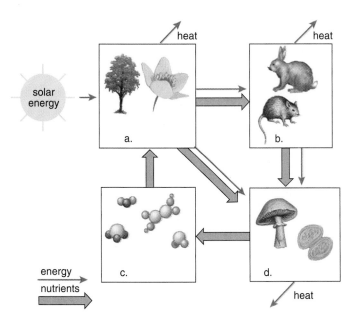

12. Match the key terms to these definitions.

a. _____ Pictorial graph of the tropic levels in a food web—from the producers to the final consumer populations.

b. _____ Sequential change in the relative dominance of species within a community; can begin on bare rock or where soil already exists.

c. _____ Remains of once living organisms that are burned to release energy, such as coal, oil, and natural gas.

d. _____ Process by which atmospheric nitrogen gas is changed to forms that plants can use.

e. _____ Formed from oxygen in the upper atmosphere, it protects the earth from ultraviolet radiation.

Understanding Key Terms

acid deposition 489
aquifer 485
autotrophic organism 480
biogeochemical cycle 484
biological magnification 491
biomass 483
carbon cycle 486
carnivore 480
chlorofluoro-
    carbon (CFC) 492
community 478
consumer 480
decomposer 480
denitrification 488
detrital food chain 483
detrital food web 483
detritus 480
ecological pyramid 483
ecosystem 477
eutrophication 491
food chain 483
food web 483
fossil fuel 486
global warming 486
grazing food chain 483

grazing food web 483
greenhouse effect 486
greenhouse gases 486
habitat 480
herbivore 480
heterotrophic organism 480
niche 480
nitrification 488
nitrogen cycle 488
nitrogen fixation 488
omnivore 480
ozone shield 492
PAN (peroxyacetylnitrate)
    489
phosphorus cycle 490
photochemical smog 489
population 478
precipitation 485
producer 480
succession 478
thermal inversion 489
transfer rate 486
trophic level 483
water (hydrologic) cycle 485

e-Learning Connection   **www.mhhe.com/biosci/genbio/maderhuman7/**

24.1 The Nature of Ecosystems

 Succession *Essential Study Partner*

 Looking at Both Sides *critical thinking activity*

24.2 Energy Flow and Chemical Cycling

 Introduction *Essential Study Partner*
Energy Flow *Essential Study Partner*

 Path of Energy Through Food Webs *art quiz*
Food Web in Cayuga Lake *art quiz*

24.3 Global Biogeochemical Cycles

 Nutrient Cycles *Essential Study Partner*
Biological Magnification *Essential Study Partner*

 Carbon Cycle *animation activity*
Nitrogen Cycle *animation activity*
Bioaccumulation *animation activity*
Global Warming *animation activity*
Acid Rain *animation activity*
The Water Cycle *art quiz*
The Carbon Cycle *art quiz*
The Nitrogen Cycle *art quiz*

 Radioactive Wastes From Medical Facilities: Where Do They Go?? *case study*

Chapter Summary

 Concept Review *Essential Study Partner*

 Key Term Flashcards *vocabulary quiz*
Chapter Quiz *objective quiz covering all chapter concepts*

Chapter 25

Conservation of Biodiversity

Chapter Concepts

Figure 25.1   Forest.
Humans don't always realize the many ways in which they disturb local ecosystems.

Susan was so excited. She and her husband Bob were moving out of the city to a house in the suburbs. Although their housing development had replaced much of a forest, enough remained to put in walking trails for the local residents.

Susan and Bob had to decide what kind of landscaping they wanted. A small brook meandered along their backyard and into the forest. Susan was thinking how nice it would be if they left the lawn natural and encouraged wildflowers to grow. An article in the paper told her what to plant in order to attract butterflies, bees, and hummingbirds. Oh, she wished she could get Bob to go along with her wishes.

Still, she thought, what would the neighbors think? She was sure that, like most suburban residents, her neighbors would be putting in grass lawns. She knew that grass lawns require frequent applications of fertilizers, herbicides, and pesticides. Aside from the extra work and expense, Susan was thinking about all those chemicals entering the brook and going into the forest. The brook joined a stream, and eventually the stream entered a river. The osprey, a bird of prey in the forest, would be taking fish from the stream to feed itself and its chicks. Susan worried that the chemicals would enter the food chain and become concentrated as they moved from minnows to larger fish to the osprey. She had read that pesticides can interfere with the ability of the osprey to reproduce. In addition, new evidence suggests that pesticides might contribute to the development of Parkinson disease in humans.

Susan thought that perhaps she should become an activist and try to convince her neighbors to put in natural lawns. Would she have enough courage to talk to people and get them to change their ways? Maybe she wouldn't be able to change their minds, but somehow she was going to find the courage to speak out.

25.1 Conservation Biology and Biodiversity

Conservation biology is a new discipline that studies all aspects of biodiversity with the goal of conserving natural resources for this generation and all future generations. Conservation biology is unique in that it is concerned with both the development of scientific concepts and the application of these concepts to the everyday world. A primary goal is the management of biodiversity for sustainable use by humans. To achieve this goal, conservation biologists are interested in, and come from, many subfields of biology that only now have been brought together into a cohesive whole:

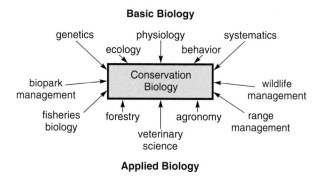

Basic Biology

genetics physiology systematics
ecology behavior
biopark management Conservation Biology wildlife management
fisheries biology forestry agronomy range management
veterinary science

Applied Biology

Like a physician, a conservation biologist must be aware of the latest findings, both theoretical and practical, and be able to use this knowledge to diagnose the source of trouble and suggest a suitable treatment. Often, it is necessary to work with government officials at both the local and federal levels.

Conservation biology is a unique science in another way. It takes a leap of faith and unabashedly supports the following ethical principles: (1) biodiversity is desirable for the biosphere and therefore for humans; (2) extinctions, due to human actions, are therefore undesirable; (3) the complex interactions in ecosystems support biodiversity and are desirable; and (4) biodiversity brought about by evolutionary change has value in and of itself, regardless of any practical benefit.

Conservation biology has emerged in response to a crisis—never before in the history of the earth are so many extinctions expected in such a short period of time. Estimates vary, but at least 10–20% of all species now living most likely will become extinct in the next 20 to 50 years unless immediate action is taken. It is urgently important, then, that all citizens understand the concept of biodiversity, the value of biodiversity, the likely causes of present-day extinc-

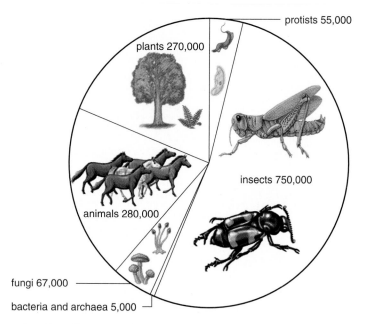

protists 55,000
plants 270,000
insects 750,000
animals 280,000
fungi 67,000
bacteria and archaea 5,000

Figure 25.2 **Number of described species.**
There are only about 1.5 million described species; insects are far more prevalent than organisms in other groups. Undescribed species probably number far more than those species that have been described.

tions, and what could possibly be done to prevent extinctions from occurring.

Biodiversity

At its simplest level, **biodiversity** is simply the variety of life on earth. It is common practice to describe biodiversity in terms of the number of species among various groups of organisms. Figure 25.2 only accounts for the species that have so far been described. The majority of estimates state that there are probably 5 to 15 million species in all; if so, many species are still to be found and described.

You can well appreciate that simply giving the number of species on earth is not very meaningful, and certainly ecologists describe biodiversity as an attribute of three other levels of biological organization: genetic diversity, community diversity, and landscape diversity.

Genetic diversity refers to variations among the members of a population. Populations with high genetic diversity are more likely to have some individuals that can survive a change in the structure of an ecosystem. For example, the 1846 potato blight in Ireland, the 1922 wheat failure in the Soviet Union, and the 1984 outbreak of citrus canker in Florida were all made worse by limited genetic variation among these crops. If its populations are quite small and isolated, a species is more likely to eventually become extinct because of a loss of genetic diversity.

Community (and therefore ecosystem) diversity is dependent on the interactions of species at a particular locale. One community's species composition can be completely

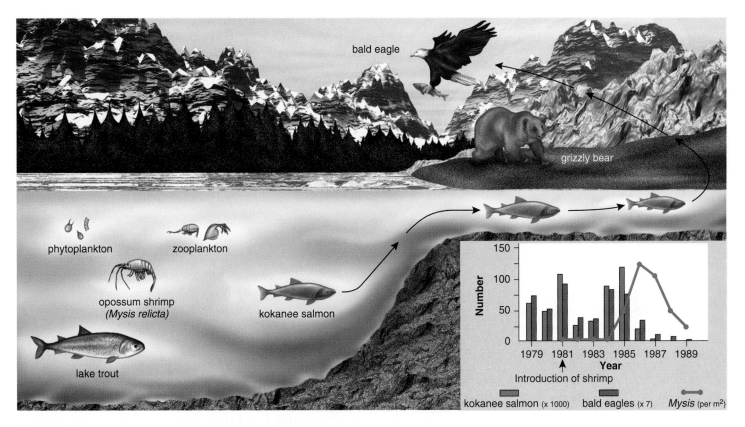

Figure 25.3   Eagles and bears feed on spawning salmon.
Humans introduced the opossum shrimp as prey for salmon. Instead, the shrimp competed with salmon for zooplankton as a food source. The salmon, eagle, and bear populations subsequently declined.

different from that of other communities. Community composition, therefore, increases the amount of biodiversity in the biosphere. Although in the past conservationists have frequently concentrated on saving particular species, such as the California condor, the black-footed ferret, or the spotted owl, this is a shortsighted approach. Saving entire communities can save many species, and the contrary is also true—disrupting a community threatens the existence of more than one species. Opossum shrimp (*Mysis relicta*) were introduced into Flathead Lake in Montana and its tributaries as food for salmon. The shrimp ate so much zooplankton that there was in the end far less food for the fish and ultimately the grizzly bears and bald eagles as well (Fig. 25.3).

Landscape diversity involves a group of interacting ecosystems; in one **landscape**, for example, there may be plains, mountains, and rivers. Any of these ecosystems can be so fragmented that they are connected by only patches (remnants) or strips of land that allow organisms to move from one ecosystem to the other.

Distribution of Diversity

Biodiversity is not evenly distributed throughout the biosphere; therefore, protecting some areas will save more species than protecting other areas. Biodiversity is highest at the tropics and declines toward each pole on land, in fresh water, and in the ocean. Also, there are more species in the coral reefs of the Indonesian archipelago than in other coral reefs as one moves westward across the Pacific.

Some regions of the world are called **biodiversity hotspots** because they contain an unusually large concentrations of species. Hotspots contain about 20% of the earth's species but cover only about half of 1% of the earth's land area. The island of Madagascar, the Cape region of South Africa, and the Great Barrier Reef of Australia are all biodiversity hotspots.

One surprise of late has been the discovery that rain forest canopies and the deep-sea benthos have many more species than formerly thought. Some conservationists refer to these two areas as biodiversity frontiers.

Conservation biology is the study and sustainable management of biodiversity for the benefit of human beings. Biodiversity is often defined as a variety of life at the species level, but biodiversity is also important at the genetic, community (ecosystem), and landscape levels.

25.2   Value of Biodiversity

Conservation biology strives to reverse the trend toward the possible extinction of thousands of plants and animals. To bring this about, it is necessary to make all people aware that biodiversity is a resource of immense value.

Direct Value

Various individual species perform services for human beings and contribute greatly to the value we should place on biodiversity. Only some of the most obvious values are discussed here and illustrated in Figure 25.4.

Medicinal Value

Most of the prescription drugs used in the United States were originally derived from living organisms. The rosy periwinkle from Madagascar is an excellent example of a tropical plant that has provided us with useful medicines. Potent chemicals from this plant are now used to treat two forms of cancer: leukemia and Hodgkin disease. Because of these drugs, the survival rate for childhood leukemia has gone from 10% to 90%, and Hodgkin disease is usually curable. Although the value of saving a life cannot be calculated, it is still sometimes easier for us to appreciate the worth of a resource if it is explained in monetary terms. Thus, researchers tell us that, judging from the success rate in the past, there are 328 more types of drugs to be found in tropical rain forests, and the value of this resource to society is probably $147 billion.

You may already know that the antibiotic penicillin is derived from a fungus and that certain species of bacteria produce the antibiotics tetracycline and streptomycin. These drugs have proven to be indispensable in the treatment of diseases, including certain sexually transmitted diseases.

Leprosy is among those diseases for which there is as yet no cure. The bacterium that causes leprosy will not grow in the laboratory, but scientists discovered that it grows naturally in the nine-banded armadillo. Having a source for the bacterium may make it possible to find a cure for leprosy. The blood of horseshoe crabs contains a substance called limulus amoebocyte lysate, which is used to ensure that medical devices such as pacemakers, surgical implants, and prosthetic devices are free of bacteria. Blood is taken from 250,000 crabs a year, and then they are returned to the sea unharmed.

Agricultural Value

Crops such as wheat, corn, and rice are derived from wild plants that have been modified to be high producers. The same high-yield, genetically similar strains tend to be grown worldwide. When rice crops in Africa were being devastated by a virus, researchers grew wild rice plants from thousands of seed samples until they found one that contained a gene for resistance to the virus. These wild plants were then used in a breeding program to transfer the gene into high-yield rice plants. If this variety of wild rice had become extinct before it could be discovered, rice cultivation in Africa might have collapsed.

Biological pest controls—natural predators and parasites—are often preferable to using chemical pesticides. When a rice pest called the brown planthopper became resistant to pesticides, farmers began to use natural brown planthopper enemies instead. The economic savings were calculated at well over $1 billion. Similarly, cotton growers in Cañete Valley, Peru, found that pesticides were no longer working against the cotton aphid because of resistance. Research identified natural predators that are now being used to an ever greater degree by cotton farmers. Again, savings have been enormous.

Most flowering plants are pollinated by animals, such as bees, wasps, butterflies, beetles, birds, and bats. The honeybee, *Apis mellifera*, has been domesticated, and it pollinates almost $10 billion worth of food crops annually in the United States. The danger of this dependency on a single species is exemplified by mites that have now wiped out more than 20% of the commercial honeybee population in the United States. Where can we get resistant bees? From the wild, of course. The value of wild pollinators to the U.S. agricultural economy has been calculated at $4.1 to $6.7 billion a year.

Consumptive Use Value

We have had much success cultivating crops, keeping domesticated animals, growing trees in plantations, and so forth. But so far, aquaculture, the growing of fish and shellfish for human consumption, has contributed only minimally to human welfare—instead, most freshwater and marine harvests depend on the catching of wild animals, such as fishes (e.g., trout, cod, tuna, and flounder) crustaceans (e.g., lobsters, shrimps, and crabs) and mammals (e.g., whales). Obviously, these aquatic organisms are an invaluable biodiversity resource.

The environment provides all sorts of other products that are sold in the marketplace worldwide, including wild fruits and vegetables, skins, fibers, beeswax, and seaweed. Also, some people obtain their meat directly from the environment. In one study, researchers calculated that the economic value of wild pig in the diet of native hunters in Sarawak, East Malaysia, was approximately $40 million per year.

Similarly, many trees are still felled in the natural environment for their wood. Researchers have calculated that a species-rich forest in the Peruvian Amazon is worth far more if the forest is used for fruit and rubber production than for timber production. Fruit and the latex needed to produce rubber can be brought to market for an unlimited number of years, whereas once the trees are gone, no more timber can be harvested.

Wild species directly provide us with all sorts of goods and services whose monetary value can sometimes be calculated.

Wild species, like the lesser long-nosed bat, *Leptonycteris curasoae*, are pollinators of agricultural and other plants.

Wild species, like the rosy periwinkle, *Catharanthus roseus*, are sources of many medicines.

Wild species, like rubber trees, *Hevea*, can provide a product indefinitely if the forest is not destroyed.

Wild species, like many marine species, provide us with food.

Wild species, like the nine-banded armadillo, *Dasypus novemcinctus*, play a role in medical research.

Wild species, like lady bugs, *Coccinella*, play a role in biological control of agricultural pests.

Figure 25.4 Direct value of wildlife.
The direct services of wild species benefit human beings immensely, and it is sometimes possible to calculate the monetary value, which is always surprisingly large.

Indirect Value

The wild species we have been discussing live in ecosystems (Fig. 25.5a). If we want to preserve them, it is more economical to save ecosystems than individual species. Ecosystems perform many services for modern humans, who increasingly live in cities (Fig. 25.5b). These services are said to be indirect because they are pervasive and not easily discernible. Even so, our very survival depends on the functions that ecosystems perform for us.

Biogeochemical Cycles

You'll recall from chapter 24 that ecosystems are characterized by energy flow and chemical cycling. The biodiversity within ecosystems contributes to the workings of the water, carbon, nitrogen, phosphorus, and other biogeochemical cycles. We are dependent on these cycles for fresh water, removal of carbon dioxide from the atmosphere, uptake of excess soil nitrogen, and provision of phosphate. When human activities upset the usual workings of biogeochemical cycles, the dire environmental consequences include excess pollutants that are harmful to us. Technology is unable to artificially contribute to or create any of the biogeochemical cycles.

Waste Disposal

Decomposers break down dead organic matter and other types of wastes to inorganic nutrients that are used by the producers within ecosystems. This function aids humans immensely because we dump millions of tons of waste material into natural ecosystems each year. If it were not for decomposition, waste would soon cover the entire surface of our planet. We can build sewage treatment plants, but they are expensive and few of them break down solid wastes completely to inorganic nutrients. It is less expensive and more efficient to water plants and trees with partially treated wastewater and let soil bacteria cleanse it completely.

Biological communities are also capable of breaking down and immobilizing pollutants, such as heavy metals and pesticides, that humans release into the environment. A review of wetland functions in Canada assigned a value of $50,000 per hectare (100 acres or 10,000 square meters) per year to the ability of natural areas to purify water and take up pollutants.

Provision of Fresh Water

Few terrestrial organisms are adapted to living in a salty environment—they need fresh water. The water cycle continually supplies fresh water to terrestrial ecosystems. Humans use fresh water in innumerable ways, including drinking it and irrigating their crops. Freshwater ecosystems like rivers and lakes also provide us with fish and other types of organisms for food.

Unlike other commodities, there is no substitute for fresh water. We can remove salt from seawater to obtain fresh water, but the cost of desalination is about four to eight times the average cost of fresh water acquired via the water cycle.

Forests and other natural ecosystems exert a "sponge effect." They soak up water and then release it at a regular rate. When rain falls in a natural area, plant foliage and dead leaves lessen its impact, and the soil slowly absorbs it, especially if the soil has been aerated by organisms. The water-holding capacity of forests reduces the possibility of flooding. The value of a marshland outside Boston, Massachusetts, has been estimated at $72,000 per hectare per year solely on its ability to reduce floods. Forests release water slowly for days or weeks after the rains have ceased. Rivers flowing through forests in West Africa release twice as much water halfway through the dry season, and between three and five times as much at the end of the dry season, as do rivers from coffee plantations.

Prevention of Soil Erosion

Intact ecosystems naturally retain soil and prevent soil erosion. The importance of this ecosystem attribute is especially observed following deforestation. In Pakistan, the world's largest dam, the Tarbela Dam, is losing its storage capacity of 12 billion cubic meters many years sooner than expected because silt is building up behind the dam due to deforestation. At one time, the Philippines were exporting $100 million worth of oysters, mussels, clams, and cockles each year. Now silt carried down rivers following deforestation is smothering the mangrove ecosystem that serves as a nursery for the sea. Most coastal ecosystems are not as bountiful as they once were because of deforestation and a myriad of other assaults.

Regulation of Climate

At the local level, trees provide shade and reduce the need for fans and air conditioners during the summer.

Globally, tropical rain forests ameliorate the climate because they take up carbon dioxide. The leaves of tropical trees use carbon dioxide when they photosynthesize, and the bodies of the trees store carbon. When trees are cut and burned, carbon dioxide is released into the atmosphere. Carbon dioxide makes a significant contribution to global warming, which is expected to be stressful for many plants and animals. Only a small percentage of wildlife will be able to move northward, where the weather will be suitable for them.

Ecotourism

Almost everyone prefers to vacation in the natural beauty of an ecosystem. In the United States, nearly 100 million people enjoy vacationing in a natural setting. To do so, they spend $4 billion each year on fees, travel, lodging, and food. Many tourists want to go sport fishing, whale watching, boat riding, hiking, birdwatching, and the like. Some want to merely immerse themselves in the beauty of a natural environment.

a.

b.

Figure 25.5   Indirect value of ecosystems.

Natural ecosystems (**a**) provide human-impacted ecosystems (**b**) with many ecological services. **c.** Research results show that the higher the biodiversity (measured by number of plant species), the greater the rate of photosynthesis of an experimental community.

c.

Biodiversity and Natural Ecosystems

Massive changes in biodiversity, such as deforestation, have a significant impact on ecosystem services like those we have been discussing. Researchers are interested in determining whether a high degree of biodiversity also helps ecosystems function more efficiently. To test the benefits of biodiversity in a Minnesota grassland habitat, researchers sowed plots with seven levels of plant diversity. Their study found that ecosystem performance improves with increasing species richness. The more diverse plots had lower concentrations of inorganic soil nitrogen, indicating a higher level of nitrate uptake. A similar study in California also showed greater overall resource use in more diverse plots.

Another group of experimenters tested the effects of an increase in diversity at four levels: producers, herbivores, parasites, and decomposers. They found that the rate of photosynthesis increased as diversity increased

(Fig. 25.5c). A computer simulation has shown that the response of a deciduous forest to elevated carbon dioxide is a function of species diversity. The more complex community, composed of nine tree species, exhibited a 30% greater amount of photosynthesis than a community composed of a single species.

More studies are needed to test whether biodiversity maximizes resource acquisition and retention within an ecosystem. Also, are more diverse ecosystems better able to withstand environmental changes and invasions by other species, including pathogens? Then, too, how does fragmentation affect the distribution of organisms within an ecosystem and the functioning of an ecosystem?

Ecosystems perform services that most likely depend on a high degree of biodiversity.

25.3 Causes of Extinction

In order to stem the tide of extinction, it is first necessary to identify its causes. Researchers examined the records of 1,880 threatened and endangered wild species in the United States and found that habitat loss was involved in 85% of the cases (Fig. 25.6a). Alien species had a hand in nearly 50%, pollution was a factor in 24%, overexploitation in 17%, and disease in 3% of cases. The percentages add up to more than 100% because most of these species are imperiled for more than one reason. Macaws are a good example that a combination of factors can lead to a species decline (Fig. 25.6b). Their habitat has been reduced by encroaching timber and mining companies. And macaws are hunted for food and collected for the pet trade.

Habitat Loss

Habitat loss has occurred in all ecosystems, but concern has now centered on tropical rain forests and coral reefs because they are particularly rich in species. A sequence of events in Brazil offers a fairly typical example of the manner in which rain forest is converted to noninhabitable land for wildlife. The construction of a major highway into the forest first provided a way to reach the interior of the forest (Fig. 25.6c). Small towns and industries sprang up along the highway, and roads branching off the main highway gave rise to even more roads.

Figure 25.6 **Habitat loss.**

a. In a study examining records of imperiled U.S. plants and animals, habitat loss emerged as the greatest threat to wildlife. **b.** Macaws that reside in South American tropical rain forests are endangered for the reasons listed in the graph in (**a**). **c.** The construction of roads opens up the rain forest and subjects it to fragmentation. **d.** The result is patches of forest and degraded land. **e.** Wildlife cannot live in destroyed portions of the forest.

The result was fragmentation of the once immense forest. The government offered subsidies to anyone willing to take up residence in the forest, and the people who came cut and burned trees in patches (Fig. 25.6d). Tropical soils contain limited nutrients, but when the trees are burned, nutrients are released that support a lush growth for the grazing of cattle for about three years. Once the land is degraded (Fig. 25.6e), the farmer and his family move on to another portion of the forest.

Loss of habitat also affects freshwater and marine biodiversity. Coastal degradation is largely due to the large concentration of people living there. Already, 60% of coral reefs have been destroyed or are on the verge of destruction; it's possible that all coral reefs may disappear during the next 40 years. Mangrove forest destruction is also a problem; Indonesia, with the most mangrove acreage, has lost 45% of its mangroves, and the percentage is even higher for other tropical countries. Wetland areas, estuaries, and seagrass beds are also being rapidly destroyed.

c.

d.

e.

b.

a.

[Graph: % Species Affected by Cause — Habitat loss ~85, Alien species ~50, Pollution ~25, Over-exploitation ~17, Disease ~3]

Figure 25.7 Alien species.
Mongooses were introduced into Hawaii to control rats, but they also prey on native birds.

Alien Species

Alien species, sometimes called exotics, are nonnative species that migrate into a new ecosystem or are deliberately or accidentally introduced by humans. Ecosystems around the globe are characterized by unique assemblages of organisms that have evolved together in one location. Migrating to a new location is not usually possible because of barriers such as oceans, deserts, mountains, and rivers. Humans, however, have introduced alien species into new ecosystems chiefly due to:

Colonization. Europeans, in particular, brought various familiar species with them when they colonized new places. For example, the pilgrims brought the dandelion to the United States as a familiar salad green.

Horticulture and agriculture. Some exotics now taking over vast tracts of land have escaped from cultivated areas. Kudzu is a vine from Japan that the Department of Agriculture thought would help prevent soil erosion. The plant now covers much landscape in the South, including even walnut, magnolia, and sweet gum trees.

Accidental transport. Global trade and travel accidentally bring many new species from one country to another. Researchers found that the ballast water released from ships into Coos Bay, Oregon, contained 367 marine species from Japan. The zebra mussel from the Caspian Sea was accidentally introduced into the Great Lakes in 1988. It now forms dense beds that squeeze out native mussels.

Alien species can disrupt food webs. As mentioned earlier in this chapter, opossum shrimp introduced into a lake in Montana led to an additional trophic level that in the end meant less food for bald eagles and grizzly bears (see Fig. 25.3).

Exotics on Islands

Islands are particularly susceptible to environmental discord caused by the introduction of alien species. Islands have unique assemblages of native species that are closely adapted to one another and cannot compete well against exotics. Myrtle trees, *Myrica faya,* introduced into the Hawaiian Islands from the Canary Islands are symbiotic with a type of bacterium that is capable of nitrogen fixation. This feature allows the species to establish itself on nutrient-poor volcanic soil, a distinct advantage in Hawaii. Once established, myrtle trees call a halt to the normal succession of native plants on volcanic soil.

The brown tree snake has been introduced onto a number of islands in the Pacific Ocean. The snake eats eggs, nestlings, and adult birds. On Guam, it has reduced 10 native bird species to the point of extinction. On the Galápagos Islands, black rats have reduced populations of giant tortoise; goats and feral pigs have changed the vegetation from highland forest to pampaslike grasslands and destroyed stands of cactus. Mongooses introduced into the Hawaiian Islands to control rats also prey on native birds (Fig. 25.7).

Pollution

In the present context, **pollution** can be defined as any environmental change that adversely affects the lives and health of living things. Pollution has been identified as the third main cause of extinction. Pollution can also weaken organisms and lead to disease, the fifth main cause of extinction. Biodiversity is particularly threatened by the following environmental concerns:

Acid deposition. Both sulfur dioxide from power plants and nitrogen oxides in automobile exhaust are converted to acids when they combine with water vapor in the atmosphere. These acids return to earth as either wet deposition (acid rain or snow) or dry deposition (sulfate and nitrate salts). Sulfur dioxide and nitrogen oxides are emitted in one locale, but deposition occurs across state and national boundaries. Acid deposition causes trees to weaken and increases their susceptibility to disease and insects. It also kills small invertebrates and decomposers so that the entire ecosystem is threatened. Many lakes in northern states are now lifeless because of the effects of acid deposition.

Eutrophication. Lakes are also under stress due to over enrichment. When lakes receive excess nutrients due to runoff from agricultural fields and wastewater from sewage treatment, algae begin to grow in abundance. An algal bloom is apparent as a green scum or excessive mats of filamentous algae. Upon death, the decomposers break down the algae, but in so doing they use up oxygen. A decreased amount of oxygen is available to fish, leading sometimes to a massive fish kill.

Ozone depletion. The ozone shield is a layer of ozone (O_3) in the stratosphere, some 50 kilometers above the earth. The ozone shield absorbs most of the wavelengths of harmful ultraviolet (UV) radiation so that they do not strike the earth. The cause of ozone depletion can be traced to chlorine atoms (Cl^-) that come from the breakdown of chlorofluorocarbons (CFCs). The best-known CFC is Freon, a heat transfer agent still found in refrigerators and air conditioners today. Severe ozone shield depletion can impair crop and tree growth and also kill plankton (microscopic plant and animal life) that sustain oceanic life. The immune system and the ability of all organisms to resist infectious diseases will most likely be weakened.

Organic chemicals. Our modern society makes use of organic chemicals in all sorts of ways. Nonylphenols are used in products ranging from pesticides to dishwashing detergents, cosmetics, plastics, and spermicides. These organic chemicals mimic the effects of hormones, and in that way most likely harm wildlife. Salmon are born in fresh water but mature in salt water. After investigators exposed young fish to nonylphenol, they found that 20–30% were unable to make the transition between fresh and salt water. Nonylphenols cause the pituitary to produce prolactin, a hormone that most likely prevents saltwater adaptation.

Global warming. The expression **global warming** refers to an expected increase in average temperature during the twenty-first century. You may recall from chapter 24 that carbon dioxide is a gas that comes from the burning of fossil fuels, and methane is a gas that comes from oil and gas wells, rice paddies, and animals. These gases are known as greenhouse gases because, just like the panes of a greenhouse, they allow solar radiation to pass through but hinder the escape of its heat back into space. Data collected around the world show a steady rise in the concentration of the various greenhouse gases. These data are used to generate computer models that predict the earth may warm to temperatures never before experienced by living things (Fig. 25.8*a*).

As the oceans warm, temperatures in the polar regions will rise to a greater degree than in other regions. The sea level will then rise because glaciers will melt and water expands as it warms. A one-meter rise in sea level in the next century could inundate 25–50% of U.S. coastal wetlands. This loss of habitat could be higher if wetlands cannot move inward because of coastal development and levees.

Global warming could very well cause many extinctions on land also. As temperatures rise, regions of suitable climate for various species will shift toward the poles and higher elevations. The present assemblages of species in ecosystems will be disrupted as some species migrate northward, leaving others behind. Plants migrate when seeds disperse and growth occurs in a new locale. For example, to remain in a favorable habitat, it's been calculated that the rate of beech tree migration would have to be 40 times faster than has ever been observed. It seems unlikely that beech or any other type of tree would be able to meet the pace required. Then, too, many species of organisms are confined to relatively small habitat patches that are surrounded by agricultural or urban areas they would not be able to cross. And even if they have the capacity to disperse to new sites, suitable habitat may not be available.

The tropics will also feel the effects of global warming. Coral growth is very dependent upon mutualistic algae that live in their walls. When the temperature rises by 4 degrees, corals expel their algae and are said to be "bleached" (Fig. 25.8*b*). Almost no growth or reproduction occurs until the algae return. Also, coral reefs prefer shallow waters, and if sea levels rise, they may "drown." Multiple assaults on coral are even now causing them to be stricken with various diseases.

Figure 25.8 Global warming.
a. Mean global temperature change is expected to rise due to the introduction of greenhouse gases into the atmosphere. Global warming has the potential to significantly affect the world's biodiversity. **b.** A temperature rise of only a few degrees causes coral reefs to "bleach" and become lifeless.

b.

Overexploitation

Overexploitation occurs when the number of individuals taken from a wild population is so great that the population becomes severely reduced in numbers. The positive feedback cycle explains overexploitation: the smaller the population, the more valuable its members, and the greater the incentive to capture the few remaining organisms.

A decorative plant and exotic pet market supports both a legal and illegal trade in wild species. Rustlers dig up rare cacti like the single-crested saguaro and sell them to gardeners. Parakeets and macaws are among the birds taken from the wild for sale to pet owners. For every bird delivered alive, many more have died in the process. The same holds true for tropical fish, which often come from the coral reefs of Indonesia and the Philippines. Divers dynamite reefs or use plastic squeeze-bottles of cyanide to stun them; in the process, many fish die.

Declining species of mammals, such as the Siberian tiger, are still hunted for their hides, tusks, horns, or bones. Because of its rarity, a single Siberian tiger is now worth more than $500,000—its bones are pulverized and used as a medicinal powder. The horns of rhinoceroses become ornate carved daggers, and their bones are ground up to sell as a medicine. The ivory of an elephant's tusk is used to make art objects, jewelry, or piano keys. The fur of a Bengal tiger sells for as much as $100,000 in Tokyo.

The U.N. Food and Agricultural organization tells us that we have now overexploited 11 of 15 major oceanic fishing areas. Fish are a renewable resource if harvesting does not exceed the ability of the fish to reproduce. But our society uses larger and more efficient fishing fleets to decimate fishing stocks. Pelagic species like tuna are captured by purse-seine fishing. A very large net surrounds a school of fish, and then the net is closed in the same manner as a drawstring purse. Dolphins that swim above schools of tuna are often captured and then killed in this type of net. Other fishing boats drag huge trawling nets, large enough to accommodate 12 jumbo jets, along the seafloor to capture bottom-dwelling fish (Fig. 25.9a). Only large fish are kept; undesirable small fish and sea turtles are discarded, dying, back into the ocean. Trawling has been called the marine equivalent of clear-cutting trees because after the net goes by, the sea bottom is devastated (Fig. 25.9b). Today's fishing practices don't allow fisheries to recover. Cod and haddock, once the most abundant bottom-dwelling fish along the northeast coast, are now often outnumbered by dogfish and skate.

A marine ecosystem can be disrupted by overfishing, as exemplified on the U.S. West Coast. When sea otters began to decline in numbers, investigators found that they were being eaten by orcas (killer whales). Usually orcas prefer seals and sea lions to sea otters, but they began eating sea otters when few seals and sea lions could be found. What caused a decline in seals and sea lions? Their preferred food source—perch and herring—were no longer plentiful due to overfishing. Ordinarily, sea otters keep the population of sea urchins, which feed on kelp, under control. But with fewer sea otters around, the sea urchin population exploded and decimated the kelp beds. Thus, overfishing set in motion a chain of events that altered the food web of an ecosystem to its detriment.

The five main causes of extinction are: habitat loss, introduction of alien species, pollution, overexploitation, and disease.

a.

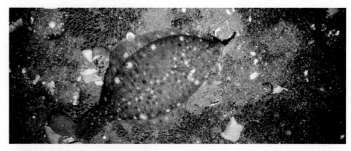

b.

Figure 25.9 Trawling.
a. These Alaskan pollock were caught by dragging a net along the seafloor. **b.** *(top)* Appearance of the seabed before and *(bottom)* after the net went by.

25.4 Conservation Techniques

Despite the value of biodiversity to our very survival, human activities are causing the extinction of thousands of species a year. Clearly, we need to reverse this trend and preserve as many species as possible. How should we go about it?

Habitat Preservation

Preservation of a species habitat is of primary concern, but first we must decide which species to preserve. As mentioned previously, the biosphere contains biodiversity hotspots, relatively small areas where there is a concentration of endemic (native) species not found any place else. In the tropical rain forests of Madagascar, 93% of the primate species, 99% of the frog species, and over 80% of the plant species are endemic to Madagascar. Preserving these forests and other hotspots will save a wide variety of organisms.

Keystone species are species that influence the viability of a community more than you would suspect from their numbers. The extinction of a keystone species can lead to other extinctions and a loss of biodiversity. For example, bats are designated a keystone species in tropical forests of the Old World. They are pollinators that also disperse the seeds of trees. When bats are killed off and their roosts destroyed, the trees fail to reproduce. The grizzly bear is a keystone species in the northwestern United States and Canada (Fig. 25.10*a*). Bears disperse the seeds of berries; as many as 7,000 seeds may be in one dung pile. Grizzlies kill the young of many hoofed animals and thereby keep their populations under control. Grizzlies are also a principal mover of soil when they dig up roots and prey upon hibernating ground squirrels and marmots.

Metapopulations

The grizzly bear population is actually a **metapopulation**—that is, a population subdivided into several small and isolated populations due to habitat fragmentation. Originally there were probably 50,000 to 100,000 grizzlies south of Canada, but this number has been reduced because communities have encroached on their home range and bears have been killed by frightened homeowners. Now there are six virtually isolated subpopulations totaling about 1,000 individuals. The Yellowstone National Park population numbers 200, but the others are even smaller.

Saving metapopulations sometimes requires determining which of the populations is a source and which are sinks. A **source population** is one that most likely lives in a favorable

a. Grizzly bear, *Ursus arctos horribilis*

Figure 25.10 **Habitat preservation.**
Other wildlife benefits when particular species are protected.
a. The Greater Yellowstone Ecosystem has been delineated in an effort to save grizzly bears, which need a very large habitat.
b. Currently, the remaining portions of old-growth forests in the Pacific Northwest are not being logged in order to save **(c)** the northern spotted owl.

b. Old-growth forest

c. Northern spotted owl, *Strix occidentalis caurina*

area, and its birthrate is most likely higher than its death rate. Individuals from source populations move into **sink populations** where the environment is not as favorable and where its birthrate equals its death rate at best. When trying to save the northern spotted owl, conservationists determined that it was best to avoid having owls move into sink habitats. The northern spotted owl reproduces successfully in old-growth rain forests of the Pacific Northwest (Fig. 25.10*b, c*) but not in nearby immature forests that are in the process of recovering from logging. Distinct boundaries that hindered the movement of owls into these sink habitats proved to be beneficial in maintaining source populations.

Landscape Dynamics

Grizzly bears inhabit a number of different types of ecosystems, including plains, mountains, and rivers. Saving any one of these particular types of ecosystems alone would not be sufficient to preserve grizzly bears. Instead, it is necessary to save diverse ecosystems that are at least connected by corridors. You will recall that a landscape encompasses different types of ecosystems. An area called the Greater Yellowstone Ecosystem, where bears are free to roam, has now been defined. It contains millions of acres in Yellowstone National Park; statelands in Montana, Idaho, and Wyoming; five different national forests; various wildlife refuges; and even private lands.

Landscape protection for one species is often beneficial for other wildlife that share the same space. The last of the contiguous 48 states' harlequin ducks, bull trout, westslope cutthroat trout, lynx, pine martens, wolverines, mountain caribou, and great gray owls are found in areas occupied by grizzlies. The recent return of gray wolves has occurred in this territory also. Then, too, grizzly range overlaps with 40% of Montana's vascular plants of special conservation concern.

The Edge Effect   When preserving landscapes, it is necessary to consider the **edge effect.** An edge reduces the amount of habitat typical of an ecosystem because the edges around a patch have a habitat slightly different from the interior of a patch. For example, forest edges are brighter, warmer, drier, and windier, with more vines, shrubs, and weeds than the interior of a forest. Also, Figure 25.11*a* shows that a small and a large patch of habitat have the same amount of edge; therefore, the effective habitat shrinks as a patch gets smaller.

The edge effect can have a serious impact on population size. Songbird populations west of the Mississippi have been declining of late, and ornithologists have noticed that the nesting success of songbirds is quite low at the edge of a forest. The cause turns out to be the brown-headed cowbird, a social parasite of songbirds. Adult cowbirds prefer to feed in open agricultural areas, and they only briefly enter the forest when searching for a host nest in which to lay their eggs (Fig. 25.11*b*). Cowbirds are therefore benefited, while songbirds are disadvantaged, by the edge effect.

Computer Analyses

Two types of computer analyses, in particular, are now available to help conservationists plan how best to protect a species.

Gap analysis is the use of the computer to find gaps in preservation—places where biodiversity is high outside of preserved areas. First computerized maps are drawn up showing the topography, vegetation, hydrology, and land ownership of a region. Then computer maps are done showing the geographic distribution of a region's animal and plant species. Once the distribution maps are superimposed onto the land-use maps, it is obvious where preserved habitats still need to be and/or could be located.

A **population viability analysis** can help researchers determine how much habitat a species requires to maintain itself. First it is necessary to calculate the minimum population size needed to prevent extinction. This size should protect the species from unforeseen events like natural catastrophes or chance swings in the birth and death rate. Another

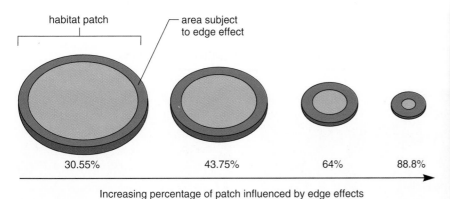

30.55%   43.75%   64%   88.8%

Increasing percentage of patch influenced by edge effects

a.

b.

Figure 25.11 Edge effect.
a. The smaller the patch, the greater the proportion that is subject to the edge effect. **b.** Cowbirds lay their eggs in the nest of songbirds. A cowbird is bigger than a warbler nestling and will be able to acquire most of the food brought by the warbler parent.

component to consider is the size needed to protect genetic diversity. This number varies according to the species. For example, analysis of red-cockaded woodpecker populations showed that an adult population of about 1,323 is needed to result in a genetically effective population of 500 because of the breeding system of the species. After you know the minimum population size, you can determine how much total acreage is needed for that size population.

All life history characteristics of organisms must be taken into account when a population viability analysis is done. For example, female grizzlies don't give birth until they are 5 or 6, and then they typically wait three years before reproducing again. After doing one of the first population viability analyses, Mark Shaffer of the Wilderness Society predicted that a total grizzly bear population of 70 to 90 individuals, each with a suitable home range of 6,000 miles, will have about a 95% chance of surviving for 100 years. But Fred Allendorf pointed out that because only a few dominant males breed, the population needs to be larger than this to protect genetic diversity. Also, to prevent inbreeding, he recommended the introduction of one or two unrelated bears each decade into populations of 100 individuals. The bottom line is that there needs to be dispersal among subpopulations to prevent inbreeding and extinction.

Habitat Restoration

Restoration ecology is a new subdiscipline of conservation biology that seeks scientific ways to return ecosystems to their former state. Three principles have so far emerged. First, it's best to begin as soon as possible before remaining fragments of the original habitat are lost. These fragments are sources of wildlife and seeds from which to restock the restored habitat. Second, once the natural history of the habitat is understood, it is best to use biological techniques that mimic natural processes to bring about restoration. This might take the form of controlled burns to bring back grassland habitats, biological pest controls to rid the area of aliens, or bioremediation techniques to clean up pollutants. Third, the goal is **sustainable development,** the ability of an ecosystem to maintain itself while providing services to human beings. We will use the Everglades ecosystem in Florida to illustrate these principles.

The Everglades

Originally, the Everglades encompassed the whole of southern Florida, from Lake Okeechobee down to Florida Bay. This ecosystem is a vast sawgrass prairie, interrupted occasionally by a cypress dome or hardwood tree island. Within these islands, both temperate and tropical evergreen trees grow amongst dense and tangled vegetation. Mangroves are found along sloughs (creeks) and at the shoreline. The prop roots of red mangroves protect over 40 different types of juvenile fishes as they grow to maturity. During the wet season, from May to November, animals disperse throughout the region, but in the dry season, from December to April, they congregate wherever pools of water are found. Alligators are famous for making "gator holes" where water collects, and fish, shrimp, crabs, birds, and a host of other living things survive until the rains come again. The Everglades once supported millions of large and beautiful birds (Fig. 25.12).

Restoration Plan A restoration plan has been developed that will sustain the Everglades ecosystem while maintaining the services society requires. The U.S. Army Corps of Engineers is to redesign local flood control measures so that the Everglades receive a more natural flow of water from Lake Okeechobee. This will require the flooding of the Everglades Agricultural Area and the growth of only crops that can tolerate these conditions, such as sugarcane and rice. This has the benefit of stopping the loss of topsoil and preventing possible residential development in the area. There will also be an extended buffer zone between an expanded Everglades and the urban areas on the east coast of Florida. The buffer zone will contain a contiguous system of interconnected marsh areas, detention reservoirs, seepage barriers, and water treatment areas. This plan is expected to stop the decline of the Everglades, while still allowing agriculture to continue and providing water and flood control to the east coast. Sustainable development will maintain the ecosystem indefinitely and still meet human needs.

Today, landscape preservation is commonly needed to protect metapopulations. Often the preserved area must be restored before sustainable development is possible.

Alligator beside his "gater hole"

White ibis in breeding coloration

Roseate spoonbill wading

Figure 25.12 Restoration of the Everglades.
If the Everglades are restored, the chances of survival for these wild animals will improve.

Bioethical Focus

Preservation of Coral Reefs

The many types of animals in a coral reef form a complex community that is admired by both snorkelers and scuba divers. The various types of fish and shellfish in a coral reef are sources of food for millions of people. Like a tropical forest, coral reefs are the most likely sources of medicines yet to be discovered. And a reef serves as a storm barrier that protects the shoreline and provides a safe harbor for ships.

Reefs around the globe are being destroyed. Tons of soil from deforested tracts of land bring nutrients that stimulate the growth of all kinds of algae. This has contributed to population explosion of the crown-of-thorn starfish that are devouring Australia's 1,200-mile-long Great Barrier Reef. Reefs are also being damaged by pollutants that seep into the sea from factories, farm fields, and sewers. These types of stress, combined with unusually warm seawater, have caused the corals to expel the symbiotic colorful algae that carry on photosynthesis and help sustain them. So-called coral bleaching has been noticed in reefs of the Pacific Ocean and Caribbean. Might worldwide global warming also contribute to coral bleaching and death?

Marine scientist Eduardo Gomez estimates that 90% of the coral reefs of the Philippines are dead or deteriorating due to pollution, but especially due to overfishing. The methods are sinister, including the use of dynamite to kill the fish so that it is easier to scoop them up, use of cyanide to stun the fish to capture them alive, and use of satellite navigation systems to hone in on areas where mature fish are spawning to reproduce. If all large herbivores are killed off, seaweed overgrows and kills the coral.

Paleobiologist Jeremy Jackson of the Smithsonian Tropical Research Institute near Panama City wonders if he is doing enough to warn the public that the reefs around the world are in

Figure 25A Coral reef.
It is wise to expend the necessary time, energy, and funds to preserve coral reefs.

danger. He estimates that we may lose 60% of all coral reefs by the year 2050.

Decide Your Opinion

1. Do you think it would be possible to make the public care about the loss of coral reefs? Explain.
2. When and under what circumstances do dire predictions help preserve the environment?
3. Considering what is causing the loss of coral reefs, would it be possible to save them? How?

Looking at Both Sides   www.mhhe.com/biosci/genbio/maderhuman7/

Every bioethical issue has at least two sides. Even if you already have an opinion, it is important to explore the opposite opinion before finalizing your position. The Online Learning Center at www.mhhe.com/biosci/genbio/maderhuman7/ will help you fine-tune your initial opinion, explore both sides, and finalize your position. You may acquire new arguments for your original opinion, or you may even change your opinion. Be sure to complete these activities in sequence:

Taking Sides   Decide your initial opinion by answering a series of questions. Then see if your opinion changes after completing the next two activities.

Further Debate   Read opposing articles that give you further information on this particular bioethical issue.

Explain Your Position   Answer another series of questions and then defend your original or changed opinion. You can e-mail your position to your instructor if he or she wishes.

Summarizing the Concepts

25.1 Conservation Biology and Biodiversity

Conservation biology is the scientific study of biodiversity and its management for sustainable human welfare. The present unequaled rate of extinctions has drawn together scientists and environmentalists in basic and applied fields to address the problem.

Biodiversity is the variety of life on earth; the exact number of species is not known, but there are many more insects than other types of organisms. Biodiversity must also be preserved at the genetic, community (ecosystem), and landscape levels of organization.

Conservationists have discovered that biodiversity is not evenly distributed in the biosphere, and therefore saving certain areas may protect more species than saving other areas.

25.2 Value of Biodiversity

The direct value of biodiversity is evidenced by the observable services of individual wild species. Wild species are our best source of new medicines to treat human ills, and they meet other medical needs as well: for example, the bacterium that causes leprosy grows naturally in armadillos, and horseshoe crab blood contains a bacteria-fighting substance.

Wild species have agricultural value. Domesticated plants and animals are derived from wild species, which also serve as a source of genes for the improvement of their phenotypes. Instead of pesticides, wild species can be used as biological controls, and most flowering plants make use of animal pollinators. Much of our food, particularly fish and shellfish, is still caught in the wild. Hardwood trees from natural forests supply us with lumber for various purposes, such as making furniture.

The indirect services provided by ecosystems are largely unseen but absolutely necessary to our well-being. These services include the workings of biogeochemical cycles, waste disposal, provision of fresh water, prevention of soil erosion, and regulation of climate. Many people enjoy vacationing in natural settings. Various studies show that more diverse ecosystems function better than less diverse systems.

25.3 Causes of Extinction

Researchers have identified the major causes of extinction. Habitat loss is the most frequent cause, followed by introduction of alien species, pollution, overexploitation, and disease. (Pollution often leads to disease, so these were discussed at the same time.) Habitat loss has occurred in all parts of the biosphere, but concern has now centered on tropical rain forests and coral reefs where biodiversity is especially high. Alien species have been introduced into foreign ecosystems because of colonization, horticulture or agriculture, and accidental transport. Among the various causes of pollution (acid rain, eutrophication, and ozone depletion), global warming is expected to cause the most instances of extinction. Overexploitation is exemplified by commercial fishing, which is so efficient that fisheries of the world are collapsing.

25.4 Conservation Techniques

To preserve species, it is necessary to preserve their habitat. Some emphasize the need to preserve biodiversity hotspots because of their richness. Often today it is necessary to save metapopulations because of past habitat fragmentation. If so, it is best to determine the source populations and save those instead of the sink populations. A keystone species like the grizzly bear requires the preservation of a landscape consisting of several types of ecosystems over millions of acres of territory. Obviously, in the process, many other species will also be preserved.

Conservation today is assisted by two types of computer analysis in particular. A gap analysis tries for a fit between biodiversity concentrations and land still available to be preserved. A population viability analysis indicates the minimum size of a population needed to prevent extinction from happening.

Since many ecosystems have been degraded, habitat restoration may be necessary before sustainable development is possible. Three principles of restoration are: (1) start before sources of wildlife and seeds are lost; (2) use simple biological techniques that mimic natural processes; and (3) aim for sustainable development so that the ecosystem fulfills the needs of humans.

Studying the Concepts

1. Explain these attributes of conservation biology: (a) both academic and applied; (b) supports ethical principles; and (c) is responding to biodiversity crisis.   498
2. Discuss the conservation of biodiversity at the species, genetic, and community levels.   498–99
3. Describe the uneven distribution of diversity in the biosphere. What is the implication of uneven distribution for conservation biologists?   499
4. List various ways in which individual wild species provide us with valuable services.   500
5. List various ways in which ecosystems provide us with indispensable services.   502
6. List and discuss the five major causes of extinction, starting with the most frequent.   504–07
7. Introduction of alien species is usually due to what events?   505
8. List and discuss four major types of pollution that particularly affect biodiversity. Of these, why is global warming considered the most significant?   505–06
9. Use the positive feedback cycle to explain why overexploitation occurs.   507
10. Using the grizzly bear population as an example, explain the terms keystone species, metapopulation, landscape preservation, and population viability analysis.   508–10
11. Explain the three principles of habitat restoration with reference to restoration of the Everglades.   510

Testing Your Knowledge of the Concepts

For questions 1–4, match the following phrases to the indirect values of biodiversity they are related to.
a. waste disposal
b. prevention of soil erosion
c. regulation of climate
d. ecotourism

_____ 1. Forests and other natural ecosystems soak up water and then release it at a regular rate.

_____ 2. People visit the Everglades to see wading birds.

_____ 3. Trees absorb carbon dioxide and keep it from entering the atmosphere.

_____ 4. Many coral reefs are covered by silt washed from the earth into the sea

In questions 5–7, indicate whether the statement is true (T) or false (F).

_____ 5. Habitat loss is the most frequent cause of extinctions today.

_____ 6. Alien species are often introduced into ecosystems by accidental transport.

_____ 7. Society doesn't always appreciate the services that wild species provide naturally and without any fanfare.

In questions 8 and 9, fill in the blanks.

8. _____ can be defined as any environmental change that adversely affects the lives and health of living things.

9. _____ is largely due to the buildup of carbon dioxide, a greenhouse gas.

10. Complete the following graph by labeling each bar with a cause of extinction, from the most influential to the least influential.

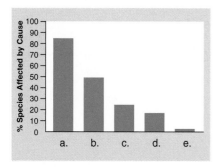

Understanding Key Terms

alien species 505
biodiversity 498
biodiversity hotspot 499
conservation biology 498
edge effect 509
gap analysis 509
global warming 506
keystone species 508
landscape 499
metapopulation 508
pollution 505
population viability analysis 509
restoration ecology 510
sink population 509
source population 508
sustainable development 510

Further Readings

Benchley, P. April 1999. Galápagos: Paradise in peril. *National Geographic* 195(4):2. The unique ecosystem of the Galápagos Islands is being threatened by tourism, development, and abnormal weather.

Chadwick, D. H. January 1999. Coral in peril. *National Geographic* 195(1):30. Coral reefs are at risk from overfishing, pollution, and disease.

Chadwick, D. H. July 1999. Humpback whales. *National Geographic* 196(1):110. Songs, travel patterns, and family life are being documented for these whales, whose numbers seem to be increasing.

Conniff, R. December 1999. Cheetahs—Ghosts of the grasslands. *National Geographic* 196(6):2. Conservationists need to move quickly to protect the cheetah.

Flather C., et al. May 1998. Threatened and endangered species geography. *BioScience* 48(5):365. Article suggests that conservation efforts should preserve an entire threatened ecosystem rather than aiding a single endangered species.

Harwell, M. A. September 1997. Ecosystem management of South Florida. *BioScience* 47(8):499. Article discusses the restoration of a sustainable natural ecosystem while maintaining the services that society requires.

Holloway, M. July 2000. The killing lakes. *Scientific American* 283(1):92. Two lakes in Cameroon may soon release a lethal gas (carbon dioxide), as they did in the 1980s; scientists are trying to prevent another tragedy.

Knapp, A., et al. January 1999. The keystone role of bison in North American tallgrass prairie. *BioScience* 49(1):39. Bison alter plant communities and ecosystem processes; grazing activities are key to conserving and restoring the prairie.

Levin, T. February/March 1999. To save a reef. *National Wildlife* 37(2):20. Article discusses the biology and ecology of the Florida Keys, and the threat to Florida coral reefs.

Long, M. E. July 1999. The shrinking world of hornbills. *National Geographic* 196(1):52. Hunting and habitat loss threaten the survival of various hornbill species.

National Geographic. February 1999. Millennium supplement: Biodiversity 195(2). The issue is devoted to biodiversity, including possible ways to preserve vanishing species.

Natural History. July/August 1998. 107(6):34–51. Preservation of Amazon rain forest diversity is discussed.

Primack, R. B. 1998. *Essentials of conservation biology.* Sunderland, Mass.: Sinauer Associates, Inc. This text for ecology majors explores the field of conservation biology.

Stone, R. June 5, 1998. Yellowstone rising again from ashes of devastating fires. *Science* 280(5369):1527. The post-fire ecosystem of Yellowstone National Park is essentially the same as before the fires.

Tangley, L. December/January 1999. How many species exist? *National Wildlife* 37(1):32. Plants and animals are vanishing before scientists can identify them.

Tattersal, I. January 2000. Once we were not alone. *Scientific American* 282(1):56. Although 20 or more types of creatures similar to modern humans have existed over the past 4 million years, only Homo sapiens remains. Article examines why.

Wong, K. April 2000. Who were the Neanderthals? *Scientific American* 282(4):99. Article discusses differences and similarities between Neanderthals and early modern Europeans.

25.1 Conservation Biology and Biodiversity

Approach *Essential Study Partner*
Issues *Essential Study Partner*
Species *Essential Study Partner*

Global Environmental Issues *readings*

25.3 Causes of Extinction

Extinction *Essential Study Partner*
Air *Essential Study Partner*
Water *Essential Study Partner*
Land *Essential Study Partner*
Concept Review *Essential Study Partner*

Greenhouse Effect *art quiz*
Biological Magnification of DDT *art quiz*

25.4 Conservation Techniques

Looking at Both Sides *critical thinking activity*

The Wolf in Yellowstone National Park *case study*

Chapter Summary

Key Term Flashcards *vocabulary quiz*
Chapter Quiz *objective quiz covering all chapter concepts*

Appendix A

Answer Key

This appendix contains the answers to the Testing Your Knowledge of the Concepts, which appear at the end of each chapter, and the Practice Problems, which appear in Chapter 20.

Chapter 1
1. d; 2. e; 3. a; 4. c; 5. T; 6. F; 7. T; 8. copy; 9. Society; 10. scientific theory

Chapter 2
1. b; 2. c; 3. a; 4. d; 5. F; 6. F; 7. T; 8. glycerol, fatty acids; 9. pentose, phosphate, base; 10. a. monomer; b. condensation; c. polymer; d. hydrolysis

Chapter 3
1. c; 2. a; 3. d; 4. b; 5. T; 6. F; 7. T; 8. electron transport system, cristae; 9. 2, 36; 10. a. nucleus—DNA specifies; b. nucleolus—produces ribosomal subunits; c. smooth endoplasmic reticulum—modifies; d. rough endoplasmic reticulum—produces; e. Golgi apparatus—further modifies

Chapter 4
1. c; 2. b; 3. a; 4. d; 5. T; 6. F; 7. F; 8. Homeostasis; 9. not striated (smooth); 10. a. columnar epithelium, lining of intestine (digestive tract), protection and absorption; b. cardiac muscle, wall of heart, pumps blood; c. compact bone, skeleton, support and protection

Chapter 5
1. d; 2. b; 3. c; 4. a; 5. T; 6. F; 7. F; 8. bile, emulsifies; 9. acid, basic; 10. a. salivary glands; b. esophagus; c. stomach; d. liver; e. gallbladder; f. pancreas; g. small intestine; h. large intestine; i. glucose and amino acids; j. lipids; k. water

Chapter 6
1. b, phagocytize; 2. a, hemoglobin; 3. d, blood clotting; 4. a; 5. F; 6. T; 7. T; 8. antibodies; 9. no; 10. a. hemoglobin; b. agglutination; c. plasma; d. platelet; e. prothrombin; 11. a. blood pressure; b. osmotic pressure; c. blood pressure; d. osmotic pressure

Chapter 7
1. b; 2. a; 3. c; 4. b; 5. F; 6. T; 7. T; 8. fat, cholesterol; 9. stroke; 10. a. diastole; b. vena cava; c. thromboembolism; d. pacemaker; e. systemic; 11. a. aorta; b. left pulmonary arteries; c. pulmonary trunk; d. left pulmonary veins; e. left atrium; f. semilunar valves; g. atrioventricular (mitral) valve; h. left ventricle; i. septum; j. inferior vena cava; k. right ventricle; l. chordae tendineae; m. atrioventricular (tricuspid) valve; n. right atrium; o. right pulmonary veins; p. right pulmonary arteries; q. superior vena cava. See also Figure 7.5, page 129.

Chapter 8
1. b; 2. c; 3. c, d; 4. d; 5. F; 6. T; 7. T; 8. tissue fluid, subclavian; 9. filter; 10. thymus; 11. The complement system; 12. plasma, memory; 13. antigen receptors; 14. cytokines; 15. APC; 16. vaccines (or antigens); 17. monoclonal; 18. memory, memory; 19. antigen-binding site; b. light chain; c. heavy chain; d. variable region; e. constant region. See also Figure 8.7a, page 152.

Chapter 9
1. b; 2. a; 3. c; 4. d; 5. F; 6. T; 7. F; 8. expanded; 9. hemoglobin; 10. a. nasal cavity; b. nostril; c. pharynx; d. epiglottis; e. glottis; f. larynx; g. trachea; h. bronchus; i. bronchiole. See also Figure 9.2, page 166.

Chapter 10
1. d; 2. b; 3. a; 4. c; 5. F; 6. T; 7. T; 8. salt; 9. Water; 10. a. glomerulus; b. efferent arteriole; c. afferent arteriole; d. proximal convoluted tubule; e. loop of the nephron; f. descending limb; g. ascending limb; h. peritubular capillary network; i. distal convoluted tubule; j. renal vein; k. renal artery; l. collecting duct.

Chapter 11
1. c; 2. d; 3. a; 4. g; 5. f; 6. e; 7. e; 8. a; 9. b; 10. c; 11. f; 12. d; 13. F; 14. T; 15. osteon; 16. red bone marrow; 17. osteoclasts; 18. sinuses; 19. pelvic girdle, rib cage, 20. a. coxal bone; b. femur; c. patella; d. tibia; e. fibula; f. metatarsals; g. phalanges; h. tarsals

Chapter 12

1. d; 2. a; 3. b; 4. c; 5. T; 6. F; 7. T; 8. creatine phosphate; 9. ACh; 10. a. T tubule; b. sarcoplasmic reticulum; c. myofibril; d. Z line; e. sarcomere; f. sarcolemma of muscle fiber

Chapter 13

1. d; 2. a; 3. b; 4. c; 5. F; 6. T; 7. F; 8. synaptic cleft; 9. sensory, motor; 10. a. sensory neuron; b. interneuron; c. motor neuron; d. receptor; e. cell body; f. dendrites; g. axon; h. nucleus of Schwann cell; i. node of Ranvier; j. effector

Chapter 14

1. b; 2. a; 3. d; 4. c; 5. F; 6. F; 7. T; 8. vestibule, semicircular canals; 9. left; 10. a. retina; b. choroid; c. sclera; d. optic nerve; e. fovea centralis; f. ciliary body; g. lens; h. iris; i. pupil; j. cornea; 11. a. ossicle; b. chemoreceptor; c. rhodopsin; d. sensation; e. spiral organ

Chapter 15

1. f; 2. a; 3. e; 4. b; 5. d; 6. c; 7. F; 8. T; 9. F; 10. T; 11. atrial natriuretic hormone; 12. negative; 13. testosterone, estrogen, progesterone; 14. a. inhibits; b. inhibits; c. releasing hormone; d. stimulating hormone; e. target gland hormone; 15. a. thyroid; b. diabetes mellitus; c. adrenocorticotropic hormone (ACTH); d. peptide hormone; e. oxytocin

Chapter 16

1. c; 2. d; 3. a; 4. b; 5. F; 6. T; 7. F; 8. blood; 9. testosterone; 10. seminal vesicles; 11. testosterone; 12. vagina; 13. follicle, endometrium; 14. estrogen, progesterone; 15. human chorionic gonadotropin (HCG); 16. laboratory glassware; 17. a. vagina; b. prostate; c. vulva; d. epididymis; e. gonad; 18. a. seminal vesicle; b. ejaculatory duct; c. prostate gland; d. bulbourethral gland; e. anus; f. vas deferens; g. epididymis; h. testis; i. scrotum; j. foreskin; k. glans penis; l. penis; m. urethra; n. vas deferens; o. urinary bladder. Path of sperm: h, g, f, n, b, m. See also Figure 16.1, page 318.

Chapter 17

1. e; 2. a; 3. c; 4. d; 5. T; 6. T; 7. T; 8. binary fission; 9. nucleic acid, protein; 10. coldsores, genital herpes; 11. rod, spherical, spiral; 12. PID; 13. condom; 14. gummas; 15. gonorrhea, chlamydia, syphilis; 16. yeast; 17. living; 18. a. Binding of the virus to the plasma membrane; b. Penetration of the virus into the cell; c. Uncoating removes capsid; d. Replication of the viral genetic material; e. Production of viral proteins; f. Assembly of new viruses; g. Budding of new viruses from the host cell. See also Figure 17.3, p. 341.

Chapter 18

1. c; 2. d; 3. b; 4. a; 5. T; 6. T; 7. T; 8. extraembryonic, amnion; 9. implant; 10. sperm, egg; 11. cleavage; 12. differentiation; 13. second; 14. placenta; 15. oxytocin; 16. a. chorion; b. amnion; c. embryo; d. allantois; e. yolk sac; f. fetal portion of placenta (chorionic villi); g. maternal portion of placenta; h. umbilical cord

Chapter 19

1. d; 2. a; 3. b; 4. c; 5. e; 6. T; 7. T; 8. F; 9. F; 10. T; 11. F; 12. chromosomes; 13. X, Y; 14. 24; 15. spermatogenesis; oogenesis; 16. fertilization; 17. a. chromatid; b. cell cycle; c. spindle; d. polar body; e. haploid; 18. Diagram to the right; 19. Homologous pairs are at equator.

Chapter 20

1. c; 2. b; 3. d; 4. a; 5. a. cystic fibrosis; b. Tay-Sachs disease; c. neurofibromatosis; d. Huntington disease; 6. F; 7. F; 8. T; 9. phenotype; 10. eeX^bY; 11. autosomal recessive

Practice Problems 1

1. *cc*; *Cc* and *Cc*; 2. 25%; 3. Heterozygous

Practice Problems 2

1. *Hh, hh*; 2. 50%, 50%; 3. Only the affected children of this couple (*Hh*) can pass on Huntington disease, because the gene (*H*) that causes the condition is dominant. Unaffected children (*hh*) do not carry the gene that causes the condition. 4. homozygous dominant or heterozygous

Practice Problems 3

1. light; 2. very light; c. baby 1 = Doe; baby 2 = Jones

Practice Problems 4

1. his mother; mother = X^HX^h, father = X^HY, son = X^hY; 2. 100%, none, 100%; 3. The husband is not the father.

Chapter 21

1. d; 2. b; 3. a; 4. c; 5. T; 6. F; 7. F; 8. RNA retrovirus; 9. copies; 10. triplet, amino acid; 11. translation; 12. transgenic; 13. a. ACU´CCU´GAA´UGC´AAA; b. UGA´GGA´CUU´ACG´UUU

Chapter 22

1. b; **2.** c; **3.** a; **4.** d; **5.** T; **6.** T; **7.** F; **8.** human papilloma; **9.** chemotherapy; **10.** metastasize; **11.** proto-oncogenes, tumor-suppressor; **12.** dividing; **13.** antibodies; **14. a.** growth factor; **b.** growth inhibitory factor; **c.** tumor-suppressor gene; **d.** proto-oncogene; **e.** signaling protein

Chapter 23

1. b; **2.** a; **3.** d; **4.** c; **5.** T; **6.** F; **7.** T; **8.** reproduce; **9.** primates; **10.** Australopithecines; **11.** Africa; **12.** erect, small; **13.** Cro-Magnon; **14.** thousands; **15.** a, b, c, d; **16.** e. See also Figure 23.8*b*, p. 471.

Chapter 24

1. d; **2.** a; **3.** b; **4.** c; **5.** T; **6.** T; **7.** F; **8.** F; **9.** nitrous oxide and nitrogen gas; **10.** respire; **11. a.** producers; **b.** consumers; **c.** inorganic nutrient pool; **d.** decomposers; **12. a.** ecological pyramid; **b.** succession; **c.** fossil fuel; **c.** nitrification; **e.** ozone shield

Chapter 25

1. c; **2.** d; **3.** a; **4.** b; **5.** T; **6.** T; **7.** T; **8.** Pollution; **9.** global warming; **10. a.** habitat loss; **b.** alien species; **c.** pollution; **d.** over-exploitation; **e.** disease

Appendix B

Metric System

Unit and Abbreviation	Metric Equivalent	Approximate English-to-Metric Equivalents	Units of Temperature
Length			
nanometer (nm)	$= 10^{-9}$ m		
micrometer (μm)	$= 10^{-6}$ m		
millimeter (mm)	$= 0.001\ (10^{-3})$ m		
centimeter (cm)	$= 0.01\ (10^{-2})$ m	1 inch = 2.54 cm	
		1 foot = 30.5 cm	
meter (m)	$= 100\ (10^{2})$ cm	1 foot = 0.30 m	
	$= 1{,}000$ mm	1 yard = 0.91 m	
kilometer (km)	$= 1{,}000\ (10^{3})$ m	1 mi = 1.6 km	
Weight (mass)			
nanogram (ng)	$= 10^{-9}$ g		
microgram (μg)	$= 10^{-6}$ g		
milligram (mg)	$= 10^{-3}$ g		
gram (g)	$= 1{,}000$ mg	1 ounce = 28.3 g	
		1 pound = 454 g	
kilogram (kg)	$= 1{,}000\ (10^{3})$ g	= 0.45 kg	
metric ton (t)	$= 1{,}000$ kg	1 ton = 0.91 t	
Volume			
microliter (μl)	$= 10^{-6}$ l $(10^{-3}$ ml)		
milliliter (ml)	$= 10^{-3}$ liter	1 tsp = 5 ml	
	$= 1$ cm^3 (cc)	1 fl oz = 30 ml	
	$= 1{,}000$ mm^3		
liter (l)	$= 1{,}000$ ml	1 pint = 0.47 liter	
		1 quart = 0.95 liter	
		1 gallon = 3.79 liter	
kiloliter (kl)	$= 1{,}000$ liter		

The temperature scale shows °F and °C columns with the following notable markings:
230–110, 212° / 210 = 100 — 100°, 160° — 160 = 70 — 71°, 134° / 131° — 130, 105.8° — 110 = 40 — 41°, 98.6° — 100 = 37°, 68.6° — 70 = 20 — 20.3°, 32° — 30 = 0 — 0°, down to −40 / −40.

°C	°F	
100	212	Water boils at standard temperature and pressure
71	160	Flash pasteurization of milk
57	134	Highest recorded temperature in the United States, Death Valley, July 10, 1913
41	105.8	Average body temperature of a marathon runner in hot weather
37	98.6	Human body temperature
20.3	68.6	Human survival is still possible at this temperature
0	32.0	Water freezes at standard temperature and pressure

To convert temperature scales:

$$°C = \frac{5(°F-32)}{9}$$

$$°F = \frac{9°C}{5} + 32$$

Appendix C

Periodic Table of the Elements

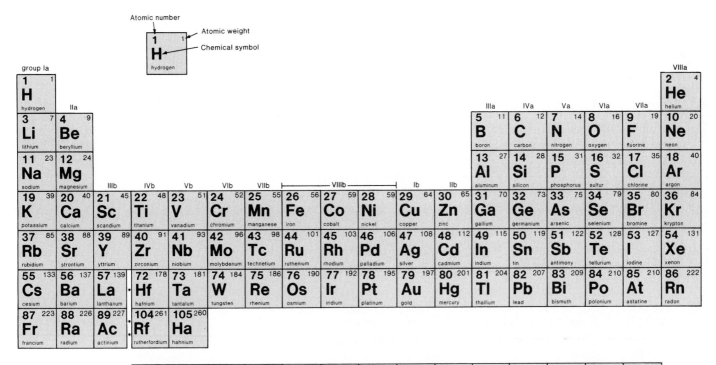

Appendix D

Drugs of Abuse

	Drugs	Often Prescribed Brand Names	Medical Uses	Potential Physical Dependence	Potential Psychological Dependence	Tolerance
Narcotics	Opium	Dover's Powder, Paregoric	Analgesic, antidiarrheal	High	High	Yes
	Morphine	Morphine	Analgesic	High	High	Yes
	Codeine	Codeine	Analgesic, antitussive	Moderate	Moderate	Yes
	Heroin	None	None	High	High	Yes
	Meperidine (Pethidine)	Demerol, Pethadol	Analgesic	High	High	Yes
	Methadone	Dolophine, Methadone, Methadose	Analgesic, heroin substitute	High	High	Yes
	Other Narcotics	Dilaudid, Leritine, Numorphan, Percodan	Analgesic, antidiarrheal, antitussive	High	High	Yes
Depressants	Chloral Hydrate	Noctec, Somnos	Hypnotic	Moderate	Moderate	Probable
	Barbiturates	Amytal, Butisol, Nembutal, Phenobarbitol, Seconal, Tuinal	Anesthetic, anticonvulsant, sedation, sleep	High	High	Yes
	Glutethimide	Doriden	Sedation, sleep	High	High	Yes
	Methaqualone	Optimil, Parest, Quaalude, Somnafac, Sopor	Sedation, sleep	High	High	Yes
	Tranquilizers	Equanil, Librium, Miltown, Serax, Tranxene, Valium	Antianxiety, muscle relaxant, sedation	Moderate	Moderate	Yes
	Other Depressants	Clonopin, Dalmane, Dormate, Noludar, Placydil, Valmid	Antianxiety, sedation, sleep	Possible	Possible	Yes
Stimulants	Cocaine*	Cocaine	Local anesthetic	Possible	High	Yes
	Amphetamines	Benzedrine, Biphetamine, Desoxyn, Dexedrine	Hyperkinesis, narcolepsy, weight control	Possible	High	Yes
	Phenmetrazine	Preludin	Weight control	Possible	High	Yes
	Methylphenidate	Ritalin	Hyperkinesis	Possible	High	Yes
	Other Stimulants	Bacarate, Cylert, Didrex, Ionamin, Plegine, Pondimin, Pro-Sate, Sanorex, Voranil	Weight control	Possible	Possible	Yes
Hallucinogens	LSD	None	None	None	Degree unknown	Yes
	Mescaline	None	None	None	Degree unknown	Yes
	Psilocybin-Psilocyn	None	None	None	Degree unknown	Yes
	MDA	None	None	None	Degree unknown	Yes
	PCP†	Sernylan	Veterinary anesthetic	None	Degree unknown	Yes
	Other Hallucinogens	None	None	None	Degree unknown	Yes
Cannabis	Marijuana, Hashish, Hashish Oil	None	Glaucoma	Degree unknown	Moderate	Yes

Source: Drugs of Abuse, produced by the Affairs in Cooperation with the Office of Public Science and Technology.

*Designated a narcotic under the Controlled Substances Act.

†Designated a depressant under the Controlled Substances Act.

Duration of Effects (in hours)	Usual Methods of Administration	Possible Effects	Effects of Overdose	Withdrawal Syndrome
3–6	Oral, smoked	Euphoria, drowsiness, respiratory depression, constricted pupils, nausea	Slow and shallow breathing, clammy skin, convulsions, coma, possible death	Watery eyes, runny nose, yawning, appetite loss, irritability, tremors, panic, chills
3–6	Injected, smoked			
3–6	Oral, injected			
3–6	Injected, sniffed			
3–6	Oral, injected			
12–24	Oral, injected			
3–6	Oral, injected			
5–8	Oral			
1–16	Oral, injected	Slurred speech, disorientation, drunken behavior without odor of alcohol	Shallow respiration, cold and clammy skin, dilated pupils, weak and rapid pulse, coma, possible death	Anxiety, insomnia, tremors, delirium, convulsions, possible death
4–8	Oral			
4–8	Oral			
4–8	Oral			
4–8	Oral			
2	Injected, sniffed	Increased alertness, excitation, euphoria, dilated pupils, increased pulse rate and blood pressure, insomnia, loss of appetite	Agitation, increased body temperature, hallucinations, convulsions, possible death	Apathy, long periods of sleep, irritability, depression, disorientation
2–4	Oral, injected			
2–4	Oral			
2–4	Oral			
2–4	Oral			
Variable	Oral	Illusions and hallucinations (with the exception of MDA), poor perception of time and distance	Longer, more intense "trip" episodes, psychosis, possible death	Withdrawal syndrome not reported
Variable	Oral, injected			
Variable	Oral			
Variable	Oral, injected, sniffed			
Variable	Oral, injected, smoked			
Variable	Oral, injected, sniffed			
2–4	Oral, smoked	Euphoria, relaxed inhibitions, increased appetite, disoriented behavior	Fatigue, paranoia, possible psychosis	Insomnia, hyperactivity, and decreased appetite reported in a limited number of individuals

Appendix E

Selected Types of Cancer

ancers are classified according to the type of tissue from which they arise. Carcinomas, the most common type, are cancers of epithelial tissues; sarcomas are cancers arising in muscle or connective tissue (especially bone or cartilage); leukemias are cancers of the blood; and lymphomas are cancers of lymphoid tissue. The chance of developing cancer in a particular tissue shows a positive correlation to the rate of cell division; new blood cells arise at a rate of 2,500,000 cells per second, and epithelial cells also reproduce at a high rate.

Five-year survival rates for the most common cancers are given in Table E.

Breast Cancer

Signs and Symptoms: Pre-clinical radiographic signs seen on a mammogram. Breast changes, such as a lump, thickening, swelling, dimpling, skin irritation, distortion, retraction, scaliness, pain, tenderness of the nipple, or nipple discharge.

Risk Factors: The risk of breast cancer increases with age. The risk is higher in the woman who has a personal or family history of breast cancer; some forms of benign breast disease; early onset of menstruation; late menopause; lengthy exposure to cyclic estrogen; never having children or having the first live birth at a later age; and higher education and socioeconomic status. International variability in breast cancer incidence rates correlates with variations in diet, especially fat intake, although a causal role for dietary factors has not been firmly established. Additional factors that may be associated with increased breast cancer risk and that are currently under study include pesticide and other chemical exposures, alcohol consumption, induced abortion, and physical inactivity. A majority of women will have one or more risk factors for breast cancer. However, most risks are at such a low level that they only partly explain the high frequency of the disease in the population. To date, knowledge about risk factors has not translated into practical ways to prevent breast cancer. Since women may not be able to alter their personal risk factors, the best opportunity at present for reducing mortality is through early detection.

Early Detection: The American Cancer Society recommends that asymptomatic women aged 40 to 49 should have a screening mammogram every one to two years; and women aged 50 and over should have a mammogram every year. In addition, a clinical breast exam is recommended every three years for women 20 to 40, and every year for women over 40. The American Cancer Society also recommends monthly breast self-exam as a routine good health habit for women 20 years or older. Most breast lumps are not cancer, but only a physician can make a diagnosis.

Mammography is recognized as a valuable diagnostic technique for women who have findings suggestive of breast cancer. When a suspicious area is identified on a mammogram, or when a woman has a suspicious lump, mammography can help determine if there are other lesions too small to be felt in the same or opposite breast. Since a small percentage of breast cancers may not be seen on a mammogram, all suspicious lumps should be biopsied for a definitive diagnosis, even when current or recent mammography findings are described as normal.

Treatment: Taking into account the medical situation and the patient's preferences, treatment may involve lumpectomy (local removal of the tumor), mastectomy (surgical removal of the breast), radiation therapy, chemotherapy, or hormone therapy. Often, two or more methods are used in combination. Patients should discuss possible options for the best management of their breast cancer with their physicians.

In recent years, new techniques have made breast reconstruction possible after mastectomy, and the cosmetic results usually are good. Reconstruction has become an important part of treatment and rehabilitation. Bone marrow transplantation is another new type of treatment that is under study.

Lung Cancer

Signs and Symptoms: Persistent cough, sputum streaked with blood, chest pain, recurring pneumonia or bronchitis.

Risk Factors: Cigarette smoking is by far the most important risk factor in the development of lung cancer. Other factors include exposure to certain industrial substances, such as arsenic; certain organic chemicals and asbestos, particularly for persons who smoke; radiation exposure from occupational, medical, and environmental sources; air pollution; tuberculosis; radon exposure, especially in cigarette smokers; and environmental tobacco smoke in nonsmokers.

Early Detection: Because symptoms often don't appear until the disease is advanced, early detection is difficult. In smokers who stop smoking when precancerous changes are found, damaged lung tissue often returns to normal. Smokers who persist in smoking may form abnormal cell growth patterns that lead to cancer. Chest X ray, analysis of the types of cells contained in sputum, and fiberoptic examination of the bronchial passages assist diagnosis.

Treatment: Determined by the type and stage of the cancer. Options include surgery, radiation therapy, and chemotherapy. For many localized cancers, surgery is usually the treatment of choice. Because the disease has usually spread by the time it is discovered, radiation therapy and

Table E Five-Year Survival Rates by Stage at Diagnosis*

Site	All Stages %	Localized %	Regional %	Distant %
Oral and pharynx	54	81	43.5	21
Colon and rectum	61	90	65	8
Pancreas	4	17	6	1
Lung and bronchus	14	48.5	20	2
Skin (melanoma)	88	96	58.5	12.5
Female breast	85	96.5	77	21
Cervix	70	91.5	49	13
Uterus	84	96	63.5	26
Ovary	50	94.6	79	28
Prostate	93	100	100	33
Bladder	81	93	49	6
Kidney and renal pelvis	61	88	61	9

*Adjusted for normal life expectancy. This chart is based on cases diagnosed from 1989 to 1996, followed through 1997.

Source: National Cancer Institute Surveillance, Epidemiology, and End Results (SEER) Cancer Statistics Review, 1973–1997.

chemotherapy are often needed in combination with surgery as the treatment of choice; on this regimen, a large percentage of patients experience remission, which in some cases is long-lasting.

Prostate Cancer

Signs and Symptoms: Weak or interrupted urine flow; inability to urinate, or difficulty starting or stopping the urine flow; the need to urinate frequently, especially at night; blood in the urine; pain or burning on urination; continuing pain in lower back, pelvis, or upper thighs. Most of these symptoms are nonspecific and may be similar to those caused by benign conditions such as infection or prostate enlargement.

Risk Factors: The incidence of prostate cancer increases with age; over 80% of all prostate cancers are diagnosed in men over age 65. African Americans have the highest prostate cancer incidence rates in the world; the disease is common in North America and Northwestern Europe and is rare in the Near East, Africa, and South America. There may be some familial tendency, but it is unclear whether this is due to genetic or environmental factors. International studies suggest that dietary fat may also be a factor.

Early Detection: Every man aged 50 and over should have a digital rectal exam as part of his regular annual physical checkup. In addition, the American Cancer Society also recommends that men aged 50 and over have an annual prostate-specific antigen blood test. Men at high risk (i.e., African American or those with a familial history of prostate cancer) should have this test starting at age 45. If either result is suspicious, further evaluation in the form of transrectal ultrasound should be performed.

Treatment: Surgery, radiation, and/or hormones and anticancer drugs are treatment options. Hormone treatment and anticancer drugs may control prostate cancer for long periods by shrinking the size of the tumor, thus relieving pain. Careful observation without immediate active treatment may be appropriate for individuals with low-grade and/or early-stage tumors.

Colon and Rectal Cancer

Signs and Symptoms: Rectal bleeding, blood in the stool, a change in bowel habits.

Risk Factors: Personal or family history of colorectal cancer or polyps, and inflammatory bowel disease have been associated with increased colorectal cancer risk. New evidence suggests cigarette smoking may cause genetic damage leading to colon cancer. Other possible risk factors include physical inactivity and high-fat and/or low-fiber diet. Recent studies have suggested that estrogen replacement therapy and nonsteroidal anti-inflammatory drugs such as aspirin may reduce colorectal cancer risk.

Early Detection: Digital rectal exam, stool blood test, and sigmoidoscopy are recommended by the American Cancer Society to detect colon or rectal cancer in asymptomatic patients. These tests offer the best opportunity for the diagnosis and removal of polyps, and hence the prevention of these cancers.

The digital rectal exam is performed by a physician during an office visit. The American Cancer Society recommends that this exam be performed annually after age 50.

The stool blood test is a simple method to test feces for hidden blood. The specimen is obtained by the patient at home and returned to the physician's office, a hospital, or a clinic for analysis. The Society recommends annual testing after age 50.

In sigmoidoscopy, the physician uses a hollow, lighted tube or a fiberoptic sigmoidoscope to inspect the rectum and lower colon. The American Cancer Society recommends sigmoidoscopy every three to five years after age 50.

If any of these tests reveal possible problems, more extensive studies, such as colonoscopy (exam of the entire colon) and barium enema (an X-ray procedure in which the intestines are viewed), may be needed.

Treatment: Surgery, at times combined with radiation, is the most effective method of treating colorectal cancer. The role of chemotherapy and immunologic agents may be beneficial in postoperative patients with cancerous lymph nodes.

Colostomy (creation of an abdominal opening for elimination of body wastes) is seldom needed for colon cancer and is infrequently required for rectal cancer. The American Cancer Society has a patient assistance program for those who do have permanent colostomies.

Bladder Cancer

Signs and Symptoms: Blood in the urine. Usually associated with increased frequency of urination.

Risk Factors: Smoking is the greatest risk factor in bladder cancer, with smokers experiencing twice the risk of nonsmokers. Smoking is estimated to be responsible for approximately 50% of the bladder cancer deaths among men and 37% among women. People living in urban areas and workers exposed to dye, rubber, or leather also are at higher risk. Chlorination in water has also been associated with a small increase in bladder cancer risk.

Early Detection: Bladder cancer is diagnosed by examination of the bladder wall with a cytoscope, a slender tube fitted with a lens and light that can be inserted into the tract through the urethra.

Treatment: Surgery, alone or in combination with other treatments, is used in over 90% of cases. Preoperative chemotherapy, alone or with radiation before cystectomy (bladder removal), has improved some treatment results.

Lymphoma

Signs and Symptoms: Enlarged lymph nodes, itching, fever, night sweats, anemia, and weight loss. Fever can come and go in periods of several days or weeks.

Risk Factors: Risk factors are largely unknown, but in part involve reduced immune function and exposure to certain infectious agents. Persons with organ transplants are at higher risk due to altered immune function. Human immunodeficiency virus (HIV) and human T lymphocyte leukemia/lymphoma virus-I (HTLV-I) are associated with increased risk of non-Hodgkin lymphoma. Burkitt lymphoma in Africa is partly caused by the Epstein-Barr herpes virus. Other possible risk factors include occupational exposures to herbicides and perhaps other chemicals.

Treatment: Hodgkin disease: chemotherapy and radiotherapy are useful for most patients. Non-Hodgkin lymphoma: early stage, localized lymph node disease can be treated with radiotherapy. Patients with later stage disease often benefit from the addition of chemotherapy. New programs using highly specific monoclonal antibodies directed at lymphoma cells and improved techniques in bone marrow preservation are under investigation in selected patients who relapse after standard treatment.

Cervical Cancer

Signs and Symptoms: Abnormal uterine bleeding or spotting; abnormal vaginal discharge. Pain and systemic symptoms are late manifestations of the disease.

Risk Factors: First intercourse at an early age, multiple sexual partners, or partners who have had multiple sexual partners are at increased risk of developing the disease. Other risk factors include genital warts and cigarette smoking.

Early Detection: The Pap test is a simple procedure that can be performed at appropriate intervals by health care professionals as part of a pelvic exam. A small sample of cells is swabbed from the cervix, transferred to a slide, and examined under a microscope. This test should be performed annually with a pelvic exam in women who are, or have been, sexually active or who have reached age 18 years. After three or more consecutive annual exams with normal findings, the Pap test may be performed less frequently at the discretion of the physician.

Treatment: Cervical cancers generally are treated by surgery or radiation, or by a combination of the two. In precancerous (in situ) stages, changes in the cervix may be treated by cryotherapy (the destruction of cells by extreme cold), by electrocoagulation (the destruction of tissue through intense heat by electric current), or by local surgery.

Skin Cancer

Signs and Symptoms: Any change in the skin, especially a change in the size or color of a mole or other darkly pigmented growth or spot. Scaliness, oozing, bleeding, or change in the appearance of a bump or nodule, the spread of pigmentation beyond its border, a change in sensation, itchiness, tenderness, or pain.

Risk Factors: Excessive exposure to ultraviolet radiation; fair complexion; occupational exposure to coal tar, pitch, creosote, arsenic compounds, or radium; family history.

Protective clothing and sunscreen should be worn. The sun's ultraviolet rays are strongest between 10 A.M. and 3 P.M.; exposure at these times should be avoided. Sunscreen comes in various strengths, ranging from those facilitating gradual tanning to those that allow practically no tanning. Because of the possible link between severe sunburns in childhood and greatly increased risk of melanoma in later life, children, in particular, should be protected from the sun.

Early Detection: Early detection is critical. Recognition of changes in skin growths or the appearance of new growths is the best way to find early skin cancer. Adults should practice skin self-exam once a month, and suspicious lesions should be evaluated promptly by a physician. Basal and squamous cell skin cancers often take the form of a pale, waxlike, pearly nodule, or a red, scaly, sharply outlined patch. A sudden or progressive change in a mole's appearance should be checked by a physician. Melanomas often start as small, molelike growths that increase in size, change color, become ulcerated, and bleed easily from a slight injury. A simple **ABCD** rule outlines the warning signals of melanoma: **A** is for asymmetry. One-half of the mole does not match the other half. **B** is for border irregularity. The edges are ragged, notched, or blurred. **C** is for color. The pigmentation is not uniform. **D** is for diameter greater than 6 mm. Any sudden or progressive increase in size should be of special concern.

Treatment: There are five methods of treatment: surgery (used in 90% of cases), radiation therapy, electrodesiccation (tissue destruction by heat), cryosurgery (tissue destruction by freezing), and laser therapy for early skin cancer. For malignant melanoma, the primary growth must be adequately excised, and it may be necessary to remove nearby lymph nodes. Removal and microscopic examination of all suspicious moles is essential. Advanced cases of melanoma are treated according to the characteristics of the case.

Oral Cavity and Pharynx Cancer

Signs and Symptoms: A sore that bleeds easily and does not heal; a lump or thickening; a red or white patch that persists. Difficulty in chewing, swallowing, or moving tongue or jaws are often late symptoms.

Risk Factors: Cigarette, cigar, or pipe smoking; use of smokeless tobacco; excess use of alcohol.

Early Detection: Cancer can affect any part of the oral cavity, including the lip, tongue, mouth, and throat. Dentists and primary care physicians have the opportunity during regular checkups to see abnormal tissue changes and to detect cancer at an early, curable stage.

Treatment: Principal methods are radiation therapy and surgery. In advanced disease, chemotherapy is being studied as an adjunct to surgery.

Leukemia

Signs and Symptoms: Fatigue, paleness, weight loss, repeated infections, bruising easily, and nosebleeds or other hemorrhages. In children, these symptoms can appear suddenly. Chronic leukemia can progress slowly and with few symptoms.

Risk Factors: Leukemia strikes both sexes and all ages. Causes of most leukemias are unknown. Persons with Down syndrome and certain other genetic abnormalities have higher than usual incidence rate of leukemia. It has also been linked to excessive exposure to ionizing radiation and to certain chemicals such as benzene, a commercially used toxic liquid that is also present in lead-free gasoline. Certain forms of leukemia and lymphoma are caused by a retrovirus, HTLV-I (human T lymphocyte leukemia/lymphoma virus-I).

Early Detection: Because symptoms often resemble those of other, less serious conditions, leukemia can be difficult to diagnose early. When a physician does suspect leukemia, diagnosis can be made using blood tests and bone marrow biopsy.

Treatment: Chemotherapy is the most effective method of treating leukemia. Various anticancer drugs are used, either in combination or as single agents. Transfusions of blood components and antibiotics are used as supportive treatments. To eliminate hidden cells, therapy of the central nervous system has become standard treatment, especially in acute lymphocytic leukemia. Under appropriate conditions, bone marrow transplantation may be useful in the treatment of certain leukemias.

Pancreas Cancer

Signs and Symptoms: Cancer of the pancreas generally occurs without symptoms until it is in advanced stages and thus is considered a "silent" disease.

Risk Factors: Very little is known about what causes the disease or how to prevent it. Risk increases after age 50, with most cases occurring between ages 60 and 80. Smoking is a risk factor; about 30% of pancreatic cancer cases are thought to result directly from cigarette smoking. Some studies have suggested associations with chronic pancreatitis, diabetes, or cirrhosis. In countries where the diet is high in fat, pancreatic cancer rates are higher.

Early Detection: At present, only surgical biopsy yields a certain diagnosis, and because of the "silent" course of the disease, the need for biopsy is likely to be obvious only after the disease has advanced. Researchers are focusing on ways to diagnose pancreatic cancer before symptoms occur. Ultrasound imaging and computerized tomography scans are under investigation.

Treatment: Surgery, radiation therapy, and anticancer drugs are treatment options, but they have had little influence on the outcome. Diagnosis is usually so late that none of these is used.

Ovarian Cancer

Signs and Symptoms: Ovarian cancer is often "silent," showing no obvious signs or symptoms until late in its development. The most common sign is enlargement of the abdomen, which is caused by the accumulation of fluid. Rarely will there be abnormal vaginal bleeding. In women over 40, vague digestive disturbances (stomach discomfort, gas, distention) that persist and cannot be explained by any other cause may indicate the need for a thorough evaluation for ovarian cancer.

Risk Factors: Risk for ovarian cancer increases with age. Women who have never had children are more likely to develop ovarian cancer than those who have. Pregnancy and the use of oral contraceptives appear to be protective against ovarian cancer. Women who have had breast cancer or have a family history of ovarian cancer are at increased risk. With the exception of Japan, industrialized countries have the highest incidence rates.

Early Detection: Periodic, thorough pelvic examinations are important. The Pap test, useful in detecting cervical cancer, only rarely uncovers ovarian cancer. Transvaginal ultrasound and a tumor marker, CA-125, may assist diagnosis. Women over the age of 40 should have a cancer-related checkup every year.

Treatment: Surgery, radiation therapy, and drug therapy are treatment options. Surgery usually includes the removal of one or both ovaries (oophorectomy), the uterus (hysterectomy), and the Fallopian tubes (salpingectomy). In some very early tumors, only the involved ovary will be removed, especially in young women. In advanced disease, an attempt is made to remove all intraabdominal disease to enhance the effect of chemotherapy.

Cancer in Children

New Cases: An estimated 12,400 new cases in 2000. As a childhood disease, cancer is rare. Common sites include the blood and bone marrow, bone, lymph nodes, brain, nervous system, kidneys, and soft tissues.

Some of the main childhood cancers are as follows:

1. Leukemia.
2. Osteogenic sarcoma is a bone cancer that may cause no pain at first; swelling in the area of the tumor is often the first sign. Ewing's sarcoma is another type of cancer that arises in bone.
3. Neuroblastoma can appear anywhere but usually in the abdomen, where a swelling occurs.
4. Rhabdomyosarcoma, the most common soft tissue sarcoma, can occur in the head and neck area, genitourinary area, trunk, and extremities.
5. Brain cancers in early stages may cause headaches, blurred or double vision, dizziness, difficulty in walking or handling objects, and nausea.
6. Lymphomas and Hodgkin disease are cancers that involve the lymph nodes but also may invade bone marrow and other organs. They may cause swelling of lymph nodes in the neck, armpit, or groin. Other symptoms may include general weakness and fever.
7. Retinoblastoma, an eye cancer, usually occurs in children under age 4. When detected early, cure is possible with appropriate treatment.
8. Wilms' tumor, a kidney cancer, may be recognized by a swelling or lump in the abdomen.

Childhood cancers can be treated by a combination of therapies. Treatment is coordinated by a team of experts including oncologic physicians, pediatric nurses, social workers, psychologists, and others who assist children and their families.

Source: From the American Cancer Society, Inc., http://www.cancer.org. Used with permission.

Appendix F

Acronyms

Acronym	Meaning
ACh	acetylcholine
AChE	acetylcholinesterase
ACTH	adrenocorticotropic hormone
AD	Alzheimer disease
ADH	antidiuretic hormone
ADP	adenosine diphosphate
AIDS	acquired immunodeficiency syndrome
ANH	atrial natriuretic hormone
APC	antigen-presenting cell
ATP	adenosine triphosphate
AV node	atrioventricular node
BP	before present
cAMP	cyclic adenosine monophosphate
CAPD	continuous ambulatory peritoneal dialysis
cDNA	complementary deoxyribonucleic acid
CFC	chlorofluorocarbon
cGMP	cyclic guanosine monophosphate
CNS	central nervous system
CVA	cardiovascular accident
DNA	deoxyribonucleic acid
DPT	diptheria–pertussis–tetanus
ECG (or EKG)	electrocardiogram
ER	endoplasmic reticulum (rough ER, smooth ER)
FAD	flavin adenine dinucleotide
FAS	fetal alcohol syndrome
FSH	follicle-stimulating hormone
GH	growth hormone
GIFT	gamete intrafallopian transfer
GIP	gastric inhibitory peptide
GMP	guanosine monophosphate
GnRH	gonadotropic-releasing hormone
HCG	human chorionic gonadotropin
HDL	high-density lipoprotein
HDN	hemolytic disease of the newborn
HHb	hemoglobin
HIV	human immunodeficiency virus
HPV	human papillomavirus
ICSH	interstitial cell-stimulating hormone
IUD	intrauterine device
IVF	in vitro fertilization
LDL	low-density lipoprotein
LH	luteinizing hormone
MHC	major histocompatibility complex
MI	myocardial infarction
MRI	magnetic resonance imaging
MSH	melanocyte-stimulating hormone
MYA	millions of years ago
NAD^+	nicotinamide adenine dinucleotide
NE	norepinephrine
NGF	nerve growth factor
NGU	nongonococcal urethritis
NO_X	nitrogen oxides
PAN	peroxyacetylnitrate
PCB	polychlorinated biphenyl
PCR	polymerase chain reaction
PG	prostaglandin
PID	pelvic inflammatory disease
PNS	peripheral nervous system
PRL	prolactin
PSA	prostate-specific antigen
PTH	parathyroid hormone
rDNA	recombinant deoxyribonucleic acid
RFLP	restriction fragment length polymorphism
RNA	ribonucleic acid
rRNA	ribosomal ribonucleic acid
SA node	sinoatrial node
SAD	seasonal affective disorder
SCID	severe combined immunodeficiency syndrome
SPF	sun protection factor
STD	sexually transmitted disease
THC	tetrahydrocannabinol
TIMP	tissue inhibitor metalloproteinase
tPA	tissue plasminogen activator
TSH	thyroid-stimulating hormone
UV	ultraviolet

Glossary

A

acetylcholine (ACh) (uh-seet-ul-KOH-leen) Neurotransmitter active in both the peripheral and central nervous systems.   251

acetylcholinesterase (AChE) (uh-SEET-ul-koh-luh-nes-tuh-rays, -rayz) Enzyme that breaks down acetylcholine bound to postsynaptic receptors within a synapse.   251

acid Molecules tending to raise the hydrogen ion concentration in a solution and to lower its pH numerically.   23

acid deposition Return to earth as rain or snow of the sulfate or nitrate salts of acids produced by commercial and industrial activities.   489

acromegaly (ak-roh-MEG-uh-lee) Condition resulting from an increase in growth hormone production after adult height has been achieved.   298

acrosome (AK-ruh-sohm) Cap at the anterior end of a sperm that partially covers the nucleus and contains enzymes that help the sperm penetrate the egg.   321

actin (AK-tin) One of two major proteins of muscle; makes up thin filaments in myofibrils of muscle fibers. See myosin.   231

action potential Electrochemical changes that take place across the axomembrane; the nerve impulse.   248

active site Region on the surface of an enzyme where the substrate binds and where the reaction occurs.   54

active transport Use of a plasma membrane carrier protein and energy to move a substance into or out of a cell from lower to higher concentration.   48

acute bronchitis (brahn-KY-tis, brahng-) Infection of the primary and secondary bronchi.   179

adaptation Organism's modification in structure, function, or behavior suitable to the environment.   463

Addison disease Condition resulting from a deficiency of adrenal cortex hormones; characterized by low blood glucose, weight loss, and weakness.   303

adenine (A) (AD-uh-neen) One of four nitrogen bases in nucleotides composing the structure of DNA and RNA.   423

adhesion junction Small region between cells in which the adjacent plasma membranes do not touch but are held together by intercellular filaments attached to buttonlike thickenings.   64

adipose tissue (AH-duh-pohs) Connective tissue in which fat is stored.   64

ADP (adenosine diphosphate) (ah-DEN-ah-seen dy-FAHS-fayt) Nucleotide with two phosphate groups that can accept another phosphate group and become ATP.   36

adrenal cortex (uh-DREE-nul KOR-teks) Outer portion of the adrenal gland; secretes mineralocorticoids such as aldosterone and glucocorticoids such as cortisol.   301

adrenal gland (uh-DREE-nul) An endocrine gland that lies atop a kidney consisting of the inner adrenal medulla and the outer adrenal cortex.   301

adrenal medulla (uh-DREE-nul muh-DUL-uh) Inner portion of the adrenal gland; secretes the hormones epinephrine and norepinephrine.   301

adrenocorticotropic hormone (ACTH) (uh-DREE-noh-kawrt-ih-koh-troh-pik) Hormone secreted by the anterior lobe of the pituitary gland that stimulates activity in the adrenal cortex.   296

afterbirth The placenta and the extraembryonic membranes, which are delivered (expelled) during the third stage of parturition.   377

agglutination (uh-gloot-un-AY-shun) Clumping of red blood cells due to a reaction between antigens on red blood cell plasma membranes and antibodies in the plasma.   120

aging Progressive changes over time, leading to loss of physiological function and eventual death.   378

agranular White blood cell that does not contain distinctive granules; 115

AIDS (acquired immunodeficiency syndrome) (im-yuh-noh-dih-FISH-un-see) Disease caused by HIV and transmitted via body fluids; characterized by failure of the immune system.   342

albumin (al-BYOO-mun) Plasma protein of the blood having transport and osmotic functions.   118

aldosterone (al-DAHS-tuh-rohn) Hormone secreted by the adrenal cortex that decreases sodium and increases potassium excretion; raises blood volume and pressure.   197, 302

alien species Nonnative species that migrate or are introduced by humans into a new ecosystem; exotics.   505

allantois (uh-LAN-toh-is) Extraembryonic membrane that contributes to the formation of umbilical blood vessels in humans.   368

allele (uh-LEEL) Alternative form of a gene—alleles occur at the same locus on homologous chromosomes.   404

allergen (AL-ur-jun) Foreign substance capable of stimulating an allergic response.   158

allergy Immune response to substances that usually are not recognized as foreign.   158

alveolus (pl., alveoli) (al-VEE-uh-lus) Air sac of a lung.   169

Alzheimer disease (AD) Brain disorder characterized by a general loss of mental abilities.   266

amino acid Organic molecule having an amino group and an acid group, that covalently bonds to produce peptide molecules.   31

amnion (AM-nee-ahn) Extraembryonic membrane that forms an enclosing, fluid-filled sac.   368

ampulla (am-POOL-uh, -PUL-uh) Base of a semicircular canal in the inner ear.   287

amygdala (uh-MIG-duh-luh) Portion of the limbic system which functions to add emotional overtones to memories.   258

anabolic steroid (a-nuh-BAHL-ik) Synthetic steroid that mimics the effect of testosterone.   306

analogous structure Structure that has a similar function in separate lineages but differs in anatomy and ancestry.   464

anaphase Mitotic phase during which daughter chromosomes move toward poles of spindle.   389

androgen (AN-druh-jun) Male sex hormone (e.g., testosterone).   306

anemia (uh-NEE-mee-uh) Inefficiency in the oxygen-carrying ability of blood due to a shortage of hemoglobin.   113

aneurysm (AN-yuh-riz-um) Ballooning of a blood vessel.   137

angina pectoris (an-JY-nuh PEK-tuh-ris) Condition characterized by thoracic pain resulting from occluded coronary arteries; precedes a heart attack.   137

angiogenesis (an-jee-oh-JEN-uh-sis) Formation of new blood vessels; one mechanism by which cancer spreads.   444

angioplasty (AN-jee-uh-plas-tee) Surgical procedure for treating clogged arteries, in which a plastic tube is threaded through a major blood vessel toward the heart and then a balloon at the end of the tube is inflated, forcing open the vessel. 138

anorexia nervosa (a-nuh-REK-see-uh nur-VOH-suh) Eating disorder characterized by a morbid fear of gaining weight. 105

anterior pituitary (pih-TOO-ih-tair-ee) Portion of the pituitary gland that is controlled by the hypothalamus and produces six types of hormones, some of which control other endocrine glands. 296

antibiotic Drug that interferes with a bacterial enzyme and in that way causes the death of a particular bacterium. 346

antibody (AN-tih-bahd-ee) Protein produced in response to the presence of an antigen; each antibody combines with a specific antigen. 115, 150

antibody-mediated immunity Specific mechanism of defense in which plasma cells derived from B cells produce antibodies that combine with antigens. 151

anticodon (an-tih-KOH-dahn) Three-base sequence in a transfer RNA molecule base that pairs with a complementary codon in mRNA. 428

antidiuretic hormone (ADH) (an-tih-dy-uh-RET-ik) Hormone secreted by the posterior pituitary that increases the permeability of the collecting ducts in a kidney. 196, 296

antigen (AN-tih-jun) Foreign substance, usually a protein or a polysaccharide, that stimulates the immune system to produce antibodies. 150

antigen receptor Receptor proteins in the plasma membrane of immune system cells whose shape allows them to combine with a specific antigen. 150

antigen-presenting cell (APC) Cell that displays the antigen to the cells of the immune system so they can defend the body against that particular antigen. 154

anus Outlet of the digestive tube. 88

aorta (ay-OR-tuh) Major systemic artery that receives blood from the left ventricle. 134

aortic body Sensory receptor in the aortic arch sensitive to the O_2, CO_2, and H^+ content of the blood. 172

apoptosis (ap-uh-TOH-sis, -ahp-) Programmed cell death involving a cascade of specific cellular events leading to death and destruction of the cell. 151, 444

appendicular skeleton (ap-un-DIK-yuh-lur) Portion of the skeleton forming the pectoral girdles and upper extremities; and the pelvic girdle and lower extremities. 216

appendix (uh-PEN-diks) Small, tubular appendage that extends outward from the cecum of the large intestine. 88

aqueous humor (AY-kwee-us, AK-wee-) Clear, watery fluid between the cornea and lens of the eye. 278

aquifer (AHK-wuh-fur) Rock layers that contain water and will release it in appreciable quantities to wells or springs. 485

arteriole (ar-TEER-ee-ohl) Vessel that takes blood from an artery to capillaries. 127

artery Vessel that takes blood away from the heart to arterioles; characteristically possessing thick elastic and muscular walls. 126

arthritis Joint inflammation. 220

articular cartilage (ar-TIK-yuh-lur) Hyaline cartilaginous covering over the articulating surface of the bones of synovial joints. 206

assisted reproductive technologies (ART) Medical techniques, sometimes performed in vitro, that are done to increase the chances of pregnancy. 332

association area One of several regions of the cerebral cortex related to memory, reasoning, judgment, and emotional feelings. 256

aster Short, radiating fibers about the centrioles at the poles of a spindle. 388

asthma (AZ-muh, as-) Condition in which bronchioles constrict and cause difficulty in breathing. 181

astigmatism (uh-STIG-muh-tiz-um) Blurred vision due to an irregular curvature of the cornea or the lens. 283

atherosclerosis (ath-uh-roh-skluh-ROH-sis) Condition in which fatty substances accumulate abnormally beneath the inner linings of the arteries. 137

atom Smallest particle of an element that displays the properties of the element. 16

ATP (adenosine triphosphate) (uh-DEN-uh-seen try-FAHS-fayt) Nucleotide with three phosphate groups. The breakdown of ATP into ADP + Ⓟ makes energy available for energy-requiring processes in cells. 36

atrial natriuretic hormone (ANH) (AY-tree-ul nay-tree-yoo-RET-ik) Hormone secreted by the heart that increases sodium excretion and therefore lowers blood volume and pressure. 197, 302

atrioventricular bundle (ay-tree-oh-ven-TRIK-yuh-lur) Group of specialized fibers that conduct impulses from the atrioventricular node to the ventricles of the heart; AV bundle. 130

atrioventricular valve Valve located between the atrium and the ventricle. 128

atrium (AY-tree-um) One of the upper chambers of either the left atrium or the right atrium which receive blood. 128

atrophy (AT-ruh-fee) Wasting away or decrease in size of an organ or tissue. 238

auditory tube Extension from the middle ear to the nasopharynx which equalizes air pressure on the eardrum. 177, 284

australopithecine (aw-stray-loh-PITH-uh-syn) Any of the first evolved hominids; classified into several species of *Australopithecus*. 468

autoimmune disease Disease that results when the immune system mistakenly attacks the body's own tissues. 160

autonomic system (awt-uh-NAHM-ik) Branch of the peripheral nervous system that has control over the internal organs; consists of the sympathetic and parasympathetic systems. 263

autosome (AW-tuh-sohm) Any chromosome other than the sex chromosomes. 395

autotrophic organism Organism that can capture energy and synthesize organic nutrients from inorganic nutrients. 480

AV (atrioventricular) node Small region of neuromuscular tissue that transmits impulses received from the SA node to the ventricles. 130

axial skeleton (AK-see-ul) Portion of the skeleton that supports and protects the organs of the head, the neck, and the trunk. 212

axon (AK-sahn) Elongated portion of a neuron that conducts nerve impulses typically from the cell body to the synapse. 246

axon bulb Small swelling at the tip of one of many endings of the axon. 251

B

B lymphocyte (LIM-fuh-syt) Lymphocyte that matures in the bone marrow and, when stimulated by the presence of a specific antigen, gives rise to antibody-producing plasma cells. 150

bacteria Microscopic prokaryotic cell that can be free living or pathogenic. 345

ball and socket joint Most freely movable type of joint (e.g., the shoulder or hip joint). 218

basal body A cytoplasmic structure that is located at the base of and may organize cilia or flagella. 54

basal nuclei (BAY-sul) Subcortical nuclei deep within the white matter that serve as relay stations for motor impulses and produce dopamine to help control skeletal muscle activities. 256

base Molecules tending to lower the hydrogen ion concentration in a solution and raise the pH numerically.   23

basement membrane Layer of nonliving material that anchors epithelial tissue to underlying connective tissue.   62

basophil (BAY-suh-fil) White blood cell with a granular cytoplasm and that is able to be stained with a basic dye.   115

bicarbonate ion Ion that participates in buffering the blood, and the form in which carbon dioxide is transported in the bloodstream.   174

bile (byl) Secretion of the liver that is temporarily stored and concentrated in the gallbladder before being released into the small intestine, where it emulsifies fat.   87

binary fission Bacterial reproduction into two daughter cells without the utilization of a mitotic spindle.   345

binomial name Scientific name of an organism, the first part of which designates the genus and second part of which designates the specific epithet.   466

biodiversity Total number of species, the variability of their genes, and the communities in which they live.   5, 498

biodiversity hotspot Region of the world that contains unusually large concentrations of species.   499

biogeochemical cycle (by-oh-jee-oh-KEM-ih-kul) Circulating pathway of elements such as carbon and nitrogen involving exchange pools, storage areas, and biotic communities.   484

biogeography Study of the geographical distribution of organisms.   464

biological evolution Change in life forms that has taken place in the past and will take place in the future; includes descent from a common ancestor and adaptation to the environment.   463

biological magnification Process by which substances become more concentrated in organisms in the higher trophic levels of a food web.   491

biomass The number of organisms multiplied by their weight.   483

biosphere (BY-oh-sfeer) Zone of air, land, and water at the surface of the earth in which living organisms are found.   2

biotechnology The use of genetic engineering and other techniques that make use of natural biological systems to produce a product or achieve a particular result desired by humans.   432

bipedalism Ability to walk upright on two feet.   467

birth-control pill Oral contraception containing estrogen and progesterone.   329

blastocyst (BLAS-tuh-sist) Early stage of embryonic development that consists of a hollow fluid-filled ball of cells.   366

blind spot Region of the retina lacking rods or cones and where the optic nerve leaves the eye.   281

blood Type of connective tissue in which cells are separated by a liquid called plasma.   66

blood pressure Force of blood pushing against the inside wall of an artery.   132

bone Connective tissue having protein fibers and a hard matrix of inorganic salts, notably calcium salts.   65

brain Enlarged superior portion of the central nervous system located in the cranial cavity of the skull.   254

brain stem Portion of the brain consisting of the medulla oblongata, pons, and midbrain.   256

bronchiole (BRAHNG-kee-ohl) Smaller air passages in the lungs that begin at the bronchi and terminate in alveoli.   169

bronchus (pl., bronchi) (BRAHNG-kus) One of two major divisions of the trachea leading to the lungs.   169

buffer Substance or group of substances that tend to resist pH changes of a solution, thus stabilizing its relative acidity and basicity.   24

bulbourethral gland (bul-boh-yoo-REE-thrul) Either of two small structures located below the prostate gland in males; each adds secretions to semen.   319

bulimia nervosa (byoo-LEE-mee-uh, -LIM-ee-, nur-VOH-suh) Eating disorder characterized by binge eating followed by purging by self-induced vomiting or use of a laxative.   104

bursa (BUR-suh) Saclike, fluid-filled structure, lined with synovial membrane, that occurs near a joint.   218

bursitis (bur-SY-tis) Inflammation of any of the friction-easing sacs called bursae within the knee joint.   218

C

calcitonin (kal-sih-TOH-nin) Hormone secreted by the thyroid gland that increases the blood calcium level.   300

calorie Amount of heat energy required to raise the temperature of 1 gram of water 1°C.   22

cancer Malignant tumor whose nondifferentiated cells exhibit loss of contact inhibition, uncontrolled growth, and the ability to invade tissue and metastasize.   444

capillary (KAP-uh-lair-ee) Microscopic vessel connecting arterioles to venules; exchange of substances between blood and tissue fluid occur across their thin walls.   62, 126

carbaminohemoglobin Hemoglobin carrying carbon dioxide.   174

carbohydrate Class of organic compounds that includes monosaccharides, disaccharides, and polysaccharides.   27

carbon cycle Continuous process by which carbon circulates in the air, water, and organisms of the biosphere.   486

carbonic anhydrase (kar-BAHN-ik an-HY-drays, -drayz) Enzyme in red blood cells that speeds the formation of carbonic acid from the reactants water and carbon dioxide.   174

carcinogen (kar-SIN-uh-jun) Environmental agent that causes mutations leading to the development of cancer.   448

carcinogenesis (kar-suh-nuh-JEN-uh-sis) Development of cancer.   444

carcinoma (kar-suh-NOH-muh) Cancer arising in epithelial tissue.   62

cardiac cycle One complete cycle of systole and diastole for all heart chambers.   130

cardiac muscle Striated, involuntary muscle found only in the heart.   67

cardiovascular system (kar-dee-oh-VAS-kyuh-lur) Organ system in which blood vessels distribute blood under the pumping action of the heart.   70

carnivore (KAR-nuh-vor) Consumer in a food chain that eats other animals.   480

carotid body (kuh-RAHT-id) Structure located at the branching of the carotid arteries and that contain chemoreceptors sensitive to the O_2, CO_2, and H^+ content in blood.   172

carrier Heterozygous individual who has no apparent abnormality but can pass on an allele for a recessively inherited genetic disorder.   409

cartilage (KAR-tul-ij, KART-lij) Connective tissue in which the cells lie within lacunae separated by a flexible proteinaceous matrix.   64, 206

cataract (KAT-uh-rakt) Opaqueness of the lens of the eye, making the lens incapable of transmitting light.   279

cecum (SEE-kum) Small pouch that lies below the entrance of the small intestine, and is the blind end of the large intestine.   88

cell Smallest unit that displays the properties of life; always contains cytoplasm surrounded by a plasma membrane.   2, 42

cell body Portion of a neuron that contains a nucleus and from which dendrites and an axon extend.   246

cell cycle Repeating sequence of cellular events that consists of interphase, mitosis, and cytokinesis.   387

cell-mediated immunity Specific mechanism of defense in which T cells destroy antigen-bearing cells. 155

cell theory One of the major theories of biology which states that all organisms are made up of cells and cells come only from pre-existing cells. 42

cellular respiration Metabolic reactions that use the energy primarily from carbohydrate but also fatty acid or amino acid breakdown to produce ATP molecules. 55

cellulose (SEL-yuh-lohs, -lohz) Polysaccharide that is the major complex carbohydrate in plant cell walls. 28

central nervous system (CNS) Portion of the nervous system consisting of the brain and spinal cord. 246

centriole (SEN-tree-ohl) Cellular structure, existing in pairs that possibly organize the mitotic spindle for chromosomal movement during mitosis and meiosis. 53

centromere (SEN-truh-meer) Constriction where sister chromatids of a chromosome are held together. 387

cerebellum (ser-uh-BEL-um) Part of the brain located posterior to the medulla oblongata and pons that coordinates skeletal muscles to produce smooth, graceful motions. 256

cerebral cortex (suh-REE-brul, SER-uh-brul KOR-teks) Outer layer of cerebral hemispheres; receives sensory information and controls motor activities. 255

cerebral hemisphere One of the large, paired structures that together constitute the cerebrum of the brain. 255

cerebrospinal fluid (sair-uh-broh-SPY-nul, suh-REE-broh-) Fluid found in the ventricles of the brain, in the central canal of the spinal cord, and in association with the meninges. 252

cerebrum (SAIR-uh-brum, suh-REE-brum) Main part of the brain consisting of two large masses, or cerebral hemispheres; the largest part of the brain in mammals. 255

cervix (SUR-viks) Narrow end of the uterus, which projects into the vagina. 322

chemical evolution Increase in the complexity of chemicals over time that could have led to the first cells. 462

chemoreceptor (kee-moh-rih-SEP-tur) Sensory receptor that is sensitive to chemical stimuli—for example, receptors for taste and smell. 272

chlamydia (kluh-MID-ee-uh) Sexually transmitted disease caused by the bacterium *Chlamydia trachomatis* that can lead to pelvic inflammatory disease. 346

chlorofluorocarbon (klor-oh-floor-oh-KAR-bun) Organic compounds containing carbon, chlorine, and fluorine atoms. CFCs such as Freon can deplete the ozone shield by releasing chlorine atoms in the upper atmosphere. 492

chondrocyte Type of cell found in the lacunae of cartilage. 206

chordae tendineae (KOR-dee TEN-din-ee-ee) Tough bands of connective tissue that attach the papillary muscles to the atrioventricular valves within the heart. 128

chorion (KOR-ee-ahn) Extraembryonic membrane that contributes to placenta formation. 368

chorionic villi (kor-ee-AHN-ik VIL-eye) Treelike extensions of the chorion which project into the maternal tissues at the placenta. 368

choroid (KOR-oyd) Vascular, pigmented middle layer of the eyeball. 278

chromatid One of the two side-by-side replicas in a duplicated chromosome. 387

chromatin (KROH-muh-tin) Network of fine threads in the nucleus which are composed of DNA and proteins. 49

chromosome (KROH-muh-som) Chromatin condensed into a compact structure. 49

chronic bronchitis Obstructive pulmonary disorder that tends to recur, marked by inflamed airways filled with mucus, and degenerative changes in the bronchi, including loss of cilia. 179

chyme (kym) Thick, semiliquid food material that passes from the stomach to the small intestine. 86

ciliary body (SIL-ee-air-ee) Structure associated with the choroid layer that contains ciliary muscle, which controls the shape of the lens of the eye. 278

cilium (pl., cilia) (SIL-ee-um) Short, hairlike projection from the plasma membrane, occurring usually in large numbers. 54, 62

circadian rhythm (sur-KAY-dee-un) Biological rhythm with a 24-hour cycle. 307

cirrhosis (sih-ROH-sis) Chronic, irreversible injury to liver tissue; commonly caused by frequent alcohol consumption. 91

cleavage furrow Indentation that begins the process of cleavage, by which human cells undergo cytokinesis. 389

clonal selection theory The concept that an antigen selects which lymphocyte will undergo clonal expansion and produce more lymphocytes bearing the same type of antigen receptor. 151

cloning Production of identical copies; can be either the production of identical individuals or in genetic engineering, the production of identical copies of a gene. 432

clotting Process of blood coagulation, usually when injury occurs. 116

cochlea (KOHK-lee-uh, KOH-klee-uh) Portion of the inner ear that resembles a snail's shell and contains the spiral organ, the sense organ for hearing. 284

cochlear canal (KOH-klee-ur) Canal within the cochlea that bears the spiral organ. 285

cochlear nerve Either of two cranial nerves that carry nerve impulses from the spiral organ to the brain; auditory nerve. 285

codominance Inheritance pattern in which both alleles of a gene are equally expressed. 413

codon Three-base sequence in messenger RNA that causes the insertion of a particular amino acid into a protein or termination of translation. 427

coelom (SEE-lum) Embryonic body cavity lying between the digestive tract and body wall that becomes the thoracic and abdominal cavities. 69

coenzyme (koh-EN-zym) Nonprotein organic molecule that aids the action of the enzyme to which it is loosely bound. 55

collagen fiber (KAHL-uh-jun) White fiber in the matrix of connective tissue, giving flexibility and strength. 64

collecting duct Duct within the kidney that receives fluid from several nephrons; the reabsorption of water occurs here. 193

colon (KOH-lun) The major portion of the large intestine and consisting of the ascending colon, the transverse colon, and the descending colon. 88

colony-stimulating factor (CSF) Protein that stimulates differentiation and maturation of white blood cells. 115

color vision Ability to detect the color of an object, dependent on three kinds of cone cells. 280

colostrum (kuh-LAHS-trum) Thin, milky fluid rich in proteins, including antibodies, that is secreted by the mammary glands a few days prior to or after delivery before true milk is secreted. 377

columnar epithelium (kuh-LUM-nur ep-uh-THEE-lee-um) Type of epithelial tissue with cylindrical cells. 62

community Assemblage of populations interacting with one another within the same environment. 478

compact bone Type of bone that contains osteons consisting of concentric layers of matrix and osteocytes in lacunae. 65, 206

complement system Series of proteins in plasma that form a nonspecific defense mechanism against a microbe invasion; it complements the antigen-antibody reaction.   150

complementary base pairing Hydrogen bonding between particular bases; in DNA, thymine (T) pairs with adenine (A) and guanine (G) pairs with cytosine (C); in RNA uracil (U) pairs with A and G pairs with C.   423

complementary DNA (cDNA) DNA that has been synthesized from mRNA by the action of reverse transcriptase.   433

conclusion Statement made following an experiment as to whether the results support or falsify the hypothesis.   8

condensation synthesis Chemical reaction resulting in a covalent bond with the accompanying loss of a water molecule.   27

condom Sheath used to cover the penis during sexual intercourse; used as a contraceptive and, if latex, to minimize the risk of transmitting infection.   330

cone cell Photoreceptor in retina of eye that responds to bright light; detects color and provides visual acuity.   280

congestive heart failure Inability of the heart to maintain adequate circulation, especially of the venous blood returned to it.   138

connective tissue Type of tissue that binds structures together, provides support and protection, fills spaces, stores fat, and forms blood cells; adipose tissue, cartilage, bone, and blood are types of connective tissue.   64

conservation biology Scientific discipline that seeks to understand the effects of human activities on diversity and seeks to develop practical approaches to preventing the extinction of species and the destruction of ecosystems.   498

constipation (kahn-stuh-PAY-shun) Delayed and difficult defecation caused by insufficient water in the feces.   89

consumer Organism that feeds on another organism in a food chain; primary consumers eat plants, and secondary consumers eat animals.   480

contraceptive (kahn-truh-SEP-tiv) Medication or device used to reduce the chance of pregnancy.   329

control group Sample that goes through all the steps of an experiment but lacks the factor or is not exposed to the factor being tested; a standard against which results of an experiment are checked.   10

cornea (KOR-nee-uh) Transparent, anterior portion of the outer layer of the eyeball.   278

coronary artery (KOR-uh-nair-ee) Artery that supplies blood to the wall of the heart.   135

corpus luteum (KOR-pus LOOT-ee-um) Yellow body that forms in the ovary from a follicle that has discharged its secondary oocyte; it secretes progesterone and some estrogen.   325

cortisol (KOR-tuh-sawl) Glucocorticoid secreted by the adrenal cortex that responds to stress on a long-term basis; reduces inflammation and promotes protein and fat metabolism.   302

covalent bond (coh-VAY-lent) Chemical bond in which atoms share one pair of electrons.   20

cranial nerve Nerve that arises from the brain.   260

creatine phosphate (KREE-uh-teen FAHS-fayt) Compound unique to muscles that contains a high-energy phosphate bond.   236

creatinine (kree-AH-tuhn-een) Nitrogenous waste, the end product of creatine phosphate metabolism.   189

cretinism (KREE-tun-iz-um) Condition resulting from improper development of the thyroid in an infant; characterized by stunted growth and mental retardation.   299

Cro-Magnon (kroh-MAG-nun) Common name for first fossils to be designated *Homo sapiens*.   472

crossing-over Exchange of segments between nonsister chromatids of a tetrad during meiosis.   390

cuboidal epithelium (kyoo-BOYD-ul) Type of epithelial tissue with cube-shaped cells.   62

culture Total pattern of human behavior; includes technology and the arts and is dependent upon the capacity to speak and transmit knowledge.   469

Cushing syndrome (KOOSH-ing) Condition resulting from hypersecretion of glucocorticoids; characterized by thin arms and legs and a "moon face," and accompanied by high blood glucose and sodium levels.   303

cyclic AMP (SY-klik, SIH-klik) ATP-related compound that acts as the second messenger in peptide hormone transduction; it initiates activity of the metabolic machinery.   309

cytokine (SY-tuh-kyn) Type of protein secreted by a T lymphocyte that stimulates cells of the immune system to perform their various functions.   154

cytokinesis (SY-tuh-kyn-EE-sus) Division of the cytoplasm following mitosis and meiosis.   389

cytoplasm (SY-tuh-plaz-um) Contents of a cell between the nucleus and the plasma membrane that contains the organelles.   44

cytosine (C) (SY-tuh-seen) One of four nitrogen bases in nucleotides composing the structure of DNA and RNA.   423

cytoskeleton Internal framework of the cell, consisting of microtubules, actin filaments, and intermediate filaments.   44

cytotoxic T cell (sy-tuh-TAHK-sik) T lymphocyte that attacks and kills antigen-bearing cells.   155

D

data Facts or pieces of information collected through observation and/or experimentation.   8

decomposer Organism, usually a bacterium or fungus, that breaks down organic matter into inorganic nutrients that can be recycled in the environment.   480

defecation (def-ih-KAY-shun) Discharge of feces from the rectum through the anus.   88

dehydrogenase Coenzyme that removes hydrogen atoms (electrons plus hydrogen ions) and carries electrons to the electron transport system in mitochondria.   55

delayed allergic response Allergic response initiated at the site of the allergen by sensitized T cells, involving macrophages and regulated by cytokines.   158

denaturation (dee-nay-chuh-RAY-shun) Loss of normal shape by an enzyme so that it no longer functions; caused by a less than optimal pH or temperature.   32

dendrite (DEN-dryt) Branched ending of a neuron which conduct signals toward the cell body.   246

denitrification Conversion of nitrate or nitrite to nitrogen gas by bacteria in soil.   488

dense fibrous connective tissue Type of connective tissue containing many collagen fibers packed together, and found in tendons and ligaments, for example.   64

dental caries (KAR-eez) Tooth decay that occurs when bacteria within the mouth metabolize sugar and give off acids that erode teeth; a cavity.   83

dermis (DUR-mus) Region of skin that lies beneath the epidermis.   72

detrital food chain (dih-TRYT-ul) Straight-line linking of organisms according to who eats whom which begins with detritus.   483

detrital food web (dih-TRYT-ul) Complex pattern of interlocking and crisscrossing food chains that begins with detritus.   483

detritus (dih-TRYT-us) Partially decomposed organic matter derived from tissues and animal wastes.   480

diabetes mellitus (dy-uh-BEE-teez MEL-ih-tus, muh-ly-tus) Condition characterized by a high blood glucose level and the appearance of glucose in the urine, due to a deficiency of insulin production and failure of cells to take up glucose. 305

diaphragm (DY-uh-fram) Dome-shaped horizontal sheet of muscle and connective tissue that divides the thoracic cavity from the abdominal cavity. Also, a birth-control device consisting of a soft rubber or latex cup that fits over the cervix.   172, 329

diarrhea (dy-uh-REE-uh) Excessively frequent bowel movements.   89

diastole (dy-AS-tuh-lee) Relaxation period of a heart chamber during the cardiac cycle.   130

diastolic pressure (dy-uh-STAHL-ik) Arterial blood pressure during the diastolic phase of the cardiac cycle. 132

diencephalon (dy-en-SEF-uh-lahn) Portion of the brain in the region of the third ventricle that includes the thalamus and hypothalamus.   256

diffusion (dih-FYOO-zhun) Movement of molecules or ions from a region of higher to lower concentration; it requires no energy and stops when the distribution is equal.   47

digestive system Organ system that includes the mouth, esophagus, stomach, small intestine, and large intestine (colon) that receives food and digests it into nutrient molecules. Also has associated organs: teeth, tongue, salivary glands, liver, gallbladder, and pancreas.   70

diploid (2n) Cell condition in which two of each type of chromosome are present in the nucleus.   386

disaccharide (dy-SAK-uh-ryd) Sugar that contains two units of a monosaccharide; e.g., maltose.   27

distal convoluted tubule Final portion of a nephron that joins with a collecting duct; associated with tubular secretion.   193

diuretic (dy-uh-RET-ik) Drug used to counteract hypertension by causing the excretion of water.   197

DNA (deoxyribonucleic acid) Nucleic acid polymer produced from covalent bonding of nucleotide

monomers that contain the sugar deoxyribose; the genetic material of nearly all organisms.   35, 422

DNA fingerprinting Using DNA fragment lengths, resulting from restriction enzyme cleavage, to identify particular individuals.   433

DNA ligase (LY-gays) Enzyme that links DNA fragments; used during production of recombinant DNA to join foreign DNA to vector DNA. 432

DNA polymerase During replication, an enzyme that joins the nucleotides complementary to a DNA template. 424

dominant allele (uh-leel) Allele that exerts its phenotypic effect in the heterozygote; it masks the expression of the recessive allele.   404

dorsal-root ganglion (GANG-glee-un) Mass of sensory neuron cell bodies located in the dorsal root of a spinal nerve.   260

duodenum (doo-uh-DEE-num) First part of the small intestine where chyme enters from the stomach.   87

dyad Duplicated chromosome having two sister chromatids.   391

E

ecological pyramid Pictorial graph based on the biomass, number of organisms, or energy content of various trophic levels in a food web—from the producer to the final consumer populations.   483

ecosystem (EK-oh-sis-tum, EE-koh-) Biological community together with the associated abiotic environment; characterized by energy flow and chemical cycling.   4, 477

ectoderm (EK-tuh-durm) Outer germ layer of the embryonic gastrula that becomes the nervous system and skin.   365

edema (ih-DEE-muh) Swelling due to tissue fluid accumulation in the intercellular spaces.   146

edge effect Edges around a patch that have a slightly different habitat than the favorable habitat in the interior of the patch.   509

egg Female gamete having the haploid number of chromosomes which is fertilized by a sperm, the male gamete.   322

elastic cartilage Type of cartilage composed of elastic fibers, allowing greater flexibility.   65

elastic fiber Yellow fiber in the matrix of connective tissue, providing flexibility.   64

electrocardiogram (ECG or EKG) (ih-lek-troh-KAR-dee-uh-gram) Recording of the electrical activity associated with the heartbeat.   131

electron Negative subatomic particle, moving about in an energy level around the nucleus of an atom.   16

electron transport system Passage of electrons along a series of membrane-bound carrier molecules from a higher to lower energy level; the energy released is used for the synthesis of ATP.   55

element Substance that cannot be broken down into substances with different properties; composed of only one type atom.   16

embolus (EM-buh-lus) Moving blood clot that is carried through the bloodstream.   137

embryo (EM-bree-oh) Immature developmental stage that is not recognizable as a human being.   365

embryonic development Period of development from the second through eighth weeks.   372

emphysema (em-fih-SEE-muh) Degenerative disorder in which the bursting of alveolar walls reduces the total surface area for gas exchange. 179

emulsification (ih-mul-suh-fuh-KAY-shun) Breaking up of fat globules into smaller droplets by the action of bile salts or any other emulsifier.   29

endocrine gland (EN-duh-krin) Ductless organ that secretes (a) hormone(s) into the bloodstream.   294

endocrine system Organ system involved in the coordination of body activities; uses hormones as chemical signals secreted into the bloodstream.   70

endoderm (EN-duh-durm) Inner germ layer of the embryonic gastrula that becomes the lining of the digestive and respiratory tract and associated organs.   366

endometriosis (en-doh-mee-tree-oh-sus) Presence of uterine tissue outside the uterus, which can contribute to infertility; possibly the result of irregular menstrual flow.   332

endometrium Mucous membrane lining the interior surface of the uterus. 323

endoplasmic reticulum (ER) (en-duh-plaz-mik reh-tik-yuh-lum) System of membranous saccules and channels in the cytoplasm, often with attached ribosomes.   50

endospore Spore formed by some bacteria within the former plasma membrane that has a thick wall and can survive unfavorable environmental conditions.   346

enzyme (en-zym) Organic catalyst, usually a protein, that speeds a reaction in cells due to its particular shape.   31

eosinophil (ee-oh-SIN-oh-fill) White blood cell containing cytoplasmic granules that stain with acidic dye. 115

epidermis (ep-uh-dur-mus) Region of skin that lies above the dermis. 71

epididymis (ep-ih-did-uh-mus) Coiled tubule next to the testes where sperm mature and may be stored for a short time. 318

epiglottis (ep-uh-glaht-us) Structure that covers the glottis during the process of swallowing. 84, 168

epinephrine (ep-uh-nef-rin) Hormone secreted by the adrenal medulla in times of stress; adrenaline. 301

episiotomy (ih-pee-zee-aht-uh-mee) Surgical procedure performed during childbirth in which the opening of the vagina is enlarged to avoid tearing. 377

episodic memory Capacity of brain to store and retrieve information with regard to persons and events. 258

epithelial tissue (ep-uh-thee-lee-ul) Type of tissue that lines hollow organs and covers surfaces; epithelium. 62

erectile dysfunction Failure of the penis to achieve erection. 319

erythropoietin (ih-rith-roh-poy-ee-tin) Hormone, produced by the kidneys, that speeds red blood cell formation. 113, 189

esophagus (ih-sahf-uh-gus) Muscular tube for moving swallowed food from the pharynx to the stomach. 84

essential amino acids Amino acids required in the human diet because the body cannot make them. 97

estrogen (es-truh-jun) Female sex hormone that helps maintain sexual organs and secondary sexual characteristics. 306, 326

eukaryotic cell Type of cell that has a membrane-bounded nucleus and membranous organelle. 463

eutrophication Overenrichment caused by human activities leading to excessive bacterial growth and oxygen depletion. 491

evolution Descent of organisms from common ancestors with the development of genetic and phenotypic changes over time that make them more suited to the environment. 2

evolutionary tree Diagram that describes the evolutionary relationship of groups of organisms; a common ancestor is presumed to have been present at points of divergence. 466

excretion Removal of metabolic wastes from the body. 187

exophthalmic goiter (ek-sahf-thal-mik) Enlargement of the thyroid gland accompanied by an abnormal protrusion of the eyes. 299

experiment Test of a hypothesis. 10

expiration (ek-spuh-ray-shun) Act of expelling air from the lungs; exhalation. 166

expiratory reserve volume (ik-spy-ruh-tor-ee) Volume of air that can be forcibly exhaled after normal exhalation. 170

external respiration Exchange of oxygen and carbon dioxide between alveoli and blood. 174

exteroceptor Sensory receptor that detects stimuli from outside the body (e.g., taste, smell, vision, hearing, and equilibrium). 272

extraembryonic membrane (ek-struh-em-bree-ahn-ik) Membrane that is not a part of the embryo but is necessary to the continued existence and health of the embryo. 368

F

facilitated transport Use of a plasma membrane carrier to move a substance into or out of a cell from higher to lower concentration: no energy required. 48

fat Organic molecule that contains glycerol and fatty acids and is found in adipose tissue. 29

fatty acid Molecule that contains a hydrocarbon chain and ends with an acid group. 29

fermentation Anaerobic breakdown of glucose that results in a gain of two ATP and end products such as alcohol and lactate. 57

fertilization Union of a sperm nucleus and an egg nucleus, which creates a zygote. 364

fetal development Period of development from the ninth week through birth. 372

fiber Structure resembling a thread; also plant material that is nondigestible. 96

fibrin (fy-brun) Insoluble protein threads formed from fibrinogen during blood clotting. 116

fibrinogen (fy-brin-uh-jun) Plasma protein that is converted into fibrin threads during blood clotting. 116

fibroblast (fy-bruh-blast) Cell in connective tissues which produces fibers and other substances. 64

fibrocartilage (fy-broh-kar-tul-ij, -kart-lij) Cartilage with a matrix of strong collagenous fibers. 65

fibrous connective tissue Tissue composed mainly of closely packed collagenous fibers and found in tendons and ligaments. 206

fimbria (fim-bree-uh) Fingerlike extension from the oviduct near the ovary. 322

flagellum (pl., flagella) (fluh-jel-um) Slender, long extension that propels a cell through a fluid medium. 54

focused Bending of light rays by the cornea, lens, and humors so that they converge and create an image on the retina. 279

follicle (fahl-ih-kul) Structure in the ovary that produces a secondary oocyte and the hormones estrogen and progesterone. 325

follicle-stimulating hormone (FSH) Hormone secreted by the anterior pituitary gland that stimulates the development of an ovarian follicle in a female or the production of sperm in a male. 321

fontanel (fahn-tun-el) Membranous region located between certain cranial bones in the skull of a fetus or infant. 212

food chain The order in which one population feeds on another in an ecosystem, from detritus (detrital food chain) or producer (grazing food chain) to final consumer. 483

food web In ecosystems, complex pattern of interlocking and crisscrossing food chains. 483

foramen magnum (fuh-ray-mun magnum) Opening in the occipital bone of the vertebrate skull through which the spinal cord passes. 212

formed element Constituent of blood that is either cellular (red blood cells and white blood cells) or at least cellular in origin (platelets). 111

fossil Any past evidence of an organism that has been preserved in the earth's crust. 464

fossil fuel Fuels such as oil, coal, and natural gas that are the result of partial decomposition of plants and animals coupled with exposure to heat and pressure for millions of years. 486

fossil record History of life recorded from remains from the past. 464

fovea centralis Region of the retina consisting of densely packed cones that is responsible for the greatest visual acuity. 279

functional group Specific cluster of atoms attached to the carbon skeleton of organic molecules that enters into reactions and behaves in a predictable way. 26

G

gallbladder Organ attached to the liver that serves to store and concentrate bile. 91

gamete (ga-meet, guh-meet) Haploid sex cell; the egg or a sperm which join in fertilization to form a zygote. 333, 386

ganglion Collection or bundle of neuron cell bodies usually outside the central nervous system. 260

gap analysis Use of computers to discover places where biodiversity is high outside of preserved areas. 509

gap junction Region between cells formed by the joining of two adjacent plasma membranes; it lends strength and allows ions, sugars, and small molecules to pass between cells.   64

gastric gland Gland within the stomach wall that secretes gastric juice.   86

gene (jeen) Unit of heredity existing as alleles on the chromosomes; typically two alleles are inherited for each trait—one from each parent.   422

gene therapy Correction of a detrimental mutation by the addition of normal DNA and its insertion in a genome.   438

genetic engineering Alteration of DNA for medical or industrial purposes.   432

genital herpes (jen-ih-tul hur-peez) Sexually transmitted disease caused by herpes simplex virus and sometimes accompanied by painful ulcers on the genitals.   343

genital warts Sexually transmitted disease caused by human papillomavirus resulting in raised growths on the external genitals.   342

genomic imprinting (jih-noh-mik, jee-nahm-ik) Inheritance pattern that differs depending upon the sex of the parent passing it on.   408

genotype (jee-nuh-typ) Genes of an individual for a particular trait or traits; often designated by letters, for example, *BB* or *Aa*.   404

gerontology (jer-un-tahl-uh-jee) Study of aging.   378

gland Epithelial cell or group of epithelial cells that are specialized to secrete a substance.   62

glaucoma (glow-koh-muh, glaw-koh-muh) Increasing loss of field of vision, caused by blockage of the ducts that drain the aqueous humor, creating pressure buildup and nerve damage.   278

global warming Predicted increase in the earth's temperature, due to human activities that promote the greenhouse effect.   486, 506

glomerular capsule (gluh-mair-yuh-lur) Double-walled cup that surrounds the glomerulus at the beginning of the nephron.   193

glomerular filtrate Filtered portion of blood contained within the glomerular capsule.   195

glomerular filtration Movement of small molecules from the glomerulus into the glomerular capsule due to the action of blood pressure.   195

glomerulus (gluh-mair-uh-lus, gloh-mair-yuh-lus) Cluster; for example, the cluster of capillaries surrounded by the glomerular capsule in a nephron, where glomerular filtration takes place.   192

glottis (glaht-us) Opening for airflow in the larynx.   84, 168

glucagon (gloo-kuh-gahn) Hormone secreted by the pancreas which causes the liver to break down glycogen and raises the blood glucose level.   304

glucocorticoid (gloo-koh-kor-tih-koyd) Type of hormone secreted by the adrenal cortex that influences carbohydrate, fat, and protein metabolism; see cortisol.   301

glucose (gloo-kohs) Six-carbon sugar that organisms degrade as a source of energy during cellular respiration.   27

glycogen (gly-koh-jun) Storage polysaccharide that is composed of glucose molecules joined in a linear fashion but having numerous branches.   28

glycolysis Anaerobic breakdown of glucose that results in a gain of two ATP.   55

Golgi apparatus (gohl-jee) Organelle, consisting of saccules and vesicles, that processes, packages, and distributes molecules about or from the cell.   50

gonad (goh-nad) Organ that produces gametes; the ovary produces eggs, and the testis produces sperm.   306

gonadotropic hormone (goh-nad-uh-trahp-ic, -troh-pic) Chemical signal secreted by anterior pituitary that regulates the activity of the ovaries and testes; principally, follicle-stimulating hormone (FSH) and luteinizing hormone (LH).   296

gonadotropin-releasing hormone (GnRH) Hormone secreted by the hypothalamus that stimulates the anterior pituitary to secrete follicle-stimulating hormone and luteinizing hormone.   321

gonorrhea (gahn-nuh-ree-uh) Sexually transmitted disease caused by the bacterium *Neisseria gonorrhoeae* that can lead to pelvic inflammatory disease.   348

granular leukocyte (gran-yuh-lur loo-kuh-syt) White blood cell with prominent granules in the cytoplasm.   115

gravitational equilibrium Maintenance of balance when the head and body are motionless.   286

gray matter Nonmyelinated axons and cell bodies in the central nervous system.   252

grazing food chain Straight-line linking of organisms according to who eats whom which begins with a producer.   483

grazing food web Complex pattern of interlocking and crisscrossing food chains that begins with populations of autotrophs serving as producers.   483

greenhouse effect Reradiation of solar heat toward the earth because gases like carbon dioxide, methane, nitrous oxide, and water vapor allow solar energy to pass through toward the earth but block the escape of heat back into space.   486

greenhouse gases Gases which are involved in the greenhouse effect.   486

growth factor Chemical signal that regulates mitosis and differentiation of cells that have receptors for it; important in such processes as fetal development, tissue maintenance and repair, and hematopoiesis; sometimes a contributing factor in cancer.   307

growth hormone (GH) Substance secreted by the anterior pituitary; controls size of individual by promoting cell division, protein synthesis, and bone growth.   296

growth plate Cartilaginous layer within an epiphysis of a long bone that permits growth of bone to occur.   208

guanine (G) (gwah-neen) One of four nitrogen-containing bases in nucleotides composing the structure of DNA and RNA; pairs with cytosine.   423

H

habitat Place where an organism lives and is able to survive and reproduce.   480

hair cell Cell with stereocilia (long microvilli) that is sensitive to mechanical stimulation; mechanoreceptor for hearing and equilibrium in the inner ear.   284

hair follicle Tubelike depression in the skin in which a hair develops.   72

haploid (n) (hap-loyd) n number of chromosomes; half the diploid number; the number characteristic of gametes which contain only one set of chromosomes.   386

hard palate (pal-it) Bony, anterior portion of the roof of the mouth.   82

heart Muscular organ located in the thoracic cavity whose rhythmic contractions maintain blood circulation.   128

heart attack Damage to the myocardium due to blocked circulation in the coronary arteries; myocardial infarction (MI).   137

heartburn Burning pain in the chest that occurs when part of the stomach contents escapes into the esophagus.   85

helper T cell T lymphocyte which secretes cytokines that stimulate all kinds of immune system cells.   155

hemodialysis (he-moh-dy-AL-uh-sus) Cleansing of blood by use of an artificial membrane that causes substances to diffuse from blood into a dialysis fluid.   200

hemoglobin (hee-muh-gloh-bun) Iron-containing pigment in red blood cells that combines with and transports oxygen.   111, 174

hemolysis (he-MAHL-uh-sus) Rupture of red blood cells accompanied by the release of hemoglobin.   113

hemophilia (he-moh-FIL-ee-uh) Genetic disorder in which the affected individual is subject to uncontrollable bleeding.   116

hemorrhoids (hem-uh-royds, hem-royds) Abnormally dilated blood vessels of the rectum.   139

hepatic portal system (hih-pat-ik) Portal system that begins at capillaries servicing the small intestine and ends at capillaries in the liver.   135

hepatic portal vein Vein leading to the liver and formed by the merging blood vessels leaving the small intestine.   135

hepatic vein Vein that runs between the liver and the inferior vena cava.   135

hepatitis (hep-uh-ty-tis) Inflammation of the liver. Viral hepatitis occurs in several forms.   91, 344

herbivore ((h)ur-buh-vor) Primary consumer in a grazing food chain; a plant eater.   480

heterotroph Organism that cannot synthesize organic molecules from inorganic nutrients and therefore must take in organic nutrients (food).   462

heterozygous Possessing unlike alleles for a particular trait.   404

hexose Six-carbon sugar.   27

hinge joint Type of joint that allows movement as a hinge does, such as the movement of the knee.   218

hippocampus (hip-uh-kam-pus) Portion of the limbic system where memories are stored.   258

histamine (his-tuh-meen, -mun) Substance, produced by basophils in blood and mast cells in connective tissue, that causes capillaries to dilate.   148

HLA (human leukocyte associated) antigen Plasma membrane protein that identifies the cell as belonging to a particular individual and acts as an antigen in other individuals.   154

homeostasis (hoh-mee-oh-stay-sis) Maintenance of normal internal conditions in a cell or an organism by means of self-regulating mechanisms.   2, 70

hominid (hahm-uh-nid) Member of the family Hominid which contains australopithecines and humans.   467

Homo erectus (hoh-moh ih-rek-tus) Hominid who used fire and migrated out of Africa to Europe and Asia.   470

Homo habilis (hoh-moh hab-uh-lus) Hominid of 2 million years ago who is believed to have been the first tool user.   469

Homo sapiens (hoh-moh say-pe-nz) Modern humans.   471

homologous chromosome (hoh-mahl-uh-gus, huh-mahl-uh-gus) Member of a pair of chromosomes that are alike and come together in synapsis during prophase of the first meiotic division.   390

homologous structure Structure that is similar in two or more species because of common ancestry.   464

homozygous dominant Possessing two identical alleles, such as *AA*, for a particular trait.   404

homozygous recessive Possessing two identical alleles, such as *aa*, for a particular trait.   404

hormone (hor-mohn) Chemical signal produced in one part of the body that controls the activity of other parts.   88, 294

host Organism that provides nourishment and/or shelter for a parasite.   340

human chorionic gonadotropin (HCG) (kor-ee-ahn-ik, goh-nad-uh-trahp-in, -troh-pin) Hormone produced by the chorion that functions to maintain the uterine lining.   328

hyaline cartilage (hy-uh-lin) Cartilage whose cells lie in lacunae separated by a white translucent matrix containing very fine collagen fibers.   64

hydrogen bond Weak bond that arises between a slightly positive hydrogen atom of one molecule and a slightly negative atom of another molecule or between parts of the same molecule.   21

hydrolysis (hy-drahl-ih-sis) Splitting of a compound by the addition of water, with the H^+ being incorporated in one fragment and the OH^- in the other.   27

hydrolytic enzyme (hy-druh-lit-ik) Enzyme that catalyzes a reaction in which the substrate is broken down by the addition of water.   92

hydrophilic (hy-druh-fil-ik) Type of molecule that interacts with water by dissolving in water and/or forming hydrogen bonds with water molecules.   21

hydrophobic (hy-druh-foh-bik) Type of molecule that does not interact with water because it is nonpolar.   21

hypertension Elevated blood pressure, particularly the diastolic pressure.   137

hypertrophy (hy-pur-truh-fee) Increase in muscle size following long-term exercise.   238

hypothalamic-inhibiting hormone (hy-poh-thuh-lah-mik) One of many hormones produced by the hypothalamus that inhibits the secretion of an anterior pituitary hormone.   296

hypothalamic-releasing hormone One of many hormones produced by the hypothalamus that stimulates the secretion of an anterior pituitary hormone.   296

hypothalamus (hy-poh-thal-uh-mus) Part of the brain located below the thalamus that helps regulate the internal environment of the body and produces releasing factors that control the anterior pituitary.   256, 296

hypothesis (hy-pahth-ih-sis) Supposition that is formulated after making an observation; it can be tested by obtaining more data, often by experimentation.   8

I

immediate allergic response Allergic response that occurs within seconds of contact with an allergen, caused by the attachment of the allergen to IgE antibodies.   158

immune system White blood cells and lymphoid organs which protect the body against foreign organisms and substances and also cancerous cells.   148

immunity Ability of the body to protect itself from foreign substances and cells, including disease-causing agents.   148

immunization (im-yuh-nuh-zay-shun) Use of a vaccine to protect the body against specific disease-causing agents.   156

immunoglobulin (Ig) (im-yuh-noh-glahb-yuh-lin, -yoo-lin) Globular plasma protein that functions as an antibody.   152

implantation Attachment and penetration of the embryo into the lining of the uterus (endometrium).   330

incomplete dominance Inheritance pattern in which the offspring has an intermediate phenotype compared to its parents; for example a normal individual and an individual with sickle-cell disease can produce a child with sickle-cell trait.   413

incus (ing-kus) Middle of three ossicles of the ear that serve to conduct vibrations from the tympanic membrane to the oval window of the inner ear.   284

induction Ability of a chemical or a tissue to influence the development of another tissue.   367

infant respiratory distress syndrome Condition in newborns, especially premature ones, in which the lungs collapse because of a lack of surfactant lining the alveoli.   169

inferior vena cava (vee-nuh kay-vuh) Large vein that enters the right atrium from below and carries blood from the trunk and lower extremities.   134

infertility Inability to have as many children as desired.   332

inflammatory reaction Tissue response to injury that is characterized by redness, swelling, pain, and heat.   148

inner ear Portion of the ear consisting of a vestibule, semicircular canals, and the cochlea where equilibrium is maintained and sound is transmitted.   284

inorganic molecule Type of molecule that is not an organic molecule; not derived from a living organism.   26

insertion End of a muscle that is attached to a movable bone.   227

inspiration (in-spuh-ray-shun) Act of taking air into the lungs; inhalation.   166

inspiratory reserve volume (in-spy-ruh-tohr-ee) Volume of air that can be forcibly inhaled after normal inhalation.   170

insulin (in-suh-lin) Hormone secreted by the pancreas that lowers the blood glucose level by promoting the uptake of glucose by cells and the conversion of glucose to glycogen by the liver and skeletal muscles.   304

integration Summing up of excitatory and inhibitory signals by a neuron or by some part of the brain.   251, 273

integumentary system (in-teg-yoo-men-tuh-ree, -men-tree) Organ system consisting of skin and various organs, such as hair, which are found in skin.   71

intercalated disks (in-tur-kuh-lay-tud) Region that holds adjacent cardiac muscle cells together and that appear as dense bands at right angles to the muscle striations.   67

interferon (in-tur-feer-ahn) Antiviral agent produced by an infected cell that blocks the infection of another cell.   150

interleukin (in-tur-loo-kun) Cytokine produced by macrophages and T lymphocytes that functions as a metabolic regulator of the immune response.   157

internal respiration Exchange of oxygen and carbon dioxide between blood and tissue fluid.   174

interneuron Neuron, located within the central nervous system, conveying messages between parts of the central nervous system.   246

interoceptor Sensory receptor that detects stimuli from inside the body (e.g., pressoreceptors, osmoreceptors, chemoreceptors).   272

interphase Cell cycle stage during which growth and DNA synthesis occur when the nucleus is not actively dividing.   387

interstitial cell (in-tur-stish-ul) Hormone-secreting cell located between the seminiferous tubules of the testes.   321

interstitial cell-stimulating hormone (ICSH) Name sometimes given to luteinizing hormone in males; controls the production of testosterone by interstitial cells.   321

intervertebral disk (in-tur-vur-tuh-brul) Layer of cartilage located between adjacent vertebrae.   214

intrauterine device (IUD) (in-truh-yoo-tur-in) Birth-control device consisting of a small piece of molded plastic inserted into the uterus, and believed to alter the uterine environment so that fertilization does not occur.   329

ion (EYE-un, -ahn) Charged particle that carries a negative or positive charge.   19

ionic bond (eye-ahn-ik) Chemical bond in which ions are attracted to one another by opposite charges.   19

iris (eye-ris) Muscular ring that surrounds the pupil and regulates the passage of light through this opening.   278

isotope (eye-suh-tohp) One of two or more atoms with the same atomic number but a different atomic mass due to the number of neutrons.   17

J

jaundice (jawn-dis) Yellowish tint to the skin caused by an abnormal amount of bilirubin (bile pigment) in the blood, indicating liver malfunction.   91

joint Articulation between two bones of a skeleton.   206

juxtaglomerular apparatus (juk-stuh-gluh-mer-yuh-lur) Structure located in the walls of arterioles near the glomerulus that regulates renal blood flow.   197

K

karyotype (kar-ee-uh-typ) Duplicated chromosomes arranged by pairs according to their size, shape, and general appearance.   395

keystone species Species whose activities have a significant role in determining community structure.   508

kidney Organ in the urinary system that produces and excretes urine.   188

kingdom One of the categories used to classify organisms; category above phylum.   2

kinin (ky-nen) Chemical mediator, released by damaged tissue cells and mast cells, which causes the capillaries to dilate and become more permeable.   148

Krebs cycle (krebz) Cycle of reactions in mitochondria that begins with citric acid; it breaks down an acetyl group as CO_2, ATP, NADH, and $FADH_2$ are given off; citric acid cycle.   55

L

lacteal (lak-tee-ul) Lymphatic vessel in an intestinal villus; it aids in the absorption of lipids.   87

lactose intolerance (LAK-tohs) Inability to digest lactose because of an enzyme deficiency.   92

lacuna (luh-koo-nuh, -kyoo-nuh) Small pit or hollow cavity, as in bone or cartilage, where a cell or cells are located.   64

landscape A number of interacting ecosystem fragments.   499

lanugo (luh-noo-goh) Short, fine hair that is present during the later portion of fetal development.   374

large intestine Last major portion of the digestive tract, extending from the small intestine to the anus and consisting of the cecum, the colon, the rectum, and the anal canal.   88

laryngitis (lar-un-jy-tis) Infection of the larynx with accompanying hoarseness.   177

larynx (lar-ingks) Cartilaginous organ located between pharynx and trachea that contains the vocal cords; voice box.   168

latency State of existing in a host without producing any signs of being present; some types of virus can go through a period of latency.   341

learning Relatively permanent change in behavior that results from practice and experience.   258

lens Clear, membranelike structure found in the eye behind the iris; brings objects into focus.   278

leptin Hormone produced by adipose tissue that acts on the hypothalamus to signal satiety.   307

leukemia (loo-kee-mee-uh) Cancer of the blood-forming tissues leading to the overproduction of abnormal white blood cells.   115, 446

ligament (lig-uh-munt) Tough cord or band of dense fibrous connective tissue that joins bone to bone at a joint.   64, 206

limbic system Association of various brain centers including the amygdala and hippocampus; governs learning and memory and various emotions such as pleasure, fear, and happiness. 257

lineage Evolutionary line of descent.   467

lipase (ly-pays, ly-payz) Fat-digesting enzyme secreted by the pancreas. 90, 92

lipid (lip-id, ly-pid) Class of organic compounds that tends to be soluble only in nonpolar solvents such as alcohol; includes fats and oils.   29

liver Large dark red internal organ that produces urea and bile, detoxifies the blood, stores glycogen, and produces the plasma proteins among other functions.   90

long-term memory Retention of information that lasts longer than a few minutes.   258

long-term potentiation (LTP) (puh-ten-shee-ay-shun) Enhanced response at synapses within the hippocampus, likely essential to memory storage. 258

loop of the nephron (nef-rahn) Portion of the nephron lying between the proximal convoluted tubule and the distal convoluted tubule that functions in water reabsorption.   193

loose fibrous connective tissue Tissue composed mainly of fibroblasts widely separated by a matrix containing collagen and elastic fibers. 64

lumen (loo-mun) Cavity inside any tubular structure, such as the lumen of the digestive tract.   85

lung cancer Malignant growth that often begins in the bronchi.   181

lungs Paired, cone-shaped organs within the thoracic cavity, functioning in internal respiration and containing moist surfaces for gas exchange. 169

luteinizing hormone (LH) Hormone, which in males, controls the production of testosterone by interstitial cells, and in females, promotes the development of the corpus luteum.   321

lymph (limf) Fluid, derived from tissue fluid, that is carried in lymphatic vessels.   119, 146

lymph node Mass of lymphoid tissue located along the course of a lymphatic vessel.   147

lymphatic system (lim-fat-ik) Organ system consisting of lymphatic vessels and lymphoid organs that transports lymph, lipids, and aids the immune system.   70, 146

lymphatic vessel Vessel that carries lymph.   146

lymphocyte (lim-fuh-syt) Specialized white blood cell that functions in specific defense; occurs in two forms—T lymphocyte and B lymphocyte.   115

lymphoid organ Organ other than a lymphatic vessel that is part of the lymphatic system; includes lymph nodes, tonsils, spleen, thymus gland, and bone marrow.   147

lysosome (ly-suh-sohm) Membrane-bounded vesicle that contains hydrolytic enzymes for digesting macromolecules.   52

M

macrophage (mak-ruh-fayj) Large phagocytic cell derived from a monocyte that ingests microbes and debris.   148

malleus (mal-ee-us) First of three ossicles of the ear that serve to conduct vibrations from the tympanic membrane to the oval window of the inner ear.   284

maltase (mahl-tays, -tayz) Enzyme produced in small intestine that breaks down maltose to two glucose molecules.   92

mast cell Cell to which antibodies, formed in response to allergens, attach, causing it to release histamine, which cause symptoms. 148

mastoiditis (mas-toyd-eye-tis) Inflammation of the mastoid sinuses of the skull.   212

matrix (may-triks) Unstructured semifluid substance that fills the space between cells in connective tissues or inside organelles.   64

matter Anything that takes up space and has mass.   16

mechanoreceptor (mek-uh-noh-rih-sep-tur) Sensory receptor that responds to mechanical stimuli, such as that from pressure, sound waves, and gravity.   272

medulla oblongata (muh-dul-uh ahb-lawng-gah-tuh) Part of the brain stem that is continuous with the spinal cord; controls heartbeat, blood pressure, breathing, and other vital functions.   256

medullary cavity (muh-DUL-uh-ree) Cavity within the diaphysis of a long bone containing marrow.   206

megakaryocyte (meg-uh-kar-ee-oh-syt, -uh-syt) Large cell that gives rise to blood platelets.   116

meiosis (my-oh-sis) Type of nuclear division that occurs as part of sexual reproduction in which the daughter cells receive the haploid number of chromosomes in varied combinations.   390

melanocyte Melanin-producing cell found in skin.   72

melanocyte-stimulating hormone (MSH) Substance that causes melanocytes to secrete melanin in lower vertebrates. 296

melatonin (mel-uh-toh-nun) Hormone, secreted by the pineal gland, that is involved in biorhythms.   307

memory Capacity of the brain to store and retrieve information about past sensations and perceptions; essential to learning.   258

meninges (sing., meninx) (muh-nin-jeez) Protective membranous coverings about the central nervous system. 69, 252

menisci (sing., meniscus) (muh-nis-ky, -kee, -sy) Cartilaginous wedges that separate the surfaces of bones in synovial joints.   218

menopause (men-uh-pawz) Termination of the ovarian and uterine cycles in older women.   328

menstruation (men-stroo-ay-shun) Loss of blood and tissue from the uterus at the end of a uterine cycle.   326

mesoderm (mez-uh-durm, MES-) Middle germ layer of embryonic gastrula that becomes muscular and connective tissue.   366

messenger RNA (mRNA) Type of RNA formed from a DNA template that bears coded information for the amino acid sequence of a polypeptide.   425

metabolism All of the chemical reactions that occur in a cell.   54

metaphase Mitotic phase during which chromosomes are aligned at the equator of the mitotic spindle.   389

metapopulation Population subdivided into several small and isolated populations due to habitat fragmentation.   508

metastasis (muh-tas-tuh-sis) Spread of cancer from the place of origin throughout the body; caused by the ability of cancer cells to migrate and invade tissues.   444

microtubule (-too-byool) Small cylindrical structure that contains thirteen rows of the protein tubulin about an empty central core; present in the cytoplasm, centrioles, cilia, and flagella.   53

microvillus (-vil-us) Cylindrical process that extends from some epithelial cell plasma membranes; serves to increase the surface area of the plasma membrane.   62

midbrain Part of the brain located below the thalamus and above the pons; contains reflex centers and tracts. 257

middle ear Portion of the ear consisting of the tympanic membrane, the oval and round windows, and the ossicles; where sound is amplified. 284

mineral Naturally occurring inorganic substance containing two or more elements; certain minerals are needed in the diet.   102

mineralocorticoid (min-ur-uh-loh-kor-tih-koyd) Type of hormone secreted by the adrenal cortex that regulates salt and water balance, leading to increases in blood volume and blood pressure.   301

mitochondrion (my-tuh-kahn-dree-un) Membrane-bounded organelle in which ATP molecules are produced during the process of cellular respiration.   52

mitosis (my-toh-sis) Type of cell division in which daughter cells receive the exact chromosomal and genetic makeup of the parent cell; occurs during growth and repair.   387, 391

molecular clock Mutational changes that accumulate at a presumed constant rate in regions of DNA not involved in adaptation to the environment.   467

molecule Union of two or more atoms of the same element; also the smallest part of a compound that retains the properties of the compound.   19

monoclonal antibody One of many antibodies produced by a clone of hybridoma cells which all bind to the same antigen.   158

monocyte (mahn-uh-syt) Type of agranular white blood cell that functions as a phagocyte and an antigen-presenting cell.   115

monosaccharide (mahn-uh-sak-uh-ryd) Simple sugar; a carbohydrate that cannot be decomposed by hydrolysis; e.g., glucose.   27

mosaic evolution Concept that human characteristics did not evolve at the same rate; e.g., some body parts are more humanlike than others in early hominids.   468

motor neuron Nerve cell that conducts nerve impulses away from the central nervous system and innervates effectors (muscle and glands).   246

motor unit Motor neuron and all the muscle fibers it innervates.   234

mucous membrane (myoo-kus) Membrane that lines a cavity or tube that opens to the outside of the body; mucosa.   69

multiple alleles (uh-leelz) Inheritance pattern in which there are more than two alleles for a particular trait; each individual has only two of all possible alleles.   412

multiregional continuity hypothesis Proposal that modern humans evolved independently in at least three different places: Asia, Africa, and Europe.   471

muscle fiber Muscle cell.   231

muscle twitch Contraction of a whole muscle in response to a single stimulus.   234

muscular (contractile) tissue Type of tissue composed of fibers that can shorten and thicken.   67

mutagen (myoo-tuh-jun) Agent, such as radiation or a chemical, that brings about a mutation.   448

mutation Alteration in chromosome structure or number and also an alteration in a gene due to a change in DNA composition.   424

myelin sheath (my-uh-lin) White, fatty material—derived from the membrane of Schwann cells that forms a covering for nerve fibers. 247

myocardium (MY-oh-kar-dee-um) Cardiac muscle in the wall of the heart.   128

myofibril (my-uh-fy-brul) Contractile portion of muscle cells which contain a linear arrangement of sarcomeres and shorten to produce muscle contraction.   231

myoglobin (my-uh-gloh-bin) Pigmented compound in muscle tissue that stores oxygen.   238

myogram Recording of a muscular contraction.   234

myosin (my-uh-sin) One of two major proteins of muscle; makes up thick filaments in myofibrils of muscle fibers. See actin.   231

myxedema (mik-sih-dee-muh) Condition resulting from a deficiency of thyroid hormone in an adult.   299

N

nasal cavity One of two canals in the nose separated by a septum.   167

nasopharynx (nay-zoh-far-ingks) Region of the pharynx associated with the nasal cavity.   84

natural killer (NK) cell Lymphocyte that causes an infected or cancerous cell to burst.   150

natural selection Mechanism resulting in adaptation to the environment.   464

Neanderthal (nee-an-dur-thahl, -tahl) Hominid with a sturdy build who lived during the last Ice Age in Europe and the Middle East; hunted large game and has left evidence of being culturally advanced.   472

negative feedback Mechanism of homeostatic response in which a stimulus initiates reactions that reduce the stimulus.   74

nephron (nef-rahn) Microscopic kidney unit that regulates blood composition by glomerular filtration, tubular reabsorption, and tubular secretion. 191

nerve Bundle of long axons outside the central nervous system.   68, 260

nerve impulse Action potential (electrochemical change) traveling along a neuron.   248

nervous system Organ system consisting of the brain, spinal cord, and associated nerves that coordinates the other organ systems of the body. 70

nervous tissue Tissue that contains nerve cells (neurons), which conduct

impulses, and neuroglial cells, which support, protect, and provide nutrients to neurons.   68

neuroglia (noo-rahg-lee-uh, noo-rohg-lee-uh) Nonconducting nerve cells that are intimately associated with neurons and function in a supportive capacity.   68, 246

neuromuscular junction Region where an axon bulb approaches a muscle fiber; contains a presynaptic membrane, a synaptic cleft, and a postsynaptic membrane.   232

neuron (noor-ahn, nyoor-) Nerve cell that characteristically has three parts: dendrites, cell body, and axon. 68, 246

neurotransmitter Chemical stored at the ends of axons that is responsible for transmission across a synapse.   251

neutron (noo-trahn) Neutral subatomic particle, located in the nucleus and having a weight of approximately one atomic mass unit.   16

neutrophil (noo-truh-fil) Granular leukocyte that is the most abundant of the white blood cells; first to respond to infection.   115

niche (nich) Role an organism plays in its community, including its habitat and its interactions with other organisms.   480

nitrification Process by which nitrogen in ammonia and organic molecules is oxidized to nitrites and nitrates by soil bacteria.   488

nitrogen cycle Continuous process by which nitrogen circulates in the air, soil, water, and organisms of the biosphere.   488

nitrogen fixation Process whereby free atmospheric nitrogen is converted into compounds, such as ammonium and nitrates, usually by bacteria.   488

node of Ranvier (rahn-vee-ay) Gap in the myelin sheath around a nerve fiber.   247

nondisjunction Failure of homologous chromosomes or daughter chromosomes to separate during meiosis I and meiosis II, respectively. 395

norepinephrine (NE) (nor-ep-uh-nef-rin) Neurotransmitter of the postganglionic fibers in the sympathetic division of the autonomic system; also a hormone produced by the adrenal medulla. 251, 301

nuclear envelope Double membrane that surrounds the nucleus and is connected to the endoplasmic reticulum; has pores that allow substances to pass between the nucleus and the cytoplasm.   49

nuclear pore Opening in the nuclear envelope which permits the passage of proteins into the nucleus and ribosomal subunits out of the nucleus.   49

nuclei (noo-klee-eye, nyoo-) Masses of cell bodies in the CNS.   257

nucleolus (noo-klee-us, nyoo-) Dark-staining, spherical body in the nucleus that produces ribosomal subunits.   49

nucleoplasm (noo-klee-uh-plaz-um) Semifluid medium of the nucleus, containing chromatin.   49

nucleotide Monomer of DNA and RNA consisting of a five-carbon sugar bonded to a nitrogen-containing base and a phosphate group.   35

nucleus (noo-klee-us, nyoo-) Membrane-bounded organelle that contains chromosomes and controls the structure and function of the cell.   44

O

obesity (oh-bee-sih-tee) Excess adipose tissue; exceeding ideal weight by more than 20%.   104

oil Substance, usually of plant origin and liquid at room temperature, formed when a glycerol molecule reacts with three fatty acid molecules.   29

oil gland Gland of the skin, associated with hair follicle, that secretes sebum; sebaceous gland.   72

olfactory cell (ahl-fak-tuh-ree, -tree, ohl-) Modified neuron that is a sensory receptor for the sense of smell.   277

omnivore (ahm-nuh-vor) Organism in a food chain that feeds on both plants and animals.   480

oncogene (ahng-koh-jeen) Cancer-causing gene.   446

oogenesis (oh-uh-jen-uh-sis) Production of an egg in females by the process of meiosis and maturation.   322, 394

opportunistic infection Infection that has an opportunity to occur because the immune system has been weakened.   342

optic nerve Either of two cranial nerves that carry nerve impulses from the retina of the eye to the brain, thereby contributing to the sense of sight.   279

organ Combination of two or more different tissues performing a common function.   2

organelle (or-guh-nel) Small membranous structure in the cytoplasm having a specific structure and function.   44

organ system Group of related organs working together.   2

organic molecule Type of molecule that contains carbon and hydrogen; it often contains oxygen also.   26

origin End of a muscle that is attached to a relatively immovable bone.   227

oscilloscope (ah-sil-uh-skohp, uh-sil-) Apparatus that records changes in voltage by graphing them on a screen; used to study the nerve impulse.   248

osmosis (ahz-moh-sis, ahs-) Diffusion of water through a selectively permeable membrane.   47

ossicle (ahs-ih-kul) One of the small bones of the middle ear—malleus, incus, stapes.   284

ossification Formation of bone tissue.   208

osteoblast Bone-forming cell.   208

osteoclast (ahs-tee-uh-klast) Cell that causes erosion of bone.   208

osteocyte (ahs-tee-uh-syt) Mature bone cell located within the lacunae of bone.   206

osteoporosis Condition in which bones break easily because calcium is removed from them faster than it is replaced.   102

otitis media (oh-ty-tis mee-dee-uh) Bacterial infection of the middle ear, characterized by pain and possibly by a sense of fullness, hearing loss, vertigo, and fever.   177

otolith (oh-tuh-lith) Calcium carbonate granule associated with ciliated cells in the utricle and the saccule.   287

outer ear Portion of ear consisting of the pinna and auditory canal.   284

out-of-Africa hypothesis Proposal that modern humans originated only in Africa; then they migrated out of Africa and supplanted populations of early *Homo* in Asia and Europe about 100,000 years ago.   471

oval window Membrane-covered opening between the stapes and the inner ear.   284

ovarian cycle (oh-vair-ee-un) Monthly follicle changes occurring in the ovary that control the level of sex hormones in the blood and the uterine cycle.   325

ovary Female gonad that produces eggs and the female sex hormones.   306, 322

oviduct (oh-vuh-dukt) Tube that transports eggs to the uterus; uterine tube.   322

ovulation (ahv-yuh-lay-shun, ohv-) Release of a secondary oocyte from the ovary; if fertilization occurs the secondary oocyte becomes an egg.   322

oxygen debt Amount of oxygen needed to metabolize lactate, a compound that accumulates during vigorous exercise.   236

oxyhemoglobin (ahk-see-hee-muh-gloh-bin) Compound formed when oxygen combines with hemoglobin.   174

oxytocin (ahk-sih-toh-sin) Hormone released by the posterior pituitary that causes contraction of uterus and milk letdown.   296

ozone shield Accumulation of O_3, formed from oxygen in the upper atmosphere; a filtering layer that protects the earth from ultraviolet radiation.   492

P

pacemaker See SA (sinoatrial) node.   130

pain receptor Sensory receptor that is sensitive to chemicals released by damaged tissues or excess stimuli of heat or pressure.   272

PAN (peroxyacetylnitrate) (puh-rahk-see-uh-see-tul) Type of noxious chemical found in photochemical smog.   489

pancreas (pang-kree-us, PAN-) Internal organ that produces digestive enzymes and the hormones insulin and glucagon.   90, 304

pancreatic amylase (pang-kree-at-ik am-uh-lays, -layz) Enzyme in the pancreas that digests starch to maltose.   90, 92

pancreatic islets (islets of Langerhans) Masses of cells that constitute the endocrine portion of the pancreas.   304

Pap test Analysis done on cervical cells for detection of cancer.   323

parasympathetic division That part of the autonomic system that is active under normal conditions; uses acetylcholine as a neurotransmitter.   263

parathyroid gland (par-uh-thy-royd) Gland embedded in the posterior surface of the thyroid gland; it produces parathyroid hormone.   300

parathyroid hormone (PTH) Hormone secreted by the four parathyroid glands that increases the blood calcium level and decreases the blood phosphate level.   300

Parkinson disease Progressive deterioration of the central nervous system due to a deficiency in the neurotransmitter dopamine.   266

parturition (par-tyoo-rish-un, par-chuh-) Processes that lead to and include birth and the expulsion of the afterbirth.   376

pathogen (path-uh-jun) Disease-causing agent.   62, 111, 340

pectoral girdle (pek-tur-ul) Portion of the skeleton that provides support and attachment for an arm; consists of a scapula and a clavicle.   216

pelvic girdle Portion of the skeleton to which the legs are attached; consists of the coxal bones.   217

pelvic inflammatory disease (PID) Disease state of the reproductive organs caused by a sexually transmitted disease that can result in scarring and infertility.   347

penis External organ in males through which the urethra passes and that serves as the organ of sexual intercourse.   319

pentose (pen-tohs, -tohz) Five-carbon sugar; deoxyribose is the pentose sugar found in DNA; ribose is a pentose sugar found in RNA.   27

pepsin (pep-sin) Enzyme secreted by gastric glands that digests proteins to peptides.   86, 92

peptidase (pep-tih-days, -dayz) Intestinal enzyme that breaks down short chains of amino acids to individual amino acids that are absorbed across the intestinal wall.   92

peptide bond Type of covalent bond that joins two amino acids.   32

peptide hormone Type of hormone that is a protein, a peptide, or is derived from an amino acid.   309

perception Mental awareness of sensory stimulation.   273

perforin (pur-for-in) Molecule secreted by a cytotoxic T cell that perforates the plasma membrane of a target cell so that water and salts enter, causing the cell to swell and burst.   155

pericardium (paier-ih-kar-dee-um) Protective serous membrane that surrounds the heart.   128

periosteum (paier-ee-ahs-tee-um) Fibrous connective tissue covering the surface of bone.   207

peripheral nervous system (PNS) (puh-rif-ur-ul) Nerves and ganglia that lie outside the central nervous system.   246

peristalsis (paier-ih-stawl-sis) Wavelike contractions that propel substances along a tubular structure like the esophagus.   84

peritonitis (paier-ih-tuh-ny-tis) Generalized infection of the lining of the abdominal cavity.   69

peritubular capillary network (paier-ih-too-byuh-lur) Capillary network that surrounds a nephron and functions in reabsorption during urine formation.   192

pH scale Measurement scale for the hydrogen ion concentration.   24

phagocytosis (fag-uh-sy-toh-sis) Process by which amoeboid-type cells engulf large substances, forming an intracellular vacuole.   115

pharynx (far-ingks) Portion of the digestive tract between the mouth and the esophagus which serves as a passageway for food and also air on its way to the trachea.   84, 167

phenotype (fee-nuh-typ) Visible expression of a genotype—for example, brown eyes or attached earlobes.   404

phlebitis (flih-by-tis) Inflammation of a vein.   139

phospholipid (fahs-foh-lip-id) Molecule that forms the bilayer of the cell's membranes; has a polar, hydrophilic head bonded to two nonpolar, hydrophobic tails.   30

phosphorus cycle Continuous process by which phosphorus circulates in the soil, water, and organisms of the biosphere.   490

photochemical smog Air pollution that contains nitrogen oxides and hydrocarbons which react to produce ozone and PAN (peroxyacetyl-nitrate).   489

photoreceptor Sensory receptor in retina that responds to light stimuli.   272

pineal gland (pin-ee-ul, py-nee-ul) Endocrine gland located in the third ventricle of the brain that produces melatonin.   307

pituitary dwarfism (pih-too-ih-taier-ee, -tyoo-) Condition in which affected individual has normal proportions but small stature, caused by inadequate growth hormone.   298

pituitary gland Endocrine gland that lies just inferior to the hypothalamus; consists of the anterior pituitary and posterior pituitary.   296

placenta (pluh-sen-tuh) Structure that forms from the chorion and the uterine wall and allows the embryo, and then the fetus, to acquire nutrients and rid itself of wastes.   328, 368

plaque (plak) Accumulation of soft masses of fatty material, particularly cholesterol, beneath the inner linings of the arteries.   98, 137

plasma (plaz-muh) Liquid portion of blood; contains nutrients, wastes, salts, and proteins.   66, 118

plasma cell Cell derived from a B lymphocyte that is specialized to mass-produce antibodies.   151

plasma membrane Membrane surrounding the cytoplasm that consists of a phospholipid bilayer with embedded proteins; functions to regulate the entrance and exit of molecules from the cell.   44

plasmid (plaz-mid) Self-replicating ring of accessory DNA in the cytoplasm of bacteria.   432

platelet (playt-lit) Component of blood that is necessary to blood clotting; thrombocyte.   66, 116

pleural membrane (ploor-ul) Serous membrane that encloses the lungs.   69, 172

pneumonectomy (noo-muh-nek-tuh-mee, nyoo-) Surgical removal of all or part of a lung.   181

pneumonia (noo-mohn-yuh, nyoo-) Infection of the lungs that causes alveoli to fill with mucus and pus.   179

polar body In oogenesis, a nonfunctional product; two to three meiotic products are of this type.   394

pollution Any environmental change that adversely affects the lives and health of living things.   505

polygenic inheritance (pahl-ee-jen-ik) Inheritance pattern in which a trait is controlled by several allelic pairs; each dominant allele contributes to the phenotype in an additive and like manner.   411

polymerase chain reaction (PCR) (pahl-uh-muh-rays, -rayz) Technique that uses the enzyme DNA polymerase to produce millions of copies of a particular piece of DNA.   433

polyp (pahl-ip) Small, abnormal growth that arises from the epithelial lining.   89

polypeptide Polymer of many amino acids linked by peptide bonds.   32

polyribosome (pahl-ih-ry-buh-sohm) String of ribosomes simultaneously translating regions of the same mRNA strand during protein synthesis.   50, 428

polysaccharide (pahl-ee-sak-uh-ryd) Polymer made from sugar monomers; the polysaccharides starch and glycogen are polymers of glucose monomers.   28

pons (pahnz) Portion of the brain stem above the medulla oblongata and below the midbrain; assists the medulla oblongata in regulating the breathing rate.   257

population Organisms of the same species occupying a certain area.   4, 478

population viability analysis Calculation of the minimum population size needed to prevent extinction.   509

positive feedback Mechanism in which the stimulus initiates reactions that lead to an increase in the stimulus.   76, 296

posterior pituitary Portion of the pituitary gland that stores and secretes oxytocin and antidiuretic hormone, which are produced by the hypothalamus.   296

precipitation Water deposited on the earth in the form of rain, snow, sleet, hail, or fog.   485

prefrontal area Association area in the frontal lobe that receives information from other association areas and uses it to reason and plan actions.   256

primary motor area Area in the frontal lobe where voluntary commands begin; each section controls a part of the body.   255

primary somatosensory area (soh-mat-uh-sens-ree, -suh-ree) Area dorsal to the central sulcus where sensory information arrives from skin and skeletal muscles.   255

primate (pry-mayt) Animal that belongs to the order Primate; includes prosimians, monkeys, apes, and humans all of whom have adaptations for living in trees.   466

principle Theory that is generally accepted by an overwhelming number of scientists; a law.   8

producer Photosynthetic organism at the start of a grazing food chain that makes its own food (e.g., green plants on land and algae in water).   480

product Substance that forms as a result of a reaction.   54

progesterone (proh-jes-tuh-rohn) Female sex hormone that helps maintain sexual organs and secondary sexual characteristics.   306, 326

prokaryotic cell Type of cell that lacks a membrane-bounded nucleus and organelles.   463

prolactin (PRL) (proh-lak-tin) Hormone secreted by the anterior pituitary that stimulates the production of milk from the mammary glands.   296

prophase (proh-fayz) Mitotic phase during which chromatin condenses so that chromosomes appear; chromosomes are scattered.   388

proprioceptor (proh-pree-oh-sep-tur) Sensory receptor in skeletal muscles and joints that assists the brain in knowing the position of the limbs.   274

prostaglandin (PG) (prahs-tuh-glan-din) Hormone that has various and powerful local effects.   307

prostate gland (prahs-tayt) Gland located around the male urethra below the urinary bladder; adds secretions to semen.   318

protein Molecule consisting of one or more polypeptides.   31

protein-first hypothesis In chemical evolution, the proposal that protein originated before other macromolecules and allowed the formation of protocells.   462

prothrombin (proh-thrahm-bin) Plasma protein that is converted to thrombin during the steps of blood clotting.   116

prothrombin activator Enzyme that catalyzes the transformation of the precursor prothrombin to the active enzyme thrombin.   116

protocell In biological evolution, a possible cell forerunner that became a cell once it could reproduce.   462

proton Positive subatomic particle, located in the nucleus and having a weight of approximately one atomic mass unit.   16

proto-oncogene (proh-toh-ahng-koh-jeen) Normal gene that can become an oncogene through mutation.   446

proximal convoluted tubule Highly coiled region of a nephron near the glomerular capsule, where tubular reabsorption takes place.   193

pulmonary artery (pool-muh-naier-ee, puul-) Blood vessel that takes blood away from the heart to the lungs.   129

pulmonary circuit Circulatory pathway that consists of the pulmonary trunk, the pulmonary arteries, and the pulmonary veins; takes O_2-poor blood from the heart to the lungs and O_2-rich blood from the lungs to the heart.   134

pulmonary fibrosis (fy-broh-sis) Accumulation of fibrous connective tissue in the lungs; caused by inhaling irritating particles, such as silica, coal dust, or asbestos.   179

pulmonary tuberculosis Tuberculosis of the lungs, caused by the tubercle bacillus.   179

pulmonary vein Blood vessel that takes blood from the lungs to the heart.   129

pulse Vibration felt in arterial walls due to expansion of the aorta following ventricle contraction.   132

Punnett square (pun-ut) Gridlike device used to calculate the expected results of simple genetic crosses.   406

pupil (pyoo-pul) Opening in the center of the iris of the eye.   278

Purkinje fibers (pur-kin-jee) Specialized muscle fibers that conduct the cardiac impulse from AV bundle into the ventricles.   130

R

reactant (re-ak-tunt) Substance that participates in a reaction.   54

recessive allele (uh-leel) Allele that exerts its phenotypic effect only in the homozygote; its expression is masked by a dominant allele.   404

recombinant DNA (rDNA) DNA that contains genes from more than one source.   432

red blood cell (erythrocyte) Formed element that contains hemoglobin and carries oxygen from the lungs to the tissues; erythrocyte.   66, 111

red bone marrow Blood cell-forming tissue located in the spaces within spongy bone.   148, 206

reduced hemoglobin (hee-muh-gloh-bun) Hemoglobin that is carrying hydrogen ions.   174

referred pain Pain perceived as having come from a site other than that of its actual origin.   275

reflex action Automatic, involuntary response of an organism to a stimulus.   84

reflex Automatic, involuntary response of an organism to a stimulus.   261

refractory period (rih-frak-tuh-ree) Time following an action potential when a neuron is unable to conduct another nerve impulse.   248

renal artery (ree-nul) Vessel that originates from the aorta and delivers blood to the kidney.   188

renal cortex (ree-nul kor-teks) Outer portion of the kidney that appears granular.   191

renal medulla (ree-nul muh-dul-uh) Inner portion of the kidney that consists of renal pyramids.   191

renal pelvis Hollow chamber in the kidney that lies inside the renal medulla and receives freshly prepared urine from the collecting ducts.   191

renal vein (ree-nul) Vessel that takes blood from the kidney to the inferior vena cava.   188

renin (ren-in) Enzyme released by kidneys that leads to the secretion of aldosterone and a rise in blood pressure.   197, 302

replication Making an exact copy, as when one copy of DNA is duplicated by a complementary base pairing process.   423

reproduce To produce a new individual of the same kind.   2

reproductive system Organ system that contains male or female organs and specializes in the production of offspring.   70

residual volume Amount of air remaining in the lungs after a forceful expiration.   170

respiratory center Group of nerve cells in the medulla oblongata that send out nerve impulses on a rhythmic basis, resulting in involuntary inspiration on an ongoing basis.   172

respiratory system Organ system consisting of the lungs and tubes that bring oxygen into the lungs and take carbon dioxide out.   70

resting potential Polarity across the plasma membrane of a resting neuron due to an unequal distribution of ions.   248

restoration ecology Subdiscipline of conservation biology that seeks ways to return ecosystems to their former state.   510

restriction enzyme Bacterial enzyme that stops viral reproduction by cleaving viral DNA; used to cut DNA at specific points during production of recombinant DNA.   432

reticular activating system (rih-tik-yuh-lur) Complex network of nerve fibers within the central nervous system that arouses the cerebrum.   257

reticular fiber (rih-tik-yuh-lur) Very thin collagen fibers in the matrix of connective tissue, highly branched and forming delicate supporting networks.   64

retina (ret-n-uh, ret-nuh) Innermost layer of the eyeball, which contains the rod cells and the cone cells.   279

retinal (ret-n-al, -awl) Light-absorbing molecule, which is a derivative of vitamin A and a component of rhodopsin. 280

rhodopsin (roh-DAHP-sun) Light-absorbing molecule in rod cells and cone cells that contains a pigment and the protein opsin. 280

rib cage Top and sides of the thoracic cavity; contains ribs and intercostal muscles. 172

ribosomal RNA (rRNA) (ry-buh-soh-mul) Type of RNA found in ribosomes where protein synthesis occurs. 425

ribosome (RY-buh-sohm) RNA and protein in two subunits; site of protein synthesis in the cytoplasm. 50

RNA (ribonucleic acid) (ry-boh-noo-klee-ik) Nucleic acid produced from covalent bonding of nucleotide monomers that contain the sugar ribose; occurs in three forms: messenger RNA, ribosomal RNA, and transfer RNA. 35, 425

RNA polymerase (pahl-uh-muh-rays) During transcription, an enzyme that joins nucleotides complementary to a DNA template. 427

RNA-first hypothesis In chemical evolution, the proposal that RNA originated before other macromolecules and allowed the formation of the first cell(s). 462

rod cell Photoreceptor in retina of eyes that responds to dim light. 280

rotational equilibrium Maintenance of balance when the head and body are suddenly moved or rotated. 286

rotator cuff Tendons that encircle and help form a socket for the humerus and help to reinforce the shoulder joint. 216

round window Membrane-covered opening between the inner ear and the middle ear. 284

S

SA (sinoatrial) node (sy-noh-ay-tree-ul) Small region of neuromuscular tissue that initiates the heartbeat; pacemaker. 130

saccule (sak-yool) Saclike cavity in the vestibule of the inner ear; contains sensory receptors for gravitational equilibrium. 287

salivary amylase (sal-uh-vair-ee am-uh-lays, -layz) Secreted from the salivary glands; the first enzyme to act on starch. 83, 92

salivary gland Gland associated with the oral cavity that secretes saliva. 82

sarcolemma (sar-kuh-lem-uh) Plasma membrane of a muscle fiber; also forms the tubules of the T system involved in muscular contraction. 231

sarcomere (sar-kuh-mir) One of many units, arranged linearly within a myofibril, whose contraction produces muscle contraction. 231

sarcoplasmic reticulum (sar-kuh-plaz-mik rih-tik-yuh-lum) Smooth endoplasmic reticulum of skeletal muscle cells; surrounds the myofibrils and stores calcium ions. 231

saturated fatty acid Fatty acid molecule that lacks double bonds between the atoms of its carbon chain. 29

Schwann cell Cell that surrounds a fiber of a peripheral nerve and forms the myelin sheath. 247

science Development of concepts about the natural world often by utilizing the scientific method. 8

scientific method Process of attaining knowledge by making observations, testing hypotheses, and coming to conclusions. 8

scientific theory Concept supported by a broad range of observations, experiments, and conclusions. 8

sclera (skleer-uh) White, fibrous, outer layer of the eyeball. 278

scrotum (skroh-tum) Pouch of skin that encloses the testes. 318

selectively permeable Having degrees of permeability; the cell is impermeable to some substances and allows others to pass through at varying rates. 47

semantic memory Capacity of the brain to store and retrieve information with regard to words or numbers and such. 258

semen (see-mun) Thick, whitish fluid consisting of sperm and secretions from several glands of the male reproductive tract. 318

semicircular canal (sem-ih-sur-kyuh-lur) One of three tubular structures within the inner ear that contain sensory receptors responsible for the sense of rotational equilibrium. 284

semilunar valve (sem-ee-loo-nur) Valve resembling a half moon located between the ventricles and their attached vessels. 128

seminal vesicle (sem-uh-nul) Convoluted, saclike structure attached to the vas deferens near the base of the urinary bladder in males; adds secretions to semen. 318

seminiferous tubule (sem-uh-nif-ur-us) Long, coiled structure contained within chambers of the testis where sperm are produced. 320

sensation Conscious awareness of a stimulus due to nerve impulse sent to the brain from a sensory receptor by way of sensory neurons. 273

sensory adaptation Phenomenon of a sensation becoming less noticeable once it has been recognized by constant repeated stimulation. 273

sensory neuron Nerve cell that transmits nerve impulses to the central nervous system after a sensory receptor has been stimulated. 246

sensory receptor Structure that receives either external or internal environmental stimuli and is a part of a sensory neuron or transmits signals to a sensory neuron. 272

serous membrane (seer-us) Membrane that covers internal organs and lines cavities without an opening to the outside of the body; serosa. 69

serum (seer-um) Light yellow liquid left after clotting of blood. 116

sex chromosome Chromosome that determines the sex of an individual; in humans, females have two X chromosomes and males have an X and Y chromosome. 395, 414

sex-influenced trait Autosomal phenotype controlled by an allele that is expressed differently in the two sexes; for example, the possibility of pattern baldness is increased by the presence of testosterone in human males. 418

sex-linked Allele that occurs on the sex chromosomes but may control a trait that has nothing to do with sexual characteristics of an individual. 414

sexually transmitted disease Illness communicated primarily or exclusively through sexual contact. 349

short-term memory Retention of information for only a few minutes, such as remembering a telephone number. 258

sickle-cell disease Genetic disorder in which the affected individual has sickle-shaped red blood cells that are subject to hemolysis. 113

simple goiter (goy-tur) Condition in which an enlarged thyroid produces low levels of thyroxine. 299

sink population Population found in an unfavorable area where at best the birthrate equals the death rate; sink populations receive new members from source populations. 509

sinus (sy-nus) Cavity or hollow space in an organ such as the skull. 212

sinusitis (sy-nuh-sy-tis) Infection of the sinuses, caused by blockage of the openings to the sinuses, and characterized by postnasal discharge and facial pain. 177

skeletal muscle Striated, voluntary muscle tissue found in muscles that move the bones. 67

skill memory Capacity of the brain to store and retrieve information necessary to perform motor activities, such as riding a bike. 258

skin Outer covering of the body; can be called the integumentary system because it contains organs such as sense organs. 71

sliding filament theory An explanation for muscle contraction based on the movement of actin filaments in relation to myosin filaments.   231

small intestine Long, tubelike chamber of the digestive tract between the stomach and large intestine.   87

smooth (visceral) muscle Nonstriated, involuntary muscle tissue found in the walls of internal organs.   67

sodium-potassium pump Carrier protein in the plasma membrane that moves sodium ions out of and potassium into cells; important in nerve and muscle cells.   248

soft palate (pal-it) Entirely muscular posterior portion of the roof of the mouth.   82

somatic cell (soh-mat-ik) A body cell; excludes cells that undergo meiosis and become a sperm or egg.   386

somatic system That portion of the peripheral nervous system containing motor neurons that control skeletal muscles.   261

source population Population that can provide members to other populations of the species because it lives in a favorable area, and the birthrate is most likely higher than the death rate.   508

sperm Male gamete having a haploid number of chromosomes and the ability to fertilize an egg, the female gamete.   321

spermatogenesis (spur-mat-uh-jen-ih-sis) Production of sperm in males by the process of meiosis and maturation.   321, 394

sphincter (sfingk-tur) Muscle that surrounds a tube and closes or opens the tube by contracting and relaxing.   85

spinal cord Part of the central nervous system; the nerve cord that is continuous with the base of the brain plus the vertebral column which protects the nerve cord.   252

spinal nerve Nerve that arises from the spinal cord.   260

spindle Microtubule structure that brings about chromosomal movement during nuclear division.   388

spiral organ Organ in the cochlear duct of the inner ear which is responsible for hearing; organ of Corti.   285

spleen Large, glandular organ located in the upper left region of the abdomen that stores and purifies blood.   147

spongy bone Porous bone found at the ends of long bones where red bone marrow is sometimes located.   65, 206

squamous epithelium (skway-mus, skwah-) Type of epithelial tissue that contains flat cells.   62

stapes (stay-peez) Last of three ossicles of the ear that serve to conduct vibrations from the tympanic membrane to the oval window of the inner ear.   284

starch Storage polysaccharide found in plants that is composed of glucose molecules joined in a linear fashion with few side chains.   28

stem cell Any undifferentiated cell that can divide and differentiate into more functionally specific cell types such as those in red bone marrow that give rise to blood cells.   113, 372

steroid (steer-oyd) Type of lipid molecule having a complex of four carbon rings; examples are cholesterol, progesterone, and testosterone.   30

steroid hormone (steer-oyd) Type of hormone that has a complex of four carbon rings but different side chains from other steroid hormones.   309

stimulus Change in the internal or external environment that a sensory receptor can detect leading to nerve impulses in sensory neurons.   272

stomach Muscular sac that mixes food with gastric juices to form chyme, which enters the small intestine.   86

striated (stry-ayt-ud) Having bands; in cardiac and skeletal muscle, alternating light and dark crossbands produced by the distribution of contractile proteins.   67

stroke Condition resulting when an arteriole in the brain bursts or becomes blocked by an embolism; cerebrovascular accident.   137

subcutaneous layer (sub-kyoo-tay-nee-us) Tissue layer that lies just beneath the skin and contains adipose tissue.   72

substrate Reactant in a reaction controlled by an enzyme.   54

succession Sequential change in the relative dominance of species within a community; primary succession begins on bare rock, and secondary succession begins where soil already exists.   478

superior vena cava (vee-nuh kay-vuh) Large vein that enters the right atrium from above and carries blood from the head, thorax, and upper limbs to the heart.   134

sustainable development Management of an ecosystem so that it maintains itself while providing services to human beings.   510

sustentacular (Sertoli) cell (sus-tun-tak-yuh-lur) Elongated cell in the wall of the seminiferous tubules to which spermatids are attached during spermatogenesis.   321

suture (soo-chur) Type of immovable joint articulation found between bones of the skull.   218

sweat gland Skin gland that secretes a fluid substance for evaporate cooling; sudoriferous gland.   72

sympathetic division That part of the autonomic system that usually promotes activities associated with emergency (fight or flight) situations; uses norepinephrine as a neurotransmitter.   263

synapse (sin-aps, si-naps) Junction between neurons consisting of the presynaptic (axon) membrane, the synaptic cleft, and the postsynaptic (usually dendrite) membrane.   251

synapsis (sih-nap-sis) Pairing of homologous chromosomes during prophase I of meiosis I.   390

synaptic cleft (sih-nap-tik) Small gap between presynaptic and postsynaptic membranes of a synapse.   251

syndrome Group of symptoms that appear together and tend to indicate the presence of a particular disorder.   386

synovial joint (sih-noh-vee-ul) Freely moving joint in which two bones are separated by a cavity.   218

synovial membrane Membrane that forms the inner lining of the capsule of a freely movable joint.   69

syphilis (sif-uh-lis) Sexually transmitted disease caused by the bacterium *Treponema pallidum* that if untreated can lead to cardiac and central nervous system disorders.   349

systemic circuit Blood vessels that transport blood from the left ventricle and back to the right atrium of the heart.   134

systole (sis-tuh-lee) Contraction period of a heart during the cardiac cycle.   130

systolic pressure (sis-tahl-ik) Arterial blood pressure during the systolic phase of the cardiac cycle.   132

T

T (transverse) tubule Membranous channel that extends inward.   231

T lymphocyte (lim-fuh-syt) Lymphocyte that matures in the thymus; cytotoxic T cells kill antigen-bearing cells outright; helper T cells release cytokines that stimulate other immune system cells.   150

taste bud Sense organ containing the receptors associated with the sense of taste.   276

tectorial membrane (tek-tor-ee-ul) Membrane that lies above and makes contact with the hair cells in the spiral organ.   285

teleological Belief that the outcome is known ahead of time, because the outcome is influenced by the need to progress in a certain direction.   465

telomere (tel-uh-meer) Tip of the end of a chromosome.   446

telophase (tel-uh-fayz) Mitotic phase during which daughter chromosomes are located at each pole.   389

template (tem-plit) Pattern or guide used to make copies; parental strand of DNA serves as a guide for the production of daughter DNA strands and DNA also serves as guide for the production of messenger RNA.   424

tendon (ten-dun) Strap of fibrous connective tissue that connects skeletal muscle to bone.   64, 206, 227

testis (tes-tis, -tus) Male gonad which produces sperm and the male sex hormones.   306, 318

testosterone (tes-tahs-tuh-rohn) Male sex hormone that helps maintain sexual organs and secondary sexual characteristics.   306, 321

tetanus (tet-n-us) Sustained muscle contraction without relaxation.   234

tetany (tet-n-ee) Severe twitching caused by involuntary contraction of the skeletal muscles due to a calcium imbalance.   300

tetrad Four chromatids that result when homologous chromosomes pair during meiosis I.   390

thalamus (thal-uh-mus) Part of the brain located in the lateral walls of the third ventricle that serves as the integrating center for sensory input; it plays a role in arousing the cerebral cortex.   256

thermal inversion Temperature inversion in which warm air traps cold air and pollutants near the earth.   489

thermoreceptor Sensory receptor that is sensitive to changes in temperature.   272

threshold Electrical potential level (voltage) at which an action potential or nerve impulse is produced.   248

thrombin (thrahm-bin) Enzyme that converts fibrinogen to fibrin threads during blood clotting.   116

thromboembolism (thrahm-boh-em-buh-liz-um) Obstruction of a blood vessel by a thrombus that has dislodged from the site of its formation.   137

thrombus (thrahm-bus) Blood clot that remains in the blood vessel where it formed.   137

thymine (T) (thy-meen) One of four nitrogen-containing bases in nucleotides composing the structure of DNA; pairs with adenine.   423

thymus gland Lymphoid organ, located along the trachea behind the sternum, involved in the maturation of T lymphocytes mature in the thymus gland. Secretes hormones called thymosins, which aid the maturation of T cells and perhaps stimulate immune cells in general.   148, 307

thyroid gland Endocrine gland in the neck that produces several important hormones, including thyroxine, triiodothyronine, and calcitonin.   299

thyroid-stimulating hormone (TSH) Substance produced by the anterior pituitary that causes the thyroid to secrete thyroxine and triiodothyromine.   296

thyroxine (T_4) (thy-rahk-sin) Hormone secreted from the thyroid gland that promotes growth and development; in general, it increases the metabolic rate in cells.   299

tidal volume Amount of air normally moved in the human body during an inspiration or expiration.   170

tight junction Region between cells where adjacent plasma membrane proteins join to form an impermeable barrier.   64

tissue Group of similar cells which perform a common function.   2, 62

tissue fluid Fluid that surrounds the body's cells; consists of dissolved substances that leave the blood capillaries by filtration and diffusion.   118

tone Continuous, partial contraction of muscle.   234

tonicity (toh-nis-ih-tee) Osmolarity of a solution compared to that of a cell; if the solution is isotonic to the cell, there is no net movement of water; if the solution is hypotonic, the cell gains water; and if the solution is hypertonic, the cell loses water.   47

tonsillectomy (tahn-suh-LEK-tuh-mee) Surgical removal of the tonsils.   177

tonsillitis Infection of the tonsils that causes inflammation, and can spread to the middle ears.   82, 177

tonsils Partially encapsulated lymph nodules located in the pharynx.   147, 177

trachea (tray-kee-uh) Passageway that conveys air from the larynx to the bronchi; windpipe.   168

tracheostomy (tray-kee-ahs-tuh-mee) Creation of an artificial airway by incision of the trachea and insertion of a tube.   168

tract Bundle of myelinated axons in the central nervous system.   252

trait Specific term for a distinguishing phenotypic feature studied in heredity.   403

transcription Process whereby a DNA strand serves as a template for the formation of mRNA.   427

transcription factor Protein that initiates transcription by RNA polymerase and thereby starts the process that results in gene expression.   431

transfer rate Amount of a substance that moves from one component of the environment to another within a specified period of time.   486

transfer RNA (tRNA) Type of RNA that transfers a particular amino acid to a ribosome during protein synthesis; at one end it binds to the amino acid and at the other end it has an anticodon that binds to an mRNA codon.   425

transgenic organisms Free-living organism in the environment that has a foreign gene in its cells.   434

translation Process whereby ribosomes use the sequence of codons in mRNA to produce a polypeptide with a particular sequence of amino acids.   428

triglyceride (trih-glis-uh-ryd) Neutral fat composed of glycerol and three fatty acids.   29

triplet code Each sequence of three nucleotide bases in the DNA of genes stands for a particular amino acid.   427

trophic level Feeding level of one or more populations in a food web.   483

tropomyosin (trahp-uh-my-uh-sin, trohp-) Protein that functions with troponin to block muscle contraction until calcium ions are present.   233

troponin (troh-puh-nin) Protein that functions with tropomyosin to block muscle contraction until calcium ions are present.   233

trypsin (trip-sin) Protein-digesting enzyme secreted by the pancreas.   90, 92

tubular reabsorption Movement of primarily nutrient molecules and water from the contents of the nephron into blood at the proximal convoluted tubule.   195

tubular secretion Movement of certain molecules from blood into the distal convoluted tubule of a nephron so that they are added to urine.   195

tumor (too-mur) Cells derived from a single mutated cell that has repeatedly undergone cell division; benign tumors remain at the site of origin, and malignant tumors metastasize.   444

tumor marker test Blood test for a substance, such as a tumor antigen, that indicates a patient has cancer.   450

tumor-suppressor gene Gene that codes for a protein that ordinarily suppresses cell division; inactivity can lead to a tumor.   446

tympanic membrane (tim-pan-ik) Located between the outer and middle ear and receives sound waves; the eardrum.   284

U

ulcer (ul-sur) Open sore in the lining of the stomach; frequently caused by bacterial infection.   86

umbilical cord Cord connecting the fetus to the placenta through which blood vessels pass.   368

unsaturated fatty acid Fatty acid molecule that has one or more double bonds between the atoms of its carbon chain.   29

uracil (U) (yoor-uh-sil) The base in RNA that replaces thymine found in DNA; pairs with adenine.   425

urea (yoo-ree-uh) Primary nitrogenous waste of humans derived from amino acid breakdown.   189

ureter (yoor-uh-tur) One of two tubes that take urine from the kidneys to the urinary bladder.   188

urethra (yoo-ree-thruh) Tubular structure that receives urine from the bladder and carries it to the outside of the body.   188, 318

uric acid (yoor-ik) Waste product of nucleotide metabolism.   189

urinary bladder Organ where urine is stored before being discharged by way of the urethra.   188

urinary system Organ system consisting of the kidneys and urinary bladder; rids body of nitrogenous wastes and helps regulate water-salt balance of the blood.   70

uterine cycle (yoo-tur-in, -tuh-ryn) Monthly occurring changes in the characteristics of the uterine lining (endometrium).   326

uterus (yoo-tur-us) Organ located in the female pelvis where the fetus develops; the womb.   322

utricle (yoo-trih-kul) Saclike cavity in the vestibule of the inner ear that contains sensory receptors for gravitational equilibrium.   287

V

vaccine Antigens prepared in such a way that they can promote active immunity without causing disease.   156

vagina Organ that leads from the uterus to the vestibule and serves as the birth canal and organ of sexual intercourse in females.   323

valve Membranous extension of a vessel or the heart wall that opens and closes, ensuring one-way flow.   127

variable Factor that can cause an observable change during the progress of the experiment.   10

varicose veins (var-ih-kohs, vair-) Irregular dilation of superficial veins, seen particularly in lower legs, due to weakened valves within the veins.   139

vas deferens (vas def-ur-unz, -uh-renz) Tube that leads from the epididymis to the urethra in males.   318

vector (vek-tur) In genetic engineering, a means to transfer foreign genetic material into a cell; e.g., a plasmid.   432

vein Vessel that takes blood to the heart from venules; characteristically having nonelastic walls.   126

ventilation Process of moving air into and out of the lungs; breathing.   172

ventricle (ven-trih-kul) Cavity in an organ, such as a lower chamber of the heart; or the ventricles of the brain.   128, 252

venule (ven-yool, veen-) Vessel that takes blood from capillaries to a vein.   127

vernix caseosa (vur-niks kay-see-oh-suh) Cheeselike substance covering the skin of the fetus.   374

vertebral column (vur-tuh-brul) Series of joined vertebrae that extends from the skull to the pelvis.   214

vertebrate (vur-tuh-brit, -brayt) An animal with a vertebral column.   2

vertigo (vur-tih-goh) Dizziness and a sense of rotation.   287

vesicle (ves-ih-kul) Small, membranous sac that stores substances within a cell.   50

vestibule (ves-tuh-byool) Space or cavity at the entrance of a canal, such as the cavity that lies between the semicircular canals and the cochlea.   284

villus (pl., villi) (vil-us) Small, fingerlike projection of the inner small intestinal wall.   87

virus Noncellular parasitic agent having an outer capsid of protein and an inner core of nucleic acid and possibly an envelope derived from the host cell plasma membrane.   340

visual accommodation Ability of the eye to focus at different distances by changing the curvature of the lens.   279

visual field Area of vision for each eye.   282

vital capacity Maximum amount of air moved in or out of the human body with each breathing cycle.   170

vitamin Essential requirement in the diet, needed in small amounts. They are often part of coenzymes.   100

vitreous humor (vit-ree-us) Clear gelatinous material between the lens of the eye and the retina.   279

vocal cord Fold of tissue within the larynx; creates vocal sounds when it vibrates.   168

vulva External genitals of the female that surround the opening of the vagina.   323

W

water (hydrologic) cycle Interdependent and continuous circulation of water from the ocean, to the atmosphere, to the land, and back to the ocean.   485

white blood cell Type of blood cell that is transparent without staining and protects the body from invasion by foreign substances and organisms; leukocyte.   66, 115

white matter Myelinated axons in the central nervous system.   252

X

X chromosome Female sex chromosome that carries genes involved in sex determination; see Y chromosome.   395

xenotransplantation (zen-uh-trans-plan-tay-shun) Use of animal organs, instead of human organs, in transplant patients.   435

X-linked Allele located on an X chromosome but may control a trait that has nothing to do with the sexual characteristics of an individual.   414

Y

Y chromosome Male sex chromosome that carries genes involved in sex determination; see X chromosome.   395

yolk sac Extraembryonic membrane that encloses the yolk of birds; in humans it is the first site of blood cell formation.   368

Z

zygote (zy-goht) Diploid cell formed by the union of sperm and egg; the product of fertilization.   322, 364

Credits

Line Art and Readings

Chapter opening text prepared by Kathryn Sergeant Brown.

Chapter 2

Page 25, Ecology Focus – Sources: Data from G. Tyler Miller, *Living in the Environment*, 1993, Wadsworth Publishing Company, Belmont, CA; and Lester R. Brown, *State of the World*, 1992, W. W. Norton & Company, Inc., New York, NY p. 29.

Chapter 4

Figure 4.5: From John W. Hole, Jr., *Human Anatomy & Physiology*, 6th edition. Copyright 1993 McGraw-Hill Companies, Inc., Dubuque, Iowa. All Rights Reserved. Reprinted by Permission. p. 727. Figure 4.13: From D. Shier et al., *Hole's Human Anatomy & Physiology*, 8th edition. Copyright 1996 McGraw-Hill Companies, Inc. All Rights Reserved. Reprinted by Permission. p. 735.

Chapter 5

Figure 5.15: Source: Data from T.T. Shintani, *Eat More, Weigh Less ™Diet*, 1983. Table 5.4: From Sizer, F. S., and Whitney, E. N. 1994. *Hamilton/Whitney's Nutrition: Concepts and Controversies*, 6th ed. Minneapolis: West Publishing Company, page 205. Health Focus p. 99: Reprinted by permission from page 374 of *Understanding Nutrition*, Fifth Edition, by W.N.Whitner et al. Copyright ©1990 by West Publishing Company. All rights reserved. Figure 5.17: Reprinted by permission from *Nutrition: Concepts and Controversies*, Sixth Edition, by Frances Sienkiewicz Sizer and Eleanor Noss Whitney. Copyright ©1994 by West Publishing Company. All Rights Reserved. Table 5.8: Source: David C. Nieman et al., *Nutrition*. Copyright © 1990 Wm. C. Brown.

Chapter 6

Health Focus, page 117: Courtesy of the American Red Cross.

Chapter 7

Figure 7.3: From Kent M. Van De Graaff and Stuart Ira Fox, *Concepts of Human Anatomy and Physiology*, 3rd edition, © 1992 The McGraw-Hill Companies, Inc. All Rights Reserved. Reprinted by permission.

Chapter 8

Figure 8.10a: Source: Approved by the Advisory Committee on Immunization Practices (ACIP), the American Academy of Pediatrics (AAP), and the American Academy of Family Physicians. Figure 8.10b: From David Shier, et al., *Hole's Human Anatomy & Physiology*, 7th Edition. Copyright © 1996 Times Mirror Higher Education Group, Inc., Dubuque, Iowa. All Rights Reserved. Reprinted by permission.

Chapter 9

Health Focus, Page 182: Source: From "The Most Often Asked Questions About Smoking Tobacco and Health and The Answers," revised July 1993. Copyright © American Cancer Society, Inc. Atlanta, GA. Used with permission.

Chapter 11

Figure 11.2: From David Shier et al., *Hole's Human Anatomy & Physiology*, 4th Edition, William C. Brown, Figure 7.8, page 192. Figure 11.6: From David Shier et al., *Hole's Human Anatomy & Physiology*, 4th Edition, William C. Brown, Figure 7.19, page 207.

Chapter 12

Figure 12.4: From Kent Van De Graaff & Stuart Ira Fox, *Concepts in Anatomy & Physiology*, 4th Edition, Figure 13.1, page 282.

Chapter 15

Health Focus Page 308: Adapted from Robert L. Sack, Alfred J. Lewy, and Mary L. Blood. *Journal of Visual Impairment and Blindness* 92: 145-161. March 1998.

Chapter 16

Figure 16.1: From John W. Hole, Jr., *Human Anatomy & Physiology*, 6th Edition, Figure 22.1A, page 807. Figure 16.2: From Kent Van De Graaff & Stuart Ira Fox, *Concepts in Anatomy & Physiology*, 4th Edition, Figure 28.22, page 857.

Chapter 17

Figures 17.4, 17.5, 17.8, 17.10, 17.12: Data from Division of STD Prevention, Sexually Transmitted Disease Surveillance, 1998. U. S. Department of Health and Human Services, Public Health Service. Atlanta: Centers for Disease Control and Prevention, September, 2000.

Chapter 18

Figure 18.5 from John W. Hole, Jr. *Human Anatomy and Physiology*, 6th Edition, ©1993 The McGraw-Hill Companies, Inc. All Rights Reserved. Reprinted by permission. Figures 18.13 and 18.15 from Kent Van De Graaff and Stuart Ira Fox, *Concepts of Human*

Anatomy and Physiology, 4th edition. Copyright ©1995 The McGraw-Hill Companies, Inc. All Rights Reserved. Reprinted by permission.

Chapter 22

Figure 22.9 from Lance A. Liotta, "Cancer Cell Invasion and Metastasis," *Scientific American*, February 1992, Illustration by Dana Burns-Pizer, p. 62; modified figure depicting tissue inhibitors of metalloproteinase. Health Focus: provided by the American Red Cross. Used by permission.

Chapter 25

Figure 25.3: Redrawn from *Shrimp Stocking, Salmon Collapse, and Eagle Displacement: Cascading Interactions in the Food Web of a Large Aquatic Ecosystem*, by Spencer, C.N., B.R. McCleeland, and J.A. Stanford, *Bioscience*, 41(1):14-21. Figure 25.8: Data from David M. Gates, *Climate Change and Its Biological Consequences*, 1993, Sinauer Associates, Inc., Sunderland, MA. p. 478.

Photographs

Chapter 1

Fig. 1.1: © John Cunningham/Visuals Unlimited; Fig. 1A(Background): © Barbara von Hoffman/Tom Stack & Associates; Fig.1A(Toucan): © Ed Reschke/Peter Arnold, Inc.; Fig. 1A(Morpho): © Kjell Sandved/Butterfly Alphabet; Fig.1A(Jaguar): © BIOS (Seitre)/Peter Arnold, Inc.; Fig. 1A(Orchid): © Max & Bea Hunn/Visuals Unlimited; Fig. 1A(Frog): © Kevin Schafer & Martha Hill/Tom Stack & Associates; Fig.1B : © George Holton/Photo Researchers, Inc.; Fig.1C : © People for the Ethical Treatment of Animals.

Chapter 2

Fig. 2.1: © The McGraw/Hill Companies, Inc./Jim Shaffer, photographer; Fig. 2.3a: © Biomed Commun./Custom Medical Stock Photos; Fig. 2.4d: © Charles M. Falco/Photo Researchers, Inc.; Fig. 2.8a: © Martin Dohrn/SPL/Photo Researchers, Inc.; Fig. 2.8b: © Comstock, Inc.; Fig. 2.8c: © Marty Cooper/Peter Arnold, Inc.; Fig. 2A(left): © Ray Pfortner/Peter Arnold, Inc.; Fig. 2A(middle): © John Miller/Stone; Fig. 2A(right)3: © Frederica Georgia/Photo Researchers, Inc.; Fig. 2.13: © Dwight Kuhn; Fig. 2.14,2.15: © Dwight Kuhn; Fig. 2.18b: © Don W. Fawcett/Photo Researchers, Inc.; Fig. 2.19b: © Richard C. Johnson/Visuals

Unlimited; Fig. 2C: © Jeff Greenberg/Unicorn Stock Photos.

Chapter 3

Fig. 3.1a: © Pascal Rondeau/Stone; Fig. 3.1b(inset): © Prof. P. Motta, Dept. of Anatomy, Univ. La Sapienza, Rome/Photo Researchers, Inc.; Fig. 3.2a: © David M. Phillips/Visuals Unlimited; Fig. 3.2b: © Robert Caughey/Visuals Unlimited; Fig. 3.2c : © Warren Rosenberg/BPS/Stone; Fig. 3.3b: © Alfred Pasieka/Photo Researchers, Inc.; Fig. 3.7(inset): Courtesy of Ron Milligan/Scripps Research Institute; Fig. 3.7(right): Courtesy of E. G. Pollock; Fig. 3.8b: © Barry F. King/Biological Photo Service; Fig. 3.9(top right): © R. Rodewald/Biological Photo Service; Fig. 3.9(lower right) : Courtesy Tim Wakefield/John Brown University; Fig. 3.10a: Courtesy of Dr. Keith Porter; Fig. 3.11a: Courtesy of Kent McDonald, University of California, Berkley; Fig. 3.12: © David M. Phillips/Photo Researchers, Inc.; Fig. 3A: © Jerry Cooke/Photo Researchers, Inc.

Chapter 4

Fig. 4.1: © Julie Houck/Corbis; Fig. 4.2(top left): © Ed Reschke/Peter Arnold, Inc.; Fig. 4.2(top right): © Ed Reschke/Peter Arnold, Inc.; Fig. 4.2(lower left): © Ed Reschke/Peter Arnold, Inc.; Fig. 4.2(lower right): © Ed Reschke/Peter Arnold, Inc., Fig. 4.4a-d: © Ed Reschke; Fig. 4.6a: © Ed Reschke; Fig. 4.6b: © Ed Reschke; Fig. 4.6c: © Ed Reschke; Fig. 4Aa: © Ken Greer/Visuals Unlimited; Fig. 4Ab: © Dr. P. Marazzi/SPL/Photo Researchers, Inc.; Fig. 4Ac: © James Stevenson/SPL/Photo Researchers, Inc.

Chapter 5

Fig. 5.4b: © Biophoto Associates/Photo Researchers, Inc.; Fig. 5.5b: © Ed Reschke/Peter Arnold, Inc.; Fig. 5.5c: © St. Bartholomew's Hospital/PL/Photo Researchers, Inc., Fig. 5.6b(right): © Manfred Kage/Peter Arnold, Inc.; Fig. 5.14: © The McGraw-Hill Companies, Inc./Bob Coyle, photographer; Fig. 5A: © F. Hache/Photo Researchers, Inc.; Fig. 5.16a: © Biophoto Associates/Photo Researchers, Inc.; Fig. 5.16b: © Ken Greer/Visuals Unlimited; Fig. 5.16c: © Biophoto Associates/Photo Researchers, Inc.; Fig. 5.18: © Carl Centineo/Medichrome/The Stock Shop, Inc.; Fig. 5.19: © Donna Day/Stone, Fig. 5.20: © Tony Freeman/PhotoEdit.

Chapter 6

Fig. 6.1: Courtesy Camp Good Days and Special Times; Fig. 6.3a: © Lennart Nilsson, "Behold Man" Little Brown and Company, Boston; Fig. 6.7(right): © Manfred Kage/Peter Arnold, Inc.; Fig. 6.10a: Courtesy of Stuart I. Fox.

Chapter 7

Fig. 7.1: © Mark Harmel/Stone; Fig. 7.2d, 7.7b, 7.7c: © Ed Reschke; Fig. 7A: © Biophoto Associates/Photo Researchers, Inc.; Fig. 7B: © Mary Kate Denny/PhotoEdit.

Chapter 8

Fig. 8.1: © Damien Lovegrove/SPL/Photo Researchers, Inc.; Fig. 8.3b(top left): © Ed Reschke/Peter Arnold, Inc.; Fig. 8.3b(top right): © Fred E. Hossler/Visuals Unlimited; Fig. 8.3b(lower left): © Ed Reschke/Peter Arnold, Inc., Fig. 8.3b(lower right): © R. Valentine/Visuals Unlimited; Fig. 8.7b: Courtesy Dr. Arthur J. Olson, Scripps Institute; Fig. 8A: © John Nuhn; Fig. 8.9a: © Boehringer Ingelheim International, photo by Lennart Nilsson; Fig. 8.10a: © Matt Meadows/Peter Arnold, Inc.; Fig. 8.11: © Estate of Ed Lettau/Peter Arnold, Inc.; Fig. 8B: © Martha Cooper/Peter Arnold, Inc.

Chapter 9

Fig. 9.1: © Seth Resnick/Stock Boston; Fig. 9.6a: © SIU/Visuals Unlimited; Fig. 9A: © Bill Aron/Photo Edit; Fig. 9B: © Oscar Burriel/SPL/Photo Researchers, Inc.; Fig. 9.13a,b: © Martin Rotker Photography.

Chapter 10

Fig. 10.1: Photodisc Image 15235 Vol. Series 15; Fig. 10.5(top left) : © R.G. Kessel and R.H. Kardon, /it/Tissues and Organs: A Text-Atlas of Scanning Electron Microscopy, 1979; Fig. 10.5b: © 1966 Academic Press, from A.B. Maunsbach, J. Ultrastruct. Res. 15:242-282; Fig. 10.6a: © Gennaro/Photo Researchers, Inc.; Fig. 10.11a: © SIU/Peter Arnold, Inc.; Fig. 10B: Courtesy National Kidney Foundation, photo by Jay Laprete.

Chapter 11

Fig. 1.1(top left): © Ed Reschke; Fig. 11.1(top right): © Ed Reschke; Fig. 11.1(lower): © Biophoto Associates/Photo Researchers, Inc.; Fig. 11.3b: © Junebug Clark/Photo Researchers, Inc.; Fig. 11A: © Royce Blair/Unicorn Stock Photos; Fig. 11Ab: © Michael Klein/Peter Arnold, Inc.; Fig. 11.6b: © The McGraw-Hill Companies, Inc./Joe DeGrandis, photographer; Fig. 11.11(top left): © Ed Reschke; Fig. 11.11(lower left): © Ed Reschke; Fig. 11.11(top right): © Ed Reschke; Fig. 11.11(right center): © Ed Reschke; Fig. 11.11(lower right): © Ed Reschke/Peter Arnold, Inc.; Fig. 11.13: © SIU/Peter Arnold, Inc.

Chapter 12

Fig. 12.1: © Jim McHugh/Corbis Outline; Fig. 12.2b: Courtesy Dr. Hugh E. Huxley; Fig. 12.5b : © Ed Reschke/Peter Arnold, Inc.; Fig.12.6: © Ed Reschke; Fig. 12.7b: © Victor B. Eichler; Fig. 12.10b: © Tim Davis/Photo Researchers, Inc.; Fig.12A: © Ronald C. Modra/Sports Illustrated/Time Inc.; Fig. 12.11b: © G.W. Willis/Biological Photo Service; Fig. 12.12: Courtesy Jan B. Kuks, Associate Professor of Neurology/University Hospital, Groningen, The Netherlands.

Chapter 13

Fig. 13.1a: © Gerhard Gscheidle/Peter Arnold, Inc.; Fig.13.3b: © M.B. Bunge/Biological Photo Service; Fig. 13.4c: © Linda Bartlett; Fig.13.5(left): Courtesy of Dr. E. R. Lewis, University of California Berkeley; Fig. 13.8c: © Manfred Kage/Peter Arnold, Inc.; Fig. 13.9b: © Colin Chumbley/Science Source/Photo Researchers, Inc.; Fig.13.14: © Marcus Raichle; Fig. 13.19: © Mary Ellen Mark/Falkland Road.

Chapter 14

Fig. 14.1: © Karen Holsinger Mullen/Unicorn Stock Photos; Fig. 14.5b: © Omikron/SPL/Photo Researchers, Inc.; Fig. 14.9a: © Lennart Nilsson, from "The Incredible Machine"; Fig. 14.10: © The McGraw-Hill Companies, Inc./Dennis Strete, photographer; Fig. 14.14b: © P. Motta/SPL/Photo Researchers, Inc., Fig. 14A: Robert S. Preston and Joseph E. Hawkins, Kresge Hearing Research Institute, University of Michigan.

Chapter 15

Fig. 15.1: © Bob Daemmrich/Stock Boston; Fig. 15.4(right): © Bob Daemmrich/Stock Boston; Fig. 15.4(left): © Ewing Galloway, Inc.; Fig. 15.5: From Clinical Pthological Conference, "Acromegaly, Diabetes, Hypermetabolism, Proteinura and Heart Failure", American Journal of Medicine 20 (1956) 133, with permission from Excerpta Medica, Inc.; Fig. 15.6: © Biophoto Associates/Photo Researchers, Inc.; Fig. 15.7: © John Paul Kay/Peter Arnold, Inc.; Fig. 15.11a: © Custom Medical Stock Photos; Fig. 15.11b: © NMSB/Custom Medical Stock Photos; Fig. 15.12a,b: "Atlas of Pediatric Physical Diagnosis," Second Edition by Zitelli & Davis, 1992. Mosby-Wolfe Europe Limited, London, UK; Fig. 15A: © James Darell/Stone; Fig. 15B: © Erika Stone/Photo Researchers, Inc.

Chapter 16

Fig. 16.3b: © Secchi-Leaque/CNRI/SPL/Photo Researchers, Inc.; Fig. 16.3c: © Ed Reschke; Fig. 16.7b: © Ed Reschke/Peter Arnold, Inc.; Fig. 16.10: © Dr. Landrum B. Shettles; Fig. 16.11a,c,d,f : © The McGraw-Hill Companies, Inc./Bob Coyle, photographer; Fig. 16.11b: © The McGraw-Hill Companies, Inc./Vincent Ho, photographer; Fig. 16.11e: © Hank Morgan/Photo Researchers, Inc.; Fig. 16.12: © CC Studio/SPL/Photo Researchers,

Inc.; Fig. 16B: © Myrleen Ferguson/
PhotoEdit.

Chapter 17

Fig. 17.1: © Paul Barton/The Stock Market;
Fig. 17.2b: © K.G. Murti/Visuals Unlimited;
Fig. 17.4: © CDC/Peter Arnold, Inc.; Fig. 17.5:
© Charles Lightdale/Photo Researchers, Inc.;
Fig. 17.6a: © David M. Phillips/Visuals
Unlimited; Fig. 17.6b: © David M.
Phillips/Visuals Unlimited; Fig. 17.6c: © R.G.
Kessel - C.Y. Shih/Visuals Unlimited; Fig.
17.8: © G.W. Willis/BPS/Stone; Fig. 17.10:
© CNRI/SPL/Photo Researchers, Inc.; Fig.
17.11: Courtesy of Dr. Ira Abrahamson; Fig.
17.12(top left): Courtesy Centers for Disease
Control; Fig. 17.12b,c: © Carroll
Weiss/Camera M.D.; Fig. 17.12d : © Science
VU/Visuals Unlimited; Fig. 17A: © Malcolm
Linton/Liaison Agency; Fig. 17.13: © David
Phillips/Photo Researchers, Inc.; Fig. 17.14: ©
David M. Phillips/Visuals Unlimited; Fig.
17.15: © Oliver Meckes/Photo Researchers,
Inc.; Fig. S.1: © NIBSC/SPL/Photo
Researchers, Inc.; Fig. S.4a-c: © Nicholas
Nixon; Fig. S.A : Courtesy Dr. Jack Shields,
from Lymphology 22:62-66 (1989).

Chapter 18

Fig. 18.1: © Joseph Nettis/Stock Boston;
Fig. 18.2a: © David M. Phillips/Photo
Researchers, Inc.; Fig. 18.9: © Petit
Format/Nestle/Photo Researchers, Inc.;
Fig. 18.10,18.11: © Lennart Nilsson, from A
CHILD IS BORN.; Fig. 18.15: © John
Cunningham/Visuals Unlimited; Fig. 18.16:
© Richard Hutchings/Photo Edit; Fig. 18B:
© Bob Daemmrich/Stock Boston.

Chapter 19

Fig. 19.4a-d: © Michael Abbey/Photo
Researchers, Inc.; Fig. 19.9(all): ©
CNRI/SPL/Photo Researchers, Inc.;
Fig. 19.11: © Jill Cannefax/EKM-Nepenthe;
Fig. 19.13a: © David M. Phillips/Visuals
Unlimited; Fig. 19.12a,b: Courtesy of G.H.
Valentine; Fig. 19.13b,c: Reprint from R.
Simensen and R. Curtis Rogers, "Fragile X
Syndrome," AMERICAN FAMILY
PHYSICIAN 39(5):186, May 1989.; Fig.
19.14a,b: Photograph by Earl Plunkett.
Courtesy of G.H. Valentine ;Fig. 19A:
© Penny Gentieu/Stone.

Chapter 20

Fig. 20.1: © Mark Gibson/Visuals
Unlimited; Fig. 20.4a: © Superstock; Fig.
20.4b: © Michael Grecco/Stock Boston;
Fig. 20.4c,d,e,f,h: © The McGraw-Hill
Companies, Inc./Bob Coyle, photographer;
Fig. 20.4g: Courtesy of Mary L. Drapeau;
Fig. 20.7: Courtesy the Cystic Fibrosis
Foundation; Fig. 20.8a,b: © Steve Uzzell; Fig.
20.11a: © Kevin Fleming/Corbis; Fig. 20A:
© Hank Morgan/Science Source/Photo
Researchers, Inc.

Chapter 21

Fig. 21.1: © Elizabeth Fulford/AP Photo
1994, Fig. 21.14a: From M.B. Roth and J.G.
Gall, "Cell Biology" 105:1047-1054, 1987.
© Rockefeller University Press; Fig. 21.17:
© Will & Deni McIntyre/Photo Researchers,
Inc.; Fig. 21A: Courtesy of Robert P. Lanza;
Fig. 21B: © William Freese.

Chapter 22

Fig. 22.1: © Corbis/Adam Smith
Productions; Fig. 22.4b: Courtesy Klaus
Hahn, Cecelia Subauste, and Gary Bokoch.
From /it/Science,/xit/ Vol. 280, April 3,
1998, p. 32; Fig. 22.5: © Bruce Roberts/Photo
Researchers, Inc.; Fig. 22.6: © Breast
Scanning Unit, Kings College Hospital,
London/SPL/Photo Researchers, Inc.; Fig.
22.10: © Joel Gordon; Fig. 22D: © Seth
Joel/SPL/Photo Researchers, Inc.

Chapter 23

Fig. 23.2: © Tom McHugh/Photo
Researchers, Inc., Fig. 23.2(inset): © C.
Seghers/Photo Researchers, Inc.; Fig. 23.5a :
© Dan Dreyfus and Associates, Fig. 23.5b:
© John Reader/Photo Researchers, Inc.; Fig.
23.7: © National Museum of Kenya; Fig.
23.9: Courtesy of The Field Museum; Fig.
23.10: Transp. #608 Courtesy Department of
Library Services, American Museum of
Natural History; Fig. 23A: ©
CABISCO/Visuals Unlimited.

Chapter 24

Fig. 24.1: © Tom Willock/Photo Researchers,
Inc.; Fig. 24.2a : © Stephen Krasemann/Peter
Arnold, Inc.; Fig. 24.2b: © Mary
Thatcher/Photo Researchers, Inc.; Fig. 24.2c:
© Paul Souders/Stone; Fig. 24.2d: © Brian
Kenny/Masterfile; Fig. 24.3a: © G.R. Roberts
Photo Library; Fig. 24.3b: © Bruce Iverson;
Fig. 24.3c: © Peter K. Ziminiski/Visuals
Unlimited; Fig. 24.3d: © Kevin Magee/Tom
Stack & Associates; Fig. 24.4(left): © Kevin
Schafer/Tom Stack & Associates; Fig.
24.4(right): © Gregory K. Scott/Photo
Researchers, Inc.; Fig. 24.15a: © John
Shaw/Tom Stack & Associates; Fig. 24.15b:
© Thomas Kitchin/Tom Stack & Associates;
Fig. 24.16c: © Bill Aron/PhotoEdit; Fig. 24A:
NASA; Fig. 24B: © Foto Luiz Claudio
Marijo/Peter Arnold, Inc.

Chapter 25

Fig. 25.1: © Mary Clay/Masterfile; Fig.
25.4(Rubber tree): © Bryn Campbell/Stone;
Fig. 25.4(Flower) : © Kevin Schafer/Peter
Arnold, Inc.; Fig. 25.4(Bat): © Merlin D.
Tuttle/Bat Conservation International; Fig.
25.4(Fisherman): © Herve Donnezan/Photo
Researchers, Inc.;Fig. 25.4(Lady bug):
© Anthony Mercieca/Photo Researchers,Inc.;
Fig. 25.4(Armadillo): © John Cancalosi/Peter
Arnold, Inc.; Fig. 25.5a: © William S. Smithey,
Jr./Masterfile; Fig. 25.5b: © Don & Pat
Valenti/Stone; Fig. 25.6b: © Gunter
Ziesler/Peter Arnold, Inc.; Fig. 25.6c,e:
Courtesy Thomas A. Stone, Woods Hole
Research Center; Fig. 25.6d: © R. O.
Bierregaard; Fig. 25.7: © Chris Johns/National
Geographic Society Image Collection, Fig.
25.8b: Courtesy Walter C. Jaap/Florida Fish &
Wildlife Conservation Commission; Fig.
25.9a: © Shane Moore/Animals,
Animals/Earth Scenes; Fig. 25.9(top &
bottom): © Peter Auster/University of
Connecticut; Fig. 25.10a: © Gerard
Lacz/Peter Arnold, Inc.; Fig. 25.10b: © Art
Wolfe, Inc.; Fig. 25.10c: © Pat & Tom
Leeson/Photo Researchers, Inc.; Fig. 25.11b:
© Jeff Foot Productions; Fig. 25.12(left):
© Bruce S. Cushing/Visuals Unlimited;
Fig. 25.12(center): © Steven J. Maka; Fig.
25.12(right): © Kim Heacox/Peter Arnold,
Inc.; Fig. 25A: © Stuart Westmorland/Stone.

Index

W

X

Y

Z

IMPORTANT:

HERE IS YOUR REGISTRATION CODE TO ACCESS
YOUR McGRAW-HILL ONLINE RESOURCES.

MCGRAW-HILL
ONLINE RESOURCES

REGISTRATION CODE

dedication-94622192

THIS CODE gives you access to the Web resources for your course. Once the code is entered, you will be able to use the Web resources for the length of your course.

If your instructor is using **WebCT** or **Blackboard**, you will also use this code to access the McGraw-Hill content within your instructor's online course.

Access is provided if you have purchased a new book. If the registration code is missing from this book, the registration screen on our Website, and within your WebCT or Blackboard course, will tell you how to obtain a new code.

Registering for McGraw-Hill Online Resources

TO gain access to your McGraw-Hill web resources simply follow the steps below:

1. USE YOUR WEB BROWSER TO GO TO: http://www.mhhe.com/biosci/genbio/maderhuman7/

2. CLICK ON **FIRST TIME USER**.

3. ENTER THE REGISTRATION CODE* PRINTED ON THE TEAR-OFF BOOKMARK ON THE RIGHT.

4. AFTER YOU HAVE ENTERED YOUR REGISTRATION CODE, CLICK **REGISTER**.

5. FOLLOW THE INSTRUCTIONS TO SET-UP YOUR PERSONAL UserID AND PASSWORD.

6. WRITE YOUR UserID AND PASSWORD DOWN FOR FUTURE REFERENCE.
   KEEP IT IN A SAFE PLACE.

TO GAIN ACCESS to the McGraw-Hill content in your instructor's **WebCT** or **Blackboard** course simply log in to the course with the UserID and Password provided by your instructor. Enter the registration code exactly as it appears in the box to the right when prompted by the system. You will only need to use the code the first time you click on McGraw-Hill content.

Thank you, and welcome to your McGraw-Hill Online Resources!

*YOUR REGISTRATION CODE CAN BE USED ONLY ONCE TO ESTABLISH ACCESS. IT IS NOT TRANSFERABLE.

0-07-232481-3 MADER: HUMAN BIOLOGY, 7E